Auxin Signaling

From Synthesis to Systems Biology

SECOND EDITION

A subject collection from *Cold Spring Harbor Perspectives in Biology*

OTHER SUBJECT COLLECTIONS FROM *COLD SPRING HARBOR PERSPECTIVES IN BIOLOGY*

T-Cell Memory
Stem Cells: From Biological Principles to Regenerative Medicine
Heart Development and Disease
Cell Survival and Cell Death, Second Edition
Calcium Signaling, Second Edition
Engineering Plants for Agriculture
Protein Homeostasis, Second Edition
Translation Mechanisms and Control
Cytokines: From Basic Mechanisms of Cellular Control to New Therapeutics
Circadian Rhythms
Immune Memory and Vaccines: Great Debates
Cell–Cell Junctions, Second Edition
Prion Biology
The Biology of the TGF-β Family
Synthetic Biology: Tools for Engineering Biological Systems
Cell Polarity
Cilia
Microbial Evolution

SUBJECT COLLECTIONS FROM *COLD SPRING HARBOR PERSPECTIVES IN MEDICINE*

Influenza: The Cutting Edge
Leukemia and Lymphoma: Molecular and Therapeutic Insights
Addiction, Second Edition
Hepatitis C Virus: The Story of a Scientific and Therapeutic Revolution
The PTEN Family
Metastasis: Mechanism to Therapy
Genetic Counseling: Clinical Practice and Ethical Considerations
Bioelectronic Medicine
Function and Dysfunction of the Cochlea: From Mechanisms to Potential Therapies
Next-Generation Sequencing in Medicine
Prostate Cancer
RAS and Cancer in the 21st Century
Enteric Hepatitis Viruses
Bone: A Regulator of Physiology
Multiple Sclerosis
Cancer Evolution
The Biology of Exercise
Prion Diseases

Auxin Signaling

From Synthesis to Systems Biology

SECOND EDITION

A subject collection from *Cold Spring Harbor Perspectives in Biology*

EDITED BY

Dolf Weijers
Wageningen University

Karin Ljung
Swedish University of Agricultural Sciences

Mark Estelle
University of California San Diego

Ottoline Leyser
University of Cambridge

COLD SPRING HARBOR LABORATORY PRESS
Cold Spring Harbor, New York • www.cshlpress.org

Auxin Signaling: From Synthesis to Systems Biology, Second Edition
A subject collection from *Cold Spring Harbor Perspectives in Biology*
Articles online at www.cshperspectives.org

Printed in the United States of America

Executive Editor	Richard Sever
Managing Editor	Maria Smit
Senior Project Manager	Barbara Acosta
Permissions Administrator	Carol Brown
Production Editor	Diane Schubach
Production Manager/Cover Designer	Denise Weiss
Publisher	John Inglis

Front cover artwork: Auxin response factors ARF2 (*top left*), ARF3 (*top right*), and ARF5 (*bottom left*) are differentially expressed in the shoot apical meristem of *Arabidopsis*. The pattern of the synthetic auxin responsive promoter DR5rev (*bottom right*) does not fully reflect the expression of ARFs. The four main panels are separated by white horizontal and vertical lines. Different colors of lines represent different microscopic axes (*x*, *y*, *z*). For additional details, see chapter by Pernisová and Vernoux. Figure created by Markéta Pernisová.

Library of Congress Cataloging-in-Publication Data

Names: Weijers, Dolf, 1954- editor. | Ljung, Karin, editor. | Estelle, Mark, editor. | Leyser, Ottoline, editor.
Title: Auxin signaling : from synthesis to systems biology : a subject collection from Cold Spring Harbor Perspectives in Biology / Dolf Weijers, Wageningen University, Karin Ljung, Swedish University of Agricultural Sciences, Mark Estelle, University of California San Diego and Ottoline Leyser, University of Cambridge.
Description: Second edition. | Cold Spring Harbor, New York : Cold Spring Harbor Laboratory Press, [2022] | Includes bibliographical references and index. | Summary: "Auxin is an important plant hormone that controls numerous aspects of development and physiology, including responses to light, tissue patterning, and organogenesis. The book examines how auxin is synthesized, the biochemical basis for its actions, and its various different roles in plant biology"-- Provided by publisher.
Identifiers: LCCN 2021024334 (print) | LCCN 2021024335 (ebook) | ISBN 9781621824008 (hardcover) | ISBN 9781621824015 (epub)
Subjects: LCSH: Auxin. | Plants--Development. | Growth (Plants)
Classification: LCC QK753.A8 A98 2021 (print) | LCC QK753.A8 (ebook) | DDC 571.7/42--dc23
LC record available at https://lccn.loc.gov/2021024334
LC ebook record available at https://lccn.loc.gov/2021024335

10 9 8 7 6 5 4 3 2 1

For a complete catalog of all Cold Spring Harbor Laboratory Press publications, visit our website at www.cshlpress.org.

Contents

Preface

IT HAS BEEN 11 YEARS SINCE THE FIRST EDITION of *Auxin Signaling: From Synthesis to Systems Biology* was published. Since then, auxin biology has continued to be a robust, dynamic, and multifaceted field. In the first edition of this collection, we noted that auxin has claimed center stage in many areas of plant biology and this continues to be the case. However, technological advances in analytical methods, imaging, genome-level studies, and computational biology have led to astounding new insights at every turn. In this volume, we once again attempt to highlight the latest exciting advances in each of the diverse areas of auxin biology.

In the first chapter, Jiří Friml reflects on the progress made in the broad auxin biology field, touching upon all other chapters in this book. He highlights 14 "stations" where exceptional progress has been made, where questions were answered, where unexpected turns have been taken, or where new questions arose. This chapter will likely convince readers that the auxin field will still be vital for decades to come.

This is followed by two chapters that explore the complex mechanisms that regulate auxin levels in both time and space. In their chapter about auxin metabolism, Casanova-Saéz et al. discuss how auxin biosynthesis, conjugation, and degradation are coordinated to achieve the spatiotemporal regulation of auxin levels that drive developmental processes. Complementing this chapter, Hammes et al. describe work on the various auxin transporters with an emphasis on their molecular mechanisms.

The next four chapters focus on mechanisms of auxin signaling. Morffy and Strader discuss the canonical auxin signaling pathway from a protein structural perspective and highlight recent work showing that formation of cytoplasmic condensates of certain ARF proteins has a regulatory function. The three next chapters address the exciting and fast-moving topic of noncanonical auxin signaling. Rapid, nontranscriptional auxin signaling has been investigated for many years. Dubey et al. discuss recent discoveries that have generated new interest in this area. Related to this topic, Napier contributes a historical perspective on auxin-binding protein 1 (ABP1), including the latest information on its potential function as an auxin receptor. McLaughlin et al. present recent work on novel auxin signaling mechanisms. One of these mechanisms involves direct binding of auxin to a transcription factor, thus affecting its interaction with binding partners, while another focuses on a small family of receptor kinases called the TMKs. Clearly there is more to learn about both canonical and noncanonical auxin signaling.

We have learned much in the last decade about the role of auxin during development. Seven chapters take a deep dive into auxin regulation of the development of different organs or tissues. Pernisová and Vernoux lead the way by describing how auxin maxima and minima are established in the shoot apical meristem and drive the development of new organs. There has been extensive progress in our understanding of auxin during root development and two chapters are devoted to this topic. Roychoudhry and Kepinski describe recent work on roots with an emphasis on root system architecture and branch angle. Cavallari et al. focus on the role of auxin in the integration of endogenous and environmental signals during lateral root development. Some of the oldest observations concerning auxin and plant development relate to the vascular system. Lavania et al. illustrate how the logic of vascular patterning is becoming clear. However, they also note that many of the molecular details are still unclear. Sticking to the theme of patterning, Verma et al. describe the profound role of auxin during embryogenesis and also point out the importance of emerging tools that may be important for future studies of auxin and embryogenesis. Of course, embryogenesis depends on the proper development of reproductive structures present on flowers, and Cucinotta et al. present the latest work on the diverse roles of auxin during flower development. They also emphasize the importance of the comparative approach since flower development is amazingly diverse across all angiosperm lineages.

One of the most complex areas of hormone biology are the interactions between hormones. Mazzoni-Putman et al. tackle this topic in the context of auxin and illustrate how auxin interacts with most, perhaps all, of the major plant hormones. In fact, one need not venture beyond auxin itself to observe complex regulatory systems. Ma et al. document the many homeostatic mechanisms that govern the regulation of auxin levels and signaling output.

Throughout the volume, the important role that auxin plays in environmental response is never far from the surface. This topic is addressed directly in three chapters. Leftley et al. discuss how auxin facilitates changes in root system architecture in stress conditions while Kunkel and Johnson focus on the important role of auxin during plant–pathogen interactions. Casal and Estevez take a different approach to this topic by considering how auxin regulation enables the plant to maximize capture of limiting resources.

During the last few years, there has been increasing interest in the evolutionary history of auxin signaling. Bowman et al. take a broad approach to this topic by addressing changes in auxin biology during the evolution of land plants. On the other hand, Suzuki et al. focus on the Bryophyta, the descendants of the first land plants and sister group to the vascular plants. Both chapters illustrate how much we can learn about auxin biology by taking an evolutionary perspective.

The volume finishes with several very forward-looking topics: synthetic biology, computational biology, and chemical biology. Wright et al. describe efforts to establish the logic and mechanisms of auxin signaling by using synthetic systems. Two chapters, Rutten et al. and Cieslak et al., describe the latest efforts to produce robust computational models for root and shoot growth, respectively. Finally, Hayashi describes the development of powerful new chemical tools to study various aspects of auxin biology.

It has been more than 140 years since Charles Darwin first described a "substance" that moved from the tip of an oat coleoptile to its base and produced a growth response. We hope this book provides an informative overview of the current state of our knowledge, and offers viewpoints that can stimulate discussion. Of course, the most important purpose of such a volume is to illustrate what we don't know, and by doing so, provoke new questions from the next generation of curious scientists. We hope that we have succeeded in this respect as well. We are grateful to Richard Sever and Barbara Acosta from Cold Spring Harbor Laboratory Press for their encouragement and patience. Finally, we sincerely thank all the authors for the time and thoughtfulness they committed to these chapters.

Dolf Weijers
Karin Ljung
Mark Estelle
Ottoline Leyser

Fourteen Stations of Auxin

Jiří Friml

Institute of Science and Technology Austria, 3400 Klosterneuburg, Austria

Correspondence: jiri.friml@ist.ac.at

Auxin has always been at the forefront of research in plant physiology and development. Since the earliest contemplations by Julius von Sachs and Charles Darwin, more than a century-long struggle has been waged to understand its function. This largely reflects the failures, successes, and inevitable progress in the entire field of plant signaling and development. Here I present 14 stations on our long and sometimes mystical journey to understand auxin. These highlights were selected to give a flavor of the field and to show the scope and limits of our current knowledge. A special focus is put on features that make auxin unique among phytohormones, such as its dynamic, directional transport network, which integrates external and internal signals, including self-organizing feedback. Accented are persistent mysteries and controversies. The unexpected discoveries related to rapid auxin responses and growth regulation recently disturbed our contentment regarding understanding of the auxin signaling mechanism. These new revelations, along with advances in technology, usher us into a new, exciting era in auxin research.

GOSPEL—UNIQUE NATURE OF AUXIN AS A SIGNALING MOLECULE

Among the most remarkable features of this chemically inconspicuous derivative of tryptophan, indole-3-acetic acid, is its ability to Influence Almost Anything (IAA) (Paque and Weijers 2016). This versatility can either highlight auxin's universal importance, or it can be presented as an argument against its action as a specific signaling molecule. While the proponents of the latter are unlikely to number among the readers of this review, one has to agree that the ability of auxin to directly or indirectly affect virtually any physiological or developmental process in plants is one of its greatest mysteries. Other plant signaling molecules also have multiple effects, but auxin is the undisputed champion. Only the list of developmental effects would be endless, ranging from gametophyte development, early embryogenesis and, throughout the whole life of the plant, all the way to senescence and programmed cell death. In decades of studies, it became obvious that auxin acts as a main developmental trigger, but also as a coordinative signal between individual cells, different tissues, different parts of the plant, and even between the plant and environment. Thus, obvious questions are (1) How can auxin accomplish this feat of regulating so many apparently unrelated processes? and (2) What are the unique features of auxin compared to other signals that make this possible?

Concerning the mechanism that allows such a staggering diversity of auxin effects, it is clear that several auxin signaling pathways are at

Cite this article as *Cold Spring Harb Perspect Biol* doi: 10.1101/cshperspect.a039859

work (Fig. 1). The most studied and presumably the major mechanism for auxin-mediated transcriptional reprogramming is based on the nuclear TRANSPORT INHIBITOR RESPONSE1/ AUXIN-SIGNALING F-BOX (TIR1/AFB) auxin receptors and potentially involves hundreds of possible combinations of downstream transcriptional repressors and activators from Aux/IAA (auxin/indole-3-acetic acid) and ARF (AUXIN RESPONSE FACTOR) families (for review, see Morffy and Strader 2021), which can be also regulated by posttranslational modifications (Orosa-Puente et al. 2018). In addition, auxin regulates transcription through at least two other mechanisms, in which transmembrane kinase (TMK) or ARF3 (ETTIN) signaling components are involved (for review, see McLaughlin et al. 2021). Furthermore, TIR1/AFB- and TMK-mediated pathways have branches involved in rapid, nontranscriptional auxin effects (for review, see Dubey et al. 2021). These distinct but likely mutually intertwined auxin perception and downstream signaling machineries provide a rich mechanistic basis for the multitude of cellular and developmental auxin effects.

When contemplating what makes auxin unique among plant signaling molecules, the obvious answer seems to be its directional intercellular transport (for review, see Hammes et al. 2021). Similar to other plant hormones, auxin (IAA) is synthesized in many tissues with different efficiency (for review, see Casanova-Sáez et al. 2021). The two-step, indole-3-pyruvic acid (IPyA)-based IAA synthesis from tryptophan emerged as the main mechanism for auxin production. The key importance of localized synthesis in several developmental contexts has already been demonstrated (for review, see Mazzoni-Putman et al. 2021); however, it is the movement from the places of biosynthesis to other cells and tissues that empowers auxin to act as a versatile coordinative signal.

It is a defining feature of hormonal-type signaling molecules to have their production and action at distinct places, which necessitates transport. Nonetheless, for auxin, transport is much more than that. Indeed, although transport and transporters of other plant hormones have been identified, the directional, regulated transport across tissues (so-called polar auxin transport [PAT]) is, based on our current state of knowledge, unique to auxin (Fig. 2). As will be described in more detail later, it is the PAT with its exceptional capacity to respond to endogenous and exogenous signals (Fig. 3) and to self-regulate and self-organize (Fig. 4) that gives auxin the decisive charm to work its universal magic.

Thus, the mysterious plethora of auxin effects is made possible by multiple, heavily branched signaling pathways and a large gang of downstream signaling components. It can be also attributed to the unique property of auxin being locally synthesized and directionally transported by an elaborated transport system that integrates many internal and external signals (for review, see Adamowski and Friml 2015). This mechanism generates auxin gradients and local maxima and minima. The resulting auxin readout then elicits a specific response in a given cell. In roots, it has been shown that the amplitude of transcriptional auxin response follows the predicted auxin gradient (Bargmann et al. 2013; for review, see Roosjen et al. 2018); nonetheless, auxin action as a bona fide morphogen (i.e., different auxin concentrations elicit different responses in the same cells) has not been demonstrated. In most contexts, it acts as a permissive rather than instructive signal, "a universal trigger of change" in a developmental program, a change that is predefined by the combination of receptors and downstream signaling components in different target cells, and ultimately leads to a multitude of very diverse responses (formulated in Vanneste and Friml 2009).

STATION 1: GET EQUIPPED FOR PILGRIMAGE—NEW TOOLS AND METHODS

As in all of science, curiosity, reasoning, and intuition accompanied by hard work and a great deal of passion have been the main prerequisites of progress in the auxin field. Nonetheless, the emergence of new technologies was the necessary ingredient and repeatedly provided a crucial push. This has been particularly evident in the last decade with a dramatic improvement in all different "omics" technologies and imaging

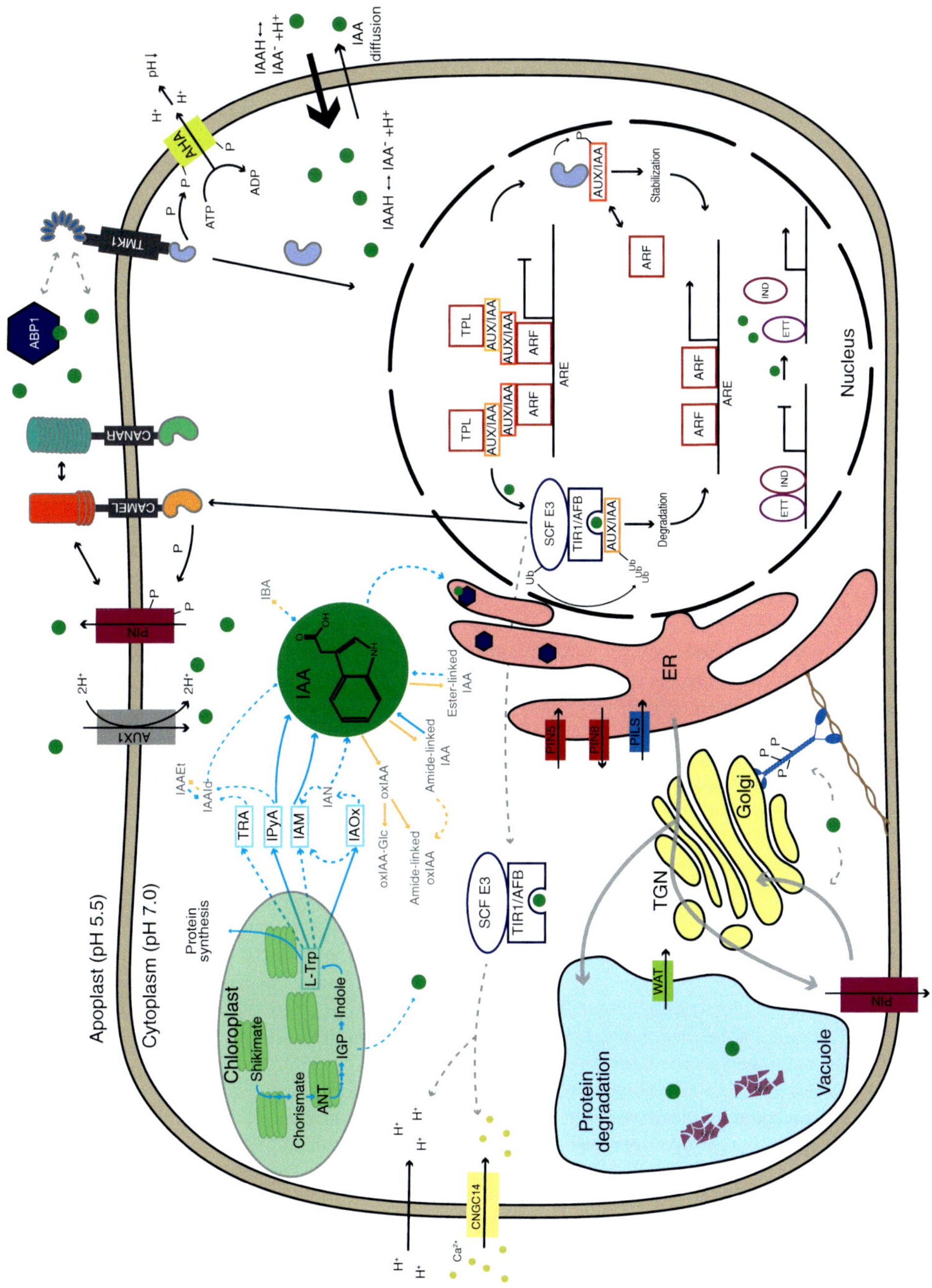

Figure 1. *(See following page for legend.)*

methods, both on the side of equipment and genetically encoded markers. Noteworthy is the development of the vertical microscope (von Wangenheim et al. 2017) coupled to a microfluidic device, which allows real-time observation of root growth and developmental processes with subcellular resolution that was instrumental to identify a nontranscriptional branch of TIR1/AFB signaling (Fendrych et al. 2018; Serre et al. 2021).

Access to next-generation sequencing technologies tremendously supported genetic studies, including the mapping of mutants (Rowan et al. 2017), GWAS (Ogura et al. 2019), and the development of new genetic models, for example, for evolutionary studies such as the morphologically simple *Klebsormidium* (Hori et al. 2014) or the more complex *Chara* (Nishiyama et al. 2018) algae. Transcriptomics, now approaching the single-cell resolution limit (Seyfferth et al. 2021), is changing the way we approach questions related to auxin functions in transcriptional reprogramming (Bargmann et al. 2013; Wendrich et al. 2020). Similar developments in proteomics and phosphoproteomics (Smith et al. 2020; Han et al. 2021) promise to advance our knowledge of nontranscriptional aspects of auxin signaling. Further improvements in metabolomics, especially allowing subcellular resolution, will boost our knowledge on compartmentalization of auxin and its metabolism (Pěnčík et al. 2018).

Chemical tools have a long tradition in the auxin field and help dissect different facets of auxin function (Ma et al. 2018). Inhibitors such as naphthylphthalamic acid (NPA) have multiple targets, including (1) a low-affinity effect on actin-mediated trafficking (Dhonukshe et al. 2008; Zhu et al. 2016), and (2) a high-affinity association with PIN-FORMED (PIN) auxin transporters (Abas et al. 2021; Teale et al. 2021). Nonetheless, these inhibitors were invaluable in establishing a versatile role for auxin transport in plant development (for review,

Figure 1. Auxin biosynthesis, homeostasis, and signaling. The main precursor for IAA (indole-3-acetic acid) biosynthesis, L-tryptophan (L-Trp), is synthesized in chloroplasts. The dominant IAA biosynthesis pathway in *Arabidopsis* is via indole-3-pyuvic acid (IPyA). Free IAA is either oxidized or conjugated to, for example, amino acids and sugars (for review, see Zhang and Peer 2017 and Casanova-Sáez et al. 2021). IAA can either enter cells through the plasma membrane (PM)-localized H^+/auxin symporter AUX1/LAX or diffuse through the PM in its protonated form (IAAH). Inside the cell, IAA is predominantly in an anionic form, necessitating export, which is mediated mainly by PIN proteins at the PM (for review, see Hammes et al. 2021). PIN proteins undergo constant endocytic cycling and are degraded in the vacuole (for review, see Adamowski and Friml 2015). Auxin influences these processes by an unclear mechanism involving myosin phosphorylation (Han et al. 2021). So-called short PINs and PILS mediate auxin in- and efflux in the endoplasmic reticulum (ER) (for review, see Abdollahi Sisi and Růžička 2020). WAT1 is an auxin transporter located in the tonoplast (Ranocha et al. 2013). In the nucleus, auxin binds to the canonical TIR1/AFB-Aux/IAA receptor complex and activates transcription via ubiquitination and degradation of Aux/IAA transcriptional repressors, which releases repressive ARF-Aux/IAA-TPL (TOPLESS) complexes (for review, see Morffy and Strader 2021). This pathway also regulates transcription of CAMEL. The CAMEL/CANAR complex at the PM interacts with and phosphorylates PINs (Hajný et al. 2020). Furthermore, auxin in the nucleus also releases the repressive complex between ETTIN/ARF3 (ETT) and other transcription factors such as INDEHISCENT (IND), causing transcriptional reprogramming required for various developmental processes. Transmembrane kinase (TMK) proteins are components of cell surface auxin signaling. Auxin can trigger the cleavage of the carboxy-terminal kinase domain of TMK1, which then translocates to the nucleus, where it regulates transcription by binding to the noncanonical AUX/IAAs (for review, see McLaughlin et al. 2021). TMK1 also mediates the auxin effect on AHA phosphorylation leading to apoplast acidification (Li et al. 2021b; Lin et al. 2021). It is unclear how auxin triggers the TMK pathway. The possible function of ABP1 as an auxin receptor in the apoplast and its role in the ER remains controversial (for review, see Napier 2021). TIR1/AFB auxin signaling, by an unknown mechanism, also mediates apoplast alkalinization and PM depolarization (Li et al. 2021b; Serre et al. 2021) and Ca^{2+} influx (Dindas et al. 2018; for review, see Dubey et al. 2021). Green-filled circles depict auxin molecules. Grey dashed arrows indicate hypothetical regulations. Blue arrows indicate IAA biosynthesis and orange arrows indicate metabolic pathways. Figure based on data in Skalický et al. (2018) and Gallei et al. (2020). (*See figure on preceding page.*)

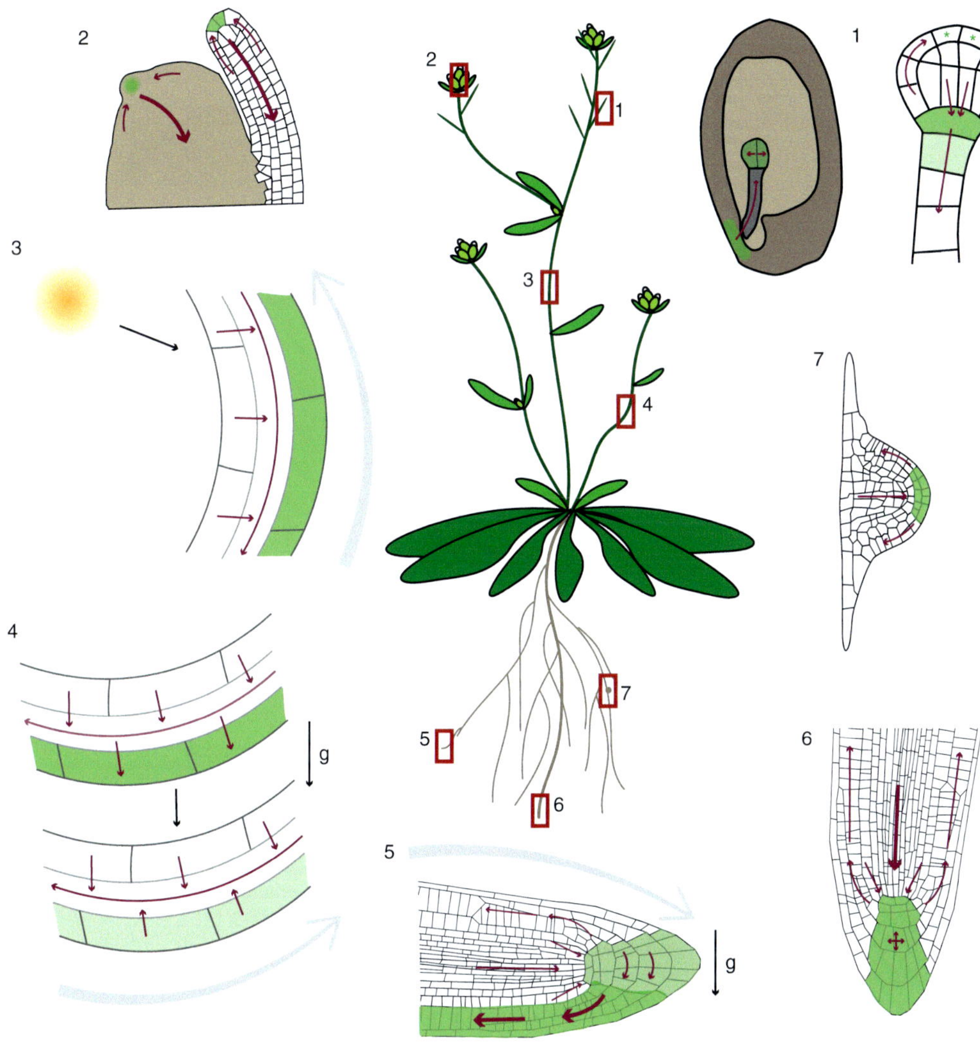

Figure 2. PIN-dependent auxin transport network generating local auxin maxima and gradients. (1) Maternally produced auxin is transported to the embryo proper, where it accumulates and regulates development (Robert et al. 2018). At early globular stage, auxin production starts at the embryo apex, auxin transport routes reverse, and a new auxin maximum is generated for the root pole specification (Robert et al. 2013; Wabnik et al. 2013). Auxin production is indicated by green stars. (2) Converging auxin fluxes in the epidermis of the shoot apical meristem generate auxin maxima for organ initiation. From the developing primordia, auxin is canalized toward preexisting vasculature (Benková et al. 2003). (3) Light diverges auxin fluxes toward the shaded side of the shoot (Ding et al. 2011), where auxin accumulates (Friml et al. 2002b) and promotes bending. It remains unclear whether this is the primary, causal mechanism. (4) Gravity stimulation diverges auxin fluxes toward the lower side of the shoot (Rakusová et al. 2011), where auxin accumulates and promotes bending. At a later stage, auxin accumulation leads to PIN relocation to the inner cell sides, thus equalizing the auxin distribution and terminating bending (Rakusová et al. 2016). (5) Gravity stimulation diverges auxin fluxes in root columella toward the lower root side, where auxin accumulates (Luschnig et al. 1998; Friml et al. 2002b). Auxin promotes its own flow along the lower root side (Baster et al. 2013) delivering auxin to the elongation zone where it inhibits growth (Fendrych et al. 2018), leading to downward bending of the root. (6) Multiple, redundant PIN fluxes converge to generate an auxin gradient in the central root meristem, which maintains cell divisions and fate specification (Friml et al. 2002a). (7) Auxin flow through the lateral root primordium generates an auxin maximum at the tip, which is partly dissipated by the flow through the epidermis (Benková et al. 2003). Gray arrows indicate direction of growth after gravity or light stimulation. Green shading shows auxin accumulation. Darker green indicates higher auxin levels than lighter green. g denotes direction of gravity.

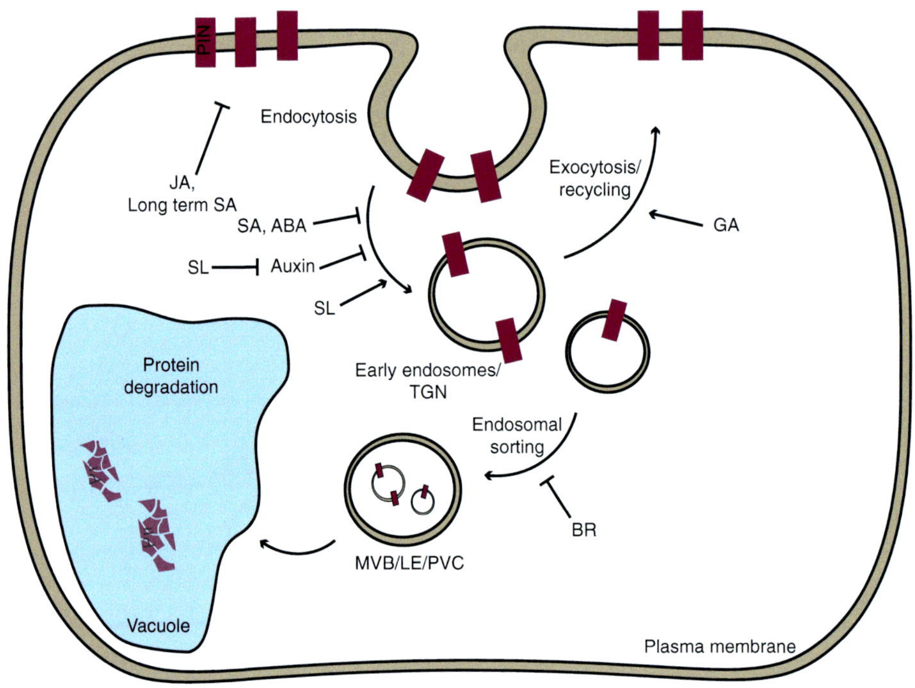

Figure 3. Hormonal regulation of PIN trafficking. PIN auxin exporters undergo constant, subcellular endocytic recycling, which assists in maintaining polar localization (Kleine-Vehn et al. 2011) and integrates various signals. PINs can as well be directed from the plasma membrane (PM) to the vacuole through the trans-Golgi network (TGN)/endosomal network for lytic degradation. Environmental and developmental signals, including hormones, can modulate various steps of PIN trafficking, thus changing PIN incidence at the PM. This eventually fine-tunes directionality and capacity of auxin transport leading to adaptation of plant growth and development. Figure based on data reviewed in Semeradova et al. (2020).

see Michniewicz et al. 2007). Inhibitors of auxin biosynthesis, as well as agonists and antagonists of specific auxin signaling mechanisms, are also widely used (for review, see Hayashi 2021). The latter ones were, for example, instrumental to confirm a TIR1/AFB-independent binding site that mediates the auxin effect on endocytic trafficking (Robert et al. 2010; De Vriese et al. 2019; Oochi et al. 2019) or TIR1/AFB involvement in rapid root growth inhibition (Fendrych et al. 2018). Also very useful is a synthetic biology tool activating specifically TIR1/AFB signaling by means of the engineered receptor (concave TIR1–ccvTIR1) and fitting ligand (convex IAA–cvxIAA), which does not bind to endogenous TIR1 (Uchida et al. 2018; for review, see Wright et al. 2021).

Mathematical modeling has aided the planning and interpretation of experimental work in the auxin field for decades (for review, see Prusinkiewicz and Runions 2012). Many models recapitulate auxin transport fluxes and local auxin maxima in different contexts such as embryos (Wabnik et al. 2013) and roots (for review, see Rutten et al. 2021). The self-organizing processes such as phyllotaxis in shoots (for review, see Cieslak et al. 2021) or leaf venation (for review, see Lavania et al. 2021) with their underlying mathematical beauty hold a special fascination.

STATION 2: TRUE AND FALSE PROPHETS—AUXIN AND ITS ANALOGS

What exactly do we mean when talking about auxin? In contrast to some other plant hormones, where a relatively diverse group of chemical compounds has similar physiological effects, by nat-

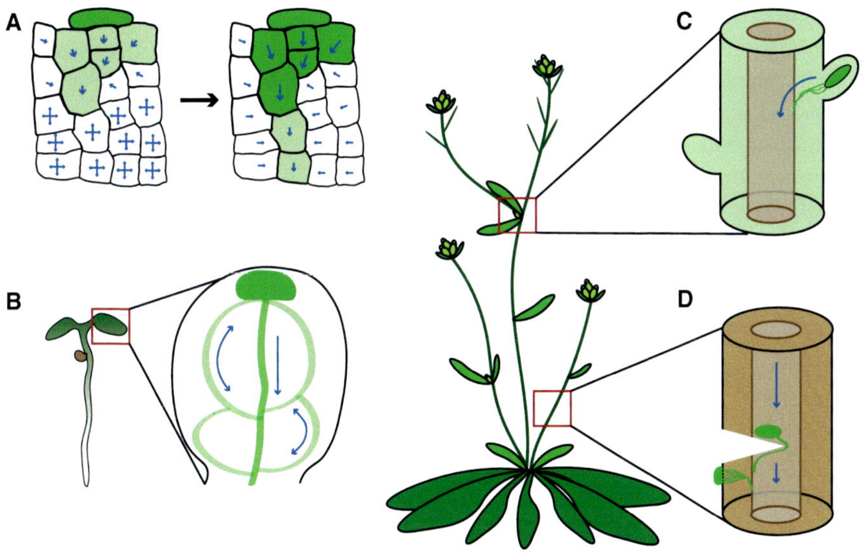

Figure 4. Canalization processes for vasculature formation and regeneration. (*A*) Auxin source (dark green color) polarizes originally homogenous cells to create a directional transport (blue arrows) away from the source. The self-organizing property of auxin transport allows to canalize auxin from an initially broad domain into a narrow channel with high auxin-transporting capacity. (*B*) Auxin maximum in a cotyledon tip drives auxin canalization in a conserved pattern, demarcating the position of future vasculature. Transport-independent patterning mechanisms are likely also involved (Verna et al. 2019; for review, see Lavania et al. 2021). (*C*) The shoot apex is a well-known source of auxin, which keeps lateral buds inhibited. Once the apex is removed, the closest lateral bud is released from the inhibition and becomes a new dominant auxin source. Auxin canalization from this bud guides vasculature formation, connecting the lateral bud to the preexisting stem vasculature (Balla et al. 2011). (*D*) Wounding of stem vasculature results in a local auxin accumulation above the wound. Subsequently, auxin is canalized around the wound to initiate reconnection of the preexisting vasculature (Sauer et al. 2006; Mazur et al. 2016). Local external auxin application to the side of the stem also triggers auxin channel formation and guides formation of vasculature (Mazur et al. 2020b). Other cases of auxin canalization are during organogenesis at the shoot apical meristem and during embryogenesis (see Fig. 2).

ural auxin we typically mean indole-3-acetic acid (IAA). In addition, 4-chloro-indole-3-acetic acid (4-Cl-IAA) and phenylacetic acid (PAA) can be found in plants. The former is found rarely and the latter appears to be much less potent than IAA (for review, see Cook 2019). Indole-3-butyric acid (IBA) has auxin-like effects when applied exogenously to plants, but it appears to be more of an IAA precursor or a storage form. The persistently debated question about IBA is, however, whether it is present at physiologically relevant concentrations in plant tissues (Novák et al. 2012; for review, see Frick and Strader 2018). Besides IAA biosynthesis (mainly from L-Trp) and irreversible oxidation, much of IAA is present in storage (IAA conjugates and methyl-IAA) and catabolite (amide-linked conjugates) forms (for review, see Casanova-Sáez et al. 2021). Widely used synthetic auxins such as 1-naphthaleneacetic acid (1-NAA) and 2,4-dichlorophenoxy acetic acid (2,4-D) are more chemically stable, but it must be taken into account that they are metabolized and transported differently compared to IAA (Simon et al. 2013).

STATION 3: PLACES OF WORSHIP—SUBCELLULAR AUXIN DISTRIBUTION

One of the biggest enigmas in the auxin field is its subcellular distribution (Fig. 1). In the future, fluorescently labeled auxins, specific anti-IAA antibodies, direct, sufficiently sensitive genetically encoded auxin sensors, and subcellular fractionations coupled to metabolomics may provide more reliable insights; but for now, we are inferring this information indirectly from the localization of auxin receptors, transporters, and metabolic enzymes. IAA, as a small molecule, would freely diffuse through the cytosol and connected nucleus, where the majority of TIR1/AFB auxin receptors reside. Another place where auxin accumulates is the vacuole as suggested by the presence of the WALLS ARE THIN1 (WAT1) auxin transporter and confirmed by metabolomics on isolated vacuoles (for review, see Skalický et al. 2018). However, what auxin does there is completely unclear.

The endoplasmic reticulum (ER) has always been on the radar of auxin researchers, originally because AUXIN-BINDING PROTEIN1 (ABP1) is localized there along with some auxin metabolic enzymes (for review, see Woodward and Bartel 2005; Kriechbaumer et al. 2016; Sanchez Carranza et al. 2016). The ER received even more prominence through the discovery that some transporters from PIN (Mravec et al. 2009) and PIN-LIKES (PILS; Barbez et al. 2012) families are localized at the ER membrane. Transport-mediated intracellular redistribution of auxin and its metabolites between compartments with different sets of metabolic enzymes and receptors would allow a rapid regulation of auxin metabolism and signaling (for review, see Abdollahi Sisi and Růžička 2020). In addition, given that receptors for ethylene (Bisson et al. 2009) and cytokinin (Antoniadi et al. 2020; Kubiasová et al. 2020) are also localized at the ER, this subcellular compartment may play an underappreciated role as a hub for hormone signaling, metabolism, and cross talk (for review, see Friml and Jones 2010).

Unanswered questions related to the subcellular auxin distribution are (1) how auxin crosses the apparently highly oxidative environment of the apoplast, and (2) the importance of auxin movement through plasmodesmata (for review, see Paterlini 2020).

STATION 4: NEW ADDITIONS TO THE OLD TESTAMENT—RAPID CELLULAR RESPONSES

Rapid cellular responses were the bread and butter of auxin signaling in the premolecular era. The discovery that auxin regulates protein expression (for review, see Abel and Theologis 2010) was a milestone that turned the focus of the field to the developmental responses mediated by transcriptional reprogramming. Given the speed of transcription and translation, any effects that take less than a few minutes can be considered nontranscriptional.

A classic rapid effect is the depolarization of the PM, which in roots is linked with apoplast alkalinization and cytosolic Ca^{2+} transients (for review, see Dubey et al. 2021). These processes have recently enjoyed a renaissance due to their recently confirmed link to root gravitropism. Given that they are nontranscriptional, these responses were proposed to require a yet unidentified auxin receptor (Monshausen et al. 2011; Shih et al. 2015; for review, see Leyser 2018). Recent findings, however, suggest that rapid, auxin-mediated root growth inhibition requires TIR1/AFB receptors (Dindas et al. 2018; Fendrych et al. 2018; Li et al. 2021b; Serre et al. 2021).

Another rapid, nontranscriptional, and indeed TIR1/AFB-independent process is the auxin effect on endocytic trafficking of PIN auxin transporters (Paciorek et al. 2005; Robert et al. 2010). As a potential mechanism involved in the auxin feedback regulation of its own transport and its polarization (Mazur et al. 2020a; Han et al. 2021), this response has received substantial attention. Recently, the original interpretation that involves auxin inhibiting endocytosis has been revised with new tools and approaches (Narasimhan et al. 2021) and it appears that auxin rather targets actin/myosin-related aspects of endocytic PIN trafficking (Arieti and Staiger 2020; Han et al. 2021).

The recent identification of an ultrafast, auxin-triggered protein phosphorylation response in *Arabidopsis* roots came as a surprise. It appears that hundreds of proteins involved in a wide variety of cellular functions, different from those targeted by transcriptional regulation, become differentially phosphorylated within 2 min of auxin treatment (Han et al. 2021). This dataset provides a rich resource to tackle the nontranscriptional auxin responses at molecular level.

 Cite this article as *Cold Spring Harb Perspect Biol* doi: 10.1101/cshperspect.a039859

STATION 5: ALPHA AND OMEGA OF AUXIN ACTION—GROWTH

The word "auxin" is derived from the Greek "auxano" = "to grow," which indicates that the effect on growth was at the very beginning of auxin research. Despite being the most classical of auxin effects, studies of the underlying mechanism still bring surprises, such as the recent identification of a noncanonical branch of TIR1/AFB auxin signaling and the implication of TMK1-based cell surface auxin signaling in growth regulation (see also stations 6 and 7).

One of the biggest enigmas of plant physiology is how shoots grow upward and roots downward. In both cases, auxin accumulates at the lower organs' side, but promotes growth in shoots and inhibits it in roots. Not only is the sign of the effect opposite but also the timing is different with a much faster reaction in roots (<30 sec) than in shoots (>10 min). How the same signal can trigger such opposite responses remains a mystery.

Plant cells have special features—a very high turgor pressure (comparable to a car tire) and rigid cell walls. Any growth regulation mechanism must therefore directly or indirectly target these two parameters. Although several alternatives for a cellular mechanism underlying growth have been proposed, such as microtubule reorientation or vacuole fragmentation, the classical acid growth hypothesis has stood the test of time and has retained its prominence to explain the auxin-mediated growth promotion in shoots (for review, see Gallei et al. 2020). It states that H^+ export across the PM increases the membrane potential, which drives secondary ion influx, ultimately leading to higher turgor pressure. At the same time, such apoplast acidification leads to the activation of pH-sensitive enzymes that loosen the cell wall. Higher turgor and lower cell wall rigidity then together result in cell expansion (for review, see Arsuffi and Braybrook 2018). Indeed, this has been nicely demonstrated in shoots, where auxin perceived by the canonical, transcriptional TIR1/AFB pathway ultimately activates the PM H^+-ATPases (AHAs), leading to apoplast acidification and growth promotion (Spartz et al. 2014; Fendrych et al. 2016; Ren et al. 2018; for review, see Dubey et al. 2021). Recently, a parallel mechanism for a direct activation of H^+ export by the cell surface kinase TMK1, which phosphorylates and activates H^+-ATPases, has been identified (Li et al. 2021b; Lin et al. 2021). This should lead to apoplast acidification and growth promotion; however, as mentioned earlier, these processes in the shoot are strictly dependent on TIR1/AFB activity and occur with a delay of about 10 min (Fendrych et al. 2016). Therefore, the functional connection between TIR/AFB and TMK pathways for a cooperative promotion of shoot growth still needs to be clarified.

In roots, auxin inhibits growth, and, as recent manipulations of external pH demonstrated, the acid growth principle also underlies this effect. The TMK1-AHA signaling module also acts in roots, mediating apoplast acidification, but this effect is overshadowed by the more dominant TIR1/AFB-mediated apoplast alkalinization, which occurs by an unknown cellular mechanism (Li et al. 2021b). These two antagonistic pathways, which converge on the regulation of apoplastic pH, thus presumably prime roots for rapid changes in growth direction important for the efficient navigating of the soil. Thus, the TMK1-AHA mechanism between shoot and root is conserved, but what causes roots to react oppositely to shoots is the addition of the nontranscriptional TIR1/AFB signaling branch mediating apoplast alkalinization.

STATION 6: NOTHING GROWS TO HEAVEN —CANONICAL AUXIN SIGNALING

A dozen years ago, the field of auxin signaling appeared to be comfortably simple. Genetic work had identified and characterized mutants such as *tir1* (*transport inhibitor response1*) and *axr1* (*auxin resistant1*), whose root growth was insensitive to auxin (for review, see Lavy and Estelle 2016; Leyser 2018) and molecular biology studies had explored the mechanism of auxin-mediated transcription. A beautiful synthesis of these two threads (Gray et al. 2001) culminated in a somewhat unexpected realization that TIR1 itself is the major part of the auxin receptor (Dharmasiri et al. 2005; Kepinski and Leyser 2005).

Downstream, the essentially unlimited combinations of Aux/IAA repressors and ARF transcription factors can remodel the transcriptomes of individual cells in myriad different ways (Bargmann et al. 2013; for review, see Morffy and Strader 2021).

In our admiration of this elegant mechanism, by association, it was tacitly assumed that this pathway also accounts for the inhibitory auxin effects on root growth, not least because this was the phenotypic trait originally used to identify the key mutants including *tir1*. This notion was further reinforced by elucidation of the TIR1/AFB-mediated transcriptional mechanism, by which auxin promotes growth in shoots (for review, see McLaughlin et al. 2021). Although the gravitropic response and auxin effects in roots are fast, the exact timing was unclear and, thus, either transcription regulation was assumed to be rapid enough to account for this, or TIR1/AFB-independent signaling mechanisms were invoked (Monshausen et al. 2011; Shih et al. 2015; for review, see Leyser 2018). The vertical microscope coupled with root microfluidics allowed a detailed evaluation of growth kinetics and revealed a visible decrease in root growth within 30 sec after auxin treatment. This, and pharmacological interference with de novo protein synthesis, showed beyond any doubt that auxin-mediated root growth inhibition is nontranscriptional. Nonetheless, the involvement of a TIR1/AFB pathway has been unequivocally confirmed using mutants and the engineered ccvTIR1/cvxIAA system. Therefore, the only interpretation reconciling these new observations is that TIR1/AFB auxin signaling has a nontranscriptional branch that regulates root growth (Fendrych et al. 2018). This unexpected notion was further reinforced by observations that very rapid cytosolic Ca^{2+} spikes also require TIR1/AFB signaling (Dindas et al. 2018). The more cytosolic-resident AFB1 member of the TIR1/AFB family has been proposed to mediate these rapid effects (Prigge et al. 2020).

The next challenge is to discover the mechanism(s) by which this elusive TIR1/AFB signaling branch alkalizes the apoplast to inhibit root growth. This would require a direct action at the PM, potentially activating some H^+ leakage mechanism that would lead to immediate apoplast alkalinization (Li et al. 2021b) and membrane depolarization (Serre et al. 2021). Whether these interdependent processes are also linked to the cytosolic Ca^{2+} transients (Shih et al. 2015), what the underlying cellular mechanism is, and how TIR1/AFB auxin receptors trigger it are the most imminent questions.

STATION 7: RISING HERESY—NONCANONICAL AUXIN SIGNALING

The overwhelming success of genetic and molecular biology approaches led to the identification of the TIR1/AFB-Aux/IAA-ARF transcriptional signaling module, which outshined other possible auxin signaling mechanisms. An unexpected twist to Aux/IAA- and ARF-mediated regulation of transcription came from studies of atypical members of these two protein families.

ARF3 (also called ETTIN) lacks the typical domain required for Aux/IAA interaction and instead harbors a so-called ETTIN-specific domain. IAA binds directly to this domain, thus disrupting the interaction of ETTIN to a number of different regulators, such as the TOPLESS (TPL) transcriptional repressor (Simonini et al. 2016; Kuhn et al. 2020). Such direct binding of a hormonal signal (here IAA) to a transcriptional regulator is reminiscent of similar mechanisms known from animals, for example, for thyroid hormones (for review, see McLaughlin et al. 2021). It will be interesting to see whether this type of direct regulation identified for ETTIN also applies to other signaling molecules and/or transcriptional regulators.

Similar to ARFs, also among *Arabidopsis* Aux/IAA transcriptional repressors, there are few nonconformists, such as IAA33, which lack domains that interact with TIR1/AFB and which, unlike canonical Aux/IAAs, are induced and/or stabilized by auxin to maintain the root stem cells (Ding and Friml 2010; Lv et al. 2020). Two other noncanonical Aux/IAAs (IAA32 and IAA34) are targeted and phosphorylated by the kinase domain of TMK1, which is cleaved in response to auxin. This auxin-mediated cleavage of TMK1 and the potential nuclear translocation of its kinase domain provides an elegant mechanism of how cell-surface TMK-based and nuclear TIR1/AFB-based auxin signaling converge on an overlapping set of transcriptional regulators, which in this case regulates the differential growth in the apical hook (Cao et al. 2019; Baral et al. 2021).

TMK signaling has been, however, mainly implicated in rapid, nontranscriptional auxin responses. TMK1 interacts with and phosphorylates H^+-ATPase pumps and activates them, which leads to apoplast acidification. This supports the transcriptional TIR1/AFB-mediated mechanism for growth promotion in shoots (Lin et al. 2021) and counteracts the nontranscriptional TIR1/AFB-mediated mechanism for growth inhibition in roots (Li et al. 2021b). The TMK pathway has been shown to play other auxin-related roles such as lateral root primordia development (Huang et al. 2019), puzzle-like shaping of leaf epidermal cells (Xu et al. 2010), and regulation of auxin biosynthesis (Wang et al. 2020). The established downstream effectors of TMKs are Rho of Plants (ROPs), small GTPases that regulate a plethora of cellular activities (for review, see Feiguelman et al. 2018). Interestingly, auxin induces sterol- and lipid-dependent nanoclustering of both TMK1 (Pan et al. 2020) and ROP6 (Platre et al. 2019), but the interdependence and relevance of these processes still remain unclear. As such, TMK has now been well established to mediate the auxin effect on multiple cellular and developmental processes and more are likely to follow. However, the key question remains: How is auxin perceived by this pathway? That was once a neat niche for a role of ABP1.

STATION 8: CRUCIFIED AND WAITING FOR RESURRECTION—ABP1

Only a few proteins in the plant field have a longer history than ABP1, which was first identified in the early 1970s as a protein that can bind auxin (Hertel et al. 1972), and none has been such a persistent subject of controversy. ABP1 was born to the field with high hopes as the first candidate for a plant hormone receptor. Excellent biochemical work, including ABP1 cocrystallization with auxin, confirmed the high affinity binding of maize ABP1 to 1-NAA (Woo et al. 2000). ABP1 further rose to prominence through the suggestion that it played an important role in auxin-sensitive endocytic trafficking of PIN auxin transporters (Robert et al. 2010) in puzzle-shaped epidermis morphology (Xu et al. 2010) and, in particular, through its association with the cell-surface localized TMK receptor-like kinase (Xu et al. 2014).

Unfortunately, many features of the ABP1 function remain unclear starting with its place of action. The majority of ABP1 can be found, as suggested by the ER retention signal, in the ER, where, however, the pH is not favorable to auxin binding. In the apoplast, pH is more suitable and ABP1 presence there would be required for the majority of its proposed roles, but localization studies have not been entirely conclusive (for review, see Napier 2021).

Genetic studies stirred even more controversy. A study in *Arabidopsis* claimed that *abp1* knockout mutants are embryo lethal, based on two independent alleles and complementation experiments (Chen et al. 2001). Although these lines were not used to generate any insight into the ABP1 function, they perpetuated a belief about the essential role of ABP1 in plant development. The first cracks in this prominence appeared in 2014 at the "Auxins and Cytokinins in Plant Development" meeting (ACPD Prague 2014; acpd.cas.cz/?page_id=42), where the failure to complement the *abp1* embryo lethal phenotype was reported. The following year, all hell broke loose for ABP1. The failed complementation was published (Grones et al. 2015), the alleged embryo lethality was explained by the mutation in a neighboring gene (Dai et al. 2015; Michalko et al. 2015), and verified *abp1* knockout alleles without any obvious growth defects were reported (Gao et al. 2015). Authors of the latter concluded: "ABP1 is not required for either auxin signaling or *Arabidopsis* development" (Gao et al. 2015). The crucifixion was complete.

Nonetheless, we might have thrown out the baby with the bathwater. There is no doubt that these critical reports are among the most important in the ABP1 field and that the originally believed essential role of ABP1 in development is not correct. In fact, a more comprehensive phenotype analysis of *abp1* mutants confirmed these notions and revealed only minor growth defects (Gelová et al. 2021). On the other hand, at least one or a few ABP1 sequences can be found in almost all available genomes of streptophytes, arguing that the ABP1 function provides a selective advantage. More importantly, independent gain-of-function lines show a wide range of cellular and development defects (Gelová et al. 2021). Some, such as in Brefeldin A-sensitive endocytic PIN trafficking and its auxin regulation (Narasimhan et al. 2021), are exactly opposite to those observed in the lines that down-regulate the ABP1 function using antisense or monoclonal anti-ABP1 antibody technologies (Tromas et al. 2009). Notably, *abp1* versions with a mutation in the auxin-binding site do not have this effect (Robert et al. 2010; Čovanová et al. 2013; Grones et al. 2015; Gelová et al. 2021). These internally consistent observations cannot be easily dismissed and support the role of ABP1 in mediating the auxin effect on endocytic trafficking (Robert et al. 2010): a feedback regulation important for auxin canalization.

It is possible that ABP1 acts redundantly with other members of the large cupin family that are then also targeted by the knockdown lines; this would be consistent with all observations. Alternatively, the phenotypes of knockdown lines are an artefact of the expression system used and the opposite phenotypes in gain-of-function are a coincidence. Only future meticulous evaluation of (1) binding of natural auxin IAA to *Arabidopsis* ABP1, and (2) test of physiological importance of this binding, along with (3) genetic analysis with verified *abp1* knockouts, will decide about the sequel of this intriguing story.

STATION 9: ALMIGHTY IN DEVELOPMENT—AUXIN AS A VERSATILE TRIGGER

Already early pharmacological experiments with synthetic auxins and inhibitors of auxin transport suggested a broad developmental importance of auxin (Liu et al. 1993). However, its rise to unrivaled prominence as a versatile developmental signal came with the establishment of a molecular link of many patterning mutants with transcriptional auxin signaling or auxin transport. The next crucial insight came with the detection of asymmetric auxin activity using DR5 reporters of transcriptional auxin signaling (Ulmasov et al. 1997), so-called auxin gradients and local auxin maxima in many developmental contexts (Benková et al. 2003; Friml et al. 2003; Heisler et al. 2005; Scarpella et al. 2006), such as embryo (for review, see Verma et al. 2021), root (for review, see Cavallari et al. 2021), shoot (for review, see Pernisová and Vernoux 2021), flower (for review, see Cucinotta et al. 2021), leaf venation (for review, see Lavania et al. 2021), and others. These reports typically include the visualization of both localized auxin response and PIN-dependent PAT in conjunction with pharmacological and genetic manipulations of PAT and downstream auxin signaling. Whereas PIN1 and PIN2 act largely nonredundantly when mediating organogenesis at the shoot apical meristem and root gravitropic response, respectively, they act redundantly in a mutually compensatory manner in most other developmental processes (Vieten et al. 2005). When PINs were shown to act as auxin exporters (Petrášek et al. 2006) and their subcellular polar localization to determine the directionality of auxin flow between cells (Friml et al. 2004; Wisniewska et al. 2006) and TIR1/AFB-Aux/IAA-ARF transcriptional auxin response was established (Dharmasiri et al. 2005; Kepinski and Leyser 2005), all necessary components were in place.

With auxin and its transport being implicated in an increasing number of developmental processes and the demonstration that local auxin maxima are sufficient to induce shoot and root organogenesis (Reinhardt et al. 2000; Dubrovsky et al. 2008), the general concept gradually crystalized into a scheme of how auxin coordinates development in vascular plants. Various PIN proteins are expressed in different parts of the plant and mediate directional auxin fluxes through their polar localization, which converge to generate auxin gradients or local auxin maxima (Fig. 2). Auxin accumulating in these cells is interpreted by transcriptional auxin signaling. Notably, different cell types in different developmental contexts remodel their transcriptomes in response to auxin differently; this variety is presumably enabled by different combinations of TIR1/AFB receptors and downstream Aux/IAA and

ARF transcriptional regulators in target cells (Vernoux et al. 2011; Calderón Villalobos et al. 2012; Bargmann et al. 2013; for review, see Ma et al. 2021). Auxin, which is distributed via a PIN-dependent network, thus acts as a versatile facilitator or trigger of predefined change in the developmental program (formulated in Benková et al. 2009; Vanneste and Friml 2009). Often in conjunction with the indispensable localized auxin biosynthesis, as shown for maternal control of embryogenesis (Robert et al. 2018), apical-basal axis establishment (Robert et al. 2013), or shoot apical meristem patterning (Brumos et al. 2018), it is nevertheless its unique directional transport system that empowers auxin to signal between different plant parts and to act as a major coordinative signal in plant development.

STATION 10: THY WILL BE DONE—INTEGRATION THROUGH A DYNAMIC AUXIN TRANSPORT NETWORK

The auxin transport network requires multiple critical molecular components (for review, see Hammes et al. 2021). Among these, polarly localized PIN auxin exporters are not only mediating flow directionality, but on account of their dynamic ability to change their localization, they also redirect auxin fluxes in response to various external and internal signals. Such dynamic repolarization is enabled by a constitutive endocytic recycling of PIN proteins (Dhonukshe et al. 2007), which allows PIN subcellular relocation after each cycle of internalization by a process called transcytosis (Kleine-Vehn et al. 2008). PIN phosphorylation, which occurs at different residues by different kinases (for review, see Barbosa et al. 2018; Tan et al. 2021), is a prominent mechanism controlling PIN polarity and activity. PIN phosphorylation is not only crucial for maintaining PINs at the specific polar domains (Glanc et al. 2021) but also for their dynamic repolarization (for review, see Rakusová et al. 2015).

One of the best-characterized examples of this concept is gravitropism, which aligns the growth of plants with a gravity vector. The gravity is perceived in seed plants by sedimentation of starch-containing amyloplasts (statoliths) in root columella or shoot starch sheath (for review, see Su et al. 2020). As shown in *Arabidopsis*, several PIN proteins are expressed in the very same cells and polarize to the bottom cell sides after gravistimulation (Friml et al. 2002b; Kleine-Vehn et al. 2010; Rakusová et al. 2011). This redirects auxin fluxes accordingly toward the lower side of the organ, where auxin promotes growth (in the shoot) or inhibits growth (in the root) (for review, see Roychoudhry and Kepinski 2021) causing the organ to bend. The mechanism, by which statolith sedimentation leads to PIN relocation, remains unclear but likely involves NEGATIVE GRAVITROPIC RESPONSE OF ROOT (NGR)/LAZY1-LIKE (LZY) proteins (Ge and Chen 2016; Furutani et al. 2020) and PIN protein phosphorylation (Grones et al. 2018). Also, during the shoot phototropic response, the same PINs gradually polarize to the cell sides away from the light (Ding et al. 2011) providing a possible mechanism for generating an auxin asymmetry during phototropic bending. Nonetheless, the causality and underlying mechanism are less clear here, but PIN phosphorylation is involved again (Ding et al. 2011; Haga et al. 2018).

Additional external signals such as nutrient availability can influence auxin biosynthesis or signaling (for review, see Casal and Estevez 2021). Nonetheless, they often have an impact on development through PIN-dependent auxin transport (for review, see Leftley et al. 2021). This is exemplified by nitrate-dependent PIN2 dephosphorylation and polarity rearrangements that modulate root growth during the forage for nitrogen resources (Ötvös et al. 2021) or phosphorus deficiency inhibiting PIN2 degradation and thus promoting root hair development (Lin et al. 2020).

Plant pathogen studies focused mainly on auxin biosynthesis and signaling (for review, see Kunkel and Johnson 2021), but it is conceivable that modification of PIN-dependent auxin transport will also be involved. Thus, the dynamic auxin distribution network serves as a unique, plant-specific mechanism to integrate different signals from the environment and translate these into auxin gradients and auxin maxima, which modify plant development.

STATION 11: FIRST AMONG EQUALS—AUXIN CROSS TALK OVER A TRANSPORT NETWORK

Complex interactions between different pathways are a hallmark of signaling in plants. This occurs on multiple levels: synthesis, transport, response, or sharing signaling components (for review, see Mazzoni-Putman et al. 2021). For auxin, a prominent way of integrating other signaling pathways is again through a PIN-dependent auxin distribution network, either by regulating PIN transcription or dominantly by targeting different aspects of PIN subcellular trafficking and polar distribution (Fig. 3; for review, see Semeradova et al. 2020).

Cytokinins (CKs) are hormones whose action in controlling cell division and differentiation has been linked to auxin. In root gravitropism (Pernisova et al. 2016) and lateral root formation (Marhavý et al. 2011), cytokinins were shown to regulate PIN degradation and polarity via canonical CYTOKININ RESPONSE1 (CRE1)/ARABIDOPSIS HISTIDINE KINASE4 (AHK4) receptors and an unclear downstream, nontranscriptional mechanism (Marhavý et al. 2011). Similarly, gibberellic acid (GA), using the canonical DELLA pathway, has a nontranscriptional inhibitory effect on PIN vacuolar delivery. Here, it targets the retromer and cytoskeleton aspects of this process (Salanenka et al. 2018). This mechanism plays a role in root gravitropism, in which an asymmetric accumulation of gibberellin at the lower root side presumably stabilizes PIN2 there and promotes the auxin flow along the lower side of the root (Löfke et al. 2013). Salicylic acid (SA) also interferes with Brefeldin A-sensitive endocytic trafficking of PINs (Du et al. 2013). In this case, the canonical signaling components are not involved but salicylic acid binds to PROTEIN PHOSPHATASE 2A (PP2A), inhibiting its activity and thereby enhancing PIN2 phosphorylation leading to its reduced polar membrane localization (Tan et al. 2020a). Structurally related aspirin and other anti-inflammatory painkillers target TWISTED DWARF1-regulated actin dynamics and auxin transport (Tan et al. 2020b). PIN trafficking is also targeted by abscisic acid (ABA) through PP2A, in this case via PP2A interaction with its canonical PYRABACTIN RESISTANCE (PYR)/PYR-LIKE (PYL) receptors (Li et al. 2020). Jasmonate (JA) (Sun et al. 2011) and brassinosteroids (BR) (Retzer et al. 2019) also influence PIN trafficking and degradation using their canonical signaling components, presumably through a nontranscriptional mechanism that has yet to be identified. In addition, strigolactones (SLs) cooperate closely with auxin in regulating shoot branching and the formation of vascular tissue and its regeneration, all of which involve the formation of PIN-expressing, auxin-transporting channels (Sauer et al. 2006; Balla et al. 2011). Indeed, strigolactones promote PIN depletion from the PM in shoots (Shinohara et al. 2013), which presumably occurs by uncoupling auxin feedback during PIN-mediated canalization. This strigolactone effect on auxin feedback provides a cellular mechanism by which strigolactones execute their role in a multitude of auxin-mediated processes (Zhang et al. 2020).

It is remarkable and certainly biased by choice of experiments that almost any tested plant hormone has some impact on the PIN-dependent auxin transport network. In many cases, it remains to be clarified how direct and specific such effects are; but available partial or full elucidation of the underlying molecular mechanisms shows that these observations provide an excellent starting point to unravel nontranscriptional mechanisms and branches of many plant hormonal signaling pathways.

STATION 12: SHOW ME THE WAY—POLARIZATION AND CANALIZATION

The prominent regulation of PIN-dependent auxin transport is carried out by auxin itself. Such feedback is a main prerequisite of the so-called auxin canalization, a hypothesis originally proposed for the flexible and self-organizing formation of the vasculature (for review, see Sachs 2000). Auxin has the unique property to promote the formation of its own polarized transporting channels from the auxin source through initially homogeneous tissues, which then guide vasculature formation (for review, see Bennett et al. 2014; Ravichandran et al. 2020; Lavania et al. 2021). Indeed, a gradual, coordinated establishment of polarized, PIN-expressing channels prior to vasculature formation has been observed (Fig. 4): (1) connecting new organs at the shoot apical meristem (Benková et al. 2003) and lateral buds released from dormancy (Balla et al. 2011) with the preexisting vasculature network; (2) during vasculature regeneration following wounding (Sauer et al. 2006; Mazur et al. 2016); and (3) in leaf venation (Scarpella et al. 2006), where an additional patterning mechanism operates (Verna et al. 2019). Some processes, such as apical-basal embryo patterning (Robert et al. 2013; Wabnik et al. 2013) or termination of tropic shoot bending (Rakusová et al. 2016), do not involve any formation of channels but still involve feedback regulation of PIN polarity by auxin.

The experimental system that enables local auxin application on the *Arabidopsis* stem confirmed that auxin is the main signal sufficient to induce channel formation (Mazur et al. 2020b), although other short-range signals are likely to be involved. It is conceptually intriguing how an initially homogenous mass of cells can collectively coordinate and propagate polarities throughout the tissue, with each individual cell positioning PINs on the correct cell side away from the auxin source and toward the sink. Although it is clear that it must somehow involve the integration of information from immediate neighbors as well as global tissue context, the underlying mechanisms remain mysterious.

Based on modeling, different mechanisms have been proposed for how information that guides PIN polarity can be transmitted between cells. The first mechanism involves an auxin effect on growth and transmission of mechanical stresses via the cell wall and microtubule orientation influencing PIN polarity (Heisler et al. 2010). Another proposes a combination of intracellular and extracellular auxin perception, which, respectively, influences PIN transcription or PIN endocytic trafficking that leads to a different PIN stabilization on opposite cell sides (Wabnik et al. 2010). Basic prerequisites of these models are fulfilled: (1) PINs are clustered at the polar domains and linked to the cell wall (Boutté et al. 2006; Feraru et al. 2011; Li et al. 2021a); and (2) auxin regulates PIN expression through TIR1/AFB signaling (Vieten et al. 2005) and modulates, by a different signaling mechanism, PIN trafficking and incidence at the PM (Paciorek et al. 2005; Robert et al. 2010; Han et al. 2021). Notably, TIR1/AFB signaling and endocytic PIN trafficking are indeed required for the formation of auxin channels from localized auxin sources (Mazur et al. 2020a).

Nonetheless, molecular insights into canalization remain limited. TIR1/AFBs signaling regulates, via the transcription factor WRKY23 (Prát et al. 2018), the expression of CANALIZATION-RELATED AUXIN-REGULATED MALECTIN-TYPE RECEPTOR-LIKE KINASE (CAMEL). CAMEL forms a complex with the CANALIZATION-RELATED RECEPTOR-LIKE KINASE (CANAR) to phosphorylate PINs and thus guide their polarity during canalization (Hajný et al. 2020). CAMEL and CANAR as receptor-like kinases at the PM represent potential receptors for the hypothetical short-range canalization signal but its identity remains elusive.

In addition to the identity of CAMEL/CANAR ligand, another mystery is how auxin enters the canalization mechanism. It is unlikely that this happens only indirectly through regulation of CAMEL expression, but whether this occurs through regulation of growth and tensions or actually through extracellular auxin perception remains an open question. An enticing possibility is the involvement of extracellularly localized ABP1, which would be in line with its proposed role in PIN endocytic trafficking (Robert et al. 2010; Grones et al. 2015).

With the identification of the first genetic components of canalization, an important step has been taken to move canalization from a more theoretical and conceptual exploration of cellular mechanisms toward tangible studies of a molecular mechanism of this fascinating and developmentally crucial process.

STATION 13: MIRACULOUS HEALINGS—AUXIN IN REGENERATION

Plants are famous for their ability to regenerate tissues, organs, or even the whole organism from a single cell. Early studies on wounded tissue regeneration, which relied on external auxin applications, implicated a role of auxin (for review, see Bloch 1941). Later it was shown that auxin biosynthesis and signaling are specifically activated in wounded tissues (e.g., in leaf explants to mediate de novo root formation) (Liu et al. 2014; Chen et al. 2016; Xu et al. 2017). These studies suggest that auxin, while not necessarily being a part of the wounding signal, is a key player in the required reprogramming of the cell fates necessary for the regeneration of lost tissues.

Also in roots, fully or partly removed meristems can be restored completely (Sena et al. 2009). Here, auxin influences stem-cell-like divisions and the subsequent acquisition of correct cell fates, which is crucial for the reestablishment of the stem cell niche after it has been removed microsurgically (Xu et al. 2006), died from cold stress (Hong et al. 2017), or by the complete removal of the root tip (Efroni et al. 2016). Insights into the cellular processes associated with wound healing have been gained through the novel combination of targeted ablation of individual root cells with a UV laser followed by an automatic tracking (for review, see Hoermayer and Friml 2019; Marhava et al. 2019).

Much of the downstream processes activated by auxin in the context of regeneration remain unclear, but single-cell transcriptomics and similar approaches will certainly bring the much-needed insights. Auxin contributes to the expression of wound-responsive genes like *ETHYLENE RESPONSE FACTOR115* (*ERF115*; Zhou et al. 2019; Canher et al. 2020; Hoermayer et al. 2020). In turn, *ERF115* expression sensitizes cells to auxin signaling (Canher et al. 2020) and increases auxin biosynthesis in cells (Matosevich et al. 2020). Hence, roots under reduced auxin levels display a decreased regeneration capacity. Auxin also influences the regeneration capacities in the root through its effect on cell expansion. Auxin, inhibiting or promoting growth, depending on the context, affects the regeneration efficiency, which leads to a lack of regeneration or overproliferation (Hoermayer et al. 2020).

Therefore, auxin, which executes its classical role in regulating cell expansion, division, and cell fate reprogramming, plays key roles in practically all stages of plant tissue and organ regeneration.

STATION 14: GENESIS—SEARCHING FOR AUXIN BEGINNINGS

We conclude our way of auxin with the search for its origins. How did it all start and how did auxin evolve to become the universal coordinative signal of vascular plants? These questions can now be approached with insights into the genomes of bryophytes (= a group of land plants, sister to vascular plants) and charophytes (= green algae that are sister to land plants). By studying the properties of auxin response and molecular and genetic properties across charophytes, bryophytes, and vascular plants, we can now begin to deduce the properties of the response system at different times in the evolution of land plants. While molecular and genetic resources are readily available in *Arabidopsis* and bryophytes, they are still limited in charophytes.

Available evidence from bryophytes suggests that the first land plants had the key components of auxin biology in place: the IPyA biosynthetic pathway, the PAT components, and TIR1/AFB-Aux/IAA-ARF transcriptional auxin response (for review, see Bowman et al. 2021; Suzuki et al. 2021). The key role of auxin at this stage is best demonstrated by a complete loss of differentiation in *M. polymorpha* mutants lacking IAA biosynthesis (Eklund et al. 2015). An interesting recent hypothesis proposes that IAA first served as an exogenous indicator of the presence of nutrient-rich decomposing organic matter and thereby stimulated the production of rhizoids in bryophytes (for review, see Sheldrake 2021). As cell death was coopted in bryophytes and vascular plants as part of development, IAA may have taken on a new role as an endogenous hormone. During land plant evolution, a simple, primordial auxin system has expanded, allowing functional diversification within the involved gene families. Examples of this are the stepwise functional innovations within the PIN family, which occurred at three distinct evolutionary milestones, the origin of (1) land plants, (2) vascular plants, and (3) flowering plants. Rapid root gravitropic response, which involved the coevolution of cell polarity, and trafficking machineries, root anatomical features, as well as auxin cellular responses are other examples of complex and recent evolutionary innovations in seed plants (Zhang et al. 2019).

The true mystery lies, nevertheless, in watery ages before plants colonized the land. Given the simplicity of IAA biosynthesis and the fact that microbes can also synthesize auxin and use it as a signal—a feat extensively exploited in plant–microbe interactions (for review, see Kunkel and Johnson 2021)—it is perhaps not unexpected that algae would make a similar choice of signaling molecule. Charophytes such as *Chara* do have a defined response to auxin (Klämbt et al. 1992), but what the auxin function was before plants acquired morphological complexity—a classic auxin playground in the land plants—is unclear. Auxin (IAA) and related metabolites can be detected in all tested algae species (Žižková et al. 2017), which is, however, no proof of its role as a signaling molecule. There is no indication that a full TIR1/AFB-Aux/IAA-ARF signaling module is present or that there is any meaningful transcriptional auxin response (Mutte et al. 2018; Nishiyama et al. 2018). Nonetheless, wherever tested, charophyte algae have PIN components of PAT (for review, see Vosolsobě et al. 2020), which most likely act as specific auxin exporters, as demonstrated for the PIN from the filamentous alga *Klebsormidium* (Skokan et al. 2019).

Why this morphologically simple algae would transport auxin to their surroundings remains entirely unclear. As a quorum-sensing signal? As a growth inhibitor that offers an advantage over the competitors that are not equipped with auxin export machinery? Or does it simply want to get rid of the unwanted and potentially toxic metabolic product? Answering these questions will be an important step on our way to understand the beginnings of the spectacular rise of auxin to become the major developmental signal of land plants.

ACKNOWLEDGMENTS

The author thanks the whole community of researchers consciously or unconsciously working on questions related to auxin, whose hard work and enthusiasm contributed to development of this exciting story. Particular thanks go to many brilliant present and past members of the Friml group and our numerous excellent collaborators, without whom my own personal journey would not be possible. The way of the cross with its 14 stations is a popular devotion among Roman Catholics and inspires them to make a spiritual pilgrimage through contemplation of Christ on his last day. Its aspects of gradual progress, struggle, passion, and revelation served as an inspiration for the formal depiction of our journey to understanding auxin as described in this review. It is in no way intended to reflect the personal beliefs of the author and readers. I am grateful to Nick Barton, Eva Benková, Lenka Caisová, Matyáš Fendrych, Lukáš Fiedler, Monika Frátriková, Jarmila Frimlová, Michelle Gallei, Jakub Hajný, Lukas Hoermayer, Alexandra Mally, Ondřej Novák, Jan Petrášek, Aleš Pěnčík, Steffen Vanneste, Tongda Xu, and Zhenbiao Yang for their valuable comments. Special thanks go to Michelle Gallei for her invaluable assistance with the figures.

REFERENCES

Reference is also in this collection.

Abas L, Kolb M, Stadlmann J, Janacek DP, Lukic K, Schwechheimer C, Sazanov LA, Mach L, Friml J, Hammes UZ. 2021. Naphthylphthalamic acid associates with and inhibits PIN auxin transporters. *Proc Natl Acad Sci* **118:** e2020857118. doi:10.1073/pnas.2020857118

Abdollahi Sisi N, Růžička K. 2020. ER-localized PIN carriers: regulators of intracellular auxin homeostasis. *Plants (Basel)* **9:** 1527. doi:10.3390/plants9111527

Abel S, Theologis A. 2010. Odyssey of auxin. *Cold Spring Harb Perspect Biol* **2:** a004572. doi:10.1101/cshperspect.a004572

Adamowski M, Friml J. 2015. PIN-dependent auxin transport: action, regulation, and evolution. *Plant Cell* **27:** 20–32. doi:10.1105/tpc.114.134874

Antoniadi I, Novák O, Gelová Z, Johnson A, Plíhal O, Simerský R, Mik V, Vain T, Mateo-Bonmatí E, Karady M, et al. 2020. Cell-surface receptors enable perception of extracellular cytokinins. *Nat Commun* **11:** 4284. doi:10.1038/s41467-020-17700-9

Arieti RS, Staiger CJ. 2020. Auxin-induced actin cytoskeleton rearrangements require AUX1. *New Phytol* **226:** 441–459. doi:10.1111/nph.16382

Arsuffi G, Braybrook SA. 2018. Acid growth: an ongoing trip. *J Exp Bot* **69:** 137–146. doi:10.1093/jxb/erx390

Balla J, Kalousek P, Reinöhl V, Friml J, Procházka S. 2011. Competitive canalization of PIN-dependent auxin flow from axillary buds controls pea bud outgrowth. *Plant J* **65:** 571–577. doi:10.1111/j.1365-313X.2010.04443.x

Baral A, Aryal B, Jonsson K, Morris E, Demes E, Takatani S, Verger S, Xu T, Bennett M, Hamant O, et al. 2021. External mechanical cues reveal a katanin-independent mechanism behind auxin-mediated tissue bending in plants. *Dev Cell* **56:** 67–80.e3. doi:10.1016/j.devcel.2020.12.008

Barbez E, Kubeš M, Rolčík J, Béziat C, Pěnčík A, Wang B, Rosquete MR, Zhu J, Dobrev PI, Lee Y, et al. 2012. A novel putative auxin carrier family regulates intracellular auxin homeostasis in plants. *Nature* **485:** 119–122. doi:10.1038/nature11001

Barbosa ICR, Hammes UZ, Schwechheimer C. 2018. Activation and polarity control of PIN-FORMED auxin transporters by phosphorylation. *Trends Plant Sci* **23:** 523–538. doi:10.1016/j.tplants.2018.03.009

Bargmann BO, Vanneste S, Krouk G, Nawy T, Efroni I, Shani E, Choe G, Friml J, Bergmann DC, Estelle M, et al. 2013. A map of cell type-specific auxin responses. *Mol Syst Biol* **9:** 688. doi:10.1038/msb.2013.40

Baster P, Robert S, Kleine-Vehn J, Vanneste S, Kania U, Grunewald W, De Rybel B, Beeckman T, Friml J. 2013. $SCF^{TIR1/AFB}$-auxin signalling regulates PIN vacuolar trafficking and auxin fluxes during root gravitropism. *EMBO J* **32:** 260–274. doi:10.1038/emboj.2012.310

Benková E, Michniewicz M, Sauer M, Teichmann T, Seifertová D, Jürgens G, Friml J. 2003. Local, efflux-dependent auxin gradients as a common module for plant organ formation. *Cell* **115:** 591–602. doi:10.1016/S0092-8674(03)00924-3

Benková E, Ivanchenko MG, Friml J, Shishkova S, Dubrovsky JG. 2009. A morphogenetic trigger: is there an emerging concept in plant developmental biology? *Trends Plant Sci* **14:** 189–193. doi:10.1016/j.tplants.2009.01.006

Bennett T, Hines G, Leyser O. 2014. Canalization: what the flux? *Trends Genet* **30:** 41–48. doi:10.1016/j.tig.2013.11.001

Bisson MM, Bleckmann A, Allekotte S, Groth G. 2009. EIN2, the central regulator of ethylene signalling, is localized at the ER membrane where it interacts with the ethylene receptor ETR1. *Biochem J* **424:** 1–6. doi:10.1042/BJ20091102

Bloch R. 1941. Wound healing in higher plants. *Botanical Review* **7:** 110–146. doi:10.1007/BF02872446

Boutté Y, Crosnier MT, Carraro N, Traas J, Satiat-Jeunemaitre B. 2006. The plasma membrane recycling pathway and cell polarity in plants: studies on PIN proteins. *J Cell Sci* **119:** 1255–1265. doi:10.1242/jcs.02847

* Bowman JL, Flores Sandoval E, Kato H. 2021. On the evolutionary origins of land plant auxin biology. *Cold Spring Harb Perspect Biol* **13:** a040048. doi:10.1101/cshperspect.a040048

Brumos J, Robles LM, Yun J, Vu TC, Jackson S, Alonso JM, Stepanova AN. 2018. Local auxin biosynthesis is a key regulator of plant development. *Dev Cell* **47:** 306–318.e5. doi:10.1016/j.devcel.2018.09.022

Calderón Villalobos LI, Lee S, De Oliveira C, Ivetac A, Brandt W, Armitage L, Sheard LB, Tan X, Parry G, Mao H, et al. 2012. A combinatorial TIR1/AFB-Aux/IAA co-receptor system for differential sensing of auxin. *Nat Chem Biol* **8:** 477–485. doi:10.1038/nchembio.926

Canher B, Heyman J, Savina M, Devendran A, Eekhout T, Vercauteren I, Prinsen E, Matosevich R, Xu J, Mironova V, et al. 2020. Rocks in the auxin stream: wound-induced auxin accumulation and *ERF115* expression synergistically drive stem cell regeneration. *Proc Natl Acad Sci* **117:** 16667–16677. doi:10.1073/pnas.2006620117

Cao M, Chen R, Li P, Yu Y, Zheng R, Ge D, Zheng W, Wang X, Gu Y, Gelová Z, et al. 2019. TMK1-mediated auxin signalling regulates differential growth of the apical hook. *Nature* **568:** 240–243. doi:10.1038/s41586-019-1069-7

* Casal JJ, Estevez JM. 2021. Auxin-environment integration in growth responses to forage for resources. *Cold Spring Harb Perspect Biol* **13:** a040030. doi:10.1101/cshperspect.a040030

* Casanova-Sáez R, Mateo-Bonmatí E, Ljung K. 2021. Auxin metabolism in plants. *Cold Spring Harb Perspect Biol* **13:** a039867. doi:10.1101/cshperspect.a039867

* Cavallari N, Artner C, Benkova E. 2021. Auxin-regulated lateral root organogenesis. *Cold Spring Harb Perspect Biol* doi:10.1101/cshperspect.a039941

Chen JG, Ullah H, Young JC, Sussman MR, Jones AM. 2001. ABP1 is required for organized cell elongation and division in *Arabidopsis* embryogenesis. *Genes Dev* **15:** 902–911. doi:10.1101/gad.866201

Chen L, Tong J, Xiao L, Ruan Y, Liu J, Zeng M, Huang H, Wang JW, Xu L. 2016. *YUCCA*-mediated auxin biogenesis is required for cell fate transition occurring during *de novo* root organogenesis in *Arabidopsis*. *J Exp Bot* **67:** 4273–4284. doi:10.1093/jxb/erw213

* Cieslak M, Owens A, Prusinkiewicz P. 2021. Computational models of auxin-driven patterning in shoots. *Cold Spring Harb Perspect Biol* doi:10.1101/cshperspect.a040097

Cook SD. 2019. An historical review of phenylacetic acid. *Plant Cell Physiol* **60:** 243–254. doi:10.1093/pcp/pcz004

Čovanová M, Sauer M, Rychtář J, Friml J, Petrášek J, Zažímalová E. 2013. Overexpression of the auxin binding protein1 modulates PIN-dependent auxin transport in tobacco cells. *PLoS ONE* **8:** e70050. doi:10.1371/journal.pone.0070050

* Cucinotta M, Cavalleri A, Chandler JW, Colombo L. 2021. Auxin and flower development: a blossoming field. *Cold Spring Harb Perspect Biol* **13:** a039974. doi:10.1101/cshperspect.a039974

Dai X, Zhang Y, Zhang D, Chen J, Gao X, Estelle M, Zhao Y. 2015. Embryonic lethality of *Arabidopsis* abp1-1 is caused by deletion of the adjacent BSM gene. *Nat Plants* **1:** 15183. doi:10.1038/nplants.2015.183

De Vriese K, Himschoot E, Dünser K, Nguyen L, Drozdzecki A, Costa A, Nowack MK, Kleine-Vehn J, Audenaert D, Beeckman T, et al. 2019. Identification of novel inhibitors of auxin-induced Ca^{2+} signaling via a plant-based chemical screen. *Plant Physiol* **180:** 480–496. doi:10.1104/pp.18.01393

Dharmasiri N, Dharmasiri S, Estelle M. 2005. The F-box protein TIR1 is an auxin receptor. *Nature* **435:** 441–445. doi:10.1038/nature03543

Dhonukshe P, Aniento F, Hwang I, Robinson DG, Mravec J, Stierhof YD, Friml J. 2007. Clathrin-mediated constitutive endocytosis of PIN auxin efflux carriers in *Arabidopsis*. *Curr Biol* **17:** 520–527. doi:10.1016/j.cub.2007.01.052

Dhonukshe P, Grigoriev I, Fischer R, Tominaga M, Robinson DG, Hasek J, Paciorek T, Petrásek J, Seifertová D, Tejos R, et al. 2008. Auxin transport inhibitors impair vesicle motility and actin cytoskeleton dynamics in diverse eukaryotes. *Proc Natl Acad Sci* **105:** 4489–4494. doi:10.1073/pnas.0711414105

Dindas J, Scherzer S, Roelfsema MRG, von Meyer K, Muller HM, Al-Rasheid KAS, Palme K, Dietrich P, Becker D, Bennett MJ, et al. 2018. AUX1-mediated root hair auxin influx governs $SCF^{TIR1/AFB}$-type Ca^{2+} signaling. *Nat Commun* **9:** 1174. doi:10.1038/s41467-018-03582-5

Ding Z, Friml J. 2010. Auxin regulates distal stem cell differentiation in *Arabidopsis* roots. *Proc Natl Acad Sci* **107:** 12046–12051. doi:10.1073/pnas.1000672107

Ding Z, Galván-Ampudia CS, Demarsy E, Łangowski Ł, Kleine-Vehn J, Fan Y, Morita MT, Tasaka M, Fankhauser C, Offringa R, et al. 2011. Light-mediated polarization of the PIN3 auxin transporter for the phototropic response in *Arabidopsis*. *Nat Cell Biol* **13:** 447–452. doi:10.1038/ncb2208

Du Y, Tejos R, Beck M, Himschoot E, Li H, Robatzek S, Vanneste S, Friml J. 2013. Salicylic acid interferes with clathrin-mediated endocytic protein trafficking. *Proc Natl Acad Sci* **110:** 7946–7951. doi:10.1073/pnas.1220205110

* Dubey SM, Serre NBC, Oulehlová D, Vittal P, Fendrych M. 2021. No time for transcription-rapid auxin responses in plants. *Cold Spring Harb Perspect Biol* doi:10.1101/cshperspect.a039891

Dubrovsky JG, Sauer M, Napsucialy-Mendivil S, Ivanchenko MG, Friml J, Shishkova S, Celenza J, Benková E. 2008. Auxin acts as a local morphogenetic trigger to specify lateral root founder cells. *Proc Natl Acad Sci* **105:** 8790–8794. doi:10.1073/pnas.0712307105

Efroni I, Mello A, Nawy T, Ip PL, Rahni R, DelRose N, Powers A, Satija R, Birnbaum KD. 2016. Root regeneration triggers an embryo-like sequence guided by hormonal interactions. *Cell* **165:** 1721–1733. doi:10.1016/j.cell.2016.04.046

Eklund DM, Ishizaki K, Flores-Sandoval E, Kikuchi S, Takebayashi Y, Tsukamoto S, Hirakawa Y, Nonomura M, Kato H, Kouno M, et al. 2015. Auxin produced by the indole-3-pyruvic acid pathway regulates development and gemmae dormancy in the liverwort *Marchantia polymorpha*. *Plant Cell* **27:** 1650–1669. doi:10.1105/tpc.15.00065

Feiguelman G, Fu Y, Yalovsky S. 2018. ROP GTPases structure-function and signaling pathways. *Plant Physiol* **176:** 57–79. doi:10.1104/pp.17.01415

Fendrych M, Leung J, Friml J. 2016. TIR1/AFB-Aux/IAA auxin perception mediates rapid cell wall acidification and growth of *Arabidopsis* hypocotyls. *eLife* **5:** e19048. doi:10.7554/eLife.19048

Fendrych M, Akhmanova M, Merrin J, Glanc M, Hagihara S, Takahashi K, Uchida N, Torii KU, Friml J. 2018. Rapid and reversible root growth inhibition by TIR1 auxin signalling. *Nat Plants* **4:** 453–459. doi:10.1038/s41477-018-0190-1

Feraru E, Feraru MI, Kleine-Vehn J, Martinière A, Mouille G, Vanneste S, Vernhettes S, Runions J, Friml J. 2011. PIN polarity maintenance by the cell wall in *Arabidopsis*. *Curr Biol* **21:** 338–343. doi:10.1016/j.cub.2011.01.036

Frick EM, Strader LC. 2018. Roles for IBA-derived auxin in plant development. *J Exp Bot* **69:** 169–177. doi:10.1093/jxb/erx298

Friml J, Jones AR. 2010. Endoplasmic reticulum: the rising compartment in auxin biology. *Plant Physiol* **154:** 458–462. doi:10.1104/pp.110.161380

Friml J, Benková E, Blilou I, Wisniewska J, Hamann T, Ljung K, Woody S, Sandberg G, Scheres B, Jürgens G, et al. 2002a. AtPIN4 mediates sink-driven auxin gradients and root patterning in *Arabidopsis*. *Cell* **108:** 661–673. doi:10.1016/S0092-8674(02)00656-6

Friml J, Wiśniewska J, Benková E, Mendgen K, Palme K. 2002b. Lateral relocation of auxin efflux regulator PIN3 mediates tropism in *Arabidopsis*. *Nature* **415:** 806–809. doi:10.1038/415806a

Friml J, Vieten A, Sauer M, Weijers D, Schwarz H, Hamann T, Offringa R, Jürgens G. 2003. Efflux-dependent auxin gradients establish the apical-basal axis of *Arabidopsis*. *Nature* **426:** 147–153. doi:10.1038/nature02085

Friml J, Yang X, Michniewicz M, Weijers D, Quint A, Tietz O, Benjamins R, Ouwerkerk PB, Ljung K, Sandberg G, et al. 2004. A PINOID-dependent binary switch in apical-basal PIN polar targeting directs auxin efflux. *Science* **306:** 862–865. doi:10.1126/science.1100618

Furutani M, Hirano Y, Nishimura T, Nakamura M, Taniguchi M, Suzuki K, Oshida R, Kondo C, Sun S, Kato K, et al. 2020. Polar recruitment of RLD by LAZY1-like protein during gravity signaling in root branch angle control. *Nat Commun* **11:** 76. doi:10.1038/s41467-019-13729-7

Gallei M, Luschnig C, Friml J. 2020. Auxin signalling in growth: Schrödinger's cat out of the bag. *Curr Opin Plant Biol* **53:** 43–49. doi:10.1016/j.pbi.2019.10.003

Gao Y, Zhang Y, Zhang D, Dai X, Estelle M, Zhao Y. 2015. Auxin binding protein 1 (ABP1) is not required for either auxin signaling or *Arabidopsis* development. *Proc Natl Acad Sci* **112:** 2275–2280. doi:10.1073/pnas.1500365112

Ge L, Chen R. 2016. Negative gravitropism in plant roots. *Nat Plants* **2:** 16155. doi:10.1038/nplants.2016.155

Gelová Z, Gallei M, Pernisová M, Brunoud G, Zhang X, Glanc M, Li L, Michalko J, Pavlovičová Z, Verstraeten I, et al. 2021. Developmental roles of auxin binding protein 1 in *Arabidopsis thaliana*. *Plant Sci* **303:** 110750. doi:10.1016/j.plantsci.2020.110750

Glanc M, Van Gelderen K, Hoermayer L, Tan S, Naramoto S, Zhang X, Domjan D, Včelařová L, Hauschild R, Johnson A, et al. 2021. AGC kinases and MAB4/MEL proteins maintain PIN polarity by limiting lateral diffusion in plant cells. *Curr Biol* **31:** 1918–1930.e1915. doi:10.1016/j.cub.2021.02.028

Gray WM, Kepinski S, Rouse D, Leyser O, Estelle M. 2001. Auxin regulates SCFTIR1-dependent degradation of Aux/IAA proteins. *Nature* **414:** 271–276. doi:10.1038/35104500

Grones P, Chen X, Simon S, Kaufmann WA, De Rycke R, Nodzyński T, Zažímalová E, Friml J. 2015. Auxin-binding pocket of ABP1 is crucial for its gain-of-function cellular and developmental roles. *J Exp Bot* **66:** 5055–5065. doi:10.1093/jxb/erv177

Grones P, Abas M, Hajný J, Jones A, Waidmann S, Kleine-Vehn J, Friml J. 2018. PID/WAG-mediated phosphorylation of the *Arabidopsis* PIN3 auxin transporter mediates polarity switches during gravitropism. *Sci Rep* **8:** 10279. doi:10.1038/s41598-018-28188-1

Haga K, Frank L, Kimura T, Schwechheimer C, Sakai T. 2018. Roles of AGCVIII kinases in the hypocotyl phototropism of *Arabidopsis* seedlings. *Plant Cell Physiol* **59:** 1060–1071. doi:10.1093/pcp/pcy048

Hajný J, Prát T, Rydza N, Rodriguez L, Tan S, Verstraeten I, Domjan D, Mazur E, Smakowska-Luzan E, Smet W, et al. 2020. Receptor kinase module targets PIN-dependent auxin transport during canalization. *Science* **370:** 550–557. doi:10.1126/science.aba3178

* Hammes UZ, Murphy AS, Schwechheimer C. 2021. Auxin transporters—a biochemical view. *Cold Spring Harb Perspect Biol* doi:10.1101/cshperspect.a039875

Han H, Verstraeten I, Roosjen M, Mazur E, Rýdza N, Hajný J, Ötvös K, Weijers D, Friml J. 2021. Rapid auxin-mediated phosphorylation of myosin regulates trafficking and polarity in *Arabidopsis.* bioRxiv 2021.2004.2013.439603

* Hayashi KI. 2021. Chemical biology in auxin research. *Cold Spring Harb Perspect Biol* **13:** a040105. doi:10.1101/cshperspect.a040105

Heisler MG, Ohno C, Das P, Sieber P, Reddy GV, Long JA, Meyerowitz EM. 2005. Patterns of auxin transport and gene expression during primordium development revealed by live imaging of the *Arabidopsis* inflorescence meristem. *Curr Biol* **15:** 1899–1911. doi:10.1016/j.cub.2005.09.052

Heisler MG, Hamant O, Krupinski P, Uyttewaal M, Ohno C, Jönsson H, Traas J, Meyerowitz EM. 2010. Alignment between PIN1 polarity and microtubule orientation in the shoot apical meristem reveals a tight coupling between morphogenesis and auxin transport. *PLoS Biol* **8:** e1000516. doi:10.1371/journal.pbio.1000516

Hertel R, Thomson KS, Russo VE. 1972. In-vitro auxin binding to particulate cell fractions from corn coleoptiles. *Planta* **107:** 325–340. doi:10.1007/BF00386394

Hoermayer L, Friml J. 2019. Targeted cell ablation-based insights into wound healing and restorative patterning. *Curr Opin Plant Biol* **52:** 124–130. doi:10.1016/j.pbi.2019.08.006

Hoermayer L, Montesinos JC, Marhava P, Benková E, Yoshida S, Friml J. 2020. Wounding-induced changes in cellular pressure and localized auxin signalling spatially coordinate restorative divisions in roots. *Proc Natl Acad Sci* **117:** 15322–15331. doi:10.1073/pnas.2003346117

Hong JH, Savina M, Du J, Devendran A, Kannivadi Ramakanth K, Tian X, Sim WS, Mironova VV, Xu J. 2017. A sacrifice-for-survival mechanism protects root stem cell niche from chilling stress. *Cell* **170:** 102–113.e14. doi:10.1016/j.cell.2017.06.002

Hori K, Maruyama F, Fujisawa T, Togashi T, Yamamoto N, Seo M, Sato S, Yamada T, Mori H, Tajima N, et al. 2014. *Klebsormidium flaccidum* genome reveals primary factors for plant terrestrial adaptation. *Nat Commun* **5:** 3978. doi:10.1038/ncomms4978

Huang R, Zheng R, He J, Zhou Z, Wang J, Xiong Y, Xu T. 2019. Noncanonical auxin signaling regulates cell division pattern during lateral root development. *Proc Natl Acad Sci* **116:** 21285–21290. doi:10.1073/pnas.1910916116

Kepinski S, Leyser O. 2005. The *Arabidopsis* F-box protein TIR1 is an auxin receptor. *Nature* **435:** 446–451. doi:10.1038/nature03542

Klämbt D, Knauth B, Dittmann I. 1992. Auxin dependent growth of rhizoids of *Chara globularis. Physiol Plant* **85:** 537–540. doi:10.1111/j.1399-3054.1992.tb05823.x

Kleine-Vehn J, Langowski L, Wisniewska J, Dhonukshe P, Brewer PB, Friml J. 2008. Cellular and molecular requirements for polar PIN targeting and transcytosis in plants. *Mol Plant* **1:** 1056–1066. doi:10.1093/mp/ssn062

Kleine-Vehn J, Ding Z, Jones AR, Tasaka M, Morita MT, Friml J. 2010. Gravity-induced PIN transcytosis for polarization of auxin fluxes in gravity-sensing root cells. *Proc Natl Acad Sci* **107:** 22344–22349. doi:10.1073/pnas.1013145107

Kleine-Vehn J, Wabnik K, Martinière A, Łangowski Ł, Willig K, Naramoto S, Leitner J, Tanaka H, Jakobs S, Robert S, et al. 2011. Recycling, clustering, and endocytosis jointly maintain PIN auxin carrier polarity at the plasma membrane. *Mol Syst Biol* **7:** 540. doi:10.1038/msb.2011.72

Kriechbaumer V, Botchway SW, Hawes C. 2016. Localization and interactions between *Arabidopsis* auxin biosynthetic enzymes in the TAA/YUC-dependent pathway. *J Exp Bot* **67:** 4195–4207. doi:10.1093/jxb/erw195

Kubiasová K, Montesinos JC, Šamajová O, Nisler J, Mik V, Semerádová H, Plíhalová L, Novák O, Marhavý P, Cavallari N, et al. 2020. Cytokinin fluoroprobe reveals multiple sites of cytokinin perception at plasma membrane and endoplasmic reticulum. *Nat Commun* **11:** 4285. doi:10.1038/s41467-020-17949-0

Kuhn A, Ramans Harborough S, McLaughlin HM, Natarajan B, Verstraeten I, Friml J, Kepinski S, Østergaard L. 2020. Direct ETTIN-auxin interaction controls chromatin states in gynoecium development. *eLife* **9:** e51787. doi:10.7554/eLife.51787

* Kunkel BN, Johnson JMB. 2021. Auxin plays multiple roles during plant–pathogen interactions. *Cold Spring Harb Perspect Biol* doi:10.1101/cshperspect.a040022

* Lavania D, Linh NM, Scarpella E. 2021. Of cells, strands, and networks: auxin and the patterned formation of the vascular system. *Cold Spring Harb Perspect Biol* **13:** a039958. doi:10.1101/cshperspect.a039958

Lavy M, Estelle M. 2016. Mechanisms of auxin signaling. *Development* **143:** 3226–3229. doi:10.1242/dev.131870

* Leftley N, Banda J, Pandey B, Bennett M, Voß U. 2021. Uncovering how auxin optimizes root systems architecture in response to environmental stresses. *Cold Spring Harb Perspect Biol* doi:10.1101/cshperspect.a040014

Leyser O. 2018. Auxin signaling. *Plant Physiol* **176:** 465–479. doi:10.1104/pp.17.00765

Li Y, Wang Y, Tan S, Li Z, Yuan Z, Glanc M, Domjan D, Wang K, Xuan W, Guo Y, et al. 2020. Root growth adaptation is mediated by PYLs ABA receptor-PP2A protein phosphatase complex. *Adv Sci (Weinh)* **7:** 1901455. doi:10.1002/advs.201901455

Li H, von Wangenheim D, Zhang X, Tan S, Darwish-Miranda N, Naramoto S, Wabnik K, De Rycke R, Kaufmann WA, Gütl D, et al. 2021a. Cellular requirements for PIN polar cargo clustering in *Arabidopsis thaliana. New Phytol* **229:** 351–369. doi:10.1111/nph.16887

Li L, Verstraeten I, Roosjen M, Takahashi K, Rodriguez L, Merrin J, Chen J, Shabala L, Smet W, Ren H, et al. 2021b. Antagonistic cell surface and intracellular auxin signalling regulate plasma membrane H^+-fluxes for root growth. *Research Square* doi:10.21203/rs.3.rs-266395/v1

Lin DL, Yao HY, Jia LH, Tan JF, Xu ZH, Zheng WM, Xue HW. 2020. Phospholipase D-derived phosphatidic acid promotes root hair development under phosphorus deficiency by suppressing vacuolar degradation of PIN-FORMED2. *New Phytol* **226:** 142–155. doi:10.1111/nph.16330

Lin W, Tang W, Takahashi K, Ren H, Pan S, Zheng H, Gray W, Xu T, Kinoshita T, Yang Z. 2021. TMK-based cell surface auxin signaling activates cell wall acidification in *Arabidopsis. Research Square* doi:10.21203/rs.3.rs-203621/v1

Liu C, Xu Z, Chua NH. 1993. Auxin polar transport is essential for the establishment of bilateral symmetry during early plant embryogenesis. *Plant Cell* **5:** 621–630. doi:10.2307/3869805

Liu J, Sheng L, Xu Y, Li J, Yang Z, Huang H, Xu L. 2014. *WOX11* and *12* are involved in the first-step cell fate transition during de novo root organogenesis in *Arabidopsis. Plant Cell* **26:** 1081–1093. doi:10.1105/tpc.114.122887

Löfke C, Zwiewka M, Heilmann I, Van Montagu MC, Teichmann T, Friml J. 2013. Asymmetric gibberellin signaling regulates vacuolar trafficking of PIN auxin transporters during root gravitropism. *Proc Natl Acad Sci* **110:** 3627–3632. doi:10.1073/pnas.1300107110

Luschnig C, Gaxiola RA, Grisafi P, Fink GR. 1998. EIR1, a root-specific protein involved in auxin transport, is required for gravitropism in *Arabidopsis thaliana. Genes Dev* **12:** 2175–2187. doi:10.1101/gad.12.14.2175

Lv B, Yu Q, Liu J, Wen X, Yan Z, Hu K, Li H, Kong X, Li C, Tian H, et al. 2020. Non-canonical AUX/IAA protein IAA33 competes with canonical AUX/IAA repressor IAA5 to negatively regulate auxin signaling. *EMBO J* **39:** e101515.

Ma Q, Grones P, Robert S. 2018. Auxin signaling: a big question to be addressed by small molecules. *J Exp Bot* **69:** 313–328. doi:10.1093/jxb/erx375

* Ma Y, Wolf S, Lohmann JU. 2021. Casting the net-connecting auxin signaling to the plant genome. *Cold Spring Harb Perspect Biol* doi:10.1101/cshperspect.a040006

Marhava P, Hoermayer L, Yoshida S, Marhavý P, Benková E, Friml J. 2019. Re-activation of stem cell pathways for pattern restoration in plant wound healing. *Cell* **177:** 957–969.e13. doi:10.1016/j.cell.2019.04.015

Marhavý P, Bielach A, Abas L, Abuzeineh A, Duclercq J, Tanaka H, Pařezová M, Petrášek J, Friml J, Kleine-Vehn J, et al. 2011. Cytokinin modulates endocytic trafficking of PIN1 auxin efflux carrier to control plant organogenesis. *Dev Cell* **21:** 796–804. doi:10.1016/j.devcel.2011.08.014

Matosevich R, Cohen I, Gil-Yarom N, Modrego A, Friedlander-Shani L, Verna C, Scarpella E, Efroni I. 2020. Local auxin biosynthesis is required for root regeneration after wounding. *Nat Plants* **6:** 1020–1030. doi:10.1038/s41477-020-0737-9

Mazur E, Benková E, Friml J. 2016. Vascular cambium regeneration and vessel formation in wounded inflorescence stems of *Arabidopsis. Sci Rep* **6:** 33754. doi:10.1038/srep33754

Mazur E, Gallei M, Adamowski M, Han H, Robert HS, Friml J. 2020a. Clathrin-mediated trafficking and PIN trafficking are required for auxin canalization and vascular tissue formation in *Arabidopsis. Plant Sci* **293:** 110414. doi:10.1016/j.plantsci.2020.110414

Mazur E, Kulik I, Hajný J, Friml J. 2020b. Auxin canalization and vascular tissue formation by TIR1/AFB-mediated auxin signaling in *Arabidopsis. New Phytol* **226:** 1375–1383. doi:10.1111/nph.16446

* Mazzoni-Putman SM, Brumos J, Zhao C, Alonso JM, Stepanova AN. 2021. Auxin interactions with other hormones in plant development. *Cold Spring Harb Perspect Biol* doi:10.1101/cshperspect.a039990

* McLaughlin HM, Ang ACH, Østergaard L. 2021. Noncanonical auxin signaling. *Cold Spring Harb Perspect Biol* **13:** a039917. doi:10.1101/cshperspect.a039917

Michalko J, Dravecká M, Bollenbach T, Friml J. 2015. Embryo-lethal phenotypes in early abp1 mutants are due to disruption of the neighboring BSM gene. *F1000Res* **4:** 1104. doi:10.12688/f1000research.7143.1

Michniewicz M, Brewer PB, Friml J. 2007. Polar auxin transport and asymmetric auxin distribution. *Arabidopsis Book* **5:** e0108.

Monshausen GB, Miller ND, Murphy AS, Gilroy S. 2011. Dynamics of auxin-dependent Ca^{2+} and pH signaling in root growth revealed by integrating high-resolution imaging with automated computer vision-based analysis. *Plant J* **65:** 309–318. doi:10.1111/j.1365-313X.2010.04423.x

* Morffy N, Strader LC. 2021. Structural aspects of auxin signaling. *Cold Spring Harb Perspect Biol* doi:10.1101/cshperspect.a039883

Mravec J, Skůpa P, Bailly A, Hoyerová K, Krecek P, Bielach A, Petrásek J, Zhang J, Gaykova V, Stierhof YD, et al. 2009. Subcellular homeostasis of phytohormone auxin is mediated by the ER-localized PIN5 transporter. *Nature* **459:** 1136–1140. doi:10.1038/nature08066

Mutte SK, Kato H, Rothfels C, Melkonian M, Wong GK, Weijers D. 2018. Origin and evolution of the nuclear auxin response system. *eLife* **7:** e33399. doi:10.7554/eLife.33399

* Napier R. 2021. The story of auxin-binding protein 1 (ABP1). *Cold Spring Harb Perspect Biol* doi:10.1101/cshperspect.a039909

Narasimhan M, Gallei M, Tan S, Johnson A, Verstraeten I, Li L, Rodriguez L, Han H, Himschoot E, Wang R, et al. 2021. Systematic analysis of specific and nonspecific auxin effects on endocytosis and trafficking. *Plant Physiol* **186:** 1122–1142. doi:10.1093/plphys/kiab134

Nishiyama T, Sakayama H, de Vries J, Buschmann H, Saint-Marcoux D, Ullrich KK, Haas FB, Vanderstraeten L, Becker D, Lang D, et al. 2018. The chara genome: secondary complexity and implications for plant terrestrialization. *Cell* **174:** 448–464.e24. doi:10.1016/j.cell.2018.06.033

Novák O, Hényková E, Sairanen I, Kowalczyk M, Pospíšil T, Ljung K. 2012. Tissue-specific profiling of the *Arabidopsis thaliana* auxin metabolome. *Plant J* **72:** 523–536. doi:10.1111/j.1365-313X.2012.05085.x

Ogura T, Goeschl C, Filiault D, Mirea M, Slovak R, Wolhrab B, Satbhai SB, Busch W. 2019. Root system depth in *Arabidopsis* is shaped by EXOCYST70A3 via the dynamic modulation of auxin transport. *Cell* **178:** 400–412.e16. doi:10.1016/j.cell.2019.06.021

Oochi A, Hajny J, Fukui K, Nakao Y, Gallei M, Quareshy M, Takahashi K, Kinoshita T, Harborough SR, Kepinski S, et al. 2019. Pinstatic acid promotes auxin transport by inhibiting PIN internalization. *Plant Physiol* **180:** 1152–1165. doi:10.1104/pp.19.00201

Orosa-Puente B, Leftley N, von Wangenheim D, Banda J, Srivastava AK, Hill K, Truskina J, Bhosale R, Morris E, Srivastava M, et al. 2018. Root branching toward water involves posttranslational modification of transcription factor ARF7. *Science* **362:** 1407–1410. doi:10.1126/science.aau3956

Ötvös K, Marconi M, Vega A, O'Brien J, Johnson A, Abualia R, Antonielli L, Montesinos JC, Zhang Y, Tan S, et al. 2021. Modulation of plant root growth by nitrogen source-defined regulation of polar auxin transport. *EMBO J* **40:** e106862. doi:10.15252/embj.2020106862

Paciorek T, Zazímalová E, Ruthardt N, Petrásek J, Stierhof YD, Kleine-Vehn J, Morris DA, Emans N, Jürgens G, Geldner N, et al. 2005. Auxin inhibits endocytosis and promotes its own efflux from cells. *Nature* **435:** 1251–1256. doi:10.1038/nature03633

Pan X, Fang L, Liu J, Senay-Aras B, Lin W, Zheng S, Zhang T, Guo J, Manor U, Van Norman J, et al. 2020. Auxin-induced signaling protein nanoclustering contributes to cell polarity formation. *Nat Commun* **11:** 3914. doi:10.1038/s41467-020-17602-w

Paque S, Weijers D. 2016. Q&A: auxin: the plant molecule that influences almost anything. *BMC Biol* **14:** 67. doi:10.1186/s12915-016-0291-0

Paterlini A. 2020. Uncharted routes: exploring the relevance of auxin movement via plasmodesmata. *Biol Open* **9:** bio055541. doi:10.1242/bio.055541

Pěnčík A, Casanova-Sáez R, Pilarová V, Žukauskaite A, Pinto R, Micol JL, Ljung K, Novák O. 2018. Ultra-rapid auxin metabolite profiling for high-throughput mutant screening in *Arabidopsis*. *J Exp Bot* **69:** 2569–2579. doi:10.1093/jxb/ery084

* Pernisová M, Vernoux T. 2021. Auxin does the SAMba: auxin signaling in the shoot apical meristem. *Cold Spring Harb Perspect Biol* doi:10.1101/cshperspect.a039925

Pernisova M, Prat T, Grones P, Harustiakova D, Matonohova M, Spichal L, Nodzynski T, Friml J, Hejatko J. 2016. Cytokinins influence root gravitropism via differential regulation of auxin transporter expression and localization in *Arabidopsis*. *New Phytol* **212:** 497–509. doi:10.1111/nph.14049

Petrášek J, Mravec J, Bouchard R, Blakeslee JJ, Abas M, Seifertová D, Wisniewska J, Tadele Z, Kubes M, Covanová M, et al. 2006. PIN proteins perform a rate-limiting function in cellular auxin efflux. *Science* **312:** 914–918. doi:10.1126/science.1123542

Platre MP, Bayle V, Armengot L, Bareille J, Marquès-Bueno MDM, Creff A, Maneta-Peyret L, Fiche JB, Nollmann M, Miège C, et al. 2019. Developmental control of plant Rho GTPase nano-organization by the lipid phosphatidylserine. *Science* **364:** 57–62. doi:10.1126/science.aav9959

Prát T, Hajný J, Grunewald W, Vasileva M, Molnár G, Tejos R, Schmid M, Sauer M, Friml J. 2018. WRKY23 is a component of the transcriptional network mediating auxin feedback on PIN polarity. *PLoS Genet* **14:** e1007177. doi:10.1371/journal.pgen.1007177

Prigge MJ, Platre M, Kadakia N, Zhang Y, Greenham K, Szutu W, Pandey BK, Bhosale RA, Bennett MJ, Busch W, et al. 2020. Genetic analysis of the *Arabidopsis* TIR1/AFB auxin receptors reveals both overlapping and specialized functions. *eLife* **9:** e54740. doi:10.7554/eLife.54740

Prusinkiewicz P, Runions A. 2012. Computational models of plant development and form. *New Phytol* **193:** 549–569. doi:10.1111/j.1469-8137.2011.04009.x

Rakusová H, Gallego-Bartolomé J, Vanstraelen M, Robert HS, Alabadí D, Blázquez MA, Benková E, Friml J. 2011. Polarization of PIN3-dependent auxin transport for hypocotyl gravitropic response in *Arabidopsis thaliana*. *Plant J* **67:** 817–826. doi:10.1111/j.1365-313X.2011.04636.x

Rakusová H, Fendrych M, Friml J. 2015. Intracellular trafficking and PIN-mediated cell polarity during tropic responses in plants. *Curr Opin Plant Biol* **23:** 116–123. doi:10.1016/j.pbi.2014.12.002

Rakusová H, Abbas M, Han H, Song S, Robert HS, Friml J. 2016. Termination of shoot gravitropic responses by auxin feedback on PIN3 polarity. *Curr Biol* **26:** 3026–3032. doi:10.1016/j.cub.2016.08.067

Ranocha P, Dima O, Nagy R, Felten J, Corratgé-Faillie C, Novák O, Morreel K, Lacombe B, Martinez Y, Pfrunder S, et al. 2013. *Arabidopsis* WAT1 is a vacuolar auxin transport facilitator required for auxin homoeostasis. *Nat Commun* **4:** 2625. doi:10.1038/ncomms3625

Ravichandran SJ, Linh NM, Scarpella E. 2020. The canalization hypothesis—challenges and alternatives. *New Phytol* **227:** 1051–1059. doi:10.1111/nph.16605

Reinhardt D, Mandel T, Kuhlemeier C. 2000. Auxin regulates the initiation and radial position of plant lateral organs. *Plant Cell* **12:** 507–518. doi:10.1105/tpc.12.4.507

Reinhardt D, Pesce ER, Stieger P, Mandel T, Baltensperger K, Bennett M, Traas J, Friml J, Kuhlemeier C. 2003. Regulation of phyllotaxis by polar auxin transport. *Nature* **426:** 255–260. doi:10.1038/nature02081

Ren H, Park MY, Spartz AK, Wong JH, Gray WM. 2018. A subset of plasma membrane-localized PP2C.D phosphatases negatively regulate SAUR-mediated cell expansion in *Arabidopsis*. *PLoS Genet* **14:** e1007455. doi:10.1371/journal.pgen.1007455

Retzer K, Akhmanova M, Konstantinova N, Malínská K, Leitner J, Petrášek J, Luschnig C. 2019. Brassinosteroid signaling delimits root gravitropism via sorting of the *Arabidopsis* PIN2 auxin transporter. *Nat Commun* **10:** 5516. doi:10.1038/s41467-019-13543-1

Robert S, Kleine-Vehn J, Barbez E, Sauer M, Paciorek T, Baster P, Vanneste S, Zhang J, Simon S, Čovanová M, et al. 2010. ABP1 mediates auxin inhibition of clathrin-dependent endocytosis in *Arabidopsis*. *Cell* **143:** 111–121. doi:10.1016/j.cell.2010.09.027

Robert HS, Grones P, Stepanova AN, Robles LM, Lokerse AS, Alonso JM, Weijers D, Friml J. 2013. Local auxin sources orient the apical-basal axis in *Arabidopsis* embryos. *Curr Biol* **23:** 2506–2512. doi:10.1016/j.cub.2013.09.039

Robert HS, Park C, Gutièrrez CL, Wójcikowska B, Pěnčík A, Novák O, Chen J, Grunewald W, Dresselhaus T, Friml J, et al. 2018. Maternal auxin supply contributes to early embryo patterning in *Arabidopsis*. *Nat Plants* **4:** 548–553. doi:10.1038/s41477-018-0204-z

Roosjen M, Paque S, Weijers D. 2018. Auxin response factors: output control in auxin biology. *J Exp Bot* **69:** 179–188. doi:10.1093/jxb/erx237

Rowan BA, Seymour DK, Chae E, Lundberg DS, Weigel D. 2017. Methods for genotyping-by-sequencing. *Methods Mol Biol* **1492:** 221–242. doi:10.1007/978-1-4939-6442-0_16

* Roychoudhry S, Kepinski S. 2021. Auxin in root development. *Cold Spring Harb Perspect Biol* doi:10.1101/cshperspect.a039933

* Rutten J, van den Berg T, ten Tusscher K. 2021. Modeling auxin signaling in roots: auxin computations. *Cold Spring Harb Perspect Biol* doi:10.1101/cshperspect.a040089

Sachs T. 2000. Integrating cellular and organismic aspects of vascular differentiation. *Plant Cell Physiol* **41:** 649–656. doi:10.1093/pcp/41.6.649

Salanenka Y, Verstraeten I, Löfke C, Tabata K, Naramoto S, Glanc M, Friml J. 2018. Gibberellin DELLA signaling targets the retromer complex to redirect protein trafficking to the plasma membrane. *Proc Natl Acad Sci* **115:** 3716–3721. doi:10.1073/pnas.1721760115

Sanchez Carranza AP, Singh A, Steinberger K, Panigrahi K, Palme K, Dovzhenko A, Dal Bosco C. 2016. Hydrolases of the ILR1-like family of *Arabidopsis thaliana* modulate auxin response by regulating auxin homeostasis in the endoplasmic reticulum. *Sci Rep* **6:** 24212. doi:10.1038/srep24212

Sauer M, Balla J, Luschnig C, Wisniewska J, Reinöhl V, Friml J, Benková E. 2006. Canalization of auxin flow by Aux/IAA-ARF-dependent feedback regulation of PIN polarity. *Genes Dev* **20:** 2902–2911. doi:10.1101/gad.390806

Scarpella E, Marcos D, Friml J, Berleth T. 2006. Control of leaf vascular patterning by polar auxin transport. *Genes Dev* **20:** 1015–1027. doi:10.1101/gad.1402406

Semeradova H, Montesinos JC, Benkova E. 2020. All roads lead to auxin: post-translational regulation of auxin transport by multiple hormonal pathways. *Plant Commun* **1:** 100048. doi:10.1016/j.xplc.2020.100048

Sena G, Wang X, Liu HY, Hofhuis H, Birnbaum KD. 2009. Organ regeneration does not require a functional stem cell niche in plants. *Nature* **457:** 1150–1153. doi:10.1038/nature07597

Serre NB, Kralík D, Yun P, Shabala S, Slouka Z, Fendrych M. 2021. The AFB1 auxin receptor controls rapid auxin signaling and root growth through membrane depolarization in *Arabidopsis thaliana*. bioRxiv 2021.2001.2005.425399

Seyfferth C, Renema J, Wendrich JR, Eekhout T, Seurinck R, Vandamme N, Blob B, Saeys Y, Helariutta Y, Birnbaum KD, et al. 2021. Advances and opportunities of single-cell transcriptomics for plant research. *Annu Rev Plant Biol* **72:** 847–866. doi:10.1146/annurev-arplant-081720-010120

Sheldrake AR. 2021. The production of auxin by dying cells. *J Exp Bot* **72:** 2288–2300. doi:10.1093/jxb/erab009

Shih HW, DePew CL, Miller ND, Monshausen GB. 2015. The cyclic nucleotide-gated channel CNGC14 regulates root gravitropism in *Arabidopsis thaliana*. *Curr Biol* **25:** 3119–3125. doi:10.1016/j.cub.2015.10.025

Shinohara N, Taylor C, Leyser O. 2013. Strigolactone can promote or inhibit shoot branching by triggering rapid depletion of the auxin efflux protein PIN1 from the plasma membrane. *PLoS Biol* **11:** e1001474. doi:10.1371/journal.pbio.1001474

Simon S, Kubeš M, Baster P, Robert S, Dobrev PI, Friml J, Petrášek J, Zažímalová E. 2013. Defining the selectivity of processes along the auxin response chain: a study using auxin analogues. *New Phytol* **200:** 1034–1048. doi:10.1111/nph.12437

Simonini S, Deb J, Moubayidin L, Stephenson P, Valluru M, Freire-Rios A, Sorefan K, Weijers D, Friml J, Østergaard L. 2016. A noncanonical auxin-sensing mechanism is required for organ morphogenesis in *Arabidopsis*. *Genes Dev* **30:** 2286–2296. doi:10.1101/gad.285361.116

Skalický V, Kubeš M, Napier R, Novák O. 2018. Auxins and cytokinins—the role of subcellular organization on homeostasis. *Int J Mol Sci* **19:** 3115. doi:10.3390/ijms19103115

Skokan R, Medvecká E, Viaene T, Vosolsobě S, Zwiewka M, Müller K, Skůpa P, Karady M, Zhang Y, Janacek DP, et al. 2019. PIN-driven auxin transport emerged early in streptophyte evolution. *Nat Plants* **5:** 1114–1119. doi:10.1038/s41477-019-0542-5

Smith S, Zhu S, Joos L, Roberts I, Nikonorova N, Vu LD, Stes E, Cho H, Larrieu A, Xuan W, et al. 2020. The CEP5 peptide promotes abiotic stress tolerance, as revealed by quantitative proteomics, and attenuates the AUX/IAA equilibrium in *Arabidopsis*. *Mol Cell Proteomics* **19:** 1248–1262. doi:10.1074/mcp.RA119.001826

Spartz AK, Ren H, Park MY, Grandt KN, Lee SH, Murphy AS, Sussman MR, Overvoorde PJ, Gray WM. 2014. SAUR inhibition of PP2C-D phosphatases activates plasma membrane H+-ATPases to promote cell expansion in *Arabidopsis*. *Plant Cell* **26:** 2129–2142. doi:10.1105/tpc.114.126037

Su SH, Keith MA, Masson PH. 2020. Gravity signaling in flowering plant roots. *Plants (Basel)* **9:** 1290. doi:10.3390/plants9101290

Sun J, Chen Q, Qi L, Jiang H, Li S, Xu Y, Liu F, Zhou W, Pan J, Li X, et al. 2011. Jasmonate modulates endocytosis and plasma membrane accumulation of the *Arabidopsis* PIN2 protein. *New Phytol* **191:** 360–375. doi:10.1111/j.1469-8137.2011.03713.x

* Suzuki H, Kohchi T, Nishihama R. 2021. Auxin biology in Bryophyta: a simple platform with versatile functions. *Cold Spring Harb Perspect Biol* **13:** a040055. doi:10.1101/cshperspect.a040055

Tan S, Abas M, Verstraeten I, Glanc M, Molnár G, Hajný J, Lasák P, Petřík I, Russinova E, Petrášek J, et al. 2020a. Salicylic acid targets protein phosphatase 2A to attenuate growth in plants. *Curr Biol* **30:** 381–395.e8. doi:10.1016/j.cub.2019.11.058

Tan S, Di Donato M, Glanc M, Zhang X, Klíma P, Liu J, Bailly A, Ferro N, Petrášek J, Geisler M, et al. 2020b. Non-steroidal anti-inflammatory drugs target TWISTED DWARF1-regulated actin dynamics and auxin transport-mediated plant development. *Cell Rep* **33:** 108463. doi:10.1016/j.celrep.2020.108463

Tan S, Luschnig C, Friml J. 2021. Pho-view of auxin: reversible protein phosphorylation in auxin biosynthesis, transport and signaling. *Mol Plant* **14:** 151–165. doi:10.1016/j.molp.2020.11.004

Teale WD, Pasternak T, Dal Bosco C, Dovzhenko A, Kratzat K, Bildl W, Schwörer M, Falk T, Ruperti B, Schaefer JV, et al. 2021. Flavonol-mediated stabilization of PIN efflux complexes regulates polar auxin transport. *EMBO J* **40:** e104416. doi:10.15252/embj.2020104416

Tromas A, Braun N, Muller P, Khodus T, Paponov IA, Palme K, Ljung K, Lee JY, Benfey P, Murray JA, et al. 2009. The AUXIN BINDING PROTEIN 1 is required for differential auxin responses mediating root growth. *PLoS ONE* **4:** e6648. doi:10.1371/journal.pone.0006648

Uchida N, Takahashi K, Iwasaki R, Yamada R, Yoshimura M, Endo TA, Kimura S, Zhang H, Nomoto M, Tada Y, et al. 2018. Chemical hijacking of auxin signaling with an engineered auxin–TIR1 pair. *Nat Chem Biol* **14:** 299–305. doi:10.1038/nchembio.2555

Ulmasov T, Murfett J, Hagen G, Guilfoyle TJ. 1997. Aux/IAA proteins repress expression of reporter genes containing natural and highly active synthetic auxin response elements. *Plant Cell* **9:** 1963–1971.

Vanneste S, Friml J. 2009. Auxin: a trigger for change in plant development. *Cell* **136:** 1005–1016. doi:10.1016/j.cell.2009.03.001

* Verma S, Attuluri VPS, Robert HS. 2021. An essential function for auxin in embryo development. *Cold Spring Harb Perspect Biol* **13:** a039966. doi:10.1101/cshperspect.a039966

Verna C, Ravichandran SJ, Sawchuk MG, Linh NM, Scarpella E. 2019. Coordination of tissue cell polarity by auxin transport and signaling. *eLife* **8:** e51061. doi:10.7554/eLife.51061

Vernoux T, Brunoud G, Farcot E, Morin V, Van den Daele H, Legrand J, Oliva M, Das P, Larrieu A, Wells D, et al. 2011. The auxin signalling network translates dynamic input into robust patterning at the shoot apex. *Mol Syst Biol* **7:** 508. doi:10.1038/msb.2011.39

Vieten A, Vanneste S, Wisniewska J, Benková E, Benjamins R, Beeckman T, Luschnig C, Friml J. 2005. Functional redundancy of PIN proteins is accompanied by auxin-dependent cross-regulation of PIN expression. *Development* **132:** 4521–4531. doi:10.1242/dev.02027

von Wangenheim D, Hauschild R, Fendrych M, Barone V, Benková E, Friml J. 2017. Live tracking of moving samples in confocal microscopy for vertically grown roots. *eLife* **6:** e26792. doi:10.7554/eLife.26792

Vosolsobě S, Skokan R, Petrášek J. 2020. The evolutionary origins of auxin transport: what we know and what we need to know. *J Exp Bot* **71:** 3287–3295. doi:10.1093/jxb/eraa169

Wabnik K, Kleine-Vehn J, Balla J, Sauer M, Naramoto S, Reinöhl V, Merks RM, Govaerts W, Friml J. 2010. Emergence of tissue polarization from synergy of intracellular and extracellular auxin signaling. *Mol Syst Biol* **6:** 447. doi:10.1038/msb.2010.103

Wabnik K, Robert HS, Smith RS, Friml J. 2013. Modeling framework for the establishment of the apical-basal embryonic axis in plants. *Curr Biol* **23:** 2513–2518. doi:10.1016/j.cub.2013.10.038

Wang Q, Qin G, Cao M, Chen R, He Y, Yang L, Zeng Z, Yu Y, Gu Y, Xing W, et al. 2020. A phosphorylation-based switch controls TAA1-mediated auxin biosynthesis in plants. *Nat Commun* **11:** 679. doi:10.1038/s41467-020-14395-w

Wendrich JR, Yang B, Vandamme N, Verstaen K, Smet W, Van de Velde C, Minne M, Wybouw B, Mor E, Arents HE, et al. 2020. Vascular transcription factors guide plant epidermal responses to limiting phosphate conditions. *Science* **370:** eaay4970. doi:10.1126/science.aay4970

Wisniewska J, Xu J, Seifertová D, Brewer PB, Ruzicka K, Blilou I, Rouquié D, Benková E, Scheres B, Friml J. 2006. Polar PIN localization directs auxin flow in plants. *Science* **312:** 883. doi:10.1126/science.1121356

Woo EJ, Bauly J, Chen JG, Marshall J, Macdonald H, Lazarus C, Goodenough P, Venis M, Napier R, Pickersgill R. 2000. Crystallization and preliminary X-ray analysis of the auxin receptor ABP1. *Acta Crystallogr D Biol Crystallogr* **56:** 1476–1478. doi:10.1107/S0907444900010714

Woodward AW, Bartel B. 2005. Auxin: regulation, action, and interaction. *Ann Bot* **95:** 707–735. doi:10.1093/aob/mci083

* Wright C, Moss BL, Nemhauser JL. 2021. The systems and synthetic biology of auxin. *Cold Spring Harb Perspect Biol* doi:10.1101/cshperspect.a040071

Xu J, Hofhuis H, Heidstra R, Sauer M, Friml J, Scheres B. 2006. A molecular framework for plant regeneration. *Science* **311:** 385–388. doi:10.1126/science.1121790

Xu T, Wen M, Nagawa S, Fu Y, Chen JG, Wu MJ, Perrot-Rechenmann C, Friml J, Jones AM, Yang Z. 2010. Cell surface- and rho GTPase-based auxin signaling controls cellular interdigitation in *Arabidopsis*. *Cell* **143:** 99–110. doi:10.1016/j.cell.2010.09.003

Xu T, Dai N, Chen J, Nagawa S, Cao M, Li H, Zhou Z, Chen X, De Rycke R, Rakusova H, et al. 2014. Cell surface ABP1-TMK auxin-sensing complex activates ROP GTPase signaling. *Science* **343:** 1025–1028. doi:10.1126/science.1245125

Xu D, Miao J, Yumoto E, Yokota T, Asahina M, Watahiki M. 2017. YUCCA9-mediated auxin biosynthesis and polar auxin transport synergistically regulate regeneration of root systems following root cutting. *Plant Cell Physiol* **58:** 1710–1723. doi:10.1093/pcp/pcx107

Zhang J, Peer WA. 2017. Auxin homeostasis: the DAO of catabolism. *J Exp Bot* **68:** 3145–3154. doi:10.1093/jxb/erx221

Zhang Y, Xiao G, Wang X, Zhang X, Friml J. 2019. Evolution of fast root gravitropism in seed plants. *Nat Commun* **10:** 3480. doi:10.1038/s41467-019-11471-8

Zhang J, Mazur E, Balla J, Gallei M, Kalousek P, Medveďová Z, Li Y, Wang Y, Prát T, Vasileva M, et al. 2020. Strigolactones inhibit auxin feedback on PIN-dependent auxin transport canalization. *Nat Commun* **11:** 3508. doi:10.1038/s41467-020-17252-y

Zhou W, Lozano-Torres JL, Blilou I, Zhang X, Zhai Q, Smant G, Li C, Scheres B. 2019. A jasmonate signaling network activates root stem cells and promotes regeneration. *Cell* **177:** 942–956.e14. doi:10.1016/j.cell.2019.03.006

Zhu J, Bailly A, Zwiewka M, Sovero V, Di Donato M, Ge P, Oehri J, Aryal B, Hao P, Linnert M, et al. 2016. TWISTED DWARF1 mediates the action of auxin transport inhibitors on actin cytoskeleton dynamics. *Plant Cell* **28:** 930–948. doi:10.1105/tpc.15.00726

Žižková E, Kubeš M, Dobrev PI, Přibyl P, Šimura J, Zahajská L, Záveská Drábková L, Novák O, Motyka V. 2017. Control of cytokinin and auxin homeostasis in cyanobacteria and algae. *Ann Bot* **119:** 151–166. doi:10.1093/aob/mcw194

Auxin Metabolism in Plants

Rubén Casanova-Sáez,[1] Eduardo Mateo-Bonmatí,[1] and Karin Ljung

Umeå Plant Science Centre, Department of Forest Genetics and Plant Physiology, Swedish University of Agricultural Sciences, SE-901 83 Umeå, Sweden

Correspondence: Karin.ljung@slu.se

The major natural auxin in plants, indole-3-acetic acid (IAA), orchestrates a plethora of developmental responses that largely depend on the formation of auxin concentration gradients within plant tissues. Together with inter- and intracellular transport, IAA metabolism—which comprises biosynthesis, conjugation, and degradation—modulates auxin gradients and is therefore critical for plant growth. It is now very well established that IAA is mainly produced from Trp and that the IPyA pathway is a major and universally conserved biosynthetic route in plants, while other redundant pathways operate in parallel. Recent findings have shown that metabolic inactivation of IAA is also redundantly performed by oxidation and conjugation processes. An exquisite spatiotemporal expression of the genes for auxin synthesis and inactivation have been shown to drive several plant developmental processes. Moreover, a group of transcription factors and epigenetic regulators controlling the expression of auxin metabolic genes have been identified in past years, which are illuminating the road to understanding the molecular mechanisms behind the coordinated responses of local auxin metabolism to specific cues. Besides transcriptional regulation, subcellular compartmentalization of the IAA metabolism and posttranslational modifications of the metabolic enzymes are emerging as important contributors to IAA homeostasis. In this review, we summarize the current knowledge on (1) the pathways for IAA biosynthesis and inactivation in plants, (2) the influence of spatiotemporally regulated IAA metabolism on auxin-mediated responses, and (3) the regulatory mechanisms that modulate IAA levels in response to external and internal cues during plant development.

The extensive search for the plant molecule responsible for tropic responses to light and gravity led to the identification of the first auxin molecule, indole-3-acetic acid (IAA), more than 80 years ago (Abel and Theologis 2010). Since then, it has been determined that auxin is not only involved in the control of tropisms, but also regulates numerous plant developmental responses that mainly rely on the spatiotemporal control of cell division, growth, and differentiation (Zhao 2018; Gallei et al. 2020). In addition to IAA, phenylacetic acid (PAA) and 4-chloro-indole-3-acetic acid (4-Cl-IAA) are naturally occurring auxins in plants. Although both are perceived by the auxin signaling machinery (Shimizu-Mitao and Kakimoto 2014; Jayasinghege et al. 2019), 4-Cl-IAA is not widespread (Lam et al. 2015) and PAA has been studied far less than IAA due to generally less potent effects (Cook 2019).

[1]These authors contributed equally to this work.

Cite this article as *Cold Spring Harb Perspect Biol* doi: 10.1101/cshperspect.a039867

Because plant responses to auxin concentrations are threshold-dependent, IAA levels must be finely regulated during plant growth in response to external and internal cues. Tuning of the IAA concentration within cells and tissues is largely performed by directional transport and localized biosynthesis (Brumos et al. 2018; Robert et al. 2018) as well as inactivation of IAA (Zheng et al. 2016; Di Mambro et al. 2019). A classical model for auxin gradient formation in plants is based on a primary synthesis of IAA in the shoot—mainly in the young leaves and cotyledons—which is then distributed throughout the plant by polar transport with a major rootward component. However, early reports already suggested a role for the root tip in auxin biosynthesis (van Raalte 1936; van Overbeek 1939; Davies and Mitchell 1972). Improvements in the sensitivity of analytical technologies that allowed quantification of IAA from minute amounts of plant tissue, along with the identification of newly synthesized auxin in these tissues using stable isotope labeling (Novák et al. 2012, 2017), have helped researchers establish that auxin biosynthesis occurs locally in different plant organs and, albeit at different rates, in every root cell type (Ljung et al. 2001a, 2005; Petersson et al. 2009). Together with advances in analytical methodologies, the availability of the *Arabidopsis* genome sequence boosted the identification of genes associated with auxin biosynthesis, conjugation, and degradation. The detailed genetic and biochemical evaluation of these genes and their regulatory networks has revealed the importance of an initially disregarded role of auxin metabolism in plant development.

AUXIN BIOSYNTHESIS IN PLANTS

Several decades of research on auxin metabolism have firmly established the aromatic amino acid L-tryptophan (Trp) as a central precursor for IAA biosynthesis in plants. Trp is produced in chloroplasts via the shikimate pathway, a route through which most living organisms—excluding animals—produce aromatic amino acids (Maeda and Dudareva 2012). Far from being linear, Trp-dependent auxin biosynthesis involves various parallel pathways converging at the production of IAA, being IAOx (indole-3-acetaldoxime), IAM (indole-3-acetamide) and IPyA (indole-3-pyruvic acid) the most common intermediates (Fig. 1). A Trp-independent pathway for auxin synthesis was proposed after finding that maize and *Arabidopsis* mutants defective in Trp biosynthesis were still producing IAA (Wright et al. 1991; Normanly et al. 1993). It was later suggested that a cytosolic indole synthase (INS) mediates Trp-independent IAA production via the conversion of indole-3-glycerol-phosphate to indole (Zhang et al. 2008; Wang et al. 2015). However, the biochemical pathway for the Trp-independent conversion of indole to IAA remains unclear (Nonhebel 2015). The following section will describe the Trp-dependent pathways for auxin synthesis.

The Main Pathway for IAA Biosynthesis

Whereas auxin production from IAOx and IAM is not yet fully understood, the IPyA pathway has been established as the prevailing route for IAA synthesis in plants. It consists of a two-step reaction in which Trp is first deaminated to IPyA by TRYPTOPHAN AMINOTRANSFERASE OF ARABIDOPSIS 1 (TAA1) and TAA1-RELATED proteins (TARs) (Stepanova et al. 2008; Tao et al. 2008; Yamada et al. 2009). IPyA is then decarboxylated to IAA in a rate-limiting and irreversible reaction catalyzed by flavin-containing monooxygenases from the YUCCA (YUC) family (Mashiguchi et al. 2011; Stepanova et al. 2011; Won et al. 2011). *TAA1* and *YUC* homologs are found across the genomes of vascular and nonvascular plants (Yue et al. 2014; Eklund et al. 2015; Poulet and Kriechbaumer 2017; Matthes et al. 2019), and functional conservation of *TAA1* and *YUC* homologs has been shown in maize (Gallavotti et al. 2008; Phillips et al. 2011; Bernardi et al. 2012), rice (Yamamoto et al. 2007; Yoshikawa et al. 2014), and *Marchantia polymorpha* (Eklund et al. 2015). Taken together, this suggests that the IPyA pathway is a universal route for IAA synthesis in land plants. Compared to single mutants in *TAA1*/*TARs* and *YUCCA* genes, which show only subtle developmental

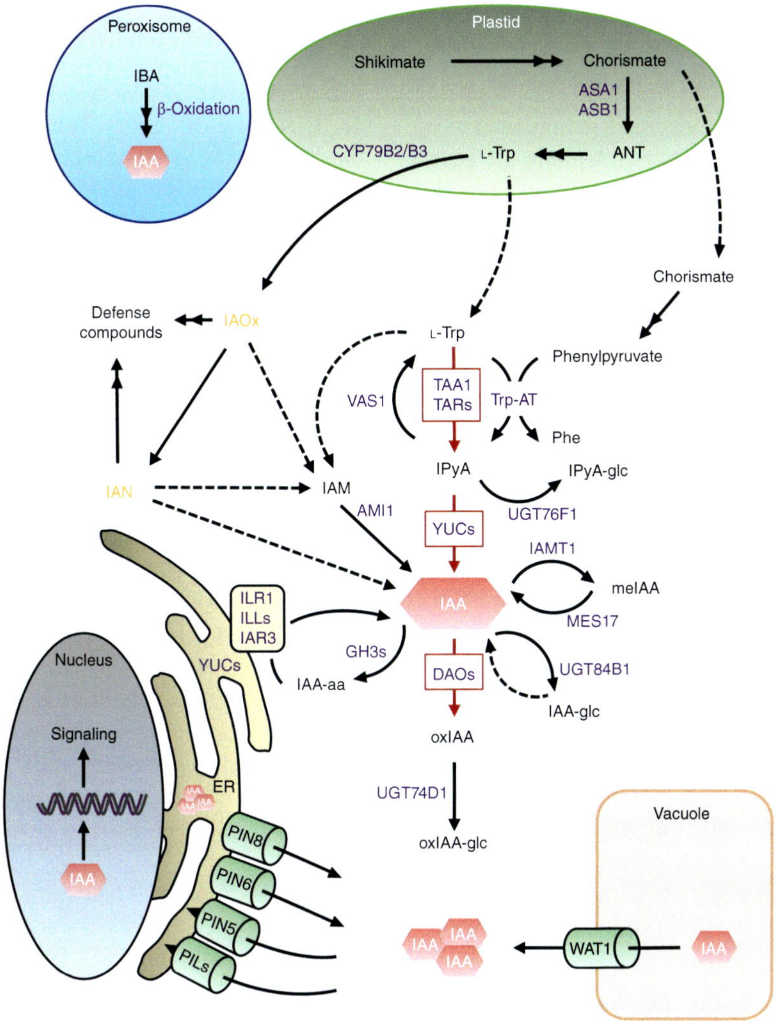

Figure 1. Main pathways for indole-3-acetic acid (IAA) metabolism in plants. Biosynthesis of the IAA precursor L-tryptophan (Trp) takes place in the plastid (green oval). Subsequent IAA biosynthesis, catabolism, and conjugation reactions commonly operate in the cytoplasm. IAA molecules are indicated as red hexagons and IAA-related metabolites are shown in black. Enzymes catalyzing the metabolic reactions are indicated in purple. Major components for IAA biosynthesis and inactivation are indicated with red arrows. Solid arrows indicate pathways in which the enzymes, genes, or intermediates are known, while dashed arrows indicate pathways that are not yet well defined. *Brassica*-specific metabolites are indicated in orange. Green cylinders indicate intracellular IAA transporters and the attached arrows indicate the IAA movement direction. The blue circle, the gray oval, and the pale orange square indicate peroxisome, nucleus, and vacuole, respectively. Endoplasmic reticulum (ER) is indicated by the dark yellow structure attached to the nucleus. Organelles are not drawn to scale. (AMI1) AMIDASE-LIKE PROTEIN 1, (ANT) anthranilate, (ASA1) ANTHRANILATE SYNTHASE α SUBUNIT 1, (ASB1) ANTHRANILATE SYNTHASE β SUBUNIT 1, (CYP79B2/B3) CYTOCHROME P450, family 79, subfamily B, polypeptides 2 and 3, (DAO) DIOXYGENASE FOR AUXIN OXIDATION, (GH3) GRETCHEN HAGEN3, (IAA-glc) IAA-glucose, (IAM) indole-3-acetamide, (IAMT1) indole-3-acetate *O*-methyltransferase 1, (IAN) indole-3-acetonitrile, (IAOx) indole-3-acetaldoxime, (IBA) indole-3-butyric acid, (IAR3) IAA-ALANINE RESISTANT3, (ILLs) ILR1-LIKE, (ILR1) IAA-LEUCINE RESISTANT1, (IPyA) indole-3-pyruvic acid, (meIAA) methylindole-3-acetic acid, (MES17) METHYLESTERASE 17, (oxIAA) 2-oxindole-3-acetic acid, (Phe) phenylalanine, (PILS) PIN-likes, (PIN) PIN-FORMED, (TAA1) TRYPTOPHAN AMINOTRANSFERASE OF ARABIDOPSIS 1, (TAR) TRYPTOPHAN AMINOTRANSFERASE-RELATED PROTEIN, (Trp-AT) TRYPTOPHAN AMINOTRANSFERASE, (UGTs) URIDINE-DIPHOSPHATE GLYCOSYLTRANSFERASE, (VAS1) REVERSAL OF SAV3 PHENOTYPE 1, (WAT1) WALLS ARE THIN1, (YUC) YUCCA flavin-containing monooxygenases.

phenotypes, higher-order *taa1/tar* and *yuc* mutants that bypass functional redundancy result in marked reductions in IAA levels and severe developmental defects in *Arabidopsis*, including abnormal embryo patterning, reduced stature, aberrant vasculature, defective root growth and gravitropic response, abnormal apical hook formation, and altered leaf and floral patterning (Cheng et al. 2006, 2007; Stepanova et al. 2008; Chen et al. 2014). Loss-of-function of the maize *vt2* gene, a TAA1 ortholog, resulted in reduced plant growth and dramatic effects in inflorescence development, which entails a severe sterility of the mutant plants (Phillips et al. 2011). Similar reproductive defects were observed in the maize *spi1* mutants, impaired in the function of a monocot-specific YUC-like protein (Gallavotti et al. 2008). In the liverwort *Marchantia*, the knock-out of its single *TAA* gene resulted in severe growth and developmental defects caused by a loss of cell and tissue differentiation (Eklund et al. 2015). The severity of the above-mentioned phenotypes, along with the seedling lethality observed in the *Arabidopsis taa1 tar1 tar2* triple mutant (Stepanova et al. 2008), points to the IPyA pathway as a major and essential route for IAA biosynthesis in plants.

Parallel Pathways for IAA Biosynthesis

The Trp derivative IAOx is an intermediate in an IAA biosynthetic route that is yet to be fully understood. The conversion of Trp to IAOx is mediated by two isozymes from the cytochrome P450 (CYP) monooxygenase family, CYP79B2 and CYP79B3 (Hull et al. 2000; Mikkelsen et al. 2000; Zhao et al. 2002). Both IAOx and *CYP79B2/3* genes have so far only been found in *Brassica* species (Sugawara et al. 2009), which suggests that this pathway is restricted to the Brassicaceae family (Fig. 1). Moreover, the *cyp79b2 cyp79b3* double mutant shows conditional auxin phenotypes when grown at high temperatures (Zhao et al. 2002; Sugawara et al. 2009), further suggesting a role for the IAOx-dependent IAA synthesis specifically during adverse conditions. IAOx is a well-known precursor of indole glucosynolates (IGs) and camalexin, which serve as defense metabolites in plants (Hull et al. 2000; Glawischnig et al. 2004; Nafisi et al. 2007). Nevertheless, an increase in IAOx levels—through either genetic disruption of the IG pathway or overexpression of genes associated with IAOx biosynthesis—results in elevated IAA and high-auxin phenotypes (Boerjan et al. 1995; King et al. 1995; Delarue et al. 1998; Barlier et al. 2000; Zhao et al. 2002; Grubb et al. 2004; Sugawara et al. 2009; Novák et al. 2012; Kong et al. 2015). A study in which isotope-labeled IAOx was fed to *Arabidopsis* seedlings revealed that IAOx, IAM, and IAN (indole-3-acetonitrile) are intermediates of IAA biosynthesis (Sugawara et al. 2009). IAN-to-IAA conversion by a family of plant nitrilases (NITs) is thought to account for the IAOx-dependent auxin biosynthesis (Lehmann et al. 2017). However, the lack of direct genetic and biochemical evidence for NIT-mediated auxin synthesis *in planta*, together with the knowledge that nitrilases participate in cyanide and glutathione detoxification (Piotrowski 2008; Niehaus et al. 2019), means that the biochemical route from IAOx to IAA remains unresolved.

IAM is a well-known auxin biosynthesis intermediate in certain plant-associated bacteria (Patten et al. 2013) in which Trp is converted to IAA through the formation of IAM. While it was demonstrated that IAM can be produced from IAOx in *Arabidopsis* (Sugawara et al. 2009), IAM has also been detected in non-*Brassica* species that lack IAOx (Pollmann et al. 2002; Sugawara et al. 2009; Novák et al. 2012). IAM application results in classical high-auxin phenotypes (Sugawara et al. 2009; Gao et al. 2020), indicating that IAM-to-IAA conversion operates *in planta*. However, the disruption of the main IAM hydrolases in *Arabidopsis*, IAMH1, and IAMH2, did not lead to substantial developmental defects or variations in IAA contents, suggesting that the IAM pathway only plays a minor role in auxin homeostasis under standard growth conditions (Gao et al. 2020).

Indole-3-butyric acid (IBA) is a compound that has been shown to stimulate an auxin response when applied to plants. However, it is very unlikely (1) that IBA itself is perceived by the plant (Strader and Bartel 2011; Uzunova et al. 2016), and (2) that it is transported via polar transport (Liu et al. 2012). Instead, the effects of IBA can be attributed to its conversion to IAA by a group of peroxisomal enzymes (Zolman et al. 2008). IBA-to-IAA conversion has been found to be relevant to plant development (Frick and Strader 2018). How plants synthesize IBA is still unknown and, as such, it is not clear whether IBA is an IAA precursor or storage molecule. Also, whether endogenous IBA is present at physiologically relevant concentrations in plants has been questioned (Novák et al. 2012) and is still under debate (Frick and Strader 2018).

AUXIN METABOLIC INACTIVATION

Together with directional transport and local biosynthesis, metabolic inactivation of IAA also modulates auxin concentrations across plant cells and tissues. Indeed, research has shown that the majority of plant IAA exists as (1) inactive conjugates and methyl ester forms that can be reversibly converted to IAA (auxin storage forms), to rapidly fine-tune auxin levels without the need for de novo synthesis; and (2) as irreversible inactive IAA (auxin catabolites), which is the result of the removal of excess auxin or a regulated response to create auxin minima. The most extensively studied inactive forms of auxin will be summarized in the following section.

IAA Storage Forms

Many different auxin storage forms have been identified in plants (Korasick et al. 2013). These forms fall into three main groups: ester-linked IAA conjugates, amide-linked IAA conjugates, and methyl IAA (meIAA). The most prevalent and abundant ester-linked auxin in plants is IAA-glucose (IAA-glc), which is present at higher levels than any other directly measured conjugate (Pěnčík et al. 2009, 2018; Brunoni et al. 2020). IAA-glc has been detected in seedling extracts from different plants (Kai et al. 2007), and is the predominant IAA metabolite throughout *Arabidopsis* tissues (Porco et al. 2016). IAA-glc and its metabolic derivative IAA-myo-inositol (IAA-Ins) are particularly abundant in plant seeds (Hall 1980; Cohen and Bandurski 1982), and are thought to be the main auxin source during early seedling establishment in vascular plants (Bartel et al. 2001; Ljung et al. 2001b). High-molecular-weight IAA-glycan and -glycoprotein conjugates have also been found in plants (Korasick et al. 2013), although their specific roles in auxin homeostasis are not yet understood. Hydrolases that release free IAA from IAA-glc and IAA-Ins were identified in maize kernels (Jakubowska and Kowalczyk 2005), and rice (TGW6; Ishimaru et al. 2013). Specific UDP-glycosyltransferases (UGTs) that produce IAA-glc have been identified in plants (Szerszen et al. 1994; Jackson et al. 2002; Ludwig-Müller et al. 2005; Liu et al. 2019). Overexpression of these *UGT*s results in increased levels of IAA-glc and reduced levels of amide-linked auxins (Jackson et al. 2002; Ludwig-Müller et al. 2005), indicating that IAA homeostasis was disturbed. The availability of knockout mutants for the known auxin UGTs, together with the discovery of additional IAA glycosylases and hydrolases, will help clarify the roles of ester-linked auxins throughout plant development.

Amide-linked auxins encompass a group of compounds in which IAA is conjugated to amino acids, small peptides and proteins, among which IAA-amino acid (IAA-aa) conjugates are the best characterized. The formation of the IAA-amide bond is catalyzed by a group of IAA acyl acid amido synthetases from the GRETCHEN HAGEN3 (GH3) family (Staswick et al. 2005; Ludwig-Müller et al. 2009). *GH3* genes (Terol et al. 2006; Okrent and Wildermuth 2011), along with different IAA-aa conjugates (Korasick et al. 2013; Záveská Drábková et al. 2015), are found all across land plants. *GH3* co-orthologs have been functionally characterized in vascular and nonvascular plants (Staswick et al. 2005; Ludwig-Müller et al. 2009; Brunoni et al. 2020). The application of various IAA-aa conjugates results in plant phenotypes that are similar to what can be observed upon the addition of exogenous IAA, which provides strong evidence that IAA-aa conjugates can serve as IAA storage forms (LeClere et al. 2002; Rampey et al. 2004). Several IAA-aa amidohydrolases, including IAA-LEUCINE RESISTANT1 (ILR1), ILR1-LIKE proteins (ILLs), and IAA-ALANINE RESISTANT3 (IAR3), were identified in screens for mutants that are insensitive to IAA-aa conjugates (Bartel and Fink 1995; Davies et al. 1999; LeClere et al. 2002). These amidohydrolases appear to be functionally conserved across plant species. IAR3 was found to be required for root architectural changes under osmotic stress in *Arabidopsis* (Kinoshita et al. 2012), and to mediate defense responses of tomato and potato plants upon infection (D'Ippolito et al. 2016). GH3s are known to mediate different responses to biotic and abiotic stress (Park et al. 2007; Zhang et al. 2007, 2009; Ding et al. 2008; Du et al. 2012; Kirungu et al. 2019). A note of caution is due here since certain GH3 enzymes that function as IAA amido synthetases, as GH3.3, GH3.5, and GH3.6, present a promiscuous conjugating activity toward different substrates like salicylic acid, jasmonic acid, and benzoic acid (Zhang et al. 2007; Gutierrez et al. 2012; Westfall et al. 2016), at least in vitro. Hence, GH3-related phenotypes might be the result of the perturbation of the homeostasis of not just IAA, but also of other phytohormones.

The methyl ester of IAA, meIAA, also serves as an auxin storage form in plants. IAA methylation at the carboxyl group is mediated by IAA CARBOXYL METHYLTRANSFERASE1 (IAMT1) (Zubieta et al. 2003; Qin et al. 2005). As an inactive form of auxin, melAA does not interfere with the auxin signaling machinery (Li et al. 2008; Abbas et al. 2018). Nevertheless, the application of meIAA results in auxin-related phenotypes in plants (Qin et al. 2005), a dynamic that can be attributed to the hydrolysis of meIAA to IAA by METHYLESTERASE 17 (MES17) and related enzymes (Yang et al. 2008). Research using knockout *iamt1* mutants revealed that auxin methylation has little impact on auxin levels, both locally or in whole seedlings (Abbas et al. 2018; Takubo et al. 2020). meIAA is a nonpolar compound that is transported in plants by both passive influx and PIN-mediated efflux and, thus, affects IAA gradients rather than IAA levels (Li et al. 2008; Abbas et al. 2018). Specific expression of *IAMT1* in the hypocotyl endodermis was shown to be important for gravitropic growth (Abbas et al. 2018). The role of auxin methylation in regulating plant development is, however, still under debate (Takubo et al. 2020).

Irreversible IAA Catabolites

The amide-linked IAA-Asp and IAA-Glu conjugates, unlike other IAA-amino acid conjugates, are not hydrolyzed back to IAA *in planta* (Östin et al. 1998; Rampey et al. 2004) and are thus considered catabolites. These two irreversible conjugates are found in plants at much higher levels than the reversible IAA-aa conjugates (Kowalczyk and Sandberg 2001; Pěnčík et al. 2009). GH3 IAA-amido synthetases show different substrate preferences for Asp and Glu (Staswick et al. 2005; Brunoni et al. 2020). GH3.17, and to a lesser extent GH3.5, is known to preferentially use Glu as a cosubstrate (Staswick et al. 2005). Accordingly, *gh3.17* plants show remarkably reduced levels of IAA-Glu in their hypocotyls (Zheng et al. 2016) and roots (Di Mambro et al. 2017), while the *gh3.1,2,3,4,5,6* sextuple mutant does not produce any IAA-Asp, although IAA-Glu production is up-regulated, and is likely supported by the still functional GH3.17 (Porco et al. 2016).

The major catabolic pathway that regulates IAA levels in plants is the irreversible oxidation of IAA to oxIAA (2-oxindole-3-acetic acid), with further glycosylation to oxIAA-glc (Östin et al. 1998; Kai et al. 2007; Kubeš et al. 2012; Novák et al. 2012; Pěnčík et al. 2013). There is extensive evidence that the levels of these oxidative catabolites rapidly increase after IAA application (Östin et al. 1998; Kubeš et al. 2012). Moreover, they are prevalent at higher levels than amide-linked catabolites at physiological conditions in algae, as well as in vascular and nonvascular land plants, which suggests that oxidation is a major pathway for IAA catabolism across the plant kingdom (Novák et al. 2012; Pěnčík et al. 2013; Záveská Drábková et al. 2015; Porco et al. 2016; Žižková et al. 2017). Conifers were found to be an exception, as conjugation and not oxidation dominates IAA homeostasis (Brunoni et al. 2020). The conversion of IAA to oxIAA is catalyzed by DIOXYGENASE FOR AUXIN OXIDATION (DAO) proteins (Zhao et al. 2013), which belong to the 2-oxoglutarate-dependent Fe(II) dioxygenase superfamily (Kawai et al. 2014; Nadi et al. 2018), while the UGT74D1 enzyme participates in the glycosylation of oxIAA to oxIAA-glc (Tanaka et al. 2014). *Arabidopsis DAO1* accounts for most de novo IAA oxidation and is widely expressed in plant tissues. However, *DAO1* loss-of-function results in only mild developmental defects. Because amide-linked catabolites are greatly increased in *dao1* mutants, while decreased in *DAO1* overexpressors, DAO and GH3 enzymes are proposed to function redundantly in regulating IAA levels (Mellor et al. 2016; Porco et al. 2016; Zhang et al. 2016). Despite this redundant IAA catabolism, multicellular modeling of auxin gradients has shown that impaired oxidation perturbs IAA levels in specific root tissues (Mellor et al. 2016), which is supported by the localized, albeit subtle, phenotypes of *dao1-1* mutant roots (Porco et al. 2016; Zhang et al. 2016).

LOCALIZED IAA METABOLISM COORDINATES PLANT DEVELOPMENT

The identification and characterization of several genes related to auxin metabolism has revealed an intricate spatiotemporal orchestration of their localized expression (exemplified in Fig. 2) that turned out to contribute to the regulation of local auxin concentrations. *YUCCA* genes represent an excellent example of specialization, as several rounds of gene duplication have resulted in multiple *YUCs* that have unique expression domains (Zhao 2018). For example, a group of shoot- and root-specific *YUCCAs* has been defined (Won et al. 2011; Chen et al. 2014). *YUC1* and *YUC4* are specifically expressed in the shoot apical meristem, along with flower and leaf primordia (Cheng et al. 2006, 2007). In the embryo, *YUC1*, *YUC4*, *YUC10*, and *YUC11* were found to be expressed in the apical cells, with each showing certain temporal changes in expression domain as embryo development progressed (Cheng et al. 2007). The quadruple *yuc1 yuc4 yuc10 yuc11* mutant was found to lack the hypophysis, a root meristem precursor cell, and thus germinated without a primary root (Cheng et al. 2007). Localized expression of *TAA1* in the developing embryo was also found to be critical for root and apical embryonic meristem specification (Stepanova et al. 2008; Robert et al. 2013). In roots, *TAA1* is specifically expressed at the quiescent center (QC), while *TAR2* expression was reported in the root provasculature (Stepanova et al. 2008). Disruption of both of these genes resulted in the complete loss of the stem cell niche and root growth abortion early after germination (Stepanova et al. 2008). This dynamic and localized expression, together with the developmental abnormalities noted for plants with loss-of-function mutations in the *TAA1*/*TAR* and *YUC* genes, supports a role for IPyA-dependent local auxin biosynthesis in embryo patterning, root meristem maintenance, gynoecium formation, and leaf and floral development (Cheng et al. 2006, 2007; Stepanova et al. 2008). Localized expression of *CYP79B2*/*B3* genes at the root meristem and lateral root primordia initiation sites additionally suggests that the IAOx pathway participates in local IAA synthesis during root development (Ljung et al. 2005).

Cooperation between local IAA biosynthesis and polar transport generates auxin concentration gradients that drive plant growth (Ikeda et al. 2009; Brumos et al. 2018). For example, the spatiotemporally coordinated expression of *TAA1* and *YUCs* in the basal and apical embryo, together with the resulting auxin-triggered PIN polarization, was shown to define the apicobasal embryo axis (Robert et al. 2013; Wabnik et al. 2013). However, auxin transport cannot always compensate for deficiencies in local synthesis. Proper root development largely depends on auxin production in the root (Bhalerao et al. 2002; Chen et al. 2014), and localized auxin biosynthesis at the root QC was found to be sufficient for root meristem maintenance in the absence of functional polar transport (Brumos et al. 2018). Recently, the capacity of the roots to regenerate their tips after wounding was shown to be highly dependent on local auxin biosynthesis mediated by TAA1 and YUCs near the cut site and in the protoxylem and

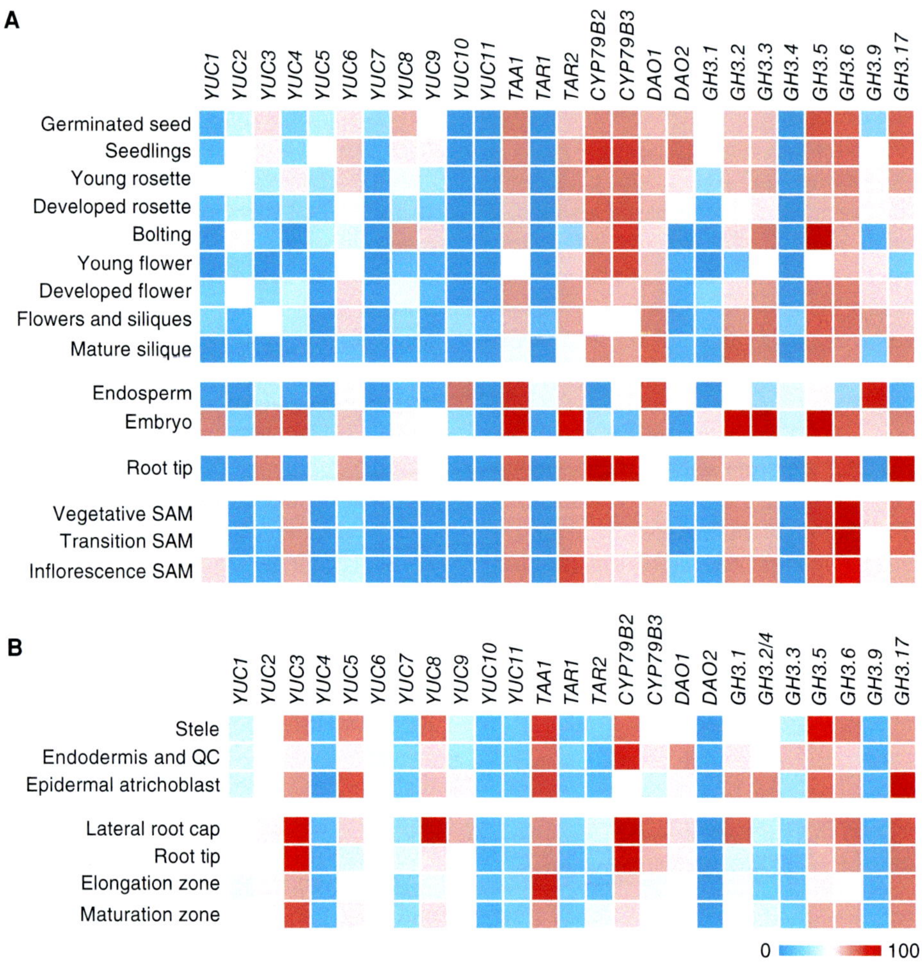

Figure 2. Relative expression level of the main genes for auxin metabolism in different developmental stages, organs, and tissues from *Arabidopsis*. Data was retrieved from Genevestigator (genevestigator.com; Hruz et al. 2008) using datasets from (*A*) RNA-seq experiments, and (*B*) Affymetrix *Arabidopsis* ATH1 Genome Array. Results are expressed in percentage of expression potential (the maximum expression a gene reaches across all experiments). *GH3.2* and *GH3.4* share array probes in the results shown in *B*. AGI codes: *YUC1* (At4G32540), *YUC2* (At4G13260), *YUC3* (At1G04610), *YUC4* (At5G11320), *YUC5* (At5G43890), *YUC6* (At5G25620), *YUC7* (At2G33230), *YUC8* (At4G28720), *YUC9* (At1G04180), *YUC10* (At1G48910), *YUC11* (At1G21430), *TAA1* (At1G70560), *TAR1* (At1G23320), *TAR2* (At4G24670), *CYP79B2* (At4G39950), *CYP79B3* (At2G22330), *DAO1* (At1G14130), *DAO2* (At1G14120), *GH3.1* (At2G14960), *GH3.2* (At4G37390), *GH3.3* (At2G23170), *GH3.4* (At1G59500), *GH3.5* (At4G27260), *GH3.6* (At5G54510), *GH3.9* (At2G47750), and *GH3.17* (At1G28130).

xylem-pole-pericycle, while PIN-mediated transport was found dispensable (Matosevich et al. 2020).

Localized auxin degradation was also found to play various roles in plant growth and development. The IAA-amido synthetase GH3.17 was shown to modulate hypocotyl elongation in response to shade and temperature independently of auxin transport and de novo biosynthesis (Zheng et al. 2016). *GH3.17*, *GH3.5*, and *GH3.6* are specifically expressed in the lateral root cap (Di Mambro et al. 2019; Pierdonati et al. 2019), with GH3.17—at the very least—participating in the IAA degradation required to create an auxin minimum at the transition zone, which influences root meristem size (Di Mambro et al. 2019).

REGULATION OF AUXIN METABOLISM

Decades of intensive research have resulted in a deep understanding of the network of interconnected and redundant pathways involved in IAA synthesis and inactivation. To ensure plastic and coordinated plant development, these pathways need to be fine-tuned by an array of regulating mechanisms that control the activation/deactivation of various synthetic and catabolic routes. However, our understanding of the regulation of this complex homeostatic machinery is mainly based on research that has been conducted during the past decade. The regulatory mechanisms underlying IAA synthesis and inactivation include (1) metabolic regulation, (2) hormonal cross talk, (3) transcriptional regulation, (4) posttranslational modifications, and (5) subcellular compartmentalization.

Metabolic Regulation of IAA Metabolism

It has been well established that IAA biosynthesis is controlled by feedback inhibitory mechanisms dependent on IAA levels (Ljung et al. 2001a). Auxin perception and signaling was shown to cause the down-regulation of genes related to the IPyA biosynthetic pathway and, consequently, to decrease endogenous IAA levels in a feedback loop (Suzuki et al. 2015; Takato et al. 2017). Likewise, the impaired conversion of IAA-aa conjugates, IBA, and meIAA to free IAA has been shown to up-regulate the expression of genes associated with IPyA-dependent auxin synthesis (Spiess et al. 2014). Moreover, the disrupted oxidative degradation of auxin in *Arabidopsis dao1* mutant plants up-regulates not only redundant GH3-mediated degradation, but also de novo auxin biosynthesis, which is a rather counterintuitive way of returning to homeostasis after an increase in auxin (Mellor et al. 2016; Porco et al. 2016).

Chorismate, the terminal product of the shikimate pathway, is a common precursor in the synthesis of the aromatic amino acids Trp, tyrosine (Tyr), and phenylalanine (Phe) (Maeda and Dudareva 2012). While the Tyr and Phe biosynthetic routes share common intermediates, Trp synthesis proceeds independently. However, a recent investigation revealed metabolic interplay between the Phe and IAA biosynthesis pathways. Phenylpyruvate, an intermediate in cytosolic Phe synthesis, can also serve as an amino acceptor in the Trp-to-IPyA conversion by Trp amino transferases and, thus, modulates IAA biosynthesis in response to Phe fluctuations (Lynch et al. 2020). Such metabolic interplay might also play a role under specific stresses, such as wounding (Lynch et al. 2020).

A screening for suppressors of the impaired shade-avoidance response observed in the *TAA1* mutant *sav3-1* identified the aminotransferase VAS1, which converts IPyA back to Trp and, thus, limits IAA synthesis (Zheng et al. 2013). VAS1 uses the ethylene intermediate methionine as a preferred amino donor, which revealed a link in the metabolic control of auxin and ethylene biosynthesis (Zheng et al. 2013). VAS1-mediated regulation is specifically, but most likely not exclusively, required for shade-induced elongation of hypocotyl and petioles (Zheng et al. 2013) and for the control of parthenocarpy in Solanaceae species (Matsuo et al. 2020).

Additional metabolic control of the IPyA pathway relies on the glycosylation of IPyA to IPyA-glc by UGT76F1, which was recently identified as a fine-tuning mechanism in the modulation of IPyA availability for IAA biosynthesis during light- and temperature-induced hypocotyl elongation (Chen et al. 2020).

Hormonal Cross Talk Modulates IAA Metabolism

The conversion of chorismate to anthranilate, which represents the first, rate-limiting step in the Trp biosynthetic pathway, is catalyzed by the anthranilate synthase complex (Niyogi and Fink 1992). Previous research has identified anthranilate production as a central hub in the modulation of IAA biosynthesis by other hormonal pathways. A screen for mutants with altered responses to ethylene identified two mutants, *wei2* and *wei7*, which harbored mutations in genes encoding anthranilate synthase α1 (ASA1) and β1 (ASB1) subunits, respectively (Stepanova et al. 2005). The *ASA1* promoter was later found to be a direct target of ETHYLENE RESPONSE FACTOR 1 (ERF1), which up-regulates *ASA1* expression in response to ethylene, resulting in increased auxin biosynthesis and root growth inhibition (Mao et al. 2016). ASA1 was also found to be at the core of jasmonate-auxin metabolic cross talk (Sun et al. 2009), in a mechanism that relies on the direct up-regulation of *ASA1* expression by the jasmonate-responsive ETHYLENE RESPONSE FACTOR 109 (ERF109) (Cai et al. 2014). ERF109 additionally binds to *YUC2* promoter to stimulate auxin biosynthesis in response to an increase in jasmonate levels (Cai et al. 2014). *YUC8* and *YUC9* might also be involved in jasmonate-regulated auxin biosynthesis, as their expression is promoted by methyl jasmonate (Hentrich et al. 2013).

The interplay between auxin and cytokinin has long been recognized as crucial to plant growth and development (Moubayidin et al. 2009). In *Arabidopsis*, cytokinin signal transduction was shown to modulate the rate of IAA biosynthesis (Jones et al. 2010). This metabolic cross talk involves the cytokinin-mediated up-regulation of *YUC1* and *YUC4* at the gynoecium primordia (Müller et al. 2017) and of *YUC8* in roots (Di et al. 2016). The cytokinin-response transcriptional effector ARABIDOPSIS RESPONSE REGULATOR 1 (ARR1) promotes auxin biosynthesis in the stem cell niche via the up-regulation of ASB1 (Moubayidin et al. 2013). While a direct molecular link for cytokinin-regulated auxin biosynthesis remains unknown, cytokinin was found to directly modulate IAA degradation. More specifically, ARR1 directly binds to *GH3.17* and activates its transcription in response to cytokinin, thus promoting auxin degradation (Di Mambro et al. 2017). This regulatory module was found to be a prerequisite for root meristem size determination (Di Mambro et al. 2017, 2019).

Transcriptional Regulation of IAA Metabolism

The coordinated response of auxin biosynthesis and inactivation to specific physiological and environmental cues is mediated by the direct modulation that specific transcription factors and components of the epigenetic machinery exert on the expression of auxin metabolic genes. Although little is known about the transcriptional control of auxin metabolism, several transcription factors and epigenetic regulators have been identified and associated to the control of auxin metabolism during specific plant developmental responses (see Tables 1 and 2).

Specific posttranslational histone modifications, also termed histone marks, are associated with an active or repressed transcriptional state. Whole-genome occupancy studies of the histone repressive mark H3K27me3 found that epigenetic mechanisms control auxin-related genes. In comparisons of dividing and differentiated cells, differential H3K27me3 modifications were observed at genes involved in auxin biosynthesis (*YUCs*, *CYPs*, *TAA1/TARs*, *SUR1*, *NITs*), inactivation (*GH3s*, *IAMT*), transport (*PINs*, *AUX/LAXs*), and signaling (*TIR1/AFBs*, *IAAs*, *ARFs*), thus revealing that this histone mark exerts a profound effect on auxin action (Lafos et al. 2011; He et al. 2012). *YUC1* and *YUC4*, for example, were found to be specifically involved in early auxin-mediated de novo root regeneration, in direct correlation with an H3K27me3 drop along their promoter regions (Chen et al. 2016). The Polycomb Repressive Complex 2 (PRC2) accessory protein LIKE HETEROCHROMATIN 1 (LHP1) (Derkacheva et al. 2013) are recruited to *YUC1*, *YUC2*, *YUC4*, *YUC5*, *YUC6*, *YUC8*, *YUC9*, and *YUC10* promoters to control their expression (Rizzardi et al. 2011). Beyond general correlations and whole-genome comparisons, specific mechanisms related not only to histone modifications —but also DNA methylation and small RNAs— have been shown to alter auxin homeostasis by regulating the transcription of *YUCs* (for review, see Mateo-Bonmatí et al. 2019).

During the first steps of flower determination—a process in which floral meristem cells stop proliferating and initiate a floral organ primordium—flower primordia formation requires auxin-driven rapid cell expansion and elongation. The required increase in auxin levels is, at least in part, achieved by the activation of

Table 1. Transcription factors that have been shown to directly modulate the expression of genes related to IAA metabolism

Target	Transcriptional regulator(s)	Structural type	Biological process(es)	References
YUC1	ARR1, ARR10, ARR12	B-ARR	Shoot stem cell niche maintenance	Meng et al. 2017
	SUP	C2H2-type zinc finger	Floral patterning	Xu et al. 2018
YUC4	LEC2	B3 domain	Embryogenesis	Stone et al. 2008
	SHI/STY1	RING-like zinc finger	Leaf and flower development	Eklund et al. 2010
	ARR1, ARR10, ARR12	B-ARR	Shoot stem cell niche maintenance	Meng et al. 2017
	LEC2, FUS3	B3 domain	Lateral root formation	Tang et al. 2017
	SUP	C2H2-type zinc finger	Floral patterning	Xu et al. 2018
	CRC	C2C2 type zinc finger	Floral determinacy	Yamaguchi et al. 2018
	AG	MADS	Floral determinacy	Yamaguchi et al. 2018
YUC5	IDD14, IDD15, IDD16	IDD	Organ morphogenesis, gravitropism	Cui et al. 2013
YUC8	PIF4	bHLH	Hypocotyl elongation at high temperature	Franklin et al. 2011
YUC9	EIN3	EIL	Aluminum-induced root growth inhibition	Liu et al. 2016b
TAA1	PIF4	bHLH	Hypocotyl elongation at high temperature	Franklin et al. 2011
	IDD14, IDD15, IDD16	IDD	Organ morphogenesis, gravitropism	Cui et al. 2013
	SPT	bHLH	Gynoecium development	Reyes-Olalde et al. 2017
	ARR1s	B-ARR	Light-induced tissue-specific IAA synthesis	Yan et al. 2017
CYP79B2	PIF4	bHLH	Hypocotyl elongation at high temperature	Franklin et al. 2011
UGT76F1	PIF4	bHLH	Hypocotyl elongation at high temperature	Chen et al. 2020
OsYUC8	OsEIL1	EIL	Ethylene-mediated primary root elongation	Qin et al. 2017
ZmGH3.2	ZmDREB2A	AP2/ERF	Longevity of maize seed	Han et al. 2020
GH3.2 GH3.6	MYB30	R2R3-MYB	Root growth	Zhao and Xue 2020

(Os) *Oryza sativa*, (Zm) *Zea mays*; otherwise *Arabidopsis thaliana*.

YUC4 by the chromatin-remodeling factors CHROMATIN REMODELLING 11 (CHR11) and CHR17, both of which are specifically recruited to the *YUC4* promoter during floral primordium formation (Yamaguchi et al. 2018). Later in floral organ development, the C2H2-type zinc-finger transcription factor *SUPERMAN* (*SUP*) actively represses *YUC1* and *YUC4* expression and, thereby, auxin biosynthesis at the boundaries between carpels and stamen primordia (Xu et al. 2018). *SUP*-mediated transcriptional silencing of *YUC1* and *YUC4* further involves the recruitment of members of the PRC2 machinery, such as CURLY LEAF (CLF) or LHP1, for the trimethylation of H3K27 (Xu et al. 2018).

Table 2. Epigenetic regulators that have been shown to directly modulate the expression of genes involved in IAA biosynthesis in a specific biological process

Target	Epigenetic regulator(s)	Biological process(es)	References
YUC1	LHP1	Floral patterning (−)	Xu et al. 2018
	CLF-PRC2		
YUC2	CMT3, DRM1, DRM2	Leaf development/growth (−)	Forgione et al. 2019
	LOCUS_77297	Ambient temperature (−)	Gyula et al. 2018
YUC4	LHP1	Floral patterning (−)	Xu et al. 2018
	CLF-PRC2		
	CHR11	Floral determinacy (+)	Yamaguchi et al. 2018
	CHR17		
YUC8	MRG2	Hypocotyl elongation (shade) (+)	Peng et al. 2018
	HDA9	Hypocotyl elongation (temperature) (+)	van der Woude et al. 2019
YUC9	ARP4	Hypocotyl elongation (shade) (−)	Lee and Seo 2017
YUC10	FIS2-PRC2	Endosperm development (−)	Figueiredo et al. 2015

(ARP4) ACTIN-RELATED PROTEIN 4, (CHR11) CHROMATIN REMODELLING 11, (CLF) CURLY LEAF, (CMT3) CHROMOMETHYLASE 3, (DRM1) DOMAINS REARRANGED 1, (FIS2) FERTILIZATION INDEPENDENT SEED 2, (HDA9) HISTONE DEACETYLASE 9, (LHP1) LIKE HETEROCHROMATIN 1, (MRG2) MORF-RELATED GENE 2, (PRC2) POLYCOMB REPRESSIVE COMPLEX 2. The (-) and (+) reflect transcriptional repression and activation of the target, respectively.

During angiosperm fertilization, auxin biosynthesis is constitutively repressed in maternal-derived tissues by the action of the FERTILIZATION-INDEPENDENT SEED-PRC2 (FIS-PRC2) complex (Figueiredo et al. 2015). Mutations in genes encoding subunits of this complex lead to premature expression of *YUC10* in the nonfertilized diploid central cell, which will result in the development of empty seeds. Recent reports indicate that additional mechanisms, driven by EMSY-like Tudor/Agenet H3K36me3 histone readers EMSY-Like protein 1 (EML1) and EML3, actively repress auxin biosynthesis, transport, and signaling during seed coat and endosperm development (Milutinovic et al. 2019).

Auxin integrates internal and external signals as sugar levels, shade, and temperature into regulated plant growth responses (Sairanen et al. 2012; for review, see Zhao 2018). A good model to exemplify such interaction is the auxin biosynthesis-driven hypocotyl elongation in response to shade or high temperature (Gray et al. 1998; Tao et al. 2008). In the hypocotyls of plants grown under normal light conditions, *YUC9* is actively repressed by the action of two AT-HOOK-CONTAINING NUCLEAR-LOCALIZED (AHL) proteins, AHL27 and AHL29 (Lee and Seo 2017). AHL29 recruits ACTIN-RELATED PROTEIN 4 (ARP4), a member of the SWI2/SNF2-RELATED1 (SWR1) chromatin-remodeling complex, to the *YUC9* regulatory region to promote the deposition of histone variant H2A.Z and, hence, block the access of RNApol II to the DNA (Lee and Seo 2017). Similar mechanisms for transcriptional regulation are driven by ARP6 under normal temperatures (Kumar and Wigge 2010). However, independent mechanisms trigger *YUC8* expression in plant hypocotyls to facilitate their elongation under shade or high temperature. The transcription factor PHYTOCHROME-INTERACTING FACTOR 7 (PIF7) and the H3K4me3/H3K36me3-binding protein Morf-Related Gene 2 (MRG2) bind to the *YUC8* promoter in response to shade, and facilitate its transcription by allowing the acetylation of H3 and H4 (Peng et al. 2018). At high temperature, HISTONE DEACETYLASE 9 (HDA9) accumulates, facilitating the H2A.Z removal from the *YUC8* locus and providing a looser chromatin environment that allows PIF4-mediated activation of *YUC8* transcription (van der Woude et al. 2019). High temperatures additionally promote auxin biosynthesis through the temperature-specific recruitment of PIF4 to the promoters of the IPyA glycosylase *UGT76F1* and the IAOx-pathway-related *CYP79B2* gene to repress and promote their transcription, respectively, by unknown epigenetic mechanisms (Chen et al. 2020).

DNA methylation of cytosines is another important epigenetic mark mostly associated with gene repression that has also been shown to participate in auxin homeostasis. In animals, DNA methylation is almost entirely restricted to CG dinucleotides while plant DNA can additionally be methylated at CHG and CHH sequence contexts (with H representing A, T, or C) (Zhang et al. 2018; Gallego-Bartolomé 2020). The cytosine methylation is mediated by RNA-directed DNA methylation (RdDM), in which 24-nt small interfering RNA (siRNA) guides the DNA methyltransferases DOMAINS REARREANGED METHYLTRANSFERASE 1 (DRM1) and DRM2 to the target region (Zhang et al. 2018). Following DNA replication, cell-specific methylation patterns are maintained by another subset of DNA methyltransferases (METHYLTRANSFERASE 1 [MET1], CHROMOMETHYLASE 2 [CMT2], and CMT3) which, in contrast to DRMs, function in a context-specific manner, targeting CG, CHH, and CHG sequences, respectively (Zhang et al. 2018). Interestingly, several phenotypes that have been linked with auxin deficiency, such as root agravitropism, aberrant embryogenesis, and vascular disorders, were also observed in a triple *drm1 drm2 cmt3* mutant (Forgione et al. 2019). Tissue-specific expression analyses in *drm1 drm2 cmt3* seedlings revealed that the auxin biosynthetic genes *YUC2* and *TAA1* were specifically up-regulated in leaves. Me-DIP (methylated DNA immunoprecipitation)-PCR experiments further confirmed a reduction of non-GC DNA methylation at the *YUC2* promoter, thus linking RdDM with the regulation of auxin biosynthesis (Forgione et al. 2019). A search for thermoresponsive regulatory RNAs identified a novel temperature-regulated 24-nt siRNA, coined Locus_77297, in the vicinity of the *YUC2* promoter (Gyula et al. 2018). The expression of this siRNA in leaves was positively correlated with CHH methylation at the *YUC2* promoter. Plants grown at high temperatures showed severe reductions in both Locus_77297 expression and CHH methylation, which triggered the up-regulation of *YUC2* (Gyula et al. 2018).

Posttranslational Regulation of IAA Metabolism

After transcriptional control, additional regulatory mechanisms also govern protein function posttranslationally. Posttranslational modifications (PTMs) refer to covalent modifications that generally modulate protein folding or activity. There is recent evidence that auxin homeostasis is also modulated by PTMs. *Arabidopsis* TAA1 phosphorylation at the Thr101 was shown to be triggered by auxin perception itself, and to serve as an on/off switch controlling the activity of the enzyme, and therefore IAA biosynthesis (Wang et al. 2020). DAO1 auxin oxidase is known to be barely induced by IAA (Porco et al. 2016). Instead, experiments in rice demonstrated that DAO activity is regulated posttranslationally by substrate-mediated multimerization, which involves the IAA-triggered formation of DAO dimers that show increased affinity for IAA (Takehara et al. 2020). Interestingly, IAA-triggered TAA1 phosphorylation also enables this enzyme to dimerize with homologous TAR enzymes, a dynamic that likely regulates IAA biosynthesis through the control of several isoenzymes (Wang et al. 2020). Whether DAO dimerization also depends on IAA-triggered phosphorylation remains unexplored, and future research might identify a common mechanism for the posttranslational control of auxin metabolism.

Subcellular Compartmentalization of the IAA Metabolism

The cellular compartmentalization of bioactive IAA, IAA metabolites, and IAA metabolic enzymes represents yet another mechanism through which intracellular levels of IAA are regulated (Skalický et al. 2018). The first compartmentalization of IAA regards its biosynthesis, as the central precursor Trp is produced in chloroplasts, and the IAOx-pathway-related CYP79B2 and CYP79B3 contain a chloroplast transit peptide (Hull et al. 2000). Nevertheless, the major routes for IAA biosynthesis and degradation are believed to take place in the cytosol, as the enzymes TAA1, YUC1, YUC2, YUC3, YUC6, YUC11, and DAO1 have been shown to share a cytoplasmic localization (Stepanova et al. 2008; Tao et al. 2008; Zhao et al. 2013; Kriechbaumer et al. 2016; Porco et al. 2016; Zhang et al. 2016). The IAA amido synthetase GH3.17 is also localized to the cytosol (Di Mambro et al. 2019). However, TAR2, YUC5, YUC7, YUC8, and YUC9 colocalize with ER-membrane markers (Kriechbaumer et al. 2016). *YUC4* is expressed as two tissue-specific splice variants, with one localized to the cytosol and the other localized to the cytosolic face of the endoplasmic reticulum (ER) membrane (Kriechbaumer et al. 2012). This suggests that part of IAA biosynthesis is compartmentalized to the ER. In line with this hypothesis, isolated ER microsomes were shown to significantly contribute to IPyA-dependent IAA biosynthesis (Kriechbaumer et al. 2015, 2016). Additionally, the IAA amidohydrolases ILR1, IAR3, and ILL2 have been shown to localize to the ER (Sanchez Carranza et al. 2016).

Together with the apparent subcellular localization of the IAA metabolism, active transport between organelles and the cytosol is a major factor determining cellular IAA homeostasis. The atypical members of the PIN family of auxin transporters PIN5, PIN6, and PIN8, together with PIN-LIKES proteins (PILS), have been detected at the ER (Mravec et al. 2009; Barbez et al. 2012; Dal Bosco et al. 2012; Ding et al. 2012; Simon et al. 2016). Genetic analyses on these PINs and PILS mutants and overexpressors suggest that ER-compartmentalization of IAA regulates auxin signaling by limiting the available cytosolic IAA that can enter to the nucleus (Mravec et al. 2009; Béziat et al. 2017; Feraru et al. 2019). Genetic manipulation of the ER transporters leads to altered levels of IAA as well as the IAA-Asp and IAA-Glu conjugates (Mravec et al. 2009; Dal Bosco et al. 2012; Ding et al. 2012; Simon et al. 2016), which further suggests that the compartmentalization of certain IAA metabolites is important to maintaining auxin homeostasis. Vacuole-associated auxin transporters that move IAA out of the vacuole have also been found (Ranocha et al. 2013; Liu et al. 2016a), pointing to a role of the vacuole in intracellular IAA regulation (Fig. 1). The presence of IAA precursors, as well as oxIAA and IAA-glc, in *Arabidopsis* vacuoles suggests that this organelle is involved in at least part of IAA metabolism (Ranocha et al. 2013). Additionally, the IBA vacuolar transporter TOB1 was shown to move IBA from the cytoplasm to the vacuole, which provides further evidence that the vacuole participates in subcellular IAA metabolism (Michniewicz et al. 2019).

CONCLUDING REMARKS AND FUTURE PERSPECTIVES

Auxin is mainly produced in plants from Trp through the formation of IPyA by the action of plant-conserved members of the TAA and YUC enzyme families. The IPyA pathway is a major and essential pathway for auxin biosynthesis, while other parallels routes for auxin production turned out to have minor or environmentally restricted roles during plant development. Irreversible IAA degradation by the redundant action of DAO oxidases and GH3 amido synthetases additionally control auxin concentrations across plant tissues. Research on when and where these genes are expressed has revealed a strict requirement of localized auxin metabolism for regulated plant growth and development. Major breakthroughs during the past decade have shed light on the transcriptional control of auxin biosynthesis, as several transcription factors and epigenetic regulators have been described to control local auxin production during specific developmental processes in response to environmental and physiological cues (Tables 1 and 2). However, we are only starting to understand the molecular basis for such regulation. As such, the transcriptional control of genes associated with auxin inactivation remains largely unexplored. Whether the so-far-known regulatory mechanisms operate in a similar way across different tissues and/or developmental times needs further exploration. Moreover, when considering the number of transcription factors that have been reported to directly control auxin biosynthesis (Table 1), it seems probable that many additional epigenetic mechanisms for the regulation of auxin metabolism will be elucidated in the coming years. The recently described posttranslational modulation of the activity of the auxin biosynthetic enzyme TAA1 (Wang et al. 2020) and the auxin oxidase DAO (Takehara et al. 2020) represents a thrilling starting point to explore similar mechanisms controlling the activity of other enzymes from the pathway.

Despite extensive efforts to unravel intracellular IAA dynamics to understand the role of organelles such as the ER and the vacuole on cellular auxin homeostasis, many critical questions remain unanswered. For example, the subcellular localization of several enzymes involved in IAA metabolism, notably, most of the GH3s, remains unexplored. Moreover, our understanding of how various IAA metabolites are distributed among different organelles, and how this compartmentalization influences intracellular auxin levels, transport, and signaling, is still at a rudimentary level. Fluorescent activated cell sorting has been employed to characterize the hormone distribution in different *Arabidopsis* root cell types (Petersson et al. 2009; Antoniadi et al. 2015). Sorting pure fractions of organelles by fluorescent activated organelle sorting (FAOS), a technique successfully used in mammalian cells (Gauthier et al. 2008), followed by high-resolution IAA metabolite profiling is expected to advance our knowledge regarding how the intracellular compartmentalization of different enzymes and metabolites influences auxin homeostasis (Novák et al. 2017; Skalický et al. 2018). Novel techniques, however, will be required to study the dynamic changes of subcellular IAA distribution and metabolism in response to different stimuli *in planta*. As most of the IAA is sensed in the nucleus, determination of IAA levels in nuclei during different responses to auxin in living plants, and the specific role of other cell compartments in modulating the available IAA to be sensed, remains an exciting challenge.

ACKNOWLEDGMENTS

We apologize to the authors who have made valuable contributions to the field, but whose research we could not include because of space constraints. Research in the laboratory of Karin Ljung is supported by grants from the Swedish Foundation for Strategic Research (Vinnova), the Knut and Alice Wallenberg Foundation (KAW), the Swedish Research Councils VR and Formas, and Carl Tryggers Stiftelse för Vetenskaplig Forskning. R.C.-S. held a postdoctoral fellowship from Kempestiftelserna (JCK-1111). E.M.-B. holds a postdoctoral fellowship from Kempestiftelserna (JCK-1811).

REFERENCES

Abbas M, Hernández-García J, Pollmann S, Samodelov SL, Kolb M, Friml J, Hammes UZ, Zurbriggen MD, Blázquez MA, Alabadí D. 2018. Auxin methylation is required for differential growth in *Arabidopsis*. *Proc Natl Acad Sci* **115:** 6864–6869. doi:10.1073/pnas.1806565115

Abel S, Theologis A. 2010. Odyssey of auxin. *CSH Perspect Biol* **2:** a004572.

Antoniadi I, Plačková L, Simonovik B, Doležal K, Turnbull C, Ljung K, Novák O. 2015. Cell-type-specific cytokinin distribution within the *Arabidopsis* primary root apex. *Plant Cell* **27:** 1955–1967. doi:10.1105/tpc.15.00176

Barbez E, Kubeš M, Rolčík J, Béziat C, Pěnčík A, Wang B, Rosquete MR, Zhu J, Dobrev PI, Lee Y, et al. 2012. A novel putative auxin carrier family regulates intracellular auxin homeostasis in plants. *Nature* **485:** 119–122. doi:10.1038/nature11001

Barlier I, Kowalczyk M, Marchant A, Ljung K, Bhalerao R, Bennett M, Sandberg G, Bellini C. 2000. The *SUR2* gene of *Arabidopsis thaliana* encodes the cytochrome P450 CYP83B1, a modulator of auxin homeostasis. *Proc Natl Acad Sci* **97:** 14819–14824. doi:10.1073/pnas.260502697

Bartel B, Fink GR. 1995. ILR1, an amidohydrolase that releases active indole-3-acetic acid from conjugates. *Science* **268:** 1745–1748. doi:10.1126/science.7792599

Bartel B, LeClere S, Magidin M, Zolman BK. 2001. Inputs to the active indole-3-acetic acid pool: de novo synthesis, conjugate hydrolysis, and indole-3-butyric acid β-oxidation. *J Plant Growth Regul* **20:** 198–216. doi:10.1007/s003440010025

Bernardi J, Lanubile A, Li QB, Kumar D, Kladnik A, Cook SD, Ross JJ, Marocco A, Chourey PS. 2012. Impaired auxin biosynthesis in the *defective endosperm18* mutant is due to mutational loss of expression in the *ZmYuc1* gene encoding endosperm-specific YUCCA1 protein in maize. *Plant Physiol* **160:** 1318–1328. doi:10.1104/pp.112.204743

Béziat C, Barbez E, Feraru MI, Lucyshyn D, Kleine-Vehn J. 2017. Light triggers PILS-dependent reduction in nuclear auxin signalling for growth transition. *Nat Plants* **3:** 17105. doi:10.1038/nplants.2017.105

Bhalerao RP, Eklöf J, Ljung K, Marchant A, Bennett M, Sandberg G. 2002. Shoot-derived auxin is essential for early lateral root emergence in *Arabidopsis* seedlings. *Plant J* **29:** 325–332. doi:10.1046/j.0960-7412.2001.01217.x

Boerjan W, Cervera MT, Delarue M, Beeckman T, Dewitte W, Bellini C, Caboche M, Van Onckelen H, Van Montagu M, Inzé D. 1995. *Superroot*, a recessive mutation in *Arabidopsis*, confers auxin overproduction. *Plant Cell* **7:** 1405–1419.

Brumos J, Robles LM, Yun J, Vu TC, Jackson S, Alonso JM, Stepanova AN. 2018. Local auxin biosynthesis is a key regulator of plant development. *Dev Cell* **47:** 306–318. e5. doi:10.1016/j.devcel.2018.09.022

Brunoni F, Collani S, Casanova-Sáez R, Šimura J, Karady M, Schmid M, Ljung K, Bellini C. 2020. Conifers exhibit a characteristic inactivation of auxin to maintain tissue homeostasis. *New Phytol* **226:** 1753–1765. doi:10.1111/nph.16463

Cai XT, Xu P, Zhao PX, Liu R, Yu LH, Xiang CB. 2014. *Arabidopsis* ERF109 mediates cross-talk between jasmonic acid and auxin biosynthesis during lateral root formation. *Nat Commun* **5:** 5833. doi:10.1038/ncomms6833

Chen Q, Dai X, De-Paoli H, Cheng Y, Takebayashi Y, Kasahara H, Kamiya Y, Zhao Y. 2014. Auxin overproduction in shoots cannot rescue auxin deficiencies in *Arabidopsis* roots. *Plant Cell Physiol* **55:** 1072–1079. doi:10.1093/pcp/pcu039

Chen L, Tong J, Xiao L, Ruan Y, Liu J, Zeng M, Huang H, Wang JW, Xu L. 2016. *YUCCA*-mediated auxin biogenesis is required for cell fate transition occurring during de novo root organogenesis in *Arabidopsis. J Exp Bot* **67:** 4273–4284. doi:10.1093/jxb/erw213

Chen L, Huang XX, Zhao SM, Xiao DW, Xiao LT, Tong JH, Wang WS, Li YJ, Ding Z, Hou BK. 2020. IPya glucosylation mediates light and temperature signaling to regulate auxin-dependent hypocotyl elongation in *Arabidopsis. Proc Natl Acad Sci* **117:** 6910–6917. doi:10.1073/pnas.2000172117

Cheng Y, Dai X, Zhao Y. 2006. Auxin biosynthesis by the YUCCA flavin monooxygenases controls the formation of floral organs and vascular tissues in *Arabidopsis. Genes Dev* **20:** 1790–1799. doi:10.1101/gad.1415106

Cheng Y, Dai X, Zhao Y. 2007. Auxin synthesized by the YUCCA flavin monooxygenases is essential for embryogenesis and leaf formation in *Arabidopsis. Plant Cell* **19:** 2430–2439. doi:10.1105/tpc.107.053009

Cohen JD, Bandurski RS. 1982. Chemistry and physiology of the bound auxins. *Annu Rev Plant Biol* **33:** 403–430. doi:10.1146/annurev.pp.33.060182.002155

Cook SD. 2019. An historical review of phenylacetic acid. *Plant Cell Physiol* **60:** 243–254. doi:10.1093/pcp/pcz004

Cui D, Zhao J, Jing Y, Fan M, Liu J, Wang Z, Xin W, Hu Y. 2013. The *Arabidopsis* IDD14, IDD15, and IDD16 cooperatively regulate lateral organ morphogenesis and gravitropism by promoting auxin biosynthesis and transport. *PLoS Genet* **9:** e1003759. doi:10.1371/journal.pgen.1003759

Dal Bosco C, Dovzhenko A, Liu X, Woerner N, Rensch T, Eismann M, Eimer S, Hegermann J, Paponov IA, Ruperti B, et al. 2012. The endoplasmic reticulum localized PIN8 is a pollen-specific auxin carrier involved in intracellular auxin homeostasis. *Plant J* **71:** 860–870. doi:10.1111/j.1365-313X.2012.05037.x

Davies PJ, Mitchell EK. 1972. Transport of indoleacetic acid in intact roots of *Phaseolus coccineus. Planta* **105:** 139–154. doi:10.1007/BF00385573

Davies RT, Goetz DH, Lasswell J, Anderson MN, Bartel B. 1999. *IAR3* encodes an auxin conjugate hydrolase from *Arabidopsis. Plant Cell* **11:** 365–376. doi:10.1105/tpc.11.3.365

Delarue M, Prinsen E, Onckelen HV, Caboche M, Bellini C. 1998. *Sur2* mutations of *Arabidopsis thaliana* define a new locus involved in the control of auxin homeostasis. *Plant J* **14:** 603–611. doi:10.1046/j.1365-313X.1998.00163.x

Derkacheva M, Steinbach Y, Wildhaber T, Mozgová I, Mahrez W, Nanni P, Bischof S, Gruissem W, Hennig L. 2013. *Arabidopsis* MSI1 connects LHP1 to PRC2 complexes. *EMBO J* **32:** 2073–2085. doi:10.1038/emboj.2013.145

Di DW, Wu L, Zhang L, An CW, Zhang TZ, Luo P, Gao HH, Kriechbaumer V, Guo GQ. 2016. Functional roles of *Arabidopsis CKRC2/YUCCA8* gene and the involvement of PIF4 in the regulation of auxin biosynthesis by cytokinin. *Sci Rep* **6:** 36866. doi:10.1038/srep36866

Di Mambro R, De Ruvo M, Pacifici E, Salvi E, Sozzani R, Benfey PN, Busch W, Novák O, Ljung K, Di Paola L, et al. 2017. Auxin minimum triggers the developmental switch from cell division to cell differentiation in the *Arabidopsis* root. *Proc Natl Acad Sci* **114:** E7641–E7649. doi:10.1073/pnas.1705833114

Di Mambro R, Svolacchia N, Dello Ioio R, Pierdonati E, Salvi E, Pedrazzini E, Vitale A, Perilli S, Sozzani R, Benfey PN, et al. 2019. The lateral root cap acts as an auxin sink that controls meristem size. *Curr Biol* **29:** 1199–1205.e4. doi:10.1016/j.cub.2019.02.022

Ding X, Cao Y, Huang L, Zhao J, Xu C, Li X, Wang S. 2008. Activation of the indole-3-acetic acid-amido synthetase GH3-8 suppresses expansin expression and promotes salicylate- and jasmonate-independent basal immunity in rice. *Plant Cell* **20:** 228–240. doi:10.1105/tpc.107.055657

Ding Z, Wang B, Moreno I, Dupláková N, Simon S, Carraro N, Reemmer J, Pěnčík A, Chen X, Tejos R, et al. 2012. ER-localized auxin transporter PIN8 regulates auxin homeostasis and male gametophyte development in *Arabidopsis. Nat Commun* **3:** 941. doi:10.1038/ncomms1941

D'Ippolito S, Vankova R, Joosten MH, Casalongué CA, Fiol DF. 2016. Knocking down expression of the auxin-amidohydrolase *IAR3* alters defense responses in Solanaceae family plants. *Plant Sci* **253:** 31–39. doi:10.1016/j.plantsci.2016.09.008

Du H, Wu N, Fu J, Wang S, Li X, Xiao J, Xiong L. 2012. A GH3 family member, OsGH3-2, modulates auxin and abscisic acid levels and differentially affects drought and cold tolerance in rice. *J Exp Bot* **63:** 6467–6480. doi:10.1093/jxb/ers300

Eklund DM, Ståldal V, Valsecchi I, Cierlik I, Eriksson C, Hiratsu K, Ohme-Takagi M, Sundström JF, Thelander M, Ezcurra I, et al. 2010. The *Arabidopsis thaliana* STYLISH1 protein acts as a transcriptional activator regulating auxin biosynthesis. *Plant Cell* **22:** 349–363. doi:10.1105/tpc.108.064816

Eklund DM, Ishizaki K, Flores-Sandoval E, Kikuchi S, Takebayashi Y, Tsukamoto S, Hirakawa Y, Nonomura M, Kato H, Kouno M, et al. 2015. Auxin produced by the indole-3-pyruvic acid pathway regulates development and gemmae dormancy in the liverwort *Marchantia polymorpha*. *Plant Cell* **27:** 1650–1669. doi:10.1105/tpc.15.00065

Feraru E, Feraru MI, Barbez E, Waidmann S, Sun L, Gaidora A, Kleine-Vehn J. 2019. PILS6 is a temperature-sensitive regulator of nuclear auxin input and organ growth in *Arabidopsis thaliana*. *Proc Natl Acad Sci* **116:** 3893–3898. doi:10.1073/pnas.1814015116

Figueiredo DD, Batista RA, Roszak PJ, Köhler C. 2015. Auxin production couples endosperm development to fertilization. *Nat Plants* **1:** 15184. doi:10.1038/nplants.2015.184

Forgione I, Wołoszynska M, Pacenza M, Chiappetta A, Greco M, Araniti F, Abenavoli MR, Van Lijsebettens M, Bitonti MB, Bruno L. 2019. Hypomethylated *drm1 drm2 cmt3* mutant phenotype of *Arabidopsis thaliana* is related to auxin pathway impairment. *Plant Sci* **280:** 383–396. doi:10.1016/j.plantsci.2018.12.029

Franklin KA, Lee SH, Patel D, Kumar SV, Spartz AK, Gu C, Ye S, Yu P, Breen G, Cohen JD, et al. 2011. Phytochrome-interacting factor 4 (PIF4) regulates auxin biosynthesis at high temperature. *Proc Natl Acad Sci* **108:** 20231–20235. doi:10.1073/pnas.1110682108

Frick EM, Strader LC. 2018. Roles for IBA-derived auxin in plant development. *J Exp Bot* **69:** 169–177. doi:10.1093/jxb/erx298

Gallavotti A, Barazesh S, Malcomber S, Hall D, Jackson D, Schmidt RJ, McSteen P. 2008. *Sparse inflorescence1* encodes a monocot-specific *YUCCA*-like gene required for vegetative and reproductive development in maize. *Proc Natl Acad Sci* **105:** 15196–15201. doi:10.1073/pnas.0805596105

Gallego-Bartolomé J. 2020. DNA methylation in plants: mechanisms and tools for targeted manipulation. *New Phytol* **227:** 38–44. doi:10.1111/nph.16529

Gallei M, Luschnig C, Friml J. 2020. Auxin signalling in growth: Schrödinger's cat out of the bag. *Curr Opin Plant Biol* **53:** 43–49. doi:10.1016/j.pbi.2019.10.003

Gao Y, Dai X, Aoi Y, Takebayashi Y, Yang L, Guo X, Zeng Q, Yu H, Kasahara H, Zhao Y. 2020. Two homologous *INDOLE-3-ACETAMIDE (IAM) HYDROLASE* genes are required for the auxin effects of IAM in *Arabidopsis*. *J Genet Genom* **47:** 157–165. doi:10.1016/j.jgg.2020.02.009

Gauthier DJ, Sobota JA, Ferraro F, Mains RE, Lazure C. 2008. Flow cytometry-assisted purification and proteomic analysis of the corticotropes dense-core secretory granules. *Proteomics* **8:** 3848–3861. doi:10.1002/pmic.200700969

Glawischnig E, Hansen BG, Olsen CE, Halkier BA. 2004. Camalexin is synthesized from indole-3-acetaldoxime, a key branching point between primary and secondary metabolism in *Arabidopsis*. *Proc Natl Acad Sci* **101:** 8245–8250. doi:10.1073/pnas.0305876101

Gray WM, Östin A, Sandberg G, Romano CP, Estelle M. 1998. High temperature promotes auxin-mediated hypocotyl elongation in *Arabidopsis*. *Proc Natl Acad Sci* **95:** 7197–7202. doi:10.1073/pnas.95.12.7197

Grubb CD, Zipp BJ, Ludwig-Müller J, Masuno MN, Molinski TF, Abel S. 2004. *Arabidopsis* glucosyltransferase UGT74B1 functions in glucosinolate biosynthesis and auxin homeostasis. *Plant J* **40:** 893–908. doi:10.1111/j.1365-313X.2004.02261.x

Gutierrez L, Mongelard G, Floková K, Păcurar DI, Novák O, Staswick P, Kowalczyk M, Păcurar M, Demailly H, Geiss G, et al. 2012. Auxin controls *Arabidopsis* adventitious root initiation by regulating jasmonic acid homeostasis. *Plant Cell* **24:** 2515–2527. doi:10.1105/tpc.112.099119

Gyula P, Baksa I, Tóth T, Mohorianu I, Dalmay T, Szittya G. 2018. Ambient temperature regulates the expression of a small set of sRNAs influencing plant development through *NF-YA2* and *YUC2*. *Plant Cell Environ* **41:** 2404–2417. doi:10.1111/pce.13355

Hall PJ. 1980. Indole-3-acetyl-myo-inositol in kernels of *Oryza sativa*. *Phytochemistry* **19:** 2121–2123. doi:10.1016/S0031-9422(00)82206-2

Han Q, Chen K, Yan D, Hao G, Qi J, Wang C, Dirk LMA, Bruce Downie A, Gong J, Wang J, et al. 2020. ZmDREB2A regulates *ZmGH3.2* and *ZmRAFS*, shifting metabolism towards seed aging tolerance over seedling growth. *Plant J* **104:** 1268–1282. doi:10.1111/TPJ.14922

He C, Chen X, Huang H, Xu L. 2012. Reprogramming of H3K27me3 is critical for acquisition of pluripotency from cultured *Arabidopsis* tissues. *PLoS Genet* **8:** e1002911. doi:10.1371/journal.pgen.1002911

Hentrich M, Böttcher C, Düchting P, Cheng Y, Zhao Y, Berkowitz O, Masle J, Medina J, Pollmann S. 2013. The jasmonic acid signaling pathway is linked to auxin homeostasis through the modulation of *YUCCA8* and *YUCCA9* gene expression. *Plant J* **74:** 626–637. doi:10.1111/tpj.12152

Hruz T, Laule O, Szabo G, Wessendorp F, Bleuler S, Oertle L, Widmayer P, Gruissem W, Zimmermann P. 2008. Genevestigator V3: a reference expression database for the meta-analysis of transcriptomes. *Adv Bioinformatics* **2008:** 420747. doi:10.1155/2008/420747

Hull AK, Vij R, Celenza JL. 2000. *Arabidopsis* cytochrome P450s that catalyze the first step of tryptophan-dependent indole-3-acetic acid biosynthesis. *Proc Natl Acad Sci* **97:** 2379–2384. doi:10.1073/pnas.040569997

Ikeda Y, Men S, Fischer U, Stepanova AN, Alonso JM, Ljung K, Grebe M. 2009. Local auxin biosynthesis modulates gradient-directed planar polarity in *Arabidopsis*. *Nat Cell Biol* **11:** 731–738. doi:10.1038/ncb1879

Ishimaru K, Hirotsu N, Madoka Y, Murakami N, Hara N, Onodera H, Kashiwagi T, Ujiie K, Shimizu B, Onishi A, et al. 2013. Loss of function of the IAA-glucose hydrolase gene *TGW6* enhances rice grain weight and increases yield. *Nat Genet* **45:** 707–711. doi:10.1038/ng.2612

Jackson RG, Kowalczyk M, Li Y, Higgins G, Ross J, Sandberg G, Bowles DJ. 2002. Over-expression of an *Arabidopsis* gene encoding a glucosyltransferase of indole-3-acetic acid: phenotypic characterisation of transgenic lines. *Plant J* **32:** 573–583. doi:10.1046/j.1365-313X.2002.01445.x

Jakubowska A, Kowalczyk S. 2005. A specific enzyme hydrolyzing 6-*O*(4-*O*)-indole-3-ylacetyl-β-D-glucose in immature kernels of *Zea mays*. *J Plant Physiol* **162:** 207–213. doi:10.1016/j.jplph.2004.05.015

Jayasinghege CPA, Ozga JA, Nadeau CD, Kaur H, Reinecke DM. 2019. TIR1 auxin receptors are implicated in the differential response to 4-Cl-IAA and IAA in developing pea fruit. *J Exp Bot* **70:** 1239–1253. doi:10.1093/jxb/ery456

Jones B, Gunnerås SA, Petersson SV, Tarkowski P, Graham N, May S, Dolezal K, Sandberg G, Ljung K. 2010. Cytokinin regulation of auxin synthesis in *Arabidopsis* involves a homeostatic feedback loop regulated via auxin and cytokinin signal transduction. *Plant Cell* **22:** 2956–2969. doi:10.1105/tpc.110.074856

Kai K, Wakasa K, Miyagawa H. 2007. Metabolism of indole-3-acetic acid in rice: identification and characterization of *N*-β-D-glucopyranosyl indole-3-acetic acid and its conjugates. *Phytochemistry* **68:** 2512–2522. doi:10.1016/j.phytochem.2007.05.040

Kawai Y, Ono E, Mizutani M. 2014. Evolution and diversity of the 2-oxoglutarate-dependent dioxygenase superfamily in plants. *Plant J* **78:** 328–343. doi:10.1111/tpj.12479

King JJ, Stimart DP, Fisher RH, Bleecker AB. 1995. A mutation altering auxin homeostasis and plant morphology in *Arabidopsis*. *Plant Cell* **7:** 2023–2037. doi:10.2307/3870148

Kinoshita N, Wang H, Kasahara H, Liu J, Macpherson C, Machida Y, Kamiya Y, Hannah MA, Chua NH. 2012. *IAA-Ala Resistant3*, an evolutionarily conserved target of miR167, mediates *Arabidopsis* root architecture changes during high osmotic stress. *Plant Cell* **24:** 3590–3602. doi:10.1105/tpc.112.097006

Kirungu JN, Magwanga RO, Lu P, Cai X, Zhou Z, Wang X, Peng R, Wang K, Liu F. 2019. Functional characterization of Gh_A08G1120 (GH3.5) gene reveal their significant role in enhancing drought and salt stress tolerance in cotton. *BMC Genet* **20:** 62. doi:10.1186/s12863-019-0756-6

Kong W, Li Y, Zhang M, Jin F, Li J. 2015. A novel *Arabidopsis* microRNA promotes IAA biosynthesis via the indole-3-acetaldoxime pathway by suppressing *SUPERROOT1*. *Plant Cell Physiol* **56:** 715–726. doi:10.1093/pcp/pcu216

Korasick DA, Enders TA, Strader LC. 2013. Auxin biosynthesis and storage forms. *J Exp Bot* **64:** 2541–2555. doi:10.1093/jxb/ert080

Kowalczyk M, Sandberg G. 2001. Quantitative analysis of indole-3-acetic acid metabolites in *Arabidopsis*. *Plant Physiol* **127:** 1845–1853. doi:10.1104/pp.010525

Kriechbaumer V, Wang P, Hawes C, Abell BM. 2012. Alternative splicing of the auxin biosynthesis gene *YUCCA4* determines its subcellular compartmentation. *Plant J* **70:** 292–302. doi:10.1111/j.1365-313X.2011.04866.x

Kriechbaumer V, Seo H, Park WJ, Hawes C. 2015. Endoplasmic reticulum localization and activity of maize auxin biosynthetic enzymes. *J Exp Bot* **66:** 6009–6020. doi:10.1093/jxb/erv314

Kriechbaumer V, Botchway SW, Hawes C. 2016. Localization and interactions between *Arabidopsis* auxin biosynthetic enzymes in the TAA/YUC-dependent pathway. *J Exp Bot* **67:** 4195–4207. doi:10.1093/jxb/erw195

Kubeš M, Yang H, Richter GL, Cheng Y, Młodzińska E, Wang X, Blakeslee JJ, Carraro N, Petrášek J, Zažímalová E, et al. 2012. The *Arabidopsis* concentration-dependent influx/efflux transporter ABCB4 regulates cellular auxin levels in the root epidermis. *Plant J* **69:** 640–654. doi:10.1111/j.1365-313X.2011.04818.x

Kumar SV, Wigge PA. 2010. H2a.Z-containing nucleosomes mediate the thermosensory response in *Arabidopsis*. *Cell* **140:** 136–147. doi:10.1016/j.cell.2009.11.006

Lafos M, Kroll P, Hohenstatt ML, Thorpe FL, Clarenz O, Schubert D. 2011. Dynamic regulation of H3K27 trimethylation during *Arabidopsis* differentiation. *PLoS Genet* **7:** e1002040. doi:10.1371/journal.pgen.1002040

Lam HK, McAdam SA, McAdam EL, Ross JJ. 2015. Evidence that chlorinated auxin is restricted to the Fabaceae but not to the Fabeae. *Plant Physiol* **168:** 798–803. doi:10.1104/pp.15.00410

LeClere S, Tellez R, Rampey RA, Matsuda SP, Bartel B. 2002. Characterization of a family of IAA-amino acid conjugate hydrolases from *Arabidopsis*. *J Biol Chem* **277:** 20446–20452. doi:10.1074/jbc.M111955200

Lee K, Seo PJ. 2017. Coordination of matrix attachment and ATP-dependent chromatin remodeling regulate auxin biosynthesis and *Arabidopsis* hypocotyl elongation. *PLoS ONE* **12:** e0181804. doi:10.1371/journal.pone.0181804

Lehmann T, Janowitz T, Sánchez-Parra B, Alonso MP, Trompetter I, Piotrowski M, Pollmann S. 2017. *Arabidopsis* NITRILASE 1 contributes to the regulation of root growth and development through modulation of auxin biosynthesis in seedlings. *Front Plant Sci* **8:** 36. doi:10.3389/fpls.2017.00036

Li L, Hou X, Tsuge T, Ding M, Aoyama T, Oka A, Gu H, Zhao Y, Qu LJ. 2008. The possible action mechanisms of indole-3-acetic acid methyl ester in *Arabidopsis*. *Plant Cell Rep* **27:** 575–584. doi:10.1007/s00299-007-0458-9

Liu X, Barkawi L, Gardner G, Cohen JD. 2012. Transport of indole-3-butyric acid and indole-3-acetic acid in *Arabidopsis* hypocotyls using stable isotope labeling. *Plant Physiol* **158:** 1988–2000. doi:10.1104/pp.111.191288

Liu F, Zhang L, Luo Y, Xu M, Fan Y, Wang L. 2016a. Interactions of *Oryza sativa* OsCONTINUOUS VASCULAR RING-LIKE 1 (OsCOLE1) and OsCOLE1-INTERACTING PROTEIN reveal a novel intracellular auxin transport mechanism. *New Phytol* **212:** 96–107. doi:10.1111/nph.14021

Liu G, Gao S, Tian H, Wu W, Robert HS, Ding Z. 2016b. Local transcriptional control of YUCCA regulates auxin promoted root-growth inhibition in response to aluminium stress in *Arabidopsis*. *PLoS Genet* **12:** e1006360. doi:10.1371/journal.pgen.1006360

Liu Q, Chen TT, Xiao DW, Zhao SM, Lin JS, Wang T, Li YJ, Hou BK. 2019. *OsIAGT1* is a glucosyltransferase gene involved in the glucose conjugation of auxins in rice. *Rice* **12:** 92. doi:10.1186/s12284-019-0357-z

Ljung K, Bhalerao RP, Sandberg G. 2001a. Sites and homeostatic control of auxin biosynthesis in *Arabidopsis* during vegetative growth. *Plant J* **28:** 465–474. doi:10.1046/j.1365-313X.2001.01173.x

Ljung K, Östin A, Lioussanne L, Sandberg G. 2001b. Developmental regulation of indole-3-acetic acid turnover in Scots pine seedlings. *Plant Physiol* **125:** 464–475. doi:10.1104/pp.125.1.464

Ljung K, Hull AK, Celenza J, Yamada M, Estelle M, Normanly J, Sandberg G. 2005. Sites and regulation of auxin biosynthesis in *Arabidopsis* roots. *Plant Cell* **17:** 1090–1104. doi:10.1105/tpc.104.029272

Ludwig-Müller J, Walz A, Slovin JP, Epstein E, Cohen JD, Dong W, Town CD. 2005. Overexpression of maize *IAGLU* in *Arabidopsis thaliana* alters plant growth and sensitivity to IAA but not IBA and 2,4-D. *J Plant Growth Regul* **24:** 127–141. doi:10.1007/s00344-004-0006-6

Ludwig-Müller J, Jülke S, Bierfreund NM, Decker EL, Reski R. 2009. Moss (*Physcomitrella patens*) GH3 proteins act in auxin homeostasis. *New Phytol* **181:** 323–338. doi:10.1111/j.1469-8137.2008.02677.x

Lynch JH, Qian Y, Guo L, Maoz I, Huang XQ, Garcia AS, Louie G, Bowman ME, Noel JP, Morgan JA, et al. 2020. Modulation of auxin formation by the cytosolic phenylalanine biosynthetic pathway. *Nat Chem Biol* **16:** 850–856. doi:10.1038/s41589-020-0519-8

Maeda H, Dudareva N. 2012. The shikimate pathway and aromatic amino acid biosynthesis in plants. *Annu Rev Plant Biol* **63:** 73–105. doi:10.1146/annurev-arplant-042811-105439

Mao JL, Miao ZQ, Wang Z, Yu LH, Cai XT, Xiang CB. 2016. *Arabidopsis* ERF1 mediates cross-talk between ethylene and auxin biosynthesis during primary root elongation by regulating *ASA1* expression. *PLoS Genet* **12:** e1005760. doi:10.1371/journal.pgen.1005760

Mashiguchi K, Tanaka K, Sakai T, Sugawara S, Kawaide H, Natsume M, Hanada A, Yaeno T, Shirasu K, Yao H, et al. 2011. The main auxin biosynthesis pathway in *Arabidopsis*. *Proc Natl Acad Sci* **108:** 18512–18517. doi:10.1073/pnas.1108434108

Mateo-Bonmatí E, Casanova-Sáez R, Ljung K. 2019. Epigenetic regulation of auxin homeostasis. *Biomolecules* **9:** 623. doi:10.3390/biom9100623

Matosevich R, Cohen I, Gil-Yarom N, Modrego A, Friedlander-Shani L, Verna C, Scarpella E, Efroni I. 2020. Local auxin biosynthesis is required for root regeneration after wounding. *Nat Plants* **6:** 1020–1030. doi:10.1038/s41477-020-0737-9

Matsuo S, Miyatake K, Endo M, Urashimo S, Kawanishi T, Negoro S, Shimakoshi S, Fukuoka H. 2020. Loss of function of the *Pad-1* aminotransferase gene, which is involved in auxin homeostasis, induces parthenocarpy in Solanaceae plants. *Proc Natl Acad Sci* **117:** 12784–12790. doi:10.1073/pnas.2001211117

Matthes MS, Best NB, Robil JM, Malcomber S, Gallavotti A, McSteen P. 2019. Auxin EvoDevo: conservation and diversification of genes regulating auxin biosynthesis, transport, and signaling. *Mol Plant* **12:** 298–320. doi:10.1016/j.molp.2018.12.012

Mellor N, Band LR, Pěnčík A, Novák O, Rashed A, Holman T, Wilson MH, Voß U, Bishopp A, King JR, et al. 2016. Dynamic regulation of auxin oxidase and conjugating enzymes *AtDAO1* and *GH3* modulates auxin homeostasis. *Proc Natl Acad Sci* **113:** 11022–11027. doi:10.1073/pnas.1604458113

Meng WJ, Cheng ZJ, Sang YL, Zhang MM, Rong XF, Wang ZW, Tang YY, Zhang XS. 2017. Type-B ARABIDOPSIS RESPONSE REGULATORs specify the shoot stem cell niche by dual regulation of *WUSCHEL*. *Plant Cell* **29:** 1357–1372. doi:10.1105/tpc.16.00640

Michniewicz M, Ho CH, Enders TA, Floro E, Damodaran S, Gunther LK, Powers SK, Frick EM, Topp CN, Frommer WB, et al. 2019. TRANSPORTER OF IBA1 links auxin and cytokinin to influence root architecture. *Dev Cell* **50:** 599–609.e4. doi:10.1016/j.devcel.2019.06.010

Mikkelsen MD, Hansen CH, Wittstock U, Halkier BA. 2000. Cytochrome P450 CYP79B2 from *Arabidopsis* catalyzes the conversion of tryptophan to indole-3-acetaldoxime, a precursor of indole glucosinolates and indole-3-acetic acid. *J Biol Chem* **275:** 33712–33717. doi:10.1074/jbc.M001667200

Milutinovic M, Lindsey BE III, Wijeratne A, Hernandez JM, Grotewold N, Fernández V, Grotewold E, Brkljacic J. 2019. *Arabidopsis* EMSY-like (EML) histone readers are necessary for post-fertilization seed development, but prevent fertilization-independent seed formation. *Plant Sci* **285:** 99–109. doi:10.1016/j.plantsci.2019.04.007

Moubayidin L, Di Mambro R, Sabatini S. 2009. Cytokinin-auxin crosstalk. *Trends Plant Sci* **14:** 557–562. doi:10.1016/j.tplants.2009.06.010

Moubayidin L, Di Mambro R, Sozzani R, Pacifici E, Salvi E, Terpstra I, Bao D, van Dijken A, Dello Ioio R, Perilli S, et al. 2013. Spatial coordination between stem cell activity and cell differentiation in the root meristem. *Dev Cell* **26:** 405–415. doi:10.1016/j.devcel.2013.06.025

Mravec J, Skůpa P, Bailly A, Hoyerová K, Křeček P, Bielach A, Petrášek J, Zhang J, Gaykova V, Stierhof YD, et al. 2009. Subcellular homeostasis of phytohormone auxin is mediated by the ER-localized PIN5 transporter. *Nature* **459:** 1136–1140. doi:10.1038/nature08066

Müller CJ, Larsson E, Spíchal L, Sundberg E. 2017. Cytokinin-auxin crosstalk in the gynoecial primordium ensures correct domain patterning. *Plant Physiol* **175:** 1144–1157. doi:10.1104/pp.17.00805

Nadi R, Mateo-Bonmatí E, Juan-Vicente L, Micol JL. 2018. The 2OGD superfamily: emerging functions in plant epigenetics and hormone metabolism. *Mol Plant* **11:** 1222–1224. doi:10.1016/j.molp.2018.09.002

Nafisi M, Goregaoker S, Botanga CJ, Glawischnig E, Olsen CE, Halkier BA, Glazebrook J. 2007. *Arabidopsis* cytochrome P450 monooxygenase 71A13 catalyzes the conversion of indole-3-acetaldoxime in camalexin synthesis. *Plant Cell* **19:** 2039–2052. doi:10.1105/tpc.107.051383

Niehaus TD, Patterson JA, Alexander DC, Folz JS, Pyc M, MacTavish BS, Bruner SD, Mullen RT, Fiehn O, Hanson AD. 2019. The metabolite repair enzyme Nit1 is a dual-targeted amidase that disposes of damaged glutathione in *Arabidopsis*. *Biochem J* **476:** 683–697. doi:10.1042/BCJ20180931

Niyogi KK, Fink GR. 1992. Two anthranilate synthase genes in *Arabidopsis*: defense-related regulation of the tryptophan pathway. *Plant Cell* **4:** 721–733.

Nonhebel HM. 2015. Tryptophan-independent indole-3-acetic acid synthesis: critical evaluation of the evidence. *Plant Physiol* **169:** 1001–1005. doi:10.1104/pp.15.01091

Normanly J, Cohen JD, Fink GR. 1993. *Arabidopsis thaliana* auxotrophs reveal a tryptophan-independent biosynthetic pathway for indole-3-acetic acid. *Proc Natl Acad Sci* **90:** 10355–10359. doi:10.1073/pnas.90.21.10355

Novák O, Hényková E, Sairanen I, Kowalczyk M, Pospíšil T, Ljung K. 2012. Tissue-specific profiling of the *Arabidopsis thaliana* auxin metabolome. *Plant J* **72:** 523–536. doi:10.1111/j.1365-313X.2012.05085.x

Novák O, Napier R, Ljung K. 2017. Zooming in on plant hormone analysis: tissue- and cell-specific approaches. *Annu Rev Plant Biol* **68:** 323–348. doi:10.1146/annurev-arplant-042916-040812

Okrent RA, Wildermuth MC. 2011. Evolutionary history of the GH3 family of acyl adenylases in rosids. *Plant Mol Biol* **76:** 489–505. doi:10.1007/s11103-011-9776-y

Östin A, Kowalyczk M, Bhalerao RP, Sandberg G. 1998. Metabolism of indole-3-acetic acid in *Arabidopsis*. *Plant Physiol* **118:** 285–296. doi:10.1104/pp.118.1.285

Park JE, Park JY, Kim YS, Staswick PE, Jeon J, Yun J, Kim SY, Kim J, Lee YH, Park CM. 2007. GH3-mediated auxin homeostasis links growth regulation with stress adaptation response in *Arabidopsis*. *J Biol Chem* **282:** 10036–10046. doi:10.1074/jbc.M610524200

Patten CL, Blakney AJ, Coulson TJ. 2013. Activity, distribution and function of indole-3-acetic acid biosynthetic pathways in bacteria. *Crit Rev Microbiol* **39:** 395–415. doi:10.3109/1040841X.2012.716819

Pěnčík A, Rolčík J, Novák O, Magnus V, Barták P, Buchtík R, Salopek-Sondi B, Strnad M. 2009. Isolation of novel indole-3-acetic acid conjugates by immunoaffinity extraction. *Talanta* **80:** 651–655. doi:10.1016/j.talanta.2009.07.043

Pěnčík A, Simonovik B, Petersson SV, Henyková E, Simon S, Greenham K, Zhang Y, Kowalczyk M, Estelle M, Zažímalová E, et al. 2013. Regulation of auxin homeostasis and gradients in *Arabidopsis* roots through the formation of the indole-3-acetic acid catabolite 2-oxindole-3-acetic acid. *Plant Cell* **25:** 3858–3870. doi:10.1105/tpc.113.114421

Pěnčík A, Casanova-Sáez R, Pilařová V, Žukauskaitė A, Pinto R, Micol JL, Ljung K, Novák O. 2018. Ultra-rapid auxin metabolite profiling for high-throughput mutant screening in *Arabidopsis*. *J Exp Bot* **69:** 2569–2579. doi:10.1093/jxb/ery084

Peng M, Li Z, Zhou N, Ma M, Jiang Y, Dong A, Shen WH, Li L. 2018. Linking PHYTOCHROME-INTERACTING FACTOR to histone modification in plant shade avoidance. *Plant Physiol* **176:** 1341–1351. doi:10.1104/pp.17.01189

Petersson SV, Johansson AI, Kowalczyk M, Makoveychuk A, Wang JY, Moritz T, Grebe M, Benfey PN, Sandberg G, Ljung K. 2009. An auxin gradient and maximum in the *Arabidopsis* root apex shown by high-resolution cell-specific analysis of IAA distribution and synthesis. *Plant Cell* **21:** 1659–1668. doi:10.1105/tpc.109.066480

Phillips KA, Skirpan AL, Liu X, Christensen A, Slewinski TL, Hudson C, Barazesh S, Cohen JD, Malcomber S, McSteen P. 2011. *Vanishing tassel2* encodes a grass-specific tryptophan aminotransferase required for vegetative and reproductive development in maize. *Plant Cell* **23:** 550–566. doi:10.1105/tpc.110.075267

Pierdonati E, Unterholzner SJ, Salvi E, Svolacchia N, Bertolotti G, Dello Ioio R, Sabatini S, Di Mambro R. 2019. Cytokinin-dependent control of GH3 group II family genes in the *Arabidopsis* root. *Plants* **8:** 94. doi:10.3390/plants8040094

Piotrowski M. 2008. Primary or secondary? Versatile nitrilases in plant metabolism. *Phytochemistry* **69:** 2655–2667. doi:10.1016/j.phytochem.2008.08.020

Pollmann S, Müller A, Piotrowski M, Weiler EW. 2002. Occurrence and formation of indole-3-acetamide in *Arabidopsis thaliana*. *Planta* **216:** 155–161. doi:10.1007/s00425-002-0868-4

Porco S, Pěnčík A, Rashed A, Voß U, Casanova-Sáez R, Bishopp A, Golebiowska A, Bhosale R, Swarup R, Swarup K, et al. 2016. Dioxygenase-encoding *AtDAO1* gene controls IAA oxidation and homeostasis in *Arabidopsis*. *Proc Natl Acad Sci* **113:** 11016–11021. doi:10.1073/pnas.1604375113

Poulet A, Kriechbaumer V. 2017. Bioinformatics analysis of phylogeny and transcription of TAA/YUC auxin biosynthetic genes. *Int J Mol Sci* **18:** 1791. doi:10.3390/ijms18081791

Qin G, Gu H, Zhao Y, Ma Z, Shi G, Yang Y, Pichersky E, Chen H, Liu M, Chen Z, et al. 2005. An indole-3-acetic acid carboxyl methyltransferase regulates *Arabidopsis* leaf development. *Plant Cell* **17:** 2693–2704. doi:10.1105/tpc.105.034959

Qin H, Zhang Z, Wang J, Chen X, Wei P, Huang R. 2017. The activation of OsEIL1 on *YUC8* transcription and auxin biosynthesis is required for ethylene-inhibited root elongation in rice early seedling development. *PLoS Genet* **13:** e1006955. doi:10.1371/journal.pgen.1006955

Rampey RA, LeClere S, Kowalczyk M, Ljung K, Sandberg G, Bartel B. 2004. A family of auxin-conjugate hydrolases that contributes to free indole-3-acetic acid levels during *Arabidopsis* germination. *Plant Physiol* **135:** 978–988. doi:10.1104/pp.104.039677

Ranocha P, Dima O, Nagy R, Felten J, Corratgé-Faillie C, Novák O, Morreel K, Lacombe B, Martinez Y, Pfrunder S, et al. 2013. *Arabidopsis* WAT1 is a vacuolar auxin transport facilitator required for auxin homoeostasis. *Nat Commun* **4:** 2625. doi:10.1038/ncomms3625

Reyes-Olalde JI, Zúñiga-Mayo VM, Serwatowska J, Chavez Montes RA, Lozano-Sotomayor P, Herrera-Ubaldo H, Gonzalez-Aguilera KL, Ballester P, Ripoll JJ, Ezquer I, et al. 2017. The bHLH transcription factor SPATULA enables cytokinin signaling, and both activate auxin biosynthesis and transport genes at the medial domain of the gynoecium. *PLoS Genet* **13:** e1006726. doi:10.1371/journal.pgen.1006726

Rizzardi K, Landberg K, Nilsson L, Ljung K, Sundås-Larsson A. 2011. *TFL2/LHP1* is involved in auxin biosynthesis through positive regulation of *YUCCA* genes. *Plant J* **65:** 897–906. doi:10.1111/j.1365-313X.2010.04470.x

Robert HS, Grones P, Stepanova AN, Robles LM, Lokerse AS, Alonso JM, Weijers D, Friml J. 2013. Local auxin sources orient the apical-basal axis in *Arabidopsis* embryos. *Curr Biol* **23:** 2506–2512. doi:10.1016/j.cub.2013.09.039

Robert HS, Park C, Gutièrrez CL, Wojcikowska B, Pěnčík A, Novák O, Chen J, Grunewald W, Dresselhaus T, Friml J, et al. 2018. Maternal auxin supply contributes to early embryo patterning in *Arabidopsis*. *Nat Plants* **4:** 548–553. doi:10.1038/s41477-018-0204-z

Sairanen I, Novák O, Pěnčík A, Ikeda Y, Jones B, Sandberg G, Ljung K. 2012. Soluble carbohydrates regulate auxin biosynthesis via PIF proteins in *Arabidopsis*. *Plant Cell* **24:** 4907–4916. doi:10.1105/tpc.112.104794

Sanchez Carranza AP, Singh A, Steinberger K, Panigrahi K, Palme K, Dovzhenko A, Dal Bosco C. 2016. Hydrolases of the ILR1-like family of *Arabidopsis thaliana* modulate auxin response by regulating auxin homeostasis in the endoplasmic reticulum. *Sci Rep* **6:** 24212. doi:10.1038/srep24212

Shimizu-Mitao Y, Kakimoto T. 2014. Auxin sensitivities of all *Arabidopsis* Aux/IAAs for degradation in the presence of every TIR1/AFB. *Plant Cell Physiol* **55:** 1450–1459. doi:10.1093/pcp/pcu077

Simon S, Skůpa P, Viaene T, Zwiewka M, Tejos R, Klima P, Čarná M, Rolčík J, De Rycke R, Moreno I, et al. 2016. PIN6 auxin transporter at endoplasmic reticulum and plasma membrane mediates auxin homeostasis and organogenesis in *Arabidopsis*. *New Phytol* **211:** 65–74. doi:10.1111/nph.14019

Skalický V, Kubeš M, Napier R, Novák O. 2018. Auxins and cytokinins—the role of subcellular organization on homeostasis. *Int J Mol Sci* **19:** 3115. doi:10.3390/ijms19103115

Spiess GM, Hausman A, Yu P, Cohen JD, Rampey RA, Zolman BK. 2014. Auxin input pathway disruptions are mitigated by changes in auxin biosynthetic gene expression in *Arabidopsis*. *Plant Physiol* **165:** 1092–1104. doi:10.1104/pp.114.236026

Staswick PE, Serban B, Rowe M, Tiryaki I, Maldonado MT, Maldonado MC, Suza W. 2005. Characterization of an *Arabidopsis* enzyme family that conjugates amino acids to indole-3-acetic acid. *Plant Cell* **17:** 616–627. doi:10.1105/tpc.104.026690

Stepanova AN, Hoyt JM, Hamilton AA, Alonso JM. 2005. A link between ethylene and auxin uncovered by the characterization of two root-specific ethylene-insensitive mutants in *Arabidopsis*. *Plant Cell* **17:** 2230–2242. doi:10.1105/tpc.105.033365

Stepanova AN, Robertson-Hoyt J, Yun J, Benavente LM, Xie DY, Doležal K, Schlereth A, Jürgens G, Alonso JM. 2008. *TAA1*-mediated auxin biosynthesis is essential for hormone crosstalk and plant development. *Cell* **133:** 177–191. doi:10.1016/j.cell.2008.01.047

Stepanova AN, Yun J, Robles LM, Novák O, He W, Guo H, Ljung K, Alonso JM. 2011. The *Arabidopsis* YUCCA1 flavin monooxygenase functions in the indole-3-pyruvic acid branch of auxin biosynthesis. *Plant Cell* **23:** 3961–3973. doi:10.1105/tpc.111.088047

Stone SL, Braybrook SA, Paula SL, Kwong LW, Meuser J, Pelletier J, Hsieh TF, Fischer RL, Goldberg RB, Harada JJ. 2008. *Arabidopsis* LEAFY COTYLEDON2 induces maturation traits and auxin activity: implications for somatic embryogenesis. *Proc Natl Acad Sci* **105:** 3151-3156. doi:10.1073/pnas.0712364105

Strader LC, Bartel B. 2011. Transport and metabolism of the endogenous auxin precursor indole-3-butyric acid. *Mol Plant* **4:** 477–486. doi:10.1093/mp/ssr006

Sugawara S, Hishiyama S, Jikumaru Y, Hanada A, Nishimura T, Koshiba T, Zhao Y, Kamiya Y, Kasahara H. 2009. Biochemical analyses of indole-3-acetaldoxime-dependent auxin biosynthesis in *Arabidopsis*. *Proc Natl Acad Sci* **106:** 5430–5435. doi:10.1073/pnas.0811226106

Sun J, Xu Y, Ye S, Jiang H, Chen Q, Liu F, Zhou W, Chen R, Li X, Tietz O, et al. 2009. *Arabidopsis ASA1* is important for jasmonate-mediated regulation of auxin biosynthesis and transport during lateral root formation. *Plant Cell* **21:** 1495–1511. doi:10.1105/tpc.108.064303

Suzuki M, Yamazaki C, Mitsui M, Kakei Y, Mitani Y, Nakamura A, Ishii T, Soeno K, Shimada Y. 2015. Transcriptional feedback regulation of *YUCCA* genes in response to auxin levels in *Arabidopsis*. *Plant Cell Rep* **34:** 1343–1352. doi:10.1007/s00299-015-1791-z

Szerszen JB, Szczyglowski K, Bandurski RS. 1994. *iaglu*, a gene from *Zea mays* involved in conjugation of growth hormone indole-3-acetic acid. *Science* **265:** 1699–1701. doi:10.1126/science.8085154

Takato S, Kakei Y, Mitsui M, Ishida Y, Suzuki M, Yamazaki C, Hayashi KI, Ishii T, Nakamura A, Soeno K, et al. 2017. Auxin signaling through $SCF^{TIR1/AFBs}$ mediates feedback regulation of IAA biosynthesis. *Biosci Biotechnol Biochem* **81:** 1320–1326. doi:10.1080/09168451.2017.1313694

Takehara S, Sakuraba S, Mikami B, Yoshida H, Yoshimura H, Itoh A, Endo M, Watanabe N, Nagae T, Matsuoka M, et al. 2020. A common allosteric mechanism regulates homeostatic inactivation of auxin and gibberellin. *Nat Commun* **11:** 2143. doi:10.1038/s41467-020-16068-0

Takubo E, Kobayashi M, Hirai S, Aoi Y, Ge C, Dai X, Fukui K, Hayashi KI, Zhao Y, Kasahara H. 2020. Role of *Arabidopsis INDOLE-3-ACETIC ACID CARBOXYL METHYLTRANSFERASE 1* in auxin metabolism. *Biochem Biophys Res Commun* **527:** 1033–1038. doi:10.1016/j.bbrc.2020.05.031

Tanaka K, Hayashi K, Natsume M, Kamiya Y, Sakakibara H, Kawaide H, Kasahara H. 2014. UGT74D1 catalyzes the glucosylation of 2-oxindole-3-acetic acid in the auxin metabolic pathway in *Arabidopsis*. *Plant Cell Physiol* **55:** 218–228. doi:10.1093/pcp/pct173

Tang LP, Zhou C, Wang SS, Yuan J, Zhang XS, Su YH. 2017. FUSCA3 interacting with LEAFY COTYLEDON2 controls lateral root formation through regulating *YUCCA4* gene expression in *Arabidopsis thaliana*. *New Phytol* **213:** 1740–1754. doi:10.1111/nph.14313

Tao Y, Ferrer JL, Ljung K, Pojer F, Hong F, Long JA, Li L, Moreno JE, Bowman ME, Ivans LJ, et al. 2008. Rapid synthesis of auxin via a new tryptophan-dependent pathway is required for shade avoidance in plants. *Cell* **133:** 164–176. doi:10.1016/j.cell.2008.01.049

Terol J, Domingo C, Talón M. 2006. The GH3 family in plants: genome wide analysis in rice and evolutionary history based on EST analysis. *Gene* **371:** 279–290. doi:10.1016/j.gene.2005.12.014

Uzunova VV, Quareshy M, Del Genio CI, Napier RM. 2016. Tomographic docking suggests the mechanism of auxin receptor TIR1 selectivity. *Open Biol* **6:** 160139. doi:10.1098/rsob.160139

van der Woude LC, Perrella G, Snoek BL, van Hoogdalem M, Novák O, van Verk MC, van Kooten HN, Zorn LE, Tonckens R, Dongus JA, et al. 2019. HISTONE DEACETYLASE 9 stimulates auxin-dependent thermomorphogenesis in *Arabidopsis thaliana* by mediating H2A.Z depletion. *Proc Natl Acad Sci* **116:** 25343–25354. doi:10.1073/pnas.1911694116

van Overbeek J. 1939. Is auxin produced in roots? *Proc Natl Acad Sci* **25:** 245–248. doi:10.1073/pnas.25.5.245

van Raalte MH. 1936. On the influence of glucose on auxin production by the root tip of *Vicia faba*. *Proc K Akad Wetensch* **39:** 261–265.

Wabnik K, Robert HS, Smith RS, Friml J. 2013. Modeling framework for the establishment of the apical-basal embryonic axis in plants. *Curr Biol* **23:** 2513–2518. doi:10.1016/j.cub.2013.10.038

Wang B, Chu J, Yu T, Xu Q, Sun X, Yuan J, Xiong G, Wang G, Wang Y, Li J. 2015. Tryptophan-independent auxin biosynthesis contributes to early embryogenesis in *Arabidopsis*. *Proc Natl Acad Sci* **112:** 4821–4826. doi:10.1073/pnas.1503998112

Wang Q, Qin G, Cao M, Chen R, He Y, Yang L, Zeng Z, Yu Y, Gu Y, Xing W, et al. 2020. A phosphorylation-based switch controls TAA1-mediated auxin biosynthesis in plants. *Nat Commun* **11:** 679. doi:10.1038/s41467-020-14395-w

Westfall CS, Sherp AM, Zubieta C, Alvarez S, Schraft E, Marcellin R, Ramirez L, Jez JM. 2016. *Arabidopsis thali-*

ana GH3.5 acyl acid amido synthetase mediates metabolic crosstalk in auxin and salicylic acid homeostasis. *Proc Natl Acad Sci* **113:** 13917–13922. doi:10.1073/pnas.1612635113

Won C, Shen X, Mashiguchi K, Zheng Z, Dai X, Cheng Y, Kasahara H, Kamiya Y, Chory J, Zhao Y. 2011. Conversion of tryptophan to indole-3-acetic acid by TRYPTOPHAN AMINOTRANSFERASES OF *ARABIDOPSIS* and YUCCAs in *Arabidopsis*. *Proc Natl Acad Sci* **108:** 18518–18523. doi:10.1073/pnas.1108436108

Wright AD, Sampson MB, Neuffer MG, Michalczuk L, Slovin JP, Cohen JD. 1991. Indole-3-acetic acid biosynthesis in the mutant maize *orange pericarp*, a tryptophan auxotroph. *Science* **254:** 998–1000. doi:10.1126/science.254.5034.998

Xu Y, Prunet N, Gan ES, Wang Y, Stewart D, Wellmer F, Huang J, Yamaguchi N, Tatsumi Y, Kojima M, et al. 2018. SUPERMAN regulates floral whorl boundaries through control of auxin biosynthesis. *EMBO J* **37:** e97499.

Yamada M, Greenham K, Prigge MJ, Jensen PJ, Estelle M. 2009. The *TRANSPORT INHIBITOR RESPONSE2* gene is required for auxin synthesis and diverse aspects of plant development. *Plant Physiol* **151:** 168–179. doi:10.1104/pp.109.138859

Yamaguchi N, Huang J, Tatsumi Y, Abe M, Sugano SS, Kojima M, Takebayashi Y, Kiba T, Yokoyama R, Nishitani K, et al. 2018. Chromatin-mediated feed-forward auxin biosynthesis in floral meristem determinacy. *Nat Commun* **9:** 5290. doi:10.1038/s41467-018-07763-0

Yamamoto Y, Kamiya N, Morinaka Y, Matsuoka M, Sazuka T. 2007. Auxin biosynthesis by the *YUCCA* genes in rice. *Plant Physiol* **143:** 1362–1371. doi:10.1104/pp.106.091561

Yan Z, Liu X, Ljung K, Li S, Zhao W, Yang F, Wang M, Tao Y. 2017. Type B response regulators act as central integrators in transcriptional control of the auxin biosynthesis enzyme TAA1. *Plant Physiol* **175:** 1438–1454. doi:10.1104/pp.17.00878

Yang Y, Xu R, Ma CJ, Vlot AC, Klessig DF, Pichersky E. 2008. Inactive methyl indole-3-acetic acid ester can be hydrolyzed and activated by several esterases belonging to the *At*MES esterase family of *Arabidopsis*. *Plant Physiol* **147:** 1034–1045. doi:10.1104/pp.108.118224

Yoshikawa T, Ito M, Sumikura T, Nakayama A, Nishimura T, Kitano H, Yamaguchi I, Koshiba T, Hibara K, Nagato Y, et al. 2014. The rice *FISH BONE* gene encodes a tryptophan aminotransferase, which affects pleiotropic auxin-related processes. *Plant J* **78:** 927–936. doi:10.1111/tpj.12517

Yue J, Hu X, Huang J. 2014. Origin of plant auxin biosynthesis. *Trends Plant Sci* **19:** 764–770. doi:10.1016/j.tplants.2014.07.004

Záveská Drábková L, Dobrev PI, Motyka V. 2015. Phytohormone profiling across the Bryophytes. *PLoS ONE* **10:** e0125411. doi:10.1371/journal.pone.0125411

Zhang Z, Li Q, Li Z, Staswick PE, Wang M, Zhu Y, He Z. 2007. Dual regulation role of *GH3.5* in salicylic acid and auxin signaling during *Arabidopsis–Pseudomonas syringae* interaction. *Plant Physiol* **145:** 450–464. doi:10.1104/pp.107.106021

Zhang R, Wang B, Ouyang J, Li J, Wang Y. 2008. *Arabidopsis* indole synthase, a homolog of tryptophan synthase α, is an enzyme involved in the Trp-independent indole-containing metabolite biosynthesis. *J Integra Plant Biol* **50:** 1070–1077. doi:10.1111/j.1744-7909.2008.00729.x

Zhang SW, Li CH, Cao J, Zhang YC, Zhang SQ, Xia YF, Sun DY, Sun Y. 2009. Altered architecture and enhanced drought tolerance in rice via the down-regulation of indole-3-acetic acid by *TLD1/OsGH3.13* activation. *Plant Physiol* **151:** 1889–1901. doi:10.1104/pp.109.146803

Zhang J, Lin JE, Harris C, Campos Mastrotti Pereira F, Wu F, Blakeslee JJ, Peer WA. 2016. DAO1 catalyzes temporal and tissue-specific oxidative inactivation of auxin in *Arabidopsis thaliana*. *Proc Natl Acad Sci* **113:** 11010–11015. doi:10.1073/pnas.1604769113

Zhang H, Lang Z, Zhu JK. 2018. Dynamics and function of DNA methylation in plants. *Nat Rev Mol Cell Biol* **19:** 489–506. doi:10.1038/s41580-018-0016-z

Zhao Y. 2018. Essential roles of local auxin biosynthesis in plant development and in adaptation to environmental changes. *Annu Rev Plant Biol* **69:** 417–435. doi:10.1146/annurev-arplant-042817-040226

Zhao CY, Xue HW. 2020. PI4Kγ2 interacts with E3 ligase MIEL1 to regulate auxin metabolism and root development. *Plant Physiol* **184:** 933–944. doi:10.1104/pp.20.00799

Zhao Y, Hull AK, Gupta NR, Goss KA, Alonso J, Ecker JR, Normanly J, Chory J, Celenza JL. 2002. Trp-dependent auxin biosynthesis in *Arabidopsis*: involvement of cytochrome P450s CYP79B2 and CYP79B3. *Genes Dev* **16:** 3100–3112. doi:10.1101/gad.1035402

Zhao Z, Zhang Y, Liu X, Zhang X, Liu S, Yu X, Ren Y, Zheng X, Zhou K, Jiang L, et al. 2013. A role for a dioxygenase in auxin metabolism and reproductive development in rice. *Dev Cell* **27:** 113–122. doi:10.1016/j.devcel.2013.09.005

Zheng Z, Guo Y, Novák O, Dai X, Zhao Y, Ljung K, Noel JP, Chory J. 2013. Coordination of auxin and ethylene biosynthesis by the aminotransferase VAS1. *Nat Chem Biol* **9:** 244–246. doi:10.1038/nchembio.1178

Zheng Z, Guo Y, Novák O, Chen W, Ljung K, Noel JP, Chory J. 2016. Local auxin metabolism regulates environment-induced hypocotyl elongation. *Nat Plants* **2:** 16025. doi:10.1038/nplants.2016.25

Žižková E, Kubeš M, Dobrev PI, Přibyl P, Šimura J, Zahajská L, Záveská Drábková L, Novák O, Motyka V. 2017. Control of cytokinin and auxin homeostasis in cyanobacteria and algae. *Ann Bot* **119:** 151–166. doi:10.1093/aob/mcw194

Zolman BK, Martinez N, Millius A, Adham AR, Bartel B. 2008. Identification and characterization of *Arabidopsis* indole-3-butyric acid response mutants defective in novel peroxisomal enzymes. *Genetics* **180:** 237–251. doi:10.1534/genetics.108.090399

Zubieta C, Ross JR, Koscheski P, Yang Y, Pichersky E, Noel JP. 2003. Structural basis for substrate recognition in the salicylic acid carboxyl methyltransferase family. *Plant Cell* **15:** 1704–1716. doi:10.1105/tpc.014548

Auxin Transporters—A Biochemical View

Ulrich Z. Hammes,[1] Angus S. Murphy,[2,3] and Claus Schwechheimer[1]

[1]Plant Systems Biology, School of Life Sciences, Technical University of Munich, 85354 Freising, Germany

[2]Department of Plant Science and Landscape Architecture; [3]Agriculture Biotechnology Center, University of Maryland, College Park, Maryland 20742, USA

Correspondence: claus.schwechheimer@wzw.tum.de

From embryogenesis to fruit formation, almost every aspect of plant development and differentiation is controlled by the cellular accumulation or depletion of auxin from cells and tissues. The respective auxin maxima and minima are generated by cell-to-cell auxin transport via transporter proteins. Differential auxin accumulation as a result of such transport processes dynamically regulates auxin distribution during differentiation. In this review, we introduce all auxin transporter (families) identified to date and discuss the knowledge on prominent family members, namely, the PIN-FORMED exporters, ATP-binding cassette B (ABCB)-type transporters, and AUX1/LAX importers. We then concentrate on the biochemical features of these transporters and their regulation by posttranslational modifications and interactors.

AUXIN TRANSPORT AND AUXIN TRANSPORTERS

The formation of auxin maxima or minima in individual cells and tissues is essential for almost every aspect of plant development, ranging from embryogenesis to fruit formation (Weijers and Wagner 2016). Auxin does not necessarily act at the sites of its synthesis but is transported within tissues and through the entire plant to regulate growth and differentiation (Petrasek and Friml 2009). Over the past two decades, the identities of many auxin transporters have been revealed, often on the basis of mutants and their defects in differentiation or tropic growth responses (Figs. 1 and 2). Additional auxin transporters were identified based on their apparent evolutionary relationship or their topological similarity to known transporters (Figs. 1 and 2).

Using functional criteria, auxin transporters can be grouped as plasma membrane transporters contributing to vectorial and long distance polar auxin streams; intercellular transporters that primarily contribute to cellular auxin homeostasis; and multifunctional transporters that predominantly mobilize other small molecules, but transport auxin under specific conditions (Fig. 1). Throughout this review, we introduce these different transporters and transporter families and discuss in detail the knowledge on the PIN-FORMED (PIN) auxin efflux carriers, the ABCB family transporters, and AUXIN-RESISTANT1 (AUX1)/LIKE-AUX1 (LAX) importers. We further concentrate on their biochemical features, their

Cite this article as *Cold Spring Harb Perspect Biol* doi: 10.1101/cshperspect.a039875

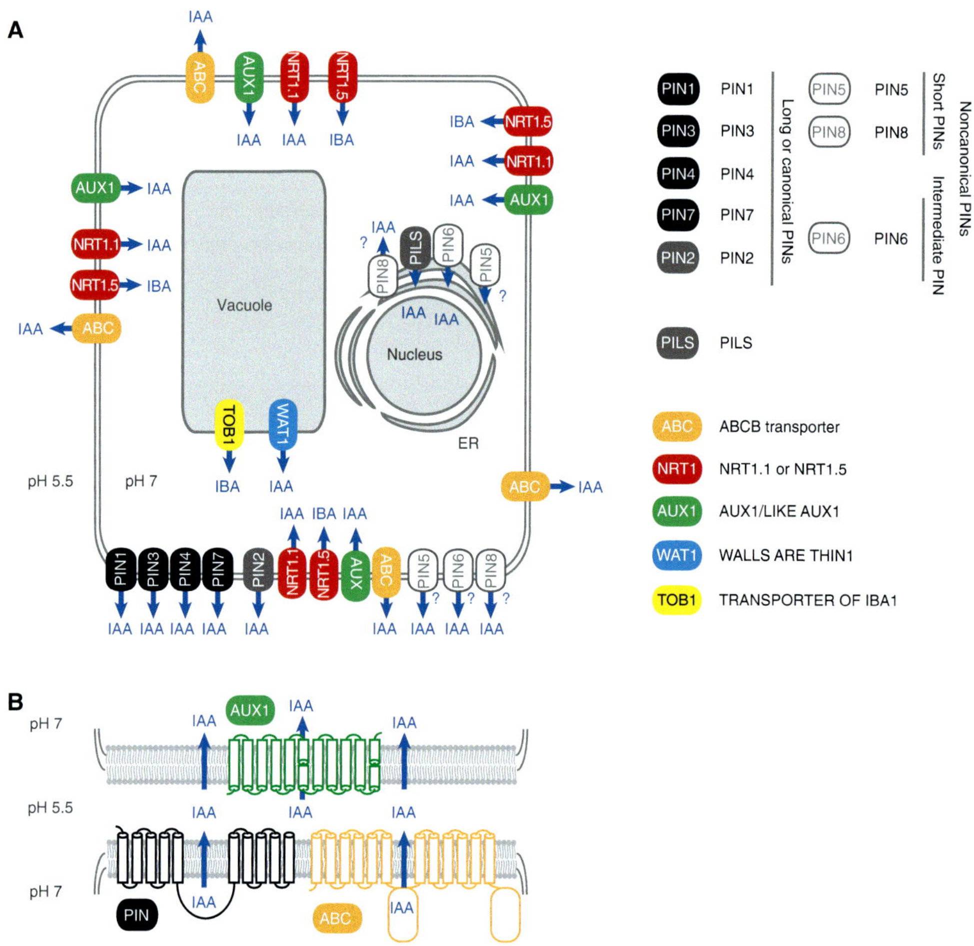

Figure 1. Auxin transporters. (*A*) Schematic representation of all auxin transporters identified to date, their distribution in the plasma membrane or in endomembranes, as well as the direction of auxin transport (arrows). (*B*) Schematic representation of the interplay between the well-studied AUX/LAX auxin importers and the PIN or ATP-binding cassette B (ABCB)-type auxin exporters at the cell–cell interface as an example for cell-to-cell auxin transport. The transporters are displayed with their assumed topology as shown in Figure 2. (ER) Endoplasmic reticulum.

regulation by posttranslational modifications, and interacting regulators. With regard to factors influencing auxin transporter cell biological behavior or transport, the reader is guided to a recent review on the topic (Semeradova et al. 2020).

PIN AUXIN EFFLUX CARRIERS

PIN proteins are auxin efflux carriers that can be subdivided into at least two transporter classes, "long" canonical and "short" or "intermediate" noncanonical PINs (Fig. 2; Bennett et al. 2014). Vascular plants typically have five or six canonical plasma membrane–resident PINs (Fig. 1; Bennett et al. 2014; Sauer and Kleine-Vehn 2019; Vosolsobě et al. 2020). In *Arabidopsis thaliana*, the canonical PINs, PIN1–PIN4, and PIN7 are distinguished from the noncanonical short PINs PIN5 and PIN8, through their long cytoplasmic loop that contains regulatory elements that are absent from the short PINs (Fig. 3). In view of this definition, the PIN6 transporter is an intermediate length PIN because it has only a subset of these regulatory elements (Figs. 2 and 3).

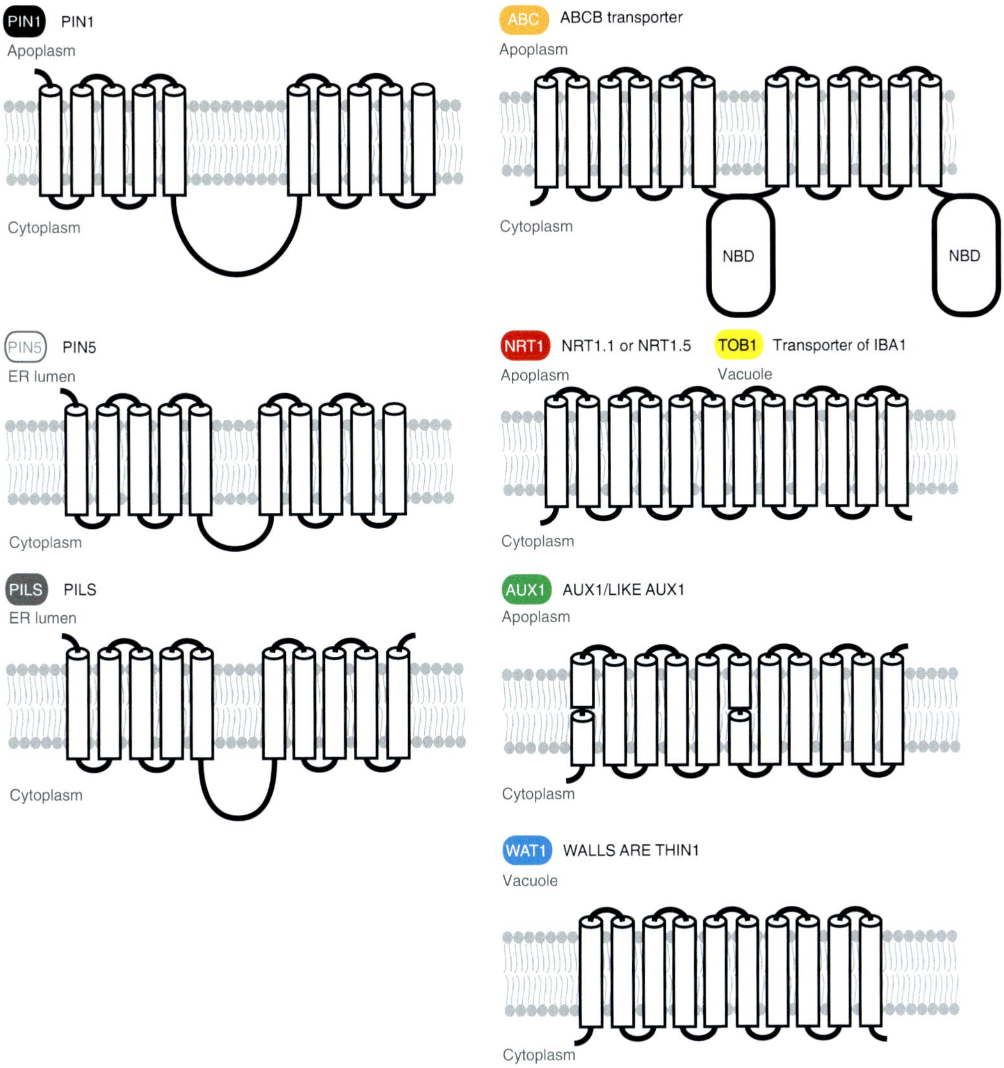

Figure 2. Schematic representation of the known auxin transporters and their minimal topology. Auxin transporters are shown with their transmembrane helices and differentially sized loops connecting the helices, as well as their cellular orientation: PIN topology (Nodzyński et al. 2016); PILS topology (Barbez et al. 2012); ABC transporter topology (Verhalen et al. 2017; Kodan et al. 2019); NRT1.1 topology (Sun et al. 2014); NRT1.5 and TOB1 topologies are derived from NRT1.1; AUX1 topology derived from the crystal structure of AAAP family member SLC38A9 (Lei et al. 2018); WAT1 topology is predicted based on the consensus from 18 topology prediction algorithms (aramemnon.botanik.uni-koeln.de). (NBD) Nucleotide-binding domain, (ER) endoplasmic reticulum.

Canonical PINs are polarly distributed in many cell types (e.g., PIN1 is basally [rootward] localized in root stele cells, PIN2 is basally localized in root cortex cells, but apically [shootward] in root epidermal cells, PIN3 is dynamically polarly distributed in hypocotyl endodermis cells, etc.) (Fig. 1; Rakusová et al. 2015; Sauer and Kleine-Vehn 2019). The polar localization of canonical PINs at the plasma membranes of many cells has inspired physiologists and cell biologists because PIN polarity allows explaining the experimentally observed or anticipated

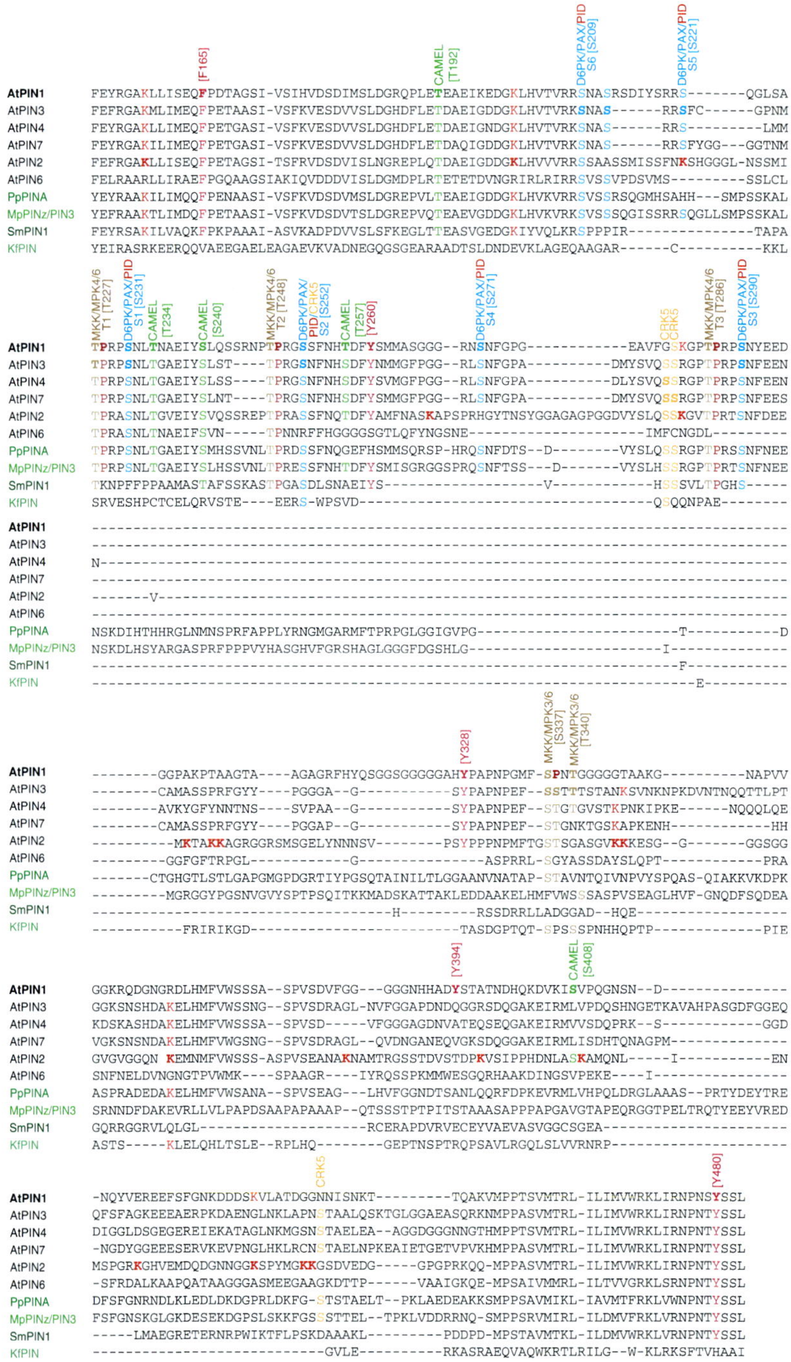

Figure 3. Alignment of the cytoplasmic loops of PIN proteins and their posttranslational modifications. Protein alignment of *Arabidopsis thaliana* canonical PINs and the intermediate PIN6, together with representative canonical PINs from *Physcomitrium patens* (PpPINA), *Marchantia polymorpha* (MpPINz/PIN3), *Selaginella moellendorffii* (SmPIN1), as well as a PIN-related protein from *Klebsomirdium flaccidum* (KfPIN). Amino acids related to protein regulation are marked in colored bold letters in the PIN where they were identified or studied, and in colored regular letters in the cases where the respective amino acids are conserved among the aligned PINs. The color code of the phosphorylation sites corresponds, where known and where graphically possible, to the protein kinases shown to regulate the sites as further specified in Table 1. In the case of serine or threonine phosphorylation, both amino acids were considered to be functionally equivalent.

directionality of cell-to-cell auxin transport through tissues and the entire plant (Bennett et al. 2014). Moreover, the development- and tissue-dependent cellular distribution of the canonical PINs, as well as their dynamic relocalization during tropic responses raised many questions about the processes regulating their polar targeting, recycling, and the identity of upstream regulators (Rakusová et al. 2015; Barbosa et al. 2018). The endoplasmic reticulum (ER)-localized noncanonical PINs may, in turn, likely have evolved to control auxin levels and distribution between intracellular compartments (Fig. 2; Bennett et al. 2014).

The importance of PIN proteins to polar auxin transport is clearly evident, and the potential of PINs to transport auxin has been demonstrated qualitatively in several expression systems; however, the biochemical properties and knowledge about the substrate specificity of PIN isoforms is still limited (Barbosa et al. 2018). Recently, two studies confirmed that PINs are direct targets of *N*-1-naphthylphthalamic acid (NPA), a well-established inhibitor of polar auxin transport, which explains the perfect alignment of *pin*-mutant phenotypes and the phenotypes of wild-type plants treated with NPA (Abas et al. 2021; Teale et al. 2021). One of these studies showed that the activation of PINs by protein kinases, which is explained below in detail, is not affected by NPA, suggesting that the compound inhibits the transporter allosterically (Abas et al. 2021).

PIN REGULATION BY PHOSPHORYLATION

Canonical PINs contain elements for activity regulation by phosphorylation that were functionally characterized in *A. thaliana* (Fig. 3; Table 1; Willige et al. 2013; Zourelidou et al. 2014; Weller et al. 2017; Marhava et al. 2018). Canonical PINs from the moss *Physcomitrium patens*, the liverwort *Marchantia polymorpha*, and the lycophyte *Selaginella moellendorffii* also contain many of these regulatory elements and are most likely subject to the same regulation as their counterparts from vascular plants (Fig. 3; Bennett et al. 2014). Although bona fide canonical PINs are absent from algae, one study describes an auxin-transporting PIN from the algae *Klebsomirdium flaccidum*, which is activated after coexpression of an *Arabidopsis* PIN-activating kinase (Skokan et al. 2019). The phosphoregulation of this algal KfPIN must, however, occur through a mechanism distinct from the mechanism described for *Arabidopsis* PINs since the respective regulatory modules are not conserved in KfPIN (Fig. 3; Skokan et al. 2019).

In vitro, canonical PINs are directly phosphorylated by AGC1 and AGC3 kinases, MITOGEN-ACTIVATED PROTEIN (MAP) kinases, Ca^{2+}/calmodulin-dependent protein kinase-related kinases (CRKs), as well as by CAMEL (CANALIZATION-RELATED AUXIN-REGULATED MALECTIN-TYPE RLK) (Fig. 3; Table 1). In the following, the currently existing knowledge about this kinase regulation is summarized.

PIN ACTIVITY AND POLARITY REGULATION BY AGCVIII KINASES

AGC kinases are evolutionarily conserved serine/threonine kinases related to mammalian cAMP-dependent kinases, cGMP-dependent kinases, and Ca^{2+}-dependent kinases (Galván-Ampudia and Offringa 2007). In *A. thaliana*, the 23-membered AGCVIII clade is comprised of AGC1, AGC2, AGC3, which are derived from the evolutionarily older phototropin blue light receptor AGC4 kinases (Galván-Ampudia and Offringa 2007). The plant-specific AGCVIII kinases have two signature motifs, a conserved DFD motif implicated in Mg^{2+}-ATP binding, that replaces the DFG motif present in other kinases, and a functionally and sequence diverse insertion domain between the conserved protein kinase subdomains VII and VIII (Galván-Ampudia and Offringa 2007). In kinases of the AGC1 clade, this insertion domain is required and sufficient for plasma membrane association and contains a positively charged polybasic motif for interactions with the negatively charged headgroups of phospholipids, as demonstrated for several AGC1 kinases but also the AGC3 kinase PINOID (PID) (Barbosa et al. 2016; Platre and Jaillais 2017).

AGC1 and AGC3 kinases phosphorylate and activate “long” PINs as shown specifically

Table 1. Regulation of PIN auxin transporter activity by posttranslational modifications

PIN transporter	Amino acid	Effect or function	References
PIN1	F165, Y260, Y328, Y394, Y480	Predicted adaptin-binding sites; individual amino acids can interact with adaptins; only F165 is essential for proper PIN trafficking and localization in vivo	Marcote et al. 2016; Sancho-Andrés et al. 2016
	T192, T234, S240, T257, S408	CAMEL phosphosites; replacement of S or T to A affects PIN1 function at the developmental and cell biological level	Hajný et al. 2020
	S252	In vitro phosphorylation site for AtCRK5	Baba et al. 2019b
	S1 (S231), S2 (S252), S3 (S290)	PID phosphosites; S to A replacement mutations induce dominant embryo and flower phenotypes and lead to constitutive basal PIN1 polarity; replacement of S1–S3 by phosphorylation-mimicking E correlates with apical polarity	Huang et al. 2010
	S1 (S231), S2 (S252), S3 (S290), S4 (S271)	AGC1 (D6PK) and AGC3 (PID/WAG) kinases in vitro phosphosites; phosphorylation detected in vivo at the basal plasma membrane; sensitive to brefeldin A (BFA) when at the basal plasma membrane but not at the apical plasma membrane; sensitive to cantharidin; the study also reveals that replacement of S1–S3 by phosphorylation-mimicking E does not lead to constitutive PIN activation and renders the protein insensitive to activating kinases	Weller et al. 2017
	T1 (T227), T2 (T248), T3 (T286)	MPK4 and MPK6 in vitro phosphosites	Dory et al. 2018
	P228, P249, P287, P338	Putative Pin1At *cis*/*trans* isomerization sites	Xi et al. 2016
	S337, T340	Phosphosites not targeted by PID; after mutagenesis of both phosphosites in PIN1 to A or D, PIN1 variants can rescue *pin1* but lines display, genetically dominant, developmental defects; PIN1-GFP with S/T to D mutations was frequently apolar or apically rather than basally localized; PIN1-HA with S/T to D mutations localizes apically in the *PIN2* expression domain	Zhang et al. 2010
	S337	Mutagenesis of S337 to A led to a dramatic reduction of phosphorylation by MKK7-MPK6; 35S::PIN1-GFP S337D but not 35S::PIN1-GFP S337A plants fail to properly polarize PIN1 in xylem parenchyma cells of inflorescence stems; in hypocotyl both forms show normal basal polarity	Jia et al. 2016
PIN2	S1 (S237), S2 (S258), S3 (S310)	Phosphosites required for polarity switch after overexpression of AGC3 kinases (PID, WAG1, WAG2)	Dhonukshe et al. 2010

Cite this article as *Cold Spring Harb Perspect Biol* doi: 10.1101/cshperspect.a039875

	K158, K201, K318, K321, K322, K361, K362, K363, K381, K497, K556, K614	Mutant PIN2 variant where K were replaced by R (collectively $pin2^{12K\text{-}R}$), cannot complement *pin2* mutant; $pin2^{12K\text{-}R}$ is not degraded in response to naphthalene acetic acid (NAA) treatment and less strongly ubiquitylated than the wild-type; transporter activity is unaffected; protein is defective in endocytosis after gravistimulation	Abas et al. 2006; Leitner et al. 2012
	K221, K273, K303, K318, K321, K322, K361, K362, K381, K401, K417, K429, K443, K457, K463, K464	Mutant PIN2 variant where K was replaced by R (collectively $pin2^{17K\text{-}R}$), cannot complement *pin2* mutant; $pin2^{17K\text{-}R}$ is defective in sorting or proteolytic turnover during early seedling development	Abas et al. 2006; Leitner et al. 2012
	C39, C560	Mutant PIN2 variant where C were replaced by A (C39,560A) can rescue a *pin2* allele but displays increased cellular accumulation and affects the distribution of the protein in the plasma membrane	Retzer et al. 2017
PIN3	S209, S212, S5 (S215), T222, S1 (S226)	Mutagenesis of at least three of the five predicted phosphosites partially impairs PIN3 function in a root hair cell assays for auxin efflux and disrupts PIN3 targeting to the plasma membrane; PIN3 phosphorylated form is not detectable after mutagenesis of all five sites but still after mutagenesis of S209, S212, and S5 (S215) (collectively 3m1)	Ganguly et al. 2012; Sasayama et al. 2013
	S209, S212, S5 (S215)	Mutagenesis of S209, S212, and S5 (S215) (collectively 3m1) to A partially impairs PIN3 function in a root hair cell assays for auxin efflux and disrupts PIN3 targeting to the plasma membrane	Sasayama et al. 2013
	S1 (S226), S2 (S243), S3 (S283)	Mutagenesis of the three serines (collectively P1) to A or D alters migration pattern in gels and renders the protein insensitive to phosphorylation changes induced by PID overexpression; phosphomimic S to D mutants exhibit a defect in PIN3 relocation after gravistimulation of roots and hypocotyls; phosphor-dead S to A mutants behave like wild-type	Grones et al. 2018
	S316, S317, S321—analogous to PIN1 S337, T340	Mutagenesis of the three sites (collectively P2) to A or D did not affect migration of the protein on gels; the phosphomimicking S to D variant was not able to fully complement the mutant phenotype; phosphomimic S to D mutations exhibit defect in PIN3 relocation after gravistimulation of roots and hypocotyls; S to A mutants behave like wild-type	Grones et al. 2018
PIN4	S271	In vitro phosphorylation site for AtCRK5	Baba et al. 2019b
PIN7	S277 or S278 and S431	In vitro phosphorylation site for AtCRK5	Baba et al. 2019b

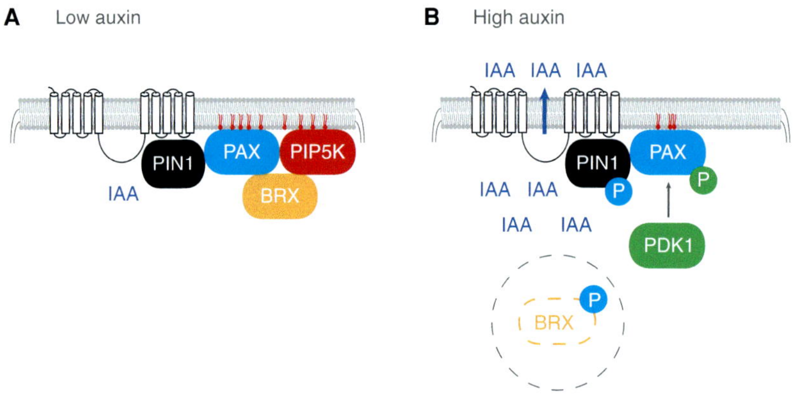

Figure 4. PAX kinase and BRX are antagonistically acting regulators of auxin transport. (*A*,*B*) Schematic representation of a model of PIN1-mediated auxin export during protophloem differentiation at low intracellular IAA levels (*A*) and at high intracellular IAA levels (*B*). Red plasma membrane components represent PI(4,5)P phospholipids that are synthesized as a result of PIP5K recruitment to the plasma membrane. The activation loop of PAX is phosphorylated by PDK1 and also at elevated IAA concentrations but it is unclear whether the two processes are linked or even identical.

for the AGC1 kinase D6PK and the related PROTEIN KINASE ASSOCIATED WITH BREVIS RADIX (PAX) as well as for the AGC3 kinases PID and WAG2 (Table 1; Barbosa et al. 2014; Zourelidou et al. 2014; Weller et al. 2017; Marhava et al. 2018). In many developmental contexts, *agc1* and *agc3* kinase mutant phenotypes align well with the phenotypes of mutants of their PIN substrates (e.g., *pid* mutants share the eponymous pin-shaped inflorescence phenotype with the *pin1* mutants), and mutants of *D6PK* and the related *D6PK-LIKE* kinases share the phototropism defect with mutants of *PIN3*, *PIN4*, and *PIN7* (Christensen et al. 2000; Benjamins et al. 2001; Zourelidou et al. 2009, 2014; Willige et al. 2013). In the case of D6PK and PID, the respective serine residues, named S1 through S5, are preceded by an arginine (R), which is followed by a variable amino acid (RXS) (Fig. 3; Zourelidou et al. 2014). Mutation of S1–S5 abolishes PIN phosphorylation in vitro and PIN activation in oocyte transport assays (Zourelidou et al. 2014). Besides the already characterized sites, also other RXS motifs are found in PIN cytoplasmic loops and these sites therefore also qualify as phosphorylation sites for these kinases but remain to be characterized (Fig. 3; Barbosa et al. 2018). Most of these sites are conserved in *A. thaliana* canonical PINs, intermediate PIN6, as well as bryophyte PINs (Fig. 3).

Although AGC1 and AGC3 kinases target the same phosphosites in canonical PINs, AGC3 kinases, unlike AGC1 kinases, do not only activate PINs but can additionally regulate PIN polar distribution (Friml et al. 2004; Michniewicz et al. 2007; Dhonukshe et al. 2010; Zhang et al. 2010). The effects of AGC3 kinases on PIN polarity, when studied at the genetic level, are prominent and antagonistically acting phosphatases have been characterized or effects of phosphatase inhibitors described (Garbers et al. 1996; Deruere et al. 1999; Rashotte et al. 2001; Shin et al. 2005; Michniewicz et al. 2007; Sukumar et al. 2009; Dai et al. 2012; Ballesteros et al. 2013; Gao et al. 2013; Guo et al. 2015; Weller et al. 2017). Nevertheless, initial models that differential PIN phosphorylation at S1–S3 by PID or WAG kinases is sufficient for PIN repolarization were questioned after it was found that PIN phosphorylation is independent of PIN plasma membrane distribution (Weller et al. 2017; Barbosa et al. 2018; Marhava et al. 2018). Thus, PIN polarity regulation by AGC3 kinases must be more complex than previously thought and the effects of AGC3 kinases on PIN polar distribution may be indirect.

Although AGC1 and AGC3 kinases phosphorylate the same serine residues, mutations of individual phosphosites have differential effects on the ability of AGC1 and AGC3 kinases to activate PINs in oocyte-based auxin transport assays. And AGC1 and AGC3 kinases phosphorylate PIN cytoplasmic loop fragments also with a differential phosphosite preference in vitro, hinting at a differential mode of PIN binding or PIN activation by the two types of kinases (Zourelidou et al. 2014; Haga et al. 2018). Strikingly, the S1–S3 phosphoserines, but not S4 and S5, reside in more complex, highly conserved TPRXS motifs, which also include putative MAPK phosphosites, as will be discussed below (Fig. 3; Table 1).

PDK1, AN UPSTREAM ACTIVATOR OF AGC1 KINASES

The *A. thaliana* genome encodes for two 3′-PHOSPHOINOSITIDE-DEPENDENT PROTEIN KINASE1 (PDK1) kinases, PDK1.1 and PDK1.2, that share >90% amino acid identity (Zegzouti et al. 2006a; Tan et al. 2020; Xiao and Offringa 2020). PDK1 kinases are conserved upstream regulators of AGC kinases in animals and the carboxy-terminal PH domain of PDK1 is activated by phosphatidylinositol (3,4,5)P_3 (PtdIns(3,4,5)P_3) (Alessi et al. 1997; Ziemba et al. 2013). PDK1 interacts with AGCVIII kinases and their conserved carboxy-terminal F(D/E)(Y/F/L)F motif, designated PDK-interacting fragment (PIF), may be required for PDK1 interactions (Balendran et al. 1999; Zegzouti et al. 2006b).

Originally, PDK1 had primarily been examined in the context of PID function but *pdk* mutants did not display phenotypes similar to *pid* mutants (Zegzouti et al. 2006a; Xiao and Offringa 2020). In contrast, a number of *pdk1* phenotypes correlated well with *agc1* kinase loss-of-function mutants, notably *d6pk d6pkl1 d6pkl2* or *pax* mutants, and detailed characterization revealed that PDK1 activates D6PK and PAX (Fig. 4; Tan et al. 2020; Xiao and Offringa 2020). Phosphorylation-mimicking mutations in the activation loop of D6PK or PAX could override the requirement for PDK1 activation (Tan et al. 2020; Xiao and Offringa 2020). At the same time, the importance of the PH domain for PDK1 function was put into question since a PH domain-deficient PDK1 variant still complemented the *pdk1.1 pdk1.2* mutant and plant species encoding PDK1 variants without a PH domain were identified (Xiao and Offringa 2020). PDK1 kinases are thus upstream regulators of auxin transport acting through AGC1 kinases.

BREVIS RADIX, A NEGATIVE REGULATOR OF PIN ACTIVATION

The AGC1 kinase PAX is critically required for protophloem differentiation, and thereby *pax* mutants share this phenotype with mutants of the protophloem regulator BREVIS RADIX (BRX), a protein devoid of any known biochemical domains (Marhava et al. 2018). BRX physically interacts with PAX and BRX attenuates PIN1 or PIN3 activation by PAX or D6PK in auxin transport experiments performed in oocytes (Fig. 4; Marhava et al. 2018). In contrast, BRX shows no effect in activation assays with PID suggesting that BRX may be specific for AGC3 kinases (Marhava et al. 2018). BRX itself responds to auxin in that elevated auxin levels result in BRX plasma membrane-to-nucleus transfer, and this auxin responsiveness of *Arabidopsis* BRX requires phosphorylation at BRX serine 228, with PAX being a likely candidate as the responsible protein kinase (Marhava et al. 2018; Koh et al. 2021). Concomitantly, auxin can induce AGC3 kinase activation loop phosphorylation kinases leading to a model where BRX and PAX dynamically and antagonistically regulate PIN-mediated auxin transport during protophloem differentiation (Fig. 4; Marhava et al. 2018). BRX also recruits phosphatidylinositol-4-phosphate 5-kinase (PIP5K) to the plasma membrane and may thereby promote PI(4,5)P synthesis at the plasma membrane and promote, for example, PAX binding (Fig. 4; Marhava et al. 2020). PAX and BRX could thereby serve as protophloem-specific regulators of auxin accumulation and export in response to cellular auxin levels (Marhava et al. 2018).

PIN REGULATION BY MAP KINASES AND HYDROXPROLINATION

Three phosphorylation sites required for PIN activation, S1–S3, each reside within TPRXS motifs in direct proximity to three threonines, named here T1–T3, which are proposed MAP kinase target sites (Fig. 3). In all canonical PINs, the TPRXS motifs are repeated three times and conserved among the family members from vascular plants and bryophytes (Fig. 3). MAP kinase regulation of PINs had initially been suggested since the MAP kinase kinase gain-of-function mutant *bushy and dwarf1* (*bud1*) displayed phenotypes that could be explained by defects in auxin transport and that could be selectively suppressed by *mpk4* and *mpk6* MAP kinase loss-of-function mutations (Dai et al. 2006). In vitro, replacement of T1–T3 by alanines abolishes PIN phosphorylation by MPK4 and MPK6 (Fig. 3; Table 1; Dory et al. 2018).

The proline residues within the TPRXS motifs, at least in the case of PIN1, may further be subject to *cis*/*trans* prolyl isomerization (Xi et al. 2016). Peptidyl-prolyl *cis*/*trans* isomerases catalyze the *cis*/*trans* isomerization of prolines (P) preceding phosphorylated S or T residues, resulting in energetically more stable conformations of the phosphorylated residues. Down-regulation of the peptidyl-prolyl *cis*/*trans* isomerase Pin1At leads to agravitropic root growth and genetically suppresses gravitropism, root meristem collapse, and PIN polarity defects of *PID* overexpression lines (Xi et al. 2016). Pin1At catalyzes *cis*/*trans* conformational changes of the MPK3/6-targeted phosphosite PIN1 S337 and, less efficiently, of the three MPK4/MPK6-targeted T1–T3 (Fig. 3; Xi et al. 2016).

REGULATION OF PIN2 PLASMA MEMBRANE DISTRIBUTION BY CRKs

A. thaliana encodes for eight CRKs (Ca^{2+}/CALMODULIN-DEPENDENT PROTEIN KINASE-RELATED KINASE), most of which localize at the plasma membrane and phosphorylate PINs in vitro (Baba et al. 2018, 2019a,b). Whereas the localization of PIN1, PIN3, PIN4, and PIN7 is not disturbed in *crk5* mutants, PIN2 is depleted from apical membranes of epidermal cells and shows basal to apical localization in the cortex in the transition zone of *crk5* mutants. Subsequent studies revealed comparable PIN2 localization defects in *Arabidopsis crk1–crk4* and *crk8* mutants (Rigó et al. 2013; Baba et al. 2018). CRK5 can phosphorylate PIN2 and the presumed defects in PIN2 phosphorylation in the *crk* mutants may lead to enhanced PIN2 instability or degradation, which may, in turn, be causal for the weak and improper PIN2 localization patterns observed in *crk* mutants (Fig. 3; Table 1; Rigó et al. 2013; Baba et al. 2019a). In line with differential PIN2 localization, all *Arabidopsis crk* mutants displayed altered root gravitropic responses after gravistimulation (Baba et al. 2018). Subsequent studies identified further in vitro phosphorylation sites in PINs that may be relevant for PIN regulation (Fig. 3; Table 1; Baba et al. 2018, 2019a,b).

PIN PHOSPHORYLATION AND REGULATION BY THE RECEPTOR KINASE COMPLEX CAMEL–CANAR

The formation of the vascular system during the patterning of leaf venation or vasculature regeneration after wounding requires feedback mechanisms between auxin response and transport. The auxin-regulated receptor CAMEL and its interactor CANALIZATION-RELATED RECEPTOR-LIKE KINASE (CANAR) are required for proper leaf venation, as well as for regeneration (Hajný et al. 2020). While CANAR is a catalytically inactive pseudokinase, CAMEL phosphorylates PINs in vitro and mutation of PIN1 phosphosites affects PIN1 function at the developmental and PIN1 polarization at the cell biological level (Fig. 3; Table 1; Hajný et al. 2020).

PIN-LIKE TRANSPORTERS (PILS)

PILS transporters are ER-resident auxin transporters phylogenetically and topologically related to PINs (Barbez et al. 2012). The PILS family is phylogenetically older than the PIN family and consists of seven members in *A. thaliana* (Feraru et al. 2012; Viaene et al. 2013). The membrane topology of PILS resembles that of the noncanonical PINs in that the transporters possess ten transmembrane α-helices and a short loop reaching into the cytoplasm (Fig. 2; Barbez et al. 2012; Viaene et al. 2013). IAA retention assays conducted in plant cells and yeast with PILS2 and PILS5 suggest an export activity, which, in view of their ER localization, may contribute to the regulation of intracellular auxin distribution (Barbez et al. 2012). This view is supported by the phenotypes of *Arabidopsis pils3 and pils6* single-mutant and a *pils2 pils5* double-mutant or *PILS* overexpressors where apical hook opening defects or organ growth phenotypes can be explained by alterations in auxin transport and/or homeostasis (Béziat et al. 2017; Feraru et al. 2019). To date, auxin transport activity has only been reported for PILS2 and PILS5 but not for any of the other five family members.

ABCB TRANSPORTERS MEDIATE AUXIN EFFLUX AND PREVENT REUPTAKE

A subset of the 21 *Arabidopsis* B subclass ATP-binding cassette B (ABCB) transporters functions in the maintenance of long distance auxin transport, with some also contributing to nonpolar auxin distribution in vegetative tissues (Noh et al. 2001; Geisler et al. 2005; Blakeslee et al. 2007; Kubeš et al. 2012; Jenness et al. 2019). In *A. thaliana*, sorghum, and maize, the loss of the prominent members of the ABCB1 subclade, *ABCB1* and *ABCB19*, results in a ∼75% decrease of long distance rootward transport in mature vegetative stems and in seedling hypocotyls, while loss of *ABCB1* and *ABCB4* in roots and *ABCB21* in leaves results in irregular auxin distribution in the epidermis (Multani et al. 2003; Blakeslee et al. 2007; Knöller et al. 2010; McLamore et al. 2010; Kubeš et al. 2012; Jenness et al. 2019). Although the structural similarity of most plant ABCB transporters to mammalian ABCB1/P-glycoprotein/multidrug resistance 1 transporters initially suggested a specificity for multiple substrates (Bernasconi et al. 1996; Sidler et al. 1998), subsequent mutational, and heterologous expression analyses suggest, however, a narrower substrate specificity for plant ABCBs (Geisler et al. 2005; Blakeslee et al. 2007). A substrate promiscuity similar to that observed with mammalian ABC transporters, in plants and fungi, is found primarily in the diverse subfamily of ABCG/PDR (PLEOTROPIC DRUG RESISTANCE) transporters (Banasiak and Jasiński 2014). At least two isoforms of these, ABCG36 and ABCG37, prevent uptake of xenobiotics and exogenously applied indole-3-butyric acid (IBA), but do not appear to contribute to polar auxin transport (Ruzicka et al. 2010).

In *A. thaliana*, paired subclades are found for most of the transporters, although a major gene duplication that predates the monocot/dicot divergence is found with *Arabidopsis* ABCB16, ABCB17, ABCB18, and ABCB22 (Knöller et al. 2010). ABCB2 and ABCD10 have a high degree of sequence similarity with ABCB1 and ABCB19 but exhibit low-energy IAA docking values, no visible mutant phenotypes, and no IAA transport in heterologous systems (Yang and Murphy 2009). In contrast, the ABCB6/ABCB20 pair exhibits slightly less sequence identity with ABCB1/ABCB19, but higher IAA docking energies, their mutants have clear auxin-related mutant phenotypes and the transporters promote auxin transport in heterologous systems (Zhang et al. 2018).

Considerable insight into the mechanism of ABCB-mediated auxin transport has been gained from the high degree of sequence and structural conservation across kingdoms, the availability of multi-conformational, high-resolution crystal structures of mammalian ABCB transporters, as well as the availability of amenable systems for single-cell functional assays (Verrier et al. 2008; Bailly et al. 2011; Verhalen et al. 2017; Kim and Chen 2018; Lefevre and Boutry 2018; Kodan et al. 2019). Initial substrate binding occurs when free-running ATP hydrolysis in a nucleotide-binding domain exposes an inner leaflet cavity that is variably hydrophobic or an outer leaflet membrane interface site (Verhalen et al. 2017). Cryoelectron microscopy revealed that ATP binding in a second nucleotide domain induces an outward-facing conformational change and substrate release from a cavity that is much shallower than surmised from earlier structural models (Kim and Chen 2018). This shallower cavity is also more consistent with secondary substrate-activated anion currents that appear to be present in *Arabi-*

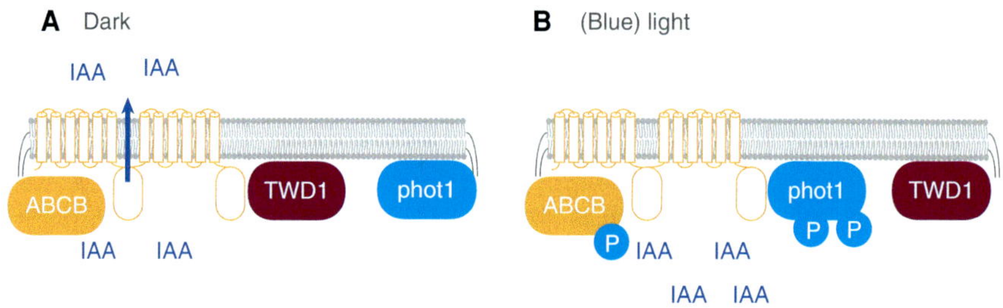

Figure 5. Blue light–activated phototropin1 (phot1) phosphorylates and inactivates auxin transport by ABCB19. (*A*) In the dark, ABCB19 mediates net cellular efflux of IAA. The active conformation of ABCB19 is maintained by the FK506-binding protein TWD1. (*B*) Upon blue light stimulation, phot1 phosphorylation of ABCB19 inactivates auxin transport. *In planta*, phot1 interactions with ABCB19 also disrupt association with TWD1. Phosphoproteomic analyses suggest that phot1-dependent phosphorylation occurs within the linker region that connects the two pseudo-halves of ABCB19. (ABCB) ATP-binding cassette B.

dopsis ABCBs and that are associated with the human cystic fibrosis transmembrane receptor (Cho et al. 2014).

Modeling and heterologous expression studies of *Arabidopsis* ABCBs suggest that substrate interactions in the inner leaflet site may be more specific than those in outer leaflet sites (Yang and Murphy 2009; Bailly et al. 2011; Hayashi et al. 2014; Jenness et al. 2019). ABCB prevention of auxin reuptake in the outer leaflet is supported by functional studies in heterologous systems and *in planta*. In yeast, mammalian, and plant cells, IAA and NAA export activities of *Arabidopsis* ABCB1, ABCB4, and ABCB19 are relatively specific, but some export of benzoic acid, an analogous compound that is not polarly transported *in planta* is observed (Geisler et al. 2005; Blakeslee et al. 2007; Yang and Murphy 2009; Kubeš et al. 2012).

CONDITIONAL AUXIN UPTAKE MEDIATED BY ABCB4 AND ABCB21

Cellular uptake of simple organic acids into guard cells by *Arabidopsis* ABCB14 and altered secondary cell wall structure in *abcb14* inflorescences is supported by inner cavity docking sites in this subclade of ABCBs (Lee et al. 2008; Yang and Murphy 2009; Kaneda et al. 2011; Jenness and Murphy 2014). The ABCB4/ABCC21 subclade features some import-enabling structural features and hydrophobic docking sites that appear to enable auxin import when cellular auxin concentrations are marginal, but canonical efflux activity when intracellular auxin concentrations are elevated (Yang and Murphy 2009; Jenness and Murphy 2014). ABCB4 regulates auxin levels in root epidermal cells and ABCB21 conditional transport regulates auxin retention in the pericycle as well as auxin distribution in leaf epidermal cells (Terasaka et al. 2005; Cho et al. 2007; Lewis et al. 2007; Kamimoto et al. 2012; Kubeš et al. 2012; Jenness et al. 2019).

REGULATION OF ABCB TRAFFICKING AND FUNCTION

Cellular trafficking of ABCBs is nondynamic but is highly dependent on insertion into sterol/sphingolipid-enriched ordered membrane domains (Roudier et al. 2010; Yang et al. 2013). The chaperone/peptidyl prolyl *cis*/*trans* isomerase activity of FK506-BINDING PROTEIN 42/TWISTED DWARF 1 (FKBP42/TWD1) is necessary for ER exit and activation of a subgroup of ABCB and two similar ABCC vacuolar transporters (Geisler et al. 2003; Wu et al. 2010; Wang and Deng 2014). TWD1 is energized by oligomerization with HSP90 and interacts with ABCBs and ABCC1/ABCC2 via interactions with the carboxy-terminal domains of the proteins (Kamphausen et al. 2002; Geisler et al. 2003; Bailly et al. 2014).

Less is known about kinase regulation of ABCB transport. Phosphoproteomic surveys show that ABCB1, ABCB4, and ABCB19 are conditionally phosphorylated (Nühse et al. 2007; Mergner et al. 2020). In a first step in seedling phototropism, ABCB19 is directly phosphorylated by the phototropin phot1 kinase to suppress auxin transport and to prime lateral auxin movement (Fig. 5; Christie et al. 2011). Although ABCB19 inactivation by blue light and phot1 occurs in heterologous systems, membrane trafficking and stability appear to also be affected, as TWD1/FKBP42 binding to ABCB19 is also decreased under very low fluence blue light conditions (Fig. 5). ABCB1 is activated by apparent phosphorylation of an internal linker region by PID in a manner that is coregulated by TWD1 interactions (Henrichs et al. 2012). ABCB4 is phosphorylated at a conserved regulatory residue after artificial stimulation of innate immunity responses (Nühse et al. 2007).

ABCG TRANSPORTERS CAN TRANSPORT INDOLE-3-BUTYRIC ACID

IBA is a naturally occurring auxin that is found in vanishingly low levels in most plant tissues (Novák et al. 2012). IBA is thought to function in seedling cell expansion and lateral root formation (Strader et al. 2010). ABCG36/PEN3/PDR8/PIS1 is a plasma membrane–localized ABCG transporter in *Arabidopsis* epidermal root cells that confers resistance to a wide range of xenobiotics and herbicides but also functions in plant defense responses (Ito and Gray 2006; Stein et al. 2006). ABCG36 functions in the net efflux of applied IBA (Ruzicka et al. 2010).

AUX/LAX Transporters

Cloning of the gene responsible for the agravitropic *aux1* (*auxin resistant1*) mutant, which is also insensitive to exogenously applied IAA or 2,4D, led to the identification of the auxin importer *AUX1* (Bennett et al. 1996). *AUX1* is a member of the *AUX/LAX* (*LIKE AUX1*) gene family, which consists of four members in *A. thaliana*, *AUX1*, *LAX1*, *LAX2*, and *LAX3* that share 75%–80% amino acid identity (Swarup and Péret 2012). AUX/LAX are AMINO ACID/AUXIN PERMEASE (AAAP) transporters and this plant-specific protein family can be traced back to the streptophyte and chlorophyte algae (Wipf et al. 2002; De Smet et al. 2011; Vosolsobě et al. 2020). The crystal structure of an AAAP family member proposes that these transporters possess 11 membrane-spanning α-helices with the amino terminus facing the cytosol (Fig. 2; Lei et al. 2018).

AUX1 function is essential for negative root gravitropism and contributes to embryo as well as female gametophyte development, apical hook formation, lateral root development, planar polarity during root hair development, leaf serration, and phyllotactic patterning (Bennett et al. 1996; Grebe et al. 2002; Marchant et al. 2002; Bainbridge et al. 2008; Vandenbussche et al. 2010; Kasprzewska et al. 2015; Panoli et al. 2015). *LAX1* acts redundantly with *AUX1* in embryo development and phyllotactic patterning (Bainbridge et al. 2008; Panoli et al. 2015). *LAX2* has been implicated in embryo development, regulation of cell division in the quiescent center, leaf serration, phyllotactic patterning, and mainly vascular patterning (Bainbridge et al. 2008; Péret et al. 2012; Zhang et al. 2013; Fàbregas et al. 2015; Kasprzewska et al. 2015; Panoli et al. 2015; Moreno-Piovano et al. 2017). The main physiological role of *LAX3* is in controlling lateral root emergence (Swarup et al. 2008).

AUX1 mediates high-affinity IAA transport in the physiological range ($K_m \approx 1$ μM) in *Xenopus laevis* oocytes (Yang et al. 2006). AUX1 transports also 2,4-D but not NAA, shows a pH optimum at apoplastic pH (pH 5.5–6), and is sensitive to the IAA influx inhibitors naphthoxy-1-acetic acid (1-NOA) and naphthoxy-2-acetic acid (2-NOA) (Yang et al. 2006). AUX1, like the amino acid transporters in the AAAP family, operates by a H^+-cotransport mechanism, which is in line with the preexisting hypothesis that auxin transport is carrier–mediated and depends on the proton motive force (Lomax et al. 1985; Dindas et al. 2018). The close phylogenetic relationship of the AUX/LAX to amino acid permease transporters raises the possibility that they evolved from amino acid transporters that are able to transport the IAA precursor tryptophan. However, AUX/LAX transporters may not possess this property and, conversely, IAA transport has not been reported for the AAAP amino acid transporters yet.

Similar properties were observed for AUX1, LAX1, and LAX3, indicating that these AUX/LAX proteins have redundant biochemical function (Swarup et al. 2008; Péret et al. 2012). However, no transport activity could be detected as of yet for LAX2 and *LAX2* expression only partially suppresses the *aux1* agravitropism defect, indicating that LAX2 may be distinct from the other AUX/LAX transporters (Péret et al. 2012). All features of the AUX1 transporter are consistent with older physiological studies, thus confirming the role of AUX1 as the primary auxin import system. The fact that AUX/LAX transporters are active when studied in heterologous systems may suggest that they import auxin independently from upstream regulators and that AUX/LAX expression and possibly subcellular localization determine auxin import.

TRANSPORTERS FOR OTHER AUXINS

IAA is the major but not the only naturally occurring auxin. The knowledge about transport systems for auxin besides IAA is currently very limited. Recently, NPF5.12 was identified as TRANSPORTER OF IBA1 (TOB1) (Michniewicz et al. 2019). TOB1 localizes to the tonoplast membrane where it may regulate the levels of IBA available for conversion to IAA in the cytosol and, thereby, root system architecture by slowing down lateral root formation (Michniewicz et al. 2019).

Transport of IBA was recently also demonstrated for another NPF family member, NPF7.3/NRT1.5, which was previously shown to play a role in low-affinity nitrate transport and K^+/H^+ antiport. NPF7.3-mediated IBA transport displayed 10-fold higher rates compared to IAA, which, unlike NPF6.3-mediated IAA transport, could not be competed for by nitrate, possibly indicating that, in this instance, the promiscuity toward the two substrates has evolved toward a specificity for the hormone (Watanabe et al. 2020). This preference is likely to be quite different in plant species where IBA is more prevalent.

TRANSPORTERS WITH "MOONLIGHTING" AUXIN TRANSPORT ACTIVITIES

IAA is also a substrate of transporters, which have another, presumably major, substrate. The most prominent example for such "moonlighting" transporters is NRT1.1/NPF6.3/CHL1 (NITRATE TRANSPORTER1.1/NITRATE TRANSPORTER1/PEPTIDE TRANSPORTER FAMILY6.3/CHLORINA1), which is known as a high-affinity nitrate importer (Krouk et al. 2010). At low nitrate concentrations, NRT1.1 can mediate IAA transport and IAA transport is inhibited by increasing nitrate concentrations, indicating that IAA and nitrate may compete for the transporter. Notably, NRT1.2, the closest relative of NRT1.1, does not seem to transport IAA, suggesting that the potential to transport these two substrates is not a common feature of nitrate transporters (Krouk et al. 2010). A competition, and thereby signal integration, between auxin and nitrate transport could occur during lateral root development where low nitrate levels could promote increased auxin transport via NRT1.1 and thereby lateral root formation (Krouk et al. 2010).

Several NPF family members transport hormones and nutrients or metabolites, and this phenomenon is also found in LHT1 (LYSINE/HISTIDIN TRANSPORTER 1), which transports the ethylene precursor ACC (1-aminocyclopropane-1-carboxylic acid), as well as amino acids. It is therefore tempting to speculate that competition for transport is a common, and possibly underrated, process to integrate hormone and nutrient transport and signaling (Chiba et al. 2015; Shin et al. 2015).

WALLS ARE THIN1 (WAT1/UmamiT5) of the UmamiT (USUALLY MULTIPLE ACIDS MOVE IN AND OUT) family is another example for an auxin transporter that is phylogenetically related to amino acid transporters (Ranocha et al. 2013). WAT1 localizes to the tonoplast and is involved in regulating cell wall thickness through an unknown, and very likely indirect mechanism. IAA transport by WAT1 was demonstrated in oocytes, but if also amino acids are substrates for the transporter is currently not known (Ranocha et al. 2013).

CONCLUDING REMARKS

Plants possess a number of transporters for the transport of auxin from cell-to-cell and auxin distribution within the cell. The study of each of these transporter (families) found and finds inspiration from different sources and angles such as the availability of respective mutants and phenotypes, insights from other, related biological and biochemical processes, relevant findings in nonplant systems, etc. As a result, and as it has become apparent in this review, the individual transporter families are understood at substantially distinct level, depth, and detail. And clearly the understanding of biochemical and biophysical properties of the transporters is still extremely limited, even in the case of the better studied PIN, ABCB, and AUX/LAX families. Knowledge about the substrate specificities and kinetic parameters of the individual proteins will be needed to fully understand the contribution of each individual transporter to auxin transport at the tissue or whole plant level. Knowledge of these parameters will be also crucial to disentangle the complexity of the phenomenon in a living organism by refining existing models and perhaps extending the models to dynamic models describing auxin distribution with time and cellular resolution during development. Ultimately, high-resolution structures of transporters will be needed to understand the precise transport mechanisms. Recent advantages in cryoelectron microscopy may yield interesting insights in these features in a foreseeable future.

The apparent proliferation of auxin transport mechanisms is consistent with the evolutionary emergence of auxin in regulating programmed and plastic growth during the evolution of vascular plants. Although auxin has been shown to function in green algae and in polar growth responses in mosses and bryophytes, the dependence of polarized growth on PIN and ABCB efflux transporters appears to have emerged and diversified within vascular plants (Thomas et al. 1983; Geisler et al. 2017; Thelander et al. 2018; Vosolsobě et al. 2020; Zhang et al. 2020). Clearly, while the "basal" polar auxin transport mechanisms appear very early, the evolutionary diversity of angiosperms, at least in part driven by pollinator interactions, may have strongly contributed to the elaboration of auxin transport mechanisms. A better understanding of the evolution of these auxin transport mechanisms in a diversity of species will help deepen our understanding of the important steps in auxin transport evolution and help aligning these with specific processes in plant development.

ACKNOWLEDGMENTS

Relevant work in the laboratory of C.S. is funded through grants from the Deutsche Forschungsgemeinschaft DFG/SCHW751/12-2, DFG/SCHW 751/14-1, and DFG/SCHW 751/15-1. Work in the laboratory of U.Z.H. is funded through DFG/HA3468/6-1 and DFG/HA3468/6-3. Work in the laboratory of A.S.M. is funded by the U.S. Department of Energy DE-FG02-13ER16405.

REFERENCES

Abas L, Benjamins R, Malenica N, Paciorek T, Wišniewska J, Moulinier-Anzola JC, Sieberer T, Friml J, Luschnig C. 2006. Intracellular trafficking and proteolysis of the *Arabidopsis* auxin-efflux facilitator PIN2 are involved in root gravitropism. *Nat Cell Biol* **8:** 249–256. doi:10.1038/ncb1369

Abas L, Kolb M, Stadlmann J, Janacek DP, Lukic K, Schwechheimer C, Sazanov LA, Mach L, Friml J, Hammes UZ. 2021. Naphthylphthalamic acid associates with and inhibits PIN auxin transporters. *Proc Natl Acad Sci* **118:** e2020857118. doi:10.1073/pnas.2020857118

Alessi DR, James SR, Downes CP, Holmes AB, Gaffney PR, Reese CB, Cohen P. 1997. Characterization of a 3-phosphoinositide-dependent protein kinase which phosphorylates and activates protein kinase Bα. *Curr Biol* **7:** 261–269. doi:10.1016/S0960-9822(06)00122-9

Baba AI, Rigó G, Ayaydin F, Rehman AU, Andrási N, Zsigmond L, Valkai I, Urbancsok J, Vass I, Pasternak T, et al. 2018. Functional analysis of the *Arabidopsis thaliana* CDPK-related kinase family: AtCRK1 regulates responses to continuous light. *Int J Mol Sci* **19:** 1282. doi:10.3390/ijms19051282

Baba AI, Andrási N, Valkai I, Gorcsa T, Koczka L, Darula Z, Medzihradszky KF, Szabados L, Fehér A, Rigó G, et al. 2019a. AtCRK5 protein kinase exhibits a regulatory role in hypocotyl hook development during skotomorphogenesis. *Int J Mol Sci* **20:** 3432. doi:10.3390/ijms20143432

Baba AI, Valkai I, Labhane NM, Koczka L, Andrási N, Klement E, Darula Z, Medzihradszky KF, Szabados L, Fehér A, et al. 2019b. CRK5 protein kinase contributes to the progression of embryogenesis of *Arabidopsis thaliana. Int J Mol Sci* **20:** 6120. doi:10.3390/ijms20246120

Bailly A, Yang H, Martinoia E, Geisler M, Murphy AS. 2011. Plant lessons: exploring ABCB functionality through structural modeling. *Front Plant Sci* **2:** 108.

Bailly A, Wang B, Zwiewka M, Pollmann S, Schenck D, Lüthen H, Schulz A, Friml J, Geisler M. 2014. Expression of TWISTED DWARF1 lacking its in-plane membrane anchor leads to increased cell elongation and hypermorphic growth. *Plant J* **77:** 108–118. doi:10.1111/tpj.12369

Bainbridge K, Guyomarc'h S, Bayer E, Swarup R, Bennett M, Mandel T, Kuhlemeier C. 2008. Auxin influx carriers stabilize phyllotactic patterning. *Genes Dev* **22:** 810–823. doi:10.1101/gad.462608

Balendran A, Casamayor A, Deak M, Paterson A, Gaffney P, Currie R, Downes CP, Alessi DR. 1999. PDK1 acquires PDK2 activity in the presence of a synthetic peptide derived from the carboxyl terminus of PRK2. *Curr Biol* **9:** 393–404. doi:10.1016/S0960-9822(99)80186-9

Ballesteros I, Domínguez T, Sauer M, Paredes P, Duprat A, Rojo E, Sanmartín M, Sánchez-Serrano JJ. 2013. Specialized functions of the PP2A subfamily II catalytic subunits PP2A-C3 and PP2A-C4 in the distribution of auxin fluxes and development in *Arabidopsis. Plant J* **73:** 862–872. doi:10.1111/tpj.12078

Banasiak J, Jasiński M. 2014. Defence, symbiosis and ABCG transporters. In *Plant ABC transporters*, pp. 163–184. Springer, Basel, Switzerland.

Barbez E, Kubeš M, Rolčík J, Béziat C, Pěnčík A, Wang B, Rosquete MR, Zhu J, Dobrev PI, Lee Y, et al. 2012. A novel putative auxin carrier family regulates intracellular auxin homeostasis in plants. *Nature* **485:** 119–122. doi:10.1038/nature11001

Barbosa IC, Zourelidou M, Willige BC, Weller B, Schwechheimer C. 2014. D6 PROTEIN KINASE activates auxin transport-dependent growth and PIN-FORMED phosphorylation at the plasma membrane. *Dev Cell* **29:** 674–685. doi:10.1016/j.devcel.2014.05.006

Barbosa IC, Shikata H, Zourelidou M, Heilmann M, Heilmann I, Schwechheimer C. 2016. Phospholipid composition and a polybasic motif determine D6 PROTEIN KINASE polar association with the plasma membrane and tropic responses. *Development* **143:** 4687–4700. doi:10.1242/dev.137117

Barbosa ICR, Hammes UZ, Schwechheimer C. 2018. Activation and polarity control of PIN-FORMED auxin transporters by phosphorylation. *Trends Plant Sci* **23:** 523–538. doi:10.1016/j.tplants.2018.03.009

Benjamins R, Quint A, Weijers D, Hooykaas P, Offringa R. 2001. The PINOID protein kinase regulates organ development in *Arabidopsis* by enhancing polar auxin transport. *Development* **128:** 4057–4067.

Bennett MJ, Marchant A, Green HG, May ST, Ward SP, Millner PA, Walker AR, Schulz B, Feldmann KA. 1996. *Arabidopsis* AUX1 gene: a permease-like regulator of root gravitropism. *Science* **273:** 948–950. doi:10.1126/science.273.5277.948

Bennett T, Brockington SF, Rothfels C, Graham SW, Stevenson D, Kutchan T, Rolf M, Thomas P, Wong GK, Leyser O, et al. 2014. Paralogous radiations of PIN proteins with multiple origins of noncanonical PIN structure. *Mol Biol Evol* **31:** 2042–2060. doi:10.1093/molbev/msu147

Bernasconi P, Patel BC, Reagan JD, Subramanian MV. 1996. The *N*-1-naphthylphthalamic acid-binding protein is an integral membrane protein. *Plant Physiol* **111:** 427–432. doi:10.1104/pp.111.2.427

Béziat C, Barbez E, Feraru MI, Lucyshyn D, Kleine-Vehn J. 2017. Light triggers PILS-dependent reduction in nuclear auxin signalling for growth transition. *Nat Plants* **3:** 17105. doi:10.1038/nplants.2017.105

Blakeslee JJ, Bandyopadhyay A, Lee OR, Mravec J, Titapiwatanakun B, Sauer M, Makam SN, Cheng Y, Bouchard R, Adamec J, et al. 2007. Interactions among PIN-FORMED and P-glycoprotein auxin transporters in *Arabidopsis. Plant Cell* **19:** 131–147. doi:10.1105/tpc.106.040782

Chiba Y, Shimizu T, Miyakawa S, Kanno Y, Koshiba T, Kamiya Y, Seo M. 2015. Identification of *Arabidopsis* thaliana NRT1/PTR FAMILY (NPF) proteins capable of transporting plant hormones. *J Plant Res* **128:** 679–686. doi:10.1007/s10265-015-0710-2

Cho M, Lee SH, Cho HT. 2007. P-glycoprotein4 displays auxin efflux transporter-like action in *Arabidopsis* root hair cells and tobacco cells. *Plant Cell* **19:** 3930–3943. doi:10.1105/tpc.107.054288

Cho M, Henry EM, Lewis DR, Wu G, Muday GK, Spalding EP. 2014. Block of ATP-binding cassette B19 ion channel activity by 5-nitro-2-(3-phenylpropylamino)-benzoic acid impairs polar auxin transport and root gravitropism. *Plant Physiol* **166:** 2091–2099. doi:10.1104/pp.114.250860

Christensen SK, Dagenais N, Chory J, Weigel D. 2000. Regulation of auxin response by the protein kinase PINOID. *Cell* **100:** 469–478. doi:10.1016/S0092-8674(00)80682-0

Christie JM, Yang H, Richter GL, Sullivan S, Thomson CE, Lin J, Titapiwatanakun B, Ennis M, Kaiserli E, Lee OR, et al. 2011. Phot1 inhibition of ABCB19 primes lateral auxin fluxes in the shoot apex required for phototropism. *PLoS Biol* **9:** e1001076. doi:10.1371/journal.pbio.1001076

Dai Y, Wang H, Li B, Huang J, Liu X, Zhou Y, Mou Z, Li J. 2006. Increased expression of MAP KINASE KINASE7 causes deficiency in polar auxin transport and leads to plant architectural abnormality in *Arabidopsis. Plant Cell* **18:** 308–320. doi:10.1105/tpc.105.037846

Dai M, Zhang C, Kania U, Chen F, Xue Q, McCray T, Li G, Qin G, Wakeley M, Terzaghi W, et al. 2012. A PP6-type phosphatase holoenzyme directly regulates PIN phosphorylation and auxin efflux in *Arabidopsis. Plant Cell* **24:** 2497–2514. doi:10.1105/tpc.112.098905

Deruere J, Jackson K, Garbers C, Soll D, Delong A. 1999. The RCN1-encoded A subunit of protein phosphatase 2A increases phosphatase activity in vivo. *Plant J* **20:** 389–399. doi:10.1046/j.1365-313x.1999.00607.x

De Smet I, Voß U, Lau S, Wilson M, Shao N, Timme RE, Swarup R, Kerr I, Hodgman C, Bock R, et al. 2011. Unraveling the evolution of auxin signaling. *Plant Physiol* **155:** 209–221. doi:10.1104/pp.110.168161

Dhonukshe P, Huang F, Galvan-Ampudia CS, Mahonen AP, Kleine-Vehn J, Xu J, Quint A, Prasad K, Friml J, Scheres B, et al. 2010. Plasma membrane-bound AGC3 kinases phosphorylate PIN auxin carriers at TPRXS(N/S) motifs to direct apical PIN recycling. *Development* **137:** 3245–3255. doi:10.1242/dev.052456

Dindas J, Scherzer S, Roelfsema MRG, von Meyer K, Müller HM, Al-Rasheid KAS, Palme K, Dietrich P, Becker D,

Bennett MJ, et al. 2018. AUX1-mediated root hair auxin influx governs $SCF^{TIR1/AFB}$-type Ca^{2+} signaling. *Nat Commun* **9:** 1174. doi:10.1038/s41467-018-03582-5

Dory M, Hatzimasoura E, Kállai BM, Nagy SK, Jäger K, Darula Z, Nádai TV, Mészáros T, López-Juez E, Barnabás B, et al. 2018. Co-evolving MAPK and PID phosphosites indicate an ancient environmental control of PIN auxin transporters in land plants. *FEBS Lett* **592:** 89–102. doi:10.1002/1873-3468.12929

Fàbregas N, Formosa-Jordan P, Confraria A, Siligato R, Alonso JM, Swarup R, Bennett MJ, Mähönen AP, Caño-Delgado AI, Ibañes M. 2015. Auxin influx carriers control vascular patterning and xylem differentiation in *Arabidopsis thaliana. PLoS Genet* **11:** e1005183. doi:10.1371/journal.pgen.1005183

Feraru E, Vosolsobe S, Feraru MI, Petrasek J, Kleine-Vehn J. 2012. Evolution and structural diversification of PILS putative auxin carriers in plants. *Front Plant Sci* **3:** 227.

Feraru E, Feraru MI, Barbez E, Waidmann S, Sun L, Gaidora A, Kleine-Vehn J. 2019. PILS6 is a temperature-sensitive regulator of nuclear auxin input and organ growth in *Arabidopsis thaliana. Proc Natl Acad Sci* **116:** 3893–3898. doi:10.1073/pnas.1814015116

Friml J, Yang X, Michniewicz M, Weijers D, Quint A, Tietz O, Benjamins R, Ouwerkerk PB, Ljung K, Sandberg G, et al. 2004. A PINOID-dependent binary switch in apical-basal PIN polar targeting directs auxin efflux. *Science* **306:** 862–865. doi:10.1126/science.1100618

Galván-Ampudia CS, Offringa R. 2007. Plant evolution: AGC kinases tell the auxin tale. *Trends Plant Sci* **12:** 541–547. doi:10.1016/j.tplants.2007.10.004

Ganguly A, Lee SH, Cho HT. 2012. Functional identification of the phosphorylation sites of *Arabidopsis* PIN-FORMED3 for its subcellular localization and biological role. *Plant J* **71:** 810–823. doi:10.1111/j.1365-313X.2012.05030.x

Gao HB, Chu YJ, Xue HW. 2013. Phosphatidic acid (PA) binds PP2AA1 to regulate PP2A activity and PIN1 polar localization. *Mol Plant* **6:** 1692–1702. doi:10.1093/mp/sst076

Garbers C, DeLong A, Deruére J, Bernasconi P, Söll D. 1996. A mutation in protein phosphatase 2A regulatory subunit A affects auxin transport in *Arabidopsis. EMBO J* **15:** 2115–2124. doi:10.1002/j.1460-2075.1996.tb00565.x

Geisler M, Kolukisaoglu HU, Bouchard R, Billion K, Berger J, Saal B, Frangne N, Koncz-Kálmán Z, Koncz C, Dudler R, et al. 2003. TWISTED DWARF1, a unique plasma membrane-anchored immunophilin-like protein, interacts with *Arabidopsis* multidrug resistance-like transporters AtPGP1 and AtPGP19. *Mol Biol Cell* **14:** 4238–4249. doi:10.1091/mbc.e02-10-0698

Geisler M, Blakeslee JJ, Bouchard R, Lee OR, Vincenzetti V, Bandyopadhyay A, Titapiwatanakun B, Peer WA, Bailly A, Richards EL, et al. 2005. Cellular efflux of auxin catalyzed by the *Arabidopsis* MDR/PGP transporter AtPGP1. *Plant J* **44:** 179–194. doi:10.1111/j.1365-313X.2005.02519.x

Geisler M, Aryal B, di Donato M, Hao P. 2017. A critical view on ABC transporters and their interacting partners in auxin transport. *Plant Cell Physiol* **58:** 1601–1614. doi:10.1093/pcp/pcx104

Grebe M, Friml J, Swarup R, Ljung K, Sandberg G, Terlou M, Palme K, Bennett MJ, Scheres B. 2002. Cell polarity signaling in *Arabidopsis* involves a BFA-sensitive auxin influx pathway. *Curr Biol* **12:** 329–334. doi:10.1016/S0960-9822(02)00654-1

Grones P, Abas M, Hajný J, Jones A, Waidmann S, Kleine-Vehn J, Friml J. 2018. PID/WAG-mediated phosphorylation of the *Arabidopsis* PIN3 auxin transporter mediates polarity switches during gravitropism. *Sci Rep* **8:** 10279. doi:10.1038/s41598-018-28188-1

Guo X, Qin Q, Yan J, Niu Y, Huang B, Guan L, Li Y, Ren D, Li J, Hou S. 2015. TYPE-ONE PROTEIN PHOSPHATASE4 regulates pavement cell interdigitation by modulating PIN-FORMED1 polarity and trafficking in *Arabidopsis. Plant Physiol* **167:** 1058–1075. doi:10.1104/pp.114.249904

Haga K, Frank L, Kimura T, Schwechheimer C, Sakai T. 2018. Roles of AGCVIII kinases in the hypocotyl phototropism of *Arabidopsis* seedlings. *Plant Cell Physiol* **59:** 1060–1071. doi:10.1093/pcp/pcy048

Hajný J, Prát T, Rydza N, Rodriguez L, Tan S, Verstraeten I, Domjan D, Mazur E, Smakowska-Luzan E, Smet W, et al. 2020. Receptor kinase module targets PIN-dependent auxin transport during canalization. *Science* **370:** 550–557. doi:10.1126/science.aba3178

Hayashi K, Nakamura S, Fukunaga S, Nishimura T, Jenness MK, Murphy AS, Motose H, Nozaki H, Furutani M, Aoyama T. 2014. Auxin transport sites are visualized in planta using fluorescent auxin analogs. *Proc Natl Acad Sci* **111:** 11557–11562. doi:10.1073/pnas.1408960111

Henrichs S, Wang B, Fukao Y, Zhu J, Charrier L, Bailly A, Oehring SC, Linnert M, Weiwad M, Endler A, et al. 2012. Regulation of ABCB1/PGP1-catalysed auxin transport by linker phosphorylation. *EMBO J* **31:** 2965–2980. doi:10.1038/emboj.2012.120

Huang F, Zago MK, Abas L, van Marion A, Galvan-Ampudia CS, Offringa R. 2010. Phosphorylation of conserved PIN motifs directs *Arabidopsis* PIN1 polarity and auxin transport. *Plant Cell* **22:** 1129–1142. doi:10.1105/tpc.109.072678

Ito H, Gray WM. 2006. A gain-of-function mutation in the *Arabidopsis* pleiotropic drug resistance transporter PDR9 confers resistance to auxinic herbicides. *Plant Physiol* **142:** 63–74. doi:10.1104/pp.106.084533

Jenness MK, Murphy AS. 2014. Evolution of transport directionality in ABCBs. In *Plant ABC transporters*, Springer, Basel, Switzerland.

Jenness MK, Carraro N, Pritchard CA, Murphy AS. 2019. The *Arabidopsis* ATP-BINDING CASSETTE transporter ABCB21 regulates auxin levels in cotyledons, the root pericycle, and leaves. *Front Plant Sci* **10:** 806. doi:10.3389/fpls.2019.00806

Jia W, Li B, Li S, Liang Y, Wu X, Ma M, Wang J, Gao J, Cai Y, Zhang Y, et al. 2016. Mitogen-activated protein kinase cascade MKK7-MPK6 plays important roles in plant development and regulates shoot branching by phosphorylating PIN1 in *Arabidopsis. PLoS Biol* **14:** e1002550. doi:10.1371/journal.pbio.1002550

Kamimoto Y, Terasaka K, Hamamoto M, Takanashi K, Fukuda S, Shitan N, Sugiyama A, Suzuki H, Shibata D, Wang B, et al. 2012. *Arabidopsis* ABCB21 is a facultative auxin importer/exporter regulated by cytoplasmic auxin concentration. *Plant Cell Physiol* **53:** 2090–2100. doi:10.1093/pcp/pcs149

Kamphausen T, Fanghänel J, Neumann D, Schulz B, Rahfeld JU. 2002. Characterization of *Arabidopsis thaliana* AtFKBP42 that is membrane-bound and interacts with Hsp90. *Plant J* **32:** 263–276. doi:10.1046/j.1365-313X.2002.01420.x

Kaneda M, Schuetz M, Lin BS, Chanis C, Hamberger B, Western TL, Ehlting J, Samuels AL. 2011. ABC transporters coordinately expressed during lignification of *Arabidopsis* stems include a set of ABCBs associated with auxin transport. *J Exp Bot* **62:** 2063–2077. doi:10.1093/jxb/erq416

Kasprzewska A, Carter R, Swarup R, Bennett M, Monk N, Hobbs JK, Fleming A. 2015. Auxin influx importers modulate serration along the leaf margin. *Plant J* **83:** 705–718. doi:10.1111/tpj.12921

Kim Y, Chen J. 2018. Molecular structure of human P-glycoprotein in the ATP-bound, outward-facing conformation. *Science* **359:** 915–919. doi:10.1126/science.aar7389

Knöller AS, Blakeslee JJ, Richards EL, Peer WA, Murphy AS. 2010. Brachytic2/ZmABCB1 functions in IAA export from intercalary meristems. *J Exp Bot* **61:** 3689–3696. doi:10.1093/jxb/erq180

Kodan A, Yamaguchi T, Nakatsu T, Matsuoka K, Kimura Y, Ueda K, Kato H. 2019. Inward- and outward-facing X-ray crystal structures of homodimeric P-glycoprotein CmABCB1. *Nat Commun* **10:** 88. doi:10.1038/s41467-018-08007-x

Koh SWH, Marhava P, Rana S, Graf A, Moret B, Bassukas AL, Zourelidou M, Kolb M, Hammes UZ, Schwechheimer C, et al. 2021. Mapping and engineering of auxin-induced plasma-membrane-dissociation in BRX family proteins. *Plant Cell* doi:10.1093/plcell/koab076

Krouk G, Lacombe B, Bielach A, Perrine-Walker F, Malinska K, Mounier E, Hoyerova K, Tillard P, Leon S, Ljung K, et al. 2010. Nitrate-regulated auxin transport by NRT1.1 defines a mechanism for nutrient sensing in plants. *Dev Cell* **18:** 927–937. doi:10.1016/j.devcel.2010.05.008

Kubeš M, Yang H, Richter GL, Cheng Y, Młodzińska E, Wang X, Blakeslee JJ, Carraro N, Petrášek J, Zažímalová E, et al. 2012. The *Arabidopsis* concentration-dependent influx/efflux transporter ABCB4 regulates cellular auxin levels in the root epidermis. *Plant J* **69:** 640–654. doi:10.1111/j.1365-313X.2011.04818.x

Lee M, Choi Y, Burla B, Kim YY, Jeon B, Maeshima M, Yoo JY, Martinoia E, Lee Y. 2008. The ABC transporter AtABCB14 is a malate importer and modulates stomatal response to CO2. *Nat Cell Biol* **10:** 1217–1223. doi:10.1038/ncb1782

Lefevre F, Boutry M. 2018. Towards identification of the substrates of ATP-binding cassette transporters. *Plant Physiol* **178:** 18–39.

Lei HT, Ma J, Sanchez Martinez S, Gonen T. 2018. Crystal structure of arginine-bound lysosomal transporter SLC38A9 in the cytosol-open state. *Nat Struct Mol Biol* **25:** 522–527. doi:10.1038/s41594-018-0072-2

Leitner J, Petrasek J, Tomanov K, Retzer K, Parezova M, Korbei B, Bachmair A, Zazimalova E, Luschnig C. 2012. Lysine63-linked ubiquitylation of PIN2 auxin carrier protein governs hormonally controlled adaptation of *Arabidopsis* root growth. *Proc Natl Acad Sci* **109:** 8322–8327. doi:10.1073/pnas.1200824109

Lewis DR, Miller ND, Splitt BL, Wu G, Spalding EP. 2007. Separating the roles of acropetal and basipetal auxin transport on gravitropism with mutations in two *Arabidopsis* multidrug resistance-like ABC transporter genes. *Plant Cell* **19:** 1838–1850. doi:10.1105/tpc.107.051599

Lomax TL, Mehlhorn RJ, Briggs WR. 1985. Active auxin uptake by zucchini membrane vesicles: quantitation using ESR volume and delta pH determinations. *Proc Natl Acad Sci* **82:** 6541–6545. doi:10.1073/pnas.82.19.6541

Marchant A, Bhalerao R, Casimiro I, Eklöf J, Casero PJ, Bennett M, Sandberg G. 2002. AUX1 promotes lateral root formation by facilitating indole-3-acetic acid distribution between sink and source tissues in the *Arabidopsis* seedling. *Plant Cell* **14:** 589–597. doi:10.1105/tpc.010354

Marcote MJ, Sancho-Andrés G, Soriano-Ortega E, Aniento F. 2016. Sorting signals for PIN1 trafficking and localization. *Plant Signal Behav* **11:** e1212801. doi:10.1080/15592324.2016.1212801

Marhava P, Bassukas AEL, Zourelidou M, Kolb M, Moret B, Fastner A, Schulze WX, Cattaneo P, Hammes UZ, Schwechheimer C, et al. 2018. A molecular rheostat adjusts auxin flux to promote root protophloem differentiation. *Nature* **558:** 297–300. doi:10.1038/s41586-018-0186-z

Marhava P, Aliaga Fandino AC, Koh SWH, Jelínková A, Kolb M, Janacek DP, Breda AS, Cattaneo P, Hammes UZ, Petrášek J, et al. 2020. Plasma membrane domain patterning and self-reinforcing polarity in *Arabidopsis*. *Dev Cell* **52:** 223–235.e5. doi:10.1016/j.devcel.2019.11.015

McLamore ES, Diggs A, Calvo Marzal P, Shi J, Blakeslee JJ, Peer WA, Murphy AS, Porterfield DM. 2010. Non-invasive quantification of endogenous root auxin transport using an integrated flux microsensor technique. *Plant J* **63:** 1004–1016. doi:10.1111/j.1365-313X.2010.04300.x

Mergner J, Frejno M, List M, Papacek M, Chen X, Chaudhary A, Samaras P, Richter S, Shikata H, Messerer M, et al. 2020. Mass-spectrometry-based draft of the *Arabidopsis* proteome. *Nature* **579:** 409–414. doi:10.1038/s41586-020-2094-2

Michniewicz M, Zago MK, Abas L, Weijers D, Schweighofer A, Meskiene I, Heisler MG, Ohno C, Zhang J, Huang F, et al. 2007. Antagonistic regulation of PIN phosphorylation by PP2A and PINOID directs auxin flux. *Cell* **130:** 1044–1056. doi:10.1016/j.cell.2007.07.033

Michniewicz M, Ho CH, Enders TA, Floro E, Damodaran S, Gunther LK, Powers SK, Frick EM, Topp CN, Frommer WB, et al. 2019. TRANSPORTER OF IBA1 links auxin and cytokinin to influence root architecture. *Dev Cell* **50:** 599–609.e4. doi:10.1016/j.devcel.2019.06.010

Moreno-Piovano GS, Moreno JE, Cabello JV, Arce AL, Otegui ME, Chan RL. 2017. A role for LAX2 in regulating xylem development and lateral-vein symmetry in the leaf. *Ann Bot* **120:** 577–590. doi:10.1093/aob/mcx091

Multani DS, Briggs SP, Chamberlin MA, Blakeslee JJ, Murphy AS, Johal GS. 2003. Loss of an MDR transporter in compact stalks of maize br2 and sorghum dw3 mutants. *Science* **302:** 81–84. doi:10.1126/science.1086072

Nodzyński T, Vanneste S, Zwiewka M, Pernisová M, Hejátko J, Friml J. 2016. Enquiry into the topology of plasma membrane-localized PIN auxin transport components. *Mol Plant* **9:** 1504–1519. doi:10.1016/j.molp.2016.08.010

Noh B, Murphy AS, Spalding EP. 2001. Multidrug resistance-like genes of *Arabidopsis* required for auxin transport and auxin-mediated development. *Plant Cell* **13:** 2441–2454.

Novák O, Hénykova E, Sairanen I, Kowalczyk M, Pospíšil T, Ljung K. 2012. Tissue-specific profiling of the *Arabidopsis thaliana* auxin metabolome. *Plant J* **72:** 523–536. doi:10.1111/j.1365-313X.2012.05085.x

Nühse TS, Bottrill AR, Jones AM, Peck SC. 2007. Quantitative phosphoproteomic analysis of plasma membrane proteins reveals regulatory mechanisms of plant innate immune responses. *Plant J* **51:** 931–940. doi:10.1111/j.1365-313X.2007.03192.x

Panoli A, Martin MV, Alandete-Saez M, Simon M, Neff C, Swarup R, Bellido A, Yuan L, Pagnussat GC, Sundaresan V. 2015. Auxin import and local auxin biosynthesis are required for mitotic divisions, cell expansion and cell specification during female gametophyte development in *Arabidopsis thaliana*. *PLoS ONE* **10:** e0126164. doi:10.1371/journal.pone.0126164

Péret B, Swarup K, Ferguson A, Seth M, Yang Y, Dhondt S, James N, Casimiro I, Perry P, Syed A, et al. 2012. AUX/LAX genes encode a family of auxin influx transporters that perform distinct functions during *Arabidopsis* development. *Plant Cell* **24:** 2874–2885. doi:10.1105/tpc.112.097766

Petrasek J, Friml J. 2009. Auxin transport routes in plant development. *Development* **136:** 2675–2688. doi:10.1242/dev.030353

Platre MP, Jaillais Y. 2017. Anionic lipids and the maintenance of membrane electrostatics in eukaryotes. *Plant Signal Behav* **12:** e1282022. doi:10.1080/15592324.2017.1282022

Rakusová H, Fendrych M, Friml J. 2015. Intracellular trafficking and PIN-mediated cell polarity during tropic responses in plants. *Curr Opin Plant Biol* **23:** 116–123. doi:10.1016/j.pbi.2014.12.002

Ranocha P, Dima O, Nagy R, Felten J, Corratgé-Faillie C, Novák O, Morreel K, Lacombe B, Martinez Y, Pfrunder S, et al. 2013. *Arabidopsis* WAT1 is a vacuolar auxin transport facilitator required for auxin homoeostasis. *Nat Commun* **4:** 2625. doi:10.1038/ncomms3625

Rashotte AM, DeLong A, Muday GK. 2001. Genetic and chemical reductions in protein phosphatase activity alter auxin transport, gravity response, and lateral root growth. *Plant Cell* **13:** 1683–1697. doi:10.1105/TPC.010158

Retzer K, Lacek J, Skokan R, Del Genio CI, Vosolsobě S, Laňková M, Malínská K, Konstantinova N, Zažímalova E, Napier RM, et al. 2017. Evolutionary conserved cysteines function as *cis*-acting regulators of *Arabidopsis* PIN-FORMED 2 distribution. *Int J Mol Sci* **18:** 2274. doi:10.3390/ijms18112274

Rigó G, Ayaydin F, Tietz O, Zsigmond L, Kovács H, Páy A, Salchert K, Darula Z, Medzihradszky KF, Szabados L, et al. 2013. Inactivation of plasma membrane-localized CDPK-RELATED KINASE5 decelerates PIN2 exocytosis and root gravitropic response in *Arabidopsis*. *Plant Cell* **25:** 1592–1608. doi:10.1105/tpc.113.110452

Roudier F, Gissot L, Beaudoin F, Haslam R, Michaelson L, Marion J, Molino D, Lima A, Bach L, Morin H, et al. 2010. Very-long-chain fatty acids are involved in polar auxin transport and developmental patterning in *Arabidopsis*. *Plant Cell* **22:** 364–375. doi:10.1105/tpc.109.071209

Ruzicka K, Strader LC, Bailly A, Yang H, Blakeslee J, Langowski L, Nejedla E, Fujita H, Itoh H, Syono K, et al. 2010. *Arabidopsis* PIS1 encodes the ABCG37 transporter of auxinic compounds including the auxin precursor indole-3-butyric acid. *Proc Natl Acad Sci* **107:** 10749–10753. doi:10.1073/pnas.1005878107

Sancho-Andrés G, Soriano-Ortega E, Gao C, Bernabé-Orts JM, Narasimhan M, Müller AO, Tejos R, Jiang L, Friml J, Aniento F, et al. 2016. Sorting motifs involved in the trafficking and localization of the PIN1 auxin efflux carrier. *Plant Physiol* **171:** 1965–1982. doi:10.1104/pp.16.00373

Sasayama D, Ganguly A, Park M, Cho HT. 2013. The M3 phosphorylation motif has been functionally conserved for intracellular trafficking of long-looped PIN-FORMEDs in the *Arabidopsis* root hair cell. *BMC Plant Biol* **13:** 189. doi:10.1186/1471-2229-13-189

Sauer M, Kleine-Vehn J. 2019. PIN-FORMED and PIN-LIKES auxin transport facilitators. *Development* **146:** dev168088. doi:10.1242/dev.168088

Semeradova H, Montesinos JC, Benkova E. 2020. All roads lead to auxin: post-translational regulation of auxin transport by multiple hormonal pathways. *Plant Commun* **1:** 100048. doi:10.1016/j.xplc.2020.100048

Shin H, Shin HS, Guo Z, Blancaflor EB, Masson PH, Chen R. 2005. Complex regulation of *Arabidopsis* AGR1/PIN2-mediated root gravitropic response and basipetal auxin transport by cantharidin-sensitive protein phosphatases. *Plant J* **42:** 188–200. doi:10.1111/j.1365-313X.2005.02369.x

Shin K, Lee S, Song WY, Lee RA, Lee I, Ha K, Koo JC, Park SK, Nam HG, Lee Y, et al. 2015. Genetic identification of ACC-RESISTANT2 reveals involvement of LYSINE HISTIDINE TRANSPORTER1 in the uptake of 1-aminocyclopropane-1-carboxylic acid in *Arabidopsis thaliana*. *Plant Cell Physiol* **56:** 572–582. doi:10.1093/pcp/pcu201

Sidler M, Hassa P, Hasan S, Ringli C, Dudler R. 1998. Involvement of an ABC transporter in a developmental pathway regulating hypocotyl cell elongation in the light. *Plant Cell* **10:** 1623–1636. doi:10.1105/tpc.10.10.1623

Skokan R, Medvecká E, Viaene T, Vosolsobě S, Zwiewka M, Müller K, Skůpa P, Karady M, Zhang Y, Janacek DP, et al. 2019. PIN-driven auxin transport emerged early in streptophyte evolution. *Nat Plants* **5:** 1114–1119. doi:10.1038/s41477-019-0542-5

Stein M, Dittgen J, Sánchez-Rodríguez C, Hou BH, Molina A, Schulze-Lefert P, Lipka V, Somerville S. 2006. *Arabidopsis* PEN3/PDR8, an ATP binding cassette transporter, contributes to nonhost resistance to inappropriate pathogens that enter by direct penetration. *Plant Cell* **18:** 731–746. doi:10.1105/tpc.105.038372

Strader LC, Chen GL, Bartel B. 2010. Ethylene directs auxin to control root cell expansion. *Plant J* **64:** 874–884. doi:10.1111/j.1365-313X.2010.04373.x

Sukumar P, Edwards KS, Rahman A, Delong A, Muday GK. 2009. PINOID kinase regulates root gravitropism through modulation of PIN2-dependent basipetal auxin transport in *Arabidopsis*. *Plant Physiol* **150:** 722–735. doi:10.1104/pp.108.131607

Sun J, Bankston JR, Payandeh J, Hinds TR, Zagotta WN, Zheng N. 2014. Crystal structure of the plant dual-affinity nitrate transporter NRT1.1. *Nature* **507:** 73–77. doi:10.1038/nature13074

Swarup R, Péret B. 2012. AUX/LAX family of auxin influx carriers—an overview. *Front Plant Sci* **3:** 225. doi:10.3389/fpls.2012.00225

Swarup K, Benková E, Swarup R, Casimiro I, Péret B, Yang Y, Parry G, Nielsen E, De Smet I, Vanneste S, et al. 2008. The auxin influx carrier LAX3 promotes lateral root emergence. *Nat Cell Biol* **10:** 946–954. doi:10.1038/ncb1754

Tan S, Zhang X, Kong W, Yang XL, Molnár G, Vondráková Z, Filepová R, Petrášek J, Friml J, Xue HW. 2020. The lipid code-dependent phosphoswitch PDK1-D6PK activates PIN-mediated auxin efflux in *Arabidopsis*. *Nat Plants* **6:** 556–569. doi:10.1038/s41477-020-0648-9

Teale WD, Pasternak T, Dal Bosco C, Dovzhenko A, Kratzat K, Bildl W, Schwörer M, Falk T, Ruperti B, Schaefer JV, et al. 2021. Flavonol-mediated stabilization of PIN efflux complexes regulates polar auxin transport. *EMBO J* **40:** e104416. doi:10.15252/embj.2020104416

Terasaka K, Blakeslee JJ, Titapiwatanakun B, Peer WA, Bandyopadhyay A, Makam SN, Lee OR, Richards EL, Murphy AS, Sato F, et al. 2005. PGP4, an ATP binding cassette P-glycoprotein, catalyzes auxin transport in *Arabidopsis thaliana* roots. *Plant Cell* **17:** 2922–2939. doi:10.1105/tpc.105.035816

Thelander M, Landberg K, Sundberg E. 2018. Auxin-mediated developmental control in the moss *Physcomitrella patens*. *J Exp Bot* **69:** 277–290. doi:10.1093/jxb/erx255

Thomas RJ, Harrison MA, Taylor J, Kaufman PB. 1983. Endogenous auxin and ethylene in *Pellia* (Bryophyta). *Plant Physiol* **73:** 395–397. doi:10.1104/pp.73.2.395

Vandenbussche F, Petrasek J, Zadnikova P, Hoyerova K, Pesek B, Raz V, Swarup R, Bennett M, Zazimalova E, Benkova E, et al. 2010. The auxin influx carriers AUX1 and LAX3 are involved in auxin-ethylene interactions during apical hook development in *Arabidopsis thaliana* seedlings. *Development* **137:** 597–606. doi:10.1242/dev.040790

Verhalen B, Dastvan R, Thangapandian S, Peskova Y, Koteiche HA, Nakamoto RK, Tajkhorshid E, McHaourab HS. 2017. Energy transduction and alternating access of the mammalian ABC transporter P-glycoprotein. *Nature* **543:** 738–741. doi:10.1038/nature21414

Verrier PJ, Bird D, Burla B, Dassa E, Forestier C, Geisler M, Klein M, Kolukisaoglu U, Lee Y, Martinoia E, et al. 2008. Plant ABC proteins—a unified nomenclature and updated inventory. *Trends Plant Sci* **13:** 151–159. doi:10.1016/j.tplants.2008.02.001

Viaene T, Delwiche CF, Rensing SA, Friml J. 2013. Origin and evolution of PIN auxin transporters in the green lineage. *Trends Plant Sci* **18:** 5–10. doi:10.1016/j.tplants.2012.08.009

Vosolsobě S, Skokan R, Petrášek J. 2020. The evolutionary origins of auxin transport: what we know and what we need to know. *J Exp Bot* **71:** 3287–3295. doi:10.1093/jxb/eraa169

Wang Y, Deng D. 2014. Molecular basis and evolutionary pattern of GA-GID1-DELLA regulatory module. *Mol Genet Genomics* **289:** 1–9. doi:10.1007/s00438-013-0797-x

Watanabe S, Takahashi N, Kanno Y, Suzuki H, Aoi Y, Takeda-Kamiya N, Toyooka K, Kasahara H, Hayashi KI, Umeda M, et al. 2020. The *Arabidopsis* NRT1/PTR FAMILY protein NPF7.3/NRT1.5 is an indole-3-butyric acid transporter involved in root gravitropism. *Proc Natl Acad Sci* **117:** 31500–31509. doi:10.1073/pnas.2013305117

Weijers D, Wagner D. 2016. Transcriptional responses to the auxin hormone. *Annu Rev Plant Biol* **67:** 539–574. doi:10.1146/annurev-arplant-043015-112122

Weller B, Zourelidou M, Frank L, Barbosa IC, Fastner A, Richter S, Jürgens G, Hammes UZ, Schwechheimer C. 2017. Dynamic PIN-FORMED auxin efflux carrier phosphorylation at the plasma membrane controls auxin efflux-dependent growth. *Proc Natl Acad Sci* **114:** E887–E896. doi:10.1073/pnas.1614380114

Willige BC, Ahlers S, Zourelidou M, Barbosa IC, Demarsy E, Trevisan M, Davis PA, Roelfsema MR, Hangarter R, Fankhauser C, et al. 2013. D6PK AGCVIII kinases are required for auxin transport and phototropic hypocotyl bending in *Arabidopsis*. *Plant Cell* **25:** 1674–1688. doi:10.1105/tpc.113.111484

Wipf D, Ludewig U, Tegeder M, Rentsch D, Koch W, Frommer WB. 2002. Conservation of amino acid transporters in fungi, plants and animals. *Trends Biochem Sci* **27:** 139–147. doi:10.1016/S0968-0004(01)02054-0

Wu G, Cameron JN, Ljung K, Spalding EP. 2010. A role for ABCB19-mediated polar auxin transport in seedling photomorphogenesis mediated by cryptochrome 1 and phytochrome B. *Plant J* **62:** 179–191. doi:10.1111/j.1365-313X.2010.04137.x

Xi W, Gong X, Yang Q, Yu H, Liou YC. 2016. Pin1At regulates PIN1 polar localization and root gravitropism. *Nat Commun* **7:** 10430. doi:10.1038/ncomms10430

Xiao Y, Offringa R. 2020. PDK1 regulates auxin transport and *Arabidopsis* vascular development through AGC1 kinase PAX. *Nat Plants* **6:** 544–555. doi:10.1038/s41477-020-0650-2

Yang H, Murphy AS. 2009. Functional expression and characterization of *Arabidopsis* ABCB, AUX 1 and PIN auxin transporters in *Schizosaccharomyces pombe*. *Plant J* **59:** 179–191. doi:10.1111/j.1365-313X.2009.03856.x

Yang Y, Hammes UZ, Taylor CG, Schachtman DP, Nielsen E. 2006. High-affinity auxin transport by the AUX1 influx carrier protein. *Curr Biol* **16:** 1123–1127. doi:10.1016/j.cub.2006.04.029

Yang H, Richter GL, Wang X, Młodzińska E, Carraro N, Ma G, Jenness M, Chao DY, Peer WA, Murphy AS. 2013. Sterols and sphingolipids differentially function in trafficking of the *Arabidopsis* ABCB19 auxin transporter. *Plant J* **74:** 37–47. doi:10.1111/tpj.12103

Zegzouti H, Anthony RG, Jahchan N, Bogre L, Christensen SK. 2006a. Phosphorylation and activation of PINOID by the phospholipid signaling kinase 3-phosphoinositide-dependent protein kinase 1 (PDK1) in *Arabidopsis*. *Proc Natl Acad Sci* **103:** 6404–6409. doi:10.1073/pnas.0510283103

Zegzouti H, Li W, Lorenz TC, Xie M, Payne CT, Smith K, Glenny S, Payne GS, Christensen SK. 2006b. Structural and functional insights into the regulation of *Arabidopsis* AGC VIIIa kinases. *J Biol Chem* **281:** 35520–35530. doi:10.1074/jbc.M605167200

Zhang J, Nodzynski T, Pencik A, Rolcik J, Friml J. 2010. PIN phosphorylation is sufficient to mediate PIN polarity and direct auxin transport. *Proc Natl Acad Sci* **107:** 918–922. doi:10.1073/pnas.0909460107

Zhang W, Swarup R, Bennett M, Schaller GE, Kieber JJ. 2013. Cytokinin induces cell division in the quiescent center of the *Arabidopsis* root apical meristem. *Curr Biol* **23:** 1979–1989. doi:10.1016/j.cub.2013.08.008

Zhang XD, Zhao KX, Yang ZM. 2018. Identification of genomic ATP binding cassette (ABC) transporter genes and Cd-responsive ABCs in *Brassica napus. Gene* **664:** 139–151. doi:10.1016/j.gene.2018.04.060

Zhang Z, Tong T, Fang Y, Zheng J, Zhang X, Niu C, Li J, Zhang X, Xue D. 2020. Genome-wide identification of barley ABC genes and their expression in response to abiotic stress treatment. *Plants (Basel)* **9:** 1281. doi:10.3390/plants9101281

Ziemba BP, Pilling C, Calleja V, Larijani B, Falke JJ. 2013. The PH domain of phosphoinositide-dependent kinase-1 exhibits a novel, phospho-regulated monomer-dimer equilibrium with important implications for kinase domain activation: single-molecule and ensemble studies. *Biochemistry* **52:** 4820–4829. doi:10.1021/bi400488f

Zourelidou M, Muller I, Willige BC, Nill C, Jikumaru Y, Li H, Schwechheimer C. 2009. The polarly localized D6 PROTEIN KINASE is required for efficient auxin transport in *Arabidopsis thaliana. Development* **136:** 627–636. doi:10.1242/dev.028365

Zourelidou M, Absmanner B, Weller B, Barbosa IC, Willige BC, Fastner A, Streit V, Port SA, Colcombet J, de la Fuente van Bentem S, et al. 2014. Auxin efflux by PIN-FORMED proteins is activated by two different protein kinases, D6 PROTEIN KINASE and PINOID. *eLife* **3:** e02860. doi:10.7554/eLife.02860

Structural Aspects of Auxin Signaling

Nicholas Morffy[1,2] and Lucia C. Strader[1,2,3]

[1]Department of Biology, Duke University, Durham, North Carolina 27708, USA

[2]Center for Science and Engineering Living Systems (CSELS), Washington University, St. Louis, Missouri 63130, USA

[3]Center for Engineering Mechanobiology, Washington University, St. Louis, Missouri 63130, USA

Correspondence: nicholas.morffy@duke.edu; lucia.strader@duke.edu

Auxin signaling regulates growth and developmental processes in plants. The core of nuclear auxin signaling relies on just three components: TIR1/AFBs, Aux/IAAs, and ARFs. Each component is itself made up of several domains, all of which contribute to the regulation of auxin signaling. Studies of the structural aspects of these three core signaling components have deepened our understanding of auxin signaling dynamics and regulation. In addition to the structured domains of these components, intrinsically disordered regions within the proteins also impact auxin signaling outcomes. New research is beginning to uncover the role intrinsic disorder plays in auxin-regulated degradation and subcellular localization. Structured and intrinsically disordered domains affect auxin perception, protein degradation dynamics, and DNA binding. Taken together, subtle differences within the domains and motifs of each class of auxin signaling component affect signaling outcomes and specificity.

Over the past three decades, scientists have elucidated the molecular and genetic mechanisms that underlie major aspects of nuclear auxin signaling. The auxin molecule has many functions, but the most well studied is its role in regulating transcription. Initial genetic studies of auxin-resistant plants uncovered the major players involved in auxin transcriptional control. In all members of the plant lineage, three protein families make up this nuclear auxin signaling pathway that affects the transcription of up to a third of all genes in the angiosperm *Arabidopsis thaliana* (Mutte et al. 2018; Powers et al. 2019). The F-box protein TRANSPORT INHIBITOR RESPONSE 1 (TIR1) and its AUXIN SIGNALING F-BOX (AFB) paralogs are auxin receptors and are responsible for regulating the abundance of the Auxin/INDOLE-3-ACETIC-ACID proteins (Aux/IAAs). The Aux/IAAs are a family of transcriptional corepressors that interact with the third family, the AUXIN RESPONSE FACTORS (ARFs). The ARFs are transcription factors with diverse functions (Fig. 1).

These three components form the core of the nuclear auxin signaling pathway. Their interactions with one another, themselves, and other proteins influence nearly every aspect of plant growth and development. Genetic studies of each family have yielded insights into the mechanisms of nuclear auxin signaling; however, genetic redundancy has prevented a com-

Cite this article as *Cold Spring Harb Perspect Biol* doi: 10.1101/cshperspect.a039883

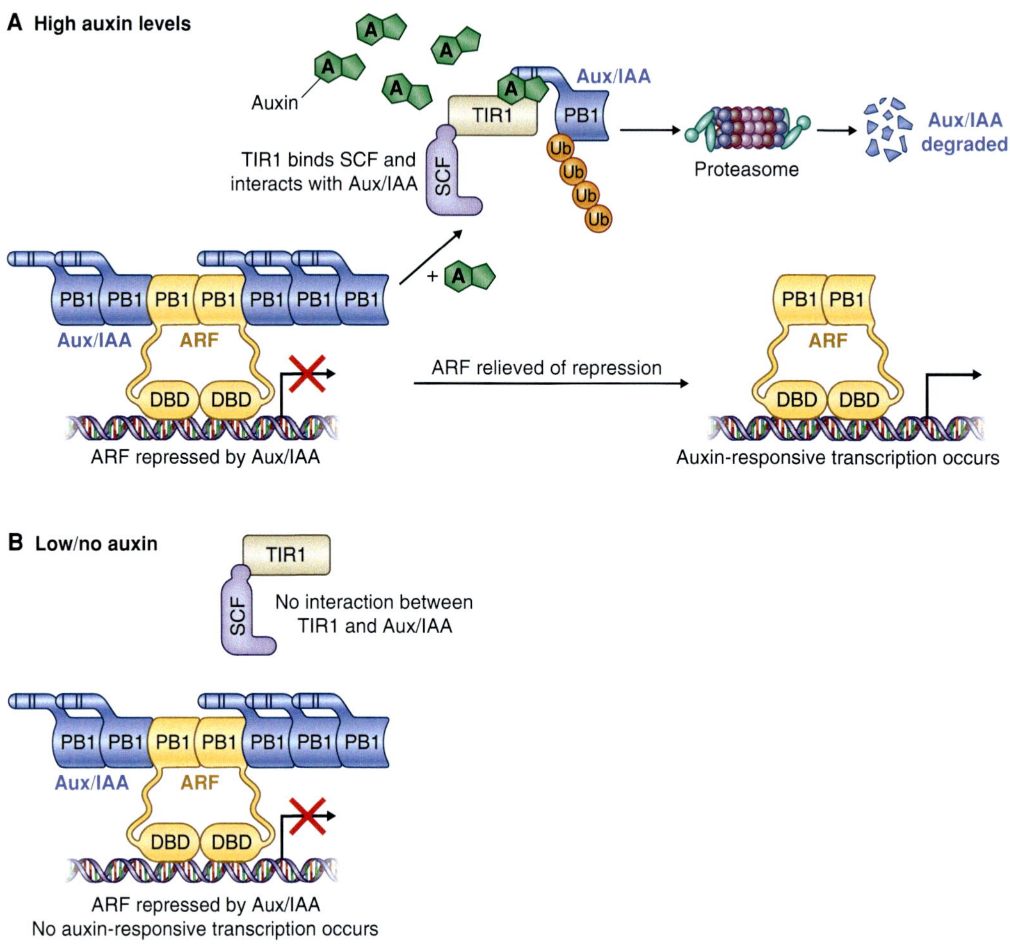

Figure 1. The canonical nuclear auxin signaling pathway. The core nuclear auxin signaling pathway is made up of three components: TIR1/AFBs, AUX/IAAs, and ARFs. (*A*) High auxin levels; (*B*) Low/no auxin. Under low auxin conditions, the AUX/IAAs interact with ARFs via the carboxy-terminal PB1 domain found in both protein families and prevent transcription of auxin-responsive genes. As auxin levels increase, the F-box protein and auxin-receptor TIR1 binds the molecule, which stabilizes interactions with its target, the AUX/IAAs. The AUX/IAAs are polyubiquitinated and degraded by the 26S proteasome, relieving the ARFs of repression and allowing auxin-responsive transcription to occur. (DBD) DNA-binding domain.

plete understanding of the function of each member. Recent work on the biochemical and structural aspects of auxin signaling has resulted in an increased understanding of the complexities of this signaling module. Here, we will discuss what is known about the structured and unstructured domains found in each family, and how these domains contribute to auxin signaling outcomes. We will also discuss new questions raised by our structural understanding of the nuclear auxin signaling pathway.

TIR1/AFBs

The Auxin Signaling F-Box Protein

The *TIR1* gene was originally identified in a mutant screen for plants with altered growth in the presence of the auxin transport inhibitor NPA (Ruegger et al. 1997, 1998). Unlike other genes identified in the initial screen, *tir1* mutants were also resistant to exogenous auxin (Ruegger et al. 1998). Genetic studies placed *TIR1* in the same pathway as other auxin signaling components, such as *AXR1* and *AXR6*, an upstream regulator and component of the ubiquitin ligase complexes, respectively (Lincoln et al. 1990; Leyser et al. 1993; del Pozo et al. 2002; Hellmann et al. 2003). Consistent with the function of AXR1 and AXR6, sequence analysis of TIR1 suggested that it is an F-box protein that links ubiquitin conjugation directly to auxin signaling (Ruegger et al. 1998; Gray et al. 1999). Like other F-box proteins, TIR1 interacts with the *Arabidopsis* SKP1 orthologs ASK1 and ASK2, via its amino-terminal F-box domain (Fig. 2A,B; Ruegger et al. 1998; Gray et al. 1999; Dharmasiri et al. 2005a,b; Kepinski and Leyser 2005; Tan et al. 2007). In addition to its F-box domain, TIR1 has several leucine-rich repeats (LRRs) (Ruegger et al. 1998; Gray et al. 1999; Tan et al. 2007). TIR1 is a member of a small sub-clade of F-box proteins that includes COI1, involved in jasmonic acid (JA) signaling, as well as the AFBs (Dharmasiri et al. 2005b; Mutte et al. 2018). Increasingly higher-order knockout mutants of the *TIR1/AFB* family display increasing severity of developmental defects, suggesting that TIR1 and the AFBs act redundantly to regulate auxin signaling (Dharmasiri et al. 2005b; Parry et al. 2009; Prigge et al. 2020).

Initial protein stability studies suggested that TIR1 directly regulates Aux/IAA levels (discussed below) by polyubiquitinating the

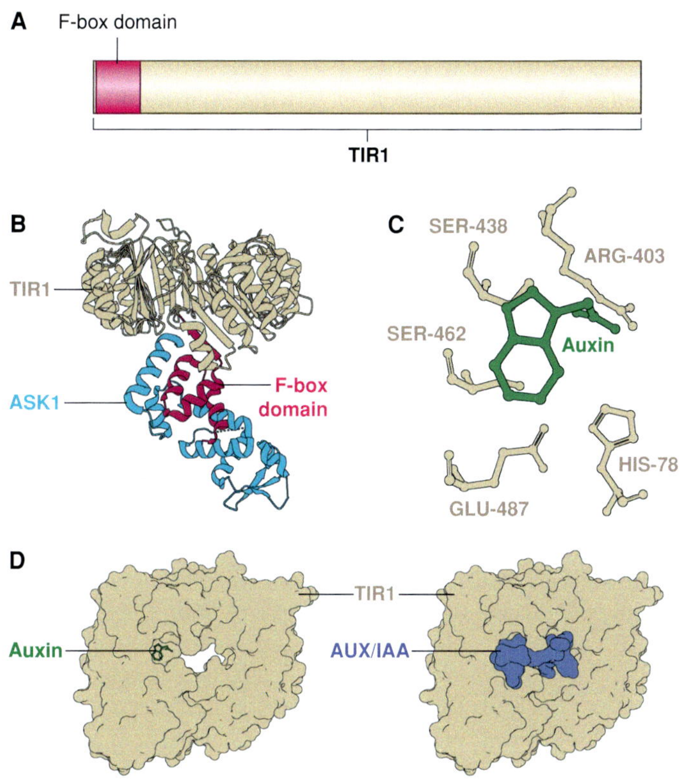

Figure 2. TIR1 is the auxin receptor. (*A*) The TIR1 protein encodes a 48-amino-acid F-box domain at its amino-terminus that is required for interactions with the SCF complex. (*B*) The TIR1 crystal structure (tan) in complex with the SCF adaptor component ASK1 (cyan) was resolved by Tan et al. (2007). The F-box domain (magenta) contacts the helices of ASK1 at several points and disturbing this interaction results in a dominant-negative form of TIR1. (*C*) Critical residues in the auxin-binding pocket of TIR1 are required for its function. This region was targeted to generate the synthetic ccvTIR1 form. (*D*) Upon auxin (green) binding, TIR1 (tan) can strongly interact with the degron (DII) of the AUX/IAA (pale blue) that seals in the auxin molecule.

Aux/IAAs in the presence of increased auxin. However, the mechanism of auxin perception was still unknown (Gray et al. 2001; Zenser et al. 2001, 2003; Tian et al. 2003; Kepinski and Leyser 2004; Yang et al. 2004). Studies of other F-box proteins raised the possibility that modification of either TIR1 or the substrate Aux/IAA in the presence of auxin promoted their direction interaction. However, subsequent biochemical and structural studies of TIR1 and the Aux/IAAs showed that TIR1 is the auxin receptor itself (Fig. 2C). Two studies published simultaneously showed that radiolabeled auxin interacts directly with both TIR1 and an Aux/IAA (Dharmasiri et al. 2005a; Kepinski and Leyser 2005). In addition, TIR1–Aux/IAA interactions occur in a cell-free system in the presence of auxin, suggesting that no further modification of either TIR1 or Aux/IAA is required for auxin-induced interaction (Dharmasiri et al. 2003, 2005a; Tan et al. 2007). Taken together, this work resulted in a model whereby auxin acts as a "molecular glue" directly interacting with both the E3 complex and its substrate (Fig. 2D; Tan et al. 2007).

Under low auxin conditions, TIR1/AFBs display low affinity for the Aux/IAA corepressors (Fig. 1; Kepinski and Leyser 2004, 2005; Yang et al. 2004; Dharmasiri et al. 2005a,b; Dreher et al. 2006; Tan et al. 2007; Maraschin Fdos et al. 2009; Calderón Villalobos et al. 2012). As auxin levels increase, the molecule binds in a pocket of the TIR1/AFB (Fig. 2C; Tan et al. 2007; Lee et al. 2014). When TIR1/AFBs bind the auxin molecule, the affinity for the Aux/IAAs increases, stabilizing the interaction between the F-box and the corepressor that results in polyubiquitination of the Aux/IAAs and their degradation (Tian et al. 2003; Kepinski and Leyser 2004, 2005; Yang et al. 2004; Dharmasiri et al. 2005a,b; Dreher et al. 2006; Tan et al. 2007; Maraschin Fdos et al. 2009; Calderón Villalobos et al. 2012). Auxin contacts both the TIR1-binding pocket and the Aux/IAA degron (DII, see below) to generate a coreceptor complex (Fig. 2D; Kepinski and Leyser 2004, 2005; Yang et al. 2004; Dharmasiri et al. 2005a,b; Dreher et al. 2006; Tan et al. 2007; Maraschin Fdos et al. 2009; Calderón Villalobos et al. 2012). Changes to either the auxin-binding surface, or the interaction surfaces between TIR1 and the Aux/IAA, result in stabilization of the Aux/IAA (Rouse et al. 1998; Nagpal et al. 2000; Worley et al. 2000; Gray et al. 2001; Tiwari et al. 2001; Dharmasiri et al. 2003, 2005a,b; Zenser et al. 2003; Kepinski and Leyser 2004, 2005; Tan et al. 2007; Maraschin Fdos et al. 2009; Calderón Villalobos et al. 2012; Uzunova et al. 2016). This coreceptor complex is the core of the nuclear auxin response pathway. The dynamics of auxin perception by TIR1 necessarily dictate the steady state of Aux/IAAs in the system, which in turn modulates auxin responses.

Consistent with other components of auxin signaling, the TIR1/AFB family is larger in angiosperms (six in *Arabidopsis*) than in bryophytes (one in *Marchantia polymorpha* and four in *Physcomitrella patens*) (Dharmasiri et al. 2005b; Parry et al. 2009; Flores-Sandoval et al. 2015; Mutte et al. 2018). The expansion of TIR1/AFBs in the flowering plants suggests subfunctionalization (Parry et al. 2009). Indeed, recent studies indicate that TIR1/AFBs from both within and between species have distinct auxin-binding affinities and different affinities for the Aux/IAAs (Calderón Villalobos et al. 2012; Havens et al. 2012; Pierre-Jerome et al. 2013; Lee et al. 2014; Moss et al. 2015; Uzunova et al. 2016; Báez et al. 2020). These small differences act as points of tunability in the auxin response. Mutating residues in the auxin-binding pocket or the interaction surface of TIR1/AFBs impairs the ability of the F-box to interact with Aux/IAAs without affecting the ability of TIR1/AFBs to interact with the SCF complex. Initial observations show that TIR1 binds full-length Aux/IAA with a higher affinity than a peptide encompassing the degron (see below) (Worley et al. 2000; Gray et al. 2001; Ramos et al. 2001; Kepinski and Leyser 2004, 2005; Yang et al. 2004; Dharmasiri et al. 2005a; Tan et al. 2007; Calderón Villalobos et al. 2012). Similar results were obtained using purified proteins and also in the yeast synthetic auxin signaling system (Havens et al. 2012; Pierre-Jerome et al. 2014; Moss et al. 2015). Maize TIR1/AFBs show differing degradation dynamics depending on the Aux/IAA tested (Báez et al. 2020). Binding pocket size and electrostatic potential affect

TIR1/AFB-binding affinities and specificities for different active auxins (Lee et al. 2014; Uzunova et al. 2016). Further studies are required to see how different auxins impact the ability of low- and high-affinity TIR1/AFB interactions to proceed in vivo.

Mutations in either the F-box domain or the auxin-binding pocket affect TIR1 function. However, the relative contributions of each domain to TIR1/AFB function result in differing phenotypes. If the F-box domain is perturbed, a dominant-negative form of TIR1 or AFBs is generated (Yu et al. 2015). It is hypothesized that signal transduction fails to occur because the Aux/IAAs are not polyubiquitinated and degraded (Yu et al. 2015). Consistent with that hypothesis, these mutations result in phenotypes similar to dominant gain-of-function Aux/IAA mutations (Rouse et al. 1998; Nagpal et al. 2000; Worley et al. 2000; Tiwari et al. 2001; Fukaki et al. 2002; Yu et al. 2015). The TIR1-auxin-Aux/IAA coreceptor is stabilized and protected from endogenous, functioning TIR1 and AFBs. Mutations in the ligand-binding pocket only affect the ability of TIR1 to interact with Aux/IAAs (Yu et al. 2013).

The vast majority of TIR1/AFBs are broadly expressed and show overlapping biological functions (Dharmasiri et al. 2005b; Parry et al. 2009; Lavy et al. 2016; Prigge et al. 2020). Consistent with its role in regulating nuclear auxin signaling, the TIR1/AFBs are nuclear localized proteins with the exception of AFB1, which is found primarily in the cytoplasm (Prigge et al. 2020). Like the other TIR1/AFBs, AFB1 can interact with Aux/IAAs in an auxin-dependent manner; however, it has lower affinity for the SCF complex than other members of the family. This suggests that AFB1 regulates auxin responses in a distinct manner. The biochemical basis for nuclear or cytoplasmic localization of the TIR1/AFB family is not understood.

Synthetic Auxin Signaling

Studies of the TIR1-auxin-Aux/IAA coreceptor module have resulted in synthetic biology approaches in several different organisms. The yeast synthetic auxin signaling system has led to new insights into auxin signaling kinetics (Pierre-Jerome et al. 2014; Báez et al. 2020). In addition, the TIR1-auxin-Aux/IAA coreceptor module has been used to modulate protein stability in other organisms. The auxin-induced degron (AID) system relies on an Aux/IAA degron fused to a protein of interest and coexpressed with TIR1 (Nishimura et al. 2009). Upon auxin application, the target protein is degraded by TIR1 resulting in controlled protein degradation and depletion. The AID system has been used in a number of different study systems from yeast to mammalian tissue culture (Kanke et al. 2011; Holland et al. 2012; Trost et al. 2016).

Synthetic approaches have also been used to generate orthogonal auxin signaling systems in *Arabidopsis*. Knowledge of the structural basis for auxin perception and signaling in the nuclear auxin-response pathway was used to develop a novel TIR1 variant (ccvTIR1) and synthetic auxin (cvxIAA) pair (Uchida et al. 2018). The auxin-binding pocket in ccvTIR1 has been enlarged and binds endogenous auxin molecules poorly, but still perceives the synthetic auxin analog, cvxIAA. This molecule has a large aromatic ring and has no bioactivity in the absence of ccvTIR1. When cvxIAA is applied, ccvTIR1 functions normally (Uchida et al. 2018; Mazur et al. 2020). Early studies relying on this system have elucidated new aspects of auxin signaling by increasing the control researchers have over their study system and allowing for examination of TIR1-mediated auxin responses. For example, a role for TIR1 in PIN repolarization was interrogated using this synthetic auxin signaling system (Mazur et al. 2020). One could imagine higher order combinations of ccvTIR1 and ccvAFBs to better identify the contributions of different TIR1/AFBs to specific aspects of auxin signaling.

AUX/IAAs

The Aux/IAA family (34 members in *Arabidopsis*) is the other major class of transcription factor involved in auxin signaling. Originally identified in screens for auxin-resistant mutants, these highly structured proteins, along with the class A ARFs, provide much of our understanding of auxin-responsive transcription (Fig. 1; for review, see Powers and Strader 2020). The Aux/IAAs are smaller than the ARFs and lack a DNA-binding domain (DBD) (Fig. 3A). Aux/IAA proteins have three major domains, DI that encodes the repressive domain, DII that encodes the Aux/IAA degron, and a type I/II Phox and Bem1 (PB1) domain that was previously termed DIII/DIV (Fig. 3A; Han et al. 2014; Korasick et al. 2014; Nanao et al. 2014).

The Aux/IAA PB1 domain mediates interactions between Aux/IAAs and between Aux/IAAs and the ARFs (Fig. 1). Under low auxin conditions, this interaction results in repression of auxin-responsive transcription via an amino-terminal ethylene-responsive element-binding factor-associated amphiphilic repression (EAR) motif (DI) (reviewed in Kagale and Rozwadowski 2011 and Powers and Strader 2020).

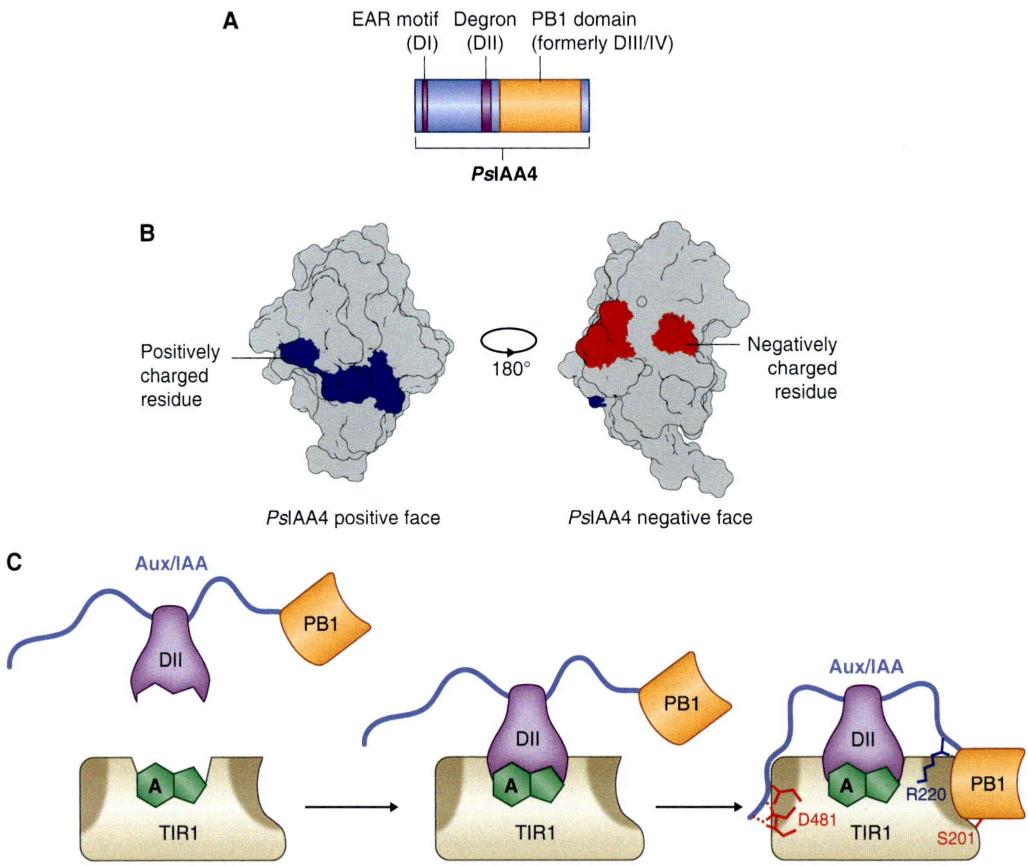

Figure 3. The AUX/IAAs are modular proteins with only a single structured domain. (*A*) An idealized motif and domain structure based on the *Pisum sativum* IAA4 protein. The only globular domain is the carboxy-terminal PB1 domain that mediates interactions with other AUX/IAAs and the ARFs. Both the degron (DII) and the EAR motif (DI) are short linear motifs that mediate interactions with TIR1 and the TPL corepressor family, respectively. (*B*) The atomic structure of the *Ps*IAA4 PB1 domain here, resolved by Dinesh et al. (2015), shows that each face of the PB1 domain is enriched for either positively (blue) or negatively (red) charged residues that facilitate head-to-tail interactions with one another. (*C*) The current model for TIR1–AUX/IAA interactions now incorporates the intrinsically disordered regions that flank the degron. Key residues on both TIR1 and the AUX/IAA help stabilize the degron-mediated interaction and ensure downstream degradation and auxin signaling as shown in Niemeyer et al. (2020).

Domain I

The amino-terminal DI region encodes a canonical EAR motif in the Aux/IAAs (Fig. 3; Ulmasov et al. 1997b; Tiwari et al. 2004). This motif is necessary to repress transcription of auxin-responsive genes in the absence of auxin (Szemenyei et al. 2008). Like other canonical EAR motifs, the Aux/IAA EAR domain recruits TOPLESS (TPL) and TPL-RELATED (TPR) corepressors to loci. TPL and TPR in turn recruit histone deacetylases, which remove marks of active transcription from histones (Long et al. 2006; Szemenyei et al. 2008). Dominant mutants of Aux/IAAs were suppressed in the absence of functional TPL and TPR proteins (Szemenyei et al. 2008). The Aux/IAAs repress transcription of auxin-responsive genes via their interactions with the ARFs (see below).

Domain II

Of the 34 Aux/IAAs found in *Arabidopsis*, 29 have a conserved degron in the DII region of the protein. DII is the most well-studied motif in the Aux/IAAs. The discovery and characterization of DII not only helped push the field of auxin signaling forward by explaining one of the core mechanisms (Worley et al. 2000; Gray et al. 2001; Ramos et al. 2001; Tiwari et al. 2001; Zenser et al. 2001), but it is also used to generate reporters that measure auxin abundance and signaling activity (Brunoud et al. 2012).

In the presence of auxin, DII interacts with both the auxin molecule and the TIR1/AFB surface, and is ultimately polyubiquitinated and degraded via the ubiquitin-proteasome pathway (Fig. 2D; Dreher et al. 2006; Tan et al. 2007; Calderón Villalobos et al. 2012). The Aux/IAAs are traditionally considered to be rapidly degrading proteins even in the absence of auxin (Abel et al. 1994). The interaction between different Aux/IAA DII motifs and members of the TIR1/AFB family affects the rate of Aux/IAA degradation and therefore auxin sensitivity in different cell types (Ouellet et al. 2001; Calderón Villalobos et al. 2012; Havens et al. 2012; Pierre-Jerome et al. 2014; Moss et al. 2015; Báez et al. 2020; Niemeyer et al. 2020).

Initial characterization of the Aux/IAAs was primarily through dominant mutations in the DII domain, stabilizing the Aux/IAAs and resulting in an auxin-resistant phenotype (Worley et al. 2000; Gray et al. 2001; Tiwari et al. 2001; Zenser et al. 2001). Since those initial observations, DII mutations have been used to investigate differences between Aux/IAA function and circumvent genetic redundancy.

The PB1 Domain

Like ARFs, Aux/IAA proteins have a carboxy-terminal type-I/II PB1 domain (Fig. 3A,B). Originally, this region was classified as two distinct conserved domains, termed domains III and IV (DIII/IV). Recent structural analyses show DIII/DIV assumes a single globular domain that shares many of the hallmarks of a type-I/II PB1 domain, including a negative and positive face with conserved lysine and aspartic acid residues, respectively (Han et al. 2014; Dinesh et al. 2015), as originally proposed by Guilfoyle and Hagen (2012) (Fig. 3B).

Amino acid substitutions in the positive face of the IAA16 PB1 domain that affect its ability to multimerize can suppress dominant DII mutations, suggesting that these interactions are necessary for Aux/IAA function in vivo (Korasick et al. 2014). In addition, mutations that disrupt IAA17 or IAA19 PB1 domain oligomerization increase the magnitude of auxin response in protoplast experiments, suggesting they are less effective repressors (Nanao et al. 2014). Conversely, mutations that disrupt IAA14 PB1 domain oligomerization fail to suppress the dominant DII mutation in vivo (Pierre-Jerome et al. 2016), suggesting that oligomerization is not required for IAA14 function. Taken together, these data suggest that some Aux/IAAs, like IAA16, may require oligomerization for full repressive activity. Alternatively, IAA16 may interact with its target ARFs along a single PB1 face, and, thus, disrupting this face of the PB1 domain abrogates its ability to bind targets in addition to disrupting oligomerization. Other Aux/IAAs, like IAA14, may interact with target ARFs along either face, which would be consistent with no single-face mutation resulting in decreased IAA14 repressive activity. Further work is necessary to determine whether Aux/IAA oligomerization is necessary for its function *in planta*. Further, these results raise the possibility that auxin signaling outcomes are regulated in part by Aux/IAA affinity for ARFs in a PB1 domain face-specific manner.

PB1 domain interactions between ARFs and Aux/IAAs display higher affinity than either ARF–ARF or Aux/IAA–Aux/IAA interactions (Han et al. 2014; Nanao et al. 2014). Thus, low levels of Aux/IAA proteins may efficiently bind members of the ARF family and repress transcription, whereas Aux/IAA–Aux/IAA interactions are less stable. Rapid turnover of the Aux/IAAs and the weak self-interactions via the PB1 domain may act to tune the thresholds required for specific auxin signaling events to occur. These weaker self-interactions may also prevent complete repression of ARF targets, including the Aux/IAAs themselves, likely maintaining auxin sensitivity in a physiologically relevant range and ensuring appropriate negative feedback regulation on the auxin signaling module.

Not all Aux/IAA–ARF interactions are equal. For example, ARF PB1 domains are not interchangeable in the *Marchantia* auxin signaling pathway. Additional structural modeling data suggests some ARF PB1 domains prevent Aux/IAA PB1 domain oligomerization, stopping both strong Aux/IAA transcriptional repression and sequestration of otherwise functional Aux/IAAs (Kato et al. 2020). In this model, the negatively charged face of the ARF PB1 domain interacts with the positively charged lysine face of the Aux/IAA PB1 domain in a manner that obscures the surface of the OPCA face of the Aux/IAA PB1 domain, preventing further interactions. Biochemical and structural analysis of TPL and TPR function suggests that tetramers are the active form of the TPL corepressor complex (Ke et al. 2015). The Aux/IAA PB1 domain, in addition to mediating interactions with the ARFs, may also provide a scaffold for higher order TPL and TPR interactions to occur through Aux/IAA multimerization. Indeed, Aux/IAA with mutated PB1 domains are less effective repressors (Korasick et al. 2014; Nanao et al. 2014; Dinesh et al. 2015). Further work is required to determine the actual interaction affinities and oligomeric state of different Aux/IAA and ARF PB1 domains for one another and each other.

The Aux/IAA PB1 domain regulates a number of biological functions either directly via protein–protein interactions, or indirectly via functional complex formation. Further studies that attempt to disentangle the contribution of Aux/IAA oligomerization from Aux/IAA–ARF interaction, as well as studies investigating interaction specificity, will be critical to understanding how the large Aux/IAA family contributes to so many distinct auxin signaling outcomes. The contribution of each face of the PB1 domain to any or all of these functions also remains a mystery.

Disorder in the Aux/IAAs

In addition to their ordered regions, the Aux/IAAs also have regions of intrinsic disorder (Niemeyer et al. 2020). These intrinsically disordered regions (IDRs) contribute to Aux/IAA–auxin–TIR1 interaction specificity and stability. The role of intrinsic disorder in protein and biological function in auxin signaling is only just beginning to be understood (see ARF intrinsic disorder below). Unlike ordered protein regions, disordered regions adopt a broad range of possible conformations. Ordered protein regions maintain a stable secondary and tertiary structure, whereas proteins with regions of intrinsic disorder lack a stable secondary structure. Protein conformation studies can present an ensemble of possible conformations, and the lack of strong predictive modeling due to the complex multifaceted nature of disordered regions makes hypothesis generation difficult (for review, see Holehouse 2019 and Uversky 2019). IDR analysis relies more on general amino acid composition and spacing than specific amino acid sequence. However, the biological and molecular functions of disordered regions can be studied via traditional genetic means.

Many Aux/IAA proteins in *Arabidopsis* have IDRs that span the region between the EAR motif (DI) and the PB1 domain (Niemeyer et al. 2020). Initial studies of the Aux/IAA degron (DII) suggested that regions flanking the conserved domain contribute to auxin signal transduction, but the mechanism was not understood (Calderón Villalobos et al. 2012; Moss et al. 2015). These flanking regions are intrinsically disordered and they contribute, along with the PB1 domain, to the interaction surface between Aux/IAAs and TIR1/AFBs (Fig. 3C; Niemeyer et al. 2020).

A model has emerged where the two IDR flanking DII contribute to the stability of the Aux/IAA–auxin–TIR1 interaction (Fig. 3C; Niemeyer et al. 2020). The regions act to clamp DII into the binding pocket on TIR1 and ensure the interaction lasts long enough to transfer ubiquitin to Aux/IAA, marking it for degradation. Interestingly, the DII flanking regions appear to be evolutionarily constrained; however, not all Aux/IAAs in *Arabidopsis* show similar amino acid compositions.

AUXIN-RESPONSE FACTORS

The ARF transcription factors are represented by large gene families in many plant species (Mutte et al. 2018). Mutant plants lacking functional ARFs have diverse developmental and growth phenotypes consistent with their roles in auxin signaling. However, in many cases, mutation of a single ARF does not result in a mutant phenotype (Sessions et al. 1997; Harper et al. 2000; Hardtke et al. 2004; Li et al. 2004; Tian et al. 2004; Ellis et al. 2005; Okushima et al. 2005; Wilmoth et al. 2005), suggesting overlapping functions. The ARFs fall into three ancient clades and have a conserved domain structure (Fig. 4A; Mutte et al. 2018). At the amino terminus lies a B3 DBD flanked by dimerization domains (Fig. 4B). A carboxy-terminal type-I/II PB1 domain promotes interactions between ARFs and the Aux/IAAs as well as between the ARFs themselves (Fig. 4C,D). The region that lies in between the amino-terminal DBD and the carboxy-terminal PB1 domain is known as the middle region (Fig. 4A). Class A ARFs, originally called activator ARFs, are defined by glutamine-rich middle regions, whereas class B and C ARFs, classically defined as repressor ARFs, are recognized by serine-rich middle regions (Ulmasov et al. 1997a, 1999a,b; Mutte et al. 2018). The middle regions of all ARFs contain varying degrees of intrinsic disorder and are implicated in transcriptional activity, subcellular localization, and protein condensation.

The ARF DNA-Binding Domain

The ARF DBD is made up of a canonical B3 DBD (Ulmasov et al. 1997a, 1999b; Boer et al. 2014; Mutte et al. 2018). Flanking the B3 DBD lie two regions that make up a dimerization domain, which contributes to ARF homodimerization and stabilizes DNA binding (Fig. 4B; Boer et al. 2014). Conserved amino acid residues along the dimerization surface and DNA-binding surface are necessary for ARF function (Boer et al. 2014; Pierre-Jerome et al. 2016; Lanctot et al. 2020). These results suggest that dimerization of ARF DBDs may dictate some aspects of ARF signaling specificity.

All studied ARFs bind to the same consensus sequence, the auxin response element (AuxRE) (Ulmasov et al. 1995, 1997a,b, 1999a, b; Boer et al. 2014; Pierre-Jerome et al. 2016; Lanctot et al. 2020). Synthetic AuxREs, like DR5, have served as auxin signaling reporters for over two decades (Ulmasov et al. 1997b). Naturally occurring AuxREs share the same core sequence TGTC followed by two variable positions. DNA affinity purification sequencing (DAP-seq) experiments have confirmed these findings for multiple *Arabidopsis* and maize ARFs (O'Malley et al. 2016; Galli et al. 2018). The initial biochemical characterization of the ARF DBD resulted in a caliper model in which ARF dimers bound DNA with differing AuxRE topologies or grammar, thus driving ARF DNA-binding specificity (Boer et al. 2014). ARF1 shows preferences for everted repeats with seven or eight base pair spacing, whereas ARF5 can bind everted repeats with five to nine base pair spacing. Similar conclusions were drawn from global binding patterns of maize and *Arabidopsis* ARFs by DAP experiments, and the yeast synthetic auxin signaling system (Boer et al. 2014; Pierre-Jerome et al. 2016; Lanctot et al. 2020). In all published studies, class A and B

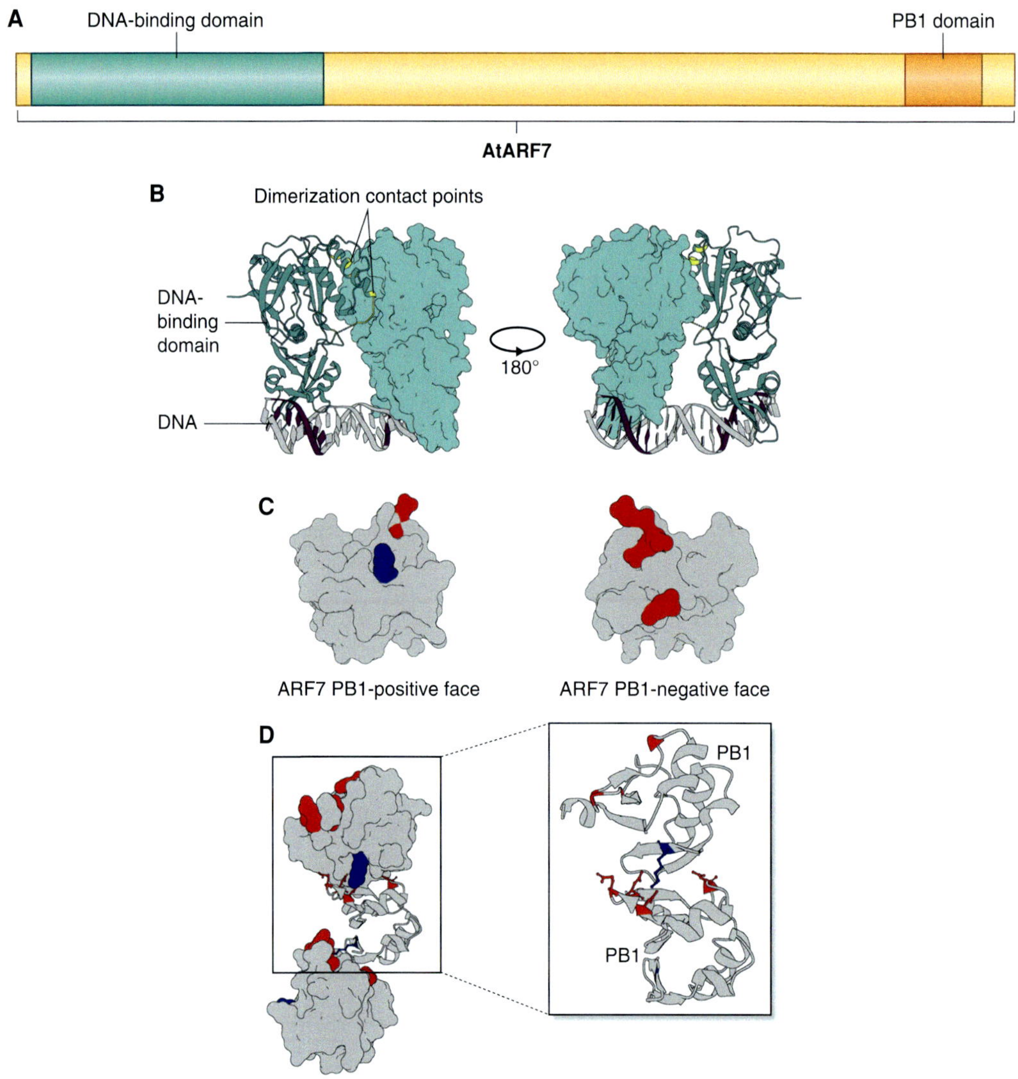

Figure 4. The ARFs are defined by three major domains. (*A*) The ARFs are defined by two globular domains, the amino-terminal DNA-binding domain (DBD) and the carboxy-terminal PB1 domain, and an intrinsically disordered middle region that lies between them, based on the *Arabidopsis* ARF7 protein model. (*B*) The crystal structure of the class B ARF1 DBD binding an ER7 AuxRE, resolved by Boer et al. (2014), shows dimerization contact points (yellow) that stabilize the interactions with TGTCTC consensus sequence (purple) in an antiparallel manner. (*C*) The ARF7 PB1 domain crystal structure resolved by Korasick et al. (2014), like the AUX/IAA PB1 domain, has both a positively and negatively charged face. (*D*) These faces interact in a head-to-tail fashion to generate strings of PB1 domains in the crystal structure. The lysine residue on the positive face interacts with the aspartic and glutamic acid residues on the negative face similar to a bar magnet. Perturbing these residues results in the loss of PB1 domain interactions along that face.

ARFs show distinct preferences for AuxRE grammar. These observations raise the possibility that the ARF DBD dictates signaling specificity by reducing binding affinity for loci with suboptimal AuxRE grammar. Further studies of class A and B ARF DNA binding suggests that inverted AuxRE repeats with eight-base-pair spacing (IR8) can recruit both classes of ARFs, whereas DR5-binding sites only recruit the class A ARFs (Freire-Rios et al. 2020). Consistent with this finding, meta-analysis of auxin transcriptional response data sets shows that IR8 motifs are found upstream of genes that show both repression and induction upon auxin signaling and DR5 motifs are only found upstream of genes with auxin-induced expression (Freire-Rios et al. 2020). Taken together, this suggests that some auxin-responsive targets can be bound by both class A and class B ARFs and others can only be bound by class A ARFs. ARF DBD swap experiments in the nonvascular plants *Physcomitrella* and *Marchantia* also suggest that ARFs compete for similar genomic targets (Lavy et al. 2016; Kato et al. 2020).

Marchantia has a single member from each ancestral class of ARF (Flores-Sandoval et al. 2015; Kato et al. 2020). Domain swap experiments showed that for MpARF1 and MpARF2, the class A and class B ARFs, respectively, the DBDs were interchangeable. Thus, replacement of the MpARF1 DBD with the same domain from MpARF2 resulted in a functional protein and could rescue the *Mparf1* loss of function mutation (Kato et al. 2020). Replacing the MpARF1 DBD with the MpARF3, the class C ARF, DBD was unable to rescue *Mparf1*. These data suggest a model in which class A and class B ARFs compete for genomic targets and class C ARFs participate in separate processes.

Differences in AuxRE-binding affinities present an appealing model for ARF DNA-binding specificity. However, experiments in the yeast synthetic auxin signaling system suggest that some ARFs may activate transcription on a single AuxRE, but dimerization between the ARFs is necessary for transcription to occur (Lanctot et al. 2020). Consistent with the yeast synthetic signaling system, enrichment for single AuxREs upstream of auxin-responsive genes is also detectable (Freire-Rios et al. 2020). The biochemical basis for differences in DNA-binding specificity to single AuxRE binding sites is not yet known.

The ARF PB1 Domain

The DBD is not the only ARF interaction domain, however. The carboxy-terminus of most ARFs contain a type-I/II PB1 domain (Mutte et al. 2018). This class of PB1 domain has two faces: a positively charged, basic face and a negatively charged, acidic face (Fig. 4C). The ARF PB1 domain, previously referred to as domains III and IV, is shared with Aux/IAAs (Guilfoyle and Hagen 2012). Like other type-I/II PB1 domains, the ARF PB1 domains allow for head-to-tail oligomerization where the positive face of a single PB1 domain will interact with the negative face of another PB1 domain (Fig. 4D; Korasick et al. 2014, 2015; Nanao et al. 2014). Oligomerization was observed both in size exclusion chromatography and protein crystallization experiments in the ARFs (Han et al. 2014; Korasick et al. 2014; Nanao et al. 2014).

The positive face is characterized by an invariant lysine residue that interacts with an array of conserved aspartic and glutamic acids (Fig. 4C). Mutations in the lysine residue disrupt interactions with the negative face preventing oligomerization (Han et al. 2014; Korasick et al. 2014; Nanao et al. 2014; Powers et al. 2019). The PB1 domain likely contributes to ARF function in a number of ways, including interactions between ARFs and Aux/IAAs, which is required for appropriate auxin signal transduction (Vernoux et al. 2011; Han et al. 2014; Nanao et al. 2014). In *Arabidopsis*, a transgenic copy of ARF19 with a mutation on the positive face that ablates oligomerization showed increases in transcription of both auxin-responsive genes and novel targets in the absence of auxin (Powers et al. 2019). This result suggests that the ARF19 PB1 mutant is acting as a constitutive auxin signaling factor likely due to its lack of interaction with its transcriptional corepressor, the Aux/IAAs. Interestingly, in vivo oligomerization data suggests that the ARF19 PB1 mutant fails to dimerize in the nucleus (Powers et al.

2019). This observation raises the possibility that the ARF PB1 domain, and not the DBD, primarily contributes to ARF homodimerization. Indeed, PB1 mutants showed reduced transcriptional activity in the yeast-synthetic auxin signaling system compared to wild-type ARFs (Pierre-Jerome et al. 2016; Lanctot et al. 2020).

Unlike the DBD, class A and class B PB1 domains seem to have differing functions in *Marchantia.* Domain swap experiments show that only the MpARF1 PB1 domain can rescue the *Mparf1* mutant (Kato et al. 2020). When the MpARF1 PB1 domain was replaced with either the MpARF2 or MpARF3 PB1 domain, the plant showed severe developmental defects and a loss of auxin responsiveness. Differences in charge distribution along the PB1 interaction surfaces may explain the differences in function; however, this has not been tested empirically. This data raises the possibility that PB1-mediated interactions, both between different ARFs and between ARFs and Aux/IAAs, drive signaling specificity by tuning auxin responsiveness.

In addition to differential Aux/IAA ARF interactions, the ARF PB1 is necessary for transcriptional control. In the synthetic yeast auxin signaling system, mutations impairing PB1 domain interactions negatively affected some, but not all, of the reporters tested (Pierre-Jerome et al. 2016; Lanctot et al. 2020). ARF19 with a mutant PB1 domain that prevents dimerization failed to activate transcription on single AuxRE targets, but is able to function on paired AuxRE targets. Unlike the PB1 domain, DBD dimerization was necessary in all cases. This data raises the possibility that the ARF PB1 domain modulates the ability of the ARFs to activate transcription on some AuxREs and serves to stabilize ARF dimerization under less ideal AuxRE grammar. All experiments testing the contribution of the ARF PB1 domain to transcriptional activity have relied on ideal AuxRE grammar with short spacing (DR5). The multimeric nature of the ARF PB1 domain also raises the possibility that AuxREs located thousands of base pairs apart from one another can be spanned by multimeric complexes of the ARFs. Further work is necessary to determine whether ARFs can span large genomic distances, and whether those multimeric complexes are biologically relevant.

Taken together, a more complex model of ARF-mediated auxin responsiveness emerges (Fig. 5B; Lavy et al. 2016; Kato et al. 2020). In tissues where class A ARF nuclear abundance is high, and therefore ARF–Aux/IAA interactions are prevalent, auxin responsiveness is high. In tissues where class B ARF abundance is high, and therefore lacking ARF–Aux/IAA interactions, auxin responsiveness is inhibited. This gradient of auxin responsiveness requires that class A and B ARFs share transcriptional targets and that Aux/IAA interact only with class A ARFs. Many of the observations of ARF function and activity are consistent between angiosperms and nonvascular plants. However, it remains to be seen whether plants with larger ARF gene families are governed by the same general principles seen in *Physcomitrella* and *Marchantia* (Lavy et al. 2016).

The ARF Middle Region

Whereas structural domains in the ARFs have been well characterized, the unstructured middle region is less understood. Initial reports on ARF function identified the middle region, making up approximately one-half of the total protein length in most ARFs, as containing activation and repressive domains in carrot protoplast experiments (Ulmasov et al. 1999a; Tiwari et al. 2003). Class A ARFs act generally as auxin-inducible activators, whereas the class B ARFs act as repressors. Interestingly, activation and repressive activity can be decoupled from auxin induction by expressing the middle region alone in a synthetic transcription factor assay in carrot protoplasts (Tiwari et al. 2003).

A conserved TPL-binding motif (RLFGV) in *Arabidopsis* class B ARFs acts as a repressor domain in vivo (Choi et al. 2018). Interestingly, some class B ARFs also encode a canonical EAR motif in addition to the RLFGV motif. In the case of ARF2, for example, both the conserved RLFGV motif and the additional EAR motif are required for ARF2 to function as a transcriptional corepressor; however, only the RLFGV

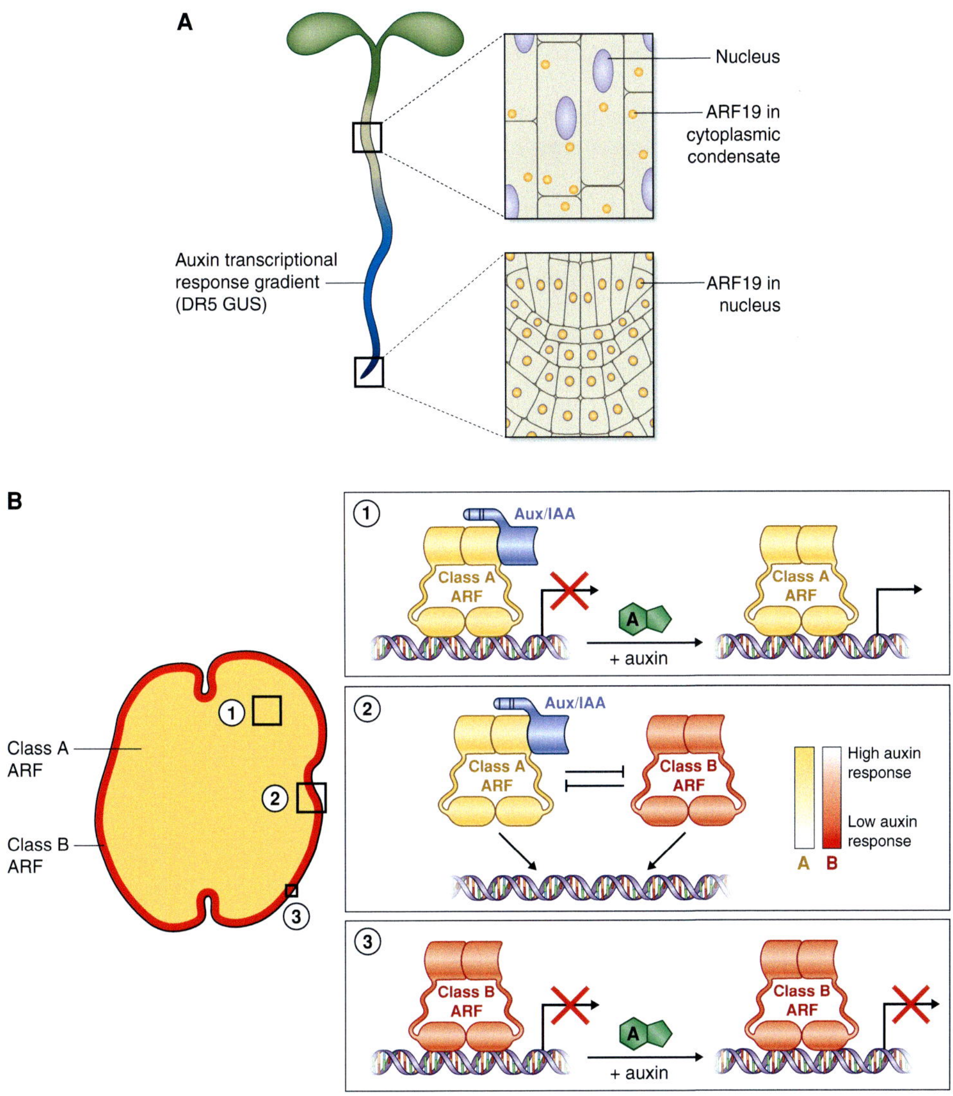

Figure 5. Auxin signaling relies on abundant, nuclear class A ARFs. (*A*) Powers et al. (2019) show that auxin signaling is absent in tissues in which ARF19 has moved out of the nucleus into cytoplasmic condensates. This effect can be eliminated if ARF19 remains nuclear. (*B*) As data in Kato et al. (2020) show, auxin signaling is present in the tissues that express class A ARF1 in *Marchantia* and absent in the tissues that express class B ARF2. Taken together, these two studies present a model in which high abundance of nuclear localized class A ARFs define auxin-responsive tissues, whereas tissues lacking class A ARFs entirely or from the nucleus are generally auxin insensitive.

motif is required for TPL interactions in yeast two-hybrid experiments (Choi et al. 2018). Taken together with data from *Marchantia*, it suggests that class B ARFs act as auxin-insensitive negative regulators of auxin-responsive genes (Fig. 5B).

Unlike ARF repressor domains, the ARF activation domain remains unknown. This is likely due in part to the intrinsic disorder found in the class A ARF middle region. Most activation domains are defined not by sequence similarity or identity, but by sequence features such as hydrophobicity and negative charge (Ravarani et al. 2018; Staller et al. 2018; Erijman et al. 2020). Further, whether class B ARFs can act as transcriptional activators at certain loci or in the presence of other unknown cofactors remains unknown.

In addition to direct activation of transcription, class A ARFs may induce transcription indirectly by recruiting the SWITCH/SUCROSE NONFERMENTING (SWI/SNF) chromatin-remodeling complex (reviewed in Clapier and Cairns 2009). The ARF5 middle region can recruit the SWI/SWF complex through interactions with BRAHMA and SPLAYED (Wu et al. 2015). These interactions are improved in the presence of auxin and increase chromatin accessibility at ARF5-binding sites (Wu et al. 2015). These results suggest a model in which ARF5, and possibly additional class A ARFs, alter nucleosome positioning to make more transcription factor-binding sites accessible (Wu et al. 2015; Weijers and Wagner 2016).

The intrinsic disorder found in the middle region of class A ARFs, and to a lesser degree in class B and C ARFs, does not dictate transcription potential alone. The middle regions of two class A ARFs that regulate lateral root development, ARF7 and ARF19, are necessary for subcellular localization (Fig. 5A; Powers et al. 2019). ARF19 displays differential cellular localization based on tissue type, with ARF19 localized to the nucleus in young, dividing root tissue and in the cytoplasm in mature root tissue. Mutating the PB1 domain alters this localization pattern, driving more ARF19 into the nucleus in mature roots compared to wild-type ARF19. The PB1 domain is necessary for this behavior but insufficient. The ARF middle region, cooperatively with the PB1 domain, drives differences in localization through protein condensation (see below). The ARF19 middle region and PB1 domain are required for protein condensation and cytoplasmic localization of ARF19 in the mature regions of the *Arabidopsis* root. If either the PB1 domain or middle region is perturbed, ARF19 localizes to the nucleus in these tissues. Interestingly, the mature zone of the root showed reduced or absent auxin-induced expression of the DR5 GUS reporter (Fig. 5A). However, plants expressing ARF19 with a mutated PB1 domain show expansion of DR5 GUS expression into these mature tissues, consistent with increased nuclear localization of ARF19. This data is supported by increased transcript abundance of both novel and known auxin-responsive genes (Powers et al. 2019). The PB1 domain likely increases the local concentration of ARF19 and that the intrinsic disorder of the middle region contributes to phase separation and protein condensation (for review, see Holehouse 2019 and Powers and Strader 2020). The relationship between ARF localization and transcriptional activity provides another possible route of regulation for the auxin signaling cascade (for review, see Lanctot and Nemhauser 2020). Understanding the mechanisms that drive ARF condensation, and the degree to which other ARFs participate in similar processes, will be crucial for our understanding of auxin signaling specificity.

Posttranslational modifications (for review, see Powers and Strader 2020) and alternative splicing of the ARFs (for review, see Lanctot and Nemhauser 2020) also contribute to ARF function. Understanding the contribution of different ARF posttranscriptional and posttranslational modifications to ARF function may further explain how a single, albeit large, transcription factor family can impact so many distinct processes.

CONCLUDING REMARKS

The structural aspects of nuclear auxin signaling have shed light on one of the main questions in auxin biology. The roles of protein–protein and protein–DNA interaction specificity, protein degradation dynamics, and subcellular localization all provide insights into how a module made up of only three components can result in such diverse signaling outcomes. Furthermore, the disordered regions of the ARFs and Aux/IAAs are opening new avenues for understanding how these proteins function in vivo. More work is required to fully understand how the ensemble of structures, modifications, and interactions in these disordered regions contribute to auxin signaling specificity and outcomes.

In addition to intrinsic disorder, there is an emerging role for protein subcellular localization in our understanding of auxin signaling potential. The ARFs appear to potentiate activity in different cell types based on nuclear and cytoplasmic localization patterns and are regulated by both posttranscriptional and posttranslational mechanisms. Uncovering the mechanisms that regulate ARF localization patterns may provide new insights into hormone cross talk as well as how plants integrate environmental and developmental signals.

ACKNOWLEDGMENTS

The authors thank members of the Strader laboratory for their critical reading of this manuscript, as well as support from the National Science Foundation Postdoctoral Research Program (IOS-1907098 to N.M.), the National Institutes of Health (R35 GM136338 to L.C.S.), and the National Science Foundation (IOS-1453750 to L.C.S.). Artwork by Debbie Maizels, Zoobotanicals Scientific Illustration.

REFERENCES

Abel S, Oeller PW, Theologis A. 1994. Early auxin-induced genes encode short-lived nuclear proteins. *Proc Natl Acad Sci* **91:** 326–330. doi:10.1073/pnas.91.1.326

Báez RR, Buckley Y, Yu H, Chen Z, Gallavotti A, Nemhauser J, Moss BL. 2020. A synthetic approach allows rapid characterization of the maize nuclear auxin response circuit. *Plant Physiol* **182:** 1713–1722.

Boer DR, Freire-Rios A, van den Berg WA, Saaki T, Manfield IW, Kepinski S, López-Vidrieo I, Franco-Zorrilla JM, de Vries SC, Solano R, et al. 2014. Structural basis for DNA binding specificity by the auxin-dependent ARF transcription factors. *Cell* **156:** 577–589. doi:10.1016/j.cell.2013.12.027

Brunoud G, Wells DM, Oliva M, Larrieu A, Mirabet V, Burrow AH, Beeckman T, Kepinski S, Traas J, Bennett MJ, et al. 2012. A novel sensor to map auxin response and distribution at high spatio-temporal resolution. *Nature* **482:** 103–106. doi:10.1038/nature10791

Calderón Villalobos LI, Lee S, De Oliveira C, Ivetac A, Brandt W, Armitage L, Sheard LB, Tan X, Parry G, Mao H, et al. 2012. A combinatorial TIR1/AFB-Aux/IAA co-receptor system for differential sensing of auxin. *Nat Chem Biol* **8:** 477–485. doi:10.1038/nchembio.926

Choi HS, Seo M, Cho HT. 2018. Two TPL-binding motifs of ARF2 are involved in repression of auxin responses. *Front Plant Sci* **9:** 372. doi:10.3389/fpls.2018.00372

Clapier CR, Cairns BR. 2009. The biology of chromatin remodeling complexes. *Annu Rev Biochem* **78:** 273–304. doi:10.1146/annurev.biochem.77.062706.153223

del Pozo JC, Dharmasiri S, Hellmann H, Walker L, Gray WM, Estelle M. 2002. AXR1-ECR1-dependent conjugation of RUB1 to the *Arabidopsis* cullin AtCUL1 is required for auxin response. *Plant Cell* **14:** 421–433. doi:10.1105/tpc.010282

Dharmasiri N, Dharmasiri S, Jones AM, Estelle M. 2003. Auxin action in a cell-free system. *Curr Biol* **13:** 1418–1422. doi:10.1016/S0960-9822(03)00536-0

Dharmasiri N, Dharmasiri S, Estelle M. 2005a. The F-box protein TIR1 is an auxin receptor. *Nature* **435:** 441–445. doi:10.1038/nature03543

Dharmasiri N, Dharmasiri S, Weijers D, Lechner E, Yamada M, Hobbie L, Ehrismann JS, Jürgens G, Estelle M. 2005b. Plant development is regulated by a family of auxin receptor F box proteins. *Dev Cell* **9:** 109–119. doi:10.1016/j.devcel.2005.05.014

Dinesh DC, Kovermann M, Gopalswamy M, Hellmuth A, Calderón Villalobos LI, Lilie H, Balbach J, Abel S. 2015. Solution structure of the PsIAA4 oligomerization domain reveals interaction modes for transcription factors in early auxin response. *Proc Natl Acad Sci* **112:** 6230–6235. doi:10.1073/pnas.1424077112

Dreher KA, Brown J, Saw RE, Callis J. 2006. The *Arabidopsis* Aux/IAA protein family has diversified in degradation and auxin responsiveness. *Plant Cell* **18:** 699–714. doi:10.1105/tpc.105.039172

Ellis CM, Nagpal P, Young JC, Hagen G, Guilfoyle TJ, Reed JW. 2005. AUXIN RESPONSE FACTOR1 and AUXIN RESPONSE FACTOR2 regulate senescence and floral organ abscission in *Arabidopsis thaliana*. *Development* **132:** 4563–4574. doi:10.1242/dev.02012

Erijman A, Kozlowski L, Sohrabi-Jahromi S, Fishburn J, Warfield L, Schreiber J, Noble WS, Söding J, Hahn S. 2020. A high-throughput screen for transcription activation domains reveals their sequence features and permits prediction by deep learning. *Mol Cell* **78:** 890. doi:10.1016/j.molcel.2020.04.020

Flores-Sandoval E, Eklund DM, Bowman JL. 2015. A simple auxin transcriptional response system regulates multiple morphogenetic processes in the liverwort *Marchantia polymorpha*. *PLoS Genet* **11:** e1005207. doi:10.1371/journal.pgen.1005207

Freire-Rios A, Tanaka K, Crespo I, Wijk Evd, Sizentsova Y, Levitsky V, Lindhoud S, Fontana M, Hohlbein J, Boer DR, et al. 2020. Architecture of DNA elements mediating ARF transcription factor binding and auxin-responsive gene expression in *Arabidopsis*. *Proc Nat Acad Sci* **117:** 24557–24566. doi:10.1073/pnas.2009554117

Fukaki H, Tameda S, Masuda H, Tasaka M. 2002. Lateral root formation is blocked by a gain-of-function mutation in the *SOLITARY-ROOT/IAA14* gene of *Arabidopsis*. *Plant J* **29:** 153–168. doi:10.1046/j.0960-7412.2001.01201.x

Galli M, Khakhar A, Lu Z, Chen Z, Sen S, Joshi T, Nemhauser JL, Schmitz RJ, Gallavotti A. 2018. The DNA binding landscape of the maize AUXIN RESPONSE FACTOR family. *Nat Commun* **9:** 4526. doi:10.1038/s41467-018-06977-6

Gray WM, del Pozo JC, Walker L, Hobbie L, Risseeuw E, Banks T, Crosby WL, Yang M, Ma H, Estelle M. 1999. Identification of an SCF ubiquitin-ligase complex required for auxin response in *Arabidopsis thaliana*. *Genes Dev* **13:** 1678–1691. doi:10.1101/gad.13.13.1678

Gray WM, Kepinski S, Rouse D, Leyser O, Estelle M. 2001. Auxin regulates SCFTIR1-dependent degradation of AUX/IAA proteins. *Nature* **414:** 271–276. doi:10.1038/35104500

Guilfoyle TJ, Hagen G. 2012. Getting a grasp on domain III/IV responsible for auxin response factor-IAA protein interactions. *Plant Sci* **190:** 82–88. doi:10.1016/j.plantsci.2012.04.003

Han M, Park Y, Kim I, Kim EH, Yu TK, Rhee S, Suh JY. 2014. Structural basis for the auxin-induced transcriptional regulation by Aux/IAA17. *Proc Natl Acad Sci* **111:** 18613–18618. doi:10.1073/pnas.1419525112

Hardtke CS, Ckurshumova W, Vidaurre DP, Singh SA, Stamatiou G, Tiwari SB, Hagen G, Guilfoyle TJ, Berleth T. 2004. Overlapping and non-redundant functions of the *Arabidopsis* auxin response factors *MONOPTEROS* and *NONPHOTOTROPIC HYPOCOTYL 4*. *Development* **131:** 1089–1100. doi:10.1242/dev.00925

Harper RM, Stowe-Evans EL, Luesse DR, Muto H, Tatematsu K, Watahiki MK, Yamamoto K, Liscum E. 2000. The *NPH4* locus encodes the auxin response factor ARF7, a conditional regulator of differential growth in aerial *Arabidopsis* tissue. *Plant Cell* **12:** 757–770. doi:10.1105/tpc.12.5.757

Havens KA, Guseman JM, Jang SS, Pierre-Jerome E, Bolten N, Klavins E, Nemhauser JL. 2012. A synthetic approach reveals extensive tunability of auxin signaling. *Plant Physiol* **160:** 135–142. doi:10.1104/pp.112.202184

Hellmann H, Hobbie L, Chapman A, Dharmasiri S, Dharmasiri N, del Pozo C, Reinhardt D, Estelle M. 2003. *Arabidopsis AXR6* encodes CUL1 implicating SCF E3 ligases in auxin regulation of embryogenesis. *EMBO J* **22:** 3314–3325. doi:10.1093/emboj/cdg335

Holehouse AS. 2019. IDPs and IDRs in biomolecular condensates. In *Intrinsically disordered proteins* (ed. Salvi N), pp. 209–255. Academic, Cambridge, MA.

Holland AJ, Fachinetti D, Han JS, Cleveland DW. 2012. Inducible, reversible system for the rapid and complete degradation of proteins in mammalian cells. *Proc Natl Acad Sci* **109:** E3350–E3357. doi:10.1073/pnas.1216880109

Kagale S, Rozwadowski K. 2011. EAR motif-mediated transcriptional repression in plants: an underlying mechanism for epigenetic regulation of gene expression. *Epigenetics* **6:** 141–146. doi:10.4161/epi.6.2.13627

Kanke M, Nishimura K, Kanemaki M, Kakimoto T, Takahashi TS, Nakagawa T, Masukata H. 2011. Auxin-inducible protein depletion system in fission yeast. *BMC Cell Biol* **12:** 8. doi:10.1186/1471-2121-12-8

Kato H, Mutte SK, Suzuki H, Crespo I, Das S, Radoeva T, Fontana M, Yoshitake Y, Hainiwa E, van den Berg W, et al. 2020. Design principles of a minimal auxin response system. *Nat Plants* **6:** 473–482. doi:10.1038/s41477-020-0662-y

Ke J, Ma H, Gu X, Thelen A, Brunzelle JS, Li J, Xu HE, Melcher K. 2015. Structural basis for recognition of diverse transcriptional repressors by the TOPLESS family of corepressors. *Sci Adv* **1:** e1500107. doi:10.1126/sciadv.1500107

Kepinski S, Leyser O. 2004. Auxin-induced SCF^{TIR1}-Aux/IAA interaction involves stable modification of the SCF^{TIR1} complex. *Proc Natl Acad Sci* **101:** 12381–12386. doi:10.1073/pnas.0402868101

Kepinski S, Leyser O. 2005. The *Arabidopsis* F-box protein TIR1 is an auxin receptor. *Nature* **435:** 446–451. doi:10.1038/nature03542

Korasick DA, Westfall CS, Lee SG, Nanao MH, Dumas R, Hagen G, Guilfoyle TJ, Jez JM, Strader LC. 2014. Molecular basis for AUXIN RESPONSE FACTOR protein interaction and the control of auxin response repression. *Proc Natl Acad Sci* **111:** 5427–5432. doi:10.1073/pnas.1400074111

Korasick DA, Chatterjee S, Tonelli M, Dashti H, Lee SG, Westfall CS, Fulton DB, Andreotti AH, Amarasinghe GK, Strader LC, et al. 2015. Defining a two-pronged structural model for PB1 (Phox/Bem1p) domain interaction in plant auxin responses. *J Biol Chem* **290:** 12868–12878. doi:10.1074/jbc.M115.648253

Lanctot A, Nemhauser JL. 2020. It's Morphin' time: how multiple signals converge on ARF transcription factors to direct development. *Curr Opin Plant Biol* **57:** 1–7. doi:10.1016/j.pbi.2020.04.008

Lanctot A, Taylor-Teeples M, Oki EA, Nemhauser J. 2020. Specificity in auxin responses is not explained by the promoter preferences of activator ARFs. *Plant Physiol* **182:** 1533–1536.

Lavy M, Prigge MJ, Tao S, Shain S, Kuo A, Kirchsteiger K, Estelle M. 2016. Constitutive auxin response in *Physcomitrella* reveals complex interactions between Aux/IAA and ARF proteins. *eLife* **5:** e13325. doi:10.7554/eLife.13325

Lee S, Sundaram S, Armitage L, Evans JP, Hawkes T, Kepinski S, Ferro N, Napier RM. 2014. Defining binding efficiency and specificity of auxins for $SCF^{TIR1/AFB}$-Aux/IAA co-receptor complex formation. *Acs Chem Biol* **9:** 673–682. doi:10.1021/cb400618m

Leyser HM, Lincoln CA, Timpte C, Lammer D, Turner J, Estelle M. 1993. *Arabidopsis* auxin-resistance gene *AXR1* encodes a protein related to ubiquitin-activating enzyme E1. *Nature* **364:** 161–164. doi:10.1038/364161a0

Li H, Johnson P, Stepanova A, Alonso JM, Ecker JR. 2004. Convergence of signaling pathways in the control of differential cell growth in *Arabidopsis*. *Dev Cell* **7:** 193–204. doi:10.1016/j.devcel.2004.07.002

Lincoln C, Britton JH, Estelle M. 1990. Growth and development of the *axr1* mutants of *Arabidopsis*. *Plant Cell* **2:** 1071–1080.

Long JA, Ohno C, Smith ZR, Meyerowitz EM. 2006. TOPLESS regulates apical embryonic fate in *Arabidopsis*. *Science* **312:** 1520–1523. doi:10.1126/science.1123841

Maraschin Fdos S, Memelink J, Offringa R. 2009. Auxin-induced, SCF^{TIR1}-mediated poly-ubiquitination marks AUX/IAA proteins for degradation. *Plant J* **59:** 100–109. doi:10.1111/j.1365-313X.2009.03854.x

Mazur E, Kulik I, Hajný J, Friml J. 2020. Auxin canalization and vascular tissue formation by TIR1/AFB-mediated auxin signaling in *Arabidopsis*. *New Phytol* **226:** 1375–1383. doi:10.1111/nph.16446

Moss BL, Mao H, Guseman JM, Hinds TR, Hellmuth A, Kovenock M, Noorassa A, Lanctot A, Villalobos LI, Zheng N, et al. 2015. Rate motifs tune auxin/indole-3-acetic acid degradation dynamics. *Plant Physiol* **169:** 803–813. doi:10.1104/pp.15.00587

Mutte SK, Kato H, Rothfels C, Melkonian M, Wong GK, Weijers D. 2018. Origin and evolution of the nuclear auxin response system. *eLife* **7:** e33399. doi:10.7554/eLife.33399

Nagpal P, Walker LM, Young JC, Sonawala A, Timpte C, Estelle M, Reed JW. 2000. *AXR2* encodes a member of the Aux/IAA protein family. *Plant Physiol* **123:** 563–574. doi:10.1104/pp.123.2.563

Nanao MH, Vinos-Poyo T, Brunoud G, Thévenon E, Mazzoleni M, Mast D, Lainé S, Wang S, Hagen G, Li H, et al. 2014. Structural basis for oligomerization of auxin transcriptional regulators. *Nat Commun* **5:** 3617. doi:10.1038/ncomms4617

Niemeyer M, Moreno Castillo E, Ihling CH, Iacobucci C, Wilde V, Hellmuth A, Hoehenwarter W, Samodelov SL, Zurbriggen MD, Kastritis PL, et al. 2020. Flexibility of intrinsically disordered degrons in AUX/IAA proteins reinforces auxin co-receptor assemblies. *Nat Commun* **11:** 2277. doi:10.1038/s41467-020-16147-2

Nishimura K, Fukagawa T, Takisawa H, Kakimoto T, Kanemaki M. 2009. An auxin-based degron system for the rapid depletion of proteins in nonplant cells. *Nat Methods* **6:** 917–922. doi:10.1038/nmeth.1401

Okushima Y, Mitina I, Quach HL, Theologis A. 2005. AUXIN RESPONSE FACTOR 2 (ARF2): a pleiotropic developmental regulator. *Plant J* **43:** 29–46. doi:10.1111/j.1365-313X.2005.02426.x

O'Malley RC, Huang SSC, Song L, Lewsey MG, Bartlett A, Nery JR, Galli M, Gallavotti A, Ecker JR. 2016. Cistrome and epicistrome features shape the regulatory DNA landscape. *Cell* **165:** 1280–1292. doi:10.1016/j.cell.2016.04.038

Ouellet F, Overvoorde PJ, Theologis A. 2001. IAA17/AXR3: biochemical insight into an auxin mutant phenotype. *Plant Cell* **13:** 829–841. doi:10.1105/tpc.13.4.829

Parry G, Calderon-Villalobos LI, Prigge M, Peret B, Dharmasiri S, Itoh H, Lechner E, Gray WM, Bennett M, Estelle M. 2009. Complex regulation of the TIR1/AFB family of auxin receptors. *Proc Natl Acad Sci* **106:** 22540–22545. doi:10.1073/pnas.0911967106

Pierre-Jerome E, Moss BL, Nemhauser JL. 2013. Tuning the auxin transcriptional response. *J Exp Bot* **64:** 2557–2563. doi:10.1093/jxb/ert100

Pierre-Jerome E, Jang SS, Havens KA, Nemhauser JL, Klavins E. 2014. Recapitulation of the forward nuclear auxin response pathway in yeast. *Proc Natl Acad Sci* **111:** 9407–9412. doi:10.1073/pnas.1324147111

Pierre-Jerome E, Moss BL, Lanctot A, Hageman A, Nemhauser JL. 2016. Functional analysis of molecular interactions in synthetic auxin response circuits. *Proc Natl Acad Sci* **113:** 11354–11359. doi:10.1073/pnas.1604379113

Powers SK, Strader LC. 2020. Regulation of auxin transcriptional responses. *Dev Dyn* **249:** 483–495. doi:10.1002/dvdy.139

Powers SK, Holehouse AS, Korasick DA, Schreiber KH, Clark NM, Jing H, Emenecker R, Han S, Tycksen E, Hwang I, et al. 2019. Nucleo-cytoplasmic partitioning of ARF proteins controls auxin responses in *Arabidopsis thaliana*. *Mol Cell* **76:** 177–190.e5. doi:10.1016/j.molcel.2019.06.044

Prigge MJ, Platre M, Kadakia N, Zhang Y, Greenham K, Szutu W, Pandey BK, Bhosale RA, Bennett MJ, Busch W, et al. 2020. Genetic analysis of the *Arabidopsis* TIR1/AFB auxin receptors reveals both overlapping and specialized functions. *eLife* **9:** e54740. doi:10.7554/eLife.54740

Ramos JA, Zenser N, Leyser O, Callis J. 2001. Rapid degradation of auxin/indoleacetic acid proteins requires conserved amino acids of domain II and is proteasome dependent. *Plant Cell* **13:** 2349–2360. doi:10.1105/tpc.010244

Ravarani CN, Erkina TY, De Baets G, Dudman DC, Erkine AM, Babu MM. 2018. High-throughput discovery of functional disordered regions: investigation of transactivation domains. *Mol Syst Biol* **14:** e8190. doi:10.15252/msb.20188190

Rouse D, Mackay P, Stirnberg P, Estelle M, Leyser O. 1998. Changes in auxin response from mutations in an *AUX/IAA* gene. *Science* **279:** 1371–1373. doi:10.1126/science.279.5355.1371

Ruegger M, Dewey E, Hobbie L, Brown D, Bernasconi P, Turner J, Muday G, Estelle M. 1997. Reduced naphthylphthalamic acid binding in the *tir3* mutant of *Arabidopsis* is associated with a reduction in polar auxin transport and diverse morphological defects. *Plant Cell* **9:** 745–757.

Ruegger M, Dewey E, Gray WM, Hobbie L, Turner J, Estelle M. 1998. The TIR1 protein of *Arabidopsis* functions in auxin response and is related to human SKP2 and yeast Grr1p. *Genes Dev* **12:** 198–207. doi:10.1101/gad.12.2.198

Sessions A, Nemhauser JL, McColl A, Roe JL, Feldmann KA, Zambryski PC. 1997. *ETTIN* patterns the *Arabidopsis* floral meristem and reproductive organs. *Development* **124:** 4481–4491.

Staller MV, Holehouse AS, Swain-Lenz D, Das RK, Pappu RV, Cohen BA. 2018. A high-throughput mutational scan of an intrinsically disordered acidic transcriptional activation domain. *Cell Syst* **6:** 444–455.e6. doi:10.1016/j.cels.2018.01.015

Szemenyei H, Hannon M, Long JA. 2008. TOPLESS mediates auxin-dependent transcriptional repression during *Arabidopsis* embryogenesis. *Science* **319:** 1384–1386. doi:10.1126/science.1151461

Tan X, Calderon-Villalobos LI, Sharon M, Zheng C, Robinson CV, Estelle M, Zheng N. 2007. Mechanism of auxin perception by the TIR1 ubiquitin ligase. *Nature* **446:** 640–645. doi:10.1038/nature05731

Tian Q, Nagpal P, Reed JW. 2003. Regulation of *Arabidopsis* SHY2/IAA3 protein turnover. *Plant J* **36:** 643–651. doi:10.1046/j.1365-313X.2003.01909.x

Tian CE, Muto H, Higuchi K, Matamura T, Tatematsu K, Koshiba T, Yamamoto KT. 2004. Disruption and overexpression of *auxin response factor 8* gene of *Arabidopsis* affect hypocotyl elongation and root growth habit, indicating its possible involvement in auxin homeostasis in light condition. *Plant J* **40:** 333–343. doi:10.1111/j.1365-313X.2004.02220.x

Tiwari SB, Wang XJ, Hagen G, Guilfoyle TJ. 2001. AUX/IAA proteins are active repressors, and their stability and activity are modulated by auxin. *Plant Cell* **13:** 2809–2822. doi:10.1105/tpc.010289

Tiwari SB, Hagen G, Guilfoyle T. 2003. The roles of auxin response factor domains in auxin-responsive transcription. *Plant Cell* **15:** 533–543. doi:10.1105/tpc.008417

Tiwari SB, Hagen G, Guilfoyle TJ. 2004. Aux/IAA proteins contain a potent transcriptional repression domain. *Plant Cell* **16:** 533–543. doi:10.1105/tpc.017384

Trost M, Blattner AC, Lehner CF. 2016. Regulated protein depletion by the auxin-inducible degradation system in *Drosophila melanogaster*. *Fly (Austin)* **10:** 35–46. doi:10.1080/19336934.2016.1168552

Uchida N, Takahashi K, Iwasaki R, Yamada R, Yoshimura M, Endo TA, Kimura S, Zhang H, Nomoto M, Tada Y, et al. 2018. Chemical hijacking of auxin signaling with an engineered auxin-TIR1 pair. *Nat Chem Biol* **14:** 299–305. doi:10.1038/nchembio.2555

Ulmasov T, Liu ZB, Hagen G, Guilfoyle TJ. 1995. Composite structure of auxin response elements. *Plant Cell* **7:** 1611–1623.

Ulmasov T, Hagen G, Guilfoyle TJ. 1997a. ARF1, a transcription factor that binds to auxin response elements. *Science* **276:** 1865–1868. doi:10.1126/science.276.5320.1865

Ulmasov T, Murfett J, Hagen G, Guilfoyle TJ. 1997b. Aux/IAA proteins repress expression of reporter genes containing natural and highly active synthetic auxin response elements. *Plant Cell* **9:** 1963–1971.

Ulmasov T, Hagen G, Guilfoyle TJ. 1999a. Activation and repression of transcription by auxin-response factors. *Proc Natl Acad Sci* **96:** 5844–5849. doi:10.1073/pnas.96.10.5844

Ulmasov T, Hagen G, Guilfoyle TJ. 1999b. Dimerization and DNA binding of auxin response factors. *Plant J* **19:** 309–319. doi:10.1046/j.1365-313X.1999.00538.x

Uversky VN. 2019. Intrinsically disordered proteins and their "mysterious" (meta)physics. *Front Phys* doi:10.3389/fphy.2019.00010

Uzunova VV, Quareshy M, Del Genio CI, Napier RM. 2016. Tomographic docking suggests the mechanism of auxin receptor TIR1 selectivity. *Open Biol* **6:** 160139. doi:10.1098/rsob.160139

Vernoux T, Brunoud G, Farcot E, Morin V, Van den Daele H, Legrand J, Oliva M, Das P, Larrieu A, Wells D, et al. 2011. The auxin signalling network translates dynamic input into robust patterning at the shoot apex. *Mol Syst Biol* **7:** 508. doi:10.1038/msb.2011.39

Weijers D, Wagner D. 2016. Transcriptional responses to the auxin hormone. *Annu Rev Plant Biol* **67:** 539–574. doi:10.1146/annurev-arplant-043015-112122

Wilmoth JC, Wang S, Tiwari SB, Joshi AD, Hagen G, Guilfoyle TJ, Alonso JM, Ecker JR, Reed JW. 2005. NPH4/ARF7 and ARF19 promote leaf expansion and auxin-induced lateral root formation. *Plant J* **43:** 118–130. doi:10.1111/j.1365-313X.2005.02432.x

Worley CK, Zenser N, Ramos J, Rouse D, Leyser O, Theologis A, Callis J. 2000. Degradation of Aux/IAA proteins is essential for normal auxin signalling. *Plant J* **21:** 553–562. doi:10.1046/j.1365-313x.2000.00703.x

Wu MF, Yamaguchi N, Xiao J, Bargmann B, Estelle M, Sang Y, Wagner D. 2015. Auxin-regulated chromatin switch directs acquisition of flower primordium founder fate. *eLife* **4:** e09269. doi:10.7554/eLife.09269

Yang X, Lee S, So JH, Dharmasiri S, Dharmasiri N, Ge L, Jensen C, Hangarter R, Hobbie L, Estelle M. 2004. The IAA1 protein is encoded by AXR5 and is a substrate of SCF^{TIR1}. *Plant J* **40:** 772–782. doi:10.1111/j.1365-313X.2004.02254.x

Yu H, Moss BL, Jang SS, Prigge M, Klavins E, Nemhauser JL, Estelle M. 2013. Mutations in the TIR1 auxin receptor that increase affinity for auxin/indole-3-acetic acid proteins result in auxin hypersensitivity. *Plant Physiol* **162:** 295–303. doi:10.1104/pp.113.215582

Yu H, Zhang Y, Moss BL, Bargmann BO, Wang R, Prigge M, Nemhauser JL, Estelle M. 2015. Untethering the TIR1 auxin receptor from the SCF complex increases its stability and inhibits auxin response. *Nat Plants* **1:** 14030. doi:10.1038/nplants.2014.30

Zenser N, Ellsmore A, Leasure C, Callis J. 2001. Auxin modulates the degradation rate of Aux/IAA proteins. *Proc Natl Acad Sci* **98:** 11795–11800. doi:10.1073/pnas.211312798

Zenser N, Dreher KA, Edwards SR, Callis J. 2003. Acceleration of Aux/IAA proteolysis is specific for auxin and independent of *AXR1*. *Plant J* **35:** 285–294. doi:10.1046/j.1365-313X.2003.01801.x

No Time for Transcription—Rapid Auxin Responses in Plants

Shiv Mani Dubey, Nelson B.C. Serre, Denisa Oulehlová, Pruthvi Vittal, and Matyáš Fendrych

Department of Experimental Plant Biology, Charles University, 128 00 Prague, Czech Republic

Correspondence: fendryc1@natur.cuni.cz

Auxin regulates the transcription of auxin-responsive genes by the TIR1/AFBs-Aux/IAA-ARF signaling pathway, and in this way facilitates plant growth and development. However, rapid, nontranscriptional responses to auxin that cannot be explained by this pathway have been reported. In this review, we focus on several examples of rapid auxin responses: (1) the triggering of changes in plasma membrane potential in various plant species and tissues, (2) inhibition of root growth, which also correlates with membrane potential changes, cytosolic Ca^{2+} spikes, and a rise of apoplastic pH, (3) the influence on endomembrane trafficking of PIN proteins and other membrane cargoes, and (4) activation of ROPs (Rho of plants) and their downstream effectors such as the cytoskeleton or vesicle trafficking. In most cases, the signaling pathway triggering the response is poorly understood. A role for the TIR1/AFBs in rapid root growth regulation is emerging, as well as the involvement of transmembrane kinases (TMKs) in the activation of ROPs. We discuss similarities and differences among these rapid responses and focus on their physiological significance, which remains an enigma in most cases.

The phytohormone auxin plays a central role in plant physiology and development. According to our current understanding, it does so by changing gene expression profiles that largely depend on the developmental context and varying auxin concentrations. There are, however, numerous reports from the premolecular auxin era as well as recent publications showing that auxin can initiate rapid responses at the level of organs, cells, and subcellular structures. The rapidity of these responses rules out the involvement of gene transcription and mRNA translation. The eukaryotic transcription rate is ~1.4 kbp per minute, and the translation rate ~5 amino acids per second (BioNumbers BNID 111158,104598; Milo et al. 2010). Thus, we can very roughly estimate that an average-length protein of 400 amino acids (Ramírez-Sánchez et al. 2016) needs at least 2 min to be transcribed and translated if all other processes (such as auxin signaling, mRNA processing, nuclear export, and protein maturation) are disregarded. The earliest measurable effect of the canonical auxin transcriptional pathway (the auxin-induced growth) typically shows a 10–20 min lag phase (Senn and Goldsmith 1988; Fendrych et al. 2016). Therefore, we arbitrarily consider responses that happen within a few minutes as rapid, nontranscriptional events that must involve rapid electrochemical re-

Cite this article as *Cold Spring Harb Perspect Biol* doi: 10.1101/cshperspect.a039891

sponses and/or posttranscriptional modifications.

During canonical auxin transcriptional signaling, when the concentration of auxin drops, the Aux/IAA proteins heterodimerize with the auxin response factors (ARFs) and inhibit auxin-responsive gene transcription (Weijers et al. 2005; Han et al. 2014). Auxin coreceptors TIR1/AFBs are F-box proteins, components of the SKP1–Cullin–F-box (SCF)–E3 ubiquitin ligase complex (Ruegger et al. 1998; Dharmasiri et al. 2005; Kepinski and Leyser 2005; Dezfulian et al. 2016). When auxin concentration increases, TIR1/AFB F-box proteins bind via auxin to Aux/IAA proteins, which results in Aux/IAA ubiquitylation and degradation (Gray et al. 2001; Tan et al. 2007), releasing the inhibition of ARFs. Aux/IAAs interact with auxin and TIR1/AFB via the degron domain (DII), which seals the ligand-binding groove in TIR1/AFB. The degron domain is conserved within the family; the dynamics of degradation mainly depends on the carboxy-terminal domain (Calderón Villalobos et al. 2012; Moss et al. 2015; Winkler et al. 2017). Recently, Niemeyer et al. (2020) and Ramans Harborough et al. (2019) revealed that the KR motif embedded in the intrinsic disorder region (IDR) upstream of the degron confers alternative auxin-independent-binding contacts with TIR1. Therefore, before the actual auxin-induced transcriptional changes, many rapid nontranscriptional events occur, which influence protein localization such as the Aux/IAA relocalization toward the proteasome, protein posttranslational modification, and abundance. This system of auxin-induced Aux/IAA degradation was exploited as a powerful tool to rapidly control protein levels in yeast and in animal cell lines (Nishimura et al. 2009, 2020; Holland et al. 2012). And possibly, some of the nontranscriptional auxin effects that we discuss in this review might be linked to these early phases of the TIR1/AFB-Aux/IAA signaling.

DYNAMICS OF AUXIN AND PROTON TRANSPORT ACROSS MEMBRANES

Polar auxin transport is essential for development (Petrášek and Friml 2009). At the cellular level, auxin fluxes depend on the combination of auxin chemical properties and the controlled activity and localization of influx and efflux auxin carriers (Bennett et al. 1996; Gälweiler et al. 1998; Petrášek et al. 2006). To understand the rapid effects of auxin, we need to discuss how auxin travels across membranes, focusing on associated ion transport processes.

Auxins, as weak acids, are partly undissociated in the acidic apoplast (pH 4.5–5.5) (Barbez et al. 2017), and the uncharged IAA can enter the cell by diffusion. At the more alkaline cytoplasmic pH, IAA releases its H^+ and is trapped in the cell. Plasma membrane (PM) permeability for the IAA^- anion is ~100 to 1000 times lower than for IAA (Raven 1975; Delbarre et al. 1996). Uptake of undissociated IAA will continue until the IAA concentration outside and inside equilibrates (Rubery and Sheldrake 1974). Therefore, the existence of H^+ gradient (acidic outside) should result in auxin accumulation in the cell even without active transporters (Fig. 1; Lomax et al. 1985).

The natural auxin IAA^- is imported with two protons (Rubery and Sheldrake 1974; Hertel et al. 1983; Lomax et al. 1985) by the AUX1/LAX family of transporters (Swarup and Péret 2012). Consequently, the proton gradient or the proton motive force fuels AUX1/LAX activity (Lomax et al. 1985). On the other hand, at higher apoplastic pH, the substrate of AUX1/LAX importer IAA^- (Carrier et al. 2008) is more abundant. Finally, AUX1/LAX affinity depends on pH, its optimum being at pH ~ 5.5 (Carrier et al. 2008). Taken together, these factors explain the nonmonotonic dependence of IAA influx on pH observed in several studies (Rubery and Sheldrake 1974; Loper and Spanswick 1991; Yang et al. 2006).

Auxin is exported out of cells by the polarly localized Pin Formed (PIN) and apolar ATP-binding cassette subfamily B (ABCBs) transporters (for review, see Petrášek and Friml 2009); some ABCBs can also import auxin (Kamimoto et al. 2012; Kubeš et al. 2012). The structure of PIN proteins and their mechanisms of action remain utterly unknown (Zwiewka et al. 2019). We are not aware of any report that PINs cotransport ions. If we assume that

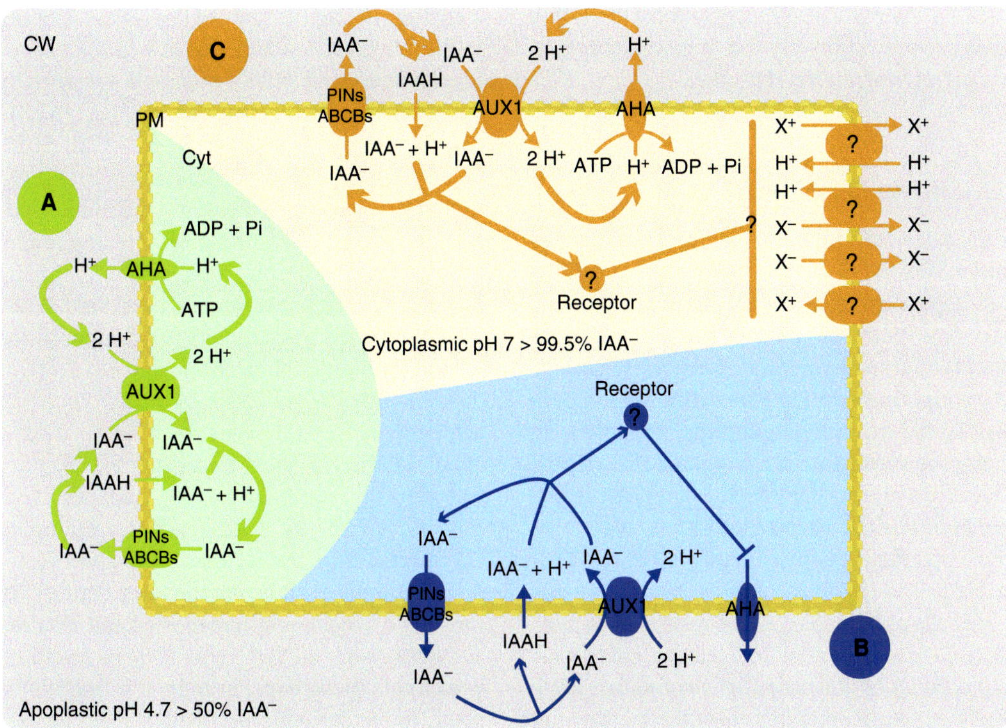

Figure 1. Possible scenarios that could explain the underlying mechanisms of auxin-induced cell depolarization. (*A*) Proton influx driven by auxin/proton symport. Import of auxin into the cell participates in the accumulation of $2H^+/1IAA^-$ by AUX1/LAX and/or $1H^+$ by diffusion (dissociation of IAAH at cytoplasmic pH 7). Such accumulation of cytoplasmic-positive charges could increase membrane potential. This could stimulate the proton pump activity, which in turn would fuel auxin import by AUX1/LAX. (*B*) Inhibition of the proton pump by an intracellular auxin receptor. Similar to *A*, auxin influx participates in the accumulation of protons in the cytoplasm. In this model, this process is enhanced by the inhibition of the proton pump by active signaling event(s) from auxin receptor(s). (*C*) Regulation of the activities of ion channels/carriers by intracellular auxin receptor(s). Inhibition of cations efflux or activation of influx and reciprocally for anions could influence membrane potential by accumulation of positive charges in the cytoplasm. The actual mechanism of depolarization is likely to be a combination of these scenarios.

PINs transport the IAA^- (at cytoplasmic pH 7.2, 99.7% of IAA is in the IAA^- form), then PINs would act as facilitators of IAA^- diffusion since, in physiological conditions, the concentration of IAA^- in the cytoplasm will always be higher than in the adjacent apoplast. A bile acid–sodium symporter was used as a template for PIN2 modeling (Retzer et al. 2017), indicating a possible relationship. This carrier harnesses the energy of the Na^+ gradient (Zhou et al. 2014); however, generalizing this mechanism to PIN proteins would be a wild speculation. Determining the electrical signature of an actively transporting PIN could help resolve the question of what, apart from IAA, PINs actually transport (Zwiewka et al. 2019).

Auxin influx, both by diffusion and carrier-mediated transport, requires a proton gradient across the PM. At the same time, the entry of IAA is accompanied by the entry of H^+, one per molecule in the case of diffusion and two per molecule in the case of AUX1/LAX-mediated import. IAA influx therefore reduces the proton motive force and is even electrogenic in the case of AUX1/LAX influx. From what we currently understand, it is impossible to conclude whether auxin efflux directly contributes to ion fluxes across the PM, but the depletion of the cytoplasmic IAA^- pool by its efflux promotes the influx carrier activity indirectly and thus contributes to AUX1/LAX-mediated H^+ influx.

AUXIN TRIGGERS RAPID CHANGES IN PLASMA MEMBRANE POTENTIAL

Application of auxin to various plant tissues causes very rapid (within seconds) concentration-dependent changes in membrane potential. This potential reflects the difference in cytoplasmic and apoplastic electrical potentials. In resting conditions, in plants, this difference is maintained by active pumping of positively charged protons into the apoplast (Sze 1985). This active process depletes the cytoplasmic potential of positive charges and leads to a membrane potential of approximately −120 to −160 mV (Sze et al. 1999). A negative membrane potential allows secondary active transport of molecules and is essential for cellular energetics. Rapid fluxes of ions in/out of the cell can drastically influence membrane potential causing accumulation of positive charges (depolarization) or the accumulation of negative charges (hyperpolarization) (Weisenseel and Meyer 1997).

Auxin rapidly depolarized PM in coleoptiles of monocots (Goering 1979; Bates and Goldsmith 1983; Felle et al. 1991), in maize (Felle et al. 1991), and white mustard (Tretyn et al. 1991) root hairs. More recently, depolarization was recorded in *Arabidopsis thaliana* root hairs (Dindas et al. 2018, 2020) and root epidermis (Paponov et al. 2019a). Surprisingly, in white mustard root hairs, picomolar IAA triggered the opposite response—a rapid hyperpolarization (Tretyn et al. 1991). Moreover, in protoplasts (cells with cell walls removed and floating in an isotonic medium), auxin triggered an outward current of positive charge in a rapid manner (~1 min) (Ruck et al. 1993).

In some tissues, the depolarization was followed by PM hyperpolarization (in 10–20 min); this response was only triggered by active auxins that triggered growth, whereas weak acids with comparable concentrations did not trigger hyperpolarization (Bates and Goldsmith 1983). The slow hyperpolarization response could result from the activation of H^+-ATPase, leading to proton extrusion and acid growth of coleoptiles and hypocotyls. This process is governed by the transcriptional activity of the TIR1/AFB1-Aux/IAA coreceptor (Takahashi et al. 2012; Fendrych et al. 2016; Uchida et al. 2018). On the other hand, instantaneous membrane potential responses constitute, without any doubt, a clear nontranscriptional reaction to auxin. What is this response caused by? What is the receptor, if any, triggering it? And, importantly, what is its biological meaning?

PM potential changes by auxin could be achieved by several scenarios (Fig. 1; Tretyn et al. 1991): (1) short-circuiting the primary transport by secondary active transport (cotransport of H^+, by AUX1/LAX, for example), (2) inactivating the primary transport (H^+-ATPase), and (3) changing the PM permeability for ions by regulating transporter and channel activities (such as Ca^{2+} and K^+). In addition, weak acids themselves, in high concentrations, could lower cytosolic pH, which leads to H^+-ATPase stimulation (Bates and Goldsmith 1983; Felle et al. 1991).

Unfortunately, reports vary as to what extent different auxins and weak acids trigger PM depolarization. Both active auxins (IAA, 1-NAA, and 2,4-D) and inactive analogs (2-NAA and 2,3-D) induced rapid PM depolarization (Bates and Goldsmith 1983; Felle et al. 1991; Tretyn et al. 1991; Dindas et al. 2018, 2020; Paponov et al. 2019a). There are indications that rapid depolarization depends on the proton–auxin symport by AUX1/LAX: First, the response is considerably diminished at high pH (Felle et al. 1991; Paponov et al. 2019a), consistent with the above-mentioned AUX1/LAX dependence on proton motive force and pH. Second, the depolarization response in root hairs is drastically diminished in the *aux1* mutant (Dindas et al. 2018). However, 1-NAA, which is not a substrate of AUX1/LAX import (Tretyn et al. 1991; Delbarre et al. 1996; Yang et al. 2006), depolarizes PM as well, as does the biologically inactive 2-NAA, at least to some extent (Felle et al. 1991; Tretyn et al. 1991; Paponov et al. 2019a; Dindas et al. 2020). In addition, auxin efflux via PIN2 contributes to PM depolarization (Dindas et al. 2018; Paponov et al. 2019a), possibly by indirectly enhancing AUX1/LAX activity by depleting the intracellular auxin. Auxin transporters are thus important actors in the auxin-induced depolarization.

The fact that picomolar IAA triggered a rapid hyperpolarization in root hairs (Tretyn et al. 1991) could hint to the existence of an auxin receptor rapidly stimulating H^+-ATPase. There is recent evidence that, in *Arabidopsis* root hairs, rapid PM depolarization depends on the TIR1/AFB-Aux/IAA auxin coreceptor. The *tir1/afb2/3* triple mutant lacking three of six receptors showed a decreased depolarization response, and the pretreatment with auxinole (a TIR1/AFB-Aux/IAA inhibitor; Hayashi et al. 2012) decreased the depolarization response (Dindas et al. 2018). However, the possibility that auxinole itself triggered depolarization, which would mask the auxin response, cannot be excluded (cf. Supplemental Fig. 4e,f in Dindas et al. 2018). The authors confirmed that the *tir1/afb2/3* phenotype was not caused by the lack of AUX1 expression, opening the possibility that the TIR1/AFB coreceptor might indeed trigger signaling that leads to depolarization response and thus be directly involved in rapid, nontranscriptional signaling.

We can therefore speculate that depolarization is triggered jointly by auxin transport processes, an active signaling response, and in cases of higher concentrations can involve the direct effect of weak acids. However, the most important question remains—what is the biological significance of auxin-induced depolarization? It is hard to imagine a situation in soil when a root hair would suddenly be flushed with micromolar IAA, even though it was shown that strains of *Bradyrhizobium japonicum* can secrete up to 20 μM IAA into the medium (Minamisawa and Fukai 1991). Root hair growth is indeed regulated by auxin, but this regulation depends on the TIR1/AFB-Aux/IAA-ARF transcriptional signaling (Mangano et al. 2017; Bhosale et al. 2018).

With the boom of molecular biology, many clearly nontranscriptional auxin effects have not been extensively studied (e.g., Marten et al. 1992) and the molecular pathways involved remain unknown. Plant electrophysiology, in contrast to animal electrophysiology, has fallen out of fashion. However, in recent years, efforts have been made to revive those techniques and it is of great importance to now create links between physiology and molecular biology as there are possibly many more rapid auxin responses to be revealed.

ROOT GROWTH INHIBITION BY AUXIN

Whereas the biological significance of auxin-induced depolarization remains a mystery, there is one rapid effect of auxin with an evident function, root growth inhibition by auxin. During root gravitropic response, auxin flows shootward to the lower root side by the coordinate action of PIN2 and AUX1 (Young et al. 1990; Band et al. 2012; Adamowski and Friml 2015) where it inhibits cell elongation. This cellular response is thus the motor of gravitropic root bending. Auxin influences cell expansion of other plant organs as well (for review, see Du et al. 2020); however, outside of the root, the effect is mediated by the canonical TIR1/AFB-Aux/IAA-mediated gene transcription (Fendrych et al. 2016; Uchida et al. 2018; Du et al. 2020).

In nanomolar and higher concentrations, IAA rapidly inhibits root growth (Evans et al. 1994; Monshausen et al. 2011; Scheitz et al. 2013); the detectable inhibition starts within 60 sec of auxin treatment (Shih et al. 2015; Fendrych et al. 2018; Prigge et al. 2020). Such rapidity effectively rules out the involvement of TIR1/AFB-Aux/IAA-ARF-mediated gene transcription. The involvement of transcription has not been formally disproven, but the translation inhibitor, cycloheximide, could not prevent the IAA-triggered growth inhibition (Fendrych et al. 2018). Interestingly and surprisingly, root growth inhibition response is quickly reversible—roots started to resume growth within 3 min after removal of IAA from the medium even after an 80-min inhibition (Fendrych et al. 2018). This indicates that this auxin response is "rapid and nontranscriptional" for a long time, and that roots can slow down and speed up rapidly, depending on the current auxin concentration in the cell.

Indeed, auxin must enter the cell to trigger growth inhibition; *Arabidopsis* roots lacking the AUX1 auxin influx carrier did not respond to low IAA concentrations, but inhibited growth rapidly when treated with the PM-permeable 1-NAA auxin analog (Fendrych et al. 2018).

The underlying molecular pathway remains unknown, but in agreement with the effect of pH on the cell wall extensibility, inhibition is correlated with alkalinization of the root surface and cell walls. In corn roots, the addition of IAA or 1-NAA led to simultaneous growth inhibition and an increase in the bathing medium pH; 2-NAA had no effect (Evans et al. 1980; Lüthen and Böttger 1988). Imaging, using pH-sensitive dye, revealed that the addition of IAA caused near-instantaneous alkalinization of the root surface in the transition and elongation zones of *Arabidopsis* root tip. The response of the *aux1* mutant to 100 nM IAA was less than the control (Monshausen et al. 2011), which is consistent with the finding that *aux1* requires higher IAA doses to trigger growth inhibition (Fendrych et al. 2018). Also, the cell wall pH, visualized by the apopHusion sensor (Gjetting et al. 2012) and a pH-sensitive dye HPTS (Barbez et al. 2017), rose in response to IAA. The alkalinization of root surface and cell wall is probably caused by proton influx into cells, as Monshausen et al. (2011) showed that the auxin-induced alkalinization is accompanied by a "small but significant" decrease of pH inside the cells.

Could the alkalization of the root apoplast be caused directly by the AUX1-mediated IAA-H^+ symport? Probably not, because even low IAA concentrations trigger apoplast alkalinization. Also, IAA and PM-permeable 1-NAA can trigger alkalinization in the *aux1* mutant. The key protein to understand the connection between apoplastic pH, auxin, and membrane potential changes is the Cyclic Nucleotide-Gated Channel 14 (CNGC14) putative calcium channel. Externally and locally applied IAA induced a rapid cytoplasmic calcium transient, as fast as the pH change along the root axis (Monshausen et al. 2011; Dindas et al. 2018). Moreover, blocking calcium channels by lanthanide impaired the calcium transient and the pH response (Monshausen et al. 2011; Shih et al. 2015). The *cngc14* mutant did not accumulate calcium in the cytoplasm after auxin treatment had impaired apoplastic alkalization and delay in rapid growth response (Shih et al. 2015; Dindas et al. 2018). These observations demonstrate the connection between rapid calcium signaling and extracellular alkalization in response to auxin.

Indeed, a massive influx of positively charged calcium into the cytoplasm could create a significant membrane depolarization and, inversely, IAA-triggered proton influx could trigger calcium transient, as pH and calcium spiking are tightly connected (Behera et al. 2018). However, cytosolic calcium concentration, measured by Shih and Dindas, returns to the resting level immediately after reaching the peak value (within 1.5 min). On the other hand, the depolarization is maintained over the course of several minutes (as measured in Dindas et al. 2018; Paponov et al. 2019a) and proton influx lasts much longer (Shih et al. 2015). One could propose that calcium is the secondary messenger in auxin-induced depolarization, and this response is maintained by a nontranscriptional activation of ion channels.

THE NONTRANSCRIPTIONAL FACE OF TIR1/AFB-AUX/IAA AUXIN RECEPTORS

It is well established that the TIR1/AFB-Aux/IAA pathway is involved in root growth inhibition, and the respective mutants show auxin insensitivity and agravitropism (e.g., Ruegger et al. 1998; Knox et al. 2003). But what about the rapid response? Based on a *tir1-1/afb1-3/afb2-3/afb3-4* quadruple mutant, Scheitz et al. (2013) showed that TIR1/AFBs are required for growth inhibition, while others claim that auxin-induced rapid pH change is unaffected in *tir1* and *tir1-1/afb2-3/afb3-4* mutants (Monshausen et al. 2011). Fendrych et al. (2018) showed that the *tir1-1/afb2-1/afb3-1* triple mutant has a decreased growth response. Using the modified ccvTIR1 receptor and cvxIAA auxin, this group also showed that the rapid growth response can be triggered directly by the ccvTIR1 protein, even though the response was slightly delayed compared to the control situation (Fendrych et al. 2018). Finally, when analyzing the phenotypes of multiple combinations of TIR1/AFB mutations, Prigge et al. (2020) discovered that each mutant combination lacking the AFB1 protein showed a defect in rapid growth inhibition. The so far enigmatic AFB1 coreceptor thus might be crucial for rapid growth response, while the other paralogs also contribute to the response. In conjunction with the impairment of membrane potential changes in the *tir1/afb* mutants (see above), this indicates that the TIR1/AFB pathway, so far studied exclusively in connection with transcriptional regulation, has a fast, nontranscriptional signaling branch (Fig. 2).

Unlike TIR1, which localizes to the nucleus, some AFBs are both cytoplasmic and nuclear, and this is most prominent in the case of AFB1 (Prigge et al. 2020). Interestingly, the AFB1 protein harbors a polymorphism relative to other members of the family (E8K), which, when introduced into TIR1, untethers TIR1 from the SCF complex, increases its levels, and reduces its ability to respond to auxin (Yu et al. 2015). Is AFB1 associated with the SCF complex to trigger root growth inhibition? Does the cytoplasmic localization of AFB1 matter for the rapid response? These exciting questions should be solved in the near future.

Finally, it remains unclear what the "real life" significance of the rapidity of auxin root growth inhibition is. Unlike organs growing in the air, roots embedded in soil cannot grow back, thus making a turn too late means no return. But regardless of the adaptive value of the

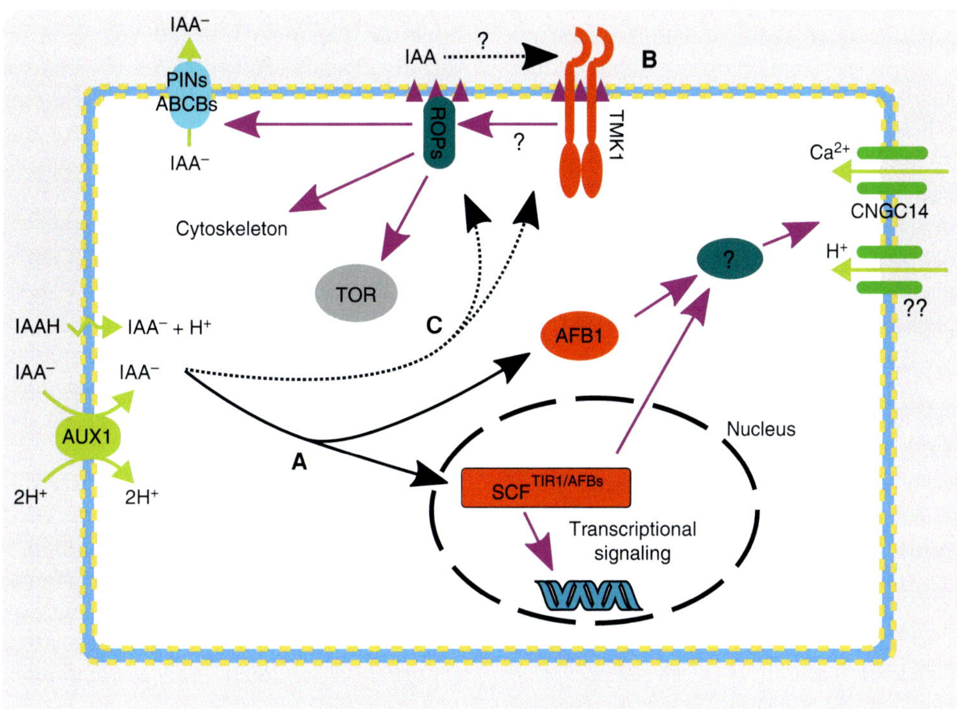

Figure 2. Schematic model of nontranscriptional auxin signaling at the cellular level. (*A*) Auxin entering the cell by influx or diffusion activates the TIR1/AFB receptors in the nucleus or possibly in the cytoplasm. Further, these receptors activate Ca^{2+} and H^+ fluxes at the plasma membrane (PM) via an unknown signaling branch and lead to root growth inhibition. (*B*) Auxin activates ROPs (Rho of plants) by unknown mechanism(s), possibly via the TMK1 receptor. Lipid composition and organization of the membrane play a key role (magenta triangles). Activation by cytoplasmic auxin has not been excluded (*C*). Activated ROPs trigger downstream events including changes in PIN endomembrane trafficking, cytoskeletal reorganization, and activation of the TOR (target of rapamycin) pathway.

response, roots do react quickly to internal auxin during gravitropism and the *cngc14* mutant shows a mild delay in gravitropic bending that corresponds to the delay in reaction to auxin added externally (Shih et al. 2015). On the other hand, the single *afb1* mutant that showed a much stronger defect in response to external auxin than *cngc14* did not have a detectable problem during gravitropism and a gravitropic defect appeared only in the higher-order mutant combinations (Prigge et al. 2020).

AUXIN AND ENDOCYTOSIS

In the last 15 years, the field of rapid auxin responses has been dominated by another nontranscriptional auxin effect: inhibition of endocytosis by auxin. The reason for this interest was that the auxin effect on endocytosis could provide a mechanistic explanation for auxin flux canalization, a symmetry-breaking process that ensures vascular connectivity during auxin-induced vascular tissue differentiation (Ravichandran et al. 2020). Inhibition of endocytosis by auxin postulated extracellular auxin perception and local PIN accumulation due to reduced PIN removal from the PM. As a result, auxin promotes its own efflux because more PINs pump more auxin toward the apoplastic auxin maximum (Paciorek et al. 2005; Robert et al. 2010). Auxin inhibits the intracellular accumulation of PINs, other membrane cargoes, and, importantly, PM-derived vesicles labeled by FM4-64. This means that the entire endocytotic flow is affected (Paciorek et al. 2005) through inhibiting clathrin-mediated endocytosis (Robert et al. 2010). Endocytosis rate was mostly approximated by using Brefeldin A (BFA), which helps visualize cargoes that undergo constitutive recycling (Geldner et al. 2001). BFA bodies accumulate both endocytosed cargoes (e.g., FM4-64) and newly synthesized proteins (Reichardt et al. 2007; Jásik et al. 2016). Consequently, Jásik et al. (2016) used PIN2 tagged with a photoconvertible fluorescent protein to monitor the endocytosis rate of specific protein pools more directly. Whereas auxins decreased PIN2 accumulation in BFA bodies as expected, neither 1-NAA nor IAA reduced PIN2 removal from the PM. Instead, the delivery of PIN2 to the PM was affected. This could be explained by an effect on endomembrane trafficking or PIN2 expression, but not by inhibition of PIN2 endocytosis. On the other hand, in the same PIN2-expressing root cells, 1-NAA triggered the detachment of clathrin light chains from the PM (Robert et al. 2010), indicating that 1-NAA somehow influences the PM pool of clathrin. Clearly, we need detailed and quantitative analysis of auxin effects on endocytosis and endomembrane trafficking to clarify these discrepancies.

The effect of auxin on cargo accumulation in BFA bodies requires high concentrations of the hormone. Importantly, the most potent inhibitor of accumulation of PINs in BFA bodies is 1-NAA, with a near-maximum effect of around 5 µM (Paciorek et al. 2005). The native auxin IAA was effective only in combination with an antioxidant, which should protect the labile IAA molecule. Paponov et al. (2019b) readdressed this discrepancy between IAA and 1-NAA and confirmed that 1-NAA inhibited PIN accumulation in BFA bodies at 10 µM, but IAA did so only at very high concentrations (100 µM). The inactive 2-NAA (with respect to physiological auxin response) triggers a similar cargo accumulation inhibition as 1-NAA, which was also shown independently by Simon et al. (2013). In the original Paciorek et al. (2005) publication, however, 2-NAA did not elicit the inhibitory response. Similarly, in tobacco cells, 1-NAA had a stronger inhibitory effect than IAA (Jelínková et al. 2015). In contrast with these results, there is evidence that genetically increased IAA auxin levels led to reduced BFA accumulation of cargoes (Paciorek et al. 2005). In addition, in a gravistimulated root, the cells on the lower side, which accumulate higher levels of IAA, showed signs of auxin-inhibited PIN internalization (Paciorek et al. 2005). The concentration of IAA in gravistimulated root cells was estimated to be approximately in the range of tens of nanomolars (Band et al. 2012). This means that a long-lasting increase of internal auxin could trigger the changes in PIN internalization. This is in sharp contrast with the transient nature of the effect when 1-NAA is applied externally: after ~2 h of treatment, the accumulation of membrane cargoes resumes and the clathrin light chain PM localization recovers (Robert et al. 2010). Once again, there are significant disagreements between publications and it is not clear to what extent the effect is relevant for the native IAA auxin and thus for the actual physiology and development of the plant.

Is this phenomenon a nontranscriptional auxin response? The effect is clearly rapid as it can be triggered by a short (<5 min) 1-NAA treatment. Pharmacological inhibition of transcription and translation did not interfere with the 1-NAA effect, which also points to the nontranscriptional nature of the response (Paciorek et al. 2005). In addition, there is evidence that the effect is not mediated by the TIR1/AFB1-Aux/IAA coreceptor or the downstream auxin-regulated gene transcription (Robert et al. 2010). However, there is one publication that claims exactly the opposite—that the response depends on TIR1/AFB-mediated auxin signaling and the downstream sterol composition of membranes (Pan et al. 2009).

The concentration ranges and spectra of auxin analogs show that the cargo accumulation inhibition is a molecularly distinct process from the pathway that leads to root growth inhibition. Furthermore, these effects do not behave similarly to the rapid membrane potential changes. The effective spectrum of concentrations and auxin analogs resembles the binding preferences of the ABP1 (Badescu and Napier 2006). Using ABP1 knockdown lines, expression of ABP1 variants and the *abp1-5* point mutation, Robert et al. (2010) showed that the process is controlled by ABP1. Later, Xu et al. (2014) showed that ABP1 activates ROPs (Rho of plants) via the interaction with the transmembrane kinase 1 (TMK1) membrane receptor, which explained the missing link between apoplastic ABP1 and effects on membrane trafficking. However, in 2015, Gao et al. demonstrated that true ABP1 knockouts do not show any major developmental phenotypes. This was a hard hit for the model because the null mutation of ABP1 was supposedly lethal (Chen et al. 2001), and thus demonstrated the importance of ABP1-mediated regulation of PIN endocytosis. Furthermore, Michalko et al. (2016) demonstrated that the phenotype of the ABP1 knockdown lines was not caused by the down-regulation of ABP1 protein. Finally, Paponov et al. (2019a) showed that in the true ABP1 knockout lines, 1-NAA still inhibits PIN accumulation in BFA bodies. In summary, it is not clear which receptor and signaling pathway participates in this process.

AUXIN AND ROPs

The existence of an apoplastic or PM-localized auxin receptor has been proposed for years, and the receptor-like kinase TMK1 is an excellent candidate. TMK1 seems to regulate both a non-TIR/AFB transcriptional pathway (through noncanonical IAAs; Cao et al. 2019) as well as a nontranscriptional pathway, in which the signal is transduced through ROP GTPases (Xu et al. 2014; Pan et al. 2020).

ROPs belong to a plant-specific, small GTPase superfamily. Like other G-proteins, ROPs act as GTPases that can cycle rapidly between active (GTP-bound) and inactive (GDP-bound) forms based on interaction with their regulatory proteins (for review, see Feiguelman et al. 2018). The first clues that G-proteins may be involved in auxin signaling came from bulk biochemical experiments, where biologically active auxins (IAA, 1-NAA, and 2,4-D) caused a rapid increase in active G-binding protein levels on vesicle membranes of rice coleoptiles (Zaina et al. 1990). On the other hand, the inhibitor of G-protein activation, GDP-β-S, blocked swelling of wheat protoplasts induced by selected hormones, including 1-NAA (Bossen et al. 1991). These data point to possible activation of any class of GTPases present in the cell fractions tested, but so far, only ROPs were reported to be auxin responsive. Tobacco ROP NtRac1 became gradually activated within minutes when seedlings were grown on 1-NAA (Tao et al. 2002), and even finer timescale experiments revealed that *Arabidopsis* ROP2 and ROP6 are activated in protoplasts after only 30 sec of 1-NAA treatment; this is, beyond doubt, too fast to be transcriptionally regulated (Xu et al. 2010). Activation of these ROPs by the natural auxin IAA was also demonstrated (Xu et al. 2014). In all three studies, the total amount of ROP proteins remained unaffected, clearly indicating a nontranscriptional modulation of their activity.

Auxin perception and signal transduction upstream of ROPs are still enigmatic. Originally, physical interaction of TMK1 with ABP1 upon auxin binding was shown to be responsible for ROP activation (Xu et al. 2014). However, recently, the same group postulated that auxin triggers sterol-dependent nanoclustering of TMK1 that is upstream of ROP6 nanoclustering and activation (Fig. 2; Pan et al. 2020). S-acylation of conserved residues within a G-domain of ROP6 is known to induce its transient activation and association with membrane lipid rafts, while the GTPase activity remains unaffected (Sorek et al. 2010), and this lipid modification is crucial for the ROP6 activation in the auxin-TMK1 pathway (Pan et al. 2020). However, regulation of ROP6 behavior at the PM seems to be more complex. In roots, another phospholipid, phosphatidylserine is required for rapid stabilization of the GTPase into immobile nanoclusters in response to auxin (quantified after 5 min 1-NAA treatment). Abolishing the lipid interaction (but not the GTPase activity) has a direct impact on ROP6 localization and auxin response (Platre et al. 2019). Furthermore, ROP6 binds to phosphatidylglycerol, which activates the GTPase in a dose-dependent manner (Han et al. 2018). Clearly, TMK1 membrane organization and correct lipid composition play a key role in the activation of ROPs. Nevertheless, we do not understand how and where auxin is perceived. Although the direct interaction of ROPs with membrane phospholipids modulate their localization and activity, the spatiotemporal regulation is the function of specific activating (RopGEFs) and inhibiting proteins (RopGAPs, RopGDIs), and we have to bear in mind that these regulatory proteins are also controlled by various interactors including fast-responding kinases (for review, see, e.g., Fehér and Lajkó 2015).

Besides TMK1, mitogen-activated protein kinases (MAPKs) rapidly and transiently respond to auxin. Although relatively high concentrations of auxin were used (20 µM), only IAA, 1-NAA, and 2,4-D were able to trigger the MAPK activation in *Arabidopsis* roots, but not the inactive auxin analog PCIB (clofibric acid) (Mockaitis and Howell 2000). The auxin-induced MAPK phosphorylation cascade involving MKK3, MPK1, and RBK1 (ROP BINDING PROTEIN KINASE 1) was then shown to regulate cotyledon expansion and root elongation in *Arabidopsis* seedlings through the activity of ROP4 and ROP6 (Enders et al. 2017).

What is the cellular and physiological role of auxin activation of ROPs? One of the effects is unexpectedly connected to the reinitiation of mRNA translation via the TOR (target of rapamycin) pathway (Fig. 2). After activation by auxin (1-NAA 100 nM), ROP2 directly interacts with TOR and promotes its activation, which becomes profound 60–120 min after the treatment. TOR is a conserved protein kinase that acts as a general growth "sensor," linked to protein synthesis. In *Arabidopsis*, auxin enhances its activity by phosphorylation through active ROP2. When activated, TOR promotes translational reinitiation of certain (µORF-containing) mRNAs. These mRNAs can be stored in a cell and, after a certain signal, they assemble into polysomes and are translated. Because of the rather slow activation process, TOR involvement in the rapid auxin response remains questionable (Schepetilnikov et al. 2013, 2017).

In planta, when ROP activity is perturbed, auxin-related phenotypes often emerge. In particular, this research has focused on how ROPs influence PM localization and trafficking of PINs that, in turn, influence auxin fluxes and levels. Polar recruitment of PINs in *Arabidopsis* seedlings was shown to be regulated by ICR1 (interactor of constitutive active ROP1), and auxin treatment (1-NAA) induced ICR1 expression within 30 min (Hazak et al. 2010). Similarly, PIN2 recycling in roots depends on ROP6 activation by its GEF-protein SPIKE1 (Lin et al. 2012). Whereas the gravitropic root bending of the *rop6* mutant is slower (Lin et al. 2012; Han et al. 2018), both overexpression and constitutively active ROP6 bend their roots faster and more than wild-type (WT) plants (Han et al. 2018; Platre et al. 2019). The auxin-ROP-PIN hypothesis has been studied and established in detail on a model system of cotyledon pavement cells, where lobe formation was postulated to be auxin regulated (for review, see Chen et al. 2015). In *Arabidopsis*, mutants with affected auxin biosynthesis (*yuc 1/2/4/6*) or ROP signaling (rop2RNAi/*rop4-1*) share similar features, such as reduced lobe number and length. Addition of exogenous auxin (1-NAA) increased the number of lobes in both WT and the auxin biosynthetic mutant, but not in the ROP signaling mutant. Furthermore, a similar phenotype of pavement cells was shared by *abp1-5*, *pin1*, and *tmk1* mutants (*tmk1234*), and none of them could be rescued by application of external auxin (1-NAA). Based on additional experiments, the authors postulated a model of a positive feedback loop where auxin gets recognized by the ABP1-TMK1 complex leading to ROP activation, cytoskeleton remodeling via its effectors and, finally, to PIN1 endocytosis, relocalization, and auxin redistribution to lobe-initiation sites (Xu et al. 2010, 2014; Nagawa et al. 2012). However, independent characterizations of the lobe phenotypes of mutants brought more doubt on the model; pavement cells of *abp1* knockouts (represented by two null alleles, *abp1-TD1* and *abp1-c1*) do not differ from WT, and they form more lobes in response to 1-NAA (Gao et al. 2015). Similarly, pavement cells of *ric4* (ROP2 effector) or *pin1-1* mutants were indistinguishable from WT. In addition, PIN1-GFP protein driven by its native promoter was not expressed in developing pavement cells. Nevertheless, other PINs (PIN3, PIN4, and PIN7) did localize to the PM of expanding pavement cells, but their mutations (either single or multiple) still resulted in reduced cell area but not lobe number, indicating their role in cell expansion but not lobe initiation (Belteton et al. 2018). Clearly, the model requires more experimental attention and deserves revision.

ROP signaling has a direct link to actin remodeling (e.g., through SCAR/WAVE, activators of ARP2/3 complex with actin-nucleation activity) (Feiguelman et al. 2018). One important question is whether microfilaments remodeling after ROP activation occurs rapidly enough to act as a messenger. Alternatively, is microfilament remodeling simply a component of the growth response? Arieti and Staiger (2020) followed actin dynamics in vivo in *Arabidopsis* roots after short-term auxin treatment. At the whole-cell level, after 20–30 min of IAA treatment, the actin network was denser, less bundled, and, interestingly, more organized. Shorter treatment of just 5 min was, however, not sufficient to trigger all these changes. An analysis of individual filaments that can detect finer modifications revealed that a 7-min treatment with 10 nM IAA had a significant impact on actin filament behavior, especially on unbundling throughout both the elongation and transition zone of a root. None of these effects were detectable in the *aux1* mutant, indicating that auxin needs to enter the cell (i.e., it cannot trigger actin remodeling from an apoplast located receptor/signal). Together with the timing of actin remodeling, it is plausible that cytoskeleton reacts to the rapid growth inhibition and associated physiological changes.

Actin, the downstream target of ROPs that orchestrates both cell morphogenesis and expansion, also dynamically responds to auxin treatment. Still, too many conclusions, especially in pavement cells studies, are based on transient biolistic experiments that may provide misleading data. We need a detailed characterization of microfilament dynamics within the first minutes following auxin treatment (preferentially IAA) in various tissues/organs to dissect the actin role in the rapid auxin responses. Despite the contradictions in the data and gaps in the models, auxin seems to activate certain ROP GTPases in a transcription-independent way, and ROPs interaction with membrane phospholipids is crucial for their activation and signal transduction (Fig. 2). Interestingly, based on one study (Pan et al. 2009), membrane phospholipid composition depends on auxin signaling via the TIR1/AFB pathway. A putative PM located receptor/receptor-like kinase has been considered so far as an upstream regulator of ROPs, but could the TIR1/AFB nontranscriptional signaling also play a role? And besides ROPs, are there any other GTPases involved?

CONCLUDING REMARKS

Auxin triggers several classes of undoubtedly rapid, nontranscriptional responses. We attempted to discuss the most prominent ones in this review. We think it is critical to clarify which of these are physiologically relevant. Do these responses occur during normal plant growth and development? Although explicitly including the concentration of hormone used in experiments might seem too meticulous, we did so to stress the differences in effectiveness of various auxinic compounds for different auxin rapid responses. Different auxin analogs should not be freely interchanged in interpretation of data and development of models. We believe it is time to shift experimental attention to the effects of lower auxin concentrations that trigger detectable physiological responses.

What do the rapid responses have in common? What are the receptors through which auxin triggers the responses? It is possible that some of the responses do not require a receptor, being triggered directly by the hormone. The weak acid effects discussed in the membrane potential section are one example. Other responses might be activated indirectly by other auxin-triggered processes, for example by the substantial physiological changes that growth inhibition causes in roots. But for others, it is important to unravel the molecular mechanisms that connect auxin perception and the execution of the response. Are there still unknown auxin perception sites in the cell walls, the PM, or the cytoplasm? How is the famous TIR1/AFB-Aux/IAA coreceptor connected to the growth and regulating processes? What activates the ROPs? These and other questions guarantee that the auxin research will not be boring in the future.

ACKNOWLEDGMENTS

The authors thank Maria Akhmanova, Jan Petrášek, and Matouš Glanc for helpful discussions and Jessica L. Barilone for editing the text. The team received support from the European Research Council (Grant No. 803048) and Charles University Primus (Grant No. PRIMUS/19/SCI/09). P.V. was supported by MMVPUK (Grant No. CZ.02.2.69/0.0/0.0/16_027/0008495).

REFERENCES

Adamowski M, Friml J. 2015. PIN-dependent auxin transport: action, regulation, and evolution. *Plant Cell* **27:** 20–32. doi:10.1105/tpc.114.134874

Arieti RS, Staiger CJ. 2020. Auxin-induced actin cytoskeleton rearrangements require AUX1. *New Phytol* **226:** 441–459. doi:10.1111/nph.16382

Badescu GO, Napier RM. 2006. Receptors for auxin: will it all end in TIRs? *Trends Plant Sci* **11:** 217–223. doi:10.1016/j.tplants.2006.03.001

Band LR, Wells DM, Larrieu A, Sun J, Middleton AM, French AP, Brunoud G, Sato EM, Wilson MH, Péret B, et al. 2012. Root gravitropism is regulated by a transient lateral auxin gradient controlled by a tipping-point mechanism. *Proc Natl Acad Sci* **109:** 4668–4673. doi:10.1073/pnas.1201498109

Barbez E, Dünser K, Gaidora A, Lendl T, Busch W. 2017. Auxin steers root cell expansion via apoplastic pH regulation in *Arabidopsis thaliana*. *Proc Natl Acad Sci* **114:** E4884–E4893. doi:10.1073/pnas.1613499114

Bates GW, Goldsmith MH. 1983. Rapid response of the plasma-membrane potential in oat coleoptiles to auxin and other weak acids. *Planta* **159:** 231–237. doi:10.1007/BF00397530

Behera S, Xu Z, Luoni L, Bonza MC, Doccula FG, De Michelis MI, Morris RJ, Schwarzländer M, Costa A. 2018. Cellular Ca^{2+} signals generate defined pH signatures in plants. *Plant Cell* **30:** 2704–2719. doi:10.1105/tpc.18.00655

Belteton SA, Sawchuk MG, Donohoe BS, Scarpella E, Szymanski DB. 2018. Reassessing the roles of PIN proteins and anticlinal microtubules during pavement cell morphogenesis. *Plant Physiol* **176:** 432–449. doi:10.1104/pp.17.01554

Bennett MJ, Marchant A, Green HG, May ST, Ward SP, Millner PA, Walker AR, Schulz B, Feldmann KA. 1996. *Arabidopsis* AUX1 gene: a permease-like regulator of root gravitropism. *Science* **273:** 948–950. doi:10.1126/science.273.5277.948

Bhosale R, Giri J, Pandey BK, Giehl RFH, Hartmann A, Traini R, Truskina J, Leftley N, Hanlon M, Swarup K, et al. 2018. A mechanistic framework for auxin dependent *Arabidopsis* root hair elongation to low external phosphate. *Nat Commun* **9:** 1409. doi:10.1038/s41467-018-03851-3

Bossen ME, Tretyn A, Kendrick RE, Vredenberg WJ. 1991. Comparison between swelling of etiolated wheat (*Triticum aestivum* L.) protoplasts induced by phytochrome and α-naphthaleneacetic acid, benzylaminopurine, gibberellic acid abscisic acid and acetylcholine. *J Plant Physiol* **137:** 706–710. doi:10.1016/S0176-1617(11)81226-5

Calderón Villalobos LIA, Lee S, De Oliveira C, Ivetac A, Brandt W, Armitage L, Sheard LB, Tan X, Parry G, Mao H, et al. 2012. A combinatorial TIR1/AFB-Aux/IAA co-receptor system for differential sensing of auxin. *Nat Chem Biol* **8:** 477–485. doi:10.1038/nchembio.926

Cao M, Chen R, Li P, Yu Y, Zheng R, Ge D, Zheng W, Wang X, Gu Y, Gelová Z, et al. 2019. TMK1-mediated auxin signalling regulates differential growth of the apical hook. *Nature* **568:** 240–243. doi:10.1038/s41586-019-1069-7

Carrier DJ, Bakar NTA, Swarup R, Callaghan R, Napier RM, Bennett MJ, Kerr ID. 2008. The binding of auxin to the *Arabidopsis* auxin influx transporter AUX1. *Plant Physiol* **148:** 529–535. doi:10.1104/pp.108.122044

Chen JG, Ullah H, Young JC, Sussman MR, Jones AM. 2001. ABP1 is required for organized cell elongation and division in *Arabidopsis* embryogenesis. *Genes Dev* **15:** 902–911. doi:10.1101/gad.866201

Chen J, Wang F, Zheng S, Xu T, Yang Z. 2015. Pavement cells: a model system for non-transcriptional auxin signalling and crosstalks. *J Exp Bot* **66:** 4957–4970. doi:10.1093/jxb/erv266

Delbarre A, Muller P, Imhoff V, Guern J. 1996. Comparison of mechanisms controlling uptake and accumulation of 2,4-dichlorophenoxy acetic acid, naphthalene-1-acetic acid, and indole-3-acetic acid in suspension-cultured tobacco cells. *Planta* **198:** 532–541. doi:10.1007/BF00262639

Dezfulian MH, Jalili E, Roberto DKA, Moss BL, Khoo K, Nemhauser JL, Crosby WL. 2016. Oligomerization of SCFTIR1 is essential for Aux/IAA degradation and auxin signaling in *Arabidopsis. PLoS Genet* **12:** e1006301. doi:10.1371/journal.pgen.1006301

Dharmasiri N, Dharmasiri S, Estelle M. 2005. The F-box protein TIR1 is an auxin receptor. *Nature* **435:** 441–445. doi:10.1038/nature03543

Dindas J, Scherzer S, Roelfsema MRG, Von Meyer K, Müller HM, Al-Rasheid KAS, Palme K, Dietrich P, Becker D, Bennett MJ, et al. 2018. AUX1-mediated root hair auxin influx governs SCFTIR1/AFB-type Ca^{2+} signaling. *Nat Commun* **9:** 1174. doi:10.1038/s41467-018-03582-5

Dindas J, Becker D, Roelfsema MRG, Scherzer S, Bennett M, Hedrich R. 2020. Pitfalls in auxin pharmacology. *New Phytol* **227:** 286–292. doi:10.1111/nph.16491

Du M, Spalding EP, Gray WM. 2020. Rapid auxin-mediated cell expansion. *Annu Rev Plant Biol* **71:** 379–402. doi:10.1146/annurev-arplant-073019-025907

Enders TA, Frick EM, Strader LC. 2017. An *Arabidopsis* kinase cascade influences auxin-responsive cell expansion. *Plant J* **92:** 68–81. doi:10.1111/tpj.13635

Evans ML, Mulkey TJ, Vesper MJ. 1980. Auxin action on proton influx in corn roots and its correlation with growth. *Planta* **148:** 510–512. doi:10.1007/BF02395322

Evans ML, Ishikawa H, Estelle MA. 1994. Responses of *Arabidopsis* roots to auxin studied with high temporal resolution: comparison of wild type and auxin-response mutants. *Planta* **194:** 215–222. doi:10.1007/BF01101680

Fehér A, Lajkó DB. 2015. Signals fly when kinases meet Rho-of-plants (ROP) small G-proteins. *Plant Sci* **237:** 93–107. doi:10.1016/j.plantsci.2015.05.007

Feiguelman G, Fu Y, Yalovsky S. 2018. ROP GTPases structure–function and signaling pathways. *Plant Physiol* **176:** 57–79. doi:10.1104/pp.17.01415

Felle H, Peters W, Palme K. 1991. The electrical response of maize to auxins. *Biochim Biophys Acta* **1064:** 199–204. doi:10.1016/0005-2736(91)90302-O

Fendrych M, Leung J, Friml J. 2016. Tir1/AFB-Aux/IAA auxin perception mediates rapid cell wall acidification and growth of *Arabidopsis* hypocotyls. *eLife* **5:** e19048. doi:10.7554/eLife.19048

Fendrych M, Akhmanova M, Merrin J, Glanc M, Hagihara S, Takahashi K, Uchida N, Torii KU, Friml J. 2018. Rapid and reversible root growth inhibition by TIR1 auxin signalling. *Nat Plants* **4:** 453–459. doi:10.1038/s41477-018-0190-1

Gälweiler L, Guan C, Müller A, Wisman E, Mendgen K, Yephremov A, Palme K. 1998. Regulation of polar auxin transport by AtPIN1 in *Arabidopsis* vascular tissue. *Science* **282:** 2226–2230. doi:10.1126/science.282.5397.2226

Gao Y, Zhang Y, Zhang D, Dai X, Estelle M, Zhao Y. 2015. Auxin binding protein 1 (ABP1) is not required for either auxin signaling or *Arabidopsis* development. *Proc Natl Acad Sci* **112:** 2275–2280. doi:10.1073/pnas.1500365112

Geldner N, Friml J, Stierhof YD, Jürgens G, Palme K. 2001. Auxin transport inhibitors block PIN1 cycling and vesicle trafficking. *Nature* **413:** 425–428. doi:10.1038/35096571

Gjetting KSK, Ytting CK, Schulz A, Fuglsang AT. 2012. Live imaging of intra-and extracellular pH in plants using pHusion, a novel genetically encoded biosensor. *J Exp Bot* **63:** 3207–3218. doi:10.1093/jxb/ers040

Goering H. 1979. Depolarization of transmembrane potential of corn and wheat coleoptiles under reduced water potential and after IAA application. *Plant Cell Physiol* **20:** 649–656.

Gray WM, Kepinski S, Rouse D, Leyser O, Estelle M. 2001. Auxin regulates SCFTIR1-dependent degradation of AUX/IAA proteins. *Nature* **414:** 271–276. doi:10.1038/35104500

Han M, Park Y, Kim I, Kim EH, Yu TK, Rhee S, Suh JY. 2014. Structural basis for the auxin-induced transcriptional regulation by Aux/IAA17. *Proc Natl Acad Sci* **111:** 18613–18618. doi:10.1073/pnas.1419525112

Han X, Shi Y, Liu G, Guo Y, Yang Y. 2018. Activation of ROP6 GTPase by phosphatidylglycerol in *Arabidopsis. Front Plant Sci* **9:** 347. doi:10.3389/fpls.2018.00347

Hayashi KI, Neve J, Hirose M, Kuboki A, Shimada Y, Kepinski S, Nozaki H. 2012. Rational design of an auxin antagonist of the SCF^{TIR1} auxin receptor complex. *ACS Chem Biol* **7:** 590–598. doi:10.1021/cb200404c

Hazak O, Bloch D, Poraty L, Sternberg H, Zhang J, Friml J, Yalovsky S. 2010. A rho scaffold integrates the secretory system with feedback mechanisms in regulation of auxin distribution. *PLoS Biol* **8:** e1000282. doi:10.1371/journal.pbio.1000282

Hertel R, Lomax TL, Briggs WR. 1983. Auxin transport in membrane vesicles from *Cucurbita pepo* L. *Planta* **157:** 193–201. doi:10.1007/BF00405182

Holland AJ, Fachinetti D, Han JS, Cleveland DW. 2012. Inducible, reversible system for the rapid and complete degradation of proteins in mammalian cells. *Proc Natl Acad Sci* **109:** E3350–E3357. doi:10.1073/pnas.1216880109

Jásik J, Bokor B, Stuchlík S, Mičieta K, Turňa J, Schmelzer E. 2016. Effects of auxins on PIN-FORMED2 (PIN2) dynamics are not mediated by inhibiting PIN2 endocytosis. *Plant Physiol* **172:** 1019–1031.

Jelínková A, Müller K, Fílová-Pařezová M, Petrášek J. 2015. NtGNL1a ARF-GEF acts in endocytosis in tobacco cells. *BMC Plant Biol* **15:** 272. doi:10.1186/s12870-015-0621-3

Kamimoto Y, Terasaka K, Hamamoto M, Takanashi K, Fukuda S, Shitan N, Sugiyama A, Suzuki H, Shibata D, Wang

B, et al. 2012. *Arabidopsis* ABCB21 is a facultative auxin importer/exporter regulated by cytoplasmic auxin concentration. *Plant Cell Physiol* **53:** 2090–2100. doi:10.1093/pcp/pcs149

Kepinski S, Leyser O. 2005. The *Arabidopsis* F-box protein TIR1 is an auxin receptor. *Nature* **435:** 446–451. doi:10.1038/nature03542

Knox K, Grierson CS, Leyser O. 2003. AXR3 and SHY2 interact to regulate root hair development. *Development* **130:** 5769–5777. doi:10.1242/dev.00659

Kubeš M, Yang H, Richter GL, Cheng Y, Młodzińska E, Wang X, Blakeslee JJ, Carraro N, Petrášek J, Zažímalová E, et al. 2012. The *Arabidopsis* concentration-dependent influx/efflux transporter ABCB4 regulates cellular auxin levels in the root epidermis. *Plant J* **69:** 640–654. doi:10.1111/j.1365-313X.2011.04818.x

Lin D, Nagawa S, Chen J, Cao L, Chen X, Xu T, Li H, Dhonukshe P, Yamamuro C, Friml J, et al. 2012. A ROP GTPase-dependent auxin signaling pathway regulates the subcellular distribution of PIN2 in *Arabidopsis* roots. *Curr Biol* **22:** 1319–1325. doi:10.1016/j.cub.2012.05.019

Lomax TL, Mehlhorn RJ, Briggs WR. 1985. Active auxin uptake by zucchini membrane vesicles: quantitation using ESR volume and delta pH determinations. *Proc Natl Acad Sci* **82:** 6541–6545. doi:10.1073/pnas.82.19.6541

Loper MT, Spanswick RM. 1991. Auxin transport in suspension-cultured soybean root cells. II: Anion effects on carrier-mediated uptake. *Plant Physiol* **96:** 192–197. doi:10.1104/pp.96.1.192

Lüthen H, Böttger M. 1988. Kinetics of proton secretion and growth in maize roots: action of various plant growth effectors. *Plant Sci* **54:** 37–43. doi:10.1016/0168-9452(88)90053-2

Mangano S, Denita-Juarez SP, Choi HS, Marzol E, Hwang Y, Ranocha P, Velasquez SM, Borassi C, Barberini ML, Aptekmann AA, et al. 2017. Molecular link between auxin and ROS-mediated polar growth. *Proc Natl Acad Sci* **114:** 5289–5294. doi:10.1073/pnas.1701536114

Marten I, Zeilinger C, Redhead C, Landry DW, Al-Awqati Q, Hedrich R. 1992. Identification and modulation of a voltage-dependent anion channel in the plasma membrane of guard cells by high-affinity ligands. *EMBO J* **11:** 3569–3575. doi:10.1002/j.1460-2075.1992.tb05440.x

Michalko J, Glanc M, Perrot-Rechenmann C, Friml J. 2016. Strong morphological defects in conditional *Arabidopsis* abp1 knock-down mutants generated in absence of functional ABP1 protein. *F1000Res* **5:** 86. doi:10.12688/f1000research.7654.1

Milo R, Jorgensen P, Moran U, Weber G, Springer M. 2010. BioNumbers—the database of key numbers in molecular and cell biology. *Nucleic Acids Res* **38:** D750–D753. doi:10.1093/nar/gkp889

Minamisawa K, Fukai K. 1991. Production of indole-3-acetic acid by *Bradyrhizobium japonicum*: a correlation with genotype grouping and rhizobitoxine production. *Plant Cell Physiol* **32:** 1–9.

Mockaitis K, Howell SH. 2000. Auxin induces mitogenic activated protein kinase (MAPK) activation in roots of *Arabidopsis* seedlings. *Plant J* **24:** 785–796. doi:10.1046/j.1365-313x.2000.00921.x

Monshausen GB, Miller ND, Murphy AS, Gilroy S. 2011. Dynamics of auxin-dependent Ca^{2+} and pH signaling in root growth revealed by integrating high-resolution imaging with automated computer vision-based analysis. *Plant J* **65:** 309–318. doi:10.1111/j.1365-313X.2010.04423.x

Moss BL, Mao H, Guseman JM, Hinds TR, Hellmuth A, Kovenock M, Noorassa A, Lanctot A, Calderón Villalobos LIA, Zheng N, et al. 2015. Rate motifs tune auxin/indole-3-acetic acid degradation dynamics. *Plant Physiol* **169:** 803–813. doi:10.1104/pp.15.00587

Nagawa S, Xu T, Lin D, Dhonukshe P, Zhang X, Friml J, Scheres B, Fu Y, Yang Z. 2012. ROP GTPase-dependent actin microfilaments promote PIN1 polarization by localized inhibition of clathrin-dependent endocytosis. *PLoS Biol* **10:** e1001299. doi:10.1371/journal.pbio.1001299

Niemeyer M, Moreno Castillo E, Ihling CH, Iacobucci C, Wilde V, Hellmuth A, Hoehenwarter W, Samodelov SL, Zurbriggen MD, Kastritis PL, et al. 2020. Flexibility of intrinsically disordered degrons in AUX/IAA proteins reinforces auxin co-receptor assemblies. *Nat Commun* **11:** 2277. doi:10.1038/s41467-020-16147-2

Nishimura K, Fukagawa T, Takisawa H, Kakimoto T, Kanemaki M. 2009. An auxin-based degron system for the rapid depletion of proteins in nonplant cells. *Nat Methods* **6:** 917–922. doi:10.1038/nmeth.1401

Nishimura K, Yamada R, Hagihara S, Iwasaki R, Uchida N, Kamura T, Takahashi K, Torii KU, Fukagawa T. 2020. A super-sensitive auxin-inducible degron system with an engineered auxin-TIR1 pair. *Nucleic Acids Res* **48:** e108. doi:10.1093/nar/gkaa748

Paciorek T, Zazímalová E, Ruthardt N, Petrášek J, Stierhof Y-D, Kleine-Vehn J, Morris DA, Emans N, Jürgens G, Geldner N, et al. 2005. Auxin inhibits endocytosis and promotes its own efflux from cells. *Nature* **435:** 1251–1256. doi:10.1038/nature03633

Pan J, Fujioka S, Peng J, Chen J, Li G, Chen R. 2009. The E3 ubiquitin ligase $SCF^{TIR1/AFB}$ and membrane sterols play key roles in auxin regulation of endocytosis, recycling, and plasma membrane accumulation of the auxin efflux transporter PIN2 in *Arabidopsis thaliana*. *Plant Cell* **21:** 568–580. doi:10.1105/tpc.108.061465

Pan X, Fang L, Liu J, Senay-Aras B, Lin W, Zheng S, Zhang T, Guo J, Manor U, Van Norman J, et al. 2020. Auxin-induced signaling protein nanoclustering contributes to cell polarity formation. *Nat Commun* **11:** 3914. doi:10.1038/s41467-020-17602-w

Paponov IA, Dindas J, Król E, Friz T, Budnyk V, Teale W, Paponov M, Hedrich R, Palme K. 2019a. Auxin-induced plasma membrane depolarization is regulated by auxin transport and not by AUXIN BINDING PROTEIN1. *Front Plant Sci* **9:** 1953. doi:10.3389/fpls.2018.01953

Paponov IA, Friz T, Budnyk V, Teale W, Wüst F, Paponov M, Al-Babili S, Palme K. 2019b. Natural auxin does not inhibit Brefeldin A induced PIN1 and PIN2 internalization in root cells. *Front Plant Sci* **10:** 574. doi:10.3389/fpls.2019.00574

Petrášek J, Friml J. 2009. Auxin transport routes in plant development. *Development* **136:** 2675–2688. doi:10.1242/dev.030353

Petrášek J, Mravec J, Bouchard R, Blakeslee JJ, Abas M, Seifertová D, Wisniewska J, Tadele Z, Kubes M, Covanová M, et al. 2006. PIN proteins perform a rate-limiting function in cellular auxin efflux. *Science* **312:** 914–918. doi:10.1126/science.1123542

Platre MP, Bayle V, Armengot L, Bareille J, del Mar Marquès-Bueno M, Creff A, Maneta-Peyret L, Fiche JB, Nollmann M, Miège C, et al. 2019. Developmental control of plant Rho GTPase nano-organization by the lipid phosphatidylserine. *Science* **364:** 57–62. doi:10.1126/science.aav9959

Prigge MJ, Platre M, Kadakia N, Zhang Y, Greenham K, Szutu W, Pandey BK, Bhosale RA, Bennett MJ, Busch W, et al. 2020. Genetic analysis of the *Arabidopsis* TIR1/AFB auxin receptors reveals both overlapping and specialized functions. *eLife* **9:** 1–28. doi:10.7554/eLife.54740

Ramans Harborough S, Kalverda A, Thompson G, Kieffer M, Kubes M, Quareshy M, Uzunova V, Prusinska J, Hayashi K, Napier R, et al. 2019. A fuzzy encounter complex precedes formation of the fully-engaged TIR1-Aux/IAA auxin co-receptor system. bioRxiv doi:10.1101/781922

Ramírez-Sánchez O, Pérez-Rodríguez P, Delaye L, Tiessen A. 2016. Plant proteins are smaller because they are encoded by fewer exons than animal proteins. *Genome Proteom Bioinf* **14:** 357–370. doi:10.1016/j.gpb.2016.06.003

Raven JA. 1975. Transport of indoleacetic acid in plant cells in relation to pH and electrical potential gradients, and its significance for polar IAA transport. *New Phytol* **74:** 163–172. doi:10.1111/j.1469-8137.1975.tb02602.x

Ravichandran SJ, Linh NM, Scarpella E. 2020. The canalization hypothesis—challenges and alternatives. *New Phytol* **227:** 1051–1059. doi:10.1111/nph.16605

Reichardt I, Stierhof Y-D, Mayer U, Richter S, Schwarz H, Schumacher K, Jürgens G. 2007. Plant cytokinesis requires de novo secretory trafficking but not endocytosis. *Curr Biol* **17:** 2047–2053. doi:10.1016/j.cub.2007.10.040

Retzer K, Lacek J, Skokan R, Del Genio CI, Vosolsobě S, Laňková M, Malínská K, Konstantinova N, Zažímalová E, Napier RM, et al. 2017. Evolutionary conserved cysteines function as *cis*-acting regulators of *Arabidopsis* PIN-FORMED 2 distribution. *Int J Mol Sci* **18:** 2274. doi:10.3390/ijms18112274

Robert S, Kleine-Vehn J, Barbez E, Sauer M, Paciorek T, Baster P, Vanneste S, Zhang J, Simon S, Čovanová M, et al. 2010. ABP1 mediates auxin inhibition of clathrin-dependent endocytosis in *Arabidopsis*. *Cell* **143:** 111–121. doi:10.1016/j.cell.2010.09.027

Rubery PH, Sheldrake AR. 1974. Carrier-mediated auxin transport. *Planta* **118:** 101–121. doi:10.1007/BF00388387

Ruck A, Palme K, Venis MA, Napier RM, Felle HH. 1993. Patch-clamp analysis establishes a role for an auxin binding protein in the auxin stimulation of plasma membrane current in *Zea mays* protoplasts. *Plant J* **4:** 41–46. doi:10.1046/j.1365-313X.1993.04010041.x

Ruegger M, Dewey E, Gray WM, Hobbie L, Turner J, Estelle M. 1998. The TIR1 protein of *Arabidopsis* functions in auxin response and is related to human SKP2 and yeast Grr1p. *Genes Dev* **12:** 198–207. doi:10.1101/gad.12.2.198

Scheitz K, Lüthen H, Schenck D. 2013. Rapid auxin-induced root growth inhibition requires the TIR and AFB auxin receptors. *Planta* **238:** 1171–1176. doi:10.1007/s00425-013-1941-x

Schepetilnikov M, Dimitrova M, Mancera-Martínez E, Geldreich A, Keller M, Ryabova LA. 2013. TOR and S6K1 promote translation reinitiation of uORF-containing mRNAs via phosphorylation of eIF3h. *EMBO J* **32:** 1087–1102. doi:10.1038/emboj.2013.61

Schepetilnikov M, Makarian J, Srour O, Geldreich A, Yang Z, Chicher J, Hammann P, Ryabova LA. 2017. GTPase ROP2 binds and promotes activation of target of rapamycin, TOR, in response to auxin. *EMBO J* **36:** 886–903. doi:10.15252/embj.201694816

Senn AP, Goldsmith MH. 1988. Regulation of electrogenic proton pumping by auxin and fusicoccin as related to the growth of *Avena* coleoptiles. *Plant Physiol* **88:** 131–138. doi:10.1104/pp.88.1.131

Shih HW, Depew CL, Miller ND, Monshausen GB. 2015. The cyclic nucleotide-gated channel CNGC14 regulates root gravitropism in *Arabidopsis thaliana*. *Curr Biol* **25:** 3119–3125. doi:10.1016/j.cub.2015.10.025

Simon S, Kubeš M, Baster P, Robert S, Dobrev PI, Friml J, Petrášek J, Zažímalová E. 2013. Defining the selectivity of processes along the auxin response chain: a study using auxin analogues. *New Phytol* **200:** 1034–1048. doi:10.1111/nph.12437

Sorek N, Segev O, Gutman O, Bar E, Richter S, Poraty L, Hirsch JA, Henis YI, Lewinsohn E, Jürgens G, et al. 2010. An S-acylation switch of conserved G domain cysteines is required for polarity signaling by ROP GTPases. *Curr Biol* **20:** 914–920. doi:10.1016/j.cub.2010.03.057

Swarup R, Péret B. 2012. AUX/LAX family of auxin influx carriers—an overview. *Front Plant Sci* **3:** 225. doi:10.3389/fpls.2012.00225

Sze H. 1985. H^+-translocating ATPases: advances using membrane vesicles. *Annu Rev Plant Physiol* **36:** 175–208. doi:10.1146/annurev.pp.36.060185.001135

Sze H, Li X, Palmgren MG. 1999. Energization of plant cell membranes by H^+-pumping ATPases: regulation and biosynthesis. *Plant Cell* **11:** 677–689.

Takahashi K, Hayashi K, Kinoshita T. 2012. Auxin activates the plasma membrane H^+-ATPase by phosphorylation during hypocotyl elongation in *Arabidopsis*. *Plant Physiol* **159:** 632–641. doi:10.1104/pp.112.196428

Tan X, Calderon-Villalobos LIA, Sharon M, Zheng C, Robinson CV, Estelle M, Zheng N. 2007. Mechanism of auxin perception by the TIR1 ubiquitin ligase. *Nature* **446:** 640–645. doi:10.1038/nature05731

Tao L, Cheung AY, Wu H. 2002. Plant Rac-like GTPases are activated by auxin and mediate auxin-responsive gene expression. *Plant Cell* **14:** 2745–2760. doi:10.1105/tpc.006320

Tretyn A, Wagner G, Felle HH. 1991. Signal transduction in *Sinapis alba* root hairs: auxins as external messengers. *J Plant Physiol* **139:** 187–193. doi:10.1016/S0176-1617(11)80606-1

Uchida N, Takahashi K, Iwasaki R, Yamada R, Yoshimura M, Endo TA, Kimura S, Zhang H, Nomoto M, Tada Y, et al. 2018. Chemical hijacking of auxin signaling with an engineered auxin-TIR1 pair. *Nat Chem Biol* **14:** 299–305. doi:10.1038/nchembio.2555

Weijers D, Benkova E, Jäger KE, Schlereth A, Hamann T, Kientz M, Wilmoth JC, Reed JW, Jürgens G. 2005. Developmental specificity of auxin response by pairs of ARF and Aux/IAA transcriptional regulators. *EMBO J* **24:** 1874–1885. doi:10.1038/sj.emboj.7600659

Weisenseel MH, Meyer AJ. 1997. Bioelectricity, gravity and plants. *Planta* **203:** S98–S106. doi:10.1007/PL00008122

Winkler M, Niemeyer M, Hellmuth A, Janitza P, Christ G, Samodelov SL, Wilde V, Majovsky P, Trujillo M, Zurbriggen MD, et al. 2017. Variation in auxin sensing guides AUX/IAA transcriptional repressor ubiquitylation and destruction. *Nat Commun* **8:** 15706. doi:10.1038/ncomms15706

Xu T, Wen M, Nagawa S, Fu Y, Chen JG, Wu MJ, Perrot-Rechenmann C, Friml J, Jones AM, Yang Z. 2010. Cell surface- and rho GTPase-based auxin signaling controls cellular interdigitation in *Arabidopsis*. *Cell* **143:** 99–110. doi:10.1016/j.cell.2010.09.003

Xu T, Dai N, Chen J, Nagawa S, Cao M, Li H, Zhou Z, Chen X, De Rycke R, Rakusová H, et al. 2014. Cell surface ABP1-TMK auxin-sensing complex activates ROP GTPase signaling. *Science* **343:** 1025–1028. doi:10.1126/science.1245125

Yang Y, Hammes UZ, Taylor CG, Schachtman DP, Nielsen E. 2006. High-affinity auxin transport by the AUX1 influx carrier protein. *Curr Biol* **16:** 1123–1127. doi:10.1016/j.cub.2006.04.029

Young LM, Evans ML, Hertel R. 1990. Correlations between gravitropic curvature and auxin movement across gravistimulated roots of *Zea mays*. *Plant Physiol* **92:** 792–796. doi:10.1104/pp.92.3.792

Yu H, Zhang Y, Moss BL, Bargmann BOR, Wang R, Prigge M, Nemhauser JL, Estelle M. 2015. Untethering the TIR1 auxin receptor from the SCF complex increases its stability and inhibits auxin response. *Nat Plants* **1:** 14030. doi:10.1038/nplants.2014.30

Zaina S, Reggiani R, Bertani A. 1990. Preliminary evidence for involvement of GTP-binding protein(s) in auxin signal transduction in rice (*Oryza sativa* L.) coleoptile. *J Plant Physiol* **136:** 653–658. doi:10.1016/S0176-1617(11)81339-8

Zhou X, Levin EJ, Pan Y, McCoy JG, Sharma R, Kloss B, Bruni R, Quick M, Zhou M. 2014. Structural basis of the alternating-access mechanism in a bile acid transporter. *Nature* **505:** 569–573. doi:10.1038/nature12811

Zwiewka M, Bilanovičová V, Seifu YW, Nodzyński T. 2019. The nuts and bolts of PIN auxin efflux carriers. *Front Plant Sci* **10:** 985. doi:10.3389/fpls.2019.00985

The Story of Auxin-Binding Protein 1 (ABP1)

Richard Napier

School of Life Sciences, University of Warwick, Coventry CV4 7AS, United Kingdom

Correspondence: Richard.napier@warwick.ac.uk

The auxin-binding protein 1 (ABP1) has endured a history of undulating prominence as a candidate receptor for this important phytohormone. Its capacity for binding auxin has not been in doubt, a feature adequately explained by its crystal structure, but any relevance of this to auxin signaling and plant development has been far more demanding to define. Over its research lifetime, it has been associated with many auxin-induced activities, including ion fluxes across the plasma membrane, rearrangement of the cytoskeleton and cell shape, and the abundance of PIN proteins at the plasma membrane via control of endocytosis, all of which required its presence in the apoplast. Yet, ABP1 has a KDEL sequence that targets it to the endoplasmic reticulum, where most of it remains. This mismatch has been more than adequately compensated for by the need for an auxin receptor to account for responses far too rapid to be executed through transcription and translation and the TIR1/AuxIAA coreceptor system. However, discoveries showing that *abp1*-null mutants are not compromised for auxin signaling or development, that TIR1 or AFB1 are necessarily involved with very rapid responses at the plasma membrane, and that these rapid responses are mediated with intracellular auxin all suggest that ABP1's auxin-binding capacity is not physiologically relevant. Nevertheless, ABP1 is ubiquitous in higher plants and throughout plant tissues. We need to complete its history by defining its function inside plant cells.

The auxin-binding protein 1 (ABP1) has had a colorful and vexatious history, being referred to as both a red herring (Hertel 1995; Habets and Offringa 2015) and an outsider (Sauer and Kleine-Vehn 2011), and with reviews ending with lists of questions rather than conclusions (Sauer and Kleine-Vehn 2011; Feng and Kim 2015). Throughout its history, its fortunes have ebbed and flowed. First discovered in 1972 using radiolabeled assays for auxin binding with preparations of *Zea* coleoptile membranes (Hertel et al. 1972), there was a rush to refine and describe its properties (for review, see Napier 1995 and Napier et al. 2002). It was then all but discounted as a receptor because it was shown to be associated with the endoplasmic reticulum (ER), not the plasma membrane (Ray 1977). Purification, using its property of binding auxin as an assay through chromatography steps (Löbler and Klämbt 1985), led to antibodies, peptide sequences, and the full primary sequence was published in 1989 (Inohara et al. 1989) starting another surge in research on ABP1.

Antibodies and sequence information offered tools with which the signaling pathway of ABP1 might be discovered. The first report claiming that an auxin-specific response was dependent on ABP1 and, hence, that it was an

Cite this article as *Cold Spring Harb Perspect Biol* doi: 10.1101/cshperspect.a039909

auxin receptor was also in 1989 (Barbier-Brygoo et al. 1989), and a series of reports extended these observations in the 1990s. However, these experiments opened again the questions about how a protein targeted to the ER might be functional at the cell surface. As details were added to the picture of receptor-like activity at the plasma membrane, the lack of genetic tools and genetic evidence to support its role in auxin signaling became ever more conspicuous.

In 2001, an *abp1*-null mutant was shown to be embryo lethal, which did support the hypothesis that ABP1 was vital to plants (Chen et al. 2001b). It also slowed further enquiries using genetics at a time when the power of genetics was starting to reveal another auxin receptor, leading to the description of the TIR1-Aux/IAA coreceptor complex in 2005 (for review, see Leyser 2018). However, a further resurgence of ABP1 research was initiated by reports that extracellular ABP1 could promote clathrin-mediated endocytosis with associated effects on the population of PIN-formed (PIN) proteins on the plasma membrane (Robert et al. 2010; Čovanová et al. 2013). As with the earlier physiological responses at the plasma membrane, this occurred rapidly, too soon after an auxin stimulus to pass through a loop of transcription and translation, and treatments to exclude transcription and translation confirmed the effects to be independent of TIR1. Once again, this resurgence grew and linked ABP1 to rearrangements of the cytoskeleton and membrane-bound kinases (Feng and Kim 2015), but all these observations were rapidly overshadowed by the description of two new *abp1*-null alleles, which were shown to have phenotypes indifferent from wild-type (Gao et al. 2015). In other words, plants were shown not to need ABP1 to interpret auxin signals, leaving all the ABP1 literature in need of reinterpretation, again.

Given its undulating history, it is no wonder that plant science is confused about ABP1. In spite of this, there are definite knowns as well as the many unknowns in its biography. In this review, I will attempt to summarize what we can be confident about (certainties) and what is less certain (uncertainties) and attempt to move our understanding toward the former. There have been thorough reviews written about ABP1 throughout its history (Napier 1995; Napier et al. 2002; Tromas et al. 2010; Sauer and Kleine-Vehn 2011; Feng and Kim 2015; Grones and Friml 2015) and I will use these where appropriate rather than list extensive original literature.

CERTAINTIES—WHAT WE KNOW ABOUT ABP1

The Biochemistry and Structure of ABP1

One aspect of ABP1 that has stood the test of time has been its name—it does bind auxin. It was identified using an auxin-binding assay using radiolabeled 1-naphthylacetic acid (1-NAA), it has been purified using auxin affinity chromatography, it has been photo-affinity labeled with an auxin derivative, and, perhaps most convincingly, it has been crystallized with auxin bound (Napier et al. 2002; Woo et al. 2002). In almost all these cases, the auxin in question was 1-NAA, or derivatives of it, although phenylacetic acid is a good affinity matrix for ABP1 purification. Kinetic measurements on the purified protein have shown that ABP1 from maize (*Zea mays* L.) binds 1-NAA with affinities varying from 6×10^{-8} M (Shimomura et al. 1986) to 2.4×10^{-7} M (Hesse et al. 1989), and from *Nicotiana* between 5 and 7×10^{-7} M (Shimomura et al. 1999). These are good affinities and are consistent with concentrations of auxins necessary for physiological activity when applied in vivo. Whether or not this binding is physiologically relevant will be discussed below.

The structure of ABP1 was solved at a resolution of 1.9 Å and with an R-free value of 0.24, suggesting a robust model for the protein (PDB 1LRH) (Woo et al. 2002). The model shows ABP1 as a dimer and with the glycan of each monomer hugging its partner monomer (Fig. 1). The structure was solved in the absence and presence of bound 1-NAA with the auxin shielded inside a tight pocket, which is the cavity of the β-barrel fold, also known as the RmlC-like jelly roll fold (Giraud et al. 2000). This barrel structure is a conserved feature for the family of proteins known as the cupins, which are numerous in plants and have diverse functions

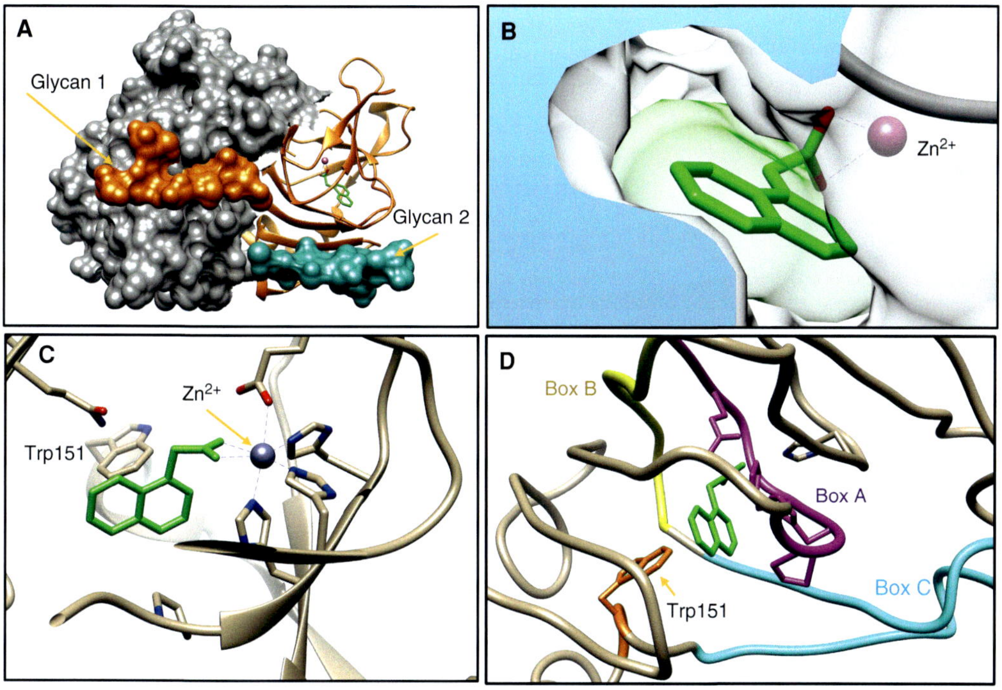

Figure 1. Structural features of auxin-binding protein 1 (ABP1). In all panels, 1-NAA is shown in green and Zn^{2+} in purple. (*A*) The dimer of ABP1 with one monomer shown as a molecular surface (gray), the other as a ribbon (orange). The glycans are both shown as surfaces, with the glycan of the orange monomer (glycan 1) folded across the gray monomer, and vice versa. (*B*) The aperture to the auxin-binding site with the surface of 1-NAA also shaded in transparent green. (*C*) The Zn^{2+} ion is coordinated by a cluster of histidine residues and an aspartate, and when 1-NAA binds a water molecule is displaced (not shown) to allow the carboxylic acid group to hydrogen bond to the Zn^{2+} (dashed lines). Tryptophan151 is seen behind the auxin at the end of an α helix (receding into the background). (*D*) The conserved sequence boxes of ABP1: box A (magenta), box B (yellow), and box C (cyan), plus the conserved Trp151 (orange). PDB 1LRH visualized using Chimera (Pettersen et al. 2004).

(Dunwell et al. 2008). A key feature is a sequence motif for coordinating a divalent metal ion, which is zinc in the case of ZmABP1 (Fig. 1B, C). This zinc ion hydrogen bonds to the acetic acid group of 1-NAA, positioning it in the binding site. The aromatic ring of the auxin is held in an edge-to-face pi stack with an adjacent tryptophan residue (Trp151).

There is a small aperture to the binding site (Fig. 1B) and no significant change in backbone carbons was reported between unbound and bound structures, which implies no auxin-induced conformational change (Woo et al. 2002). A rationale was put forward for crystal constraints promoting crystallization only of the bound structure, alongside a hypothetical model for conformational movement of the carboxy-terminal α helix upon auxin binding. At one end of this helix lies Trp151 (Fig. 1C), and the activation model suggested that as auxin binds, this Trp is displaced, pushing away the helix like a piston. This model remains unproven, although it is consistent with evidence of an auxin-induced conformational change detected using antibodies mapped to the carboxyl-terminus of the protein (Napier and Venis 1990).

Molecular dynamics simulations of ABP1 have variously supported changes in the arrangement of the carboxyl-terminus induced by auxin with general agreement that it unfolds in the absence of auxin (Bertoša et al. 2008; Grandits and Oostenbrink 2014; Tesser da Costa et al. 2017). Simulations of the glycosylated protein for *Arabidopsis* as well as maize also suggested small differences between dicot and monocot proteins, which may be linked to different auxin-binding specificities (Tesser da Costa et al. 2017).

The Sequence of ABP1

At the protein level, ABP1 is highly conserved throughout the angiosperms, and residue identity remains high with possible homologs in the lycophytes and mosses (Tromas et al. 2010). A series of very highly conserved motifs were defined as boxes A–C (Napier et al. 2002), with the cluster of zinc-binding residues being box A (Fig. 1) and the other boxes forming the walls of the barrel around the auxin-binding pocket in ABP1 (Fig. 1D).

Two additional key motifs that are conserved across higher plants are the carboxy-terminal KDEL ER retention sequence and two cysteine residues, one close to the amino-terminus of the mature ABP1 and the second just carboxy-terminal to Trp151. These cysteines are disulfide bonded, locking the two ends of the protein together, an important structural constraint (Woo et al. 2002). Both of these features are missing in lycophyte and moss sequences. The absence of these two structural features raises doubts about whether lower plant sequences are really of ABP1 homologs. The analysis by Tromas and colleagues also discussed possible homologs further back in plant ancestry, and it is certainly true that there are sequences from some algae with high similarity across the boxes but, again, they lack other key features.

Expression Profiles of ABP1

The early work of Hertel and Ray screened plant tissues for auxin-binding activity, which resulted in maize seedlings becoming the material of choice for early work. They remain the tissue with highest levels of extractable ABP1. In most plants and most tissues, ABP1 expression is modest. In *Arabidopsis*, expression appears highest in stylar tissue and seeds (Fig. 2, eFP browser for AT4G02980; Winter et al. 2007; Waese et al. 2017). Expression is elevated during early- to mid-seed maturation and in germinating endosperm, as well as in the root meristem and stele in trichomes and transiently in response to heat (38°C). Its transcript levels do not respond to hormone treatments (including auxin); there is some increase in response to some pathogen challenges although nothing consistent across different biotic stresses. Transcript levels change little with photoperiod and not at all to photoreceptor-linked light treatments. The eFP browser summaries agree in general with results from *pABP1*:GUS analyses, which suggested elevated expression in young growing leaves, the root meristem, vascular tissue, and hydathodes (Klode et al. 2011).

Expression maps for maize suggest somewhat more variation over tissues and lifetime (Fig. 2), with the highest expression in floral and embryo tissues, which is consistent with the picture from *Arabidopsis*, plus the young seedling as used for biochemical extractions.

The expression data from *Arabidopsis* are also consistent with earlier results from *Zea*, which showed that both ABP1 protein and its mRNA were remarkably stable with long half-lives, which did not respond to hormone or stress treatments, except heat, which led to rapid loss of ABP1 protein (Oliver et al. 1995). There is also an interesting report of ABP1 expression being highly enhanced in active cambium of the tree *Eucommia*, and poorly expressed in dormant cambium (Hou et al. 2006). However, if we summarize all these observations on expression, we learn little that helps explain or define the role of ABP1. Its expression profiles are not distinctive, although the highlights like trichomes and hydathodes, stele and cambium, and stylar tissues should provide clues if only we could decipher them.

Genetic Analyses of ABP1 Function

Genes for ABP1 are ubiquitous across the higher plants, with ABP1-like sequences in lower plants, as noted above (Tromas et al. 2010). In many plant species, there is a small family of *ABP1*-like genes, although there is a single-copy *ABP1* gene in *Arabidopsis*. The observation that *ABP1* pervades higher plant genomes suggests that ABP1 is essential, and this presump-

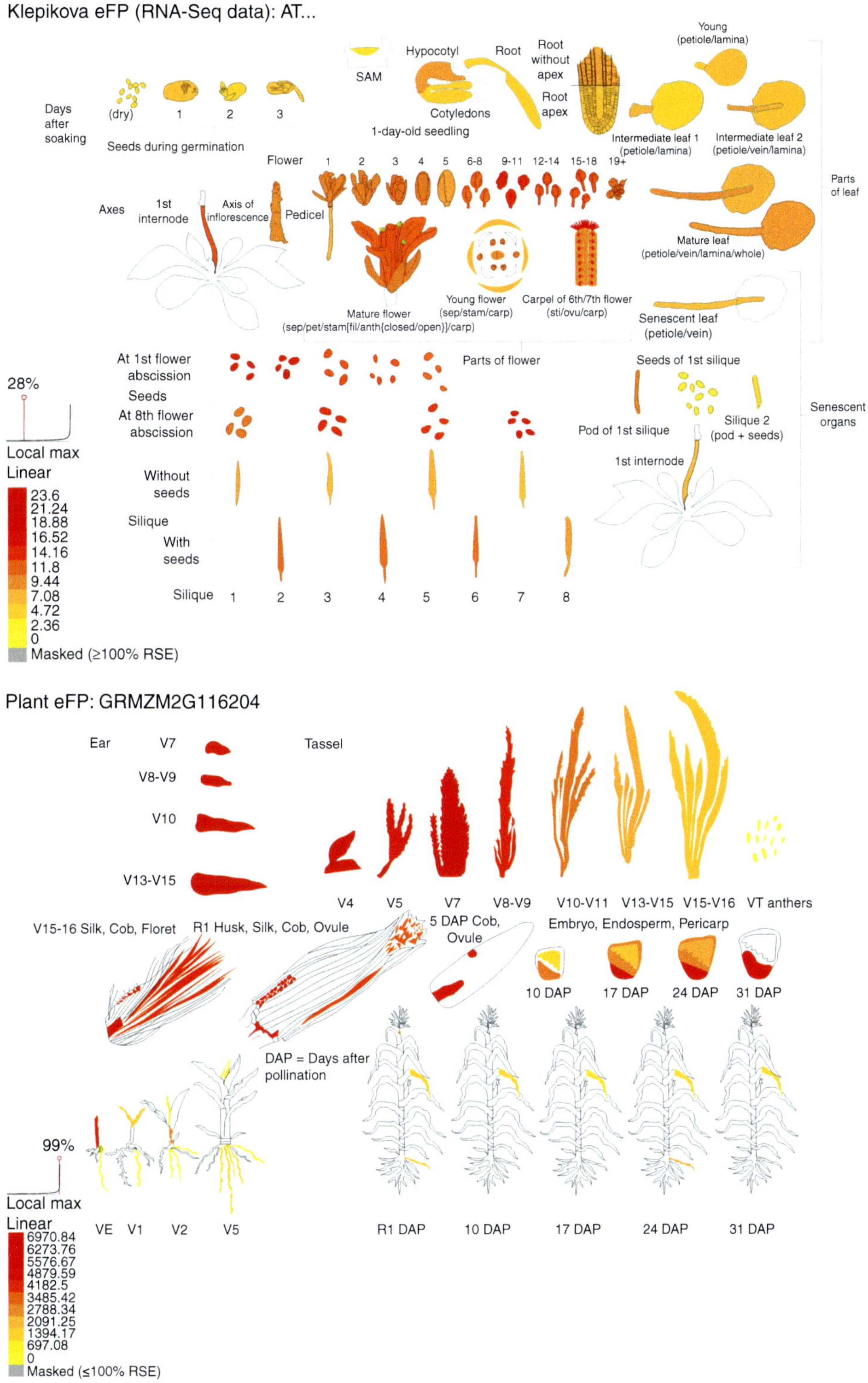

Figure 2. Expression maps for auxin-binding protein 1 (ABP1) in *Arabidopsis* (*top*, AT4G02980) and maize (*bottom*, GRMZM2FG116204). The *Arabidopsis* image was generated from data from Klepikova et al. (2016). The maize image is based on data from Downs et al. (2013). Both images were prepared by bar.utoronto.ca/eplant (Waese et al. 2017). (RSE) Relative standard error.

tion was kindled when a reverse-engineered *Arabidopsis abp1-1* knockout line was reported to be embryo lethal (Chen et al. 2001b). This awkward genetic system slowed follow-up experimentation for over 10 years. Then, when attempts to complement the *abp1-1* mutation by expressing wild-type *ABP1* failed (Gao et al. 2015), additional work revealed that the embryo lethality in *abp1-1* was caused by disruption of an adjacent gene (Dai et al. 2015; Michalko et al. 2015). Further, two new verified *abp1* knockout lines showed that plant development and auxin signaling were not dependent on ABP1 (Gao et al. 2015). Thus, genetic analysis of ABP1 function was both freed up and appeared to separate ABP1 from auxin action.

A number of alternative genetic tools had been developed to help validate proposed roles of ABP1 in auxin signaling and these were rapidly reevaluated after the paper by Gao et al. (2015). A partial loss of function line, *abp1-5*, was shown to be unreliable after second site mutations were recorded (Enders et al. 2015) and strong phenotypes were observed in the absence of ABP1 in a set of conditional *ABP1* knockdown lines (Michalko et al. 2016). Taken together, it became clear that another certainty about ABP1 was that it was not involved in the bulk of developmental and signaling roles driven by auxin. Indeed, it had long been accepted that ABP1 was not involved with the TIR1-Aux/IAA coreceptor complex nor in the genetic control of auxin action through this gene derepression pathway (Leyser 2018).

Antisense and Overexpression of ABP1

The observations on *abp1* mutant lines do not necessarily disqualify data generated using antisense and overexpression genetic strategies. Antisense data have been reported little, probably due to a lack of marked phenotypes consistent with data from the knockout lines noted above (Gao et al. 2015). Overexpression has been reported a number of times and in most cases has resulted in subtle developmental phenotypes such as increased leaf epidermal cell size and interdigitation of epidermal cell lobes (Jones et al. 1998; Grones et al. 2015), increased sensitivity of K^+ influx to applied auxin (Bauly et al. 2000), and positive regulation of endocytosis of PIN efflux carriers (Čovanová et al. 2013; Grones et al. 2015), probably through promotion of clathrin recruitment to the plasma membrane (Robert et al. 2010; Chen et al. 2014; Grones et al. 2015). Thus, there remain a cluster of phenotypes associated with overexpression of ABP1, which are linked to activities at the plasma membrane.

UNCERTAINTIES—REPORTS AND ASSOCIATIONS OF ABP1 WITH ACTIVITIES THAT NEED MORE EXPLANATION

ABP1 and Responses at the Plasma Membrane

Once antibodies and the sequence of ABP1 were available, there was a series of reports that suggested that ABP1 was active at the outer face of the plasma membrane and that this was the site of perception for auxin. The first was the demonstration that anti-ABP1 antibodies could selectively repress plasma membrane depolarization in response to added auxin in tobacco protoplasts (Barbier-Brygoo et al. 1989). There followed many reports using antibodies, purified ABP1 protein, and peptides based on parts of the ABP1 protein-associating ABP1 with the cell surface (Thiel et al. 1993; Napier et al. 2002; Scherer 2011). In addition to linking ABP1 with auxin perception on the plasma membrane, all these responses were rapid with response times of seconds or minutes, linked to ion flux across the membrane and, crucially, faster than gene-activation responses engaged by the TIR-Aux/IAA coreceptor complex (Badescu and Napier 2006). All these reports have been brought into question by subsequent findings.

ABP1 and Ion Transfer across the Plasma Membrane

One of the key drivers for seeking a TIR1-independent site for auxin perception has been the need to explain rapid responses. Many of these were recorded in ex vivo systems, but two very rapid responses to auxin have been reported in intact *Arabidopsis* tissues, the inhibition of primary root growth (Fendrych et al. 2018), and plasma membrane depolarization in root hairs (Dindas et al. 2018). In both cases, the response was maximal within ~30 sec, and in both cases the use of mutants showed that the auxin effect depended on both the auxin uptake carrier AUX1 and the receptor TIR1, although later work showed that auxin F-Box1 (AFB1) from the TIR1 family dominated in these rapid responses (Prigge et al. 2020). How AFB1 transduces the auxin signal remains to be determined but the work shows that ABP1 is not necessary for rapid, plasma-membrane-associated auxin signaling. Indeed, the *abp1*-null lines (Gao et al. 2015) have been used to show that membrane depolarization and inhibition of endocytosis are independent of ABP1 (Paponov et al. 2019).

How then may we explain the many publications implicating ABP1 in auxin perception at the plasma membrane? It seems likely that the responses were associated with off-site activity of antibodies and peptides, perhaps because of features shared between ABP1 and other cupins. The conserved box sequences of ABP1 (Fig. 1) are accessible on the surface of what is a cupin fold (Fig. 1), making it difficult to discount antibody cross-reactivity. Such speculation is easy and not easy to prove, but these observations might also spur on some to explore the wider roles of cupins in cell surface signaling.

ABP1 as an Apoplastic Protein

All reports associating ABP1 with a receptor-like function rely on it being active at the outer face of the plasma membrane. Other cupins are certainly apoplastic (Dunwell et al. 2008), including *Gossipium* ABP19 (e.g., Pei et al. 2019) and *Arabidopsis* germin-like protein 4 (GLP4) (Yin et al. 2009). Peach ABP19 and *At*GLP4 have also been shown to bind auxin (Ohmiya et al. 1998; Yin et al. 2009), albeit at pH 7, and so the claims for auxin binding by extracellular proteins are not exclusive. However, ABP1 (and not these other proteins) displays an ER retention sequence, KDEL at its carboxyl-terminus, and all studies of its subcellular localization place the majority of ABP1 in the ER (Tian et al. 1995; Napier et al. 2002; Xu et al. 2014).

Studies that have not relied on antibodies (which could cross-react with other apoplastic cupins, as noted above) to establish the subcellular localization of ABP1 are few. Quantification of glycans characteristic of post-Golgi compartments estimated that <15% of ABP1 ever reaches the Golgi apparatus and <2% passes further down the secretory pathway in maize (Henderson et al. 1997). Immunoelectron microscopy of *Arabidopsis* root cells measured 22% at the plasma membrane (Xu et al. 2014). Evidence of antibody specificity was given using western blotting but given sequence and structural similarities to other cupins, it is challenging to confirm that this holds for non-denatured proteins in microscopy, and rigorous evidence for a physical presence of ABP1 at the cell surface is lacking.

Auxin Binding and pH

One argument that has often been used to vindicate APB1's role as a receptor for auxin at the cell surface is its pH optimum for binding auxin. Many studies showed this to be between pH 5 and 5.5 (Napier 1995), which is within the pH range of the apoplast, whereas the pH in the ER is considered to be similar to that of the cytoplasm, between pH 7 and 7.5. Hence, ABP1 could work well as an ABP in the apoplast, not in the ER, and a set of careful studies confirmed that ABP1 does not bind auxin in the ER (Tian et al. 1995). Yet, most ABP1 is in the ER. Thus, one of our certainties, that ABP1 binds auxin, needs qualification: ABP1 binds auxin in vitro. In living cells, most of it does not. Of course, if any does reach the apoplast, binding conditions are favorable.

Auxin-Binding Selectivity

A further question that has often arisen over the activity of ABP1 as a receptor candidate is over its higher affinity for binding the synthetic auxin 1-NAA than for IAA. Where comparative measurements have been made, the affinity is poorer for IAA than for 1-NAA by one or two orders of magnitude (Shimomura et al. 1986; Hesse et al. 1989). This preference was also seen with unpurified ABP1 in *Zea* microsomes (Ray et al. 1977) and in an antibody-based assay for a binding-induced conformational change in purified ABP1 (Napier and Venis 1990). Molecular dynamic simulations of ABP1 have consistently found 1-NAA to bind with higher affinity than IAA, and better able to stabilize both *Arabidopsis* and maize ABP1s (Grandits and Oostenbrink 2014; Tesser da Costa et al. 2017). Yet in vivo, IAA is among the most active of auxins and this lack of correspondence of binding affinity with physiological activity formed part of the red herring argument that questioned the validity of the hypothesis that ABP1 is a receptor (Hertel 1995). Synthetic ligands often outperform natural ligands. This lack of correspondence does not undermine arguments for or against the functionality of ABP1.

The primary location of ABP1 in the ER corresponds to the growing awareness of the ER as an important contributor to auxin homeostasis (Skalický et al. 2018). It is a compartment where IAA conjugates accumulate and where there are auxin amidohydrolases to release IAA from such conjugates, for AUX1 (as well as the plasma membrane) (Dharmasiri et al. 2006) and PIN5 and PIN8 (Mravec et al. 2009; Ding et al. 2012). However, there is no evidence of ABP1 having a function related to auxin signaling in the ER.

A Binding Partner for ABP1 at the Plasma Membrane, the Docking Protein

As soon as there was evidence for ABP1 being a receptor candidate at the cell surface, it was recognized that this soluble glycoprotein would need help to transduce a signal back across the plasma membrane and a docking protein was posited (Barbier-Brygoo et al. 1991). This docking protein would act as an anchor in the membrane and help transfer the auxin signal into the cell. A glycosylphosphatidylinositol (GPI)-anchored protein known as carboxy-terminal peptide-binding protein 1 (CBP1) was shown to interact with the carboxyl-terminus of ABP1 (Shimomura 2006), but no further work has linked ABP1 with CBP1 or the related SKU5-like GPI-anchored proteins (Zhou 2019). Likewise, yeast two-hybrid work identified an *Arabidopsis* ortholog of human RING membrane-anchor E3 Ub ligase (AtRMA2) as an interactor for ABP1 (Son et al. 2010), but there has been no subsequent confirmation.

ABP1 Overexpression, Auxin-Mediated Control of Endocytosis, and TMKs

Intracellular cycling of the Rho Of Plants (ROPs) small GTPases has been linked to auxin control of cytoskeletal rearrangements and cell shape, all under the control of extracellular ABP1 (Xu et al. 2010). As noted above, some of these findings are uncertain because of the use of the *abp1-5* mutant line, but the results were supported by antisense experiments, which gave phenotypes consistent with those of the mutant. A development in the story came with the discovery that a transmembrane receptor-like kinase (TMK1) formed a complex with ABP1 at the plasma membrane and that this was the auxin-sensing unit for activating the intracellular ROPs (Xu et al. 2014). The association was demonstrated by a series of coimmunoprecipitation experiments, notably with the extracellular domain of TMK1, which showed an auxin-driven dose-dependent interaction with ABP1 (for review, see Feng and Kim 2015).

The story of the ABP1/TMK complex became uncertain when the *abp1* null lines showed no measurable changes in plant phenotype (Gao et al. 2015). More recent work links auxin activation of TMK1 (and TMK4) with downstream intracellular kinase cascades without the mention of what protein binds and perceives the IAA (Huang et al. 2019). In separate work on auxin-induced rapid cytoskeletal rearrangements, it was shown that the response required the uptake carrier AUX1, hence, intracellular auxin (Arieti and Staiger 2020), reducing the need for ABP1 as an auxin sensor in the apoplast. Further, calcium influx has been linked to rapid auxin-induced plasma membrane depolarization (Dindas et al. 2018), and calcium influx plus auxin-induced calcium-binding interactors of ROPs have been shown to be involved in root growth (Hazak et al. 2019). The need for an extracellular auxin sensor is diminishing, making it less certain that ABP1 is such a protein.

SYNTHESIS

The discussion above has served to highlight that there remain more uncertainties about ABP1 than certainties when considering auxin perception. There is little doubt that it lives up to its name in vitro, but when this is put into the context of its primary cellular destination (another certainty), the ER, the certainty is dissipated. The greatest challenge remaining is to assign a role for ABP1 in the ER—where it does not bind auxin.

ABP1 is certainly a cupin, and this has also brought uncertainty to our analysis of many of the reports on ABP1 because it is not possible to be certain the tools used were sufficiently specific. The cupin fold and key sequence motifs, most of which are accessible on the surface of the protein (Fig. 1), are shared, and many cupins are apoplastic. On these grounds, it is reasonable to question the interpretation of past research undertaken with antibodies and peptides, although this is not meant to question the validity of the work at the time. This leads us to question many of the reports linking ABP1 at the cell surface with signaling. Such uncertainty has been compounded by research suggesting that very rapid responses in intact tissues are dependent on the uptake carrier AUX1 and the intracellular receptors TIR1 and AFB1, albeit in a nontranscriptional mode of action. Hence, there is no longer a receptor missing for rapid auxin-mediated events at the plasma membrane and, hence, no longer a need for an extracellular auxin sensor.

An intriguing part of the story is the consistency of some phenotypes reported from use of techniques, such as antisense and overexpression, and the relationship of many such phenotypes to elements of auxin physiology (Jones et al. 1998; Bauly et al. 2000; Chen et al. 2001a; Čovanová et al. 2013). Another is the fact that the *ABP1* gene family is ubiquitous in higher plants. There is much to be learned about cupins and their functions in subcellular compartments.

CONCLUDING REMARKS

We have learned a lot about ABP1 and understand very little. Its expression patterns are starting to tell a story, perhaps about sugars and secretion, but the plot remains obscure. We have probably now learned enough to concur with the statement that, "Auxin-binding protein 1 (ABP1) is not required for either auxin signaling or *Arabidopsis* development" (Gao et al. 2015). Consequently, we should separate ABP1 from auxin physiology. Auxin signaling probably no longer needs an extracellular receptor, nor a second class of receptor for rapid responses. However, until we understand what function ABP1 does play in plant biology, its storybook remains open.

ACKNOWLEDGMENTS

The author thanks members of the ABP1 community for enthusiasm and support over many years.

REFERENCES

Arieti RS, Staiger CJ. 2020. Auxin-induced actin cytoskeleton rearrangements require AUX1. *New Phyt* **226:** 441–459. doi:10.1111/nph.16382

Badescu GO, Napier RM. 2006. Receptors for auxin: will it all end in TIRs? *Trends Plant Sci* **11:** 217–223. doi:10.1016/j.tplants.2006.03.001

Barbier-Brygoo H, Ephritikhine G, Klämbt D, Ghislain M, Guern J. 1989. Functional evidence for an auxin receptor at the plasmalemma of tobacco mesophyll protoplasts. *Proc Natl Acad Sci* **86:** 891–895. doi:10.1073/pnas.86.3.891

Barbier-Brygoo H, Ephritikhine G, Klämbt D, Maurel C, Palme K, Schell J, Guern J. 1991. Perception of the auxin signal at the plasma membrane of tobacco mesophyll protoplasts. *Plant J* **1:** 83–93. doi:10.1111/j.1365-313X.1991.00083.x

Bauly JM, Sealy IM, Macdonald H, Brearley J, Dröge S, Hillmer S, Robinson DG, Venis MA, Blatt MR, Lazarus CM, et al. 2000. Overexpression of auxin-binding protein enhances the sensitivity of guard cells to auxin. *Plant Physiol* **12:** 1299–1238.

Bertoša B, Kojić-Prodić B, Wade RC, Tomić S. 2008. Mechanism of auxin interaction with auxin binding protein (ABP1): a molecular dynamics simulation study. *Biophys J* **94:** 27–37. doi:10.1529/biophysj.107.109025

Chen JG, Shimomura S, Sitbon F, Sandberg G, Jones AM. 2001a. The role of auxin-binding protein 1 in the expansion of tobacco leaf cells. *Plant J* **28:** 607–617. doi:10.1046/j.1365-313x.2001.01152.x

Chen JG, Ullah H, Young JC, Sussman MR, Jones AM. 2001b. ABP1 is required for organized cell elongation and division in *Arabidopsis* embryogenesis. *Genes Dev* **15:** 902–911. doi:10.1101/gad.866201

Chen X, Grandont L, Li H, Hauschild R, Paque S, Abuzeineh A, Rakusová H, Benkova E, Perrot-Rechenmann C, Friml J. 2014. Inhibition of cell expansion by rapid ABP1-mediated auxin effect on microtubules. *Nature* **516:** 90–93. doi:10.1038/nature13889

Čovanová M, Sauer M, Rychtář J, Friml J, Petrášek J, Zažímalová E. 2013. Overexpression of the auxin binding protein1 modulates PIN-dependent auxin transport in tobacco cells. *PLoS ONE* **8:** e70050. doi:10.1371/journal.pone.0070050

Dai X, Zhang Y, Zhang D, Chen J, Gao X, Estelle M, Zhao Y. 2015. Embryonic lethality of *Arabidopsis* abp1-1 is caused by deletion of the adjacent BSM gene. *Nat Plants* **1:** 3–6.

Dharmasiri S, Swarup R, Mockaitis K, Dharmasiri N, Singh SK, Kowalchyk M, Marchant A, Mills S, Sandberg G, Bennett MJ, et al. 2006. AXR4 is required for localization of the auxin influx facilitator AUX1. *Science* **312:** 1218–1220. doi:10.1126/science.1122847

Dindas J, Scherzer S, Roelfsema MRG, von Meyer K, Müller HM, Al-Rasheid KAS, Palme K, Dietrich P, Becker D, Bennett MJ, et al. 2018. AUX1-mediated root hair auxin influx governs SCFTIR1/AFB-type Ca^{2+} signaling. *Nat Commun* **9:** 1174. doi:10.1038/s41467-018-03582-5

Ding Z, Wang B, Moreno I, Dupláková N, Simon S, Carraro N, Reemmer J, Pěnčík A, Chen X, Tejos R, et al. 2012. ER-localized auxin transporter PIN8 regulates auxin homeostasis and male gametophyte development in *Arabidopsis*. *Nat Commun* **3:** 941. doi:10.1038/ncomms1941

Downs GS, Bi YM, Colasanti J, Wu W, Chen X, Zhu T, Rothstein SJ, Lukens LN. 2013. A developmental transcriptional network for maize defines coexpression modules. *Plant Physiol* **161:** 1830–1843. doi:10.1104/pp.112.213231

Dunwell JM, Gibbings JG, Mahmood T, Naqvi SMS. 2008. Germin and germin-like proteins: evolution, structure, and function. *Crit Rev Plant Sci* **27:** 342–375. doi:10.1080/07352680802333938

Enders TA, Oh S, Yang Z, Montgomery BL, Strader LC. 2015. Genome sequencing of *Arabidopsis* abp1-5 reveals second-site mutations that may affect phenotypes. *Plant Cell* **27:** 1820–1826. doi:10.1105/tpc.15.00214

Fendrych M, Akhmanova M, Merrin J, Glanc M, Hagihara S, Takahashi K, Uchida N, Torii KU, Friml J. 2018. Rapid and reversible root growth inhibition by TIR1 auxin signalling. *Nat Plants* **4:** 453–459. doi:10.1038/s41477-018-0190-1

Feng M, Kim JY. 2015. Revisiting apoplastic auxin signaling mediated by AUXIN BINDING PROTEIN 1. *Mol Cells* **38:** 829–835. doi:10.14348/molcells.2015.0205

Gao Y, Zhang Y, Zhang D, Dai X, Estelle M, Zhao Y. 2015. Auxin binding protein 1 (ABP1) is not required for either auxin signaling or *Arabidopsis* development. *Proc Nat Acad Sci* **112:** 2275–2280. doi:10.1073/pnas.1500365112

Giraud M-F, Leonard GA, Field RA, Berlind C, Naismith JH. 2000. RmlC, the third enzyme of dTDP-L-rhamnose pathway, is a new class of epimerase. *Nat Struct Biol* **7:** 398–402. doi:10.1038/75178

Grandits M, Oostenbrink C. 2014. Molecular dynamics simulations of the auxin-binding protein 1 in complex with indole-3-acetic acid and naphthalen-1-acetic acid. *Proteins* **82:** 2744–2755. doi:10.1002/prot.24639

Grones P, Friml J. 2015. Auxin transporters and binding proteins at a glance. *J Cell Sci* **128:** 1–7. doi:10.1242/jcs.159418

Grones P, Chen X, Simon S, Kaufmann WA, De Rycke R, Nodzyński T, Zažímalová E, Friml J. 2015. Auxin-binding pocket of ABP1 is crucial for its gain-of-function cellular and developmental roles. *J Exp Bot* **66:** 5055–5065. doi:10.1093/jxb/erv177

Habets MEJ, Offringa R. 2015. Auxin binding protein 1: a red herring after all? *Mol Plant* **8:** 1131–1134. doi:10.1016/j.molp.2015.04.010

Hazak O, Mamon E, Lavy M, Sternberg H, Behera S, Schmitz-Thom I, Bloch D, Dementiev O, Gutman I, Danziger T, et al. 2019. A novel Ca^{2+}-binding protein that can rapidly transduce auxin responses during root growth. *PLoS Biol* **17:** e3000085. doi:10.1371/journal.pbio.3000085

Henderson J, Bauly JM, Ashford DA, Oliver SC, Hawes CR, Lazarus CM, Venis MA, Napier RM. 1997. Retention of maize auxin-binding protein in the endoplasmic reticulum: quantifying escape and the role of auxin. *Planta* **202:** 313–323. doi:10.1007/s004250050133

Hertel R. 1995. Auxin binding protein 1 is a red herring. *J Exp Bot* **46:** 461–462. doi:10.1093/jxb/46.5.461-a

Hertel R, Thomson KS, Russo VE. 1972. In-vitro auxin binding to particulate cell fractions from corn coleoptiles. *Planta* **107:** 325–340. doi:10.1007/BF00386394

Hesse T, Feldwisch J, Balshusemann D, Bauw G, Puype M, Vandekerckhove J, Lobler M, Klambt D, Schell J, Palme K. 1989. Molecular cloning and structural analysis of a gene from *Zea mays* (L.) coding for a putative receptor for the plant hormone auxin. *EMBO J* **8:** 2453–2461. doi:10.1002/j.1460-2075.1989.tb08380.x

Hou HW, Zhou YT, Mwange KN, Li WF, He XQ, Cui KM. 2006. ABP1 expression regulated by IAA and ABA is associated with the cambium periodicity in *Eucommia ulmoides* Oliv. *J Exp Bot* **57:** 3857–3867. doi:10.1093/jxb/erl150

Huang R, Zheng R, He J, Zhou Z, Wang J, Xiong Y, Xu T. 2019. Noncanonical auxin signaling regulates cell division pattern during lateral root development. *Proc Natl Acad Sci* **116:** 21285–21290. doi:10.1073/pnas.1910916116

Inohara N, Shimomura S, Fukui T, Futai M. 1989. Auxin-binding protein located in the endoplasmic reticulum of maize shoots: molecular cloning and complete primary structure. *Proc Natl Acad Sci* **86:** 3564–3568. doi:10.1073/pnas.86.10.3564

Jones AM, Im KH, Savka MA, Wu MJ, DeWitt NG, Shillito R, Binns AN. 1998. Auxin-dependent cell expansion mediated by overexpressed auxin-binding protein 1. *Science* **282:** 1114–1117. doi:10.1126/science.282.5391.1114

Klepikova AV, Kasianov AS, Gerasimov ES, Logacheva MD, Penin AA. 2016. A high resolution map of the *Arabidopsis thaliana* developmental transcriptome based on RNA-seq profiling. *Plant J* **88:** 1058–1070. doi:10.1111/tpj.13312

Klode M, Dahlke RI, Sauter M, Steffens B. 2011. Expression and subcellular localization of *Arabidopsis thaliana* auxin-binding protein 1 (ABP1). *J Plant Growth Reg* **30:** 416–424. doi:10.1007/s00344-011-9203-2

Leyser O. 2018. Auxin signaling. *Plant Physiol* **176:** 465–479. doi:10.1104/pp.17.00765

Löbler M, Klämbt D. 1985. Auxin-binding protein from coleoptile membranes of corn (*Zea mays* L.). I: Purification by immunological methods and characterization. *J Biol Chem* **260:** 9848–9853. doi:10.1016/S0021-9258(17)39314-6

Michalko J, Dravecká M, Bollenbach T, Friml J. 2015. Embryo-lethal phenotypes in early *abp1* mutants are due to disruption of the neighboring BSM gene. *F1000 Res* **4:** 1104. doi:10.12688/f1000research.7143.1

Michalko J, Glanc M, Perrot-Rechenmann C, Friml J. 2016. Strong morphological defects in conditional *Arabidopsis* abp1 knock-down mutants generated in absence of functional ABP1 protein. *F1000Res* **5:** 86. doi:10.12688/f1000research.7654.1

Mravec J, Skůpa P, Bailly A, Hoyerová K, Krecek P, Bielach A, Petrásek J, Zhang J, Gaykova V, Stierhof YD, et al. 2009. Subcellular homeostasis of phytohormone auxin is mediated by the ER-localized PIN5 transporter. *Nature* **459:** 1136–1140. doi:10.1038/nature08066

Napier RM. 1995. Towards an understanding of ABP1. *J Exp Bot* **46:** 1787–1795. doi:10.1093/jxb/46.12.1787

Napier RM, Venis MA. 1990. Monoclonal antibodies detect an auxin-induced conformational change in the maize auxin-binding protein. *Planta* **182:** 313–318. doi:10.1007/BF00197128

Napier RM, David KM, Perrot-Rechenmann C. 2002. A short history of auxin-binding proteins. *Plant Mol Biol* **49:** 339–348. doi:10.1023/A:1015259130955

Ohmiya A, Tanaka Y, Kadowaki KI, Hayashi T. 1998. Cloning of genes encoding auxin-binding proteins (ABP19/20) from peach: significant peptide sequence similarity with germin-like proteins. *Plant Cell Physiol* **39:** 492–499. doi:10.1093/oxfordjournals.pcp.a029396

Oliver SC, Venis MA, Freedman RB, Napier RM. 1995. Regulation of synthesis and turnover of maize auxin-binding protein and observations on its passage to the plasma membrane: comparisons to maize immunoglobulin-binding protein cognate. *Planta* **197:** 465–474. doi:10.1007/BF00196668

Paponov IA, Dindas J, Król E, Friz T, Budnyk V, Teale W, Paponov M, Hedrich R, Palme K. 2019. Auxin-induced plasma membrane depolarization is regulated by auxin transport and not by auxin binding protein1. *Front Plant Sci* **9:** 1953. doi:10.3389/fpls.2018.01953

Pei Y, Li X, Zhu Y, Ge X, Sun Y, Liu N, Jia Y, Li F, Hou Y. 2019. GhABP19, a novel germin-like protein from *Gossypium hirsutum* plays an important role in the regulation of resistance to verticillium and fusarium wilt pathogens. *Front Plant Sci* **10:** 583. doi:10.3389/fpls.2019.00583

Pettersen EF, Goddard TD, Huang CC, Couch GS, Greenblatt DM, Meng EC, Ferrin TE. 2004. UCSF Chimera—a visualization system for exploratory research and analysis. *J Comput Chem* **25:** 1605–1612. doi:10.1002/jcc.20084

Prigge MJ, Platre M, Kadakia N, Zhang Y, Greenham K, Szutu W, Kumar Pandey B, Bhosale RA, Bennett MJ, Busch W, et al. 2020. Genetic analysis of the *Arabidopsis* TIR1/AFB auxin receptors reveals both overlapping and specialized functions. *eLife* **9:** e54740. doi:10.7554/eLife.54740

Ray PM. 1977. Auxin-binding sites of maize coleoptiles are localized on membranes of the endoplasmic reticulum. *Plant Physiol* **59:** 594–599. doi:10.1104/pp.59.4.594

Ray PM, Dohrmann U, Hertel R. 1977. Specificity of auxin-binding sites on maize coleoptile membranes as possible receptor sites for auxin action. *Plant Physiol* **60:** 585–591. doi:10.1104/pp.60.4.585

Robert S, Kleine-Vehn J, Barbez E, Sauer M, Paciorek T, Baster P, Vanneste S, Zhang J, Simon S, Čovanová M, et al. 2010. ABP1 mediates auxin inhibition of clathrin-dependent endocytosis in *Arabidopsis*. *Cell* **143:** 111–121. doi:10.1016/j.cell.2010.09.027

Sauer M, Kleine-Vehn J. 2011. Auxin binding protein1: the outsider. *Plant Cell* **23:** 2033–2043. doi:10.1105/tpc.111.087064

Scherer GFE. 2011. AUXIN-BINDING-PROTEIN1, the second auxin receptor: what is the significance of a two-receptor concept in plant signal transduction? *J Exp Bot* **62:** 3339–3357. doi:10.1093/jxb/err033

Shimomura S. 2006. Identification of a glycosylphosphatidylinositol-anchored plasma membrane protein interacting with the C-terminus of auxin-binding protein 1: a photoaffinity crosslinking study. *Plant Mol Biol* **60:** 663–677. doi:10.1007/s11103-005-5471-1

Shimomura S, Sotobayashi T, Futai M, Fukui T. 1986. Purification and properties of an auxin-binding protein from maize shoot membranes. *J Biochem* **99:** 1513–1524. doi:10.1093/oxfordjournals.jbchem.a135621

Shimomura S, Watanabe S, Ichikawa H. 1999. Characterization of auxin-binding protein 1 from tobacco: content, localization and auxin-binding activity. *Planta* **209:** 118–125. doi:10.1007/s004250050613

Skalický V, Kubeš M, Napier R, Novák O. 2018. Auxins and cytokinins—the role of subcellular organization on homeostasis. *Int J Mol Sci* **19:** 3115. doi:10.3390/ijms19103115

Son O, Cho SK, Kim SJ, Kim WT. 2010. In vitro and in vivo interaction of AtRma2 E3 ubiquitin ligase and auxin binding protein 1. *Biochem Biophys Res Commun* **393:** 492–497. doi:10.1016/j.bbrc.2010.02.032

Tesser da Costa C, Pedebos C, Verli H, Fett-Neto AG. 2017. The role of Zn^{2+}, dimerization and *N*-glycosylation in the interaction of auxin-binding protein 1 (ABP1) with different auxins. *Glycobiology* **27:** 1109–1119. doi:10.1093/glycob/cwx080

Thiel G, Blatt MR, Fricker MD, White IR, Millner P. 1993. Modulation of K^+ channels in *Vicia* stomatal guard cells by peptide homologs to the auxin-binding protein C terminus. *Proc Nat Acad Sci* **90:** 11493–11497. doi:10.1073/pnas.90.24.11493

Tian H, Klämbt D, Jones AM. 1995. Auxin-binding protein 1 does not bind auxin within the endoplasmic reticulum despite this being the predominant subcellular location for this hormone receptor. *J Biol Chem* **270:** 26962–26969. doi:10.1074/jbc.270.45.26962

Tromas A, Paponov I, Perrot-Rechenmann C. 2010. Auxin binding protein 1: functional and evolutionary aspects. *Trends Plant Sci* **15:** 436–446. doi:10.1016/j.tplants.2010.05.001

Waese J, Fan J, Pasha A, Yu H, Fucile G, Shi R, Cumming M, Kelley LA, Sternberg MJ, Krishnakumar V, et al. 2017. ePlant: visualizing and exploring multiple levels of data for hypothesis generation in plant biology. *Plant Cell* **29:** 1806–1821. doi:10.1105/tpc.17.00073

Winter D, Vinegar B, Nahal H, Ammar R, Wilson GV, Provart NJ. 2007. An "electronic fluorescent pictograph" browser for exploring and analyzing large-scale biological data sets. *PLoS ONE* **2:** e718. doi:10.1371/journal.pone.0000718

Woo EJ, Marshall J, Bauly J, Chen JG, Venis M, Napier RM, Pickersgill RW. 2002. Crystal structure of auxin-binding protein 1 in complex with auxin. *EMBO J* **21:** 2877–2885. doi:10.1093/emboj/cdf291

Xu T, Wen M, Nagawa S, Fu Y, Chen JG, Wu MJ, Perrot-Rechenmann C, Friml J, Jones AM, Yang Z. 2010. Cell surface- and rho GTPase-based auxin signaling controls cellular interdigitation in *Arabidopsis*. *Cell* **143:** 99–110. doi:10.1016/j.cell.2010.09.003

Xu T, Dai N, Chen J, Nagawa S, Cao M, Li H, Zhou Z, Chen X, De Rycke R, Rakusová H, et al. 2014. Cell surface ABP1-TMK auxin-sensing complex activates ROP GTPase signaling. *Science* **343:** 1025–1028. doi:10.1126/science.1245125

Yin K, Han X, Xu Z, Xue H. 2009. *Arabidopsis* GLP4 is localized to the Golgi and binds auxin in vitro. *Acta Biochim Biophys Sin* **41:** 478–487. doi:10.1093/abbs/gmp036

Zhou K. 2019. GPI-anchored SKS proteins regulate root development through controlling cell polar expansion and cell wall synthesis. *Biochem Biophys Res Comm* **509:** 119–124. doi:10.1016/j.bbrc.2018.12.081

Noncanonical Auxin Signaling

Heather Marie McLaughlin, Aaron Chun Hou Ang, and Lars Østergaard

Crop Genetics Department, John Innes Centre, Norwich Research Park, Norwich NR4 7UH, United Kingdom

Correspondence: Lars.Ostergaard@jic.ac.uk

Auxin influences all aspects of plant growth and development and exerts its function at scales ranging from the subcellular to the whole-organism level. A canonical mechanism for auxin signaling has been elucidated, which is based on derepression of downstream genes via ubiquitin-mediated degradation of transcriptional repressors. While the combinatorial nature of this canonical pathway provides great potential for specificity in the auxin response, alternative noncanonical signaling pathways required to mediate certain processes have been identified. One such pathway affects gene regulation in a manner that is reminiscent of mechanisms employed in animal hormone signaling, while another triggers transcriptional changes through auxin perception at the plasma membrane and the stabilization of transcriptional repressors. In some cases, the exact perception mechanisms and the nature of the receptors involved are yet to be revealed. In this review, we describe and discuss current knowledge on noncanonical auxin signaling and highlight unresolved questions surrounding auxin biology.

The effects of auxin during plant development were first documented in the 1800s, when it was postulated that the movement of so-called "growth regulators" between tissues influenced both morphogenesis and responses to abiotic stimuli, such as gravity and light (for review, see Enders and Strader 2015). However, indole 3-acetic acid (IAA), the natural auxin predominantly responsible for mediating these effects, was not isolated from plants until much later (Haagen-Smit et al. 1941). Auxin has morphogenic properties and is implicated in signaling over both short and long distances, as well as in rapid responses to environmental fluctuations and long-term developmental processes (for review, see Bhalerao and Bennett 2003; Cho et al. 2007; Mroue et al. 2018; Zhao 2018).

Auxin has a profound effect on gene expression. The exogenous application of auxin to a variety of tissues has demonstrated its genome-wide effects, which include the modulation of hundreds to thousands of genes depending on cellular context (Paponov et al. 2008; Simonini et al. 2017; Kalve et al. 2020). An elegant molecular mechanism for how auxin controls gene expression has been elucidated (for review, see Salehin et al. 2015; Weijers and Wagner 2016; Leyser 2018). Central to this mechanism are proteins of the TRANSPORT INHIBITOR RESPONSE1/AUXIN SIGNALING F-BOX (TIR1/AFB) family, which act as auxin receptors. In the absence of auxin, repression is maintained at target loci through the Aux/IAA-mediated recruitment of TOPLESS/TOPLESS-RELATED

Cite this article as *Cold Spring Harb Perspect Biol* doi: 10.1101/cshperspect.a039917

(TPL/TPR) corepressors (Fig. 1; Szemenyei et al. 2008). Upon auxin binding, TIR1/AFBs increase their affinity toward Aux/IAA transcriptional repressors to facilitate their degradation, and thereby relieve the repression of transcription factors belonging to the AUXIN RESPONSE FACTOR (ARF) family (Dharmasiri et al. 2005; Kepinski and Leyser 2005). Plants with mutations in components of the TIR1/AFB pathway have severe developmental defects, demonstrating the prominent role of this pathway in relaying the auxin signal (Berleth and Jürgens 1993; Leyser et al. 1993; Hamann et al. 1999; Prigge et al. 2020). In this review, we will refer to this TIR1/AFB-based mechanism as the canonical auxin signaling pathway or simply the canonical pathway.

An open question in the field of plant biology is how canonical auxin signaling can facilitate the vast array of processes involving auxin. The duplication and neofunctionalization of pathway components are likely to contribute to this versatility. For example, the *Arabidopsis* genome encodes six TIR1/AFB receptors, 29 Aux/IAAs, and 23 ARFs operating in different combinations in different tissues (Salehin et al. 2015). However, several observations have at some point seemed incompatible with signaling through the canonical pathway either due to their distinct temporal or spatial occurrences, or because they were thought to be regulated by distinct receptors. These observations are (1) acid growth (Hager et al. 1971; for review, see Arsuffi and Braybrook 2018; Gallei et al. 2020), (2) auxin-induced cytoskeletal rearrangements (Bergfeld et al. 1988; for review, see Sauer and Kleine-Vehn 2011; Zhu and Geisler 2015), (3) rapid root-growth inhibition (Fendrych et al. 2018), (4) the rapid influx of Ca^{2+} during gravitropism (Monshausen et al. 2011), (5) TIR1/AFB2-independent inhibition of PIN cycling (Paciorek et al. 2005; Robert et al. 2010), (6) localization of canonical pathway components to the cytosol (Wang et al. 2016; Powers et al. 2019), and (7) the limited interaction between repressive ARFs and Aux/IAA repressors (Vernoux et al. 2011). While some of these phenomena have since been shown to require canonical pathway components (Dindas et al. 2018; Fendrych et al. 2018; Uchida et al. 2018), or to occur independently of auxin in the case of microtubule rearrangements (Adamowski et al. 2019), the mechanisms underlying others have not yet been elucidated. Evidence for alternative auxin signaling mechanisms has gradually accumulated, and such mechanisms may contribute to the versatility of the auxin response throughout development. Here, we provide a summary of recent findings in this area and aim to highlight critical questions that must be answered to obtain a more comprehensive understanding of auxin biology.

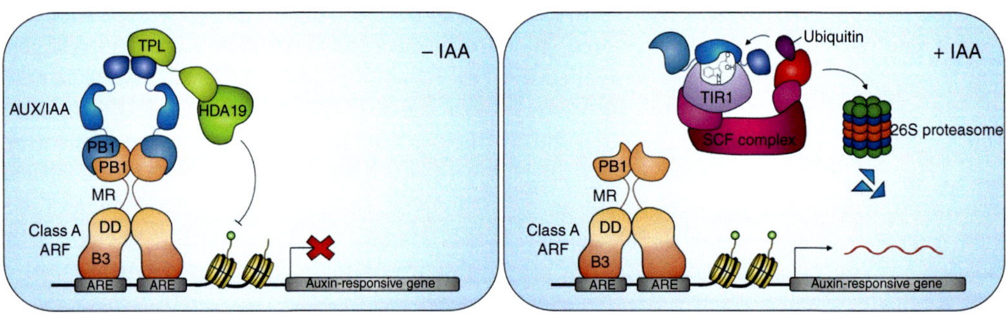

Figure 1. The canonical auxin signaling pathway. In the absence of auxin, AUXIN RESPONSE FACTORS (ARFs) are bound by Aux/IAA repressor proteins, which recruit transcriptional corepressors and prevent the expression of auxin-responsive genes. Auxin increases the affinity between the TIR1/AFB auxin receptor complex and Aux/IAAs, which are subsequently ubiquitinated and degraded. ARFs are then free to activate the expression of auxin-responsive genes. (ARE) Auxin-responsive element, (B3) B3 DNA-binding domain, (DD) dimerization domain, (MR) middle region, (PB1) Phox and Bem1 domain.

INCOMPATIBILITIES BETWEEN AUXIN EFFECTS AND THE CANONICAL PATHWAY

The acid growth hypothesis was proposed around 50 years ago (Hager et al. 1971) and is one of the many phenomena that prompted the idea of noncanonical auxin signaling. In acid growth, auxin activates the movement of protons into the apoplast through the plasma membrane (PM) proton-pumping ATPase (PM H^+-ATPase). The acidification of the apoplast upon proton accumulation leads to alterations in the properties of the cell wall, promoting hyperpolarization of the PM to allow turgor pressure-mediated cell expansion (Kutschera 1994; Hager 2003; Takahashi et al. 2012; Arsuffi and Braybrook 2018). The activity of PM H^+-ATPases depends on their phosphorylation status, and phosphorylation was observed in *tir1 afb2* double mutants implying that auxin receptors other than TIR1 and AFB2 are responsible for regulating ATPase activity (Takahashi et al. 2012).

Initially, it was proposed that AUXIN-BINDING PROTEIN1 (ABP1) was responsible for mediating nontranscriptional, rapid auxin responses such as cell expansion, cytoskeletal rearrangement, and the inhibition of endocytic PIN cycling (Chen et al. 2001, 2014; Robert et al. 2010) and could therefore be considered a candidate for the non-TIR1/AFB receptor in acid growth (Sauer and Kleine-Vehn 2011). However, it has since been shown that *abp1* loss-of-function does not cause developmental defects in *Arabidopsis*, and its importance in auxin biology is now disputed (Dünser and Kleine-Vehn 2015; Gao et al. 2015; Habets and Offringa 2015).

Later, it was discovered that a subset of the *SMALL AUXIN UP-RNA* (*SAUR*) gene family are induced through the canonical pathway and inhibit the activity of PP2C-D protein phosphatases to promote PM H^+-ATPase phosphorylation, thus directly linking the activity of the TIR1/AFB pathway to acid growth (Fig. 2; Spartz et al. 2014). A synthetic auxin response system has since been developed, using a convex auxin ligand (cvxIAA), which hijacks the downstream auxin signaling machinery in the presence of a concave TIR1 (ccvTIR1) receptor (Uchida et al. 2018). When expressing this synthetic ccvTIR1 receptor in the *tir1 afb2* double mutant under control of the endogenous *TIR1* promoter, treatment with cvxIAA-induced *SAUR19* expression, acid growth, and PM H^+-ATPase phosphorylation, providing conclusive evidence that the canonical pathway regulates acid growth (Uchida et al. 2018). Consistent with Takahashi et al. (2012), treatment of the *tir1 afb2* mutant with natural IAA also induced H^+-ATPase phosphorylation. Moreover, rapid cell elongation in response to natural IAA was still observed in the *tir1 afb2* ccvTIR1 line, suggesting that other auxin receptors contribute to this response. Whether these receptors include other members of the AFB family remains to be established (Uchida et al. 2018).

A second phenomenon that suggested a role for noncanonical signaling during growth and development is auxin-mediated root growth inhibition. Using microfluidics and live imaging, Fendrych et al. (2018) showed that root growth was rapidly inhibited upon the exogenous application of auxin on a time scale that is incompatible with a transcriptional response. Equally, removal of auxin led to the rapid reestablishment of root growth, suggesting a rapid and reversible mechanism of auxin-mediated growth regulation. Interestingly, the response still depends on TIR1/AFB, which suggests the existence of a nontranscriptional output from the canonical pathway (Fendrych et al. 2018; Dubey et al. 2020).

Root growth inhibition on the lower side of the root forms the basis of gravitropism (Shih et al. 2015), and gravistimulation is followed by the redistribution of auxin in the root, PM depolarization, and the influx of Ca^{2+}. Although auxin-mediated calcium signaling was initially thought to occur independently of the canonical receptors (Monshausen et al. 2011), the influx of Ca^{2+} depends on the intracellular perception of auxin through the canonical pathway and on the CNGC14 Ca^{2+} channel (Shih et al. 2015; Dindas et al. 2018).

The involvement of the canonical receptors in these rapid responses raises questions on how the nuclear-localized TIR1/AFBs can mediate nontranscriptional signaling and alterations in signaling components at the PM (Retzer et al.

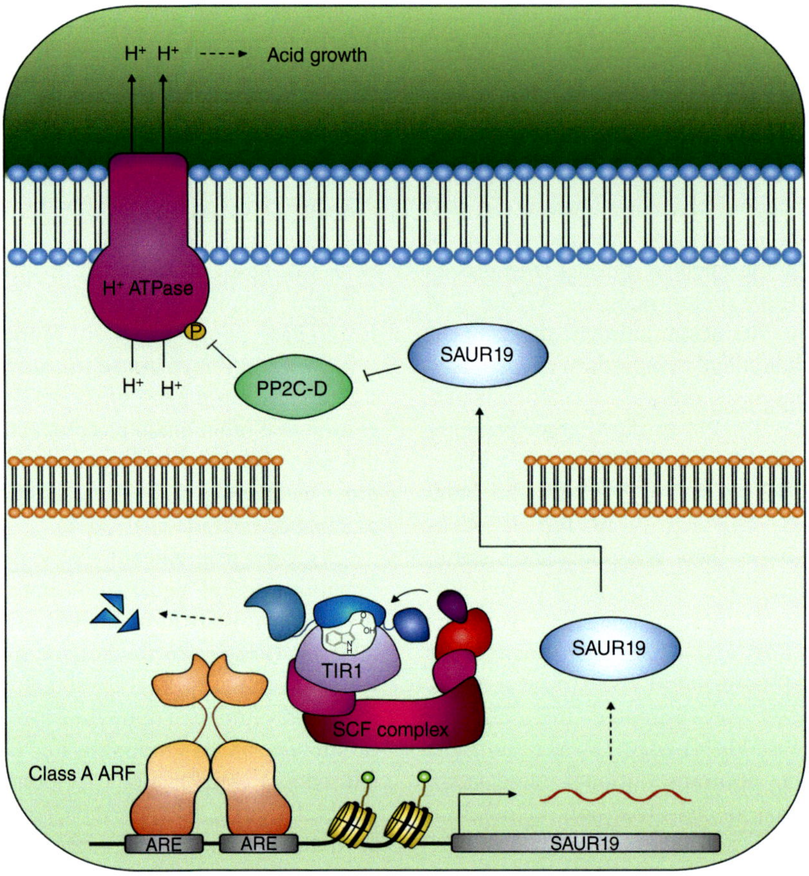

Figure 2. Auxin triggers the up-regulation of SAUR19 through the canonical signaling pathway. SAUR19 inhibits PP2C-D activity leading to the phosphorylation and activation of H^+-ATPases, the acidification of the apoplast, and acid growth.

2018). Both Fendrych et al. (2018) and Dindas et al. (2018) speculated that canonical pathway components may be translocated out of the nucleus to mediate their effects at the cell periphery. TIR1/AFBs have been observed in the cytosol, and upon their stabilization, when mutations in the receptors release them from their associated E3 ubiquitin ligases, auxin-resistant phenotypes were observed (Yu et al. 2015; for review, see Kubeš and Napier 2019). Prigge et al. (2020) observed that fluorescently tagged TIR1 exhibited a predominantly nuclear localization, whereas AFB1 was primarily cytosolic and AFB2-5 partitioned between the nucleus and the cytoplasm (Prigge et al. 2020). Furthermore, TIR1/AFBs are chaperoned by cytoplasmic Heat-Shock Protein 90s (HSP90s), which appears to stabilize the receptors to promote their nuclear localization and temperature-dependent auxin responses (Wang et al. 2016). It remains unclear whether localization of the canonical receptors to the cytosol has physiological consequences; however, other canonical pathway components have been observed in the cytosol. ARF7 and ARF19 were found to form cytoplasmic assemblies with physiological consequences on auxin sensitivity (Powers et al. 2019), although this nucleocytoplasmic partitioning has not to date been linked to HSP90 activity.

Another open question relating to ARF function in the canonical pathway is the role of ARF repressors in gene regulation. ARF proteins are B3-type transcription factors that have been divided into specific classes based on phylogeny and ability to activate/repress gene expression (Tiwari et al. 2003; Finet et al. 2013; Mutte et al. 2018). The ARF family in *Arabidopsis* has 23 members with a general domain structure where the DNA-binding domain is situated in the amino-terminal half followed by a middle region (MR) and a carboxy-terminal Phox and Bem1 (PB1) domain. The amino acid sequence of the MR dictates the function of the ARF as either activating (A-ARFs) or repressive (B-ARFs and C-ARFs) (Tiwari et al. 2003). At the carboxy-terminus, most ARFs contain a PB1 domain that facilitates Aux/IAA interactions, repressing ARF activity in the absence of auxin (Salehin et al. 2015).

The canonical model of auxin signaling proposes that the lifting of Aux/IAA-mediated repression in the presence of auxin leads to the expression of auxin-responsive genes. This model fits well with the activity of A-ARFs, but the role of B-ARFs and C-ARFs in this model was less clear until recently. Through a study in the liverwort, *Marchantia polymorpha*, which contains one ARF of each class, it was demonstrated that the B-ARFs compete with A-ARFs to bind auxin response elements (AREs) in the promoters of common target genes independently of auxin (Kato et al. 2020). Whether B-ARFs function in a similar way in *Arabidopsis* where the ARF clade has undergone expansion is unknown; however, it has been shown that B- and C-ARFs have a limited capacity to interact with the Aux/IAA repressors despite the presence of a carboxy-terminal PB1 domain, suggesting that they may also function as auxin-independent transcriptional repressors in *Arabidopsis* (Vernoux et al. 2011; Piya et al. 2014).

ETT-MEDIATED AUXIN SIGNALING

ETT, ARF13, ARF17, and ARF23 are atypical ARFs that entirely lack the PB1 domain required for interactions with Aux/IAAs (Guilfoyle and Hagen 2007; Lokerse and Weijers 2009). While the carboxyl termini of ARF13, ARF17, and ARF23 are simply truncated, ETT contains a region at its carboxy terminus, which does not share homology with other eukaryotic proteins. This domain was named the ETTIN-specific (ES) domain (Simonini et al. 2016). Given their atypical domain structures, these ARFs do not fit into the canonical model of signaling. So far, no biological function has been assigned to ARF13 or ARF23, but ARF17 is important for pollen development and anther dehiscence (Yang et al. 2013; Wang et al. 2017; Xu et al. 2019), although the molecular mechanism is unknown.

ett loss-of-function mutants have a range of auxin-related phenotypes most prominently during gynoecium development (Sessions and Zambryski 1995; Sessions et al. 1997). In addition, defects in ovule integument development, lateral root emergence, and primary branch formation have been reported (Sessions and Zambryski 1995; Sessions et al. 1997; Garcia et al. 2006; Marin et al. 2010; Kelley et al. 2012; Simonini et al. 2016). The auxin-related developmental defects of *ett* mutants led to the hypothesis that ETT can mediate auxin signaling independently of the canonical pathway. It was suggested that ETT translates local auxin concentrations to developmental outputs in the gynoecium, but the molecular mechanisms underpinning this were unknown (Nemhauser et al. 2000; Pekker et al. 2005; Simonini et al. 2016). Since then, data supporting a mechanism that is fundamentally different from canonical auxin signaling are emerging. For example, it has been established that several protein–protein interactions involving ETT are sensitive to the natural auxin, IAA, but not other auxinic compounds (Simonini et al. 2016). Several of these interactions are with transcription factors belonging to a range of different families and it has been hypothesized that such interactions are relevant for auxin responsiveness of specific tissues or cell types during development (Simonini et al. 2016). As auxin levels rise, interactions between ETT and its protein partners are broken. Chromatin immunoprecipitation (ChIP) experiments have shown that ETT remains bound to the DNA of its target loci both in the presence and absence of auxin (Simonini et al. 2016). Moreover, a genome-wide analysis of ETT-mediated gene regulation showed that the transcriptome regulated by ETT changes dynamically depending on the auxin levels (Simonini et al. 2016, 2017). For a subset of targets that were analyzed in further detail, ETT-dependent, auxin-responsive gene expression was shown to be independent of the canonical pathway (Kuhn et al. 2020). Biophysical data have shown that ETT is able to directly interact with the auxin molecule via the ES domain suggesting that binding of auxin disrupts the interaction between ETT and its partners (Kuhn et al. 2020).

Like animal genomes, plant genomes encode corepressors of the conserved Tup1/GROUCHO/TLE family, and their functions have been described in both monocots and dicots (Long et al. 2006; Liu et al. 2019). In *Arabidopsis*, this family is comprised of TOPLESS (TPL) and TPL-related (TPR) proteins and has been shown to associate with transcription factors affecting most aspects of plant development (Causier et al. 2012). The molecular function of TPL was first described in the context of auxin signaling where TPL was found to interact with Aux/IAAs when auxin levels are low to repress gene expression. As auxin levels increase and Aux/IAAs dissociate from ARFs, TPL also dissociates from the locus to allow derepression (Szemenyei et al. 2008). TPL-mediated repression occurs through the recruitment of histone deacetylases (HDACs) such as HDA19, which maintain a repressed state at target loci that can subsequently be reversed upon Aux/IAA degradation and TPL dissociation (Krogan et al. 2012).

Members of the TPL/TPR family also play a central role in ETT-mediated auxin signaling. Under low auxin conditions, ETT can interact directly with TPL/TPR proteins, but this interaction is disrupted when auxin concentrations rise (Fig. 3A). The interaction occurs via a repressive motif in the ES domain of ETT with the amino acid sequence RLFG found as a common motif in proteins interacting with TPL/TPR (Causier et al. 2012; Liu et al. 2019; Kuhn et al. 2020).

Generally, hormone signaling in plants involves the ubiquitination and subsequent degradation of transcriptional repressors upon ligand perception and facilitates the activation of gene expression (for review, see Santner and Estelle 2010; Kelley and Estelle 2012). The canonical auxin signaling pathway provides one such example (Salehin et al. 2015; Weijers and Wagner 2016).

In contrast, hormone signaling in animals often involves direct binding of the hormone ligand (as in thyroid hormone [TH] signaling) or an effector protein (as in Wingless [Wnt]/β-catenin signaling) to transcription factors. For instance, in the absence of TH, the thyroid hormone receptor (THR) binds its target genes in complex with corepressors that attract HDACs to the locus to maintain the chromatin in a repressed state. When TH levels increase, TH binds to the DNA-bound THR resulting in a conformational change and the exchange of corepressors for coactivators, triggering the recruitment of histone acetyl transferases (HATs) and the activation of target-gene expression (Fig. 3B; Tsai and O'Malley 1994).

In Wnt/β-catenin signaling the perception of the Wnt ligand at the PM triggers the release of β-catenin from the degradasome (Fig. 3C; Gammons and Bienz 2018). Components of the degradasome are sequestered at the PM forming the signalosome, allowing β-catenin to accumulate and enter the nucleus. Here, β-catenin associates with components of the enhanceosome, including the DNA-bound TCF transcription factor. Wnt association triggers a conformational change and the dissociation of GROUCHO corepressors and activates Wnt-responsive gene expression.

The mechanism by which ETT relays changes in auxin levels is reminiscent of the TH and Wnt/β-catenin pathways in animals. Like THR and TCF, ETT is constitutively bound to at least a subset of its targets; however, it remains to be established whether ETT is involved in the formation of an activating complex upon ligand perception as in the case of THR during TH perception (Fig. 3). A fundamental difference between the ETT-mediated and canonical models of auxin signaling is that the former does not in its core mechanism invoke the requirement for protein degradation to activate gene expression. To reinstate repressive conditions in the canonical pathway, a new pool of Aux/IAA repressors must first be synthesized. In con-

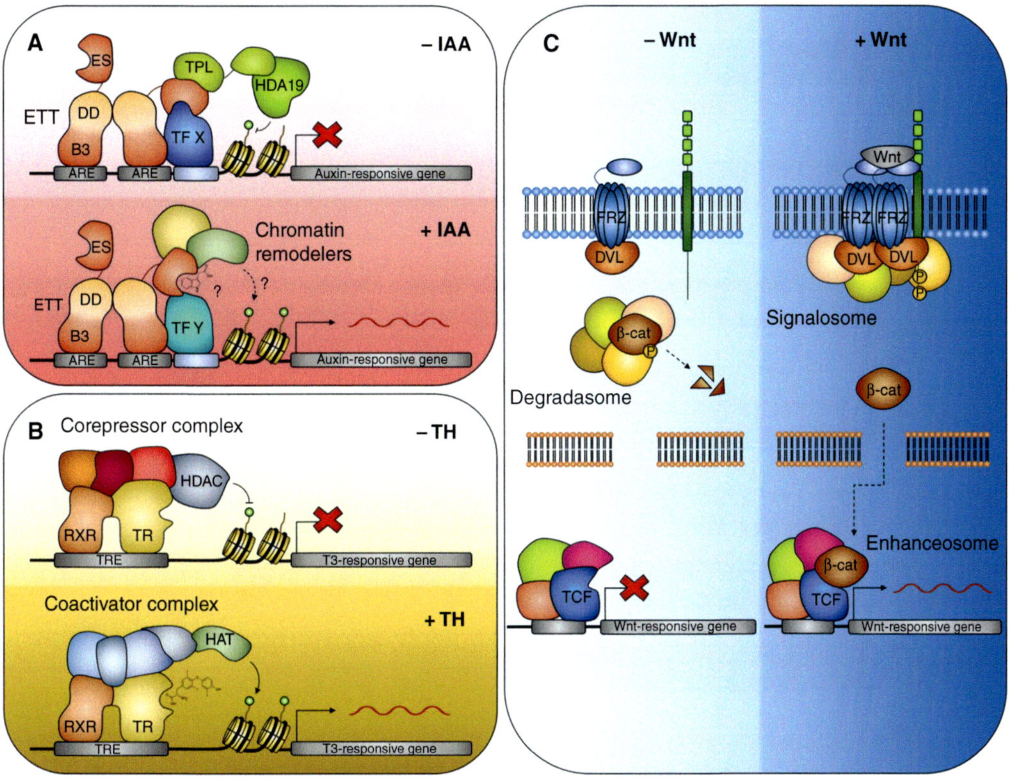

Figure 3. Ligand-mediated switch in the activity of transcription factor complexes. (*A*) The ETT-mediated auxin signaling pathway. In the absence of auxin, ETT recruits the corepressor TOPLESS (TPL) to target loci. TPL, in turn, recruits HDA19, which deacetylates histones and represses target gene expression. Auxin binds to ETT directly through the ETT-specific (ES) domain and triggers the dissociation of the repressive complex. Histone acetylation occurs, potentially through the recruitment of coactivators by ETT, and target genes are derepressed. (*B*) Thyroid hormone (TH) signaling pathway. In the absence of TH, the DNA-bound thyroid hormone receptor (THR) is associated with corepressors and histone deacetylases (HDACs), which repress target gene expression. In the presence of TH, the corepressors are exchanged for coactivators, and histone acetyltransferases (HATs) are recruited, activating target genes. (*C*) Wnt/β-catenin signaling. In the absence of Wnt, β-catenin is degraded by the degradasome. When the Wnt ligand binds to the Frizzled (FRZ) receptor, degradasome components are recruited to the plasma membrane by the active receptors, leading to the formation of the signalosome, and the accumulation of β-catenin. β-Catenin enters the nucleus and associates with the TCF transcription factor and other components of the enhanceosome, the Groucho corepressors dissociate, and Wnt-responsive genes are expressed.

trast, the ETT-based mechanism is immediately reversible, thus providing an ability to function in specialized processes where a fast and reversible response is important.

The evolutionary origin of the ETT-mediated auxin signaling pathway is so far unknown. While the canonical pathway is conserved at least back to the extant bryophytes such as *M. polymorpha* (Mutte et al. 2018; Kato et al. 2020), homologs of ETT do not appear outside the angiosperm phylum and ETT-mediated signaling may therefore be absent in nonflowering plants. The domains and residues in ETT-ES known to be important for auxin perception and TPL interaction are highly conserved throughout the angiosperms (Finet et al. 2010; Kuhn et al. 2020). ETT proteins in the basal angiosperms, such as *Amborella trichopoda*, contain a carboxy-terminal PB1 domain, as well as the conserved RLFG motif required for interaction with TPL/TPR, and this

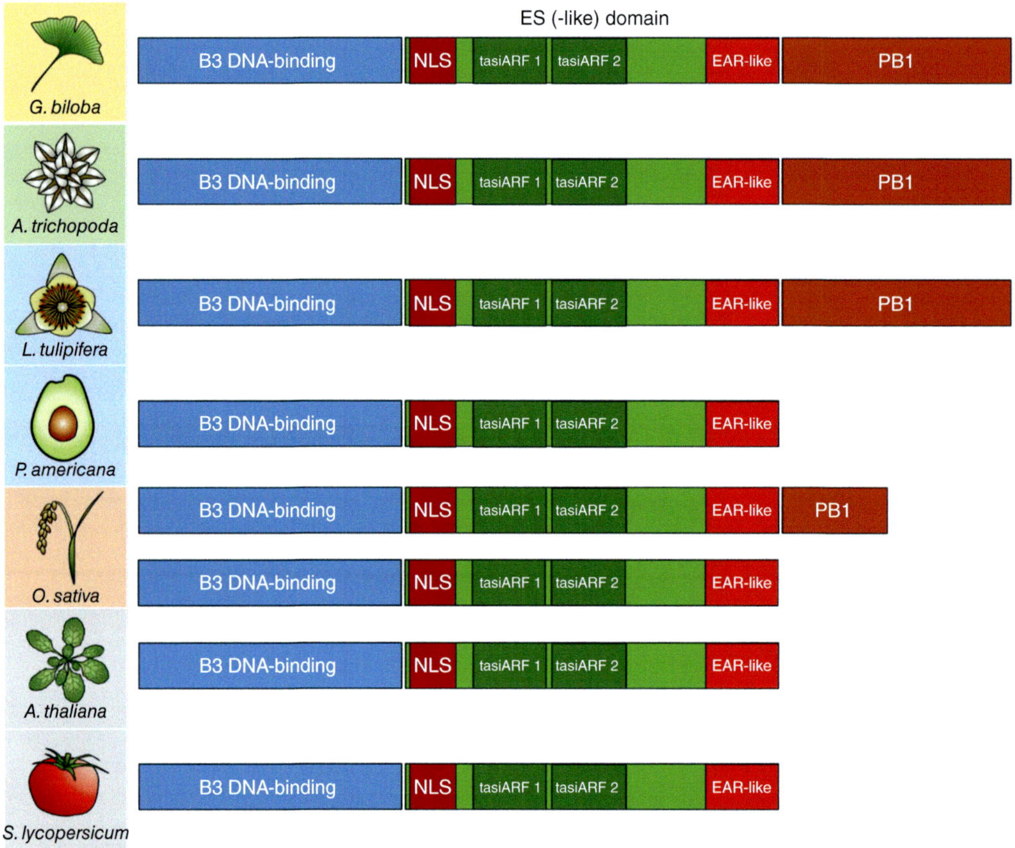

Figure 4. ETT domains and their conservation across the seed plants. ETT in the eudicot *Arabidopsis thaliana* and *Solanum lycopersicum* contains a highly conserved B3 DNA-binding domain, a nuclear localization signal (NLS), two tasi-ARF-binding sites, and an EAR-like TPL-interacting domain. The ARF3/4-like ortholog of the gymnosperm *Ginkgo biloba* and the ETT of the basal angiosperm *Amborella trichopoda* share these domains but also contain a full-length PB1 domain. Within the magnoliids, the PB1 domain is retained in the *Liriodendron tulipifera* ortholog but has been lost in the *Persea americana* ortholog. The monocot *Oryza sativa* has two clades of ETT paralogs with either a partial or total loss of the PB1 domain.

is likely to represent the ancestral state of ETT (Fig. 4; Finet et al. 2010, 2013; Causier et al. 2012; Kuhn et al. 2020). The PB1 domain was then lost independently in several lineages of extant angiosperms, including in the magnoliids and monocots (Finet et al. 2013).

The closest homolog of ETT in *Arabidopsis* is another B-ARF, ARF4 (Pekker et al. 2005). A phylogenetic study strongly suggested that ETT and ARF4 were generated by a duplication event that took place after angiosperms diverged from gymnosperms, but before the last common ancestor of extant angiosperms (Finet et al. 2010). In contrast to ETT, ARF4 contains a carboxy-terminal PB1 domain and was previously shown to interact with Aux/IAA repressors, indicating its regulation via canonical auxin signaling in *Arabidopsis* (Vernoux et al. 2011). ETT and ARF4 have overlapping functions and *arf4* loss-of-function mutants severely enhance defects of *ett* mutants (Pekker et al. 2005). Intriguingly, ARF4 is also truncated through alternative splicing in several angiosperm lineages, suggesting that the truncation of ETT and ARF4 proteins generally has a functional relevance (Finet et al. 2013). Experiments involving chimeric

ETT and ARF4 constructs suggested that the unique carboxy-terminal domain of ETT is important for its function, and that ARF4 function similarly depends on the presence of the PB1 domain (Finet et al. 2010). Further domain swap experiments using ETT and ARF4 domains from species across the angiosperm lineage will be required to fully understand how and when ETT and ARF4 adopted distinct functions, and to reveal the evolutionary origins of ETT-mediated auxin signaling.

Several mutant screens carried out in *Arabidopsis* to identify genetic factors critical for development or auxin responsiveness led to the discovery of genes encoding components of the canonical pathway (Berleth and Jürgens 1993; Leyser et al. 1993; Hardtke and Berleth 1998; Hamann et al. 1999). This clearly reflects the predominance of this pathway and its importance throughout the plant life cycle. The ETT-based mechanism is not employed to such an extent; however, its immediate reversibility suggests that it is particularly well-suited for certain processes such as the changing of tissue polarity (Simonini et al. 2016). Finally, given the mechanistic conservation between ETT-based auxin signaling and certain animal hormone pathways, it is possible that direct binding of plant hormones or other small molecules to transcription factors is a more common mechanism of regulating gene expression in plants than previously expected. Approaches involving highly sensitive proteomics and biophysics techniques will be essential to elucidate other such mechanisms.

NONCANONICAL SIGNALING THROUGH PROTEIN KINASES

Auxin gradients underpin the formative cell division and expansion patterns that are required for appropriate organ morphogenesis and the formation of the plant body plan (for review, see Bhalerao and Bennett 2003; Friml 2003). While auxin-mediated cell expansion has been widely studied, and many of the mechanisms relating to asymmetric cell growth have been elucidated (for review, see Majda and Robert 2018; Du et al. 2020), processes involved in cell division patterning are more poorly understood. The Trans-Membrane Kinase 1 (TMK1) pathway has been implicated in regulating both differential cell elongation and the establishment of cell division patterns in response to auxin.

There are four kinases in the TMK subfamily in *Arabidopsis*, all of which contain an extracellular leucine-rich-repeat (LRR) motif linked to an intracellular kinase domain through a single transmembrane region (Dai et al. 2013). The TMK subfamily was first linked to auxin signal transduction when the phenotypes of *tmk* mutant combinations were analyzed. While single mutants showed no obvious phenotypic abnormalities, double, triple, and quadruple mutants showed cell expansion and proliferation defects, miniaturized organs, infertility, and a reduced sensitivity to exogenously applied auxin (Dai et al. 2013). TMK1 has since been implicated in apical hook maintenance, lateral root development (along with TMK4), and pavement cell morphogenesis (Xu et al. 2014; Cao et al. 2019; Huang et al. 2019). The roles of other TMK family members in auxin signaling have not yet been elucidated.

TMK1 was initially thought to regulate auxin-mediated signaling through its interaction with Auxin-Binding Protein 1 (ABP1) at the PM (Xu et al. 2010, 2014). It was observed that the auxin-mediated activation of the TMK1 pathway triggered Rho-like GTPases from plants (ROPs) at the PM, leading to alterations in the cytoskeleton and the distribution of PINs that seemed to be important for lobe development in the pavement cells of the leaves (Xu et al. 2010, 2014). However, the significance of ABP1 in plant biology has since been put into question (Gao et al. 2015; Habets and Offringa 2015), and so the receptor responsible for extracellular auxin binding and subsequent TMK1 activation remains elusive.

The TMK1 pathway was discovered using the apical hook as a model system (Cao et al. 2019; for review, see Gallei et al. 2020). The apical hook is a structure produced after germination in dicots that protect the shoot apical meristem and cotyledons from damage as the hypocotyl elongates through the soil in search of light. Its formation is facilitated through asymmetric growth on either side and this process is regulated by auxin (for review, see Abbas et al. 2013). The production of the apical hook occurs in three stages: formation, maintenance, and opening. During the formation phase, auxin is concentrated at the concave (inner) side of the apical hook, inhibiting cell elongation, while the convex (outer) side of the hook elongates. This difference in growth leads to hook bending.

Auxin biosynthesis mutants show variable phenotypes relating to the abolition of differential growth in the apical hook (Cao et al. 2019). Ectopic auxin production in the *yuc1-D* mutant represses growth at the convex side of the apical hook, while auxin deficiency in the *wei8-3 tar2-1* mutant promoted growth at the concave side, highlighting that the asymmetric distribution of auxin is vital for differential growth and apical hook formation. The *tmk1* mutant also exhibited growth at the concave side of the apical hook. However, the exogenous application of auxin, which rescued defects in the *wei8-3 tar2-1* apical hook, did not rescue the *tmk1* phenotype, implying that *tmk1* was impaired in auxin perception and/or signaling.

The carboxyl-terminus of TMK1 is cleaved during the maintenance phase at the concave side of the apical hook, in response to the relatively high concentrations of auxin present in this region of the hypocotyl (Fig. 5A; Cao et al. 2019). The carboxy-terminal TMK1 fragment is then translocated to the nucleus, where it interacts with and phosphorylates the Aux/IAA repressors, IAA32 and IAA34, leading to their stabilization and ultimately the repression of growth at the concave side of the apical hook. Unlike Aux/IAA proteins in the canonical pathway, IAA32 and IAA34 lack the degron domain required for interaction with TIR1/AFB receptors and are stabilized in the presence of auxin by TMK1-mediated phosphorylation (Calderon-Villalobos et al. 2010; Cao et al. 2019). Under low auxin concentrations, TMK1 is inactive and IAA32/34 are destabilized, allowing auxin-responsive gene expression and growth (Fig. 5B).

The TMK1-mediated auxin signaling pathway has also been implicated in the control of mitogen-activated protein kinase (MAPK) signaling (Fig. 5C; Huang et al. 2019). Lateral root emergence requires tight regulation of cell division and expansion. Auxin maxima form at the locations where lateral roots begin to initiate, and disruption of auxin distribution (De Smet et al. 2007) or of MAPK signaling (Zhu et al. 2019) caused defects in lateral root development. It was previously observed that the application of exogenous auxin triggered the activation of MAPK signaling in the root, although the mechanism by which this occurs is unknown (Mockaitis and Howell 2000).

Huang et al. (2019) showed that auxin triggers the activation of TMK1 and TMK4, which directly interact with and phosphorylate the MAPK-kinases (MAPKKs) 4/5 through their carboxy-terminal intracellular kinase domains. MAPK3/6 were reported to act downstream of MAPKK4/5 in response to several stimuli and are coexpressed with TMK1 during lateral root emergence. The *tmk1 tmk4* double mutant and RNAi knockdown lines of *mapkk*4/5 and *mapk*3/6 all showed the same defects in lateral root emergence and aberrant cell division patterning in the emerging lateral root primordia, suggesting a shared role in regulating these processes (Huang et al. 2019).

In wild-type plants, the application of auxin triggers lateral root emergence. In auxin-deficient mutants, which produce no lateral roots, lateral root development can be restored by the application of auxin (Huang et al. 2019). The lateral root phenotypes in the *tmk1 tmk4* double mutant, and *mapkk*4/5 and *mapk*3/6 RNAi knockdown lines, could not be rescued in this manner implying that the TMK1/4-MAPKK4/5-MAPK3/6 module transduces the auxin signal during lateral root development. Furthermore, the level of MAPK3/6 phosphorylation directly correlated with the levels of auxin in the lateral root, and MAPK3/6 phosphorylation was abolished in the *tmk1 tmk4* but not in *tir1 afb2 afb3* mutant lines, showing that this TMK1/4-dependent pathway is independent of canonical auxin signaling.

Auxin was previously shown to regulate lateral root emergence through the IDA-HAE/HSL2 pathway, which is also upstream of the MAPKK4/5-MAPK3/6 module (Kumpf et al. 2013; Zhu et al. 2019). This raises questions about

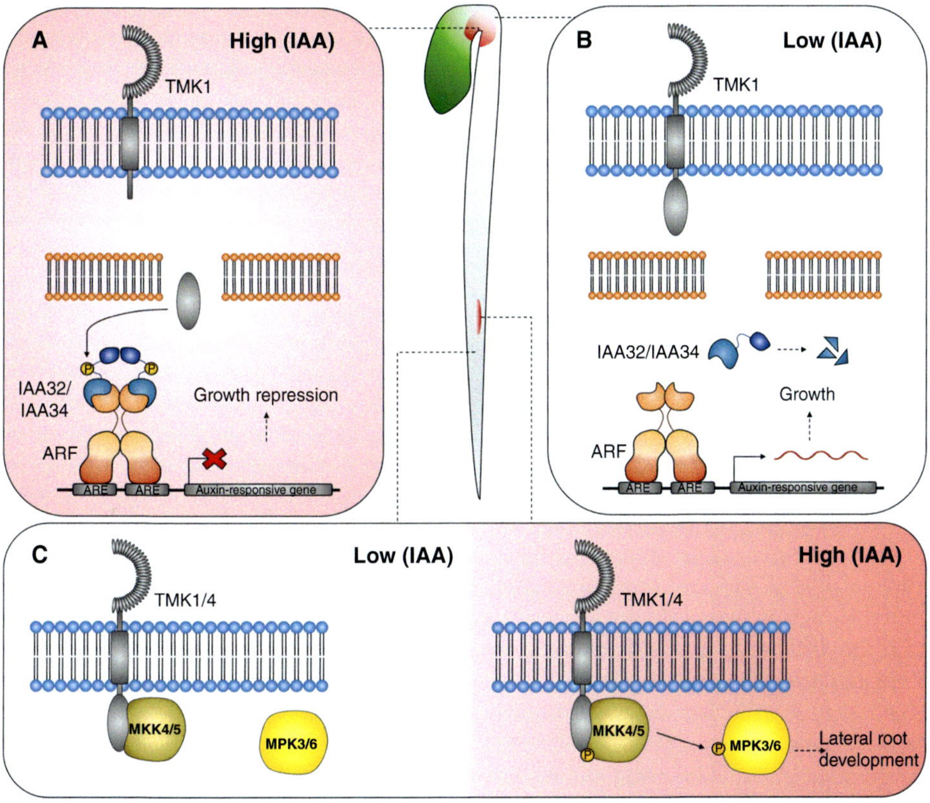

Figure 5. Trans-Membrane Kinase 1 (TMK1)-mediated auxin signaling during apical hook maintenance and lateral root emergence. (*A*) Relatively high auxin concentrations at the concave side of the apical hook activate the cleavage of the TMK1 carboxyl terminus. (*B*) At the convex side of the apical hook, relatively low auxin concentrations promote growth due to the destabilization of IAA32/34. The carboxyl-terminus of TMK1 is translocated to the nucleus, where it phosphorylates and stabilizes IAA32/34 to repress gene expression and inhibit growth. (*C*) Under high auxin conditions, TMK1 phosphorylates MKK4/5, triggering the downstream phosphorylation of MPK3/6 to regulate lateral root emergence.

how peptide and phytohormone signaling modules undergo cross talk to regulate developmental processes, and how specificity in signaling is established through shared signaling modules. The canonical pathway also has important roles during lateral root development, and the exact relationships between all of the auxin-related pathways will now need to be dissected to get a global view of auxin signaling and its regulation of the lateral roots (for review, see He and Meng 2020). Lateral root emergence phenotypes have also been reported in both *ett-3* (Simonini et al. 2016), and canonical receptor mutants (Prigge et al. 2020), and lateral roots may therefore provide an excellent model system in which to dissect interactions between all currently known auxin-mediated signaling pathways.

Other noncanonical Aux/IAAs that lack the degron domain have been implicated in growth and development, including IAA20, IAA30, and IAA33. The *iaa20 iaa30* double mutant produced ectopic protoxylem implying at least a partially redundant role for these Aux/IAAs in xylem patterning (Müller et al. 2016). The expression of both *IAA20* and *IAA30* is induced by auxin through the binding of MONOPTEROS (MP)/ARF5 to their promoters, these Aux/IAAs then bind to MP repressing its activity in a feedback loop (Sato and Yamamoto 2008; Krogan et al. 2014; Müller et al. 2016). The expression levels of these noncanonical Aux/IAAs and MP are further modulated by the HD-ZIP transcription factor PHABULOSA (PHB), which stabilizes the auxin response to allow proper xylem development (Müller et al. 2016). It remains unclear whether IAA20 and IAA30 are phosphorylated to modulate their stability, or whether they function downstream of a kinase-mediated auxin signaling pathway such as the TMK1 pathway.

IAA33 maintains stem cell identity in the root by competing with the canonical Aux/IAA IAA5 to bind ARF10 and ARF16, thereby maintaining the repression of these ARFs in the presence of auxin (Lv et al. 2020). IAA33 is stabilized in response to auxin through its phosphorylation by MAPK14, and this stabilization is required for normal stem cell identity specification in the distal root. It is still unclear how auxin triggers the activation of MAPK14, although it is plausible that it could occur through a TMK1-like pathway.

Phosphorylation is a very rapid and reversible change and is an appealing alternative mechanism to explain how the auxin signal can be transduced so swiftly; however, the TMK1 pathway is yet to be directly linked to any of these rapid auxin responses, and many of them have since been shown to be TIR1/AFB-dependent (Fendrych et al. 2018; Uchida et al. 2018; Dubey et al. 2020).

CONCLUDING REMARKS

While the combinatorial nature of the TIR1/AFB-Aux/IAA-ARF pathway of auxin signaling provides great potential to explain diverse effects of auxin on plant development, there is mounting evidence for the existence of alternative noncanonical signaling pathways to mediate certain processes. A common theme in noncanonical auxin signaling is the loss of key domains from canonical pathway components that then allows them to be recruited as novel signaling modules. For instance, the loss of the degron domain from IAA32/34 in the TMK1-mediated pathway and the PB1 domain from ETT have allowed these pathways to regulate specific developmental processes independently of the canonical pathway. Moreover, canonical auxin receptors have now been linked to nontranscriptional mechanisms and it seems likely that canonical and noncanonical pathways are connected into a wider auxin signaling network that is carefully integrated to regulate a wide range of processes.

The existence of different pathways by which auxin relays its effect opens the door for potential interactions between these mechanisms that would increase the potential for specificity in the auxin response. Understanding such interactions will therefore be crucial to fully comprehend the versatility of auxin function in plant biology. Auxin has been intensely studied by plant scientists for decades and there is little doubt that this hugely important plant hormone will continue to provide scientific excitement and fascination for a long time to come.

ACKNOWLEDGMENTS

We are grateful to Yang Dong, André Kuhn, Bhavani Natarajan, and Nicola Stacey for providing critical comments to the manuscript. H.M.M. was funded by the John Innes Foundation, A.C.H.A. by Sainsbury PhD studentship, and L.Ø. by an Institute Strategic Programme Grant from the UKRI Biotechnological and Biological Sciences Research Council (BBSRC) to the John Innes Centre (BB/P013511/1).

REFERENCES

*Reference is also in this collection.

Abbas M, Alabadí D, Blázquez MA. 2013. Differential growth at the apical hook: all roads lead to auxin. *Front Plant Sci* **4:** 441. doi:10.3389/fpls.2013.00441

Adamowski M, Li L, Friml J. 2019. Reorientation of cortical microtubule arrays in the hypocotyl of *Arabidopsis thaliana* is induced by the cell growth process and independent of auxin signaling. *Int J Mol Sci* **20:** 3337. doi:10.3390/ijms20133337

Arsuffi G, Braybrook SA. 2018. Acid growth: an ongoing trip. *J Exp Botany* **69:** 137–146. doi:10.1093/jxb/erx390

Bergfeld R, Speth V, Schopfer P. 1988. Reorientation of microfibrils and microtubules at the outer epidermal wall of maize coleoptiles during auxin-mediated growth. *Bot Acta* **101:** 57–67. doi:10.1111/j.1438-8677.1988.tb00012.x

Berleth T, Jürgens G. 1993. The role of the *monopteros* gene in organizing the basal body region of the *Arabidopsis* embryo. *Development* **118:** 575–587.

Bhalerao RP, Bennett MJ. 2003. The case for morphogens in plants. *Nat Cell Biol* **5:** 939–943. doi:10.1038/ncb1103-939

Calderon-Villalobos LI, Tan X, Zheng N, Estelle M. 2010. Auxin perception—structural insights. *Cold Spring Harb Perspect Biol* **2:** a005546. doi:10.1101/cshperspect.a005546

Cao M, Chen R, Li P, Yu Y, Zheng R, Ge D, Zheng W, Wang X, Gu Y, Gelová Z, et al. 2019. TMK1-mediated auxin signalling regulates differential growth of the apical hook. *Nature* **568:** 240–243. doi:10.1038/s41586-019-1069-7

Causier B, Ashworth M, Guo W, Davies B. 2012. The TOPLESS interactome: a framework for gene repression in *Arabidopsis*. *Plant Physiol* **158:** 423–438. doi:10.1104/pp.111.186999

Chen JG, Shimomura S, Sitbon F, Sandberg G, Jones AM. 2001. The role of auxin-binding protein 1 in the expansion of tobacco leaf cells. *Plant J* **28:** 607–617. doi:10.1046/j.1365-313x.2001.01152.x

Chen X, Grandont L, Li H, Hauschild R, Paque S, Abuzeineh A, Rakusová H, Benkova E, Perrot-Rechenmann C, Friml J. 2014. Inhibition of cell expansion by rapid ABP1-mediated auxin effect on microtubules. *Nature* **516:** 90–93. doi:10.1038/nature13889

Cho M, Lee OR, Ganguly A, Cho HT. 2007. Auxin-signaling: short and long. *J Plant Biol* **50:** 79–89. doi:10.1007/BF03030615

Dai N, Wang W, Patterson SE, Bleecker AB. 2013. The TMK subfamily of receptor-like kinases in *Arabidopsis* display an essential role in growth and a reduced sensitivity to auxin. *PLoS ONE* **8:** e60990. doi:10.1371/journal.pone.0060990

De Smet I, Tetsumura T, De Rybel B, Frei dit Frey N, Laplaze L, Casimiro I, Swarup R, Naudts M, Vanneste S, Audenaert D, et al. 2007. Auxin-dependent regulation of lateral root positioning in the basal meristem of *Arabidopsis*. *Development* **134:** 681–690. doi:10.1242/dev.02753

Dharmasiri N, Dharmasiri S, Estelle M. 2005. The F-box protein TIR1 is an auxin receptor. *Nature* **435:** 441–445. doi:10.1038/nature03543

Dindas J, Scherzer S, Roelfsema MRG, von Meyer K, Müller HM, Al-Rasheid KAS, Palme K, Dietrich P, Becker D, Bennett MJ, et al. 2018. AUX1-mediated root hair auxin influx governs $SCF^{TIR1/AFB}$-type Ca^{2+} signaling. *Nat Commun* **9:** 1174. doi:10.1038/s41467-018-03582-5

Du M, Spalding EP, Gray WM. 2020. Rapid auxin-mediated cell expansion. *Ann Rev Plant Biol* **71:** 379–402. doi:10.1146/annurev-arplant-073019-025907

* Dubey SM, Serre NBC, Oulehlova D, Vittal P, Fendrych M. 2020. No time for transcription—rapid auxin responses in plants. *Cold Spring Harb Perspect Biol* doi:10.1101/cshperspect.a039891

Dünser K, Kleine-Vehn J. 2015. Differential growth regulation in plants—the acid growth balloon theory. *Curr Opin Plant Biol* **28:** 55–59. doi:10.1016/j.pbi.2015.08.009

Enders TA, Strader LC. 2015. Auxin activity: past, present, and future. *Am J Bot* **102:** 180–196. doi:10.3732/ajb.1400285

Fendrych M, Akhmanova M, Merrin J, Glanc M, Hagihara S, Takahashi K, Uchida N, Torii KU, Friml J. 2018. Rapid and reversible root growth inhibition by TIR1 auxin signalling. *Nat Plants* **4:** 453–459. doi:10.1038/s41477-018-0190-1

Finet C, Fourquin C, Vinauger M, Berne-Dedieu A, Chambrier P, Paindavoine S, Scutt CP. 2010. Parallel structural evolution of auxin response factors in the angiosperms. *Plant J* **63:** 952–959. doi:10.1111/j.1365-313X.2010.04292.x

Finet C, Berne-Dedieu A, Scutt CP, Marlétaz F. 2013. Evolution of the ARF gene family in land plants: old domains, new tricks. *Mol Biol Evol* **30:** 45–56. doi:10.1093/molbev/mss220

Friml J. 2003. Auxin transport—shaping the plant. *Curr Opin Plant Biol* **6:** 7–12. doi:10.1016/S1369526602000031

Gallei M, Luschnig C, Friml J. 2020. Auxin signalling in growth: Schrödinger's cat out of the bag. *Curr Opin Plant Biol* **22:** 43–49.

Gammons M, Bienz M. 2018. Multiprotein complexes governing Wnt signal transduction. *Curr Opin Cell Biol* **51:** 42–49. doi:10.1016/j.ceb.2017.10.008

Gao Y, Zhang Y, Zhang D, Dai X, Estelle M, Zhao Y. 2015. Auxin binding protein 1 (ABP1) is not required for either auxin signaling or *Arabidopsis* development. *Proc Natl Acad Sci* **112:** 2275–2280. doi:10.1073/pnas.1500365112

Garcia D, Collier SA, Byrne ME, Martienssen RA. 2006. Specification of leaf polarity in *Arabidopsis* via the *trans*-acting siRNA pathway. *Curr Biol* **16:** 933–938. doi:10.1016/j.cub.2006.03.064

Guilfoyle TJ, Hagen G. 2007. Auxin response factors. *Curr Opin Plant Biol* **10:** 453–460. doi:10.1016/j.pbi.2007.08.014

Haagen-Smit AJ, Leech WD, Bergen WR. 1941. Estimation, isolation and identification of auxins in plant material. *Science* **93:** 624–625. doi:10.1126/science.93.2426.624

Habets ME, Offringa R. 2015. Auxin binding protein 1: a red herring after all? *Mol Plant* **8:** 1131–1134. doi:10.1016/j.molp.2015.04.010

Hager A. 2003. Role of the plasma membrane H^+-ATPase in auxin-induced elongation growth: historical and new aspects. *J Plant Res* **116:** 483–505. doi:10.1007/s10265-003-0110-x

Hager A, Menzel H, Krauss A. 1971. Versuche und hypothese zur primarwirkung des auxins beim Streckungswachstum [Experiments and hypothesis concerning the primary action of auxin in elongation growth]. *Planta* **100:** 47–75. doi:10.1007/BF00386886

Hamann T, Mayer U, Jürgens G. 1999. The auxin-insensitive *bodenlos* mutation affects primary root formation and apical-basal patterning in the *Arabidopsis* embryo. *Development* **126:** 1387–1395.

Hardtke CS, Berleth T. 1998. The *Arabidopsis* gene *MONOPTEROS* encodes a transcription factor mediating embryo axis formation and vascular development. *EMBO J* **17:** 1405–1411. doi:10.1093/emboj/17.5.1405

He Y, Meng X. 2020. MAPK signaling: emerging roles in lateral root formation. *Trends Plant Sci* **25:** 126–129. doi:10.1016/j.tplants.2019.11.006

Huang R, Zheng R, He J, Zhou Z, Wang J, Xiong Y, Xu T. 2019. Noncanonical auxin signaling regulates cell division pattern during lateral root development. *Proc Natl Acad Sci* **116:** 21285–21290. doi:10.1073/pnas.1910916116

Kalve S, Sizani BL, Markakis MN, Helsmoortel C, Vandeweyer G, Laukens K, Sommen M, Naulaerts S, Vissenberg K, Prinsen E, et al. 2020. Osmotic stress inhibits leaf growth of *Arabidopsis thaliana* by enhancing ARF-mediated auxin responses. *New Phytologist* **226:** 1766–1780. doi:10.1111/nph.16490

Kato H, Mutte SK, Suzuki H, Crespo I, Das S, Radoeva T, Fontana M, Yoshitake Y, Hainiwa E, van den Berg W, et al. 2020. Design principles of a minimal auxin response system. *Nat Plants* **6:** 473–482. doi:10.1038/s41477-020-0662-y

Kelley DR, Estelle M. 2012. Ubiquitin-mediated control of plant hormone signaling. *Plant Physiol* **160:** 47–55. doi:10.1104/pp.112.200527

Kelley DR, Arreola A, Gallagher TL, Gasser CS. 2012. ETTIN (ARF3) physically interacts with KANADI proteins to form a functional complex essential for integument development and polarity determination in *Arabidopsis*. *Development* **139:** 1105–1109. doi:10.1242/dev.067918

Kepinski S, Leyser HMO. 2005. The *Arabidopsis* F-box protein TIR1 is an auxin receptor. *Nature* **435:** 446–451. doi:10.1038/nature03542

Krogan NT, Hogan K, Long JA. 2012. APETALA2 negatively regulates multiple floral organ identity genes in *Arabidopsis* by recruiting the co-repressor TOPLESS and the histone deacetylase HDA19. *Development* **139:** 4180–4190. doi:10.1242/dev.085407

Krogan NT, Yin X, Ckurshumova W, Berleth T. 2014. Distinct subclades of Aux/IAA genes are direct targets of ARF5/MP transcriptional regulation. *New Phytol* **204:** 474–483. doi:10.1111/nph.12994

Kubeš M, Napier R. 2019. Non-canonical auxin signalling: fast and curious. *J Exp Bot* **70:** 2609–2614. doi:10.1093/jxb/erz111

Kuhn A, Ramans Harborough S, McLaughlin HM, Natarajan B, Verstraeten I, Friml J, Kepinski S, Østergaard L. 2020. Direct ETTIN-auxin interaction controls chromatin states in gynoecium development. *eLife* **9:** e51787. doi:10.7554/eLife.51787

Kumpf RP, Shi CL, Larrieu A, Stø IM, Butenko MA, Péret B, Riiser ES, Bennett MJ, Aalen RB. 2013. Floral organ abscission peptide IDA and its HAE/HSL2 receptors control cell separation during lateral root emergence. *Proc Natl Acad Sci* **110:** 5235–5240. doi:10.1073/pnas.1210835110

Kutschera U. 1994. The current status of the acid-growth hypothesis. *New Phytologist* **126:** 549–569. doi:10.1111/j.1469-8137.1994.tb02951.x

Leyser O. 2018. Auxin signaling. *Plant Physiol* **176:** 465–479. doi:10.1104/pp.17.00765

Leyser HM, Lincoln CA, Timpte C, Lammer D, Turner J, Estelle M. 1993. *Arabidopsis* auxin-resistance gene *AXR1* encodes a protein related to ubiquitin-activating enzyme E1. *Nature* **364:** 161–164. doi:10.1038/364161a0

Liu X, Galli M, Camehl I, Gallavotti A. 2019. RAMOSA1 ENHANCER LOCUS2-mediated transcriptional repression regulates vegetative and reproductive architecture. *Plant Physiol* **179:** 348–363. doi:10.1104/pp.18.00913

Lokerse A, Weijers D. 2009. Auxin enters the matrix—assembly of response machineries for specific outputs. *Curr Opin Plant Biol* **12:** 520–526. doi:10.1016/j.pbi.2009.07.007

Long JA, Ohno C, Smith ZR, Meyerowitz EM. 2006. TOPLESS regulates apical embryonic fate in *Arabidopsis*. *Science* **312:** 1520–1523. doi:10.1126/science.1123841

Lv B, Yu Q, Liu J, Wen X, Yan Z, Hu K, Li H, Kong X, Li C, Tian H, et al. 2020. Non-canonical AUX/IAA protein IAA33 competes with canonical AUX/IAA repressor IAA5 to negatively regulate auxin signaling. *EMBO J* **39:** e101515.

Majda M, Robert S. 2018. The role of auxin in cell wall expansion. *Int J Mol Sci* **19:** 951. doi:10.3390/ijms19040951

Marin E, Jouannet V, Herz A, Lokerse AS, Weijers D, Vaucheret H, Nussaume L, Crespi MD, Maizel A. 2010. Mir390, *Arabidopsis TAS3* tasiRNAs, and their *AUXIN RESPONSE FACTOR* targets define an autoregulatory network quantitatively regulating lateral root growth. *Plant Cell* **22:** 1104–1117. doi:10.1105/tpc.109.072553

Mockaitis K, Howell SH. 2000. Auxin induces mitogenic activated protein kinase (MAPK) activation in roots of *Arabidopsis* seedlings. *Plant J* **24:** 785–796. doi:10.1046/j.1365-313x.2000.00921.x

Monshausen GB, Miller ND, Murphy AS, Gilroy S. 2011. Dynamics of auxin-dependent Ca^{2+} and pH signaling in root growth revealed by integrating high-resolution imaging with automated computer vision-based analysis. *Plant J* **65:** 309–318. doi:10.1111/j.1365-313X.2010.04423.x

Mroue S, Simeunovic A, Robert HS. 2018. Auxin production as an integrator of environmental cues for developmental growth regulation. *J Exp Bot* **69:** 201–212. doi:10.1093/jxb/erx259

Müller CL, Valdés AE, Wang G, Ramachandran P, Beste L, Uddenberg D, Carlsbecker A. 2016. PHABULOSA mediates an auxin signaling loop to regulate vascular patterning in *Arabidopsis*. *Plant Physiol* **170:** 956–970. doi:10.1104/pp.15.01204

Mutte SK, Kato H, Rothfels C, Melkonian M, Wong GK, Weijers D. 2018. Origin and evolution of the nuclear auxin response system. *eLife* **7:** e33399. doi:10.7554/elife.33399

Nemhauser J, Feldmann LJ, Zambryski PC. 2000. Auxin and ETTIN in *Arabidopsis* gynoecium morphogenesis. *Development* **127:** 3877–3888.

Paciorek T, Zažímalová E, Ruthardt N, Petrášek J, Stierhof YD, Kleine-Vehn J, Morris DA, Emans N, Jürgens G, Geldner N, et al. 2005. Auxin inhibits endocytosis and promotes its own efflux from cells. *Nature* **435:** 1251–1256. doi:10.1038/nature03633

Paponov IA, Paponov M, Teale W, Menges M, Chakrabortee S, Murray JA, Palme K. 2008. Comprehensive transcriptome analysis of auxin responses in *Arabidopsis*. *Mol Plant* **1:** 321–337. doi:10.1093/mp/ssm021

Pekker I, Alvarez JP, Eshed Y. 2005. Auxin response factors mediate *Arabidopsis* organ asymmetry via modulation of KANADI activity. *Plant Cell* **17:** 2899–2910. doi:10.1105/tpc.105.034876

Piya S, Shrestha SK, Binder B, Stewart CN Jr, Hewezi T. 2014. Protein–protein interaction and gene co-expression maps of ARFs and Aux/IAAs in *Arabidopsis*. *Front Plant Sci* **5:** 744. doi:10.3389/fpls.2014.00744

Powers SK, Holehouse AS, Korasick DA, Schreiber KH, Clark NM, Jing H, Emenecker R, Han S, Tycksen E, Hwang I, et al. 2019. Nucleo-cytoplasmic partitioning of ARF proteins controls auxin responses in *Arabidopsis thaliana*. *Mol Cell* **76:** 177–190.e5. doi:10.1016/j.molcel.2019.06.044

Prigge M, Platre M, Kadakia N, Zhang Y, Greenham K, Szutu W, Pandey BK, Bhosale RA, Bennett MJ, Busch W, et al. 2020. Genetic analysis of the *Arabidopsis* TIR1/AFB auxin receptors reveals both overlapping and specialized functions. *eLife* **9:** e54740. doi:10.7554/eLife.54740

Retzer K, Singh G, Napier RM. 2018. It starts with TIRs. *Nat Plants* **4:** 410–411. doi:10.1038/s41477-018-0196-8

Robert S, Kleine-Vehn J, Barbez E, Sauer M, Paciorek T, Baster P, Vanneste S, Zhang J, Simon S, Čovanová M, et al. 2010. ABP1 mediates auxin inhibition of clathrin-dependent endocytosis in *Arabidopsis*. *Cell* **143:** 111–121. doi:10.1016/j.cell.2010.09.027

Salehin M, Bagchi R, Estelle M. 2015. $SCF^{TIR1/AFB}$-based auxin perception: mechanism and role in plant growth and development. *Plant Cell* **27:** 9–19. doi:10.1105/tpc.114.133744

Santner A, Estelle M. 2010. The ubiquitin–proteasome system regulates plant hormone signaling. *Plant J* **61:** 1029–1040. doi:10.1111/j.1365-313X.2010.04112.x

Sato A, Yamamoto KT. 2008. Overexpression of the noncanonical Aux/IAA genes causes auxin-related aberrant phenotypes in *Arabidopsis*. *Physiol Plant* **133:** 397–405. doi:10.1111/j.1399-3054.2008.01055.x

Sauer M, Kleine-Vehn J. 2011. AUXIN BINDING PROTEIN1: the outsider. *Plant Cell* **23:** 2033–2043. doi:10.1105/tpc.111.087064

Sessions RA, Zambryski PC. 1995. *Arabidopsis gynoecium* structure in the wild and in *ettin* mutants. *Development* **121:** 1519–1532.

Sessions A, Nemhauser JL, McColl A, Roe JL, Feldmann KA, Zambryski PC. 1997. ETTIN patterns the *Arabidopsis* floral meristem and reproductive organs. *Development* **124:** 4481–4491.

Shih HW, DePew CL, Miller ND, Monshausen GB. 2015. The cyclic nucleotide-gated channel CNGC14 regulates root gravitropism in *Arabidopsis thaliana*. *Curr Biol* **25:** 3119–3125. doi:10.1016/j.cub.2015.10.025

Simonini S, Deb J, Moubayidin L, Stephenson P, Valluru M, Freire-Rios A, Sorefan K, Weijers D, Friml J, Østergaard L. 2016. A noncanonical auxin-sensing mechanism is required for organ morphogenesis in *Arabidopsis*. *Genes Dev* **30:** 2286–2296. doi:10.1101/gad.285361.116

Simonini S, Bencivenga S, Trick M, Østergaard L. 2017. Auxin-induced modulation of ETTIN activity orchestrates gene expression in *Arabidopsis*. *Plant Cell* **29:** 1864–1882. doi:10.1105/tpc.17.00389

Spartz AK, Ren H, Park MY, Grandt KN, Lee SH, Murphy AS, Sussman MR, Overvoorde PJ, Gray WM. 2014. SAUR inhibition of PP2C-D phosphatases activates plasma membrane H^+-ATPases to promote cell expansion in *Arabidopsis*. *Plant Cell* **26:** 2129–2142. doi:10.1105/tpc.114.126037

Szemenyei H, Hannon M, Long J. 2008. TOPLESS mediates auxin-dependent transcriptional repression during *Arabidopsis* embryogenesis. *Science* **319:** 1384–1386. doi:10.1126/science.1151461

Takahashi K, Hayashi K, Kinoshita T. 2012. Auxin activates the plasma membrane H^+-ATPase by phosphorylation during hypocotyl elongation in *Arabidopsis*. *Plant Physiol* **159:** 632–641. doi:10.1104/pp.112.196428

Tiwari SB, Hagen G, Guilfoyle T. 2003. The roles of auxin response factor domains in auxin-responsive transcription. *Plant Cell* **15:** 533–543. doi:10.1105/tpc.008417

Tsai MJ, O'Malley BW. 1994. Molecular mechanisms of action of steroid/thyroid receptor superfamily members. *Ann Rev Biochem* **63:** 451–486. doi:10.1146/annurev.bi.63.070194.002315

Uchida N, Takahashi K, Iwasaki R, Yamada R, Yoshimura M, Endo TA, Kimura S, Zhang H, Nomoto M, Tada Y, et al. 2018. Chemical hijacking of auxin signaling with an engineered auxin-TIR1 pair. *Nat Chem Biol* **14:** 299–305. doi:10.1038/nchembio.2555

Vernoux T, Brunoud G, Farcot E, Movin V, Van den Daele H, Legrand J, Oliva M, Das P, Larrieu A, Wells D, et al. 2011. The auxin signalling network translates dynamic input into robust patterning at the shoot apex. *Mol Syst Biol* **7:** 508. doi:10.1038/msb.2011.39

Wang R, Zhang Y, Kieffer M, Yu H, Kepinski S, Estelle M. 2016. HSP90 regulates temperature-dependent seedling growth in *Arabidopsis* by stabilizing the auxin co-receptor F-box protein TIR1. *Nat Commun* **7:** 10269. doi:10.1038/ncomms10269

Wang B, Xue JS, Yu YH, Liu SQ, Zhang JX, Yao XZ, Liu ZX, Xu XF, Yang ZN. 2017. Fine regulation of ARF17 for anther development and pollen formation. *BMC Plant Biol* **17:** 243. doi:10.1186/s12870-017-1185-1

Weijers D, Wagner D. 2016. Transcriptional responses to the auxin hormone. *Ann Rev Plant Biol* **67:** 539–574. doi:10.1146/annurev-arplant-043015-112122

Xu T, Wen M, Nagawa S, Fu Y, Chen JG, Wu MJ, Perrot-Rechenmann C, Friml J, Jones AM, Yang Z. 2010. Cell surface- and rho GTPase-based auxin signaling controls cellular interdigitation in *Arabidopsis*. *Cell* **143:** 99–110. doi:10.1016/j.cell.2010.09.003

Xu T, Dai N, Chen J, Nagawa S, Cao M, Li H, Zhou Z, Chen X, De Rycke R, Rakusová H, et al. 2014. Cell surface ABP1-TMK auxin-sensing complex activates ROP GTPase signaling. *Science* **343:** 1025–1028. doi:10.1126/science.1245125

Xu X, Wang B, Feng YF, Xue JS, Qian XX, Liu SQ, Zhou J, Yu YH, Yang NY, Xu P, et al. 2019. AUXIN RESPONSE FACTOR17 directly regulates *MYB108* for anther dehiscence. *Plant Physiol* **181:** 645–655. doi:10.1104/pp.19.00576

Yang J, Tian L, Sun MX, Huang XY, Zhu J, Guan YF, Jia QS, Yang ZN. 2013. AUXIN RESPONSE FACTOR17 is essential for pollen wall pattern formation in *Arabidop-*

sis. Plant Physiol **162:** 720–731. doi:10.1104/pp.113.214940

Yu H, Zhang Y, Moss BL, Bargmann BO, Wang R, Prigge M, Nemhauser JL, Estelle M. 2015. Untethering the TIR1 auxin receptor from the SCF complex increases its stability and inhibits auxin response. *Nat Plants* **1:** 14030. doi:10.1038/nplants.2014.30

Zhao Y. 2018. Essential roles of local auxin biosynthesis in plant development and in adaptation to environmental changes. *Ann Rev Plant Biol* **69:** 417–435. doi:10.1146/annurev-arplant-042817-040226

Zhu J, Geisler M. 2015. Keeping it all together: auxin-actin crosstalk in plant development. *J Exp Bot* **66:** 4983–4998. doi:10.1093/jxb/erv308

Zhu Q, Shao Y, Ge S, Zhang M, Zhang T, Hu X, Liu Y, Walker J, Zhang S, Xu J. 2019. A MAPK cascade downstream of IDA-HAE/HSL2 ligand-receptor pair in lateral root emergence. *Nat Plants* **5:** 414–423. doi:10.1038/s41477-019-0396-x

Auxin Does the SAMba: Auxin Signaling in the Shoot Apical Meristem

Markéta Pernisová[1,2] and Teva Vernoux[1]

[1]Laboratoire Reproduction et Développement des Plantes, University at Lyon, ENS de Lyon, UCB Lyon 1, CNRS, INRAE, 69342 Lyon, France

[2]Functional Genomics and Proteomics, National Centre for Biomolecula Research, Faculty of Science, Masaryk University and CEITEC MU, 62500 Brno, Czech Republic

Correspondence: pernisov@sci.muni.cz; teva.vernoux@ens-lyon.fr

Plants, in contrast to animals, are unique in their capacity to postembryonically develop new organs due to the activity of stem cell populations, located in specialized tissues called meristems. Above ground, the shoot apical meristem generates aerial organs and tissues throughout plant life. It is well established that auxin plays a central role in the functioning of the shoot apical meristem. Auxin distribution in the meristem is not uniform and depends on the interplay between biosynthesis, transport, and degradation. Auxin maxima and minima are created, and result in transcriptional outputs that drive the development of new organs and contribute to meristem maintenance. To uncover and understand complex signaling networks such as the one regulating auxin responses in the shoot apical meristem remains a challenge. Here, we will discuss our current understanding and point to important research directions for the future.

The establishment and maintenance of precise developmental patterns to construct a body architecture is a common characteristic of all multicellular living organisms. Understanding the spatiotemporal regulation of patterning is central to the comprehension of development processes. A unique characteristic of plants is their ability to continuously create structures postembryonically. The dynamic building of the architecture of the aerial part of the plant is controlled in the shoot apical meristem and regulated by a combination of both developmental and environmental factors. The shoot apical meristem is a highly organized structure containing a central zone, where the stem cells are located (Fig. 1). Below the central zone, the organizing center produces signals to regulate stem cells and is itself regulated by signals originating from the central zone as we will see below. New organs including leaves and flowers are initiated in the organogenic peripheral zone that surrounds the central zone.

Among the factors controlling shoot apical meristem function, phytohormones play a key role and compelling evidence show that auxin is one of the main regulators of both stem cell activity and organogenesis. Several recent reviews have discussed how auxin transport generates dynamic auxin distribution in the shoot apical meristem and the central role of polar

Cite this article as *Cold Spring Harb Perspect Biol* doi: 10.1101/cshperspect.a039925

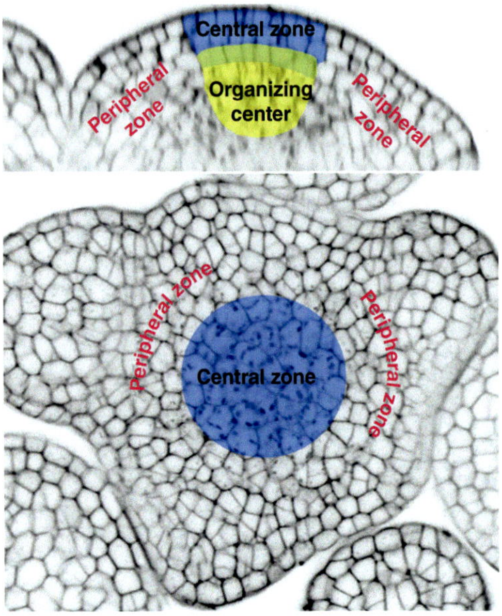

Figure 1. Shoot apical meristem structure in *Arabidopsis.*

auxin transport and auxin carriers in this regulation (Galvan-Ampudia et al. 2016; Kuhlemeier 2017; Bhatia and Heisler 2018; Shi and Vernoux 2019), and we will not cover those aspects here. We will focus on auxin signaling and on how it participates to regulate the function of the shoot apical meristem. In this introduction, we will first briefly review the different mechanisms mediating the perception and transduction of the auxin signal in plant cells.

Auxin acts through different signaling pathways, known as the canonical and noncanonical pathways. The canonical pathway also called the nuclear auxin pathway (NAP) involves three families of proteins (the numbers of members are given for *Arabidopsis*): 6 TRANSPORT INHIBITOR RESPONSE1 (TIR1)/AUXIN SIGNALING F-BOX (AFB) F-box proteins acting as auxin coreceptors, 29 AUXIN/INDOLE 3-ACETIC ACID INDUCIBLE (Aux/IAA) repressor proteins, and 23 AUXIN RESPONSE FACTOR (ARF) transcription factors. ARFs can be divided into three classes, A, B, and C, based on their evolutionary origin (Finet et al. 2013; Mutte et al. 2018). Class A ARFs are described to have a transcriptional activation activity and classes B and C as repressors (reviewed in Israeli et al. 2020). Individual class B ARFs likely compete with class A ARFs by binding to the same DNA sites (Boer et al. 2014; Kato et al. 2020). In the NAP, the presence of auxin is perceived by TIR1/AFB and triggers changes in the expression of target genes, which then induce tissue-specific biological responses. When auxin is low or absent, Aux/IAAs interact with ARF transcription factors through the shared PB1 domain and repress gene transcription and auxin responses. Auxin binding to TIR1/AFB coreceptors induces the proteasome-dependent degradation of Aux/IAA proteins by targeting them for ubiquitination, thus releasing the repression on ARFs and allowing transcriptional auxin responses. Several Aux/IAAs are themselves ARF targets thus establishing a negative feedback loop (reviewed in Gallei et al. 2020).

Differential expression of regulators of the NAP, as can be seen with the variety of expression patterns of particular ARFs in the shoot apical meristem of *Arabidopsis* (Fig. 2; Vernoux et al. 2011; Truskina et al. 2020), suggests different possible regulatory outputs arising from the auxin signaling cascade and an implication in all aspects of shoot apical meristem function. Many auxin targets, regulated by ARFs and involved in shoot apical meristem maintenance and activity, have been identified notably in *Arabidopsis*, especially targets regulated by ARF5/MONOPTEROS (ARF5/MP) (Hardtke and Berleth 1998): ARF5 itself (Lau et al. 2011), TARGET OF MONOPTEROSs (TMOs) (Schlereth et al. 2010), ARABIDOPSIS HISTIDINE PHOSPHOTRANSFER PROTEIN 6 (AHP6) (Besnard et al. 2014), LEAFY (LFY) (Blázquez et al. 1997; Yamaguchi et al. 2013), AINTEGUMENTA (ANT) (Yamaguchi et al. 2013), ANT-LIKE6/PLETHORA3 (AIL6/PLT3) (Yamaguchi et al. 2013), ARABIDOPSIS RESPONSE REGULATOR 7 and ARR15 (Zhao et al. 2010), and DORNRÖSCHEN/ENHANCER OF SHOOT REGENERATION (DRN/ESR1) (Cole et al. 2009). We will discuss in this review the role of individual ARFs and their targets in the shoot apical meristem.

There are currently two noncanonical auxin pathways that have been identified. The first noncanonical auxin response pathway involves

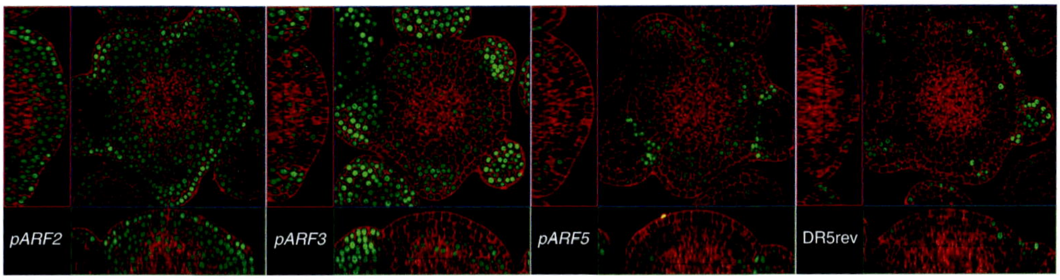

Figure 2. ARFs are differentially expressed in the shoot apical meristem of *Arabidopsis. proARF2, proARF3, proARF5* (Rademacher et al. 2011), and DR5rev (Wabnik et al. 2013) (from the *left* to the *right*) reporters show possible combinatorial actions of ARFs in the regulation of auxin transcriptional response, which might lead to different responses in the shoot apical meristem. Note that the expression of the synthetic auxin responsive promoter DR5rev (last image on the *right*) does not fully reflect the expression of ARFs.

the receptor-like kinase TRANSMEMBRANE KINASE 1 (TMK1) (Dai et al. 2013). TMK1 interacts with AUXIN-BINDING PROTEIN 1 (ABP1) and activates Rho-like guanosine triphosphatases (GTPase) from plants (ROP2/ROP6). The interaction between TMK1 and ABP1 was shown to occur in an auxin-dependent manner (Xu et al. 2014). Moreover, after auxin sensing, TMK1 phosphorylates and stabilizes IAA32 and IAA34, thus regulating ARF transcription factors (Cao et al. 2019). The role for ABP1 in this auxin signaling cascade is largely debated due to the absence of clear phenotypes in loss-of-function mutants (Gao et al. 2015; Gelová et al. 2021). An alternative noncanonical auxin pathway has been proposed and involves auxin binding to the atypical ARF3/ETTIN (ETT), resulting in modulation of chromatin and interaction with other transcriptional regulators (Simonini et al. 2016). Direct auxin binding to ARF3 leads to dissociation from corepressors of the TOPLESS/TOPLESS-RELATED family followed by histone acetylation and induction of gene expression and enables switching between repressive and derepressive chromatin states in a reversible manner (Kuhn et al. 2020).

In the next two sections, we will discuss how the canonical and noncanonical auxin signaling pathways act in the meristem to highlight the crucial role for auxin signaling not only in the dynamic initiation of organs but also in the maintenance of the shoot apical meristem.

AUXIN SIGNALING IN THE CENTRAL ZONE: A FUNCTION IN STEM CELL MAINTENANCE

Shoot apical meristem maintenance depends on the coordinated action of signals in the central zone and the organizing center, as well as on interactions with the peripheral zone. In *Arabidopsis*, the localization of the stem cell niche is primarily controlled by expression of the homeodomain transcription factor WUSCHEL (WUS) in the organizing center (Mayer et al. 1998; Schoof et al. 2000). WUS function is essential for proper shoot apical meristem development as *wus* mutants fail to organize a shoot apical meristem in the embryo. Postembryonically, an aberrant flattened shoot apical meristem is initiated and terminated prematurely (Laux et al. 1996). When produced, WUS protein migrates from the organizing center upward into the central zone and activates the synthesis of the secreted peptide CLAVATA3 (CLV3) (Fletcher et al. 1999; Yadav et al. 2011). CLV3 in turn negatively regulates *WUS* expression via the CLV1 receptor that delimits *WUS* expression to the organizing center (Yadav et al. 2011; Knauer et al. 2013; Daum et al. 2014). Thus, the WUS–CLV negative feedback loop serves as a self-controlling mechanism that maintains a stem cell niche of constant size (Brand 2000; Schoof et al. 2000; Somssich et al. 2016; Soyars et al. 2016). In addition, a plethora of other genes have been identified that are involved in stem cell maintenance alongside the WUS-CLV loop, likely contributing to provide the robust control mechanism that regulates the activity of the stem cell niche (for review, see, e.g., Kitagawa and Jackson 2019; Lee et al. 2019; Han et al. 2020; Jha et al. 2020).

The function of WUS in shoot apical meristem maintenance is intricately linked with cytokinin signaling. WUS represses the transcription of two ARR type-A genes, ARR7 and ARR15 (Leibfried et al. 2005; Zhao et al. 2010), which encode negative regulators of the cytokinin signaling pathway. WUS thus provides a positive input for cytokinin signaling in the center of the shoot apical meristem. Further, the cytokinin signaling pathway increases WUS levels and regulates the stem cell niche (Gordon et al. 2009). Moreover, WUS is stabilized by the transcription factor ARR1, a type-B response regulator that is part of the cytokinin pathway (Snipes et al. 2018), thus creating a positive feedback between WUS and cytokinins. Cytokinins produced in the L1 layer (epidermis) by the cytokinin biosynthetic enzyme LONELY GUY 4 (LOG4), along with CLV signaling, can act as positional cues for patterning the WUS domain within the stem cell niche (Chickarmane et al. 2012). These regulations are likely conserved in other plants, although with variations. For example, in maize, a mutation in *ZmLOG7* causes meristem termination (Knauer et al. 2019), which is not the case in *Arabidopsis* where mutations in single *LOG* genes result in either a minor or no phenotype (Kuroha et al. 2009). Also, *ZmWUS1* expression is not conserved within the vegetative shoot apical meristem (Nardmann and Werr 2006). The maize CLV1 ortholog, thick tassel dwarf1, is expressed in leaf primordia (Nardmann and Werr 2007). Moreover, it is ZmFCP1 and not the CLV3 ortholog that specifically signifies the shoot apical meristem in the corresponding organizing center region (Knauer et al. 2019). This indicates a conservation of the WUS-CLV signaling logic but also the existence of diversifications of the components of the circuit (Kellogg 2019; Rodriguez-Leal et al. 2019).

In maize, KNOTTED1 (KN1) (Vollbrecht et al. 1991) acts as a master regulator of meristem activity in the shoot apical meristem. Consistent with that, KN1 binds a substantial number of transcription factors acting in meristematic tissues. Moreover, genes acting in auxin signaling are overrepresented among KN1 targets (Bolduc et al. 2012; Knauer et al. 2019), providing a functional link between meristematic activity and the regulation of auxin signaling. More information on the function of auxin in meristem maintenance has come from extensive studies in *Arabidopsis*. Among the effectors involved in auxin signaling in *Arabidopsis*, nine ARFs and eight Aux/IAAs are expressed at relatively higher levels at the periphery and at comparatively lower levels in the center of the meristem (Vernoux et al. 2011; Truskina et al. 2020). The nine ARFs comprise four class A ARFs but also class B ARFs, indicating that ARFs acting as activators as well as repressors of auxin signaling show significant coexpression in the shoot apical meristem. ARF and Aux/IAA genes are expressed at low levels in the central zone and at relatively higher levels in the peripheral zone. Predictive mathematical modeling arising from a large-scale analysis of AUX/IAA-ARF interaction networks could correctly predict that the variation in expression levels between ARF activators and repressors results in different sensitivity to auxin between the central zone (low sensitivity) and the peripheral zone (high sensitivity). Furthermore, the relative expression ratios of ARF activators and repressors can buffer against input fluctuations, thus maintaining a stable regulatory response (Vernoux et al. 2011).

In a recent study, the auxin signaling input sensor R2D2, and the output signaling reporter DR5v2 (Liao et al. 2015) confirmed previous findings and identified WUS as a key regulator of the reduced sensitivity of the central zone cells to auxin (Ma et al. 2019). The auxin signaling minimum in the meristem center depends on stem cell fate. WUS binds to the promoter region of target genes and activates or represses their expression (Busch et al. 2010; Yadav et al. 2013). However, Ma et al. (2019) reported that WUS acts by triggering histone acetylation at target loci, including regulators of the NAP (e.g., TIR1, ARF5, IAA9) but also a large number of genes acting downstream of auxin (Ma et al. 2019). Histone acetylation is thought to lead to repression of gene expression, suggesting that auxin signaling and response capacity are rheostatically controlled by WUS, restricting auxin signaling to the stem cell population, while at the same time allowing instructive low levels of signaling output. Simultaneously, low levels of auxin signaling are required for stem cell maintenance (Ma et al. 2019). A yeast one-hybrid screen recently identified a network of transcription factors regulating the expression of *Arabidopsis* class A ARFs (Truskina et al. 2020). This work identified WUS as part of this network and further demonstrated that a vast majority of class A ARF regulators are transcriptional repressors. A mutation in one of these transcriptional repressors, WRKY38, leads to increased *ARF7* expression in the center of the shoot apical meristem (Truskina et al. 2020). While a more global analysis of the function of these genes in the shoot apical meristem is required, this suggests that lower ARFs expression (and thus auxin sensitivity) in the central zone implicates transcriptional repressors acting in combination with WUS in regulating ARFs. This complex regulation of signaling output of auxin together with the regulation of cytokinin signaling results in the maintenance of a constant population of stem cells in the central zone, allowing for continuous replenishment of the peripheral zone. Recent evidence suggests that the regulation of stem cell activity by auxin might be an ancestral mechanism. Indeed, in moss, stem cell activity depends on the inability of stem cells to sense the auxin they produce while stem cell progeny responds to auxin (Landberg et al. 2021). Thus, a low level of auxin signaling appears to be required for the maintenance of shoot stem cells and suggests a requirement for certain ARFs to drive this process.

Expression patterns revealed by in situ hybridization or green fluorescent protein (GFP) fusion of single *Arabidopsis* ARFs pinpointed several candidates that might have a regulatory function in the central zone (Fig. 3A). ARF2, a possible repressor of auxin signaling, seems to be expressed throughout the shoot apical meristem, and seems to be one of the few ARFs present in the central zone (Fig. 2; Vernoux et al. 2011). *arf2* mutants display a severe shoot phenotype with late flowering and a very long, thick, and wavy inflorescence stem, suggesting impaired timing of the vegetative-to-generative switch and growth determination. High numbers of undeveloped or retarded siliques point to pleiotropic defects in shoot apical meristem regulation (Okushima et al. 2005a). Weak homogenous expression in the shoot apical meristem was also detected for ARF9 and ARF10 (Vernoux et al. 2011) but their possible function in the central zone is still unknown. In the inner core of the meristem, in the organizing center, ARF4 (Vernoux et al. 2011) as well as ARF7 and ARF19 (Truskina et al. 2020) are expressed at relatively high levels. Mature *arf7* mutant plants (*nph4-1*, *arf7-1*, and *msg1-2/nph4-102*) do not show any prominent developmental defects connected with the shoot apical meristem, except for slightly shorter inflorescence stems compared to wild-type plants. Finally, *arf19* mutants are identical to wild-type, while an ARF19 overexpressor has severely impaired shoot development as well as apical dominance (Okushima et al. 2005b). So far, evidence for a role for ARFs in meristem maintenance in other species than *Arabidopsis* is missing (Matthes et al. 2019). Whether specific ARFs are regulating shoot apical meristem maintenance and stem cell niche activity is thus still largely unknown, which calls for further investigation.

INITIATION AND OUTGROWTH OF PRIMORDIA IN PERIPHERAL ZONE (PHYLLOTAXIS)—"ARF5 RULES THEM ALL"

Lateral organ primordia are initiated in the peripheral zone, following a regular spatiotemporal pattern called phyllotaxis. The most widely accepted model for phyllotaxis proposes that the spatiotemporal patterns of organ initiation emerge from a self-organizing process that involves morphogen-based inhibitory fields generated by preexisting organ primordia (Mitchison 1977; Douady and Couder 1992). While inhibitory fields block organ initiation in the vicinity of existing organs, new primordia are initiated in the position where the sum of inhibitory effects is the lowest outside of the center of the shoot apical meristem, where initiation is also inhibited. The spatiotemporal patterns of organ initiation then emerge thanks to these inhibitory fields as a result of growth. Over the last two decades,

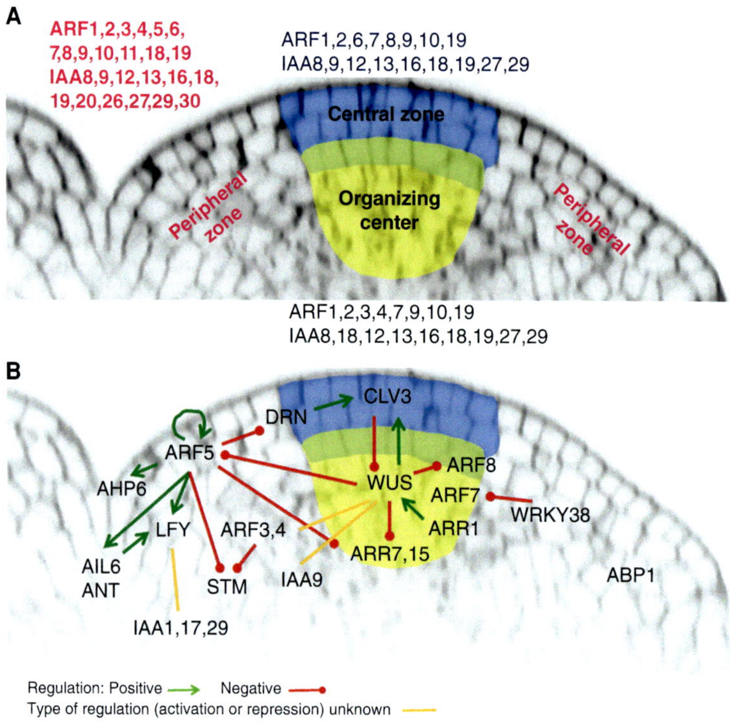

Figure 3. Auxin signaling in the shoot apical meristem of *Arabidopsis*. (*A*) ARFs and IAAs expression overview, and (*B*) ARF-dependent regulatory network.

auxin has been demonstrated to be the central regulator of phyllotaxis (for review, see Shi and Vernoux 2019; Godin et al. 2020; Véron et al. 2020). Local auxin maxima trigger organ formation while auxin depletion around organs generates inhibitory fields. Auxin spatiotemporal distribution is generated by PIN1-dependent auxin transport in cooperation with auxin influx and efflux carriers and local auxin biosynthesis (Benková et al. 2003; Reinhardt et al. 2003; Heisler et al. 2005; Jonsson et al. 2006; Smith et al. 2006; Bainbridge et al. 2008; Stoma et al. 2008; Vernoux et al. 2011; Bhatia et al. 2016; Galvan-Ampudia et al. 2020).

As discussed previously, the role of the NAP (and also of the ARF3-mediated noncanonical pathway) in organ initiation is highlighted by the expression of ARF transcription factors and Aux/IAAs. In *Arabidopsis*, ARF1, 2, 5, 7, 8, 18 and IAA12, 13, 18, 27 are expressed homogeneously throughout the periphery of the shoot apical meristem and ARF3, 4, 6 and IAA8, 9 show a stronger expression in organ primordia (Fig. 3; Vernoux et al. 2011; Truskina et al. 2020). This suggests the involvement of a large network of ARFs and Aux/IAAs in controlling sensitivity to auxin in the peripheral zone. Among these, the class A ARF5/MP, an activator of transcription, seems to work as a central regulator. The ARF5 mutant *mp* exhibits a pin-like phenotype (Przemeck et al. 1996)—a naked inflorescence stem without lateral organs—similarly to the auxin transport mutant *pin1* (Galweiler et al. 1998), even though leaves are still formed in both. But unlike the *pin1* mutant, exogenous auxin application does not rescue the defect in flower primordia initiation of the *arf5* mutant (Hardtke and Berleth 1998; Reinhardt et al. 2003; Zhao et al. 2010), demonstrating a critical role for auxin signaling in organ initiation in the inflorescence meristem. In the shoot apical meristem, ARF5 not only mediates gene activation in response to auxin as we will see below, but its activity is also determinant for regulating the dynamics of auxin distribution (Bhatia et al. 2016). ARF5 expression is required for regulating polarization of PIN1, with cell polarity patterns following ARF5 expression, which in turn follows auxin distribution. In the epidermal layer as well as in subepidermal tissues, high levels of ARF5-mediated auxin signaling are sufficient to instruct PIN1 polarity directions non-cell-autonomously. ARF5 in the shoot apical meristem has been shown to impact a chromatin-state switch by promoting histone acetylation (Wu et al. 2015). In the presence of auxin, ARF5 recruits SWI/SNF chromatin-remodeling ATPases and increases DNA accessibility through the regulation of histone acetylation in regions of chromatin that contain genes involved in flower formation, thus driving the flower primordium fate at the periphery of the shoot apical meristem. In the absence of auxin, Aux/IAA proteins bind to ARF5 preventing recruitment of SWI/SNF, in addition to recruiting a corepressor/histone deacetylase complex (Long 2006; Wu et al. 2015). This auxin-mediated chromatin state switch is key for iterative primordia initiation in the shoot apical meristem and could coordinate additional auxin-regulated cell fate transitions. Interestingly, the promotion of histone acetylation by ARF5 suggests that ARF5 might be able to counteract the histone deacetylation promoted by WUS, thus providing a possible mechanism by which auxin could contribute to the transition of stem cells to the peripheral zone.

Although playing a dominant role, ARF5 regulates organ initiation in the shoot apical meristem together with ARF3 and ARF4. A recent work demonstrates that these ARFs stimulate flower initiation by histone-deacetylation-mediated transcriptional silencing of the SHOOT-MERISTEMLESS gene (Chung et al. 2019), a key meristem identity gene, and the ortholog of maize KN1, *Oryza sativa homeobox15* (*OSH15*) or *Tomato Kn2* (*TKn2*)/*LeT6* (Hay and Tsiantis 2010). The *arf3*/*ett* loss-of function mutant displays perturbations in phyllotaxis, indicating that this noncanonical pathway is required for stable spatiotemporal patterns of organogenesis (Simonini et al. 2017; Galvan-Ampudia et al. 2020). A recent live imaging analysis has also provided important new insights on the properties of the NAP and of the ARF3-mediated noncanonical pathway in the shoot apical meristem. By comparing the dynamics of auxin accumulation and of transcription activation by auxin and exogenous auxin treatments, the authors could demonstrate that the NAP has time-integration properties, allowing to "count" cell exposure to auxin to trigger organ initiation (Galvan-Ampudia et al. 2020). The molecular mechanisms involved are still to be identified but ARF3 was shown to be involved, as the *arf3* mutant loses the capacity to integrate the auxin signal. The temporal integration capacity could then rely on perception of auxin by ARF3 but possibly also by the NAP. Whether the ARF3-dependent noncanonical pathway is functionally conserved in species other than *Arabidopsis* is unknown (Matthes et al. 2019). Among the other ARFs, ARF6 and ARF8 mutants showed no other shoot phenotype except one in flower maturation (Nagpal 2005), although aerial tissues are dwarfed in the *arf6 arf8* double mutant (Okushima et al. 2005b). Apart from ARF5, information about other ARFs and their function in organ initiation at the shoot apical meristem is fragmented and detailed genetic studies (on appropriate multiple mutants) are needed (Ori 2019).

In addition to ARFs, a role for several Aux/IAAs in the shoot apical meristem was uncovered in maize. BARREN INFLORESCENCE1 (BIF1/IAA27) and BARREN INFLORESCENCE4 (BIF4/IAA20) regulate the early steps required for inflorescence formation, particularly proper formation of new axillary meristems. The corresponding gain-of-function mutants produce fewer spikelets and flowers or even naked ears without any grains. These phenotypes are the result of defects in primordia initiation at the peripheral zone of the inflorescence meristem (Galli et al. 2015).

In *Arabidopsis*, a significant number of genes regulated by ARF5 have been identified, and some of them are crucial players in organ initiation and development (Fig. 3). The transcription factor LFY acts as a master coordinator of the floral network (Moyroud et al. 2010; Siriwardana and Lamb 2012), and is activated by auxin directly via ARF5 (Blázquez and Weigel 2000; Vernoux et al. 2000; Li et al. 2013;

Yamaguchi et al. 2013). LFY specifies the floral fate of meristems and its loss-of-function mutation causes partial transformation of flowers into inflorescence shoots in *Arabidopsis*. Whereas auxin regulates LFY expression by directly binding to the ARF5 promoter region (Yamaguchi et al. 2013), LFY feeds back to the auxin pathway by modulating auxin transport, signaling, or biosynthesis. LFY targets a key regulator of auxin transport, PINOID (PID) (Friml et al. 2004) as seen in the *pid* mutant that produces a naked inflorescence like *arf5/mp* or *pin1* (Yamaguchi et al. 2013). Moreover, LFY positively regulates auxin signaling by affecting IAA1, IAA17, and IAA29 expression and suppressing several auxin biosynthetic genes (Li et al. 2013). Therefore, floral meristem formation is under the feedback control of both auxin and the master transcription factor LFY. In addition to LFY, two other transcription factors play important roles in mediating auxin responses during floral meristem specification. ANT and AIL6/PLT3 are also direct targets of ARF5 (Yamaguchi et al. 2013), and they further activate LFY expression (Yamaguchi et al. 2016). Taken together, auxin (mainly through ARF5) activates LFY and other key regulators to define primordium founder cells. Last, but not least, ARF5 regulates its own expression by binding to its promoter region (Lau et al. 2011), which could potentiate its activation during organ initiation (Bhatia et al. 2016). In addition, ARF5 also directly represses the transcription of DRN/ESR1 in the shoot apical meristem, and DRN is involved in the activation of CLV3 in the stem cells (Luo et al. 2018). It seems that ARF5 restricts DRN expression in the meristem center thus aiding CLV3 activation. Although the *drn* mutant shows no obvious phenotype in shoot apical meristem development, overexpression of DRN results in specific defects in meristem maintenance and lateral organ formation (Kirch et al. 2003) and enhances shoot regeneration efficiency in tissue cultures (Banno et al. 2001).

Interaction of auxin with cytokinin was shown to be crucial for organ initiation in the peripheral zone as well. The cytokinin signaling inhibitor AHP6 as a direct target of ARF5 was shown to regulate the timing of organ initiation (Besnard et al. 2014). During flower initiation, AHP6 is a direct target of ARF5 and is activated one plastochrone (the time difference between two successive organ initiation, which is close to 12 h in *Arabidopsis*) (Besnard et al. 2014) after the induction of the auxin responsive reporter DR5. Sharp gradients of AHP6 centered on primordia and extending beyond its boundaries indicate that this protein creates fields around organs thus making additional, secondary inhibitory fields. Other than AHP6, ARF5 transcriptionally represses ARR7 and ARR15 (both negative regulators of cytokinin response) in the peripheral zone (Zhao et al. 2010). As ARR7 and ARR15 negatively regulate shoot apical meristem size, auxin signaling could thus promote shoot apical meristem activity. This demonstrates cooperation between both hormones in organ initiation and that cytokinin signaling acts in parallel or downstream of auxin to regulate new meristem initiation in the shoot.

CONCLUSION AND PERSPECTIVES

Insights about shoot apical meristem maintenance and patterning have increased at a steady pace over the last two decades. Although the role of auxin transport in the shoot apical meristem has been studied in great detail using mutants in auxin influx and efflux transporters (for review, see Bhatia and Heisler 2018), data regarding auxin action via transcriptional regulation (activation or repression) still remains fragmented, both in the central zone and in organ initiation. Moreover, the role for the noncanonical auxin pathway involving TMK1 in the shoot apical meristem remains unclear, as well as the function of ABP1 in auxin signaling. A recent paper uncovered a negative role for ABP1 in the initiation of leaf primordia using an ABP1 overexpressor line that was observed to produce a smaller number of rosette leaves (Gelová et al. 2021). While this raises the possibility that an ABP1-dependent mechanism acts in organ initiation in the shoot apical meristem, the absence of phenotypes in aerial organ initiation in *abp1* mutants does not allow a conclusion at the moment. Large-scale gene-functional studies will likely be critical for shedding light on the role of the effectors of auxin signaling in the shoot apical meristem, allowing to decipher the role of individual transcriptional regulators (or groups of them) in response to auxin signaling. As functional redundancy between closely related genes complicates genetic studies of single members, the availability of CRISPR/Cas9 gene-editing tools should help generate higher order mutants and overcome this redundancy limitation or alternatively gain-of-function lines could be employed (Ori 2019; Israeli et al. 2020).

We foresee that quantitative imaging will be key to advancing our knowledge on auxin signaling function in the shoot apical meristem. More sensitive auxin signaling reporters could certainly help analyze auxin-dependent transcription and understanding the role of the low auxin signaling activity in the central zone and the organizing center of the shoot apical meristem. More generally, reporters and sensors with higher sensitivity (e.g., FRET-based biosensors for detecting auxin with subcellular resolution) (Herud-Sikimic et al. 2020), together with microscopes equipped with the latest-generation highly sensitive detectors, will be important to visualize and analyze auxin distribution and signaling activity in the internal layers of the shoot apical meristem. Most imaging studies we have discussed in this review focus on the epidermal layer, leaving many questions open on how auxin signaling contribute to shoot apical meristem activity in the rest of the meristem. Obtaining quantitative data on auxin distribution and auxin signaling activity in 3D over time in the entire shoot apical meristem will certainly be key to understanding how the auxin signal is processed and how it is translated into specific cellular responses such as fate changes, cell expansion, or cell division and ultimately in morphogenesis. A recent study demonstrated that cells integrate the auxin signal over time to trigger organ initiation in the shoot apical meristem, providing a striking example of how such approaches can expand our knowledge on the role of auxin signaling in the shoot apical meristem (Galvan-Ampudia et al. 2020).

Future approaches will also need to link genome-scale transcriptional changes to the activity of the auxin signaling pathways. Recent works have begun to use mathematical approaches to integrate large-scale data sets in predictive models (for review, see Mjolsness 2019). Such a cell-based model of the *Arabidopsis* shoot apical meristem was recently developed (Banwarth-Kuhn et al. 2019). The model incorporates individual cell behaviors such as cell–cell interactions, anisotropy in cell growth, cell division, differentiation, and biochemical signaling events to quantify the impact of individual cell processes on overall tissue shape, size, and function. Such approaches are flexible enough to model diverse biological processes such as intercellular chemical kinetics, intercellular signaling, cell differentiation, and motion of cells. More importantly, they can be used to perform simulations that can capture the underlying properties of the real biological system and help pinpoint other candidates as shoot apical meristem regulators mediating auxin signaling activity.

Finally, expanding the analysis of the role of auxin signaling in the shoot apical meristem to nonmodel plant meristems that develop different shoot architecture (Wang et al. 2018a,b; Kitagawa and Jackson 2019), as for example maize given the extensive amount of genetic and expression data available or tomato (Knauer et al. 2019; Matthes et al. 2019), will undoubtedly provide valuable complementary information as highlighted in this review for maize Aux/IAA mutants (Galli et al. 2015). It will also allow the identification of conserved function(s) for auxin signaling in shoot apical meristem function.

ACKNOWLEDGMENTS

We acknowledge funding sources, particularly the European Structural and Investment Funds, Operational Programme Research, Development and Education, Project MSCAfellow@MUNI (CZ.02.2.69/0.0/0.0/17_050/0008496) to M.P., and ANR-18-CE12-0014-02 (ChromAuxi) to T.V.

REFERENCES

Bainbridge K, Guyomarc'h S, Bayer E, Swarup R, Bennett M, Mandel T, Kuhlemeier C. 2008. Auxin influx carriers stabilize phyllotactic patterning. *Genes Dev* **22:** 810–823. doi:10.1101/gad.462608

Banno H, Ikeda Y, Niu QW, Chua NH. 2001. Overexpression of *Arabidopsis ESR1* induces initiation of shoot regeneration. *Plant Cell* **13:** 2609–2618. doi:10.1105/tpc.010234

Banwarth-Kuhn M, Nematbakhsh A, Rodriguez KW, Snipes S, Rasmussen CG, Reddy GV, Alber M. 2019. Cell-based model of the generation and maintenance of the shape and structure of the multilayered shoot apical meristem of *Arabidopsis thaliana. Bull Mathematical Biol* **81:** 3245–3281. doi:10.1007/s11538-018-00547-z

Benková E, Michniewicz M, Sauer M, Teichmann T, Seifertová D, Jürgens G, Friml J. 2003. Local, efflux-dependent auxin gradients as a common module for plant organ formation. *Cell* **115:** 591–602. doi:10.1016/S0092-8674(03)00924-3

Besnard F, Refahi Y, Morin V, Marteaux B, Brunoud G, Chambrier P, Rozier F, Mirabet V, Legrand J, Lainé S, et al. 2014. Cytokinin signalling inhibitory fields provide robustness to phyllotaxis. *Nature* **505:** 417–421. doi:10.1038/nature12791

Bhatia N, Heisler MG. 2018. Self-organizing periodicity in development: organ positioning in plants. *Development* **145:** dev149336. doi:10.1242/dev.149336

Bhatia N, Bozorg B, Larsson A, Ohno C, Jönsson H, Heisler MG. 2016. Auxin acts through MONOPTEROS to regulate plant cell polarity and pattern phyllotaxis. *Curr Biol* **26:** 3202–3208. doi:10.1016/j.cub.2016.09.044

Blázquez MA, Weigel D. 2000. Integration of floral inductive signals in *Arabidopsis. Nature* **404:** 889–892. doi:10.1038/35009125

Blázquez MA, Soowal LN, Lee I, Weigel D. 1997. LEAFY expression and flower initiation in *Arabidopsis. Development* **124:** 3835–3844.

Boer DR, Freire-Rios A, van den Berg WAM, Saaki T, Manfield IW, Kepinski S, López-Vidrieo I, Franco-Zorrilla JM, de Vries SC, Solano R, et al. 2014. Structural basis for DNA binding specificity by the auxin-dependent ARF transcription factors. *Cell* **156:** 577–589. doi:10.1016/j.cell.2013.12.027

Bolduc N, Yilmaz A, Mejia-Guerra MK, Morohashi K, O'Connor D, Grotewold E, Hake S. 2012. Unraveling the KNOTTED1 regulatory network in maize meristems. *Genes Dev* **26:** 1685–1690. doi:10.1101/gad.193433.112

Brand U. 2000. Dependence of stem cell fate in *Arabidopsis* on a feedback loop regulated by CLV3 activity. *Science* **289:** 617–619. doi:10.1126/science.289.5479.617

Busch W, Miotk A, Ariel FD, Zhao Z, Forner J, Daum G, Suzaki T, Schuster C, Schultheiss SJ, Leibfried A, et al. 2010. Transcriptional control of a plant stem cell niche. *Dev Cell* **18:** 841–853. doi:10.1016/j.devcel.2010.03.012

Cao M, Chen R, Li P, Yu Y, Zheng R, Ge D, Zheng W, Wang X, Gu Y, Gelová Z, et al. 2019. TMK1-mediated auxin signalling regulates differential growth of the apical hook. *Nature* **568:** 240–243. doi:10.1038/s41586-019-1069-7

Chickarmane VS, Gordon SP, Tarr PT, Heisler MG, Meyerowitz EM. 2012. Cytokinin signaling as a positional cue for patterning the apical-basal axis of the growing *Arabidopsis* shoot meristem. *Proc Natl Acad Sci* **109:** 4002–4007. doi:10.1073/pnas.1200636109

Chung Y, Zhu Y, Wu MF, Simonini S, Kuhn A, Armenta-Medina A, Jin R, Østergaard L, Gillmor CS, Wagner D. 2019. Auxin response factors promote organogenesis by chromatin-mediated repression of the pluripotency gene SHOOTMERISTEMLESS. *Nat Commun* **10:** 886.

Cole M, Chandler J, Weijers D, Jacobs B, Comelli P, Werr W. 2009. DORNROSCHEN is a direct target of the auxin response factor MONOPTEROS in the *Arabidopsis* embryo. *Development* **136:** 1643–1651. doi:10.1242/dev.032177

Dai N, Wang W, Patterson SE, Bleecker AB. 2013. The TMK subfamily of receptor-like kinases in *Arabidopsis* display an essential role in growth and a reduced sensitivity to auxin. *PLoS ONE* **8:** e60990. doi:10.1371/journal.pone.0060990

Daum G, Medzihradszky A, Suzaki T, Lohmann JU. 2014. A mechanistic framework for noncell autonomous stem cell induction in *Arabidopsis. Proc Natl Acad Sci* **111:** 14619–14624. doi:10.1073/pnas.1406446111

Douady S, Couder Y. 1992. Phyllotaxis as a physical self-organized growth process. *Phys Rev Lett* **68:** 2098–2101. doi:10.1103/PhysRevLett.68.2098

Finet C, Berne-Dedieu A, Scutt CP, Marlétaz F. 2013. Evolution of the ARF gene family in land plants: old domains, new tricks. *Mol Biol Evol* **30:** 45–56. doi:10.1093/molbev/mss220

Fletcher JC, Brand U, Running MP, Simon R, Meyerowitz EM. 1999. Signaling of cell fate decisions by CLAVATA3 in *Arabidopsis* shoot meristems. *Science* **283:** 1911–1914. doi:10.1126/science.283.5409.1911

Friml J, Yang X, Michniewicz M, Weijers D, Quint A, Tietz O, Benjamins R, Ouwerkerk PB, Ljung K, Sandberg G, et al. 2004. A PINOID-dependent binary switch in apical-basal PIN polar targeting directs auxin efflux. *Science* **306:** 862–865. doi:10.1126/science.1100618

Gallei M, Luschnig C, Friml J. 2020. Auxin signalling in growth: Schrödinger's cat out of the bag. *Curr Opin Plant Biol* **53:** 43–49. doi:10.1016/j.pbi.2019.10.003

Galli M, Liu Q, Moss BL, Malcomber S, Li W, Gaines C, Federici S, Roshkovan J, Meeley R, Nemhauser JL, et al. 2015. Auxin signaling modules regulate maize inflorescence architecture. *Proc Natl Acad Sci* **112:** 13372–13377. doi:10.1073/pnas.1516473112

Galvan-Ampudia CS, Chaumeret AM, Godin C, Vernoux T. 2016. Phyllotaxis: from patterns of organogenesis at the meristem to shoot architecture. *Wiley Interdiscip Rev Dev Biol* **5:** 460–473. doi:10.1002/wdev.231

Galvan-Ampudia CS, Cerutti G, Legrand J, Brunoud G, Martin-Arevalillo R, Azais R, Bayle V, Moussu S, Wenzl C, Jaillais Y, et al. 2020. Temporal integration of auxin information for the regulation of patterning. *eLife* **9:** e55832. doi:10.7554/eLife.55832

Galweiler L, Guan C, Muller A, Wisman E, Mendgen K, Yephremov A, Palme K. 1998. Regulation of polar auxin transport by AtPIN1 in *Arabidopsis* vascular tissue. *Science* **282:** 2226–2230. doi:10.1126/science.282.5397.2226

Gao Y, Zhang Y, Zhang D, Dai X, Estelle M, Zhao Y. 2015. Auxin binding protein 1 (ABP1) is not required for either auxin signaling or *Arabidopsis* development. *Proc Natl Acad Sci* **112:** 2275–2280. doi:10.1073/pnas.1500365112

Gelová Z, Gallei M, Pernisová M, Brunoud G, Zhang X, Glanc M, Li L, Michalko J, Pavlovičová Z, Verstraeten I, et al. 2021. Developmental roles of auxin binding protein 1 in *Arabidopsis thaliana*. *Plant Sci* **303:** 110750. doi:10.1016/j.plantsci.2020.110750

Godin C, Golé C, Douady S. 2020. Phyllotaxis as geometric canalization during plant development. *Development* **147:** dev165878. doi:10.1242/dev.165878

Gordon SP, Chickarmane VS, Ohno C, Meyerowitz EM. 2009. Multiple feedback loops through cytokinin signaling control stem cell number within the *Arabidopsis* shoot meristem. *Proc Natl Acad Sci* **106:** 16529–16534. doi:10.1073/pnas.0908122106

Han H, Liu X, Zhou Y. 2020. Transcriptional circuits in control of shoot stem cell homeostasis. *Curr Opin Plant Biol* **53:** 50–56. doi:10.1016/j.pbi.2019.10.004

Hardtke CS, Berleth T. 1998. The *Arabidopsis* gene MONOPTEROS encodes a transcription factor mediating embryo axis formation and vascular development. *EMBO J* **17:** 1405–1411. doi:10.1093/emboj/17.5.1405

Hay A, Tsiantis M. 2010. KNOX genes: versatile regulators of plant development and diversity. *Development* **137:** 3153–3165. doi:10.1242/dev.030049

Heisler MG, Ohno C, Das P, Sieber P, Reddy GV, Long JA, Meyerowitz EM. 2005. Patterns of auxin transport and gene expression during primordium development revealed by live imaging of the *Arabidopsis* inflorescence meristem. *Curr Biol* **15:** 1899–1911. doi:10.1016/j.cub.2005.09.052

Herud-Sikimic O, Stiel AC, Ortega-Perez M, Shanmugaratnam S, Höcker B, Jürgens G. 2020. Design of a biosensor for direct visualisation of auxin. *bioRxiv* doi:10.1101/2020.01.19.911735

Israeli A, Reed JW, Ori N. 2020. Genetic dissection of the auxin response network. *Nat Plants* **6:** 1082–1090. doi:10.1038/s41477-020-0739-7

Jha P, Ochatt SJ, Kumar V. 2020. WUSCHEL: a master regulator in plant growth signaling. *Plant Cell Rep* **39:** 431–444. doi:10.1007/s00299-020-02511-5

Jonsson H, Heisler MG, Shapiro BE, Meyerowitz EM, Mjolsness E. 2006. An auxin-driven polarized transport model for phyllotaxis. *Proc Natl Acad Sci* **103:** 1633–1638. doi:10.1073/pnas.0509839103

Kato H, Mutte SK, Suzuki H, Crespo I, Das S, Radoeva T, Fontana M, Yoshitake Y, Hainiwa E, van den Berg W, et al. 2020. Design principles of a minimal auxin response system. *Nat Plants* **6:** 473–482. doi:10.1038/s41477-020-0662-y

Kellogg EA. 2019. Different ways to be redundant. *Nat Genet* **51:** 770–771. doi:10.1038/s41588-019-0406-y

Kirch T, Simon R, Grünewald M, Werr W. 2003. The *DORNRÖSCHEN/ENHANCER OF SHOOT REGENERATION1* gene of *Arabidopsis* acts in the control of meristem cell fate and lateral organ development. *Plant Cell* **15:** 694–705. doi:10.1105/tpc.009480

Kitagawa M, Jackson D. 2019. Control of meristem size. *Ann Rev Plant Biol* **70:** 269–291. doi:10.1146/annurev-arplant-042817-040549

Knauer S, Holt AL, Rubio-Somoza I, Tucker EJ, Hinze A, Pisch M, Javelle M, Timmermans MC, Tucker MR, Laux T. 2013. A protodermal miR394 signal defines a region of stem cell competence in the *Arabidopsis* shoot meristem. *Dev Cell* **24:** 125–132. doi:10.1016/j.devcel.2012.12.009

Knauer S, Javelle M, Li L, Li X, Ma X, Wimalanathan K, Kumari S, Johnston R, Leiboff S, Meeley R, et al. 2019. A high-resolution gene expression atlas links dedicated meristem genes to key architectural traits. *Genome Res* **29:** 1962–1973. doi:10.1101/gr.250878.119

Kuhlemeier C. 2017. Phyllotaxis. *Curr Biol* **27:** R882–R887. doi:10.1016/j.cub.2017.05.069

Kuhn A, Ramans Harborough S, McLaughlin HM, Natarajan B, Verstraeten I, Friml J, Kepinski S, Østergaard L. 2020. Direct ETTIN-auxin interaction controls chromatin states in gynoecium development. *eLife* **9:** e51787. doi:10.7554/eLife.51787

Kuroha T, Tokunaga H, Kojima M, Ueda N, Ishida T, Nagawa S, Fukuda H, Sugimoto K, Sakakibara H. 2009. Functional analyses of *LONELY GUY* cytokinin-activating enzymes reveal the importance of the direct activation pathway in *Arabidopsis*. *Plant Cell* **21:** 3152–3169. doi:10.1105/tpc.109.068676

Landberg K, Šimura J, Ljung K, Sundberg E, Thelander M. 2021. Studies of moss reproductive development indicate that auxin biosynthesis in apical stem cells may constitute an ancestral function for focal growth control. *New Phytol* **229:** 845–860. doi:10.1111/nph.16914

Lau S, Smet ID, Kolb M, Meinhardt H, Jürgens G. 2011. Auxin triggers a genetic switch. *Nat Cell Biol* **13:** 611–615. doi:10.1038/ncb2212

Laux T, Mayer KF, Berger J, Jurgens G. 1996. The WUSCHEL gene is required for shoot and floral meristem integrity in *Arabidopsis*. *Development* **122:** 87–96.

Lee ZH, Hirakawa T, Yamaguchi N, Ito T. 2019. The roles of plant hormones and their interactions with regulatory genes in determining meristem activity. *Int J Mol Sci* **20:** 4065. doi:10.3390/ijms20164065

Leibfried A, To JP, Busch W, Stehling S, Kehle A, Demar M, Kieber JJ, Lohmann JU. 2005. WUSCHEL controls meristem function by direct regulation of cytokinin-inducible response regulators. *Nature* **438:** 1172–1175. doi:10.1038/nature04270

Li W, Zhou Y, Liu X, Yu P, Cohen JD, Meyerowitz EM. 2013. LEAFY controls auxin response pathways in floral primordium formation. *Sci Signal* **6:** ra23.

Liao CY, Smet W, Brunoud G, Yoshida S, Vernoux T, Weijers D. 2015. Reporters for sensitive and quantitative measurement of auxin response. *Nat Methods* **12:** 207–210. doi:10.1038/nmeth.3279

Long JA. 2006. TOPLESS regulates apical embryonic fate in *Arabidopsis*. *Science* **312:** 1520–1523. doi:10.1126/science.1123841

Luo L, Zeng J, Wu H, Tian Z, Zhao Z. 2018. A molecular framework for auxin-controlled homeostasis of shoot stem cells in *Arabidopsis*. *Mol Plant* **11:** 899–913. doi:10.1016/j.molp.2018.04.006

Ma Y, Miotk A, Šutiković Z, Ermakova O, Wenzl C, Medzihradszky A, Gaillochet C, Forner J, Utan G, Brackmann K, et al. 2019. WUSCHEL acts as an auxin response rheostat to maintain apical stem cells in *Arabidopsis*. *Nat Commun* **10:** 5093. doi:10.1038/s41467-019-13074-9

Matthes MS, Best NB, Robil JM, Malcomber S, Gallavotti A, McSteen P. 2019. Auxin EvoDevo: conservation and diversification of genes regulating auxin biosynthesis, transport, and signaling. *Mol Plant* **12:** 298–320. doi:10.1016/j.molp.2018.12.012

Mayer KF, Schoof H, Haecker A, Lenhard M, Jürgens G, Laux T. 1998. Role of WUSCHEL in regulating stem cell fate in the *Arabidopsis* shoot meristem. *Cell* **95:** 805–815. doi:10.1016/S0092-8674(00)81703-1

Mitchison GJ. 1977. Phyllotaxis and the Fibonacci series. *Science* **196:** 270–275. doi:10.1126/science.196.4287.270

Mjolsness E. 2019. Prospects for declarative mathematical modeling of complex biological systems. *Bull Math Biol* **81:** 3385–3420. doi:10.1007/s11538-019-00628-7

Moyroud E, Kusters E, Monniaux M, Koes R, Parcy F. 2010. LEAFY blossoms. *Trends Plant Sci* **15:** 346–352. doi:10.1016/j.tplants.2010.03.007

Mutte SK, Kato H, Rothfels C, Melkonian M, Wong GK-S, Weijers D. 2018. Origin and evolution of the nuclear auxin response system. *eLife* **7:** e33399.doi:10.7554/eLife.33399

Nagpal P. 2005. Auxin response factors ARF6 and ARF8 promote jasmonic acid production and flower maturation. *Development* **132:** 4107–4118. doi:10.1242/dev.01955

Nardmann J, Werr W. 2006. The shoot stem cell niche in angiosperms: expression patterns of WUS orthologues in rice and maize imply major modifications in the course of mono- and dicot evolution. *Mol Biol Evol* **23:** 2492–2504. doi:10.1093/molbev/msl125

Nardmann J, Werr W. 2007. The evolution of plant regulatory networks: what *Arabidopsis* cannot say for itself. *Curr Opin Plant Biol* **10:** 653–659. doi:10.1016/j.pbi.2007.07.009

Okushima Y, Mitina I, Quach HL, Theologis A. 2005a. AUXIN RESPONSE FACTOR 2 (ARF2): a pleiotropic developmental regulator: ARF transcription factors. *Plant J* **43:** 29–46. doi:10.1111/j.1365-313X.2005.02426.x

Okushima Y, Overvoorde PJ, Arima K, Alonso JM, Chan A, Chang C, Ecker JR, Hughes B, Lui A, Nguyen D, et al. 2005b. Functional genomic analysis of the *AUXIN RESPONSE FACTOR* gene family members in *Arabidopsis thaliana*: unique and overlapping functions of *ARF7* and *ARF19*. *Plant Cell* **17:** 444–463. doi:10.1105/tpc.104.028316

Ori N. 2019. Dissecting the biological functions of ARF and Aux/IAA genes. *Plant Cell* **31:** 1210–1211. doi:10.1105/tpc.19.00330

Przemeck GH, Mattsson J, Hardtke C, Sung ZR, Berleth T. 1996. Studies on the role of the *Arabidopsis* gene MONOPTEROS in vascular development and plant cell axialization. *Planta* **200:** 229–237. doi:10.1007/BF00208313

Rademacher EH, Möller B, Lokerse AS, Llavata-Peris CI, van den Berg W, Weijers D. 2011. A cellular expression map of the *Arabidopsis* AUXIN RESPONSE FACTOR gene family: a cellular expression map of ARF gene expression. *Plant J* **68:** 597–606. doi:10.1111/j.1365-313X.2011.04710.x

Reinhardt D, Pesce ER, Stieger P, Mandel T, Baltensperger K, Bennett M, Traas J, Friml J, Kuhlemeier C. 2003. Regulation of phyllotaxis by polar auxin transport. *Nature* **426:** 255–260. doi:10.1038/nature02081

Rodriguez-Leal D, Xu C, Kwon C-T, Soyars C, Demesa-Arevalo E, Man J, Liu L, Lemmon ZH, Jones DS, Van Eck J, et al. 2019. Evolution of buffering in a genetic circuit controlling plant stem cell proliferation. *Nat Genet* **51:** 786–792. doi:10.1038/s41588-019-0389-8

Schlereth A, Möller B, Liu W, Kientz M, Flipse J, Rademacher EH, Schmid M, Jürgens G, Weijers D. 2010. MONOPTEROS controls embryonic root initiation by regulating a mobile transcription factor. *Nature* **464:** 913–916. doi:10.1038/nature08836

Schoof H, Lenhard M, Haecker A, Mayer KF, Jürgens G, Laux T. 2000. The stem cell population of *Arabidopsis* shoot meristems is maintained by a regulatory loop between the CLAVATA and WUSCHEL genes. *Cell* **100:** 635–644. doi:10.1016/S0092-8674(00)80700-X

Shi B, Vernoux T. 2019. Patterning at the shoot apical meristem and phyllotaxis. *Curr Topic Dev Biol* **131:** 81–107. doi:10.1016/bs.ctdb.2018.10.003

Simonini S, Deb J, Moubayidin L, Stephenson P, Valluru M, Freire-Rios A, Sorefan K, Weijers D, Friml J, Østergaard L. 2016. A noncanonical auxin-sensing mechanism is required for organ morphogenesis in *Arabidopsis*. *Genes Dev* **30:** 2286–2296. doi:10.1101/gad.285361.116

Simonini S, Bencivenga S, Trick M, Østergaard L. 2017. Auxin-induced modulation of ETTIN activity orchestrates gene expression in *Arabidopsis*. *Plant Cell* **29:** 1864–1882. doi:10.1105/tpc.17.00389

Siriwardana NS, Lamb RS. 2012. The poetry of reproduction: the role of LEAFY in *Arabidopsis thaliana* flower formation. *Int J Dev Biol* **56:** 207–221. doi:10.1387/ijdb.113450ns

Smith RS, Guyomarc'h S, Mandel T, Reinhardt D, Kuhlemeier C, Prusinkiewicz P. 2006. A plausible model of phyllotaxis. *Proc Natl Acad Sci* **103:** 1301–1306. doi:10.1073/pnas.0510457103

Snipes SA, Rodriguez K, DeVries AE, Miyawaki KN, Perales M, Xie M, Reddy GV. 2018. Cytokinin stabilizes WUSCHEL by acting on the protein domains required for nuclear enrichment and transcription. *PLoS Genet* **14:** e1007351. doi:10.1371/journal.pgen.1007351

Somssich M, Je BI, Simon R, Jackson D. 2016. CLAVATA-WUSCHEL signaling in the shoot meristem. *Development* **143:** 3238–3248. doi:10.1242/dev.133645

Soyars CL, James SR, Nimchuk ZL. 2016. Ready, aim, shoot: stem cell regulation of the shoot apical meristem. *Curr Opin Plant Biol* **29:** 163–168. doi:10.1016/j.pbi.2015.12.002

Stoma S, Lucas M, Chopard J, Schaedel M, Traas J, Godin C. 2008. Flux-based transport enhancement as a plausible unifying mechanism for auxin transport in meristem development. *PLoS Comput Biol* **4:** e1000207. doi:10.1371/journal.pcbi.1000207

Truskina J, Han J, Chrysanthou E, Galvan-Ampudia CS, Lainé S, Brunoud G, Macé J, Bellows S, Legrand J, Bågman AM, et al. 2020. A network of transcriptional repressors modulates auxin responses. *Nature* **589:** 116–119.

Vernoux T, Kronenberger J, Grandjean O, Laufs P, Traas J. 2000. PIN-FORMED 1 regulates cell fate at the periphery of the shoot apical meristem. *Development* **127:** 5157–5165.

Vernoux T, Brunoud G, Farcot E, Morin V, Van den Daele H, Legrand J, Oliva M, Das P, Larrieu A, Wells D, et al. 2011. The auxin signalling network translates dynamic input into robust patterning at the shoot apex. *Mol Syst Biol* **7:** 508. doi:10.1038/msb.2011.39

Véron E, Vernoux T, Coudert Y. 2020. Phyllotaxis from a single apical cell. *Trends Plant Sci* **26:** 124–131. doi:10.1016/j.tplants.2020.09.014

Vollbrecht E, Veit B, Sinha N, Hake S. 1991. The developmental gene Knotted-1 is a member of a maize homeobox gene family. *Nature* **350:** 241–243. doi:10.1038/350241a0

Wabnik K, Robert HS, Smith RS, Friml J. 2013. Modeling framework for the establishment of the apical-basal embryonic axis in plants. *Curr Biol* **23:** 2513–2518. doi:10.1016/j.cub.2013.10.038

Wang B, Smith SM, Li J. 2018a. Genetic regulation of shoot architecture. *Ann Rev Plant Biol* **69:** 437–468. doi:10.1146/annurev-arplant-042817-040422

Wang Y, Zhang T, Wang R, Zhao Y. 2018b. Recent advances in auxin research in rice and their implications for crop improvement. *J Exp Botany* **69:** 255–263. doi:10.1093/jxb/erx228

Wu M-F, Yamaguchi N, Xiao J, Bargmann B, Estelle M, Sang Y, Wagner D. 2015. Auxin-regulated chromatin switch directs acquisition of flower primordium founder fate. *eLife* **4:** e09269. doi:10.7554/eLife.09269

Xu T, Dai N, Chen J, Nagawa S, Cao M, Li H, Zhou Z, Chen X, De Rycke R, Rakusova H, et al. 2014. Cell surface ABP1-TMK auxin-sensing complex activates ROP GTPase signaling. *Science* **343:** 1025–1028. doi:10.1126/science.1245125

Yadav RK, Perales M, Gruel J, Girke T, Jonsson H, Reddy GV. 2011. WUSCHEL protein movement mediates stem cell homeostasis in the *Arabidopsis* shoot apex. *Genes Dev* **25:** 2025–2030. doi:10.1101/gad.17258511

Yadav RK, Perales M, Gruel J, Ohno C, Heisler M, Girke T, Jönsson H, Reddy GV. 2013. Plant stem cell maintenance involves direct transcriptional repression of differentiation program. *Mol Syst Biol* **9:** 654. doi:10.1038/msb.2013.8

Yamaguchi N, Wu MF, Winter CM, Berns MC, Nole-Wilson S, Yamaguchi A, Coupland G, Krizek BA, Wagner D. 2013. A molecular framework for auxin-mediated initiation of flower primordia. *Dev Cell* **24:** 271–282. doi:10.1016/j.devcel.2012.12.017

Yamaguchi N, Jeong CW, Nole-Wilson S, Krizek BA, Wagner D. 2016. AINTEGUMENTA and AINTEGUMENTA-LIKE6/PLETHORA3 induce *LEAFY* expression in response to auxin to promote the onset of flower formation in *Arabidopsis*. *Plant Physiol* **170:** 283–293. doi:10.1104/pp.15.00969

Zhao Z, Andersen SU, Ljung K, Dolezal K, Miotk A, Schultheiss SJ, Lohmann JU. 2010. Hormonal control of the shoot stem-cell niche. *Nature* **465:** 1089–1092. doi:10.1038/nature09126

Auxin in Root Development

Suruchi Roychoudhry and Stefan Kepinski

Centre for Plant Sciences, Faculty of Biological Sciences, University of Leeds, Leeds LS2 9JT, United Kingdom
Correspondence: s.roychoudhry@leeds.ac.uk; s.kepinski@leeds.ac.uk

Root system architecture is an important determinant of below-ground resource capture and hence overall plant fitness. The plant hormone auxin plays a central role in almost every facet of root development from the cellular to the whole-root-system level. Here, using *Arabidopsis* as a model, we review the multiple gene signaling networks regulated by auxin biosynthesis, conjugation, and transport that underpin primary and lateral root development. We describe the role of auxin in establishing the root apical meristem and discuss how the tight spatiotemporal regulation of auxin distribution controls transitions between cell division, cell growth, and differentiation. This includes the localized reestablishment of mitotic activity required to elaborate the root system via the production of lateral roots. We also summarize recent discoveries on the effects of auxin and auxin signaling and transport on the control of lateral root gravitropic setpoint angle (GSA), a critical determinant of the overall shape of the root system. Finally, we discuss how environmental conditions influence root developmental plasticity by modulation of auxin biosynthesis, transport, and the canonical auxin signaling pathway.

In vascular plants, root systems form the interface between the plant body and the soil. They play a crucial role in basic functions such as anchorage and nutrient and water uptake, as well as in species-specific functions such as vegetative reproduction, minimizing underground competition with adjacent individuals, and maintaining abiotic and biotic interactions within the rhizosphere (Morris et al. 2017). Plants encounter myriad subterranean macro- and microenvironmental stresses during their lifetimes, including biotic stresses, such as pathogen attack, and abiotic stresses such as nutrient and water scarcity. Root systems must therefore retain the ability to perceive and respond to these stresses at multiple levels, ranging from individual cells to tissues and eventually whole organs. The ability of root systems to maintain a high degree of plasticity is therefore key to ensuring the survival, fitness, and adaptation of individual plants (Khan et al. 2016).

Roots can have one of two distinct origins: the embryo (embryonic, i.e., primary or seminal roots) or other roots and/or non-root tissues (postembryonic). Roots that arise from preexisting roots are usually secondary or lateral roots (LRs), while those that arise from nonroot tissues are adventitious or basal roots (for review, see Du and Scheres 2018). In dicot species, including *Arabidopsis*, the primary embryonic root is dominant and produces a number of secondary LRs. Over time, these LRs reiterate the formation of new LRs, which leads to the production of a highly ordered and spatially dis-

Cite this article as *Cold Spring Harb Perspect Biol* doi: 10.1101/cshperspect.a039933

tinct root system termed as the "taproot" or allorhizic root system. In contrast, monocots form "homorhizic" root systems comprised of numerous adventitious roots that in turn produce additional secondary or LRs (Osmont et al. 2007; Bellini et al. 2014). An important distinction between the two types of root systems lies in the life span of the primary embryonic root: in taproot systems, the primary root remains dominant during the entire life span of the plant, while in homorhizic systems, embryonic roots are often short-lived and play a role in the establishment of the root systems at the seedling stage, before the development of adventitious roots, which dominate the root system (Bellini et al. 2014).

Root system architecture (RSA) describes the spatial distribution of the root system, capturing the specific deployment of root axes in the soil matrix or substrate upon which the plant grows (Smith and De Smet 2012; Khan et al. 2016). Macro RSA, the large-scale overall form of the root system, comprises three major parameters: length, degree of branching, and angle of growth. At a smaller scale, the "micro" parameters of RSA that influence the extent of root surface area include root diameter and the formation of root hairs. These macro- and micro-scale components of RSA are a critical determinant of the plant's ability to capture the water and nutrients that are typically distributed heterogeneously within the soil (Smith and De Smet 2012; Morris et al. 2017; for review, see Rellán-Álvarez et al. 2016). RSA parameters can be modulated by multiple genetic, physiological, and environmental factors, allowing for an efficient capture of resources from the rhizosphere (for review, see Rellán-Álvarez et al. 2016). Understanding the molecular mechanisms that regulate RSA and its plasticity is therefore key to enhancing both the productivity and sustainability of agriculture through developing crop varieties that are better able to take up water and nutrients, reducing the need for supplementary fertilizer application and irrigation (Lynch 2013).

The plant hormone auxin is the master regulator of plant architecture (for review, see Benjamins and Scheres 2008; Overoode et al. 2010). Auxin is a small molecule that is dynamically distributed via a system of polar auxin transport. The resulting gradients, minima and maxima of auxin concentration that are formed to act to pattern, trigger developmental transitions, and control growth. Auxin can enter cells passively, but is also actively imported into cells by the influx carriers AUX1 and LIKE AUX1 (LAX) 1-3 (Swarup et al. 2005; Péret et al. 2012). Auxin must be exported from cells by the PIN-FORMED family of efflux transporter proteins as well as the ABCB and TWISTED DWARF1 (TWD1) proteins (for review, see Geisler et al. 2017; Zwiewka et al. 2019; Hammes et al. 2021). It is this requirement for active auxin efflux from cells that forms the basis of the directional control of auxin movement. *Arabidopsis* contains eight PIN proteins, of which PINs 1–4 and 7 are canonical "long PINs" targeted to the plasma membrane, while PINs 5 and 6 are short PINs targeted to the endoplasmic reticulum (ER). Although each of the long PINs have specific expression patterns and play distinct roles in root and shoot development, recent work has shown that all five canonical PINs share equivalent capacity to generate auxin maxima and regulate different aspects of root (and shoot) development (Zhang et al. 2020). Thus, coordinated regulation of the subcellular distribution of PIN proteins within the field of cells allows flows of auxin to be generated (Blilou et al. 2005). Spatiotemporal variation in auxin concentration is then translated into developmental control principally via the TIR1/AFB-Aux/IAA-ARF system of auxin-dependent transcriptional regulation (see Morffy and Strader 2021). Briefly, in this system auxin acts to regulate the abundance of Aux/IAA transcriptional corepressor proteins, which are targeted to auxin-regulated genes via their interaction with DNA-binding transcription factors called auxin response factors (ARFs) (for review, see Leyser 2018). Auxin promotes the formation of an auxin coreceptor complex comprising an F-box protein of the TIR1/AFB family and an Aux/IAA protein. This auxin-enhanced interaction catalyzes to ubiquitin-mediated proteolysis of the Aux/IAA, thereby de-repressing the ARF-bound loci previously transcriptionally silenced by the presence of the Aux/IAA (Gray et al. 2001; Dharmasiri et al. 2005; Kepinski and Leyser

2005). Here, we review the complex role of auxin in root development and its impact on RSA.

AUXIN IN PRIMARY ROOT DEVELOPMENT

In the *Arabidopsis* root, tissues originate through the activity of the root meristem. Just above the columella, which constitutes the root tip, is the stem cell niche, which is composed of a group of slow-dividing cells that constitute the organizing center, also called quiescent center (QC), and the surrounding long-term stem cells (for review, see García-Gómez et al. 2021). These stem cells divide asymmetrically to short-term stem cells, which undergo multiple cell divisions with little expansion, in the proximal meristem (PM). The PM is enveloped by the lateral root cap (LRC), an important tissue for auxin transport and homeostasis in the root meristem. Cells are displaced away from the stem cell niche by subsequent cell divisions, until they reach the transition zone (TZ), where they stop dividing and start to differentiate. Cells then enter the elongation zone (EZ) within which cells undergo vast increases in length (up to 300% in 3 h) with virtually no division. The EZ is followed by the differentiation zone (DZ), where cells stop elongating and complete their differentiation gaining cell-type-specific features and functions (Verbelen et al. 2006; Markakis et al. 2012). Research has shown that some of the signaling processes that mediate differentiation begin while cells are still elongating (see below) (Datta et al. 2015). For this reason, the EZ and DZ have sometimes been referred to collectively as the elongation–differentiation zone (EDZ), a name that better captures the overlap in the processes of cell elongation and cell differentiation (Takatsuka and Umeda 2014).

The patterning and coordination of the processes that maintain the constant self-renewal and growth of the root is complex. At first sight, the extent of auxin's involvement in so many of them seems quite remarkable. First, a localized, auxin maximum in the root apex acts to position and maintain the QC and stem cell niche (Sabatini et al. 1999; Grieneisen et al. 2007). This auxin maximum is specified during embryogenesis in the hypophyseal cell at the base of the embryo, where it first establishes the population of root stem cells. In both the embryo and postembryonic root, the formation of these auxin maxima is principally dependent on the expression and subcellular localization of members of the PIN family of auxin efflux transporters (see Fig. 1 and legend for details; Blilou et al. 2005; Michniewicz et al. 2007; Zourelidou et al. 2014). Auxin maintains stem cell organization by regulating the differentiation of distal stem cells (the cells that give rise to the columella) through a signaling pathway involving the transcription factors PLETHORA (PLTs) and the homeodomain transcription factor WUSCHEL RELATED HOMEOBOX 5 (WOX5) (Sabatini et al. 2003; for review, see Drisch and Stahl 2015). Through an ARF10/16-IAA17/AXR3 signaling pathway, local auxin levels are tightly regulated in the root tip by biosynthesis and transport, in turn repressing the expression of WOX5 and restricting it to the QC where it is required for the expression of PLTs (Ding and Friml 2010).

Next, a complex network of hormone signaling pathways specify the boundaries of mitotic activity in the primary root apex, positioning the TZ and maintaining a meristem of sufficient size to sustain continuous root growth. At the TZ, auxin and cytokinin act antagonistically to regulate the expression of the PIN efflux transporters through modulation of the expression of the auxin signaling corepressor SHY2/IAA3. This, in turn, promotes auxin redistribution leading to cell differentiation. Conversely, auxin promotes SHY2 degradation thereby promoting cell division (Dello Ioio et al. 2008). Further insight into auxin-dependent TZ maintenance came from a recent study that combined computational and genetic approaches (Di Mambro et al. 2017). This work showed that cytokinin-dependent degradation of auxin, via conjugation by the enzymes of the GRETCHEN HAGEN (GH) family, in the LRC is required for the specification of an auxin minimum, which coincides with the TZ. The authors predicted, using mathematical modeling, that perturbation of this auxin minimum would lead to changes in TZ positioning and root meristem size and, conversely, showed that manipulation of GH expression lev-

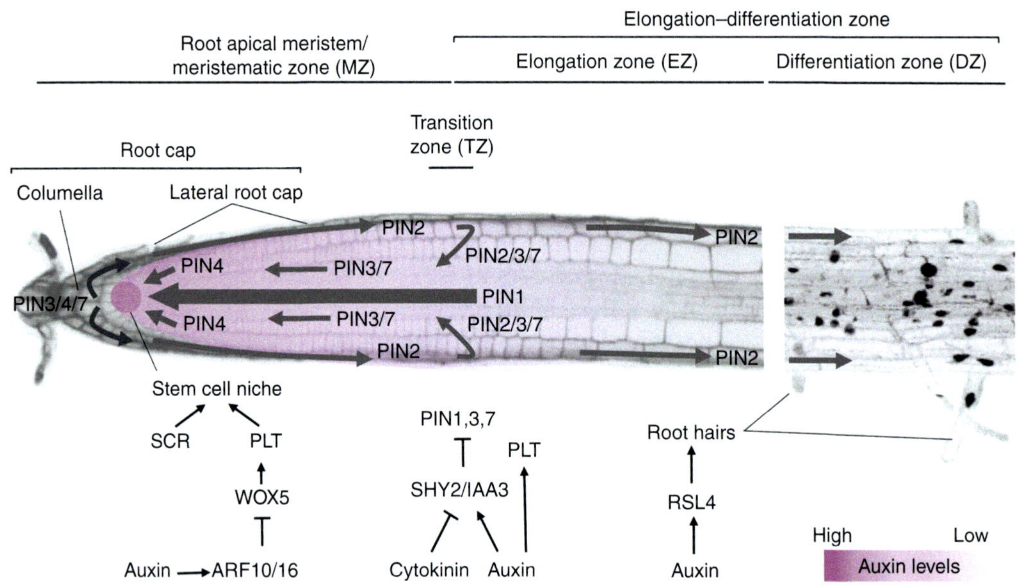

Figure 1. Organization of the primary root in *Arabidopsis*. In the root apex, individual PINs are localized to specific cell membranes to generate a pattern of auxin distribution in the root tip. In the stele, several PINs but principally PIN1 are localized to rootward cell membranes leading to a strong tipward flux of auxin. In the columella cells of the root cap, PIN3, PIN4, and PIN7 distribute auxin laterally toward the lateral root cap, from where it is transported shootward through the epidermis by AUX/LAX proteins and PIN2 (Hu et al. 2021). At the transition zone, some auxin is also refluxed back into the rootward stream in the stele via lateral, inward transport by PIN2, PIN3, and PIN7 (Blilou et al. 2005). This topology of PINs proteins in the root apex creates both the maximum of auxin that positions the stem cell niche and through the gravity-responsive, asymmetric lateral redistribution of auxin from the columella, a means to steer the growth of the root with reference to gravity (for review, see García-Gómez et al. 2021). The identity of the stem cell niche is also dependent on the high levels of *PLT* expression, tightly regulated by complex coordinated hormone signaling pathways. Further shootward, antagonistic interaction between auxin and cytokinin maintain the identity of the meristematic and transition zones, partially through the regulation of levels of the Aux/IAA, SHY2/IAA3. At the transition zone *PLT* levels are intermediate and become further reduced in the elongation–differentiation zone, where cells begin to stop elongating and differentiate to acquire specific functions, such as root hairs.

el resulted in variations in root meristem size (Di Mambro et al. 2017). In addition, auxin also controls TZ positioning by regulating the graded distribution of PLTs (Aida et al. 2004; Mähönen et al. 2014; Durgaprasad et al. 2019). PLTs are transcribed in a narrower domain around the stem cell niche, and PLT proteins form a gradient through a mechanism that involves coordinated cell divisions and cell-to-cell transport. Auxin regulates PLT levels and distribution by both inducing PLT expression around the stem cell niche and by regulating cell division in the PM. PLT controls root zonation in a dose-dependent manner: high levels of PLT in the QC are necessary to maintain slow mitotic stem cell identity, intermediate PLT levels in the meristem induce rapid cell divisions, and low PLT levels at the TZ allow for the onset of cell expansion and differentiation (Galinha et al. 2007; Mähönen et al. 2014). A recent systems biology approach further revealed the complex interplay between auxin-cytokinin-PLT signaling pathways in root development. This study showed that PLT2 levels drop in the root TZ, leading to ARR12 activation. The resulting PLT2-ARR12 antagonism controls the growth of the root meristem in the initial phases of primary root growth. Subsequent activation of ARR1 further reduces PLT2 levels through induction of the cell-cycle repressor KRP2, thus setting final meristem size (Salvi et al. 2020).

Proximal to the meristem, auxin concentrations decrease as daughter cells repeatedly divide before they differentiate. Crucially, auxin concentrations rise again after the TZ, where it is needed for appropriate cell differentiation. In the DZ, auxin is an essential driver of developmental events in both proximodistal and radial axes (Marhava et al. 2018). For example, root protophloem sieve element (PPSE) cells accumulate higher levels of auxin and begin the process of differentiation while their neighbouring cells are still meristematic. A recent study has demonstrated the novel role of two plasma membrane proteins, BRX (BREVIS RADIX) and PAX (a D6PK-like protein kinase, member of the AGC protein family) in regulating this pathway. This work showed that BRX and PAX coordinately act at the rootward membrane of PPSE cells to inhibit PIN1-dependent auxin efflux. This causes a rise in auxin levels, which triggers PPSE differentiation. Following the localized increase in auxin levels, BRX is displaced from the membrane, which in turn activates PAX-dependent auxin efflux and leads to localized reduction in auxin levels. The authors propose that BRX and PAX act as a "molecular rheostat" to control the timing of PPSE differentiation in a multicellular context (Marhava et al. 2018).

Surprisingly, the role of auxin homeostasis and signaling networks in regulating root elongation still remains poorly understood. However, a large number of studies have demonstrated that the exogenous application of both natural and synthetic auxins inhibits cellular elongation and, therefore, root growth. Interestingly, recent work also has demonstrated that tissue-specific biosynthesis of auxin effects root growth. In this study, Hu et al. (2021) expressed the YUCCA and TAA1 auxin biosynthesis genes under the control of promoters that restricted their expression to specific root tissues: *pWER*—epidermis, *pSCR*—endodermis, *pSHR*—stele, *pAPL*—phloem (protophloem, companion cells, and metaphloem sieve elements), and *pWOX5*—QC. The authors showed that the ectopic expression of YUCCA-TAA1 in the epidermis, endodermis, and stele caused very short roots, while YUCCA-TAA1 expression in the phloem and QC had relatively mild effects (Hu et al. 2021).

The inhibition of root growth by auxin has been linked at least partially to auxin-dependent disruption of the auxin cytoskeleton (Rahman et al. 2007; for review, see Zhu and Geisler 2015). Auxin homeostasis during overall root development was also recently found to be dependent on the interplay between the dynamic regulation of the auxin oxidase AtDAO1 (DIOXYGENASE FOR AUXIN OXIDATION 1) and the conjugation enzyme GH3. Mellor et al. (2016) used a combination of mathematical modeling, biochemical, and physiological approaches to show that at lower concentrations of auxin, *AtDAO1* was expressed at higher levels to oxidize endogenous auxin, while higher levels of auxin led to expression of the conjugating enzyme *GH3*. Further work by Zhang et al. (2016) showed that *Arabidopsis* contains two distinct DAO genes, *AtDAO1* and *AtDAO2*, and loss-of-function *dao* mutants had elongated primary roots, increased LR density, and longer root hairs (Porco et al. 2016a; Zhang et al. 2016) compared to wild-type (WT) seedlings. All of these phenotypes are attributed to elevated IAA levels in *dao* mutants.

Auxin also regulates epidermal cell patterning. The *Arabidopsis* root epidermis is alternately spaced with files consisting of smaller trichoblast or hair cells, which generate root hair, and longer atrichoblast or non-hair cells (Dolan 2001; Löfke et al. 2013). Interestingly, at least two auxin transporters are known to be differentially expressed within these distinct cell files. First, the activity of the auxin influx carrier AUX1 mediates auxin transport through atrichoblast cells and is necessary to maintain the identity of root hair cells, and promote root hair elongation. Surprisingly, AUX1 itself is not present in root hair cells, suggesting that auxin is transported longitudinally through canals in epidermal tissue (Jones et al. 2009). Second, PIN2 is also targeted to the plasma membrane of non-hair cells at higher levels than the hair cell files, likely by differential rates of trafficking to lytic vacuoles within these files (Löfke et al. 2015). Taken together, these data suggest that distinctly different downstream auxin-dependent processes occur in these two cell types.

Downstream of PIN/AUX1 auxin transport-dependent patterning, auxin regulates hair cell growth by modulating the expression of a basic helix-loop-helix transcription factor RSL4 (ROOT HAIR DEFECTIVE 6 LIKE 4). RSL4 is sufficient to promote postmitotic hair cell elongation and, accordingly, *rsl4* mutants have very short root hairs (Yi et al. 2010). *RSL4* is an auxin-inducible gene (Yi et al. 2010) and more recently it has been shown that root hair elongation is directly proportional to the amplitude of a pulse of *RSL4* expression that precedes formation of the root hairs in the DZ (Datta et al. 2015). The intensity of this transient increase in *RSL4* expression can be modulated by external environmental signals such as low phosphate, providing another mechanism for the integration of environmental cues to modulate plant development (Datta et al. 2015).

AUXIN IN LATERAL ROOT DEVELOPMENT

In *Arabidopsis*, LRs arise from single-layered pericycle cells. Radially, the pericycle is surrounded by the endodermal, cortical, and epidermal cell layers, and itself surrounds the central root vasculature. The pericycle is composed of two distinct cell types: phloem pole pericycle (PPP) and xylem pole pericycle (XPP). LRs arise exclusively from XPP cells, also known as pericycle founder cells. Crucially, these cells are thought to retain semi-meristematic properties based on their ultrastructure (small vacuoles, dense cytoplasm, and ribosomes) and maintain the ability to undergo cell divisions upon activation by remaining in the G2 phase of the cell cycle for long periods.

LR development occurs in distinct steps, including (1) pericycle priming and specification of founder cells; (2) initiation, defined as the first round of anticlinal divisions of the founder cells; (3) primordium formation, defined as the sum of divisions leading to the formation of a dome-shaped primordium; and (4) emergence, defined as the protrusion of the newly formed LR through overlying ground and epidermal tissue (Fig. 2; covered in detail in Cavallari et al. 2021).

Priming

Early patterning events that determine the regular spacing of LRs coincide with the oscillation of auxin-dependent transcriptional mechanism. In *Arabidopsis*, LRs emerge from the DZ of the primary root at regularly spaced intervals. When roots are observed in a 2D growth system, such as a petri dish, LRs emerge specifically at the outside edges of waves or curves. An auxin-dependent root clock transcriptional signal is one of the earliest drivers of early LR priming and, therefore, patterning as confirmed by several studies (De Smet et al. 2007; Laskowski et al. 2008; Moreno-Risueno et al. 2010; Xuan et al. 2015, 2016). Crucially, these studies first demonstrated the oscillatory expression of the auxin-responsive promoter DR5 driving the expression of two different markers, β-glucuronidase (GUS) and luciferase (LUC). The oscillations have an amplitude of ~6 h and occur within the basal meristem and elongation zones, together termed the oscillation zone (OZ). However, the periodicity of the root clock can be modified by environmental signals or auxin treatment (Moreno-Risueno et al. 2010; Kircher and Schopfer 2016). In standard conditions, however, over time, a portion of cells within the OZ maintained stable high expressions of DR5:LUC signal. These cells then began to initiate LRs as they reached the DZ in the elongating root and were defined as LR pre-branch sites. Interestingly, the authors found large numbers of genes that oscillate both in phase (2000) and in antiphase (1500) of DR5:LUC. Also, the application of exogenous auxin was not sufficient to induce LR priming. Taken together, these data suggest that priming is a result of fluctuations not only in auxin levels, but in the expression of a significant number of genes.

Studies using of loss-of-function mutants have reiterated the importance of auxin perception and signaling in LR priming and pre-branch site formation. The *arf7*, *tir1*, and *afb2* mutants all display aberrant DR5 oscillatory patterns and have markedly lower numbers of LR pre-branch sites (for review, see Du and Scheres 2018). Recently, ARF7 and its auxin-sensitive inhibitor POTENT/IAA18 were found to regulate gene

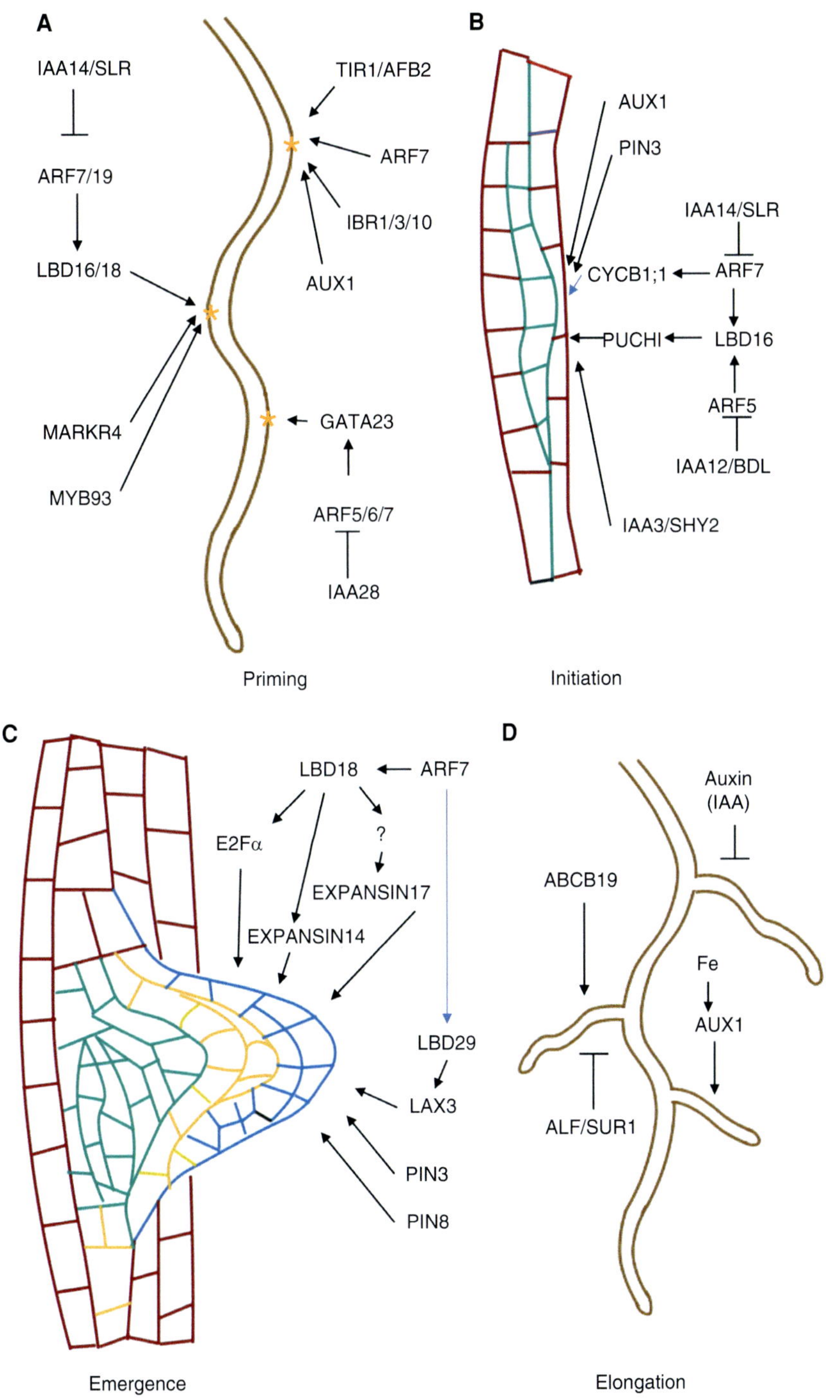

Figure 2. Auxin signaling networks in lateral root development. Visual summary of genes and networks acting both upstream and downstream of the canonical TIR1/AFB-Aux/IAA-ARF auxin signaling pathway during the multiple stages (*A*) priming, (*B*) initiation, (*C*) emergence, and (*D*) elongation of lateral root development in *Arabidopsis*. Yellow asterisks in *A* denote priming sites. Arrowheads denote up-regulation of expression of target genes, while bars denote target down-regulation.

oscillation in the OZ required for the formation of LR pre-branch sites by developing a negative regulatory loop circuit that was predicted to be influenced by environmental cues (Perianez-Rodriguez et al. 2021). Interestingly, an indole-3-butyric acid (IBA) to indole acetic acid ([IAA], the most common form of active auxin) conversion pathway acting specifically within the LRC seems to play a significant role in the generation of stable auxin fluxes required for pre-branch site establishment. Mutations in three INDOLE-3-BUTYRIC ACID RESPONSE genes (IBR1, IBR3, and IBR10) lead to reduced frequency of pre-branch sites (Fig. 2A; Xuan et al. 2015). In a follow-up study, the authors demonstrated the significance of local fluctuations in auxin levels in pre-branch site establishment (Xuan et al. 2016). LRC cells reaching the TZ displayed a gradual reduction in DR5:VENUS signal just prior to their apoptosis and act as a source of auxin for neighboring stele cells, consistent with DR5 activity oscillation. This work also showed that AUX1 expression in the LRC was required for correct DR5 oscillation amplitude and LR patterning.

Interestingly, several studies have demonstrated that not all pre-branch sites showing static DR5 signal go on to become lateral root founder cells (LRFCs) or, indeed, form LRs. One of the earliest molecular markers for LRFC formation is the auxin regulatory GATA23 transcription factor. GATA23 is activated at specific pre-branch sites through an IAA28-ARF 6/7/8/19-dependent pathway in the basal meristem (Fig. 2A). XPP cells leaving the basal meristem displays broad peaks in GATA23 expression and both *gata23* gain- and loss-of-function mutants show altered numbers and spacing of LRs (De Rybel et al. 2010). Moreover, LBD16/ASL18 and other redundantly expressed LBDs, expressed in the adjacent cells of the XPPs undergoing asymmetric cell division downstream of an SLR14-ARF7/19 signaling module, are required for founder cell specification (Fig. 2A; Goh et al. 2012). Additional molecular genetic studies also identified *MARKR4* as another candidate gene in LRFC specification. *markr4* mutants show fewer LRs, but have unaltered numbers of pre-branch sites, suggesting that *MARKR4* is required for the conversion of pre-branch sites into LRFCs (Fig. 2A). *MARKR4* was identified as a downstream molecular component of the IBA to IAA conversion pathway (Xuan et al. 2016). Another potential candidate regulator is *MYB93*, a member of the R2R3 MYB transcription factor family, which is strongly induced by auxin in the basal meristem (Gibbs et al. 2014).

A known positional cue that tightly regulates LR initiation is curvature, either induced mechanically or by gravity. Several studies have convincingly demonstrated that LRs preferentially originate from the convex side of a curved root. Several in silico and mechanotransduction studies have attempted to explain this phenomenon by considering factors such as the differences in geometry of XPP cells on convex and concave sides of the bend, and heterogenous changes in of Ca^{2+} levels in stretched XPP cells. However, experimental work recently showed that curvature alone is not sufficient to specify positioning of the LR, as even if curvature frequency increases, the frequency of LR initiation remains the same (Kircher and Shopfer 2016). Thus, while further work is required to fully understand the signaling pathways that regulate unilateral LR initiation after LRFC specification (Laskowski et al. 2008; Richter et al. 2009), it is evident that auxin transport from the LRC, via the influx transporter AUX1, also plays a role in specifying LR initiation sites. Roots of the *Arabidopsis aux1* mutant display a coiled growth phenotype, with LRs emerging mainly from the convex side of the bend, in complete contrast to the normal wavy growth patterns of WT roots (Marchant et al. 2002; De Smet et al. 2007). Further, *aux1* mutants also have fewer LRs and display lower levels of DR5 oscillatory activity in the basal meristem. However, targeted expression of *AUX1* to the LRC and epidermal cells restores both the agravitropic root and aberrant LR branching patterns in the *aux1* mutant (Swarup et al. 2005).

Studies using a heat-shock-inducible system to generate clonal sectors of XPP cells with local high IAA levels (iaaM) were sufficient to enhance LR production in pericycle cells (Dubrovsky et al. 2008). In silico approaches proposed that existing LRFCs were able to inhibit new

LRFCs in the distal region of the root, by depleting auxin levels in the surrounding cells (Laskowski et al. 2008) These local auxin gradients are formed by the action of influx and efflux carriers. Conversely, molecular genetic studies showed that roots of polar auxin transport mutants (e.g., *pin2pin3pin7* and weak alleles of *gnom*) and loss- (e.g., *mp/arf5*) or gain-of-function (e.g., *shy2*, *bdl*) signaling mutants show altered branching patterns, with closely grouped, fused, or fewer LRP/LRs (Laskowski et al. 2008; Geldner 2009; De Smet et al. 2010; Goh et al. 2012; Okumura et al. 2013).

Other regulatory genes/modules also interact with auxin to induce and regulate LRFC patterning. For example, the *PLT* triple loss-of-function *plt357* mutant has several clustered LR primordia (Hofhuis et al. 2013). Finally, auxin-responsive small terminal peptide RALF34, which is expressed around the flanks of the pericycle cells undergoing divisions to form LRPs, seems to be required for spatial distribution of LRs. Indeed, two loss-of-function insertion mutants of *RAFLF4* showed a threefold increase in clustered LRPs (Murphy et al. 2016). Another auxin-repressed peptide signaling module CEP5/CEPR has a similar phenotype in loss-of-function background (Roberts et al. 2016).

LR Initiation and Primordium Formation

The first step in LR initiation involves the formation of an auxin maximum in specified LRFCs (for review, see Torres-Martínez et al. 2019). This maximum is dependent on the influx carrier AUX1 and the activation of *YUCCA4*-mediated auxin biosynthesis by plant-specific B3 transcription factors *FUSCA3* and *LEC2* (Tang et al. 2017). During primordium formation, the first cell division occurs anticlinally to form a series of cells that undergo further periclinal division (for review, see Du and Scheres 2018). This step is preceded by nuclear migration within the XPP cells toward the center of the LR-initiation site. This is followed by asymmetric cell division, which is again regulated by *BDL-ARF5*- and *SLR14-ARF7*-dependent *LBD16* expression in the XPP cells (Fig. 2B; Goh et al. 2012; Lee et al. 2015). *LBD16* is required for LR initiation and subsequent cell fate re-specification after the first round of asymmetric divisions (Goh et al. 2012). LBD16 in turn up-regulates *PUCHI*, an auxin-inducible transcription factor expressed within newly developed LRPs following the first round of asymmetric divisions (Fig. 2A; Goh et al. 2019). Premature induction of *PUCHI* during the preinitiation phase disrupts LR primordium formation.

The divisions necessary to initiate LR formation require pericycle cells to reenter the cell cycle and start dividing actively. This first series of XPP divisions depends on mechanical feedback from the overlying endodermal cells. A *SHY2/IAA3*-dependent signal triggers loss of endodermal cell volume and turgor pressure by relinquishing their tight junction-like diffusion barriers (Vermeer et al. 2014). These biophysical changes trigger *CYCB1;1* expression in the XPP cells, which regulates G2-M transition. Again, the *SLR-ARF7/19* signaling module seems to be required for this pathway and *slr* and *arf7/19* mutants have almost no LRs and show reduced expression of *CYCB1;1*, likely through an *LBD16/LBD18* signaling module (Fig. 2B; Vanneste et al. 2005; Okushima et al. 2007). In addition to triggering changes in endodermal cell shape, auxin signaling in pericyle cells is also required to specify the plane of division. *tir1/afb* mutants show periclinal XPP cell divisions that are not competent to form primordia. This auxin-dependent input into cell plane division specification relies on a dynamic microtubule cytoskeleton (for review, see Van Norman et al. 2013).

PIN3 is transiently expressed in the endodermal cells overlying the LRFC, until just after the first round of anticlinal divisions. Interestingly, *pin3* mutants have increased numbers of founder cells but reduced numbers of LRPs, suggesting that transient endodermal *PIN3* expression is required for the transition from founder cell to primordium stage. PIN3 is initially distributed in a nonpolar manner in the endodermal cells in the DZ. However, during the course of LRP formation, PIN3 becomes polarized to the inner cell wall (inner lateralization) (Marhavý et al. 2013, 2016) facilitating auxin efflux into the LRFCs for the first round of anticlinal divisions.

LR Emergence

LRs originate from pericycle cells deep within the parental root tissue and must emerge through three distinct cell layers (endodermis, cortex, and epidermis). These cell layers undergo cell wall loosening and hydraulic separation via multiple auxin-dependent signaling modules. Particularly within the endodermis, the dynamic opening and resealing of lignified and suberized barriers are tightly regulated for LR emergence, and to prevent leakage of nutrients and pathogen entry from the soil. This endodermal remodeling is also critical for root system plasticity (Barberon et al. 2016). A new transcriptomic study identified specific members of GDSL-type esterases/lipases (GELPs), proteins that are expressed sequentially downstream of SHY2 and regulate auxin-dependent desuberization during LR emergence, and consequently resuberization of the endodermal barrier post-LR emergence (Ursache et al. 2021). A study from a few years ago, which used a combination of mathematical modeling and cutting-edge light sheet fluorescence microscopy, showed that while the first division is tightly regulated, subsequent divisions do not occur in a rigid sequence as the primordium center grows, but instead their orientations (anticlinal or periclinal) depend on cell geometry and follow the "shortest wall" rule. The orientations of these division planes, the apical growth of the primordium due to cell proliferation, and accommodation from the overlying cell layers are all key factors that determine the emergence of the characteristic layers that make up the dome-shaped primordium (von Wangenheim et al. 2016).

Interestingly, recent data has suggested that the *Arabidopsis* noncanonical ARF *ETTIN/ARF3* (*ETT*), which lacks the PB1 DNA-binding domain, may play an important role in LR emergence (Stoeckle et al. 2018). Loss-of-function and auxin-insensitive mutants have increased numbers of emerged LRs, suggesting that auxin can act directly through ETT to negatively regulate LR emergence (Simonini et al. 2017).

Downstream of an *ARF7/19* module, LBD transcription factors play multiple roles in LR initiation and emergence. An LBD18/33 dimer are other direct downstream targets of this module and themselves regulate the expression of E2Fα, which encodes a transcriptional activator of cell-cycle genes (Berckmans et al. 2011). Further, LBD18 also directly up-regulates *EXPANSIN14*, and indirectly regulates *EXPANSIN17*, required for loosening of the cell wall during LR emergence (Fig. 2C).

The auxin influx transporters, *AUX1* and *LAX3*, play a crucial role in regulating spatial separation of adjacent overlying cell files as the newly formed LRP emerges. Initially, AUX1 facilitates the loading of shoot-derived IAA into the vascular system to ensure that the auxin required for cell wall loosening is channeled to the overlying tissue via the LRPs (Marchant et al. 2002; Swarup et al. 2005; Péret et al. 2013). At later developmental stages, the primordium itself becomes the source of auxin required in the overlying tissue layers to trigger the expression of cell wall remodeling enzymes required for emergence. Several in silico and in vivo studies have also demonstrated that sequential expression of *PIN3* and *LAX3* in turn in the two adjacent cell files overlying the LRP is crucial to ensure that the auxin-dependent separation of cells occurs solely along the shared walls. *LAX3* is a downstream target of *LBD29*, which is a direct target of *ARF7* (Fig. 2C; Porco et al. 2016a,b).

The intracellular auxin homeostasis required during LR emergence is maintained by the short PIN, *PIN8*. *PIN8* is expressed in the phloem cells adjected to the primordium at a later stage of development. Although the mechanisms by which PIN8 promotes LR emergence is not fully understood, *pin8* mutants show significantly reduced numbers of emerged LRs that are arrested at the primordium stage and have reduced expression levels of *GATA23* and *LBD* genes (Lee et al. 2020).

LR Elongation

Surprisingly, there is little information available on the effect of auxin and the role of auxin signaling networks on LR elongation. In *Arabidopsis*, at least, several studies have reported that auxin inhibits the growth of LRs in young seedlings. Interestingly, however, in tomato, Muday and Haworth (1994) reported that auxin promoted LR growth, suggesting that the growth responses of LRs to auxin are likely to be species specific (for review, see Waidmann et al. 2020).

Gain- and loss-of-function studies using *Arabidopsis* mutants have identified a small number of additional genes that regulate LR elongation. Celenza et al. (1995) reported that the constitutively active allele of *alf1/sur* displays hyperproliferation of LRs (Fig. 2D). Whereas Boerjan et al. (1995) established that this mutation leads to an overproduction of endogenous IAA, the *ALF* gene is yet to be mapped. Moreover, Wu et al. (2007) demonstrated that the loss-of-function mutants of the *ABCB14/PGP19/MDR1* gene show arrested and/or impaired LR growth rates and a 40% reduction of acropetal auxin transport within the LRs. In contrast, no differences in LR growth rates have been reported between WT and *aux1* mutants in normal conditions (Linkohr et al. 2002; Giehl et al. 2012). However, localized iron supply was found to promote LR elongation via altering AUX1-mediated auxin distribution, suggesting that environmental triggers are able to influence LR growth rates via the modulation of auxin transport and/or signaling components (Fig. 2D; see the "Environmental Regulation" section).

GRAVITROPIC SETPOINT ANGLE (GSA)

Perhaps the most significant parameter that determines overall root architecture is root angle. As described previously, secondary roots in most plant species grow at oblique angles in the soil. These angles, and indeed all angles that plant organs may grow at are known as gravitropic setpoint angles (GSAs). GSAs are defined as the angle at which a plant organ grows to maintain equilibrium with gravity (Digby and Firn 1995). Spatiotemporal developmental and environmental cues regulate the setting and maintenance of these root angles within the soil, and can therefore shape root architectures that can contrastingly range from shallow, widely distributed roots, to narrow deeper rooting systems. The distribution of the root system within the soil is key to maximizing the capture of heterogenously distributed soil minerals and water. For example, phosphate and nitrate are often found while water is found in the deeper layers.

In *Arabidopsis*, as described above, LRs emerge from the pericycle layer and emerge through overlying tissues at a near horizontal orientation (stage I). Following emergence, LRs undergo a brief period of downward growth, characterized by asymmetric distribution of auxin (as visualized by the reporter DR5v2) as well as the efflux transporter PIN3 (stage II) (Rosquete et al. 2013). Subsequently, LRs undergo stable angled growth, progressively transitioning through increasingly vertical states to adopt near vertical GSAs (stages III–V). Importantly, LRs actively maintain GSAs from stage II onward (i.e., they retain the capacity to reorient themselves both downward and upward) when moved from the GSA.

Several studies have recently sought to explain the molecular mechanisms that underpin the active setting and maintenance of these GSAs, most often using *Arabidopsis* roots as a model system. One set of studies proposes that LRs grow nonvertically due to the onset of sequential expressions of *PIN3*, *PIN7*, and *PIN4* in the root columella cells of stage II, III, and IV LRs, respectively. This is in direct contrast to primary root tips where PIN3, 4, and 7 are highly expressed within the columella initials (Guyomarc'h et al. 2012; Rosquete et al. 2013; Ruiz Rosquete et al. 2018; Ogura et al. 2019). In this model, LRs grow at nonvertical angles because they lack the molecular machinery required for full gravitropic competence. However, both these studies do not explain how LRs not only grow obliquely, but also maintain the capacity to bend upward to return to their original GSAs upon reorientation. Other work proposes that nonvertical GSAs are a result of balanced auxin fluxes across the upper and lower sides of the root that act antagonistically. In this model, the authors define a gravitropic auxin flux that increases in magnitude when an LR is reoriented above its GSA, to promote downward bending. This increase in downward auxin flux occurs in an angle-dependent manner: the larger the displacement above GSA, the greater the increase in the magnitude of downward auxin flux. However, when the LR is reoriented below its GSA, the magnitude of this gravitropic auxin flux decreases. In this scenario, the relative magnitude of the antagonistic "antigravitropic offset" flux increases, and the root bends upward toward its original GSA (for review, see Roychoudhry and Kepinski 2015). Interestingly, another recent study discovered that the cytokinin reporter TCSn:GFP is expressed asymmetrically in higher levels toward the upper flank of stage II LRs. The authors proposed that asymmetric cytokinin signaling inhibits downward bending at the upper flank of LRs, and cytokinin therefore functions as an "antigravitropic" signal (Waidmann et al. 2019).

In the last few years, several studies have described the role of multiple members of the *LAZY* gene family in regulating gravitropism and root (and indeed shoot) angles in multiple plant species. Uga et al. first described *DEEPER ROOTING 1* (*DRO1*), a QTL that controlled root angle in rice. Rice accessions that contained a truncated version of the *DRO1* allele, termed DRO1-NIL, had shallow rooting angles and demonstrated impaired root gravitropism in young seedlings. *DRO1* was found to be expressed in the root EZ and was negatively regulated by auxin (Uga et al. 2011). Subsequently, *DRO1* was found to be orthologous to *LAZY4* in *Arabidopsis* (Hollender and Dardick 2015). Subsequent studies showed that *LAZY4* was part of a larger gene family in *Arabidopsis*, termed as the IGT family, which also contains *TAC1* (Guseman et al. 2017). LAZY proteins are expressed in the root tip and EZ, and share a conserved 14 amino acid carboxyl terminus (termed as CCL domain), which seems to be important for directional graviresponse in primary roots (Taniguchi et al. 2017). Several single and multiple loss-of-function mutants of *LAZYs* have multiple phenotypes including shallow LR growth angles and delayed primary root gravitropism. Interestingly, a recent study showed that primary roots of a *lazy234* multiple mutant are negatively gravitropic and bend upward (Ge and Chen 2016). Whether this phenotype is the result of an actively maintained GSA, or loss of the capacity to respond to gravity, remains to be investigated.

While the precise roles of *LAZYs* in regulating root angle are not yet fully understood, what is evident is that *PIN3* polarity is altered in LR columella cells of *lazy* mutants. Several lines of evidence suggest that the shallow LR GSA phenotype in *lazy* loss-of-function mutants is due to increased localization of PIN3 to the upper membrane of LR columella cells (Taniguchi et al. 2017; Furutani et al. 2020). Another recent study showed that LAZY proteins may themselves be polarized in LR columella cells and likely recruit PIN3 to the plasma membrane by interacting with the BRX domain of Regulator of Chromosome Condensation 1 (RCC1)-Like proteins (RLD proteins), identified as novel regulators of PIN polarity (Furutani et al. 2020).

ENVIRONMENTAL REGULATION

Vascular plants have evolved complex 3D root architectures to maximize the volume of soil they can explore and interact with. This is to acquire numerous highly heterogeneously distributed resources within multiple soil strata. The ability of the root system to adapt its genetically determined developmental parameters to multiple environmental signals, such as nutrient and water availability, has been a key factor determining the success of land plants (for review, see Morris et al. 2017). This adaptive behavior is known as developmental plasticity.

Phosphate Availability

Diverse environmental signals can have dramatic and often diametrically opposite effects on RSA. Phosphates and nitrates are both examples of plant macronutrients that are often limiting in soil (Morgan and Connolly 2013). Phosphorus, specifically, is crucial in the building of nucleic acids, ATP, and other molecules, and is acquired by the plant in its inorganic forms of phosphates. Phosphates are highly immobile in soil and often form complexes with soil components or can be converted to organic phosphate, leading to reductions in the availability of soil Pi for the plants (Brady and Weil 2008). In *Arabidopsis*, phosphate deficiency leads to PR length inhibition, increases in LR length, and deeper root angle (Fig. 3). These effects are due to a signaling pathway that enhances sensitivity to

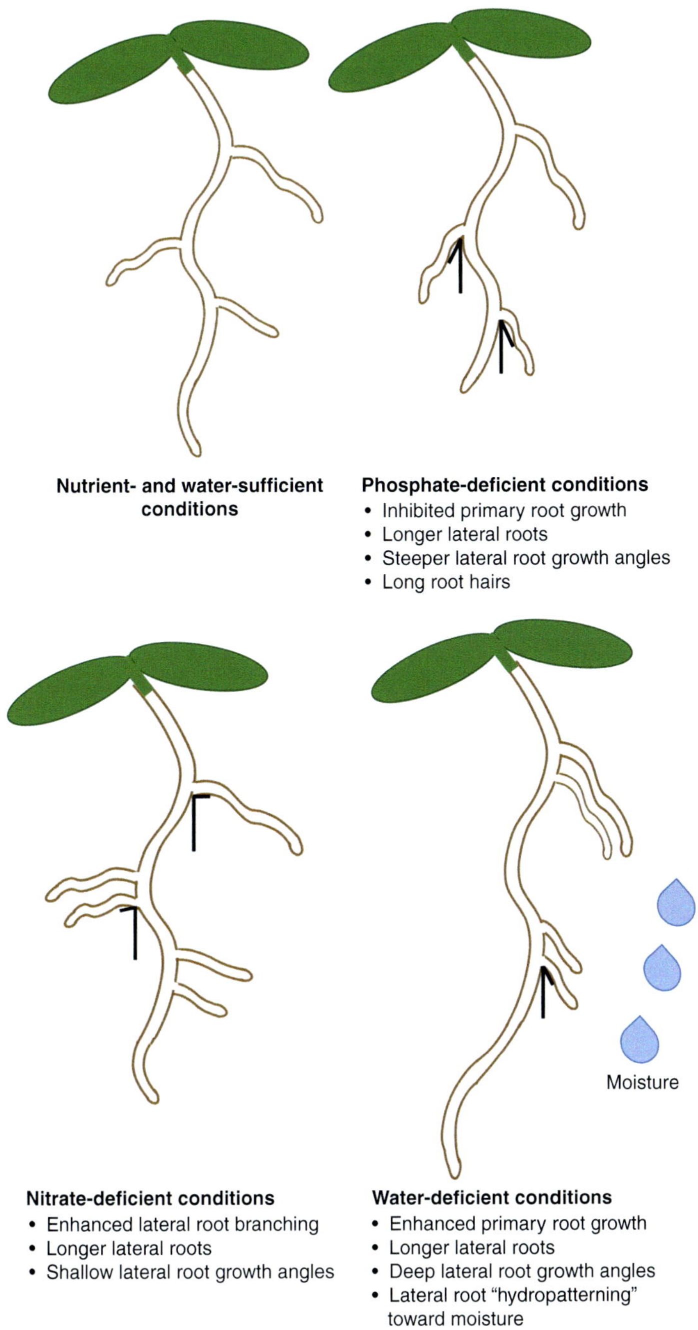

Figure 3. Environmental regulation of root system architecture (RSA) plasticity. Effect of phosphate, nitrate, and water deficiency on different parameters of RSA. Phosphate deficiency leads to inhibition of primary root growth, with contrasting increases in lateral root and root hair length along with steeper lateral root growth angles in *Arabidopsis*. In contrast, nitrate deficiency has little to no effect on primary root length, but leads to increases in lateral root length, and shallower growth angles. Water deficiency broadly enhances primary and lateral root growth, causes steeper lateral root growth angles and stimulates "hydropatterning" of lateral roots (i.e., induces enhanced lateral root branching toward the side of the primary root that is in contact with higher moisture levels).

auxin via up-regulation of the *TIR1* auxin receptor (López-Bucio et al. 2002; Ma et al. 2003; Pérez-Torres et al. 2008; Roychoudhry et al. 2017). Interestingly, these effects of low phosphate appear to be root class specific. In bean, growth under phosphorus deficiency had been demonstrated to induce shallower root angles in basal (adventitious) roots in contrast to the effect of low phosphorus on *Arabidopsis* LR growth angle (Bonser et al. 1996). Further examination of the response to low phosphate in bean showed that, as in *Arabidopsis*, bean LRs become more vertical while, as previously shown, basal roots adopt a shallower growth habit (Roychoudhry et al. 2017). The physiological basis of this root class-specific response is not yet understood.

Nitrate Availability

Nitrogen, a crucial macronutrient required for the production of amino acids and proteins, is primarily acquired by the root system in its inorganic form from nitrates and ammonia, supplied as fertilizers. In smaller quantities, nitrogen may also be taken up in its organic forms (urea, amino acids, and peptides). Depending on its source, nitrogen availability may have different effects on RSA (Fig. 3). For example, ammonium application limits PR growth (Liu et al. 2013). However, local application of ammonium enhances LR length and branching, likely due to increased expression of the ammonium transporter *AMT (AMMONIUM TRANSPORTER) 1;3* in LRs, which enhances nitrogen uptake (Lima et al. 2010). In contrast to these findings, several studies have demonstrated that localized nitrogen supply either in the form of nitrate or L-glutamate inhibits LR elongation, while nitrate deficiency enhances LR elongation and branching (Zhang and Forde 1998; Zhang et al. 1999; Linkohr et al. 2002; Tian et al. 2009). Nitrogen availability or deprivation seems to have a relatively small effect on PR length (for review, see Waidmann et al. 2020). Thus, depending on the source of nitrogen, the root system has distinct responses in terms of LR branching and elongation, suggesting that there are several additional mechanisms/hormone-signaling pathways that fine-tune the uptake of N. A recent study has also demonstrated that nitrate deficiency resulted in shallow LR growth angles in *Arabidopsis* (Roychoudhry et al. 2017). However, the effect of localized nitrate supply on LR growth angle has not yet been explored in detail.

Light

The effect of light on RSA has remained largely unexplored, as the root systems of most terrestrial plants remain in darkness, underground. Interestingly, however, several studies have demonstrated that multiple photoreceptors are expressed in root tissues, enabling them to detect ambient light and launch appropriate tropic and developmental regimes (for review, see van Gelderen et al. 2018). Broadly, total darkness inhibits root growth because of the reduction in basipetal auxin transport, due to depletion of PIN proteins from the plasma membrane (Laxmi et al. 2008). However, when only root systems are placed in darkness, root growth is largely enhanced in *Arabidopsis* (Zhang et al. 2019). In contrast, light exposure limits root growth and promotes negative phototropism, via a signaling pathway that depends on the photoreceptors *PHOT1* and *PHOT2*, with receptors *CRY1/CRY2* and *PHYA* playing minor roles. Downstream of the direction of light perception, polar auxin transport is required for directional root growth. The coordinated rootward polarities of PIN1 in the stele and PIN2 in root epidermal cells as well as the lateral relocalization of PIN3 in the columella cells are required for the formation of the auxin gradient (Wang et al. 2012; Zhang et al. 2013). During gradient formation, PIN1 polarity in the stele is also influenced by flavonols, which were found to accumulate on the side of the root exposed to light and promote cell elongation at this side (Buer and Muday 2004; Silva-Navas et al. 2015). Another recent study demonstrated that UV-B light inhibits LR growth through a signaling pathway that involved the UV-B receptor *UVR8*, which antagonistically regulates auxin-dependent gene expression. Upon UV-B irradiation, UVR8 interacts with the MYB73 transcription factor within the nucleus, thereby inhibiting its DNA-binding activity and resulting in direct repression of target auxin-responsive genes (Yang et al. 2020).

Water

The effect of water on overall RSA has not been extensively studied to date (for review, see Dietrich 2018). Root systems with deeper, soil-penetrating architectures have enhanced resistance to drought, because of their ability to capture available water from deep within the soil. In this light, it is not surprising perhaps that a recent study showed that water deficit promoted vertical growth in *Arabidopsis* LRs (Fig. 3; Rellán-Álvarez et al. 2016). Interestingly, it was found that in addition to regulating growth angles, water availability around the circumferential axis of the root also acts as a positional cue to specify the patterning of LRs. This phenomenon was shown to occur before the initiation step and was defined as "hydropatterning." It seems that high water levels promote patterning through a pathway that creates a local LR inductive signal through the biosynthesis (dependent on *TRYPTOPHAN AMINOTRANSFERASE1*) and transport (*PIN3*-dependent) of auxin (Bao et al. 2014). In addition, it has been shown that ARF7 promotes LR initiation by triggering asymmetric transcription of *LBD16* at the circumference of the root that is in contact with water. This asymmetric induction of *LBD16* was regulated by the preferential posttranslational modification of ARF7 by the small ubiquitin-like modifier (SUMO) protein. SUMOylation of ARF7 leads to an increased association with the repressor IAA3/SHY2, presumably at the "dry" circumference of the root, and consequent inhibition of *LBD16* transcription (Orosa-Puente et al. 2018).

Similar to LRs, primary roots in several species also develop hydrotropic curvatures in response to water availability. Initially, several lines of evidence suggested that this hydrotropic response was antagonistic to gravitropic bending (for review, see Takahashi et al. 2009), but this response appears to be species-specific. For instance, in cucumber, the root cap is required for normal hydrotropism, suggesting that the interactions between the two tropisms may be more complex. In *Arabidopsis*, while auxin transport is not required for the hydrotropic response, a functional auxin response seems to be needed. For example, studies showed that treatment with auxin response inhibitors led to contrasting results. Whereas treatment with *p*-chloro-phenoxy-isobutyl acetic acid (PCIB), an inhibitor of Aux/IAA protein degradation (Oono et al. 2003), produced a decrease in hydrotropic response, addition of auxinole or α-(phenylethyl-2-oxo)-indole acetic acid (PEO-IAA), a TIR1 antagonist (Hayashi et al. 2012), accelerated the response (Kaneyasu et al. 2007; Shkolnik et al. 2016). These contrasting results could be due to different modes of action and target specificities of inhibitors. Thus, currently, additional experiments with multiple auxin-responsive mutants are required to confirm the effect of auxin signaling on the hydrotropic response.

CONCLUSIONS

The number of processes governed by auxin during root, and indeed most aspects of plant development seems extraordinary. This versatility rests in part on the fact that auxin can be moved within tissues to create and maintain gradients of concentration. Equally important is capacity of the auxin signaling system to generate specific outputs relevant to distinct and diverse developmental phenomena. These context-specific responses to auxin are a property of the existence of different TIR1/AFB-Aux/IAA-ARF signaling modules, which can then be wired to downstream genes to effect appropriate modes of developmental control. Nevertheless, even with this rationalization of how so much of development can, often simultaneously, depend on one simple molecule, the sheer robustness of auxin-regulated development is remarkable. A conspicuous example of this robustness is seen in the control of LR formation. This is a process with multiple, distinct, auxin-regulated steps and yet, in the face of high concentrations of auxin, the result is simply an increase in the number of LR primordia, each initiating into perfectly formed LRs. Here, the fact that the development of individual LRs is so well buffered against large changes in auxin concentrations may well stem from the fact that successive Aux/IAA-ARF signaling modules are activated consecutively in time and space. Understanding this molecular genetic basis of auxin's capacity for self-organization is one of the frontiers of auxin research and one that is at the interface of genetics, biophysics, and mathematical modeling.

This review has been very focused on the dicot model *Arabidopsis* and within that system, principally on auxin's role in primary and LR growth. This leaves out the unsurprisingly central role of auxin in adventitious roots and also in the regulation of root development in monocot species. Nevertheless, many of the paradigms established in *Arabidopsis* have been found to be translatable to cereal species such as maize, rice, and wheat in particular, albeit with interesting differences (for review, see McSteen 2010; Balzan et al. 2014; Steffens and Rasmussen 2016; Waidmann et al. 2020). Understanding these differences and maximizing the potential application of our understanding of auxin's control of root biology will be essential if we are to make rapid progress in developing crop varieties that perform better in more sustainable farming systems in which inputs such as water, nitrogen, and phosphorous must be reduced. Indeed, there was an extent to which root performance as a trait became a casualty of post-Green Revolution breeding programs based on high levels of agricultural inputs. While this was an appropriate technological response to alleviating food insecurity at that point in history, we are now way beyond the stage at which food production needs to be radically decarbonized. Reducing the use of nitrogen fertilizer derived from the environmentally damaging Haber–Bosch process and conserving the world's finite supply of phosphorous are high on the list of targets for crop improvement. It is here that our deep understanding of the networks of auxin response controlling root growth and system architecture can be brought to bear in a way that can accelerate the development of new varieties in time to meet our commitments to limiting global warming and feeding a global population predicted to peak at 9–10 billion people just 30 years from now.

Whereas overall tremendous progress has been made in understanding the role of auxin and its signaling networks in regulating multiple facets of root and, indeed, overall plant development, many questions still remain unanswered. Whereas a large number of gene networks that regulate different steps of both primary and LR development have now been studied, at least partially, many still remain to be discovered. Specific examples of these include exceptionally rapid auxin-dependent signaling pathways that regulate root elongation (Fendrych et al. 2016, 2018) and those that underpin the mechanisms of LR growth angle setting and maintenance. Following on from this, how multiple environmental signals can fine tune these myriad auxin-dependent developmental responses remains to be fully understood. Perhaps the most basic question of all relates to the evolution of such complex systems of auxin response and how the proliferation and subfunctionalization of the three gene families in auxin signaling, including the development of specific preferential interactions between pairs and multimers of different Aux/IAAs, ARFs, and TIR1/AFBs, may have underpinned the evolution of increasingly sophisticated developmental phenomena. Interdisciplinary advances in the multiple fields of biology, physics, and mathematical modeling will enable researchers to answer these and other fundamental, yet exciting, questions in the coming decade.

ACKNOWLEDGMENTS

We apologize to those authors whose work we could not cover in this review. We thank Marta Del Bianco for critical reading of the manuscript and The Leverhulme Trust (RPG-2018-137) and BBSRC (BB/N010124/1) for their support of relevant projects in the Kepinski laboratory.

REFERENCES

**Reference is also in this collection.*

Aida M, Beis D, Heidstra R, Willemsen V, Blilou I, Galinha C, Nussaume L, Noh YS, Amasino R, Scheres B. 2004. The

PLETHORA genes mediate patterning of the *Arabidopsis* root stem cell niche. *Cell* **119:** 109–120. doi:10.1016/j.cell.2004.09.018

Balzan S, Johal GS, Carraro N. 2014. The role of auxin transporters in monocots development. *Front Plant Sci* **5:** 393. doi:10.3389/fpls.2014.00393

Bao Y, Aggarwal P, Robbins NE 2nd, Sturrock CJ, Thompson MC, Tan HQ, Tham C, Duan L, Rodriguez PL, Vernoux T, et al. 2014. Plant roots use a patterning mechanism to position lateral root branches toward available water. *Proc Natl Acad Sci* **111:** 9319–9324. doi:10.1073/pnas.1400966111

Barberon M, Vermeer JE, De Bellis D, Wang P, Naseer S, Andersen TG, Humbel BM, Nawrath C, Takano J, Salt DE, et al. 2016. Adaptation of root function by nutrient-induced plasticity of endodermal differentiation. *Cell* **164:** 447–459. doi:10.1016/j.cell.2015.12.021

Bellini C, Pacurar DI, Perrone I. 2014. Adventitious roots and lateral roots: similarities and differences. *Annu Rev Plant Biol* **65:** 639–666. doi:10.1146/annurev-arplant-050213-035645

Benjamins R, Scheres B. 2008. Auxin: the looping star in plant development. *Annu Rev Plant Biol* **59:** 443–465. doi:10.1146/annurev.arplant.58.032806.103805

Berckmans B, Vassileva V, Schmid SP, Maes S, Parizot B, Naramoto S, Magyar Z, Alvim Kamei CL, Koncz C, Bögre L, et al. 2011. Auxin-dependent cell cycle reactivation through transcriptional regulation of *Arabidopsis E2Fa* by lateral organ boundary proteins. *Plant Cell* **23:** 3671–3683. doi:10.1105/tpc.111.088377

Blilou I, Xu J, Wildwater M, Willemsen V, Paponov I, Friml J, Heidstra R, Aida M, Palme K, Scheres B. 2005. The PIN auxin efflux facilitator network controls growth and patterning in *Arabidopsis* roots. *Nature* **433:** 39–44. doi:10.1038/nature03184

Boerjan W, Cervera MT, Delarue M, Beeckman T, Dewitte W, Bellini C, Caboche M, Van Onckelen H, Van Montagu M, Inzé D. 1995. Superroot, a recessive mutation in *Arabidopsis*, confers auxin overproduction. *Plant Cell* **7:** 1405–1419.

Bonser A, Lynch J, Snapp S. 1996. Effect of phosphorus deficiency on growth angle of basal roots in *Phaseolus vulgaris*. *New Phytol* **132:** 281–288. doi:10.1111/j.1469-8137.1996.tb01847.x

Brady NC, Weil RR. 2008. *The nature and properties of soil*, 14th ed. Prentice Hall, Upper Saddle River, NJ.

Buer CS, Muday GK. 2004. The *transparent testa4* mutation prevents flavonoid synthesis and alters auxin transport and the response of *Arabidopsis* roots to gravity and light. *Plant Cell* **16:** 1191–1205. doi:10.1105/tpc.020313

* Cavallari N, Artner C, Benkova E. 2021. Auxin-regulated lateral root organogenesis. *Cold Spring Harb Perspect Biol* doi: 10.1101/cshperspect.a039941

Celenza JL Jr, Grisafi PL, Fink GR. 1995. A pathway for lateral root formation in *Arabidopsis thaliana*. *Genes Dev* **9:** 2131–2142. doi:10.1101/gad.9.17.2131

Datta S, Prescott H, Dolan L. 2015. Intensity of a pulse of RSL4 transcription factor synthesis determines *Arabidopsis* root hair cell size. *Nat Plants* **1:** 15138. doi:10.1038/nplants.2015.138

Dello Ioio R, Nakamura K, Moubayidin L, Perilli S, Taniguchi M, Morita MT, Aoyama T, Costantino P, Sabatini S. 2008. A genetic framework for the control of cell division and differentiation in the root meristem. *Science* **322:** 1380–1384. doi:10.1126/science.1164147

De Rybel B, Vassileva V, Parizot B, Demeulenaere M, Grunewald W, Audenaert D, Van Campenhout J, Overvoorde P, Jansen L, Vanneste S, et al. 2010. A novel aux/IAA28 signaling cascade activates GATA23-dependent specification of lateral root founder cell identity. *Curr Biol* **20:** 1697–1706. doi:10.1016/j.cub.2010.09.007

De Smet I, Tetsumura T, De Rybel B, Frei Dit Frey N, Laplaze L, Casimiro I, Swarup R, Naudts M, Vanneste S, Audenaert D, et al. 2007. Auxin-dependent regulation of lateral root positioning in the basal meristem of *Arabidopsis*. *Development* **134:** 681–690. doi:10.1242/dev.02753

De Smet I, Lau S, Voss U, Vanneste S, Benjamins R, Rademacher EH, Schlereth A, De Rybel B, Vassileva V, Grunewald W, et al. 2010. Bimodular auxin response controls organogenesis in *Arabidopsis*. *Proc Natl Acad Sci* **107:** 2705–2710. doi:10.1073/pnas.0915001107

Dharmasiri N, Dharmasiri S, Weijers D, Lechner E, Yamada M, Hobbie L, Ehrismann JS, Jürgens G, Estelle M. 2005. Plant development is regulated by a family of auxin receptor F box proteins. *Dev Cell* **9:** 109–119. doi:10.1016/j.devcel.2005.05.014

Dietrich D. 2018. Hydrotropism: how roots search for water. *J Exp Bot* **69:** 2759–2771. doi:10.1093/jxb/ery034

Digby J, Firn RD. 1995. The gravitropic set-point angle (GSA): the identification of an important developmentally controlled variable governing plant architecture. *Plant Cell Environ* **18:** 1434–1440. doi:10.1111/j.1365-3040.1995.tb00205.x

Di Mambro R, De Ruvo M, Pacifici E, Salvi E, Sozzani R, Benfey P, Busch W, Novak O, Ljung K, De Paola L, et al. 2017. Auxin minimum triggers the developmental switch from cell division to cell differentiation in the *Arabidopsis* root. *Proc Natl Acad Sci* **114:** E7641–E7649. doi:10.1073/pnas.1705833114

Ding Z, Friml J. 2010. Auxin regulates distal stem cell differentiation in *Arabidopsis* roots. *Proc Natl Acad Sci* **107:** 12046–12051. doi:10.1073/pnas.1000672107

Dolan L. 2001. How and where to build a root hair. *Curr Opin Plant Biol* **4:** 550–554. doi:10.1016/S1369-5266(00)00214-4

Drisch RC, Stahl Y. 2015. Function and regulation of transcription factors involved in root apical meristem and stem cell maintenance. *Front Plant Sci* **6:** 505. doi:10.3389/fpls.2015.00505

Du Y, Scheres B. 2018. Lateral root formation and the multiple roles of auxin. *J Exp Bot* **69:** 155–167. doi:10.1093/jxb/erx223

Dubrovsky JG, Sauer M, Napsucialy-Mendivil S, Ivanchenko MG, Friml J, Shishkova S, Celenza J, Benková E. 2008. Auxin acts as a local morphogenetic trigger to specify lateral root founder cells. *Proc Natl Acad Sci* **105:** 8790–8794. doi:10.1073/pnas.0712307105

Durgaprasad K, Roy MV, Venugopal A, Kareem A, Raj K, Willemsen V, Mähönen AP, Scheres B, Prasad K. 2019. Gradient expression of transcription factor imposes a boundary on organ regeneration potential in plants. *Cell Rep* **29:** 453–463.e3. doi:10.1016/j.celrep.2019.08.099

Fendrych M, Leung J, Friml J. 2016. TIR1/AFB-Aux/IAA auxin perception mediates rapid cell wall acidification and growth of *Arabidopsis* hypocotyls. *eLife* **5:** e19048. doi:10.7554/eLife.19048

Fendrych M, Akhmanova M, Merrin J, Glanc M, Hagihara S, Takahashi K, Uchida N, Torii KU, Friml J. 2018. Rapid and reversible root growth inhibition by TIR1 auxin signalling. *Nat Plants* **4:** 453–459. doi:10.1038/s41477-018-0190-1

Furutani M, Hirano Y, Nishimura T, Nakamura M, Taniguchi M, Suzuki K, Oshida R, Kondo C, Sun S, Kato K, et al. 2020. Polar recruitment of RLD by LAZY1-like protein during gravity signaling in root branch angle control. *Nat Commun* **11:** 76. doi:10.1038/s41467-019-13729-7

Galinha C, Hofhuis H, Luijten M, Willemsen V, Blilou I, Heidstra R, Scheres B. 2007. PLETHORA proteins as dose-dependent master regulators of *Arabidopsis* root development. *Nature* **449:** 1053–1057. doi:10.1038/nature06206

García-Gómez ML, Garay-Arroyo A, García-Ponce B, Sánchez MP, Álvarez-Buylla ER. 2021. Hormonal regulation of stem cell proliferation at the *Arabidopsis thaliana* root stem cell niche. *Front Plant Sci* **12:** 628491. doi: 10.3389/fpls.2021.628491

Ge L, Chen R. 2016. Negative gravitropism in plant roots. *Nat Plants* **2:** 16155. doi:10.1038/nplants.2016.155

Geisler M, Aryal B, di Donato M, Hao P. 2017. A critical view on ABC transporters and their interacting partners in auxin transport. *Plant Cell Physiol* **58:** 1601–1614. doi:10.1093/pcp/pcx104

Geldner N. 2009. Cell polarity in plants: a PARspective on PINs. *Curr Opin Plant Biol* **12:** 42–48. doi:10.1016/j.pbi.2008.09.009

Gibbs DJ, Voß U, Harding SA, Fannon J, Moody LA, Yamada E, Swarup K, Nibau C, Bassel GW, Choudhary A, et al. 2014. *At*MYB93 is a novel negative regulator of lateral root development in *Arabidopsis*. *New Phytol* **203:** 1194–1207. doi:10.1111/nph.12879

Giehl RF, Lima JE, von Wirén N. 2012. Localized iron supply triggers lateral root elongation in *Arabidopsis* by altering the AUX1-mediated auxin distribution. *Plant Cell* **24:** 33–49. doi:10.1105/tpc.111.092973

Goh T, Joi S, Mimura T, Fukaki H. 2012. The establishment of asymmetry in *Arabidopsis* lateral root founder cells is regulated by LBD16/ASL18 and related LBD/ASL proteins. *Development* **139:** 883–893. doi:10.1242/dev.071928

Goh T, Toyokura K, Yamaguchi N, Okamoto Y, Uehara T, Kaneko S, Takebayashi Y, Kasahara H, Ikeyama Y, Okushima Y, et al. 2019. Lateral root initiation requires the sequential induction of transcription factors LBD16 and PUCHI in *Arabidopsis thaliana*. *New Phytol* **224:** 749–760. doi:10.1111/nph.16065

Gray WM, Kepinski S, Rouse D, Leyser O, Estelle M. 2001. Auxin regulates SCF^{TIR1}-dependent degradation of AUX/IAA proteins. *Nature* **414:** 271–276. doi:10.1038/35104500

Grieneisen VA, Xu J, Marée AF, Hogeweg P, Scheres B. 2007. Auxin transport is sufficient to generate a maximum and gradient guiding root growth. *Nature* **449:** 1008–1013. doi:10.1038/nature06215

Guseman JM, Webb K, Srinivasan C, Dardick C. 2017. *DRO1* influences root system architecture in *Arabidopsis* and *Prunus* species. *Plant J* **89:** 1093–1105. doi:10.1111/tpj.13470

Guyomarc'h S, Léran S, Auzon-Cape M, Perrine-Walker F, Lucas M, Laplaze L. 2012. Early development and gravitropic response of lateral roots in *Arabidopsis thaliana*. *Philos Trans R Soc Lond B Biol Sci* **367:** 1509–1516. doi:10.1098/rstb.2011.0231

* Hammes UZ, Murphy AS, Schwechheimer C. 2021. Auxin transporters—a biochemical view. *Cold Spring Harb Perspect Biol* doi: 10.1101/cshperspect.a039875

Hayashi K, Neve J, Hirose M, Kuboki A, Shimada Y, Kepinski S, Nozaki H. 2012. Rational design of an auxin antagonist of the SCF^{TIR1} auxin receptor complex. *ACS Chem Biol* **17:** 590–598.

Hofhuis H, Laskowski M, Du Y, Prasad K, Grigg S, Pinon V, Scheres B. 2013. Phyllotaxis and rhizotaxis in *Arabidopsis* are modified by three PLETHORA transcription factors. *Curr Biol* **23:** 956–962. doi:10.1016/j.cub.2013.04.048

Hollender CA, Dardick C. 2015. Molecular basis of angiosperm tree architecture. *New Phytol* **206:** 541–556. doi:10.1111/nph.13204

Hu Y, Omary M, Hu Y, Doron O, Hoermayer L, Chen Q, Megides O, Chekli O, Ding Z, Friml J, et al. 2021. Cell kinetics of auxin transport and activity in *Arabidopsis* root growth and skewing. *Nat Commun* **12:** 1657. doi:10.1038/s41467-021-21802-3

Jones AR, Kramer EM, Knox K, Swarup R, Bennett MJ, Lazarus CM, Leyser HM, Grierson CS. 2009. Auxin transport through non-hair cells sustains root-hair development. *Nat Cell Biol* **11:** 78–84. doi:10.1038/ncb1815

Kaneyasu T, Kobayashi A, Nakayama M, Fujii N, Takahashi H, Miyazawa Y. 2007. Auxin response, but not its polar transport, plays a role in hydrotropism of *Arabidopsis* roots. *J Exp Bot* **58:** 1143–1150. doi:10.1093/jxb/erl274

Kepinski S, Leyser O. 2005. The *Arabidopsis* F-box protein TIR1 is an auxin receptor. *Nature* **435:** 446–451. doi:10.1038/nature03542

Khan MA, Gemenet DC, Villordon A. 2016. Root system architecture and abiotic stress tolerance: current knowledge in root and tuber crops. *Front Plant Sci* **7:** 1584.

Kircher S, Schopfer P. 2016. Priming and positioning of lateral roots in *Arabidopsis*. An approach for an integrating concept. *J Exp Bot* **67:** 1411–1420. doi:10.1093/jxb/erv541

Laskowski M, Grieneisen VA, Hofhuis H, Hove CA, Hogeweg P, Marée AF, Scheres B. 2008. Root system architecture from coupling cell shape to auxin transport. *PLoS Biol* **6:** e307. doi:10.1371/journal.pbio.0060307

Laxmi A, Pan J, Morsy M, Chen R. 2008. Light plays an essential role in intracellular distribution of auxin efflux carrier PIN2 in *Arabidopsis thaliana*. *PLoS ONE* **3:** e1510. doi:10.1371/journal.pone.0001510

Lee HW, Cho C, Kim J. 2015. *Lateral Organ Boundaries Domain16* and *18* Act downstream of the AUXIN1 and LIKE-AUXIN3 auxin influx carriers to control lateral root development in *Arabidopsis*. *Plant Physiol* **168:** 1792–1806. doi: 10.1104/pp.15.00578

Lee H, Ganguly A, Lee RD, Park M, Cho HT. 2020. Intracellularly localized PIN-FORMED8 promotes lateral root emergence in *Arabidopsis. Front Plant Sci* **10:** 1808. doi:10.3389/fpls.2019.01808

Leyser O. 2018. Auxin signaling. *Plant Physiol* **176:** 465–479. doi:10.1104/pp.17.00765

Lima JE, Kojima S, Takahashi H, von Wirén N. 2010. Ammonium triggers lateral root branching in *Arabidopsis* in an AMMONIUM TRANSPORTER1;3-dependent manner. *Plant Cell* **22:** 3621–3633. doi:10.1105/tpc.110.076216

Linkohr BI, Williamson LC, Fitter AH, Leyser HM. 2002. Nitrate and phosphate availability and distribution have different effects on root system architecture of *Arabidopsis. Plant J* **29:** 751–760. doi:10.1046/j.1365-313X.2002.01251.x

Liu Y, Lai N, Gao K, Chen F, Yuan L, Mi G. 2013. Ammonium inhibits primary root growth by reducing the length of meristem and elongation zone and decreasing elemental expansion rate in the root apex in *Arabidopsis thaliana. PLoS ONE* **8:** e61031. doi:10.1371/journal.pone.0061031

Löfke C, Dünser K, Kleine-Vehn J. 2013. Epidermal patterning genes impose non-cell autonomous cell size determination and have additional roles in root meristem size control. *J Int Plant Biol* **55:** 864–875. doi:10.1111/jipb.12097

Löfke C, Scheuring D, Dünser K, Schöller M, Luschnig C, Kleine-Vehn J. 2015. Tricho- and atrichoblast cell files show distinct PIN2 auxin efflux carrier exploitations and are jointly required for defined auxin-dependent root organ growth. *J Exp Bot* **66:** 5103–5112. doi:10.1093/jxb/erv282

López-Bucio J, Hernández-Abreu E, Sánchez-Calderón L, Nieto-Jacobo MF, Simpson J, Herrera-Estrella L. 2002. Phosphate availability alters architecture and causes changes in hormone sensitivity in the *Arabidopsis* root system. *Plant Physiol* **129:** 244–256. doi:10.1104/pp.010934

Lynch JP. 2013. Steep, cheap and deep: an ideotype to optimize water and N acquisition by maize root systems. *Ann Bot* **112:** 347–357. doi:10.1093/aob/mcs293

Ma Z, Baskin TI, Brown KM, Lynch JP. 2003. Regulation of root elongation under phosphorus stress involves changes in ethylene responsiveness. *Plant Physiol* **131:** 1381–1390. doi:10.1104/pp.012161

Mähönen AP, Ten Tusscher K, Siligato R, Smetana O, Díaz-Triviño S, Salojärvi J, Wachsman G, Prasad K, Heidstra R, Scheres B. 2014. PLETHORA gradient formation mechanism separates auxin responses. *Nature* **515:** 125–129. doi:10.1038/nature13663

Marchant A, Bhalerao R, Casimiro I, Eklöf J, Casero PJ, Bennett M, Sandberg G. 2002. AUX1 promotes lateral root formation by facilitating indole-3-acetic acid distribution between sink and source tissues in the *Arabidopsis* seedling. *Plant Cell* **14:** 589–597. doi:10.1105/tpc.010354

Marhava P, Bassukas AEL, Zourelidou M, Kolb M, Moret B, Fastner A, Schulze WX, Cattaneo P, Hammes UZ, Schwechcheimer C, et al. 2018. A molecular rheostat adjusts auxin flux to promote root protophloem differentiation. *Nature* **558:** 297–300. doi:10.1038/s41586-018-0186-z

Marhavý P, Vanstraelen M, De Rybel B, Zhaojun D, Bennett MJ, Beeckman T, Benková E. 2013. Auxin reflux between the endodermis and pericycle promotes lateral root initiation. *EMBO J* **32:** 149–158. doi:10.1038/emboj.2012.303

Marhavý P, Montesinos JC, Abuzeineh A, Van Damme D, Vermeer JE, Duclercq J, Rakusová H, Nováková P, Friml J, Geldner N, et al. 2016. Targeted cell elimination reveals an auxin-guided biphasic mode of lateral root initiation. *Genes Dev* **30:** 471–483. doi:10.1101/gad.276964.115

Markakis MN, De Cnodder T, Lewandowski M, Simon D, Boron A, Balcerowicz D, Doubbo T, Taconnat L, Renou JP, Höfte H, et al. 2012. Identification of genes involved in the ACC-mediated control of root cell elongation in *Arabidopsis thaliana. BMC Plant Biol* **12:** 208. doi:10.1186/1471-2229-12-208

McSteen P. 2010. Auxin and monocot development. *Cold Spring Harb Perspect Biol* **2:** a001479. doi:10.1101/cshperspect.a001479

Mellor N, Band LR, Pěnčík A, Novák O, Rashed A, Holman T, Wilson MH, Voß U, Bishopp A, King RJ, et al. 2016. Dynamic regulation of auxin oxidase and conjugating enzymes *AtDAO1* and *GH3* modulates auxin homeostasis. *Proc Natl Acad Sci* **113:** 11022–11027. doi:10.1073/pnas.1604458113

Michniewicz M, Zago MK, Abas L, Weijers D, Schweighofer A, Meskiene I, Heisler MG, Ohno C, Zhang J, Huang F, et al. 2007. Antagonistic regulation of PIN phosphorylation by PP2A and PINOID directs auxin flux. *Cell* **130:** 1044–1056. doi:10.1016/j.cell.2007.07.033

Moreno-Risueno MA, Van Norman JM, Moreno A, Zhang J, Ahnert SE, Benfey PN. 2010. Oscillating gene expression determines competence for periodic *Arabidopsis* root branching. *Science* **329:** 1306–1311. doi:10.1126/science.1191937

* Morffy N, Strader LC. 2021. Structural aspects of auxin signaling. *Cold Spring Harb Perspect Biol* doi: 10.1101/cshperspect.a039883

Morgan JB, Connolly EL. 2013. Plant–soil interactions: nutrient uptake. *Nature Education Knowledge* **4:** 2.

Morris EC, Griffiths M, Golebiowska A, Mairhofer S, Burr-Hersey J, Goh T, von Wangenheim D, Atkinson B, Sturrock CJ, Lynch JP, et al. 2017. Shaping 3D root system architecture. *Curr Biol* **27:** R919–R930. doi:10.1016/j.cub.2017.06.043

Muday GK, Haworth P. 1994. Tomato root growth, gravitropism, and lateral development: correlation with auxin transport. *Plant Physiol Biochem* **32:** 193–203.

Murphy E, Vu LD, Van den Broeck L, Lin Z, Ramakrishna P, van de Cotte B, Gaudinier A, Goh T, Slane D, Beeckman T, et al. 2016. RALFL34 regulates formative cell divisions in *Arabidopsis* pericycle during lateral root initiation. *J Exp Bot* **67:** 4863–4875. doi:10.1093/jxb/erw281

Ogura T, Goeschl C, Filiault D, Mirea M, Slovak R, Wolhrab B, Satbhai SB, Busch W. 2019. Root system depth in *Arabidopsis* is shaped by EXOCYST70A3 via the dynamic modulation of auxin transport. *Cell* **178:** 400–412.e16. doi:10.1016/j.cell.2019.06.021

Okumura K, Goh T, Toyokura K, Kasahara H, Takebayashi Y, Mimura T, Kamiya Y, Fukaki H. 2013. GNOM/FEWER ROOTS is required for the establishment of an auxin response maximum for *Arabidopsis* lateral root initiation. *Plant Cell Physiol* **54:** 406–417. doi:10.1093/pcp/pct018

Okushima Y, Fukaki H, Onoda M, Theologis A, Tasaka M. 2007. ARF7 and ARF19 regulate lateral root formation via direct activation of *LBD/ASL* genes in *Arabidopsis. Plant Cell* **19:** 118–130. doi:10.1105/tpc.106.047761

Oono Y, Ooura C, Rahman A, Aspuria ET, Hayashi K, Tanaka A, Uchimiya H. 2003. *p*-Chlorophenoxyisobutyric acid impairs auxin response in *Arabidopsis* root. *Plant Physiol* **133:** 1135–1147. doi:10.1104/pp.103.027847

Orosa-Puente B, Leftley N, von Wangenheim D, Banda J, Srivastava AK, Hill K, Truskina J, Bhosale R, Morris E, Srivastava M, et al. 2018. Root branching toward water involves posttranslational modification of transcription factor ARF7. *Science* **362:** 1407–1410. doi:10.1126/science.aau3956

Osmont KS, Sibout R, Hardtke CS. 2007. Hidden branches: developments in root system architecture. *Ann Rev Plant Biol* **58:** 93–113. doi:10.1146/annurev.arplant.58.032806.104006

Overoode P, Fukaki H, Beeckman T. 2010. Auxin control of root development. *Cold Spring Harb Perspect Biol* **2:** a001537.

Péret B, Swarup K, Ferguson A, Seth M, Yang Y, Dhondt S, James N, Casimiro I, Perry P, Syed A, et al. 2012. *AUX/LAX* genes encode a family of auxin influx transporters that perform distinct functions during *Arabidopsis* development. *Plant Cell* **24:** 2874–2885. doi:10.1105/tpc.112.097766

Péret B, Middleton AM, French AP, Larrieu A, Bishopp A, Njo M, Wells DM, Porco S, Mellor N, Band LR, et al. 2013. Sequential induction of auxin efflux and influx carriers regulates lateral root emergence. *Mol Syst Biol* **9:** 699. doi:10.1038/msb.2013.43

Pérez-Torres CA, López-Bucio J, Cruz-Ramírez A, Ibarra-Laclette E, Dharmasiri S, Estelle M, Herrera-Estrella L. 2008. Phosphate availability alters lateral root development in *Arabidopsis* by modulating auxin sensitivity via a mechanism involving the TIR1 auxin receptor. *Plant Cell* **12:** 3258–3272.

Perianez-Rodriguez J, Rodriguez M, Marconi M, Bustillo-Avendaño E, Wachsman G, Sanchez-Corrionero A, De Gernier H, Cabrera J, Perez-Garcia P, Gude I, et al. 2021. An auxin-regulable oscillatory circuit drives the root clock in *Arabidopsis. Sci Adv* **7:** eabd4722. doi:10.1126/sciadv.abd4722

Porco S, Pěnčík A, Rashed A, Voß U, Casanova-Sáez R, Bishopp A, Golboeweska A, Bhosale R, Swarup R, Swarup K, et al. 2016a. Dioxygenase-encoding *AtDAO1* gene controls IAA oxidation and homeostasis in *Arabidopsis. Proc Natl Acad Sci* **113:** 11016–11021. doi:10.1073/pnas.1604375113

Porco S, Larrieu A, Du Y, Gaudinier A, Goh T, Swarup K, Swarup R, Kuempers B, Bishopp A, Lavenus J, et al. 2016b. Lateral root emergence in *Arabidopsis* is dependent on transcription factor LBD29 regulation of auxin influx carrier *LAX3. Development* **143:** 3340–3349.

Rahman A, Bannigan A, Sulaman W, Pechter P, Blancaflor EB, Baskin TI. 2007. Auxin, actin and growth of the *Arabidopsis thaliana* primary root. *Plant J* **50:** 514–528. doi:10.1111/j.1365-313X.2007.03068.x

Rellán-Álvarez R, Lobet G, Dinneny J. 2016. Environmental control of root system biology. *Ann Rev Plant Biol* **67:** 619–642. doi:10.1146/annurev-arplant-043015-111848

Richter GL, Monshausen GB, Krol A, Gilroy S. 2009. Mechanical stimuli modulate lateral root organogenesis. *Plant Physiol* **151:** 1855–1866.

Roberts I, Smith S, Stes E, De Rybel B, Staes A, van de Cotte B, Njo MF, Dedeyne L, Demol H, Lavenus J, et al. 2016. CEP5 and XIP1/CEPR1 regulate lateral root initiation in *Arabidopsis. J Exp Bot* **67:** 4889–4899. doi:10.1093/jxb/erw231

Rosquete MR, von Wangenheim D, Marhavý P, Barbez E, Stelzer EH, Benková E, Maizel A, Kleine-Vehn J. 2013. An auxin transport mechanism restricts positive orthogravitropism in lateral roots. *Curr Biol* **23:** 817–822. doi:10.1016/j.cub.2013.03.064

Roychoudhry S, Kepinski S. 2015. Shoot and root branch growth angle control—the wonderfulness of lateralness. *Curr Opin Plant Biol* **23:** 124–131. doi:10.1016/j.pbi.2014.12.004

Roychoudhry S, Kieffer M, Del Bianco M, Liao CY, Weijers D, Kepinski S. 2017. The developmental and environmental regulation of gravitropic setpoint angle in *Arabidopsis* and bean. *Sci Rep* **7:** 42664. doi: 10.1038/srep42664

Ruiz Rosquete M, Waidmann S, Kleine-Vehn J. 2018. PIN7 auxin carrier has a preferential role in terminating radial root expansion in *Arabidopsis thaliana. Int J Mol Sci* **19:** 1238. doi:10.3390/ijms19041238

Sabatini S, Beis D, Wolkenfelt H, Murfett J, Guilfoyle T, Malamy J, Benfey P, Leyser O, Bechtold N, Weisbeek P, et al. 1999. An auxin-dependent distal organizer of pattern and polarity in the *Arabidopsis* root. *Cell* **99:** 463–472. doi:10.1016/S0092-8674(00)81535-4

Sabatini S, Heidstra R, Wildwater M, Scheres B. 2003. SCARECROW is involved in positioning the stem cell niche in the *Arabidopsis* root meristem. *Genes Dev* **17:** 354–358. doi:10.1101/gad.252503

Salvi E, Rutten JP, Di Mambro R, Polverari L, Licursi V, Negri R, Dello Ioio R, Sabatini S, Ten Tusscher K. 2020. A self-organized PLT/auxin/ARR-B network controls the dynamics of root zonation development in *Arabidopsis thaliana. Dev Cell* **53:** 431–443.e23. doi:10.1016/j.devcel.2020.04.004

Shkolnik D, Krieger G, Nuriel R, Fromm H. 2016. Hydrotropism: root bending does not require auxin redistribution. *Mol Plant* **9:** 757–759. doi:10.1016/j.molp.2016.02.001

Silva-Navas J, Moreno-Risueno MA, Manzano C, Pallero-Baena M, Navarro-Neila S, Téllez-Robledo B, Garcia-Mina JM, Baigorri R, Gallego FJ, del Pozo JC. 2015. D-Root: A system for cultivating plants with the roots in darkness or under different light conditions. *Plant J* **84:** 244–255. doi:10.1111/tpj.12998

Simonini S, Bencivenga S, Trick M, Østergaard L. 2017. Auxin-induced modulation of ETTIN activity orchestrates gene expression in *Arabidopsis. Plant Cell* **29:** 1864–1882. doi:10.1105/tpc.17.00389

Smith S, De Smet I. 2012. Root system architecture: insights from *Arabidopsis* and cereal crops. *Phil Trans R Soc B* **367:** 1441–1452. doi:10.1098/rstb.2011.0234

Steffens B, Rasmussen A. 2016. The physiology of adventitious roots. *Plant Phys* **170:** 603–617. doi:10.1104/pp.15.01360

Stoeckle D, Thellmann M, Vermeer JE. 2018. Breakout—lateral root emergence in *Arabidopsis thaliana*. *Curr Opin Plant Biol* **41:** 67–72. doi:10.1016/j.pbi.2017.09.005

Swarup R, Kramer EM, Perry P, Knox K, Leyser HM, Haseloff J, Beemster GT, Bhalerao R, Bennett MJ. 2005. Root gravitropism requires lateral root cap and epidermal cells for transport and response to a mobile auxin signal. *Nat Cell Biol* **7:** 1057–1065. doi:10.1038/ncb1316

Takahashi H, Miyazawa Y, Fujii N. 2009. Hormonal interactions during root tropic growth: hydrotropism versus gravitropism. *Plant Mol Biol* **69:** 489–502. doi:10.1007/s11103-008-9438-x

Takatsuka H, Umeda M. 2014. Hormonal control of cell division and elongation along differentiation trajectories in roots. *J Exp Bot* **65:** 2633–2643. doi:10.1093/jxb/ert485

Tang LP, Zhou C, Wang SS, Yuan J, Zhang XS, Su YH. 2017. FUSCA3 interacting with LEAFY COTYLEDON2 controls lateral root formation through regulating *YUCCA4* gene expression in *Arabidopsis thaliana*. *New Phyt* **213:** 1740–1754. doi:10.1111/nph.14313

Taniguchi M, Furutani M, Nishimura T, Nakamura M, Fushita T, Iijima K, Baba K, Tanaka H, Toyota M, Tasaka M, et al. 2017. The *Arabidopsis* LAZY1 family plays a key role in gravity signaling within statocytes and in branch angle control of roots and shoots. *Plant Cell* **29:** 1984–1999. doi:10.1105/tpc.16.00575

Tian QY, Sun P, Zhang WH. 2009. Ethylene is involved in nitrate-dependent root growth and branching in *Arabidopsis thaliana*. *New Phytol* **184:** 918–931. doi:10.1111/j.1469-8137.2009.03004.x

Torres-Martínez HH, Rodríguez-Alonso G, Shishkova S, Dubrovsky JG. 2019. Lateral root primordium morphogenesis in angiosperms. *Front Plant Sci* **10:** 206. doi: 10.3389/fpls.2019.00206

Uga Y, Okuno K, Yano M. 2011. *Dro1*, a major QTL involved in deep rooting of rice under upland field conditions. *J Exp Bot* **62:** 2485–2494. doi:10.1093/jxb/erq429

Ursache R, De Jesus Vieira Teixeira C, Dénervaud Tendon V, Gully K, De Bellis D, Schmid-Siegert E, Grube Andersen T, Shekhar V, Calderon S, Pradervand S, et al. 2021. GDSL-domain proteins have key roles in suberin polymerization and degradation. *Nat Plants* **7:** 353–364. doi:10.1038/s41477-021-00862-9

van Gelderen K, Kang C, Pierik R. 2018. Light signaling, root development, and plasticity. *Plant Physiol* **176:** 1049–1060. doi:10.1104/pp.17.01079

Vanneste S, De Rybel B, Beemster GT, Ljung K, De Smet I, Van Isterdael G, Naudts M, Iida R, Gruissem W, Tasaka M, et al. 2005. Cell cycle progression in the pericycle is not sufficient for SOLITARY ROOT/IAA14-mediated lateral root initiation in *Arabidopsis thaliana*. *Plant Cell* **17:** 3035–3050. doi:10.1105/tpc.105.035493

Van Norman JM, Xuan W, Beeckman T, Benfey PN. 2013. To branch or not to branch: the role of pre-patterning in lateral root formation. *Development* **140:** 4301–4310. doi:10.1242/dev.090548

Verbelen JP, Cnodder TDe, Le J, Vissenberg K, Baluška F. 2006. The root apex of *Arabidopsis thaliana* consists of four distinct zones of growth activities: meristematic zone, transition zone, fast elongation zone and growth terminating zone. *Plant Signal Behav* **1:** 296–304. doi:10.4161/psb.1.6.3511

Vermeer JE, von Wangenheim D, Barberon M, Lee Y, Stelzer EH, Maizel A, Geldner N. 2014. A spatial accommodation by neighboring cells is required for organ initiation in *Arabidopsis*. *Science* **343:** 178–183. doi:10.1126/science.1245871

von Wangenheim D, Fangerau J, Schmitz A, Smith RS, Leitte H, Stelzer EH, Maizel A. 2016. Rules and self-organizing properties of post-embryonic plant organ cell division patterns. *Curr Biol* **26:** 439–449. doi:10.1016/j.cub.2015.12.047

Waidmann S, Ruiz Rosquete M, Schöller M, Sarkel E, Lindner H, LaRue T, Petřík I, Dünser K, Martopawiro S, Sasidharan R, et al. 2019. Cytokinin functions as an asymmetric and anti-gravitropic signal in lateral roots. *Nat Commun* **10:** 3540. doi:10.1038/s41467-019-11483-4

Waidmann S, Sarkel E, Kleine-Vehn J. 2020. Same same, but different: growth responses of primary and lateral roots. *J Exp Bot* **71:** 2397–2411. doi:10.1093/jxb/eraa027

Wang Y, Deng D, Shi Y, Miao N, Bian Y, Yin Z. 2012. Diversification, phylogeny and evolution of *auxin response factor* (*ARF*) family: insights gained from analyzing maize *ARF* genes. *Mol Biol Rep* **39:** 2401–2415. doi:10.1007/s11033-011-0991-z

Wu G, Lewis DR, Spalding EP. 2007. Mutations in *Arabidopsis* multidrug resistance-like ABC transporters separate the roles of acropetal and basipetal auxin transport in lateral root development. *Plant Cell* **19:** 1826–1837. doi:10.1105/tpc.106.048777

Xuan W, Audenaert D, Parizot B, Möller BK, Njo MF, De Rybel B, De Rop G, Van Isterdael G, Mähönen AP, Vanneste S, et al. 2015. Root cap-derived auxin pre-patterns the longitudinal axis of the *Arabidopsis* root. *Curr Biol* **25:** 1381–1388. doi:10.1016/j.cub.2015.03.046

Xuan W, Band LR, Kumpf RP, Van Damme D, Parizot B, De Rop G, Opdenacker D, Moller BK, Skorzinki N, Njo MF, et al. 2016. Cyclic programmed cell death stimulates hormone signaling and root development in *Arabidopsis*. *Science* **351:** 384–387. doi:10.1126/science.aad2776

Yang Y, Zhang L, Chen P, Liang L, Li X, Liu H. 2020. UV-B photoreceptor UVR8 interacts with MYB73/MYB77 to regulate auxin responses and lateral root development. *EMBO J* **39:** e101928.

Yi K, Menand B, Bell E, Dolan L. 2010. A basic helix-loop-helix transcription factor controls cell growth and size in root hairs. *Nat Genet* **42:** 64–267.

Zhang H, Forde BG. 1998. An *Arabidopsis* MADS box gene that controls nutrient-induced changes in root architecture. *Science* **279:** 407–409. doi:10.1126/science.279.5349.407

Zhang H, Jennings A, Barlow PW, Forde BG. 1999. Dual pathways for regulation of root branching by nitrate. *Proc Natl Acad Sci* **96:** 6529–6534. doi:10.1073/pnas.96.11.6529

Zhang KX, Xu HH, Yuan TT, Zhang L, Lu YT. 2013. Blue-light-induced PIN3 polarization for root negative phototropic response in *Arabidopsis*. *Plant J* **76:** 308–321.

Zhang J, Lin JE, Harris C, Pereira FCM, Wu F, Blakeslee JJ, Peer WA. 2016. DAO1 catalyzes temporal and tissue-specific oxidative inactivation of auxin in *Arabidopsis thali-*

ana. Proc Natl Acad Sci **113:** 11010–11015. doi:10.1073/pnas.1604769113

Zhang Y, Wang C, Xu H, Shi X, Zhen W, Hu Z, Huang J, Zheng Y, Huang P, Zhang KX, et al. 2019. HY5 contributes to light-regulated root system architecture under a root-covered culture system. *Front Plant Sci.* **10:** 1490. doi:10.3389/fpls.2019.01490

Zhang Y, Rodriguez R, Li L, Zhang X, Friml J. 2020. Functional innovations of PIN auxin transporters mark crucial evolutionary transitions during rise of flowering plants. *Sci Adv* **6:** eabc8895. doi:10.1126/sciadv.abc8895

Zhu J, Geisler M. 2015. Keeping it all together: auxin-actin crosstalk in plant development. *J Exp Bot* **66:** 4983–4998. doi:10.1093/jxb/erv308

Zourelidou M, Absmanner B, Weller B, Barbosa IC, Willige BC, Fastner A, Streit V, Port SA, Colcombet J, de la Fuente van Bentem S, et al. 2014. Auxin efflux by PIN-FORMED proteins is activated by two different protein kinases, D6 PROTEIN KINASE and PINOID. *eLife* **3:** e02860. doi:10.7554/eLife.02860

Zwiewka M, Bilanovičová V, Seifu YW, Nodzyński T. 2019. The nuts and bolts of PIN auxin efflux carriers. *Front Plant Sci* **10:** 985. doi:10.3389/fpls.2019.00985

Auxin-Regulated Lateral Root Organogenesis

Nicola Cavallari, Christina Artner, and Eva Benkova

Institute of Science and Technology Austria, 3400 Klosterneuburg, Austria

Correspondence: eva.benkova@ist.ac.at

Plant fitness is largely dependent on the root, the underground organ, which, besides its anchoring function, supplies the plant body with water and all nutrients necessary for growth and development. To exploit the soil effectively, roots must constantly integrate environmental signals and react through adjustment of growth and development. Important components of the root management strategy involve a rapid modulation of the root growth kinetics and growth direction, as well as an increase of the root system radius through formation of lateral roots (LRs). At the molecular level, such a fascinating growth and developmental flexibility of root organ requires regulatory networks that guarantee stability of the developmental program but also allows integration of various environmental inputs. The plant hormone auxin is one of the principal endogenous regulators of root system architecture by controlling primary root growth and formation of LR. In this review, we discuss recent progress in understanding molecular networks where auxin is one of the main players shaping the root system and acting as mediator between endogenous cues and environmental factors.

The evolution of land plants is characterized by the appearance of specialized organs such as leaves, flowers, and roots. In particular, roots have developed to anchor the plant body to the ground, mediate absorption of water and nutrients from the soil, and serve as storage organs for water and photo-assimilates (Raven and Edwards 2001; Osmont et al. 2007; Pires and Dolan 2012; Bellini et al. 2014).

Contrary to animals, where organ formation occurs primarily during embryogenesis, plants can produce new organs along their life cycle. Hence, plant roots are either of embryonic origin, such as primary roots (PRs), or formed postembryonically from parental roots or non-root tissues (i.e., lateral roots [LRs] or adventitious roots [ARs], respectively). The root system in dicotyledonous plants is mainly based on the PR that sprouts to numerous lateral branches, which is known as a taproot or allorhizic system. Unlike dicots, monocotyledonous plants form fibrous or homorhizic root systems consisting of a dense mass of ARs that arise from the stem (Atkinson et al. 2014).

Arabidopsis thaliana, with one of the simplest and therefore the best anatomically described root systems within dicotyledonous plant species, became the preferred model to study molecular mechanisms regulating root branching. In the PR of *Arabidopsis*, concentric single layers of epidermis, cortex, and endodermis surround the pericycle, the tissue giving rise to LRs (Dolan et al. 1993). The pericycle consists of two cell populations, the xylem- and the phloem-pole pericycle cells (XPP and PPP, respectively). Typically, cells adjacent to xylem can

Cite this article as *Cold Spring Harb Perspect Biol* doi: 10.1101/cshperspect.a039941

acquire characteristics of founder cells (FCs) and reenter the cell cycle necessary for the initiation of LR primordia (LRP) organogenesis (Beeckman et al. 2001; Casimiro et al. 2003; Himanen et al. 2004; Parizot et al. 2008). LRPs, formed through a series of anti- and periclinal divisions classified in seven developmental stages, traverse through surrounding tissues until they emerge from the parental root (Malamy and Benfey 1997). Ultimately, the apical meristem of LRs, functionally analogous to that of the PR, is established. This simple, well-described root system allowed genetic and molecular screens for key factors controlling LR organogenesis. As a result, core genetic networks that regulate different phases of LR development have been identified and they are constantly updated for new components.

The plant hormone auxin is one of the principal regulators of LR organogenesis. Research focused on the mechanisms underlying root branching has corroborated the role of auxin as a trigger and a major coordinator of this postembryonic developmental process and provided molecular insights into auxin-orchestrated pathways (Torrey 1950; Blakely and Evans 1979; Blakely et al. 1988; Laskowski et al. 1995; Reed et al. 1998; Dubrovsky et al. 2008; Du and Scheres 2018; Israeli et al. 2020; Xuan et al. 2020). Intriguingly, besides its role as activator and regulator steering specific phases of LR organogenesis, auxin acts as an integrator of various endo- and exogenous inputs, thereby providing developmental plasticity and flexibility of the root system to adapt to fluctuating environmental conditions (Sun et al. 2009; Kapulnik et al. 2011; Duan et al. 2013).

Here we discuss recent advances in understanding pathways that, under the coordination of auxin, control different aspects of LR organogenesis and we cover views on the role of auxin as an integrator of environmental inputs to coordinate adaptive root responses.

ENDOGENOUS MECHANISMS DETERMINING RHIZOTACTIC PATTERNING

In *Arabidopsis*, a zone above the root apical meristem from the transition zone (TZ) to elongation zone (EZ) spatially defines a developmental window where LR initiation takes place (Casimiro et al. 2001; Dubrovsky et al. 2006). This region, also named the oscillation zone (OZ), is characterized by recurrent pulses of the DIRECT REPEAT5 (DR5) auxin-sensitive reporter and oscillatory expression of numerous genes (Ulmasov et al. 1997; De Smet et al. 2007; Moreno-Risueno et al. 2010). The oscillating auxin maxima and coregulated genes such as MADS-box transcription factors *SHATTERPROOF1* (*SHP1*), *SHP2*, and *SEEDSTICK* (*STK*) define prebranch sites, where XPP cells are specified to FCs and subsequently LRs are initiated (Moreno-Risueno et al. 2010; Van Norman et al. 2013). In the XPP cells, auxin, through the signaling module encompassing AUXIN/INDOLE-3-ACETIC ACID repressor IAA28 and transcriptional activator AUXIN RESPONSE FACTOR7 (ARF7), triggers expression of the *GATA23* transcription factor that controls FC identity (Fig. 1; De Rybel et al. 2010). Although the molecular principle of this endogenous mechanism working as a "root clock" is not fully understood, several recent studies provide important insights into pathways that contribute to regulation of the oscillation-based rhizotactic patterning.

Inhibition of auxin transport by *N*-1-naphthylphthalamic acid (NPA) dims the oscillation of auxin response and the initiation of LRs, thus hinting at a role of polar auxin transport (PAT) in control of LR positioning (Xuan et al. 2016). Analyses of mutants defective in AUXIN RESISTANT1 (AUX1)/LIKE AUX1 (LAXs) auxin influx- and PIN-FORMED (PIN) or ABCB/PGP auxin efflux transporters further corroborate contributions of PAT to LR initiation (Fig. 1B; Marchant et al. 2002; Benková et al. 2003; Laskowski et al. 2008; Mravec et al. 2008). Intriguingly, targeted expression of AUX1 in the LR cap (LRC) and epidermal cells restored the aberrant LR initiation in the *aux1* mutant, suggesting that the LRC-derived auxin might act as a trigger of LR organogenesis (De Smet et al. 2007; Xuan et al. 2016). A role of the LRC as a source of auxin essential for root branching is further supported by Naxillin, a non-auxin-like molecule that promotes LRC-specific conversion of the

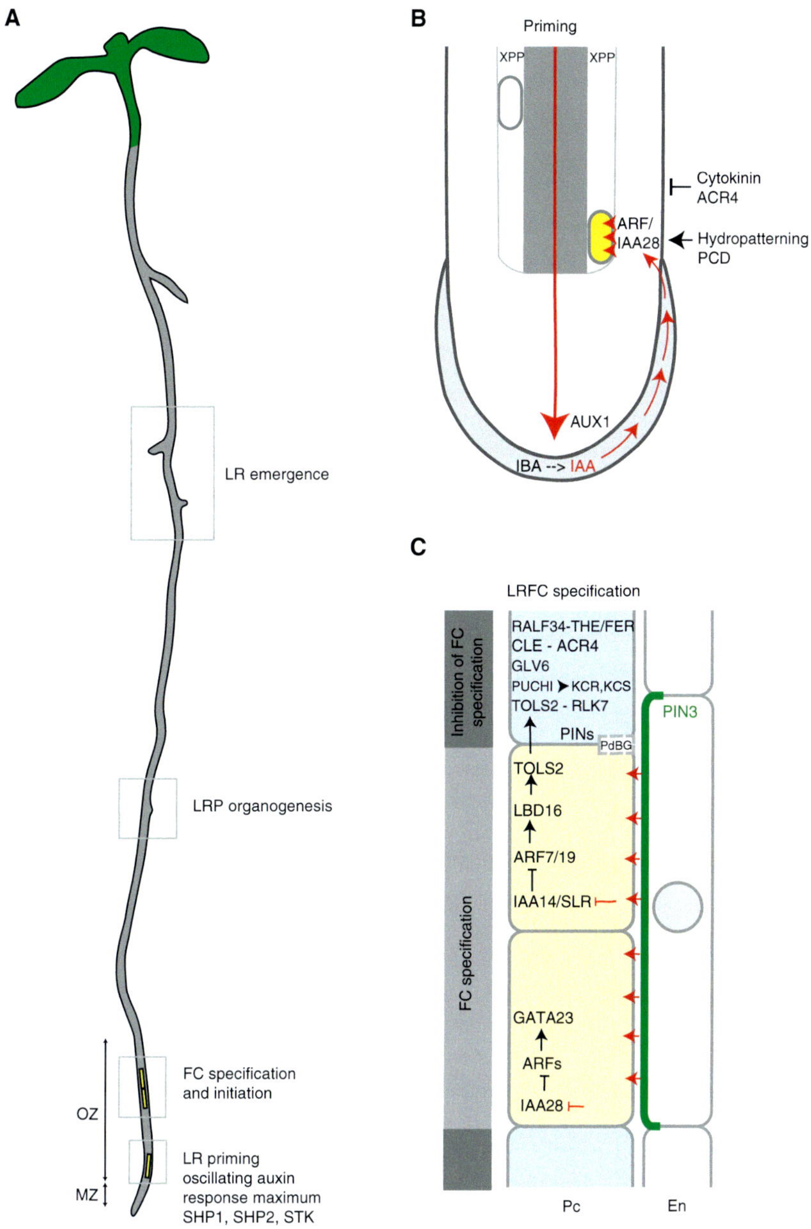

Figure 1. Lateral root priming and founder cell specification. (*A*) Schematic representation of the *Arabidopsis* seedling. Specific steps during lateral root (LR) development including priming, LR primordia (LRP) specification and initiation, and LR emergence are highlighted along the primary root (PR) in gray boxes. Pericycle cells involved in early stages of LR development are depicted in yellow. (*B*) Priming of pericycle founder cells (FCs) starts in the basal meristem and is driven by auxin reflux from the LR cap (LRC) (red arrows) to epidermal root cells. In the LRC, auxin flow is reinforced by the conversion of indole-3-butyric (IBA) to indole-3-acetic acid (IAA) (black dashed line). Factors promoting or repressing LR priming are included in the scheme (black arrows and bar-headed lines, respectively). Auxin priming of a xylem-pole pericycle (XPP) cell is highlighted in yellow. (*C*) Specification of FCs triggers the inhibition of FC establishment and LR initiation in flanking XPP cells. Polar localization of PIN3 (marked in green) regulates active auxin flow from the endodermis to the FC. Auxin signaling promoting or repressing expression of the indicated modules is given in red (arrows or bar-headed lines, respectively). Factors involved in FC specification and inhibition in the neighboring cells and their signaling pathways are depicted (black arrows and bar-headed lines). Auxin accumulates in the FC (yellow), while the neighboring FC flanking cells have a higher cytokinin content (blue). (Pc) Pericycle, (En) endodermis, (MZ) meristematic zone, (OZ) oscillation zone.

auxin precursor indole-3-butyric acid (IBA) to the active indole-3-acetic acid (IAA) and thereby stimulates LR formation. Accordingly, the *ibr1ibr3ibr10* and *ech2ibr1ibr3ibr10* mutants defective in enzymes of the IBA-to-IAA conversion pathway have a reduced number of initiated LRs (De Rybel et al. 2012; Xuan et al. 2015). Importantly, the LRC is not the only source of auxin required for establishment of prebranch sites, but it might accommodate a pathway that steers oscillation and thereby the spatiotemporal pattern of LR initiation. Programmed cell death (PCD) of the cells at the most distal end of the LRC displays oscillations and, consequently, auxin pulses are released to surrounding root tissues and define the LR spacing along the main root (Xuan et al. 2016; van den Berg and ten Tusscher 2018).

To maintain the acropetal pattern of root branching (i.e., younger LRs are positioned distally from older LRs), initiation of LRs must be restricted to a developmental window. Several processes contribute to LR spacing by suppressing ectopic initiation and thereby reinforcing and stabilizing the rhizotactic pattern. Enhanced expression of the cytokinin-sensitive two-component signaling sensor (TCS) reporter in pericycle cells adjacent to FCs and LRPs, suggests that cytokinin forms an inhibitory field that restricts the initiation of ectopic LRPs in proximity to existing ones (Fig. 1B). Consistently, inhibition of cytokinin biosynthesis or signaling resulted in LRP clustering (Bielach et al. 2012; Chang et al. 2015). In parallel, ARABIDOPSIS CRINKLY4 (ACR4), encoding a plasma membrane (PM)-localized receptor-like protein kinase (RLK), contributes to regulation of LR spacing. Defective rhizotactic patterning and accumulation of ectopic LRPs in *acr4* mutants is in line with the inhibitory function of ACR4 during the initiation phase (De Smet et al. 2008; Chang et al. 2015). Putative ligands of ACR4 are signaling peptides of the CLAVATA3/ESR-RELATED (CLE) family and in particular CLE40 has been proposed to act as a direct ligand for ACR4 in the root (Stahl et al. 2013; Berckmans et al. 2020). However, quantitative binding assays revealed that ACR4 cannot directly interact with CLE40, and it may require additional components, like the ACR4 peptide coreceptor CLAVATA1 (CLV1) in the signaling complex (Okuda et al. 2020).

Recently, a peptide-signaling cascade involving TARGET OF LBD SIXTEEN (TOLS2) and its receptor RLK7 was identified as a local inhibitory mechanism controlling LR spacing. TOLS2 peptide, whose expression is auxin-dependent in FCs, moves to neighboring cells where it inhibits FC specification in ectopic positions through interaction with the RLK7 receptor (Toyokura et al. 2019). Activation of RLK7 promotes expression of *PUCHI*, a member of the AP2/EREBP transcription factor family, previously shown to control cell proliferation during LRP formation. Typically, loss of PUCHI activity results in abnormally enlarged flank cells and clustering of LRPs (Hirota et al. 2007). Interestingly, PUCHI was found to target transcription of genes involved in the synthesis of very long chain fatty acids (VLCFAs) such as a *3-Ketoacyl-CoA Synthase* (*KCS*), a *3-Ketoacyl-CoA Reductase* (*KCR*), and *PASTICCINO2* (*PAS2*) (Fig. 1C; Trinh et al. 2019). Mutants perturbed in VLCFA biosynthesis showed similar LR initiation defects as observed in the *puchi* mutant, which suggests that PUCHI-regulated VLCFA biosynthesis is involved in control of spacing between LRPs. How VLCFAs suppress the initiation of LRs awaits further study. In addition, expression of RAPID ALKALINISATION FACTOR34 (RALF34) peptide was detected in pericycle cells flanking LRPs (Fig. 1C). RALF34, through activation of the THESEUS1 (THE1) and related FERONIA (FER) receptors, controls pathways involved in repression of LR initiation (Murphy et al. 2016; Gonneau et al. 2018). The downstream mechanism by which the RALF34-THE1/FER module regulates LR initiation has not yet been uncovered. Because swelling of the FCs is crucial for the onset of LR initiation (Vermeer et al. 2014; Ramakrishna et al. 2019; Jourquin et al. 2020), a similar mechanism to the known RALF1-FER signaling in PRs may involve the activity of the proton exporters, preventing acidification of the apoplast and triggering cell wall loosening and cell expansion (Fendrych et al. 2016; Haruta et al. 2018).

 Cite this article as *Cold Spring Harb Perspect Biol* doi: 10.1101/cshperspect.a039941

It has become evident that a rhizotactic pattern is controlled by multiple, mutually coordinated pathways. The root clock mechanism that drives oscillatory gene expression and steers the regulatory circuits to specify prebranch sites and LRP initiation is complemented by pathways that suppress LR formation outside of this developmental window.

AUXIN COORDINATES THE ESTABLISHMENT OF A DEVELOPMENTAL NICHE FOR LR INITIATION

Typically, the LR developmental program at the prebranching sites is triggered by auxin accumulating in two longitudinally adjacent FCs (i.e., two per cell file). Besides the longitudinal bicellular type of initiation, the longitudinal unicellular type has been reported in which a single FC in a cell file gives rise to the entire LRP. Long-term live-imaging experiments demonstrated that after specification of single FC its immediate neighboring pericycle cells are recruited and participate in LRP formation (von Wangenheim et al. 2016; Torres-Martínez et al. 2020).

The PIN3 auxin efflux carrier reinforces reflux of auxin from surrounding endodermal to XPP cells (Fig. 1C) and thereby allows accumulation of auxin above a threshold level required for FC specification (Dubrovsky et al. 2008; Marhavý et al. 2013). Radial swelling of pairs of FCs and the migration of their nuclei toward the shared cell wall followed by asymmetric anticlinal divisions (perpendicular to the PR growth axis) are hallmarks of initiated LR organogenesis (Malamy and Benfey 1997; Dubrovsky et al. 2000). During the early phases of LR initiation, FCs and adjacent endodermal cells undergo complex morphocytological changes that encompass modulation of actin and microtubule cytoskeleton dynamics (Vilches Barro et al. 2019), changes of biomechanical cell wall properties (Swarup et al. 2008; Vermeer et al. 2014; Ramakrishna et al. 2019), and specification of symplastic fields of cell-to-cell connectivity (Fig. 2A; Hernández-Hernández et al. 2019). These processes, essential for the proper LR developmental program, are controlled through a few auxin-regulated signaling modules, such as SOLITARY ROOT(SLR)/IAA14-ARF7 and downstream acting transcription factors LATERAL ORGAN BOUNDARIES-DOMAIN16 (LBD16) and LBD18, BODENLOS (BDL)/IAA12-ARF5, or IAA3-ARF7 (Fukaki et al. 2002; Okushima et al. 2005, 2007; Vanneste et al. 2005; De Smet et al. 2007, 2010; Orosa-Puente et al. 2018). Importantly in FCs, the cell cycle is reactivated through transcriptional regulation of the E2Fa transcription factor by auxin-responsive IAA14-ARF7-LBD16 and LBD18 modules (Fig. 2A; Berckmans et al. 2011; Sanz et al. 2011; Lee et al. 2015). It is noteworthy that attempts to trigger LR initiation through either tissue-specific activation of cell-cycle machinery in XPP cells (Vanneste et al. 2005), or release of mechanical constraints imposed on XPP cells by endodermal cell ablation (Marhavý et al. 2016) demonstrate that a simple transition of XPP cells to a meristematically active stage is not sufficient to launch an LR developmental program. Hence, auxin, through spatiotemporal coordination of all the above-discussed pathways, establishes a developmental niche for LR initiation.

New molecular players involved in the regulation of LR initiation and contributing to the plasticity of root branching have been identified recently. AtHB23, a homeodomain-leucine zipper (HD-Zip) I transcription factor, represses the expression of *LBD16* and thereby suppresses the initiation of LRPs (Fig. 2A). Interestingly, expression of this TF differs in LRPs of secondary and tertiary order, suggesting that AtHB23 might act in the upstream pathway that fine-tunes the core "LR initiation module" as the root system branches and expands (Perotti et al. 2019). Similarly, SHI-RELATED SEQUENCE5 (SRS5), a member of the SHORT-INTERNODES family, represses the transcription of *LBD16*. SRS5 is under negative control of ARF7 and ARF19 and might act in a negative feedback loop repressing LR initiation (Fig. 2B; Yuan et al. 2020). Intriguingly, the activity of the IAA14-ARF7-LBD16 regulatory module that controls asymmetric cell divisions of FCs is complemented by GOLVEN/ROOT GROWTH FACTOR/CLE-like (GLV/RGF/CLEL) signaling peptides (Fig. 2A). Two members of the

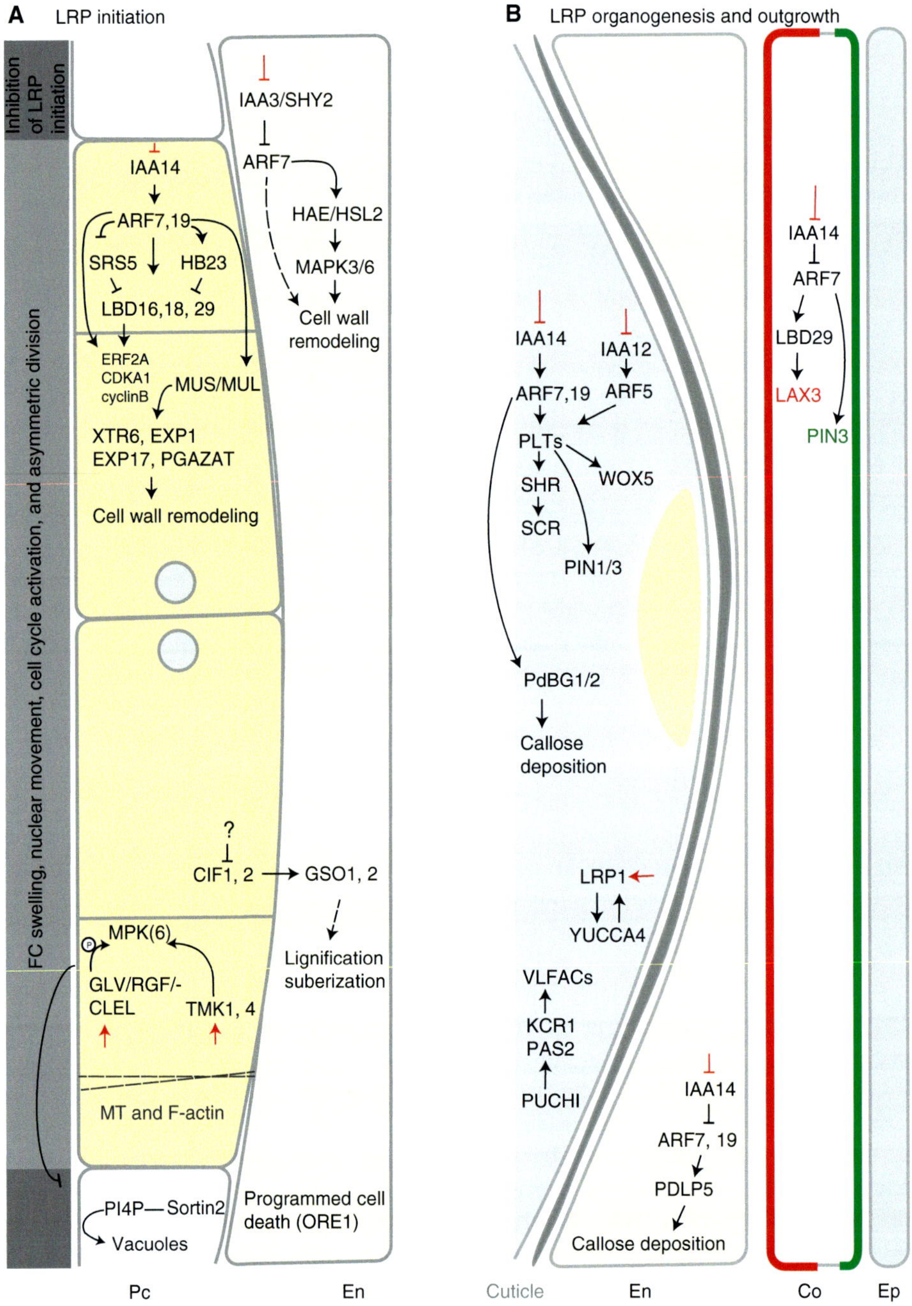

Figure 2. Lateral root primordia development: from initiation to outgrowth. (*A*) Swelling of founder cells (FCs), nuclear movement, cytoskeleton rearrangements, and first asymmetric cell divisions defines the onset of lateral root primordia (LRP) development. Auxin signaling (red bar-headed lines) promotes the degradation of Aux/IAA repressing factors such as IAA14 and IAA3 and thereby stimulates the expression of downstream ARF, LBD transcription factors, and cell cycle and cell wall remodeling genes. Different factors are acting in the neighboring cells in which ectopic LRP initiation is inhibited. The initial cells of LRP accumulate higher auxin concentrations and are reported in yellow. (*B*) LRP organogenesis and outgrowth is regulated through several auxin signaling modules. The auxin signaling cascade and expression of specific factors involved are indicated in the relevant cell layers. At the surface of the developing LRP, a cuticle layer is formed to prevent adhesion when penetrating the overlaying tissues. In the cortex, LAX3 and PIN3 localization are marked in red and green, respectively. (MT) Microtubule, (Pc) pericycle, (En) endodermis, (Co) cortex, (Ep) epidermis.

GLV gene family, *GLV6* and *GLV10*, act as inhibitors of asymmetric cell divisions and signal through RGF1 INSENSITIVE receptors and MITOGEN-ACTIVATED PROTEIN KINASE6 (MPK6) to restrict the number of initial asymmetric cell divisions that take place during LR initiation (Fernandez et al. 2020).

ROLE OF CELL-TO-CELL TRANSPORT IN LRP ORGANOGENESIS

The LR developmental program starts with the asymmetric division of FCs. Although the subsequent rounds of divisions are less deterministic and do not follow a strictly stereotypical pattern, this division process results in a highly conserved tissue patterning and shape of developing LRP (Malamy and Benfey 1997; Lucas et al. 2013; von Wangenheim et al. 2016). LRPs originate from XPP cells embedded deeply within the parental PR, implying that significant parts of LR organogenesis take place while the primordium is surrounded by several tissue layers, each of them exhibiting specific biomechanical properties (Malamy 2005; Richter et al. 2009; Naseer et al. 2012; Vermeer et al. 2014; Banda et al. 2019). Hence, LR formation is governed by a tight cross talk between regulatory circuits acting in the body of the primordium and adjacent tissues. Remarkably, auxin and its distribution driven by PAT plays an instructive role in the proper formation of LRP as well as in the coordination of its emergence through surrounding tissues. In the LRP, PIN, and AUX1/LAX auxin transporters (Marchant et al. 2002; Benková et al. 2003; Marhavý et al. 2013; Péret et al. 2013) occupy distinct expression domains and polarize to specific membranes, thereby determining a graded auxin distribution with maximum at the tip of the primordium (Figs. 1C and 2B). Hence, numerous factors and pathways that control expression or polar membrane localization of the auxin transporters, including regulators of vesicular trafficking such as ARF GTPases and ARF guanine-nucleotide exchange factors (ARF-GEFs) (Geldner 2003), and hormones such as cytokinin (Marhavý et al. 2011, 2014), gibberellins (Löfke et al. 2013), and salicylic acid (Tan et al. 2020), play an important role in steering LRP organogenesis.

Auxin released from LRP to adjacent tissues activates pathways that promote cell separation and thereby facilitate emergence of the growing primordium (Swarup et al. 2008; Banda et al. 2019). Directional transport of auxin from the tip of the LRP toward neighboring tissues is driven by coordinated activities of LAX3 and PIN3 influx and efflux transporters, respectively, acting in few cells in direct contact with the emerging primordia (Fig. 2B). The remarkably specific spatiotemporal expression pattern of both transporters is the result of tightly coordinated consecutive expression. Auxin, through an ARF7-LBD29 transcriptional cascade, induces expression of *LAX3*, followed by enhancement of PIN3-mediated efflux (Péret et al. 2013; Porco et al. 2016).

PAT appears to play an essential role in auxin distribution during LRP organogenesis and, recently, novel factors that fine-tune the distribution of auxin at the subcellular level have been identified (Mravec et al. 2009; Barbez et al. 2012). Among them TOB1, a vacuolar transporter of the auxin precursor IBA, plays a role in LRP organogenesis. *TOB1* expression is cytokinin-responsive and expands from LRP-flanking regions to the whole primordia and outer layers of the parental root. Mutations in *TOB1* result in enhanced LRP development and while detailed mechanisms underlying TOB1 function await further investigation, TOB1 appears to act as an integrator of cytokinin signaling and auxin homeostasis in the regulation of LRP organogenesis (Michniewicz et al. 2019). The role of vacuoles in LR formation is also supported by a recent study of Sortin2, a chemical identified in screening for molecules affecting protein trafficking (Norambuena et al. 2008). Sortin2, through phosphatidylinositol 4-kinases PI4-KIIIb1 and PI4KIIIb2, affects the abundance of phosphatidylinositol 4-phosphate (PI4P) at the PM and endosomal compartments, thereby promoting protein trafficking to the vacuoles. Intriguingly, the effect of Sortin2 on subcellular trafficking correlates with enhanced LR organogenesis independently of the canonical $SCF^{TIR1/AFB}$ auxin-mediated signaling. Whether Sortin2-mediated protein trafficking to the vacuoles modulates the activity of factors acting downstream of auxin signaling or involves different auxin-independent programs still needs to be resolved (Pérez-Henríquez et al. 2012; Rubilar-Hernández et al. 2019).

In parallel to cell-to-cell transport of auxin driven by PM transporters, symplastic plasmodesmata (PD)-dependent transport of molecules is critical for proper LRP organogenesis. Two related callose-degrading enzymes plasmodesmal-localized β-1,3 glucanase1 (PdBG1) and PdBG2, both transcriptionally regulated by auxin in an SLR/IAA14-dependent manner, control callose deposition in LRPs (Fig. 2B). Mutations in these genes result in defective primordia, in line with a role of symplastic transport in the regulation of LRP organogenesis (Benitez-Alfonso et al. 2013). Interestingly, in cells overlying LRPs, the auxin-inducible PLASMODESMATA-LOCATED PROTEIN5 (PDLP5) governs a temporary symplastic isolation of primordia via reversible callose accumulation. Interference with PDLP5 affects the outgrowth of LRP (Fig. 2B; Sager et al. 2020). Thus, auxin-controlled PD-dependent movement of molecules in both primordia and adjacent tissues contributes to fine-tuning of LRP organogenesis and outgrowth (Benitez-Alfonso et al. 2013; Sager et al. 2020).

AUXIN ORCHESTRATES GENETIC NETWORKS IN LRP AND ADJACENT TISSUES TO REGULATE LRP DEVELOPMENT

Auxin in LRP and adjacent cells triggers and coordinates essential pathways for the proper organogenesis and emergence. Importantly, recent works show that tissues overlaying LRPs are not just biomechanical barriers that need to be crossed during emergence from the parental root, but they play an active role in shaping and pattering the LRP (Lucas et al. 2008; Vermeer et al. 2014).

In the primordium, auxin signals transduced through specific modules, including SLR/IAA14-ARF7, ARF19, and BDL/IAA12-ARF5, activate multiple downstream pathways. Among them, the auxin-controlled PLETHORA (PLT) transcription factors, which further define expression of *PIN1* and *PIN3* transporters, as well as the *SHORT ROOT* (*SHR*), *SCARECROW* (*SCR*), and *WUSCHEL-RELATED HOMEOBOX5* (*WOX5*) transcription factors steer the cell differentiation and patterning of LRPs (Du and Scheres 2017). Recently, LATERAL ROOT PRIMORDIA1 (LRP1), a member of the SHORT INTERNODES/STYLISH (SHI/STY) family, whose expression is regulated by histone deacetylation in an auxin-dependent manner, was recognized as a novel negative regulator of LRP organogenesis. LRP1 promotes auxin biosynthesis through the regulation of *YUCCA4* (*YUC4*) and thus contributes to LRP formation (Fig. 2B; Singh et al. 2020).

Strikingly, in parallel to the well-established Aux/IAA-ARF-mediated signaling, a new noncanonical auxin transduction pathway acting through transmembrane kinases (TMKs) has been proposed. TMKs, in response to auxin, activate the mitogen-activated protein kinase (MAPK) pathway and thereby control the cell-division patterns during LRP organogenesis (Huang et al. 2019).

In outer tissues, auxin, through signaling modules such as SHORT HYPOCOTYL2 (SHY2)/IAA3-ARF7 in the endodermis and SLR/IAA14-ARF7 and ARF19 in the cortex and epidermis, regulates distinct pathways that enable separation of cells with different cell wall properties. For example, in the walls of endodermal cells, Casparian strips, enriched for suberin and lignin, form an extracellular hydrophobic barrier between the outer and inner root tissues to control water and nutrient uptake (Doblas et al. 2017; Nakayama et al. 2017). Interestingly, the expression of the CASPARIAN STRIP INTEGRITY FACTOR1 (CIF1) and CIF2, peptides, which through SCHENGEN3/GASSHO1 (SGN3/GSO1) and GSO2 leucine-rich repeat (LRR)-RLK receptors regulate lignification and suberization of the endodermal tissues, is suppressed in cells adjacent to LRPs (Fig. 2A). Although the mechanisms that control local down-regulation of CIFs and the role of auxin in this process need to be examined, the findings support the concept of a local, tissue-specific pathway engaged in cell separation during LRP outgrowth (Ghorbani et al. 2015; Nakayama et al. 2017). Degradation of pectins, which are major components of plant cell walls, is facilitated by the INFLORESCENCE DEFICIENT IN ABSCISSION (IDA) peptide. IDA interacts with the HAESA (HAE)/HAESA-LIKE2 (HSL2) LRR-RLK and activates the MKK4/MKK5–MPK3/MPK6 cascade, which promotes expression of a specific set of the cell wall remodeling enzymes and thereby enhances LR emergence (Fig. 2A; Kumpf et al. 2013; Zhu et al. 2019b).

Whereas initially major attention was given to mechanisms associated with separation of cells overlying LRP, current studies point out that in LRPs, auxin-controlled cell wall remodeling also has an important patterning function. *MUSTACHE* (*MUS*) and *MUSTACHES-LIKE* (*MUL*) genes, encoding inactive LRR-RLKs, are expressed in early-stage LRPs in an auxin/ARF7-ARF19-dependent manner. RNA-seq data suggest that MUS and MUL control LRP development, likely via regulating cell wall biosynthesis and remodeling genes such as *XYLOGLUCAN ENDOTRANSGLYCOSYLASE6* (*XTR6*), *EXPANSIN1* (*EXP1*), *EXP17*, and *POLYGALACTURONASE ABSCISSION ZONE A. THALIANA* (*PGAZAT*) (Xun et al. 2020).

In addition, recent findings hint at PCD as a parallel regulatory pathway to cell wall remodeling–driven cell separation during LR emergence. A subset of LRP-overlying cells display induction of PCD marker genes and their cell death was detected by electron and other microscopic techniques. Cell-death-deficient mutants lacking the positive cell death regulator ORESARA1/ANAC092 (ORE1) exhibit a delayed outgrowth of LRPs and these defects were restored by physical or genetic elimination of cells overlying the LRPs (Escamez et al. 2020).

Tight contact of developing LRPs and surrounding tissues raises an important question about the mechanisms that prevent adhesion of newly formed organs and parental root tissues. Intriguingly, a layer comprised of an insoluble polyester (cutin) and soluble lipids (waxes), called cuticle, was detected from the early stages on the surface of developing LRPs (Berhin et al. 2019). Mutants defective in cuticle formation exhibit deformations of LRPs, suggesting that, like in the shoot apical meristem and other plant organs, cuticle forms a separating layer, which prevents adhesion of adjacent tissues and thereby facilitates penetration of LRP toward the root surface (Fig. 2B; Nawrath et al. 2013; Ingram and Nawrath 2017; Berhin et al. 2019).

DISTINCT REGULATORY PATHWAYS ARE INVOLVED IN THE REGULATION OF LR DEVELOPMENT

The outgrowth of LRs and their maturation encompasses a number of events including the establishment of a connection of the vascular system with that of the parental root and the formation of an apical meristem at the proximal end of the LR. The apical meristem, analogous to that of the PR, is essential for production of new cells and to sustain growth of LRs. Although many regulatory pathways are shared with PRs, some genetic programs are specific to LRs. An obvious example of such a nonshared pathway is the SLR/IAA14-ARF7-ARF19 signaling module. Mutations in components of this module severely affect LR formation, while the PR remains mostly unaffected (Fukaki et al. 2002). Similarly, expression and functional analyses of *AtHB23* indicate that specialized pathways have developed to control PRs and different-order LRs (Perotti et al. 2019). Genetically distinct pathways controlling roots of either embryonic or postembryonic origin provide crucial advantages for plant survival under conditions that limit growth of the PR. Furthermore, radial expansion of the root system, essential for maximizing soil exploration, relies on diversification of gravity-sensing mechanisms in LR and PR. Whereas PRs are highly responsive to gravity and they rapidly realign with changed gravity vector (Friml et al. 2002; Zhu et al. 2019a), attenuated gravity sensing in LRs allows their horizontal growth (Rosquete et al. 2013; Waidmann and Kleine-Vehn 2020). Reduced gravity-driven bending of LRs is the result of partially suppressed activity of the PAT machinery, which mediates root gravity response through rapid relocation of auxin to the lower gravi-stimulated side of the root (Rosquete et al. 2013; Ogura et al. 2019). Moreover, in LRs, auxin-driven gravity sensing is counterbalanced by a cytokinin-de-

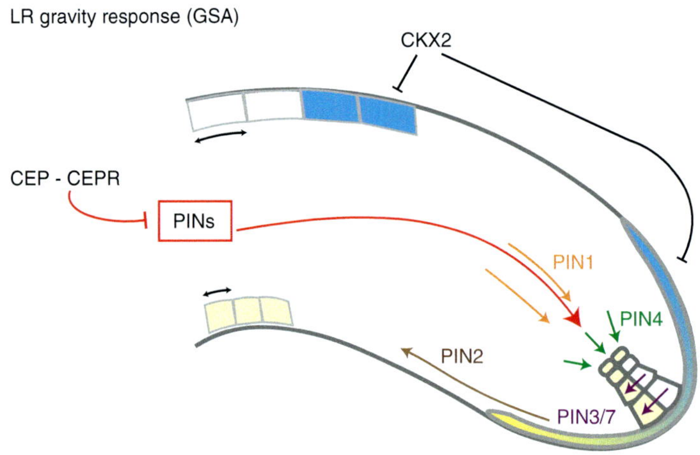

Figure 3. Auxin and cytokinin define the gravitropic set point angle of the lateral root (LR) upon emergence. PIN-driven auxin flow to LR columella cells (yellow) is depicted as a red *central* arrow. Asymmetric cytokinin distribution in the LR cap (LRC) cell and in the *upper* side of LR is highlighted in blue. Interactions among factors interfering with the gravitropic angle are indicated (bar-headed lines). Differential cell elongation between the upper and lower side of LR is represented as double-headed black arrows.

pendent mechanism. Local increase of cytokinin signaling at the upper LR flank leads to growth inhibition and attenuation of downward gravitropic bending of LRs (Fig. 3). Interestingly, a genome-wide association study revealed that activity of Cytokinin Oxidase 2 (CKX2) which catalyzes the degradation of cytokinins and thereby controls levels of cytokinins at the tip of LR, is an important factor determining gravity set point angle (GSA) of LRs (Fig. 3; Waidmann et al. 2019). Notably, regulation of the LR GSA is not fully root-autonomous and can be fine-tuned by systemic signals that coordinate the development of root and shoot organs. Recently, interaction of CARBOXY-TERMINALLY ENCODED PEPTIDE (CEP) with the CEP RECEPTOR1 (CEPR1) in shoot was shown to control the GSA of LRs by reducing rootward auxin transport and/or by altering auxin levels in shoots (Chapman et al. 2020).

ROOT SYSTEM ARCHITECTURE SHAPING BY ENVIRONMENTAL CUES

In heterogeneous soil environments, the root system is exposed to numerous cues such as mechanical obstacles, the availability of water and patches of nutrients, the level of oxygen, and light passing through the soil. The root continuously perceives these variables and translates them to adaptive developmental responses, including modulation of the branching pattern. Bending of PRs triggered either by gravity or deflection from obstacles are examples of an environmentally driven mechanism that significantly contribute to modulation of the LR initiation pattern (Ditengou et al. 2008; Lucas et al. 2008). Although the molecular details remain unclear, in silico models suggest that mechanical deformation of tissues during curve formation leads to local increases of auxin activity, which stimulates the initiation of LR (Laskowski et al. 2008). Alternatively, root bending could trigger a transient increase of Ca^{2+} within the pericycle that might act as a signal to translate mechanical forces to initiate LR development (Richter et al. 2009).

Light is one of the environmental factors essential for plant survival. It affects not only the development and physiology of the shoot, but also the underground root system. Importantly, root cells also possess photoreceptors that can be activated by light. Hence, roots not only receive information about light conditions through signaling molecules traveling from the shoot, but they also directly perceive light passing through the soil (Lee et al. 2016; Yang et al. 2020). For example, light perceived by the photoreceptor PHYTOCHROME B (PHYB) in roots activates the ELONGATED HYPOCOTYL5 (HY5) transcription factor and thereby modulates growth and gravitropic responses of the PR (Lee et al. 2016). Recently, another root-based mechanism of light-regulated responses has been described. Ultraviolet B (UV-B) light is an inherent component of sunlight that markedly affects root branching. At the molecular level, UV-B light activates the UVB-RESISTANCE8 (UVR8) receptor and triggers its relocation to the nucleus, where it inhibits the DNA-binding activities of MYB73/MYB77 transcription factors. As a result, transcription of the auxin-responsive genes is suppressed and consequently LR development is inhibited (Fig. 4; Yang et al. 2020).

The soil environment displays temporary oxygen deprivations due to seasonal rainfalls, flooding, or snow melting. Plants suffering from this drop in the soil oxygen content react by adaptive responses including inhibition of LR development. In-depth analyses revealed that *in planta*, the lack of active circulation may lead to the formation of internal hypoxic areas. These hypoxic niches, characterized by the expression of hypoxia marker genes, including a subgroup of ERF-VII transcription factors, were detected at specific stages of LRP development both in aerobic and enhanced at low oxygen conditions. ERF-VII transcription factors bind to promoters of the auxin-induced genes *LBD16*, *LBD18*, and *PUCHI* and suppress their expression, resulting in inhibition of LRP development (Shukla et al. 2019). Further insights into mechanisms associated with the regulation of oxygen levels and their impact on the root system were obtained from rice. In rice, lysigenous aerenchyma, a gas space created by cortical cell death, is established constitutively under aerobic conditions to enhance internal oxygen transport and its formation increases under oxygen-deficient conditions. In the dominant-negative mutant of the IAA13 auxin signaling repressor, reduced aerenchyma formation correlated with defects in LR formation. This hints at a role for auxin signaling in the coordination of both processes. Consistently, *LBD1-8*, a transcriptional target of IAA13 and ARF19, is expressed in cortex and LRP. Restoration of *LBD1-8* expression recovered aerenchyma and LR formation in the *iaa13* background, implying that auxin signaling coregulates aerenchyma and LR formation (Yamauchi et al. 2019).

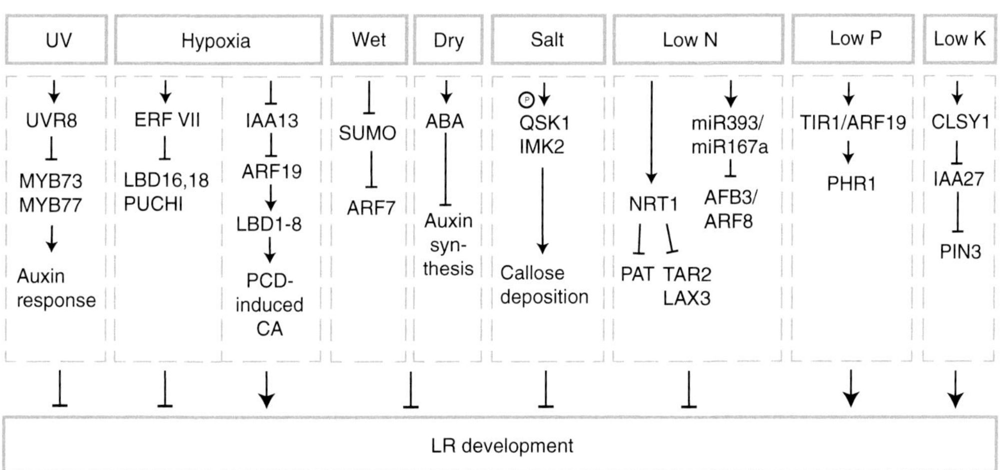

Figure 4. Environmental factors regulating lateral root (LR) development. Schematic representation of environmental cues, signaling pathways, and molecular mechanism regulating LR organogenesis. Factors promoting or repressing LR developments are indicated as black arrows or bar-headed lines, respectively.

Roots, as they grow through heterogenic soil environments, are exposed to wet soil aggregates or airy pores, where they experience a local transient water deficit. Recent advances in bioimaging techniques, which allowed monitoring of root growth in the natural soil environments, revealed that, while LR organogenesis is enhanced at the root side exposed to moisture, it is suppressed in airy microenvironments (Bao et al. 2014; Orman-Ligeza et al. 2018; Orosa-Puente et al. 2018; von Wangenheim et al. 2020). Intriguingly, both processes, hydro-patterning (promoting effect of moisture on LR organogenesis) and xero-branching (LR organogenesis inhibition in airy microenvironments) appear to be regulated by specific molecular pathways. Hydro-patterning is primarily driven through differential posttranslational control of the sumoylation pattern of ARF7 and its impact on auxin signaling. ARF7 is sumoylated on the airy side of the root, resulting in an interaction with the IAA3 repressor that inhibits LR initiation. On the water-exposed side of the root, ARF7 is not sumoylated, enabling the transcription factor to activate expression of genes involved in LR initiation (Orosa-Puente et al. 2018). Intriguingly, local heterogeneity in moisture also affects later phases of LRP organogenesis, namely the LR outgrowth angle. Typically, LR initiate from multiple adjacent XPP files, where the central cell file forms the tip of the primordium and contributes most of the cell mass to developing LRP (von Wangenheim et al. 2017). Exposure to water gradient might change proportion with which the individual XPP files contribute to formation of LRP and promote leaning of LR toward the water reach area (von Wangenheim et al. 2020).

Studies of xero-branching in cereals and *Arabidopsis* point at an indispensable role of abscisic acid (ABA) signaling when roots are not in contact with water. Endogenous ABA accumulation in the root correlates with attenuated auxin biosynthesis and consequently decreased LRP initiation (Deak and Malamy 2005; Orman-Ligeza et al. 2018).

When roots are exposed to osmotic stress in the presence of high soil salinity, the PM-located LRR-RLKs Qian Shou kinase (QSK1) and inflorescence meristem kinase2 (IMK2) rapidly relocate to the PD and through the regulation of callose-deposition fine-tune LRP organogenesis (Grison et al. 2019).

LATERAL ROOT DEVELOPMENT AS A FORAGING STRATEGY

The root system is highly responsive to nutrient availability and distribution within the soil. This is particularly true for major nutrients like nitrogen (N), phosphorus (P), and potassium (K), which are limiting nutrients for plant growth (Kellermeier et al. 2013; Guan 2017; Crombez et al. 2019; Vidal et al. 2020).

Nitrogen is an essential element that plants can acquire from various inorganic or organic sources in the soil. Fluctuations in both concentrations and the form of N sources available in the soil have prominent effects on root systems. Deficiency in N severely interferes with root growth and development, whereas low availability of nitrogen enhances root branching to promote the exploitation of this macronutrient. High levels of N might trigger production of stress hormones such as ethylene that lead to inhibition of PR and LR elongation growth (Tian et al. 2009; Gruber et al. 2013; Jia and von Wirén 2020). Interestingly, the local nitrate-rich zones enhance LR outgrowth (Remans et al. 2006). These complex adaptive responses of roots to N sources and heterogeneity in availability are regulated by a combination of systemic and local signaling (Ruffel 2018). The impact of available N on the root system is closely interconnected with the activity of plant hormones including auxin. A number of studies have demonstrated that auxin biosynthesis and transport are altered in response to different N regimes in various plant species (Gutiérrez et al. 2007; Ma et al. 2014; Krouk 2016; Maghiaoui et al. 2020). At the level of auxin signaling, auxin receptor *AFB3* and auxin response factor *AUXIN RESPONSE FACTOR 8* (*ARF8*) with associated *miR393* and *miRNA167s*, respectively, were identified as N-responsive modules that control root system architecture in response to N availability in *Arabidopsis* (Gifford et al. 2008; Vidal et al. 2010). The NITRATE TRANSPORTER1.1

(NRT1.1), a principal component of nitrate transport and sensing, acts as an important integrator of nitrate and auxin pathways in *Arabidopsis* (Krouk et al. 2010; Bouguyon et al. 2015; Sun et al. 2017; Naz et al. 2019; Vidal et al. 2020). NRT1.1 is a nitrate transceptor with a dual transport and sensing function, which in the absence or at low nitrate availability displays auxin transport activity. The ability of NRT1.1 to transport auxin has a critical role in adjustment of LRP development to nitrate levels (Fig. 4). At low nitrate, NRT1 negatively affects LRP development by promoting auxin transport out of the primordia tip, whereas at the optimal level of this macronutrient, accumulation of auxin promotes LRP development (Krouk et al. 2010; Bouguyon et al. 2015). Intriguingly, NRT1, whose abundance is tightly controlled by nitrate, not only fine-tunes flow of auxin in LRPs but also controls both auxin biosynthesis by repressing expression of *TRYPTOPHAN AMINOTRANSFERASE-RELATED2* (*TAR2*), and auxin influx into tissues overlaying LRP by attenuating *LAX3* expression (Maghiaoui et al. 2020). Hence, NRT1.1 acts as a regulatory hub that senses and controls nitrate levels in the plant body and through multiple pathways controls activity of auxin, thereby translating N status of the plant into adaptive developmental responses including the root system architecture.

Although P is a major component in soil nutrients, only a small fraction is bioavailable for plants in the form of phosphate (Pi). Pi starvation triggers a series of adaptive responses in plants (Crombez et al. 2019). A Pi shortage fosters the formation of a shallow root system by attenuating PR growth and promoting branching (Naumann et al. 2019). Increased LR formation in Pi-deprived *Arabidopsis* seedlings is concomitant with increased expression of the *TIR1* auxin receptor and *ARF19* (Pérez-Torres et al. 2008). Both ARF7 and ARF19 positively regulate *PHOSPHATE STARVATION RESPONSE1* (*PHR1*), an MYB-CC transcription factor, which is one of the main positive regulators of Pi-response genes (Bustos et al. 2010; Huang et al. 2018). Expression of *PHR1* can partially rescue the LR-lacking phenotype of *arf7/arf19* mutants, suggesting a compensatory feedback mechanism acting independently of *LBD16* and *LBD29* in response to low Pi (Huang et al. 2018). Other factors like light, ethylene, and strigolactones can coordinate *PHR1* expression and the phosphate response, suggesting a high degree of cross talk among hormone signaling pathways (Sun et al. 2014; Liu et al. 2017; Crombez et al. 2019). In particular, PHR1 activation leads to a reduction in nitrate uptake and represents a direct link between phosphorus and nitrogen regulation (Maeda et al. 2018).

In contrast to Pi, potassium is highly soluble and K-deficiency inhibits LR organogenesis by interfering with auxin signaling and the expression of *PIN3* in LRP (Kellermeier et al. 2013; Shahzad et al. 2020). Characteristic regulation of LR development in low K is achieved by RNA-directed DNA methylation (RdDM) (Matzke and Mosher 2014; Shahzad et al. 2020). The chromatin remodeling factor CLASSY (CLSY1) is a component of the RdDM complex, which is involved in silencing of a specific subset of transposable elements (TEs) and genes (Zhou et al. 2018). The promoter region of *IAA27*, an auxin signaling repressor and negative regulator of root branching, contains several of these TEs, and accordingly displays hypomethylation in RdDM mutants. Hence, a DNA-methylation-mediated transcriptional repression by CLSY1 has been proposed as an alternative pathway promoting LR development in low K by silencing *IAA27* expression when the auxin-dependent degradation is impaired (Shahzad et al. 2020).

CONCLUSION AND PERSPECTIVES

Roots are underground organs that anchor and provide the plant body with water and nutrients essential for sustained growth and development. Developmental plasticity of the root system, based on the ability of genetic programs that control LR organogenesis to integrate environmental cues, has become a successful strategy supporting plant survival in very different habitats. Current research is providing exciting insights into mechanisms that act in the core of these unique root system features. Particularly, advances in monitoring root growth in real time and in natural soil environments have brought attention to some, so far, neglected factors that significantly impact the root system (Orman-Ligeza et al. 2018; von Wangenheim et al. 2020). Heterogeneity of soils with diverse microenvironments differing in water and oxygen content, nutrient composition, temperature, or light forms a complex universe that continuously shapes the root system and determines the fitness of the whole plant. In this dynamic soil environment, our view on auxin as a key endogenous orchestrator of root branching gains new perspectives and is challenged by intriguing open questions. How would relatively few auxin signaling modules integrate such a variety of environmental signals and translate them to coordinate diverse processes during LR organogenesis: the regulation of the cell cycle and divisions, actin and microtubule dynamics, cell wall remodeling, tissue patterning, and others. What are the molecular principles underlying specificity of the auxin action? Identification of a novel noncanonical TMK-mediated auxin transduction pathway (Huang et al. 2019), acting in parallel to well-established auxin signaling modules in LRP formation represents an exciting possibility of parallel pathways acting in the context of varying environmental conditions. *Arabidopsis* is a suitable model for investigation of basic developmental and molecular principles in root system architecture, but anatomical studies on a broader range of plant species point at significant diversities in LRP organogenesis (Xiao et al. 2019). This demonstrates that studying the molecular mechanisms controlling root branching in other plant species may provide important insights into evolution of root system architecture and environmental responses. Moreover, the rapidly progressing area of soil microbiota research and interaction with the plant's roots will reveal more genetic pathways that feed into modules controlling LR organogenesis.

In conclusion, the root system architecture is the result of a fine balance between the endogenous plant developmental program and exogenous inputs perceived from the environment. In this scenario, auxin is standing as a master regulator of LR organogenesis and future research will provide more elements to fully understand its primary role in shaping root architecture.

ACKNOWLEDGMENTS

We apologize to all the authors whose scientific work could not be cited and discussed because of space restrictions. We thank Dr. Inge Verstraeten (IST Austria) and Dr. Juan Carlos Montesinos-Lopez (ETH Zürich) for helpful suggestions. This work was supported by the DOC Fellowship Programme of the Austrian Academy of Sciences (25008) to C.A.

REFERENCES

Atkinson JA, Rasmussen A, Traini R, Voss U, Sturrock C, Mooney SJ, Wells DM, Bennett MJ. 2014. Branching out in roots: uncovering form, function, and regulation. *Plant Physiol* **166:** 538–550. doi:10.1104/pp.114.245423

Banda J, Bellande K, von Wangenheim D, Goh T, Guyomarc'h S, Laplaze L, Bennett MJ. 2019. Lateral root formation in *Arabidopsis*: a well-ordered LRexit. *Trends Plant Sci* **24:** 826–839. doi:10.1016/j.tplants.2019.06.015

Bao Y, Aggarwal P, Robbins NE, Sturrock CJ, Thompson MC, Tan HQ, Tham C, Duan L, Rodriguez PL, Vernoux T, et al. 2014. Plant roots use a patterning mechanism to position lateral root branches toward available water. *Proc Natl Acad Sci* **111:** 9319–9324. doi:10.1073/pnas.1400966111

Barbez E, Kubeš M, Rolčík J, Béziat C, Pěnčík A, Wang B, Rosquete MR, Zhu J, Dobrev PI, Lee Y, et al. 2012. A novel putative auxin carrier family regulates intracellular auxin homeostasis in plants. *Nature* **485:** 119–122. doi:10.1038/nature11001

Beeckman T, Burssens S, Inzé D. 2001. The peri-cell-cycle in *Arabidopsis*. *J Exp Bot* **52:** 403–411.

Bellini C, Pacurar DI, Perrone I. 2014. Adventitious roots and lateral roots: similarities and differences. *Annu Rev Plant Biol* **65:** 639–666. doi:10.1146/annurev-arplant-050213-035645

Benitez-Alfonso Y, Faulkner C, Pendle A, Miyashima S, Helariutta Y, Maule A. 2013. Symplastic intercellular connectivity regulates lateral root patterning. *Dev Cell* **26:** 136–147. doi:10.1016/j.devcel.2013.06.010

Benková E, Michniewicz M, Sauer M, Teichmann T, Seifertová D, Jürgens G, Friml J. 2003. Local, efflux-dependent auxin gradients as a common module for plant organ formation. *Cell* **115:** 591–602. doi:10.1016/S0092-8674(03)00924-3

Berckmans B, Vassileva V, Schmid SPC, Maes S, Parizot B, Naramoto S, Magyar Z, Kamei CLA, Koncz C, Bögre L, et al. 2011. Auxin-dependent cell cycle reactivation through transcriptional regulation of *Arabidopsis E2Fa* by lateral organ boundary proteins. *Plant Cell* **23:** 3671–3683. doi:10.1105/tpc.111.088377

Berckmans B, Kirschner G, Gerlitz N, Stadler R, Simon R. 2020. CLE40 signaling regulates root stem cell fate. *Plant Physiol* **182:** 1776–1792. doi:10.1104/pp.19.00914

Berhin A, de Bellis D, Franke RB, Buono RA, Nowack MK, Nawrath C. 2019. The root cap cuticle: a cell wall structure for seedling establishment and lateral root formation. *Cell* **176:** 1367–1378.e8. doi:10.1016/j.cell.2019.01.005

Bielach A, Podlesáková K, Marhavý P, Duclercq J, Cuesta C, Müller B, Grunewald W, Tarkowski P, Benková E. 2012. Spatiotemporal regulation of lateral root organogenesis in *Arabidopsis* by cytokinin. *Plant Cell* **24:** 3967–3981. doi:10.1105/tpc.112.103044

Blakely L, Evans TA. 1979. Cell dynamics studies on the pericycle of radish seedling roots. *Plant Sci Lett* **14:** 79–83. doi:10.1016/0304-4211(79)90158-5

Blakely LM, Blakely RM, Colowit PM, Elliott DS. 1988. Experimental studies on lateral root formation in radish seedling roots. II: Analysis of the dose-response to exogenous auxin. *Plant Physiol* **87:** 414–419. doi:10.1104/pp.87.2.414

Bouguyon E, Brun F, Meynard D, Kubeš M, Pervent M, Leran S, Lacombe B, Krouk G, Guiderdoni E, Zažímalová E, et al. 2015. Multiple mechanisms of nitrate sensing by *Arabidopsis* nitrate transceptor NRT1.1. *Nat Plants* **1:** 15015. doi:10.1038/nplants.2015.15

Bustos R, Castrillo G, Linhares F, Puga MI, Rubio V, Pérez-Pérez J, Solano R, Leyva A, Paz-Ares J. 2010. A central regulatory system largely controls transcriptional activation and repression responses to phosphate starvation in *Arabidopsis. PLoS Genet* **6:** e1001102. doi:10.1371/journal.pgen.1001102

Casimiro I, Marchant A, Bhalerao RP, Beeckman T, Dhooge S, Swarup R, Graham N, Inzé D, Sandberg G, Casero PJ, et al. 2001. Auxin transport promotes *Arabidopsis* lateral root initiation. *Plant Cell* **13:** 843–852. doi:10.1105/tpc.13.4.843

Casimiro I, Beeckman T, Graham N, Bhalerao R, Zhang H, Casero P, Sandberg G, Bennett MJ. 2003. Dissecting *Arabidopsis* lateral root development. *Trends Plant Sci* **8:** 165–171. doi:10.1016/S1360-1385(03)00051-7

Chang L, Ramireddy E, Schmülling T. 2015. Cytokinin as a positional cue regulating lateral root spacing in *Arabidopsis. EXBOTJ* **66:** 4759–4768. doi:10.1093/jxb/erv252

Chapman K, Ivanovici A, Taleski M, Sturrock CJ, Ng JLP, Mohd-Radzman NA, Frugier F, Bennett MJ, Mathesius U, Djordjevic MA. 2020. CEP receptor signalling controls root system architecture in *Arabidopsis* and *Medicago. New Phytol* **226:** 1809–1821. doi:10.1111/nph.16483

Crombez H, Motte H, Beeckman T. 2019. Tackling plant phosphate starvation by the roots. *Dev Cell* **48:** 599–615. doi:10.1016/j.devcel.2019.01.002

Deak KI, Malamy J. 2005. Osmotic regulation of root system architecture: osmotic regulation of lateral root formation. *Plant J* **43:** 17–28. doi:10.1111/j.1365-313X.2005.02425.x

De Rybel B, Vassileva V, Parizot B, Demeulenaere M, Grunewald W, Audenaert D, Van Campenhout J, Overvoorde P, Jansen L, Vanneste S, et al. 2010. A novel Aux/IAA28 signaling cascade activates GATA23-dependent specification of lateral root founder cell identity. *Curr Biol* **20:** 1697–1706. doi:10.1016/j.cub.2010.09.007

De Rybel B, Audenaert D, Xuan W, Overvoorde P, Strader LC, Kepinski S, Hoye R, Brisbois R, Parizot B, Vanneste S, et al. 2012. A role for the root cap in root branching revealed by the non-auxin probe naxillin. *Nat Chem Biol* **8:** 798–805. doi:10.1038/nchembio.1044

De Smet I, Tetsumura T, De Rybel B, Frey NF, Laplaze L, Casimiro I, Swarup R, Naudts M, Vanneste S, Audenaert D, et al. 2007. Auxin-dependent regulation of lateral root positioning in the basal meristem of *Arabidopsis. Development* **134:** 681–690. doi:10.1242/dev.02753

De Smet I, Vassileva V, De Rybel B, Levesque MP, Grunewald W, Van Damme D, Van Noorden G, Naudts M, Van Isterdael G, De Clercq R, et al. 2008. Receptor-like kinase ACR4 restricts formative cell divisions in the *Arabidopsis* root. *Science* **322:** 594–597. doi:10.1126/science.1160158

De Smet I, Lau S, Voss U, Vanneste S, Benjamins R, Rademacher EH, Schlereth A, De Rybel B, Vassileva V, Grunewald W, et al. 2010. Bimodular auxin response controls organogenesis in *Arabidopsis. Proc Natl Acad Sci* **107:** 2705–2710. doi:10.1073/pnas.0915001107

Ditengou FA, Teale WD, Kochersperger P, Flittner KA, Kneuper I, van der Graaff E, Nziengui H, Pinosa F, Li X, Nitschke R, et al. 2008. Mechanical induction of lateral root initiation in *Arabidopsis thaliana. Proc Natl Acad Sci* **105:** 18818–18823. doi:10.1073/pnas.0807814105

Doblas VG, Geldner N, Barberon M. 2017. The endodermis, a tightly controlled barrier for nutrients. *Curr Opin Plant Biol* **39:** 136–143. doi:10.1016/j.pbi.2017.06.010

Dolan L, Janmaat K, Willemsen V, Linstead P, Poethig S, Roberts K, Scheres B. 1993. Cellular organisation of the *Arabidopsis thaliana* root. *Development* **119:** 71.

Du Y, Scheres B. 2017. PLETHORA transcription factors orchestrate de novo organ patterning during *Arabidopsis* lateral root outgrowth. *Proc Natl Acad Sci* **114:** 11709–11714. doi:10.1073/pnas.1714410114

Du Y, Scheres B. 2018. Lateral root formation and the multiple roles of auxin. *J Exp Bot* **69:** 155–167. doi:10.1093/jxb/erx223

Duan L, Dietrich D, Ng CH, Chan PMY, Bhalerao R, Bennett MJ, Dinneny JR. 2013. Endodermal ABA signaling promotes lateral root quiescence during salt stress in *Arabidopsis* seedlings. *Plant Cell* **25:** 324–341. doi:10.1105/tpc.112.107227

Dubrovsky JG, Doerner PW, Colón-Carmona A, Rost TL. 2000. Pericycle cell proliferation and lateral root initiation in *Arabidopsis. Plant Physiol* **124:** 1648–1657. doi:10.1104/pp.124.4.1648

Dubrovsky JG, Gambetta GA, Hernández-Barrera A, Shishkova S, González I. 2006. Lateral root initiation in *Arabidopsis*: developmental window, spatial patterning, density and predictability. *Ann Bot* **97:** 903–915. doi:10.1093/aob/mcj604

Dubrovsky JG, Sauer M, Napsucialy-Mendivil S, Ivanchenko MG, Friml J, Shishkova S, Celenza J, Benkova E. 2008. Auxin acts as a local morphogenetic trigger to specify lateral root founder cells. *Proc Natl Acad Sci* **105:** 8790–8794. doi:10.1073/pnas.0712307105

Escamez S, André D, Sztojka B, Bollhöner B, Hall H, Berthet B, Voß U, Lers A, Maizel A, Andersson M, et al. 2020. Cell death in cells overlying lateral root primordia facilitates organ growth in *Arabidopsis. Curr Biol* **30:** 455–464.e7. doi:10.1016/j.cub.2019.11.078

Fendrych M, Leung J, Friml J. 2016. TIR1/AFB-Aux/IAA auxin perception mediates rapid cell wall acidification and growth of *Arabidopsis* hypocotyls. *eLife* **5:** e19048. doi:10.7554/eLife.19048

Fernandez AI, Vangheluwe N, Xu K, Jourquin J, Claus LAN, Morales-Herrera S, Parizot B, De Gernier H, Yu Q, Drozdzecki A, et al. 2020. GOLVEN peptide signalling through RGI receptors and MPK6 restricts asymmetric cell division during lateral root initiation. *Nat Plants* **6:** 533–543. doi:10.1038/s41477-020-0645-z

Friml J, Wiśniewska J, Benková E, Mendgen K, Palme K. 2002. Lateral relocation of auxin efflux regulator PIN3 mediates tropism in *Arabidopsis*. *Nature* **415:** 806–809. doi:10.1038/415806a

Fukaki H, Tameda S, Masuda H, Tasaka M. 2002. Lateral root formation is blocked by a gain-of-function mutation in the *SOLITARY-ROOT/IAA14* gene of *Arabidopsis*. *Plant J* **29:** 153–168. doi:10.1046/j.0960-7412.2001.01201.x

Geldner N. 2003. Partial loss-of-function alleles reveal a role for GNOM in auxin transport-related, post-embryonic development of *Arabidopsis*. *Development* **131:** 389–400. doi:10.1242/dev.00926

Ghorbani S, Lin YC, Parizot B, Fernandez A, Njo MF, Van de Peer Y, Beeckman T, Hilson P. 2015. Expanding the repertoire of secretory peptides controlling root development with comparative genome analysis and functional assays. *EXBOTJ* **66:** 5257–5269. doi:10.1093/jxb/erv346

Gifford ML, Dean A, Gutierrez RA, Coruzzi GM, Birnbaum KD. 2008. Cell-specific nitrogen responses mediate developmental plasticity. *Proc Natl Acad Sci* **105:** 803–808. doi:10.1073/pnas.0709559105

Gonneau M, Desprez T, Martin M, Doblas VG, Bacete L, Miart F, Sormani R, Hématy K, Renou J, Landrein B, et al. 2018. Receptor kinase THESEUS1 is a rapid alkalinization factor 34 receptor in *Arabidopsis*. *Curr Biol* **28:** 2452–2458.e4. doi:10.1016/j.cub.2018.05.075

Grison MS, Kirk P, Brault ML, Wu XN, Schulze WX, Benitez-Alfonso Y, Immel F, Bayer EM. 2019. Plasma membrane-associated receptor-like kinases relocalize to plasmodesmata in response to osmotic stress. *Plant Physiol* **181:** 142–160. doi:10.1104/pp.19.00473

Gruber BD, Giehl RFH, Friedel S, von Wirén N. 2013. Plasticity of the *Arabidopsis* root system under nutrient deficiencies. *Plant Physiol* **163:** 161–179. doi:10.1104/pp.113.218453

Guan P. 2017. Dancing with hormones: a current perspective of nitrate signaling and regulation in *Arabidopsis*. *Front Plant Sci* **8:** 1697. doi:10.3389/fpls.2017.01697

Gutiérrez RA, Lejay LV, Dean A, Chiaromonte F, Shasha DE, Coruzzi GM. 2007. Qualitative network models and genome-wide expression data define carbon/nitrogen-responsive molecular machines in *Arabidopsis*. *Genome Biol* **8:** R7. doi:10.1186/gb-2007-8-1-r7

Haruta M, Tan LX, Bushey DB, Swanson SJ, Sussman MR. 2018. Environmental and genetic factors regulating localization of the plant plasma membrane H^+-ATPase. *Plant Physiol* **176:** 364–377. doi:10.1104/pp.17.01126

Hernández-Hernández V, Benítez M, Boudaoud A. 2019. Interplay between turgor pressure and plasmodesmata during plant development. *J Exp Bot* **71:** 768–777. doi:10.1093/jxb/erz434

Himanen K, Vuylsteke M, Vanneste S, Vercruysse S, Boucheron E, Alard P, Chriqui D, Van Montagu M, Inzé D, Beeckman T. 2004. Transcript profiling of early lateral root initiation. *Proc Natl Acad Sci* **101:** 5146–5151. doi:10.1073/pnas.0308702101

Hirota A, Kato T, Fukaki H, Aida M, Tasaka M. 2007. The auxin-regulated AP2/EREBP gene *PUCHI* is required for morphogenesis in the early lateral root primordium of *Arabidopsis*. *Plant Cell* **19:** 2156–2168. doi:10.1105/tpc.107.050674

Huang KL, Ma GJ, Zhang ML, Xiong H, Wu H, Zhao CZ, Liu CS, Jia HX, Chen L, Kjorven JO, et al. 2018. The ARF7 and ARF19 transcription factors positively regulate *PHOSPHATE STARVATION RESPONSE1* in *Arabidopsis* roots. *Plant Physiol* **178:** 413–427. doi:10.1104/pp.17.01713

Huang R, Zheng R, He J, Zhou Z, Wang J, Xiong Y, Xu T. 2019. Noncanonical auxin signaling regulates cell division pattern during lateral root development. *Proc Natl Acad Sci* **116:** 21285–21290. doi:10.1073/pnas.1910916116

Ingram G, Nawrath C. 2017. The roles of the cuticle in plant development: organ adhesions and beyond. *J Exp Bot* **68:** 5307–5321. doi:10.1093/jxb/erx313

Israeli A, Reed JW, Ori N. 2020. Genetic dissection of the auxin response network. *Nat Plants* **6:** 1082–1090. doi:10.1038/s41477-020-0739-7

Jia Z, von Wirén N. 2020. Signaling pathways underlying nitrogen-dependent changes in root system architecture: from model to crop species. *J Exp Bot* **71:** 4393–4404. doi:10.1093/jxb/eraa033

Jourquin J, Fukaki H, Beeckman T. 2020. Peptide-receptor signaling controls lateral root development. *Plant Physiol* **182:** 1645–1656. doi:10.1104/pp.19.01317

Kapulnik Y, Delaux PM, Resnick N, Mayzlish-Gati E, Wininger S, Bhattacharya C, Séjalon-Delmas N, Combier JP, Bécard G, Belausov E, et al. 2011. Strigolactones affect lateral root formation and root-hair elongation in *Arabidopsis*. *Planta* **233:** 209–216. doi:10.1007/s00425-010-1310-y

Kellermeier F, Chardon F, Amtmann A. 2013. Natural variation of *Arabidopsis* root architecture reveals complementing adaptive strategies to potassium starvation. *Plant Physiol* **161:** 1421–1432. doi:10.1104/pp.112.211144

Krouk G. 2016. Hormones and nitrate: a two-way connection. *Plant Mol Biol* **91:** 599–606. doi:10.1007/s11103-016-0463-x

Krouk G, Lacombe B, Bielach A, Perrine-Walker F, Malinska K, Mounier E, Hoyerova K, Tillard P, Leon S, Ljung K, et al. 2010. Nitrate-regulated auxin transport by NRT1.1 defines a mechanism for nutrient sensing in plants. *Dev Cell* **18:** 927–937. doi:10.1016/j.devcel.2010.05.008

Kumpf RP, Shi CL, Larrieu A, Sto IM, Butenko MA, Peret B, Riiser ES, Bennett MJ, Aalen RB. 2013. Floral organ abscission peptide IDA and its HAE/HSL2 receptors control cell separation during lateral root emergence. *Proc Natl Acad Sci* **110:** 5235–5240. doi:10.1073/pnas.1210835110

Laskowski MJ, Williams ME, Nusbaum HC, Sussex IM. 1995. Formation of lateral root meristems is a two-stage process. *Development* **121:** 3303.

Laskowski M, Grieneisen VA, Hofhuis H, ten Hove CA, Hogeweg P, Marée AFM, Scheres B. 2008. Root system architecture from coupling cell shape to auxin transport. *PLoS Biol* **6:** e307. doi:10.1371/journal.pbio.0060307

Lee HW, Cho C, Kim J. 2015. *Lateral organ boundaries Domain16* and *18* act downstream of the AUXIN1 and LIKE-AUXIN3 auxin influx carriers to control lateral root development in *Arabidopsis*. *Plant Physiol* **168:** 1792–1806. doi:10.1104/pp.15.00578

Lee HJ, Ha JH, Kim SG, Choi HK, Kim ZH, Han YJ, Kim JI, Oh Y, Fragoso V, Shin K, et al. 2016. Stem-piped light activates phytochrome B to trigger light responses in *Arabidopsis thaliana* roots. *Sci Signal* **9:** ra106. doi:10.1126/scisignal.aaf6530

Liu Y, Xie Y, Wang H, Ma X, Yao W, Wang H. 2017. Light and ethylene coordinately regulate the phosphate starvation response through transcriptional regulation of *PHOSPHATE STARVATION RESPONSE1*. *Plant Cell* **29:** 2269–2284. doi:10.1105/tpc.17.00268

Löfke C, Zwiewka M, Heilmann I, Van Montagu MCE, Teichmann T, Friml J. 2013. Asymmetric gibberellin signaling regulates vacuolar trafficking of PIN auxin transporters during root gravitropism. *Proc Natl Acad Sci* **110:** 3627–3632. doi:10.1073/pnas.1300107110

Lucas M, Godin C, Jay-Allemand C, Laplaze L. 2008. Auxin fluxes in the root apex co-regulate gravitropism and lateral root initiation. *J Exp Bot* **59:** 55–66. doi:10.1093/jxb/erm171

Lucas M, Kenobi K, von Wangenheim D, Voss U, Swarup K, De Smet I, Van Damme D, Lawrence T, Peret B, Moscardi E, et al. 2013. Lateral root morphogenesis is dependent on the mechanical properties of the overlaying tissues. *Proc Natl Acad Sci* **110:** 5229–5234. doi:10.1073/pnas.1210807110

Ma W, Li J, Qu B, He X, Zhao X, Li B, Fu X, Tong Y. 2014. Auxin biosynthetic gene *TAR2* is involved in low nitrogen-mediated reprogramming of root architecture in *Arabidopsis*. *Plant J* **78:** 70–79. doi:10.1111/tpj.12448

Maeda Y, Konishi M, Kiba T, Sakuraba Y, Sawaki N, Kurai T, Ueda Y, Sakakibara H, Yanagisawa S. 2018. A NIGT1-centred transcriptional cascade regulates nitrate signalling and incorporates phosphorus starvation signals in *Arabidopsis*. *Nat Commun* **9:** 1376. doi:10.1038/s41467-018-03832-6

Maghiaoui A, Bouguyon E, Cuesta C, Perrine-Walker F, Alcon C, Krouk G, Benkova E, Nacry P, Gojon A, Bach L. 2020. The *Arabidopsis* NRT1.1 transceptor coordinately controls auxin biosynthesis and transport to regulate root branching in response to nitrate. *J Exp Bot* **71:** 4480–4494. doi:10.1093/jxb/eraa242

Malamy JE. 2005. Intrinsic and environmental response pathways that regulate root system architecture. *Plant Cell Environ* **28:** 67–77. doi:10.1111/j.1365-3040.2005.01306.x

Malamy JE, Benfey PN. 1997. Organization and cell differentiation in lateral roots of *Arabidopsis thaliana*. *Development* **124:** 33–44.

Marchant A, Bhalerao R, Casimiro I, Eklöf J, Casero PJ, Bennett M, Sandberg G. 2002. AUX1 promotes lateral root formation by facilitating indole-3-acetic acid distribution between sink and source tissues in the *Arabidopsis* seedling. *Plant Cell* **14:** 589–597. doi:10.1105/tpc.010354

Marhavý P, Bielach A, Abas L, Abuzeineh A, Duclercq J, Tanaka H, Pařezová M, Petrášek J, Friml J, Kleine-Vehn J, et al. 2011. Cytokinin modulates endocytic trafficking of PIN1 auxin efflux carrier to control plant organogenesis. *Dev Cell* **21:** 796–804. doi:10.1016/j.devcel.2011.08.014

Marhavý P, Vanstraelen M, De Rybel B, Zhaojun D, Bennett MJ, Beeckman T, Benková E. 2013. Auxin reflux between the endodermis and pericycle promotes lateral root initiation. *EMBO J* **32:** 149–158. doi:10.1038/emboj.2012.303

Marhavý P, Duclercq J, Weller B, Feraru E, Bielach A, Offringa R, Friml J, Schwechheimer C, Murphy A, Benková E. 2014. Cytokinin controls polarity of PIN1-dependent auxin transport during lateral root organogenesis. *Curr Biol* **24:** 1031–1037. doi:10.1016/j.cub.2014.04.002

Marhavý P, Montesinos JC, Abuzeineh A, Van Damme D, Vermeer JEM, Duclercq J, Rakusová H, Nováková P, Friml J, Geldner N, et al. 2016. Targeted cell elimination reveals an auxin-guided biphasic mode of lateral root initiation. *Genes Dev* **30:** 471–483. doi:10.1101/gad.276964.115

Matzke MA, Mosher RA. 2014. RNA-directed DNA methylation: an epigenetic pathway of increasing complexity. *Nat Rev Genet* **15:** 394–408. doi:10.1038/nrg3683

Michniewicz M, Ho CH, Enders TA, Floro E, Damodaran S, Gunther LK, Powers SK, Frick EM, Topp CN, Frommer WB, et al. 2019. TRANSPORTER of IBA1 links auxin and cytokinin to influence root architecture. *Dev Cell* **50:** 599–609.e4. doi:10.1016/j.devcel.2019.06.010

Moreno-Risueno MA, Van Norman JM, Moreno A, Zhang J, Ahnert SE, Benfey PN. 2010. Oscillating gene expression determines competence for periodic *Arabidopsis* root branching. *Science* **329:** 1306–1311. doi:10.1126/science.1191937

Mravec J, Kubes M, Bielach A, Gaykova V, Petrasek J, Skupa P, Chand S, Benkova E, Zazimalova E, Friml J. 2008. Interaction of PIN and PGP transport mechanisms in auxin distribution-dependent development. *Development* **135:** 3345–3354. doi:10.1242/dev.021071

Mravec J, Skůpa P, Bailly A, Hoyerová K, Křeček P, Bielach A, Petrášek J, Zhang J, Gaykova V, Stierhof YD, et al. 2009. Subcellular homeostasis of phytohormone auxin is mediated by the ER-localized PIN5 transporter. *Nature* **459:** 1136–1140. doi:10.1038/nature08066

Murphy E, Vu LD, Van den Broeck L, Lin Z, Ramakrishna P, van de Cotte B, Gaudinier A, Goh T, Slane D, Beeckman T, et al. 2016. RALFL34 regulates formative cell divisions in *Arabidopsis* pericycle during lateral root initiation. *J Exp Bot* **67:** 4863–4875. doi:10.1093/jxb/erw281

Nakayama T, Shinohara H, Tanaka M, Baba K, Ogawa-Ohnishi M, Matsubayashi Y. 2017. A peptide hormone required for Casparian strip diffusion barrier formation in *Arabidopsis* roots. *Science* **355:** 284–286. doi:10.1126/science.aai9057

Naseer S, Lee Y, Lapierre C, Franke R, Nawrath C, Geldner N. 2012. Casparian strip diffusion barrier in *Arabidopsis* is made of a lignin polymer without suberin. *Proc Natl Acad Sci* **109:** 10101–10106. doi:10.1073/pnas.1205726109

Naumann C, Müller J, Sakhonwasee S, Wieghaus A, Hause G, Heisters M, Bürstenbinder K, Abel S. 2019. The local phosphate deficiency response activates endoplasmic reticulum stress-dependent autophagy. *Plant Physiol* **179:** 460–476. doi:10.1104/pp.18.01379

Nawrath C, Schreiber L, Franke RB, Geldner N, Reina-Pinto JJ, Kunst L. 2013. Apoplastic diffusion barriers in *Arabidopsis*. *The Arabidopsis Book* **11:** e0167. doi:10.1199/tab.0167

Naz L, Guo L, Chen F. 2019. Overexpression of nitrate transporter OsNRT2.1 enhances nitrate-dependent root elongation. *Genes (Basel)* **10:** 290. doi:10.3390/genes10040290

Norambuena L, Zouhar J, Hicks GR, Raikhel NV. 2008. Identification of cellular pathways affected by Sortin2, a synthetic compound that affects protein targeting to the vacuole in *Saccharomyces cerevisiae*. *BMC Chem Biol* **8:** 1. doi:10.1186/1472-6769-8-1

Ogura T, Goeschl C, Filiault D, Mirea M, Slovak R, Wolhrab B, Satbhai SB, Busch W. 2019. Root system depth in *Arabidopsis* is shaped by EXOCYST70A3 via the dynamic modulation of auxin transport. *Cell* **178:** 400–412.e16. doi:10.1016/j.cell.2019.06.021

Okuda S, Hothorn LA, Hothorn M. 2020. Crystal structures of *Arabidopsis* and *Physcomitrella* CR4 reveal the molecular architecture of CRINKLY4 receptor kinases. bioRxiv doi:10.1101/2020.08.10.245050

Okushima Y, Overvoorde PJ, Arima K, Alonso JM, Chan A, Chang C, Ecker JR, Hughes B, Lui A, Nguyen D, et al. 2005. Functional genomic analysis of the *AUXIN RESPONSE FACTOR* gene family members in *Arabidopsis thaliana*: unique and overlapping functions of *ARF7* and *ARF19*. *Plant Cell* **17:** 444–463. doi:10.1105/tpc.104.028316

Okushima Y, Fukaki H, Onoda M, Theologis A, Tasaka M. 2007. ARF7 and ARF19 regulate lateral root formation via direct activation of *LBD/ASL* genes in *Arabidopsis*. *Plant Cell* **19:** 118–130. doi:10.1105/tpc.106.047761

Orman-Ligeza B, Morris EC, Parizot B, Lavigne T, Babé A, Ligeza A, Klein S, Sturrock C, Xuan W, Novák O, et al. 2018. The xerobranching response represses lateral root formation when roots are not in contact with water. *Curr Biol* **28:** 3165–3173.e5. doi:10.1016/j.cub.2018.07.074

Orosa-Puente B, Leftley N, von Wangenheim D, Banda J, Srivastava AK, Hill K, Truskina J, Bhosale R, Morris E, Srivastava M, et al. 2018. Root branching toward water involves posttranslational modification of transcription factor ARF7. *Science* **362:** 1407–1410. doi:10.1126/science.aau3956

Osmont KS, Sibout R, Hardtke CS. 2007. Hidden branches: developments in root system architecture. *Annu Rev Plant Biol* **58:** 93–113. doi:10.1146/annurev.arplant.58.032806.104006

Parizot B, Laplaze L, Ricaud L, Boucheron-Dubuisson E, Bayle V, Bonke M, De Smet I, Poethig SR, Helariutta Y, Haseloff J, et al. 2008. Diarch symmetry of the vascular bundle in *Arabidopsis* root encompasses the pericycle and is reflected in distich lateral root initiation. *Plant Physiol* **146:** 140–148. doi:10.1104/pp.107.107870

Péret B, Middleton AM, French AP, Larrieu A, Bishopp A, Njo M, Wells DM, Porco S, Mellor N, Band LR, et al. 2013. Sequential induction of auxin efflux and influx carriers regulates lateral root emergence. *Mol Syst Biol* **9:** 699. doi:10.1038/msb.2013.43

Pérez-Henríquez P, Raikhel NV, Norambuena L. 2012. Endocytic trafficking towards the vacuole plays a key role in the auxin receptor SCFTIR-independent mechanism of lateral root formation in *A. thaliana*. *Mol Plant* **5:** 1195–1209. doi:10.1093/mp/sss066

Pérez-Torres CA, López-Bucio J, Cruz-Ramírez A, Ibarra-Laclette E, Dharmasiri S, Estelle M, Herrera-Estrella L. 2008. Phosphate availability alters lateral root development in *Arabidopsis* by modulating auxin sensitivity via a mechanism involving the TIR1 auxin receptor. *Plant Cell* **20:** 3258–3272. doi:10.1105/tpc.108.058719

Perotti MF, Ribone PA, Cabello JV, Ariel FD, Chan RL. 2019. AtHB23 participates in the gene regulatory network controlling root branching, and reveals differences between secondary and tertiary roots. *Plant J* **100:** 1224–1236. doi:10.1111/tpj.14511

Pires ND, Dolan L. 2012. Morphological evolution in land plants: new designs with old genes. *Phil Trans R Soc B* **367:** 508–518. doi:10.1098/rstb.2011.0252

Porco S, Larrieu A, Du Y, Gaudinier A, Goh T, Swarup K, Swarup R, Kuempers B, Bishopp A, Lavenus J, et al. 2016. Lateral root emergence in *Arabidopsis* is dependent on transcription factor LBD29 regulation of auxin influx carrier *LAX3*. *Development* **143:** 3340–3349. doi:10.1242/dev.136283

Ramakrishna P, Ruiz Duarte P, Rance GA, Schubert M, Vordermaier V, Vu LD, Murphy E, Vilches Barro A, Swarup K, Moirangthem K, et al. 2019. EXPANSIN A1-mediated radial swelling of pericycle cells positions anticlinal cell divisions during lateral root initiation. *Proc Natl Acad Sci* **116:** 8597–8602. doi:10.1073/pnas.1820882116

Raven JA, Edwards D. 2001. Roots: evolutionary origins and biogeochemical significance. *J Exp Bot* **52:** 381–401. doi:10.1093/jxb/52.suppl_1.381

Reed RC, Brady SR, Muday GK. 1998. Inhibition of auxin movement from the shoot into the root inhibits lateral root development in *Arabidopsis*. *Plant Physiol* **118:** 1369–1378. doi:10.1104/pp.118.4.1369

Remans T, Nacry P, Pervent M, Filleur S, Diatloff E, Mounier E, Tillard P, Forde BG, Gojon A. 2006. The *Arabidopsis* NRT1.1 transporter participates in the signaling pathway triggering root colonization of nitrate-rich patches. *Proc Natl Acad Sci* **103:** 19206–19211. doi:10.1073/pnas.0605275103

Richter GL, Monshausen GB, Krol A, Gilroy S. 2009. Mechanical stimuli modulate lateral root organogenesis. *Plant Physiol* **151:** 1855–1866. doi:10.1104/pp.109.142448

Rosquete MR, von Wangenheim D, Marhavý P, Barbez E, Stelzer EHK, Benková E, Maizel A, Kleine-Vehn J. 2013. An auxin transport mechanism restricts positive orthogravitropism in lateral roots. *Curr Biol* **23:** 817–822. doi:10.1016/j.cub.2013.03.064

Rubilar-Hernández C, Osorio-Navarro C, Cabello F, Norambuena L. 2019. PI4KIIIβ activity regulates lateral root formation driven by endocytic trafficking to the vacuole. *Plant Physiol* **181:** 112–126. doi:10.1104/pp.19.00695

Ruffel S. 2018. Nutrient-related long-distance signals: common players and possible cross-talk. *Plant Cell Physiol* **59:** 1723–1732. doi:10.1093/pcp/pcy152

Sager R, Wang X, Hill K, Yoo BC, Caplan J, Nedo A, Tran T, Bennett MJ, Lee JY. 2020. Auxin-dependent control of a plasmodesmal regulator creates a negative feedback loop modulating lateral root emergence. *Nat Commun* **11:** 364. doi:10.1038/s41467-019-14226-7

Sanz L, Dewitte W, Forzani C, Patell F, Nieuwland J, Wen B, Quelhas P, De Jager S, Titmus C, Campilho A, et al. 2011. The *Arabidopsis* D-type cyclin CYCD2;1 and the inhibitor ICK2/KRP2 modulate auxin-induced lateral root formation. *Plant Cell* **23:** 641–660. doi:10.1105/tpc.110.080002

Shahzad Z, Eaglesfield R, Carr C, Amtmann A. 2020. Cryptic variation in RNA-directed DNA-methylation controls lateral root development when auxin signalling is perturbed. *Nat Commun* **11:** 218. doi:10.1038/s41467-019-13927-3

Shukla V, Lombardi L, Iacopino S, Pencik A, Novak O, Perata P, Giuntoli B, Licausi F. 2019. Endogenous hypoxia in lateral root primordia controls root architecture by antagonizing auxin signaling in *Arabidopsis*. *Mol Plant* **12:** 538–551. doi:10.1016/j.molp.2019.01.007

Singh S, Yadav S, Singh A, Mahima M, Singh A, Gautam V, Sarkar AK. 2020. Auxin signaling modulates *LATERAL ROOT PRIMORDIUM 1* (*LRP 1*) expression during lateral root development in *Arabidopsis*. *Plant J* **101:** 87–100. doi:10.1111/tpj.14520

Stahl Y, Grabowski S, Bleckmann A, Kühnemuth R, Weidtkamp-Peters S, Pinto KG, Kirschner GK, Schmid JB, Wink RH, Hülsewede A, et al. 2013. Moderation of *Arabidopsis* root stemness by CLAVATA1 and ARABIDOPSIS CRINKLY4 receptor kinase complexes. *Curr Biol* **23:** 362–371. doi:10.1016/j.cub.2013.01.045

Sun J, Xu Y, Ye S, Jiang H, Chen Q, Liu F, Zhou W, Chen R, Li X, Tietz O, et al. 2009. *Arabidopsis ASA1* is important for jasmonate-mediated regulation of auxin biosynthesis and transport during lateral root formation. *Plant Cell* **21:** 1495–1511. doi:10.1105/tpc.108.064303

Sun H, Tao J, Liu S, Huang S, Chen S, Xie X, Yoneyama K, Zhang Y, Xu G. 2014. Strigolactones are involved in phosphate- and nitrate-deficiency-induced root development and auxin transport in rice. *J Exp Bot* **65:** 6735–6746. doi:10.1093/jxb/eru029

Sun CH, Yu JQ, Hu DG. 2017. Nitrate: a crucial signal during lateral roots development. *Front Plant Sci* **8:** 485.

Swarup K, Benková E, Swarup R, Casimiro I, Péret B, Yang Y, Parry G, Nielsen E, De Smet I, Vanneste S, et al. 2008. The auxin influx carrier LAX3 promotes lateral root emergence. *Nat Cell Biol* **10:** 946–954. doi:10.1038/ncb1754

Tan S, Abas M, Verstraeten I, Glanc M, Molnár G, Hajný J, Lasák P, Petřík I, Russinova E, Petrášek J, et al. 2020. Salicylic acid targets protein phosphatase 2A to attenuate growth in plants. *Curr Biol* **30:** 381–395.e8. doi:10.1016/j.cub.2019.11.058

Tian QY, Sun P, Zhang WH. 2009. Ethylene is involved in nitrate-dependent root growth and branching in *Arabidopsis thaliana*. *New Phytol* **184:** 918–931. doi:10.1111/j.1469-8137.2009.03004.x

Torres-Martínez HH, Hernández-Herrera P, Corkidi G, Dubrovsky JG. 2020. From one cell to many: morphogenetic field of lateral root founder cells in *Arabidopsis thaliana* is built by gradual recruitment. *Proc Natl Acad Sci* **117:** 20943–20949. doi:10.1073/pnas.2006387117

Torrey JG. 1950. The induction of lateral roots by indoleacetic acid and root decapitation. *Am J Bot* **37:** 257–264. doi:10.1002/j.1537-2197.1950.tb12192.x

Toyokura K, Goh T, Shinohara H, Shinoda A, Kondo Y, Okamoto Y, Uehara T, Fujimoto K, Okushima Y, Ikeyama Y, et al. 2019. Lateral inhibition by a peptide hormone-receptor cascade during *Arabidopsis* lateral root founder cell formation. *Dev Cell* **48:** 64–75.e5. doi:10.1016/j.devcel.2018.11.031

Trinh D-C, Lavenus J, Goh T, Boutté Y, Drogue Q, Vaissayre V, Tellier F, Lucas M, Voß U, Gantet P, et al. 2019. PUCHI regulates very long chain fatty acid biosynthesis during lateral root and callus formation. *Proc Natl Acad Sci* **116:** 14325–14330. doi:10.1073/pnas.1906300116

Ulmasov T, Murfett J, Hagen G, Guilfoyle TJ. 1997. Aux/IAA proteins repress expression of reporter genes containing natural and highly active synthetic auxin response elements. *Plant Cell* **9:** 1963–1971.

van den Berg T, ten Tusscher KH. 2018. Lateral root priming synergystically arises from root growth and auxin transport dynamics. bioRxiv doi:10.1101/361709

Vanneste S, De Rybel B, Beemster GTS, Ljung K, De Smet I, Van Isterdael G, Naudts M, Iida R, Gruissem W, Tasaka M, et al. 2005. Cell cycle progression in the pericycle is not sufficient for SOLITARY ROOT/IAA14-mediated lateral root initiation in *Arabidopsis thaliana*. *Plant Cell* **17:** 3035–3050. doi:10.1105/tpc.105.035493

Van Norman JM, Xuan W, Beeckman T, Benfey PN. 2013. To branch or not to branch: the role of pre-patterning in lateral root formation. *Development* **140:** 4301–4310. doi:10.1242/dev.090548

Vermeer JEM, von Wangenheim D, Barberon M, Lee Y, Stelzer EHK, Maizel A, Geldner N. 2014. A spatial accommodation by neighboring cells is required for organ initiation in *Arabidopsis*. *Science* **343:** 178–183. doi:10.1126/science.1245871

Vidal EA, Araus V, Lu C, Parry G, Green PJ, Coruzzi GM, Gutierrez RA. 2010. Nitrate-responsive miR393/AFB3 regulatory module controls root system architecture in *Arabidopsis thaliana*. *Proc Natl Acad Sci* **107:** 4477–4482. doi:10.1073/pnas.0909571107

Vidal EA, Alvarez JM, Araus V, Riveras E, Brooks M, Krouk G, Ruffel S, Lejay L, Crawford N, Coruzzi GM, et al. 2020. Nitrate 2020: thirty years from transport to signaling networks. *Plant Cell* **32:** 2094–2119. doi:10.1105/tpc.19.00748

Vilches Barro A, Stöckle D, Thellmann M, Ruiz-Duarte P, Bald L, Louveaux M, von Born P, Denninger P, Goh T, Fukaki H, et al. 2019. Cytoskeleton dynamics are necessary for early events of lateral root initiation in *Arabidopsis*. *Curr Biol* **29:** 2443–2454.e5. doi:10.1016/j.cub.2019.06.039

von Wangenheim D, Fangerau J, Schmitz A, Smith RS, Leitte H, Stelzer EHK, Maizel A. 2016. Rules and self-organizing properties of post-embryonic plant organ cell division patterns. *Curr Biol* **26:** 439–449. doi:10.1016/j.cub.2015.12.047

von Wangenheim D, Hauschild R, Friml J. 2017. Light sheet fluorescence microscopy of plant roots growing on the surface of a gel. *J Vis Exp* **119:** 55044.

von Wangenheim D, Banda J, Schmitz A, Boland J, Bishopp A, Maizel A, Stelzer EHK, Bennett M. 2020. Early developmental plasticity of lateral roots in response to asymmetric water availability. *Nat Plants* **6:** 73–77. doi:10.1038/s41477-019-0580-z

Waidmann S, Kleine-Vehn J. 2020. Asymmetric cytokinin signaling opposes gravitropism in roots. *J Integr Plant Biol* **62:** 882–886. doi:10.1111/jipb.12929

Waidmann S, Ruiz Rosquete M, Schöller M, Sarkel E, Lindner H, LaRue T, Petřík I, Dünser K, Martopawiro S, Sasidharan R, et al. 2019. Cytokinin functions as an asymmetric and anti-gravitropic signal in lateral roots. *Nat Commun* **10:** 3540. doi:10.1038/s41467-019-11483-4

Xiao TT, van Velzen R, Kulikova O, Franken C, Bisseling T. 2019. Lateral root formation involving cell division in both pericycle, cortex and endodermis is a common and ancestral trait in seed plants. *Development* **146:** dev182592. doi:10.1242/dev.182592

Xuan W, Audenaert D, Parizot B, Möller BK, Njo MF, De Rybel B, De Rop G, Van Isterdael G, Mähönen AP, Vanneste S, et al. 2015. Root cap-derived auxin pre-patterns the longitudinal axis of the *Arabidopsis* root. *Curr Biol* **25:** 1381–1388. doi:10.1016/j.cub.2015.03.046

Xuan W, Band LR, Kumpf RP, Van Damme D, Parizot B, De Rop G, Opdenacker D, Moller BK, Skorzinski N, Njo MF, et al. 2016. Cyclic programmed cell death stimulates hormone signaling and root development in *Arabidopsis*. *Science* **351:** 384–387. doi:10.1126/science.aad2776

Xuan W, De Gernier H, Beeckman T. 2020. The dynamic nature and regulation of the root clock. *Development* **147:** dev181446. doi:10.1242/dev.181446

Xun Q, Wu Y, Li H, Chang J, Ou Y, He K, Gou X, Tax FE, Li J. 2020. Two receptor-like protein kinases, MUSTACHES and MUSTACHES-LIKE, regulate lateral root development in *Arabidopsis thaliana*. *New Phytol* **227:** 1157–1173. doi:10.1111/nph.16599

Yamauchi T, Tanaka A, Inahashi H, Nishizawa NK, Tsutsumi N, Inukai Y, Nakazono M. 2019. Fine control of aerenchyma and lateral root development through AUX/IAA- and ARF-dependent auxin signaling. *Proc Natl Acad Sci* **116:** 20770–20775. doi:10.1073/pnas.1907181116

Yang Y, Zhang L, Chen P, Liang T, Li X, Liu H. 2020. UV-b photoreceptor UVR8 interacts with MYB73/MYB77 to regulate auxin responses and lateral root development. *EMBO J* **39:** e101928. doi:10.15252/embj.2019101928

Yuan T, Xu H, Li J, Lu Y. 2020. Auxin abolishes SHI-RELATED SEQUENCE 5-mediated inhibition of lateral root development in *Arabidopsis*. *New Phytol* **225:** 297–309. doi:10.1111/nph.16115

Zhou M, Palanca AMS, Law JA. 2018. Locus-specific control of the de novo DNA methylation pathway in *Arabidopsis* by the CLASSY family. *Nat Genet* **50:** 865–873. doi:10.1038/s41588-018-0115-y

Zhu Q, Gallemí M, Pospíšil J, Žádníková P, Strnad M, Benková E. 2019a. Root gravity response module guides differential growth determining both root bending and apical hook formation in *Arabidopsis*. *Development* **146:** dev175919. doi:10.1242/dev.175919

Zhu Q, Shao Y, Ge S, Zhang M, Zhang T, Hu X, Liu Y, Walker J, Zhang S, Xu J. 2019b. A MAPK cascade downstream of IDA–HAE/HSL2 ligand–receptor pair in lateral root emergence. *Nat Plants* **5:** 414–423. doi:10.1038/s41477-019-0396-x

Of Cells, Strands, and Networks: Auxin and the Patterned Formation of the Vascular System

Dhruv Lavania,[1] Nguyen Manh Linh,[1] and Enrico Scarpella

Department of Biological Sciences, University of Alberta, Edmonton, Alberta T6G 2E9, Canada

Correspondence: enrico.scarpella@ualberta.ca

Throughout plant development, vascular cells continually form from within a population of seemingly equivalent cells. Vascular cells connect end to end to form continuous strands, and vascular strands connect at both or either end to form networks of exquisite complexity and mesmerizing beauty. Here we argue that experimental evidence gained over the past few decades implicates the plant hormone auxin—its production, transport, perception, and response—in all the steps that lead to the patterned formation of the plant vascular system, from the formation of vascular cells to their connection into vascular networks. We emphasize the organizing principles of the cell- and tissue-patterning process, rather than its molecular subtleties. In the picture that emerges, cells compete for an auxin-dependent, cell-polarizing signal; positive feedback between cell polarization and cell-to-cell movement of the polarizing signal leads to gradual selection of cell files; and selected cell files differentiate into vascular strands that drain the polarizing signal from the neighboring cells. Although the logic of the patterning process has become increasingly clear, the molecular details remain blurry; the future challenge will be to bring them into razor-sharp focus.

AUXIN AND THE PLANT VASCULAR SYSTEM

In most multicellular organisms, water, nutrients, and signals are transported by tissue networks. In animals, this essential transport function is distributed over separate tissue networks—for example, the nervous, circulatory, and respiratory systems. By contrast, in plants, all transport functions are performed by the only tissue network: the vascular system.

The plant vascular system is a network of vascular strands that connect the different parts of an organ and the different organs of a plant (Fig. 1; Esau 1965). Vascular strands are bundles of files of vascular cells that are elongated along the file and connected at their ends (Fig. 1). Vascular strands are named differently in different organs: veins in flat organs like leaves, petals, sepals, and cotyledons; vascular bundles in the stem; and vascular cylinder or stele in the root.

The relation between the plant hormone auxin and vascular development is old (e.g., Kraus et al. 1936) and varied (e.g., Snow 1935; Camus 1949; Jacobs 1952; Wangermann 1967). Here we discuss evidence that suggests that auxin controls the patterned formation of the vascular system at all the system's organization levels: the cells', the strands', and the network's.

[1]These authors contributed equally to this work.

Cite this article as *Cold Spring Harb Perspect Biol* doi: 10.1101/cshperspect.a039958

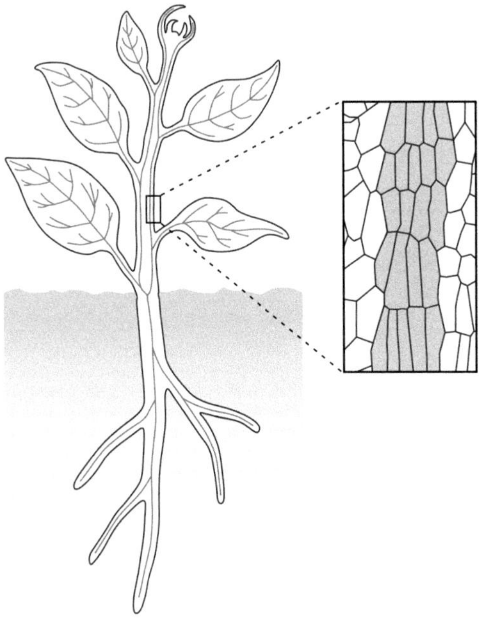

Figure 1. The plant vascular system. (*Left*) The plant vascular system is a network of vascular strands (gray lines) that connect the different parts of an organ and the different organs of a plant. (*Right*) Vascular strands are bundles of files of vascular cells (gray fill) that are elongated along the file and connected at their ends.

We will emphasize the "syntax" of such control, rather than its "semantics." This latter has been reviewed recently and comprehensively elsewhere (e.g., Jouannet et al. 2015; De Rybel et al. 2016; Etchells et al. 2016; Cho et al. 2017; Ramachandran et al. 2017; Ruonala et al. 2017; Anne and Hardtke 2018; Fischer et al. 2019; Fukuda and Ohashi-Ito 2019).

AUXIN SIGNALING AND THE FORMATION OF THE FIRST VASCULAR CELLS

Much of a seedling can be seen as a cylinder with a vascular strand in its center (Fig. 2A). The formation of this cylinder coincides with the formation of the first vascular cells in the *Arabidopsis* globular embryo (Fig. 2A; Scheres et al. 1994; De Rybel et al. 2014; Yoshida et al. 2014; Gooh et al. 2015). However, expression of vascular-specific markers suggests that the identity of these first vascular cells had been specified earlier, at the dermatogen stage (Fig. 2B; Smit et al. 2020).

The *Arabidopsis* dermatogen embryo is composed of 16 cells: eight outer cells, which are the precursors of the epidermis, and eight inner cells, which are the precursors of all the other tissue types (Fig. 2B; Mansfield and Briarty 1991; De Rybel et al. 2014; Yoshida et al. 2014; Gooh et al. 2015). These eight inner cells will divide periclinally and asymmetrically, and the resulting four larger, innermost cells in the basal half of the globular embryo will become the first vascular cells (Fig. 2B; Esau 1965).

Available evidence suggests that the formation of the seedling cylinder and the vascular strand in its center depend on auxin signaling. Indeed, dermatogen embryos of auxin signaling mutants express vascular-specific markers abnormally—if at all—and the eight inner cells in these mutant embryos fail to divide periclinally and asymmetrically and to form the first vascular cells in globular embryos (Fig. 2C; Berleth and Jurgens 1993; Hardtke and Berleth 1998; Hamann et al. 1999, 2002; Hobbie et al. 2000; Hellmann et al. 2003; Dharmasiri et al. 2005; Yoshida et al. 2014; Smit et al. 2020). Most of these mutant embryos develop into seedlings in which the vascularized cylinder is replaced by a vascularless cone (Fig. 2C). This defect characterizes mutants in auxin perception or response but also mutants in auxin production and embryos developed in the presence of auxin antagonists (Hadfi et al. 1998; Dharmasiri et al. 2003, 2007; Cheng et al. 2007; Stepanova et al. 2008; Thomas et al. 2009).

Among such mutants, defects are most severe in mutants lacking the function of the *MONOPTEROS/AUXIN RESPONSE FACTOR5* (*MP* hereafter) gene of *Arabidopsis*, which encodes a transcription factor that regulates auxin-responsive gene expression (Berleth and Jurgens 1993; Ulmasov et al. 1997, 1999; Hardtke and Berleth 1998; Mattsson et al. 2003), and in mutants with a stabilized variant of the otherwise short-lived MP-inhibitor INDOLE-3-ACETIC ACID12/BODENLOS (Hamann et al. 1999, 2002; Hardtke et al. 2004; Dharmasiri et al. 2005; Weijers et al. 2005, 2006; Schlereth et al.

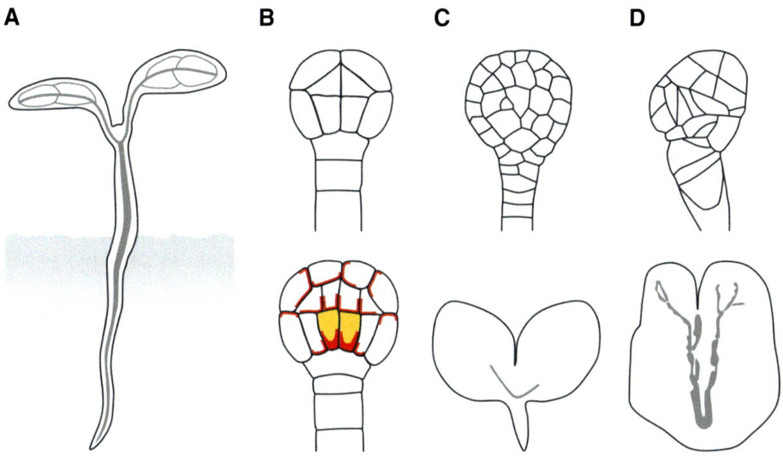

Figure 2. Auxin signaling and the formation of the first vascular cells. (*A*) Much of the seedling is a cylinder with a vascular strand (gray line) in its center. (*B*) The localization of PIN1 (red) is polarized in the first vascular cells (yellow fill) of the globular embryo (*bottom*), which originate from the periclinal, asymmetric division of the lower inner cells in the dermatogen embryo (*top*). (*C*) The lower inner cells in the dermatogen embryos of auxin signaling mutants fail to divide periclinally and asymmetrically, leading to vascularless globular embryos (*top*) and seedlings in which the vascularized cylinder is replaced by a vascularless cone (*bottom*). (*D*) Auxin-signaling-unrelated mutants whose cells divide along random planes (*top*) form vascular strands nonetheless (*bottom*).

2010; Lau et al. 2011; Garrett et al. 2012; Krogan et al. 2012). Similar defects also characterize mutants in the *EMBRYO DEFECTIVE30/GNOM* (*GN* hereafter) gene of *Arabidopsis* (Mayer et al. 1993; Shevell et al. 1994; Busch et al. 1996; Yoshida et al. 2014), which encodes a regulator of membrane trafficking that controls auxin signaling through unknown pathways (Mayer et al. 1993; Wolters et al. 2011; Verna et al. 2019).

That these mutants fail to form the first vascular cells seems to result from the inability of the lower inner cells of the dermatogen embryo to divide periclinally and asymmetrically, suggesting that auxin signaling promotes such asymmetric cell division (Hamann et al. 1999; Yoshida et al. 2014). However, vascular cells—although abnormally shaped—still form in mutant embryos in which cells divide along random planes (Fig. 2D; Torres-Ruiz and Jurgens 1994; Yoshida et al. 2014)—random planes that presumably include those which in auxin signaling mutants are correlated with failure to form vascular cells. Furthermore, auxin application to various tissues induces the formation of vascular strands even in the absence of cell division (Fig. 3A; see, e.g., Sinnott and Bloch 1944, 1945; Sachs 1969). Therefore, the vascular-differentiation-promoting influence of auxin does not seem to necessarily depend on its cell-division-orienting activity.

AUXIN TRANSPORT AND VASCULAR STRAND FORMATION

The functional unit of the vascular system is the vascular strand, whose formation seems to depend on auxin transport. This conclusion is suggested by the result of experiments in which auxin is applied to mature plant tissues: Auxin application leads to the formation of continuous files of vascular cells that connect the applied auxin to the preexisting vascular strands basal to the auxin application site (Fig. 3A; Kraus et al. 1936; Jost 1942; Jacobs 1952; Sachs 1968; Linh et al. 2018), and this polar formation of vascular strands in response to auxin application is prevented by inhibitors of polar auxin transport (Roberts 1960; Thompson and Jacobs 1966), suggesting that it depends

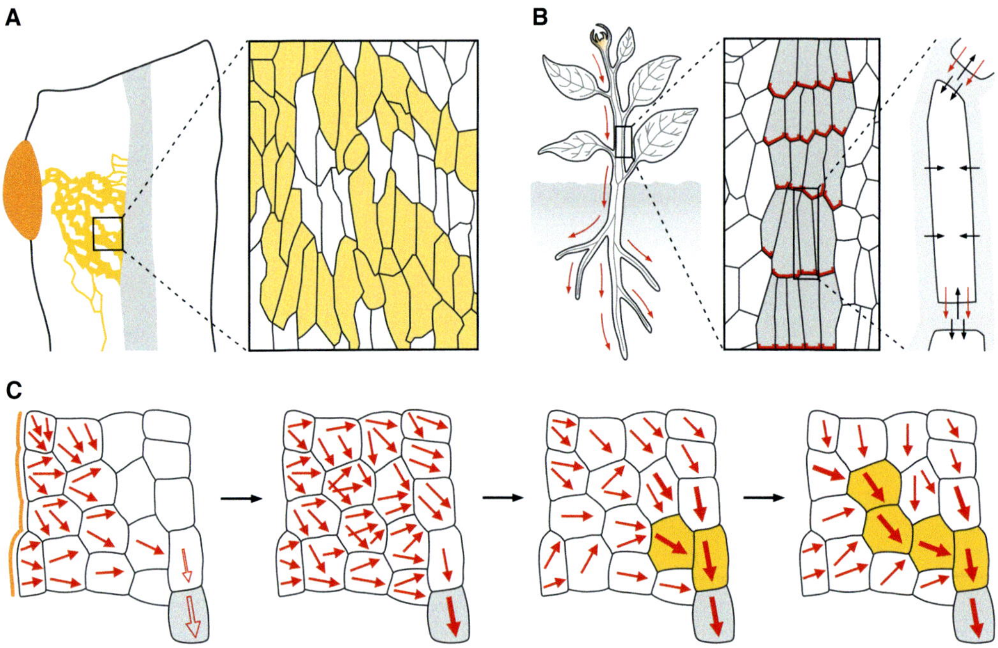

Figure 3. Auxin transport and vascular strand formation. (*A*) (*Left*) The application of auxin (orange) to mature plant tissues induces the differentiation of vascular cells in continuous files to form vascular strands (yellow lines) that connect the applied auxin to the preexisting vascular strands (gray fill) basal to the auxin application site. (*Right*) In the vascular strands formed in response to auxin application, vascular cells (yellow fill) are not aligned along the file like in Figure 1, but along the axis of the tissue. (*B*) (*Left*) Auxin (orange) is mainly produced in immature shoot organs and transported (red arrows) toward the roots through vascular strands (gray lines). (*Middle*) Apicobasal, polar auxin transport results from the localization of auxin efflux carriers of the PIN family (red) at the basal plasma membrane of vascular cells (gray fill). (*Right*) Auxin efflux carriers are required for auxin to leave the cell (red arrows) because auxin is mostly negatively charged inside the cell; by contrast, in the extracellular space, auxin is to a larger extent uncharged and can thus diffuse into the cell (black arrows). (*C*) The auxin canalization hypothesis assumes that auxin diffuses from the auxin application site (orange line) toward the preexisting vascular strands in the organ (gray fill). The positive feedback between auxin movement (red arrows) and localization of auxin efflux carriers to the site where auxin leaves the cell would gradually polarize auxin transport (thicker arrows). This would occur first in the cells connected to the preexisting vasculature (gray fill), which are still polarized along the original, apicobasal polarity of the tissue (empty red arrows) and thus orient auxin movement toward themselves. Increased auxin transport polarity, capacity, or velocity in the selected cells would lead to vascular differentiation (yellow fill) and drain auxin from neighboring cells, thus inhibiting their differentiation. The process would continue until a vascular strand formed that connected the applied auxin to the preexisting vascular strands basal to the auxin application site. Black arrows connect successive stages of auxin-induced vascular-strand formation.

on the ability of the responding tissue to transport auxin polarly.

Indeed, although auxin is mainly produced in immature shoot organs such as leaf and flower primordia (Thimann and Skoog 1934; Avery 1935), it is transported toward the roots through vascular strands (Fig. 3B; Went 1928; Wangermann 1974). This apicobasal, polar auxin transport is thought to result from the localization of auxin efflux carriers at the basal end of auxin-transporting cells (Fig. 3B; Rubery and Sheldrake 1974; Raven 1975). Indeed, the weak acid auxin is mostly negatively charged inside the cell and can thus efficiently leave the cell only through plasma-membrane-localized auxin efflux carriers (Fig. 3B). Although the mechanism of action is still unclear (for review, see Barbosa et al. 2018), available evidence suggests that these auxin efflux carriers are encoded by *PIN-FORMED* (*PIN*) genes (for review, see Adamowski and Friml 2015). Computational simulations of this model suggest that it can account for both polar auxin transport (Mitchison 1980b) and the polar formation of vascular strands in response to auxin application, provided that auxin movement through a cell positively feeds back on the localization of auxin efflux carriers to the site where auxin leaves the cell, as proposed by the auxin canalization hypothesis (Sachs 1981, 1991, 2000).

The auxin canalization hypothesis assumes that the applied auxin—or an auxin-dependent signal that is experimentally indistinguishable from auxin—initially diffuses from the auxin application site toward the preexisting vascular strands in the organ (Fig. 3C; Sachs 1981, 1991, 2000). In the cells between the application site and the preexisting vascular strands, auxin efflux carriers would initially be localized to all the different sections of the plasma membrane (i.e., nonpolarly). By contrast, in the preexisting vascular strands, auxin efflux carriers would be localized along the original apicobasal, auxin transport polarity of the tissue. Therefore, the preexisting vascular strands would efficiently remove any auxin that reached them by diffusion from the auxin application site, thereby acting as auxin sinks that orient toward themselves auxin movement in the neighboring cells and polarize the localization of auxin efflux carriers in those same cells (Fig. 3C). The resulting induction of polar auxin transport in the cells neighboring the preexisting vascular strands would be gradually enhanced by the positive feedback between auxin movement and localization of auxin efflux carriers proposed by the auxin canalization hypothesis. By draining auxin increasingly more efficiently and polarly, the cells neighboring the preexisting vascular strands would in turn induce polar localization of auxin efflux carriers and polar auxin transport in the cells above them and inhibit the same processes in the lateral neighbors (Fig. 3C). The repetition of these steps would eventually lead to preferential auxin transport through single cell files—the "canals" the hypothesis refers to—that would connect the applied auxin to the preexisting vascular strands basal to the auxin application site and that would differentiate into vascular strands (Fig. 3C). During this process, any random polarization in the localization of auxin efflux carriers would be stabilized by the positive feedback between auxin movement and localization of auxin efflux carriers proposed by the auxin canalization hypothesis and lead to random deviations in the course of the selected cell files and from the shortest route for auxin transport.

The plasma membrane localization of PIN1 auxin efflux carriers of *Arabidopsis* marks the presumed auxin efflux side of cells (Petrasek et al. 2006; Wisniewska et al. 2006). Therefore, the polarity of auxin transport can be inferred from the localization of PIN1 to the plasma membrane. Auxin application to mature plant tissues induces the formation of broad expression domains of nonpolarly localized PIN1 that connect the applied auxin to the preexisting vascular strands (Sauer et al. 2006; Mazur et al. 2016). Over time, the broad PIN1-expression domains (PEDs hereafter) become restricted to sites of auxin-induced vascular strand formation, and PIN1 localization becomes polarized: In the cells along the PEDs' midline, PIN1 localization becomes polarized toward the preexisting vascular strands basal to the site of auxin application; in the cells flanking the PEDs' midline, PIN1 localization becomes polarized toward the domains' midline. Eventually, narrow PEDs mark the sites of auxin-induced vascular strand formation, and polar PIN1 localization in the vascular strands formed in response to auxin application suggests auxin transport away from the applied auxin and toward the preexisting vascular strands basal to the auxin application site. Both the restriction of broad PEDs and the polarization of PIN1 localization during auxin-induced vascular strand formation initiate and proceed away from the preexisting vascular strands and are consistent with predictions of the auxin canalization hypothesis.

Consistent with predictions of the auxin canalization hypothesis is also the observation that auxin application fails to induce vascular strand formation in *pin1* mutants or plants treated with auxin transport inhibitors (Mazur et al. 2020). This observation is, however, unexpected because auxin application to shoot tips of *pin1* mutants or wild-type plants grown in the presence of auxin transport inhibitors leads to the formation of whole leaves, including their veins (Reinhardt et al. 2000, 2003). This finding is not the only apparent inconsistency between assumptions or predictions of the auxin canalization hypothesis and experimental observations (Bennett et al. 2014; Runions et al. 2014; Ravichandran et al. 2020). For example, the auxin canalization hypothesis assumes that auxin readily diffuses out of the cells (Sachs 1981), but auxin diffusion out of the cell is unfavored over diffusion into the cell by nearly two orders of magnitude (Runions et al. 2014).

The auxin canalization hypothesis also assumes that cells can measure net auxin transport across their plasma membranes (Sachs 1969). No biological mechanism is known that allows cells to measure net auxin transport directly (Mitchison 1980a, 1981; Kramer 2009; Cieslak et al. 2015). However, cells could use the concentration of auxin, of auxin-bound auxin carriers, or of substances produced or consumed during auxin transport to measure auxin influx and efflux separately, which would indirectly provide a measure of net auxin transport (Mitchison 1980a; Coen et al. 2004; Cieslak et al. 2015). Alternatively, cells could measure intracellular auxin gradients, which depend on auxin transport: Auxin concentration would be higher where auxin enters the cell and lower near auxin efflux carriers, where auxin leaves the cell (Mitchison 1981; Kramer 2009). Even though these mechanisms await experimental testing, they are both plausible; instead, that auxin controls polarization of PIN1 localization by inhibiting PIN1 deallocation from the section of the plasma membrane from which auxin leaves the cell can be ruled out (Paponov et al. 2019).

Finally, the auxin canalization hypothesis assumes that developing vascular strands effortlessly connect to preexisting vascular strands, an assumption that seems to be justified by experimental observations (Sachs 1968) but that cannot be easily reproduced by computational simulations (Bayer et al. 2009; Smith and Bayer 2009). This limitation is overcome in plants that have two PIN1 proteins: one that broadly polarizes auxin transport toward preexisting vascular strands, and the other that restricts the broad auxin transport domains to narrow sites of vascular strand formation (O'Connor et al. 2014, 2017). In plants like *Arabidopsis* that have only one PIN1 protein, instead, a hypothetical substance has been proposed to diffuse from preexisting vascular strands and polarize PIN1 localization in developing vascular strands toward preexisting vascular strands (Bayer et al. 2009; Smith and Bayer 2009).

Despite the apparent limitations, vascular strand formation in response to auxin application seems to be an expression of auxin transport whose essence is captured by the auxin canalization hypothesis; can we say the same of the vascular strands that form during normal development?

AUXIN TRANSPORT AND VEIN FORMATION

Just like auxin application to mature plant tissues, auxin application to developing leaves leads to the formation of continuous files of vascular cells that connect the applied auxin to the preexisting veins basal to the auxin application site (Scarpella et al. 2006; Sawchuk et al. 2007; Verna et al. 2019). During both auxin-induced and normally occurring vein formation, PIN1 expression is initiated in broad domains of leaf inner cells connected to preexisting veins (Fig. 4A; Scarpella et al. 2006; Wenzel et al. 2007; Marcos and Berleth 2014). In the cells of those broad PEDs, PIN1 is initially nonpolarly localized (Carraro et al. 2006; Scarpella et al. 2006; Wenzel et al. 2007; Lee et al. 2009; Marcos and Berleth 2014; Johnston et al. 2015). Over time, however, the broad PEDs become restricted to sites of vein formation, and PIN1 localization becomes polarized: In the cells along the PEDs' midline, PIN1 localization becomes polarized toward preexisting veins; in the cells flanking the PEDs' midline, PIN1 localization becomes polarized toward the domains' midline (Fig. 4A). Both the restriction of broad PEDs and the polarization of PIN1 localization initiate and proceed away from preexisting veins and are delayed by auxin transport inhibition. And both auxin transport inhibition and higher aux-

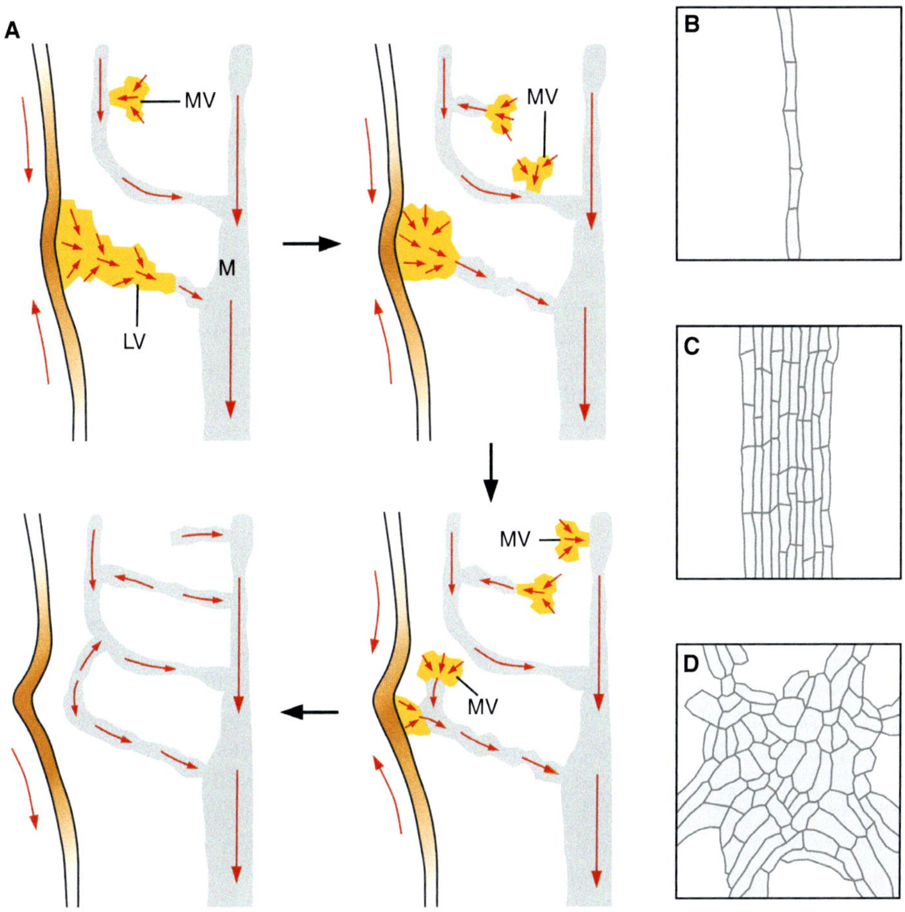

Figure 4. Auxin and vein formation. (*A*) In rounded leaves, PIN1 localization in epidermal cells of the leaf margin (red arrows) becomes polarized toward sites of leaf lateral growth (orange). Epidermal convergence points of PIN1 polarity are associated with sites of leaf lateral growth and broad PEDs in the inner tissue that become restricted to sites of lateral vein (LV) formation (yellow). Over time, PIN1 localization becomes polarized (red arrows) in the cells of these broad domains: In the cells along the PEDs' midline, PIN1 localization becomes polarized toward preexisting veins (gray); in the cells flanking PEDs' midline, PIN1 localization becomes polarized toward the domains' midline. Both the restriction of broad PEDs and the polarization of PIN1 localization initiate and proceed away from preexisting veins (gray), in which PIN1 localization is polarized (red arrows). Broad PEDs in the inner tissue that become restricted to sites of minor vein (MV) formation (yellow) branch from preexisting veins (gray). Over time, PIN1 localization becomes polarized (red arrows) in the cells of these broad domains: In the cells along the PEDs' midline, PIN1 localization becomes polarized toward preexisting veins (gray); in the cells flanking PEDs' midline, PIN1 localization becomes polarized toward the domains' midline. Both the restriction of broad PEDs and the polarization of PIN1 localization initiate and proceed away from preexisting veins, in which PIN1 localization is polarized (red arrows). Broad MV-associated PEDs can gradually disappear instead of becoming restricted and polarized. Initially, MV-associated PEDs connect to preexisting veins at one end only ("open" PEDs), but they can extend to connect to preexisting veins at both ends ("closed" PEDs). Open PEDs have a single polarity; closed PEDs have two opposite polarities, which are connected by a "bipolar" cell: a cell in which PIN1 localization is polarized at both ends (see also Fig. 6F). Each vein loop forms from the fusion of LV-associated PEDs and MV-associated closed PEDs. Black arrows connect successive stages of leaf development. (*B–D*) In wild-type (*B*) and auxin-transport-inhibited (*C*) leaves, vascular cells are connected end-to-end and aligned along the vein. By contrast, in leaves in which both auxin transport and auxin signaling are inhibited (*D*), veins are replaced by clusters of vascular cells that are randomly oriented.

in levels lead to the formation of broader PEDs, but given time even these broader PEDs eventually become restricted to sites of vein formation (Aloni et al. 2003; Mattsson et al. 2003; Hay et al. 2006; Scarpella et al. 2006; Wenzel et al. 2007).

Many of these observations can be accounted for by the positive feedback between auxin movement and localization of auxin efflux carriers proposed by the auxin canalization hypothesis (Mitchison 1980a, 1981; Sachs 1981, 1991, 2000; Rolland-Lagan and Prusinkiewicz 2005). But if the auxin canalization hypothesis were truly to account for vein formation in response to auxin, leaves of *pin* mutant plants or wild-type plants grown in the presence of auxin transport inhibitors or lacking PIN proteins should form no veins (Rolland-Lagan and Prusinkiewicz 2005). However, this prediction seems to be unsupported by experimental evidence: Leaves of wild-type plants grown in the presence of auxin transport inhibitors do form veins; and as in normally grown plants, the veins of auxin-transport-inhibited plants are oriented along the apicobasal axis of the leaf, and their vascular cells are elongated along the vein and connected at their ends (Fig. 4B,C; Sieburth 1999; Verna et al. 2019). Veins with these same features also form in leaves of *Arabidopsis* plants lacking all the PIN proteins with vascular function (Verna et al. 2019). Therefore, that auxin-induced vascular-strand formation depends on auxin transport (i.e., the core prediction of the auxin canalization hypothesis) seems to be unsupported by experimental evidence. Moreover, unlike auxin application to mature plant tissues —but like auxin application to shoot apices— *pin* mutant leaves can still respond to auxin application by forming veins that connect to preexisting veins basal to the auxin application site (Verna et al. 2019). This observation seems to suggest that vein formation in *pin* mutants still depends on auxin, but how?

AUXIN SIGNALING AND VEIN FORMATION

The residual vein-formation activity in auxin-transport-inhibited leaves turns out to depend, at least in part, on auxin signaling. That auxin signaling controls vein formation has long been known. Indeed, in auxin-signaling-inhibited leaves, fewer veins form, and veins are often incompletely differentiated; yet in those veins, vascular cells are still elongated along the vein and connected at their ends (Przemeck et al. 1996; Hardtke and Berleth 1998; Candela et al. 1999; Alonso-Peral et al. 2006; Strader et al. 2008; Esteve-Bruna et al. 2013; Verna et al. 2019; Mazur et al. 2020). Instead, inhibition of auxin signaling, either because of growth in the presence of auxin signaling inhibitors or because of mutation in auxin receptors or their regulators, leads to entirely new vein defects in auxin-transport-inhibited leaves (Verna et al. 2019).

In the leaves of plants in which both auxin transport and signaling are inhibited, the end-to-end alignment of vascular cells oriented along the vein is replaced by the formation of clusters of randomly oriented vascular cells (Fig. 4D; Verna et al. 2019). Although hypotheses have been proposed (for a recent review, see Ravichandran et al. 2020), it is currently unclear how auxin signaling, which takes place in the nucleus and is thus inherently nonpolar (for review, see Leyser 2018), would control, in the absence of polar auxin transport, the polar propagation of the auxin signal.

AUXIN AND THE FORMATION OF CONTINUOUS VASCULAR STRANDS

If vascular strand formation depended on the cell-to-cell transport of an auxin-dependent signal, vascular strands would always be "continuous" (i.e., they would connect to other vascular strands at least at one of their two ends). Yet vascular strands that fail to satisfy this requirement—vascular "fragments"—have been observed in leaves of wild types and mutants (e.g., Pray 1955a,b; Lersten 1965; Herbst 1971; Berleth and Jurgens 1993; Carland et al. 1999; Deyholos et al. 2000; Koizumi et al. 2000; Steynen and Schultz 2003; Sawa et al. 2005). Such vascular fragments are of two types.

Vascular fragments of the first type, including those observed in auxin signaling mutants, are composed of files of mature vascular cells that are connected by files of immature vascular cells (Fig. 5A; Pray 1955a,b; Lersten 1965; Herbst

1972; Przemeck et al. 1996; Hardtke and Berleth 1998; Mähönen et al. 2006; Scacchi et al. 2010; Truernit et al. 2012; Ruiz Sola et al. 2017). Because the identification of immature vascular cells can be problematic (Esau 1943), vascular strands of this type have often been interpreted as fragmented when they are instead incompletely differentiated.

Vascular fragments of the second type are composed of files of vascular cells that are separated by nonvascular cells (Fig. 5B; Herbst 1972; Carland et al. 1999; Deyholos et al. 2000; Koizumi et al. 2000). Vascular fragments of this type originate from continuous PEDs within which some cells terminate PIN1 expression and differentiate into nonvascular cells (Scarpella et al. 2006; Naramoto et al. 2009). Therefore, both types of vascular fragments are continuous, at least at formative stages, and thus compatible with a vein-formation mechanism that depends on the cell-to-cell transport of an auxin-dependent signal.

Vascular fragments of the first type have been observed in the leaf and the seedling cylinder (Pray 1955a,b; Lersten 1965; Herbst 1972; Przemeck et al. 1996; Mähönen et al. 2006; Scacchi et al. 2010; Truernit et al. 2012; Rodriguez-Villalon et al. 2014, 2015; Ruiz Sola et al. 2017; Marhava et al. 2018). However, no vascular fragments of the second type have ever been observed in the seedling cylinder, even in those mutants with such vascular fragments in their leaves. This observation suggests that the function of the genes that control the formation of continuous veins in the leaf is not required for the continuity of the vascular strand in the seedling cylinder. But how could that be?

PIN1 localization is already polarized in the first vascular cells that form in the globular embryo (Steinmann et al. 1999). These first vascular cells are stem cells, and as such they continually divide into cells with unequal fates: One cell will maintain the stem cell population; the other will extend the vascular strand in the cylinder of the embryo during embryogenesis and of the seedling root upon germination (Scheres et al. 1994; van den Berg et al. 1997; Aida et al. 2004; Yoshida et al. 2014). Unlike in the embryo, in the leaf PIN1 localization is initially nonpolar, and PIN1-expressing cells do not behave like vascular stem cells. In developing leaves of both wild type and mutants with vascular fragments of the second type, PEDs are initially continuous. In wild type, over time, PIN1 localization becomes polarized (Scarpella et al. 2006; Wenzel et al. 2007; Marcos and Berleth 2014). By contrast, in mutants with vascular fragments of the second type, PIN1 expression is terminated in some of the cells in a PED before PIN1 localization becomes polarized in any of the cells in the domain (Scarpella et al. 2006; Naramoto et al. 2009; Verna et al. 2019). There-

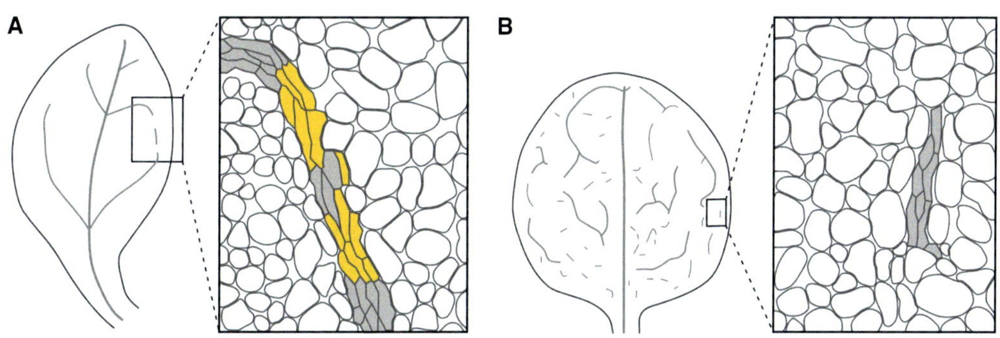

Figure 5. Auxin and the formation of continuous veins. (*A*) Vein fragments of the first type are composed of files of mature vascular cells (gray fill) that are connected by files of immature vascular cells (yellow fill). (*B*) Vein fragments of the second type are composed of files of vascular cells (gray fill) that are separated by nonvascular cells (white fill). Vein fragments of this type originate from continuous PEDs within which some cells terminate PIN1 expression and differentiate into nonvascular cells.

fore, it is possible that the function of the genes that control the formation of continuous veins in the leaf is required for the polarization of PIN1 localization, and that it is the lack of such polarization that leads to PED fragmentation. If so, the function of the genes that control the continuity of the veins in the leaf would not be required during the extension of the vascular strand in the embryo and seedling cylinder because polar PIN1 localization would need to be maintained and propagated during such extension and not established like in leaf vein formation. In support of this interpretation, available evidence suggests that distinct mechanisms control the polarization of PIN1 localization and the maintenance of such polar localization (Kleine-Vehn et al. 2011; Łangowski et al. 2016).

AUXIN AND VASCULAR NETWORK FORMATION

Leaf Vein Networks

Just like a seedling can be seen as a vascularized cylinder, early, leafless plants can be seen as two systems of branched vascularized cylinders: one above ground and one below ground (Fig. 6A; Fairon-Demaret and Li 1993). And even flat organs such as leaves can be seen, at least at early stages of their development, as vascularized cylinders (Mattsson et al. 1999; Kang and Dengler 2004; Scarpella et al. 2004). However, during their development, flat-organ primordia soon lose their cylindrical shape by expanding laterally to acquire their distinctive flattened shape, a process that coincides with the formation of branched systems of veins that largely mirror the shape of the leaf (von Ettinghausen 1861; Ash et al. 1999; Dengler and Kang 2001). These vein networks are said to be "open" if all the veins connect to other veins at only one end or "closed" if at least some veins connect to other veins at both ends (Fig. 6B; Roth-Nebelsick et al. 2001; Verna et al. 2015).

In rounded leaves like those of *Arabidopsis*, lateral veins branch from a central midvein and connect to distal veins to form vein loops; minor veins branch from midvein and loops, and either end freely or connect to other veins to form a mesh; and loops and minor veins bend near the leaf edge to give the vein network a scalloped outline (Fig. 6C; Troll 1937; Gifford and Foster 1989; Nelson and Dengler 1997). In elongated leaves of grasses like maize, vein loops are compressed laterally and stretched along the leaf, such that midvein and lateral veins seem to run parallel to one another (Fig. 6D; Troll 1937; Gifford and Foster 1989; Nelson and Dengler 1997).

Auxin and the Formation of Open Vein Networks

In both rounded and elongated leaves, localization of PIN1 proteins at the plasma membrane of epidermal cells at the shoot tip becomes polarized toward sites of leaf primordium formation (Fig. 6E; Benková et al. 2003; Reinhardt et al. 2003; Carraro et al. 2006; Scarpella et al. 2006; Bayer et al. 2009; Lee et al. 2009; Johnston et al. 2015). The resulting epidermal "convergence points" of PIN1 polarity are associated with the appearance of broad PEDs in the inner tissue that will become restricted to sites of midvein formation. Likewise, epidermal convergence points of PIN1 polarity at the leaf edge are associated with both sites of leaf lateral growth and positions of broad PEDs in the inner tissue that become restricted to sites of lateral vein formation (Fig. 4A; Hay et al. 2006; Scarpella et al. 2006; Wenzel et al. 2007). Both auxin transport inhibition or higher auxin levels reduce the distance between successive epidermal convergence points of PIN1 polarity and leaf primordia and between midvein and lateral veins (Okada et al. 1991; Bennett et al. 1995; Mattsson et al. 1999; Sieburth 1999; Reinhardt et al. 2000; Scarpella et al. 2006; Guenot et al. 2012; Sawchuk et al. 2013; Verna et al. 2019). However, the relation between convergence points of epidermal PIN1 polarity and positioning of midvein and lateral veins seems to be correlative, rather than causal. Indeed, mutants that lack such convergence points form normal vein networks (Bilsborough et al. 2011), and epidermal expression of PIN1 is neither required nor sufficient for auxin-transport-dependent vein positioning (Govindaraju et

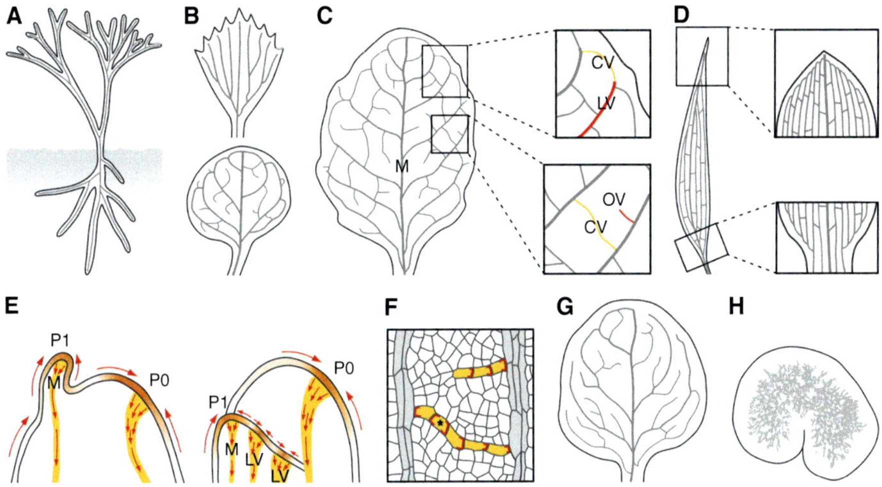

Figure 6. Auxin and vein network formation. (*A*) Early, leafless plants are systems of branched cylinders with a vascular strand (gray line) in their center. (*B*) Leaves of extant plants have "open" (*top*) or "closed" (*bottom*) vein networks. (*C*) In rounded leaves, lateral veins (LVs) branch from a single midvein (M); minor veins branch from M and LVs, and either end freely ("open" veins [OVs]; red) or connect to other veins ("closed" veins [CVs]; yellow); and vein loops form from the fusion of LVs (red) and closed minor veins (yellow) (*E*). (*D*) In elongated leaves, vein loops are compressed laterally and stretched along the leaf, such that M and LVs seem to run parallel to one another; and M and LVs are connected laterally by minor veins. (*E*) In epidermal cells at the shoot tip of plants with either rounded (*left*) or elongated (*right*) leaves, PIN1 localization becomes polarized (red arrows) toward sites of leaf primordium formation (orange). These epidermal convergence points of PIN1 polarity are associated with broad PEDs in the inner tissue that become restricted to sites of M formation (yellow). Over time, PIN1 localization becomes polarized (red arrows) in the cells of these broad domains: In the cells along the PEDs' midline, PIN1 localization becomes polarized toward preexisting veins; in the cells flanking PEDs' midline, PIN1 localization becomes polarized toward the domains' midline. Both the restriction of broad PEDs and the polarization of PIN1 localization initiate and proceed away from preexisting veins, in which PIN1 localization is polarized. P0 and P1 are successive stages of leaf primordium development. (*F*) The localization of PIN1 (red) in files of vascular cells (yellow) is polarized toward preexisting veins (gray; for simplicity, PIN1 expression in preexisting veins is not shown). In OVs, a single PIN1 localization polarity exists; in CVs, the two opposite PIN1 localization polarities are connected by a bipolar cell (asterisk): a cell with PIN1 at both ends. (*G*,*H*) Progressive reduction in the ability to polarize PIN1 localization during vein network formation leads to mutant vein networks with very few CVs (*G*); mutant vein networks in which vein fragments form along paths defined by initially continuous PEDs (Fig. 5B); or clusters of randomly oriented vascular cells, as in *gn* mutant leaves (*H*).

al. 2020). These observations suggest that the mechanism that controls the positioning of midvein and lateral veins depends on auxin levels and transport but is independent of epidermal convergence points of PIN1 polarity.

Auxin and the Formation of Closed Vein Networks

Unlike the broad PEDs that become restricted to sites of formation of midvein and lateral veins, those that become restricted to sites of minor vein formation in rounded leaves are not associated with epidermal convergence points of PIN1 polarity; instead, the broad PEDs that become restricted to sites of minor vein formation branch from preexisting veins (Fig. 4A; Scarpella et al. 2006; Wenzel et al. 2007; Marcos and Berleth 2014). Initially, these broad PEDs connect to preexisting veins at only one of their two ends ("open" PEDs), but they can extend to connect to other veins at both their ends ("closed" PEDs) (Fig. 4A; Scarpella et al. 2006; Wenzel et al. 2007; Marcos and Berleth 2014). The formation of such closed PEDs is promoted by auxin transport inhibition (Mattsson et al. 1999; Sieburth

1999; Verna et al. 2015). Because auxin transport inhibition delays restriction of broad PEDs and polarization of PIN1 localization, only PEDs that have yet to differentiate into highly efficient auxin transport paths can connect to preexisting veins at both ends.

But not all broad PEDs become restricted and polarized: Some gradually disappear (Fig. 4A; Marcos and Berleth 2014), suggesting that PEDs compete for a limiting amount of auxin (Sachs 2003). Therefore, how many PEDs will form at any given stage of leaf tissue development, where exactly they will form, how broad they will be, what shape they will have, and which ones will become restricted and polarized and which ones will instead gradually disappear will not only depend on positive feedback between polarization of PIN1 localization and polar auxin transport but on random variations in the local production of auxin and on the number, shape, size, position, and polarization of preexisting PEDs.

Within narrow PEDs—whether open or closed—PIN1 localization is polarized toward preexisting veins (Scarpella et al. 2006; Wenzel et al. 2007; Marcos and Berleth 2014). Therefore, open PEDs have a single PIN1 localization polarity (Fig. 6F). By contrast, closed PEDs are composed of two segments, each with a single PIN1 localization polarity, opposite to that of the other (Fig. 6F). The two opposite polarities are connected by a "bipolar" cell, a cell in which PIN1 localization is polarized at both ends (Fig. 6F).

In rounded leaves, each loop forms from the fusion of a lateral-vein-associated PED and a minor-vein-associated closed PED (Fig. 4A; Scarpella et al. 2006; Wenzel et al. 2007). Whether in elongated leaves PIN1 is expressed and localized during the formation of minor veins and loops as it is in rounded leaves is currently unknown; however, computational simulations suggest that the same vein-formation mechanism embedded in different leaf growth patterns can account for the different vein networks of elongated and rounded leaves (Runions et al. 2005; Fujita and Mochizuki 2006b).

Therefore, available evidence suggests that like vascular strand formation in response to auxin application and vein formation during normal leaf development, vein network formation is an expression of auxin transport, many aspects of which are consistent with the auxin canalization hypothesis (Mitchison 1980a, 1981; Sachs 1991, 2000; Feugier et al. 2005; Rolland-Lagan and Prusinkiewicz 2005; Feugier and Iwasa 2006; Fujita and Mochizuki 2006a; Stoma et al. 2008; Bayer et al. 2009; Smith and Bayer 2009; Alim and Frey 2010; Wabnik et al. 2010; Walker et al. 2013; Lee et al. 2014; Cieslak et al. 2015; Abley et al. 2016). Nevertheless, the auxin canalization hypothesis predicts the formation of networks of "open" veins (i.e., veins that connect to veins at only one of their two ends) (Sachs 1975; Mitchison 1980a, 1981; Rolland-Lagan and Prusinkiewicz 2005). The formation of "closed" veins (i.e., veins that connect to other veins at both their ends) has thus been proposed to result from the diffusion of a hypothetical substance from preexisting veins that allows approaching veins to connect to preexisting veins (Feugier and Iwasa 2006). This hypothesis predicts that in closed PEDs, PIN1 is polarized *away* from the preexisting veins to which the closed PEDs connect, when in fact in closed PEDs, PIN1 is polarized *toward* the preexisting veins to which the closed PEDs connect (Fig. 6F; Scarpella et al. 2006; Wenzel et al. 2007; Marcos and Berleth 2014). Therefore, this hypothesis seems to be unsupported by experimental evidence.

Alternatively, closed veins could form from veins meeting at points of peak auxin levels (Dimitrov and Zucker 2006) or from localized auxin production at precisely defined times and places (Sachs 1975, 1989; Mitchison 1980a; Aloni et al. 2003; Rolland-Lagan and Prusinkiewicz 2005; Runions et al. 2005). Both hypotheses are consistent with the observation of bipolar cells and predict peak auxin levels in these cells; however, this prediction remains to be tested experimentally.

The formation of bipolar cells seems to be very sensitive to even the partial loss of function of the auxin-signaling- and auxin-transport-dependent pathway that controls the formation of continuous veins: Mutants partially lacking the function of this pathway often fail to form bipolar cells and thus to polarize PIN1 localization along closed PEDs (Naramoto et al. 2009; Hou et al. 2010). This reduced ability to polarize PIN1 localization along closed PEDs often leads to their "opening" and thus to vein networks with very few closed veins (Fig. 6G; Steynen and Schultz 2003; Naramoto et al. 2009). Nevertheless, these mutants are still able to polarize PIN1 localization along open PEDs (Naramoto et al. 2009; Hou et al. 2010).

More severe loss of function of the pathway that controls the formation of continuous veins leads to the inability to polarize PIN1 localization even along open PEDs (Carland et al. 1999; Deyholos et al. 2000; Koizumi et al. 2000, 2005; Willemsen et al. 2003; Carland and Nelson 2004, 2009; Scarpella et al. 2006; Sieburth et al. 2006; Naramoto et al. 2009). The inability of these mutants to polarize PIN1 localization along PEDs leads to termination of PIN1 expression in some of the cells in a PED before PIN1 localization becomes polarized in any of the cells in the domain (Scarpella et al. 2006; Naramoto et al. 2009). The cells that terminate PIN1 expression differentiate into nonvascular cells, while the remaining PED fragments differentiate into vascular fragments of the second type (Fig. 5B; Carland et al. 1999; Deyholos et al. 2000; Koizumi et al. 2000, 2005; Willemsen et al. 2003; Carland and Nelson 2004, 2009; Scarpella et al. 2006; Sieburth et al. 2006; Naramoto et al. 2009; Tan et al. 2020; Xiao and Offringa 2020). Nevertheless, these vascular fragments still form along the paths where continuous veins would form in wild type, a vestige that these fragments were once connected.

Defects in vein network formation are most severe in *gn* mutants. In *gn* cotyledons and leaves, individual cells can still localize PIN1 polarly—although to a lesser extent—but they seem to have almost entirely lost the ability to coordinate between them the polarity of PIN1 localization (Steinmann et al. 1999; Verna et al. 2019). The result of this inability is the formation of clusters of randomly oriented vascular cells (Fig. 6H; Mayer et al. 1993; Steinmann et al. 1999; Geldner et al. 2004; Verna et al. 2019; Amalraj et al. 2020).

Based on their biochemical function and cellular localization, proteins in the pathway that controls the formation of continuous veins have been proposed to localize PIN1 to or retain it in the plasma membrane, polarize PIN1 localization, or maintain such polar localization (e.g., Geldner et al. 2004; Koizumi et al. 2005; Sieburth et al. 2006; Kleine-Vehn et al. 2008; Carland and Nelson 2009; Naramoto et al. 2009, 2010; Prabhakaran Mariyamma et al. 2018). However, the function of these proteins seems to entail more than the control of auxin transport and to include at least the control of auxin signaling. Indeed, defects in mutants in the pathway that controls the formation of continuous veins cannot be phenocopied by mutation in *PIN* genes or growth in the presence of auxin transport inhibitors; instead, those defects are reproduced in plants in which both auxin transport and signaling are compromised (Verna et al. 2019). Although it is currently unclear how proteins in the pathway that control the formation of continuous veins control auxin signaling, the most parsimonious account is that such proteins control the polar localization of proteins with vein formation function that are produced in response to auxin signaling.

CONCLUDING REMARKS

The past 20 years have witnessed unprecedented advances in our understanding of the role of auxin in the patterned formation of the vascular system; however, the very same research that has led to such advancement has also exposed unexpected gaps in our current knowledge. For example, a role for auxin signaling in vein positioning had been unsuspected because the fewer and incompletely differentiated veins of auxin signaling mutants still form in the same positions as they would in wild type. It now turns out that functions of auxin signaling in vein positioning had been obscured by nonhomologous redundancy with auxin transport. Furthermore, it also turns out that auxin signaling and auxin transport had been eclipsing each other's functions in the end-to-end alignment of vascular cells oriented along the vein. However, how precisely auxin transport and signaling control all those processes remains unclear. These and other questions will have to be addressed in future research, and as past research has taught us, even more surprises are awaiting ahead.

ACKNOWLEDGMENTS

We thank Przemek Prusinkiewicz for helpful comments on the manuscript. We apologize to colleagues whose work could not be included because of space constraints. Our research is supported by Discovery Grants of the Natural Science and Engineering Research Council of Canada (Grants NSERC RGPIN-2016-04736 and NSERC RGPAS 492872-2016).

REFERENCES

Abley K, Sauret-Güeto S, Marée AF, Coen E. 2016. Formation of polarity convergences underlying shoot outgrowths. *eLife* **5:** e18165. doi:10.7554/eLife.18165

Adamowski M, Friml J. 2015. PIN-dependent auxin transport: action, regulation, and evolution. *Plant Cell* **27:** 20–32. doi:10.1105/tpc.114.134874

Aida M, Beis D, Heidstra R, Willemsen V, Blilou I, Galinha C, Nussaume L, Noh YS, Amasino R, Scheres B. 2004. The *PLETHORA* genes mediate patterning of the *Arabidopsis* root stem cell niche. *Cell* **119:** 109–120. doi:10.1016/j.cell.2004.09.018

Alim K, Frey E. 2010. Quantitative predictions on auxin-induced polar distribution of PIN proteins during vein formation in leaves. *Eur Phys J E Soft Matter* **33:** 165–173. doi:10.1140/epje/i2010-10604-5

Aloni R, Schwalm K, Langhans M, Ullrich CI. 2003. Gradual shifts in sites of free-auxin production during leaf-primordium development and their role in vascular differentiation and leaf morphogenesis in *Arabidopsis*. *Planta* **216:** 841–853. doi:10.1007/s00425-002-0937-8

Alonso-Peral MM, Candela H, del Pozo JC, Martinez-Laborda A, Ponce MR, Micol JL. 2006. The *HVE/CAND1* gene is required for the early patterning of leaf venation in *Arabidopsis*. *Development* **133:** 3755–3766. doi:10.1242/dev.02554

Amalraj B, Govindaraju P, Krishna A, Lavania D, Linh NM, Ravichandran SJ, Scarpella E. 2020. GAL4/GFP enhancer-trap lines for identification and manipulation of cells and tissues in developing *Arabidopsis* leaves. *Dev Dyn* **249:** 1127–1146. doi:10.1002/dvdy.181

Anne P, Hardtke CS. 2018. Phloem function and development—biophysics meets genetics. *Curr Opin Plant Biol* **43:** 22–28. doi:10.1016/j.pbi.2017.12.005

Ash A, Ellis B, Hickey LJ, Johnson K, Wilf P, Wing S. 1999. *Manual of leaf architecture*. Leaf Architecture Working Group, Smithsonian Institution, Washington, DC.

Avery GS Jr. 1935. Differential distribution of a phytohormone in the developing leaf of nicotiana, and its relation to polarized growth. *Bull Torrey Bot Club* **62:** 313–330. doi:10.2307/2480818

Barbosa ICR, Hammes UZ, Schwechheimer C. 2018. Activation and polarity control of PIN-FORMED auxin transporters by phosphorylation. *Trends Plant Sci* **23:** 523–538. doi:10.1016/j.tplants.2018.03.009

Bayer EM, Smith RS, Mandel T, Nakayama N, Sauer M, Prusinkiewicz P, Kuhlemeier C. 2009. Integration of transport-based models for phyllotaxis and midvein formation. *Genes Dev* **23:** 373–384. doi:10.1101/gad.497009

Benková E, Michniewicz M, Sauer M, Teichmann T, Seifertová D, Jürgens G, Friml J. 2003. Local, efflux-dependent auxin gradients as a common module for plant organ formation. *Cell* **115:** 591–602. doi:10.1016/S0092-8674(03)00924-3

Bennett SRM, Alvarez J, Bossinger G, Smyth DR. 1995. Morphogenesis in *pinoid* mutants of *Arabidopsis thaliana*. *Plant J* **8:** 505–520. doi:10.1046/j.1365-313X.1995.8040505.x

Bennett T, Hines G, Leyser O. 2014. Canalization: what the flux. *Trends Genet* **30:** 41–48. doi:10.1016/j.tig.2013.11.001

Berleth T, Jurgens G. 1993. The role of the *monopteros* gene in organizing the basal body region of the *Arabidopsis* embryo. *Development* **118:** 575–587.

Bilsborough GD, Runions A, Barkoulas M, Jenkins HW, Hasson A, Galinha C, Laufs P, Hay A, Prusinkiewicz P, Tsiantis M. 2011. Model for the regulation of *Arabidopsis thaliana* leaf margin development. *Proc Natl Acad Sci* **108:** 3424–3429. doi:10.1073/pnas.1015162108

Busch M, Mayer U, Jurgens G. 1996. Molecular analysis of the *Arabidopsis* pattern formation of gene *GNOM*: gene structure and intragenic complementation. *Mol Gen Genet* **250:** 681–691.

Camus G. 1949. Recherches sur le role des bourgeons dans les phenomenes de morphogenese [Research on the role of buds in the phenomenon of morphogenesis]. *Rev Cytol Biol Veg* **11:** 1–199.

Candela H, Martı´nez-Laborda A, Micol JL. 1999. Venation pattern formation in *Arabidopsis thaliana* vegetative leaves. *Dev Biol* **205:** 205–216. doi:10.1006/dbio.1998.9111

Carland FM, Nelson T. 2004. *Cotyledon vascular pattern2*-mediated inositol (1,4,5) triphosphate signal transduction is essential for closed venation patterns of *Arabidopsis* foliar organs. *Plant Cell* **16:** 1263–1275. doi:10.1105/tpc.021030

Carland F, Nelson T. 2009. CVP2- and CVL1-mediated phosphoinositide signaling as a regulator of the ARF GAP SFC/VAN3 in establishment of foliar vein patterns. *Plant J* **59:** 895–907. doi:10.1111/j.1365-313X.2009.03920.x

Carland FM, Berg BL, FitzGerald JN, Jinamornphongs S, Nelson T, Keith B. 1999. Genetic regulation of vascular tissue patterning in *Arabidopsis*. *Plant Cell* **11:** 2123–2137. doi:10.1105/tpc.11.11.2123

Carraro N, Forestan C, Canova S, Traas J, Varotto S. 2006. *ZmPIN1a* and *ZmPIN1b* encode two novel putative candidates for polar auxin transport and plant architecture determination of maize. *Plant Physiol* **142:** 254–264. doi:10.1104/pp.106.080119

Cheng Y, Dai X, Zhao Y. 2007. Auxin synthesized by the YUCCA flavin monooxygenases is essential for embryogenesis and leaf formation in *Arabidopsis*. *Plant Cell* **19:** 2430–2439. doi:10.1105/tpc.107.053009

Cho H, Dang TV, Hwang I. 2017. Emergence of plant vascular system: roles of hormonal and non-hormonal regulatory networks. *Curr Opin Plant Biol* **35:** 91–97. doi:10.1016/j.pbi.2016.11.013

Cieslak M, Runions A, Prusinkiewicz P. 2015. Auxin-driven patterning with unidirectional fluxes. *J Exp Bot* **66:** 5083–5102. doi:10.1093/jxb/erv262

Coen E, Rolland-Lagan AG, Matthews M, Bangham JA, Prusinkiewicz P. 2004. The genetics of geometry. *Proc Natl Acad Sci* **101:** 4728–4735. doi:10.1073/pnas.0306308101

Dengler N, Kang J. 2001. Vascular patterning and leaf shape. *Curr Opin Plant Biol* **4:** 50–56. doi:10.1016/S1369-5266(00)00135-7

De Rybel B, Adibi M, Breda AS, Wendrich JR, Smit ME, Novák O, Yamaguchi N, Yoshida S, Van Isterdael G, Palovaara J, et al. 2014. Plant development. Integration of growth and patterning during vascular tissue formation in *Arabidopsis*. *Science* **345:** 1255215. doi:10.1126/science.1255215

De Rybel B, Mähönen AP, Helariutta Y, Weijers D. 2016. Plant vascular development: from early specification to differentiation. *Nat Rev Mol Cell Biol* **17:** 30–40. doi:10.1038/nrm.2015.6

Deyholos MK, Cordner G, Beebe D, Sieburth LE. 2000. The *SCARFACE* gene is required for cotyledon and leaf vein patterning. *Development* **127:** 3205–3213.

Dharmasiri S, Dharmasiri N, Hellmann H, Estelle M. 2003. The RUB/Nedd8 conjugation pathway is required for early development in *Arabidopsis*. *EMBO J* **22:** 1762–1770. doi:10.1093/emboj/cdg190

Dharmasiri N, Dharmasiri S, Weijers D, Lechner E, Yamada M, Hobbie L, Ehrismann JS, Jürgens G, Estelle M. 2005. Plant development is regulated by a family of auxin receptor F Box proteins. *Dev Cell* **9:** 109–119. doi:10.1016/j.devcel.2005.05.014

Dharmasiri N, Dharmasiri S, Weijers D, Karunarathna N, Jurgens G, Estelle M. 2007. AXL and AXR1 have redundant functions in RUB conjugation and growth and development in *Arabidopsis*. *Plant J* **52:** 114–123. doi:10.1111/j.1365-313X.2007.03211.x

Dimitrov P, Zucker SW. 2006. A constant production hypothesis guides leaf venation patterning. *Proc Natl Acad Sci* **103:** 9363–9368. doi:10.1073/pnas.0603559103

Esau K. 1943. Origin and development of primary vascular tissues in seed plants. *Botanical Review* **9:** 125–206. doi:10.1007/BF02872466

Esau K. 1965. *Plant anatomy*. Wiley, New York.

Esteve-Bruna D, Pérez-Pérez JM, Ponce MR, Micol JL. 2013. *Incurvata13*, a novel allele of *AUXIN RESISTANT6*, reveals a specific role for auxin and the SCF complex in *Arabidopsis* embryogenesis, vascular specification, and leaf flatness. *Plant Physiol* **161:** 1303–1320. doi:10.1104/pp.112.207779

Etchells JP, Smit ME, Gaudinier A, Williams CJ, Brady SM. 2016. A brief history of the TDIF-PXY signalling module: balancing meristem identity and differentiation during vascular development. *New Phytol* **209:** 474–484. doi:10.1111/nph.13642

Fairon-Demaret M, Li CS. 1993. *Lorophyton goense* gen. et sp. nov. from the Lower Givetian of Belgium and a discussion of the Middle Devonian Cladoxylopsida. *Rev Palaeobot Palynol* **77:** 1–22. doi:10.1016/0034-6667(93)90052-V

Feugier FG, Iwasa Y. 2006. How canalization can make loops: a new model of reticulated leaf vascular pattern formation. *J Theor Biol* **243:** 235–244. doi:10.1016/j.jtbi.2006.05.022

Feugier FG, Mochizuki A, Iwasa Y. 2005. Self-organization of the vascular system in plant leaves: inter-dependent dynamics of auxin flux and carrier proteins. *J Theor Biol* **236:** 366–375. doi:10.1016/j.jtbi.2005.03.017

Fischer U, Kucukoglu M, Helariutta Y, Bhalerao RP. 2019. The dynamics of cambial stem cell activity. *Annu Rev Plant Biol* **70:** 293–319. doi:10.1146/annurev-arplant-050718-100402

Fujita H, Mochizuki A. 2006a. Pattern formation of leaf veins by the positive feedback regulation between auxin flow and auxin efflux carrier. *J Theor Biol* **241:** 541–551. doi:10.1016/j.jtbi.2005.12.016

Fujita H, Mochizuki A. 2006b. The origin of the diversity of leaf venation pattern. *Dev Dyn* **235:** 2710–2721. doi:10.1002/dvdy.20908

Fukuda H, Ohashi-Ito K. 2019. Vascular tissue development in plants. *Curr Top Dev Biol* **131:** 141–160. doi:10.1016/bs.ctdb.2018.10.005

Garrett JJ, Meents MJ, Blackshaw MT, Blackshaw LC, Hou H, Styranko DM, Kohalmi SE, Schultz EA. 2012. A novel, semi-dominant allele of MONOPTEROS provides insight into leaf initiation and vein pattern formation. *Planta* **236:** 297–312. doi:10.1007/s00425-012-1607-0

Geldner N, Richter S, Vieten A, Marquardt S, Torres-Ruiz RA, Mayer U, Jurgens G. 2004. Partial loss-of-function alleles reveal a role for GNOM in auxin transport-related, post-embryonic development of *Arabidopsis*. *Development* **131:** 389–400. doi:10.1242/dev.00926

Gifford EM, Foster AS. 1989. *Morphology and evolution of vascular plants*. W.H. Freeman, New York.

Gooh K, Ueda M, Aruga K, Park J, Arata H, Higashiyama T, Kurihara D. 2015. Live-cell imaging and optical manipulation of *Arabidopsis* early embryogenesis. *Dev Cell* **34:** 242–251. doi:10.1016/j.devcel.2015.06.008

Govindaraju P, Verna C, Zhu T, Scarpella E. 2020. Vein patterning by tissue-specific auxin transport. *Development* **147:** dev187666. doi:10.1242/dev.187666

Guenot B, Bayer E, Kierzkowski D, Smith RS, Mandel T, Žádníková P, Benková E, Kuhlemeier C. 2012. PIN1-independent leaf initiation in *Arabidopsis*. *Plant Physiol* **159:** 1501–1510. doi:10.1104/pp.112.200402

Hadfi K, Speth V, Neuhaus G. 1998. Auxin-induced developmental patterns in *Brassica juncea* embryos. *Development* **125:** 879–887.

Hamann T, Mayer U, Jurgens G. 1999. The auxin-insensitive *bodenlos* mutation affects primary root formation and apical-basal patterning in the *Arabidopsis* embryo. *Development* **126:** 1387–1395.

Hamann T, Benkova E, Baurle I, Kientz M, Jurgens G. 2002. The *Arabidopsis BODENLOS* gene encodes an auxin response protein inhibiting MONOPTEROS-mediated embryo patterning. *Genes Dev* **16:** 1610–1615. doi:10.1101/gad.229402

Hardtke CS, Berleth T. 1998. The *Arabidopsis* gene *MONOPTEROS* encodes a transcription factor mediating embryo axis formation and vascular development. *EMBO J* **17:** 1405–1411. doi:10.1093/emboj/17.5.1405

Hardtke CS, Ckurshumova W, Vidaurre DP, Singh SA, Stamatiou G, Tiwari SB, Hagen G, Guilfoyle TJ, Berleth T. 2004. Overlapping and non-redundant functions of the *Arabidopsis* auxin response factors MONOPTEROS and NONPHOTOTROPIC HYPOCOTYL 4. *Development* **131:** 1089–1100. doi:10.1242/dev.00925

Hay A, Barkoulas M, Tsiantis M. 2006. ASYMMETRIC LEAVES1 and auxin activities converge to repress BREVIPEDICELLUS expression and promote leaf development in *Arabidopsis*. *Development* **133:** 3955–3961. doi:10.1242/dev.02545

Hellmann H, Hobbie L, Chapman A, Dharmasiri S, Dharmasiri N, del Pozo C, Reinhardt D, Estelle M. 2003. *Arabidopsis* AXR6 encodes CUL1 implicating SCF E3 ligases in auxin regulation of embryogenesis. *EMBO J* **22:** 3314–3325. doi:10.1093/emboj/cdg335

Herbst D. 1971. Disjunct foliar veins in Hawaiian Euphorbias. *Science* **171:** 1247–1248. doi:10.1126/science.171.3977.1247

Herbst D. 1972. Ontogeny of foliar venation in *Euphorbia forbesii*. *Am J Bot* **59:** 843–850. doi:10.1002/j.1537-2197.1972.tb10159.x

Hobbie L, McGovern M, Hurwitz LR, Pierro A, Liu NY, Bandyopadhyay A, Estelle M. 2000. The *axr6* mutants of *Arabidopsis thaliana* define a gene involved in auxin response and early development. *Development* **127:** 23–32.

Hou H, Erickson J, Meservy J, Schultz EA. 2010. FORKED1 encodes a PH domain protein that is required for PIN1 localization in developing leaf veins. *Plant J* **63:** 960–973. doi:10.1111/j.1365-313X.2010.04291.x

Jacobs WP. 1952. The role of auxin in differentiation of xylem around a wound. *Am J Bot* **39:** 301–309. doi:10.1002/j.1537-2197.1952.tb14277.x

Johnston R, Leiboff S, Scanlon MJ. 2015. Ontogeny of the sheathing leaf base in maize (*Zea mays*). *New Phytol* **205:** 306–315. doi:10.1111/nph.13010

Jost L. 1942. Über gefässbrücken. *Zeitsch Bot* **38:** 161–215.

Jouannet V, Brackmann K, Greb T. 2015. (Pro)cambium formation and proliferation: Two sides of the same coin? *Curr Opin Plant Biol* **23:** 54–60. doi:10.1016/j.pbi.2014.10.010

Kang J, Dengler N. 2004. Vein pattern development in adult leaves of *Arabidopsis thaliana*. *Int J Plant Sci* **165:** 231–242. doi:10.1086/382794

Kleine-Vehn J, Dhonukshe P, Sauer M, Brewer PB, Wiśniewska J, Paciorek T, Benková E, Friml J. 2008. ARF GEF-dependent transcytosis and polar delivery of PIN auxin carriers in *Arabidopsis*. *Curr Biol* **18:** 526–531. doi:10.1016/j.cub.2008.03.021

Kleine-Vehn J, Wabnik K, Martinière A, Łangowski L, Willig K, Naramoto S, Leitner J, Tanaka H, Jakobs S, Robert S, et al. 2011. Recycling, clustering, and endocytosis jointly maintain PIN auxin carrier polarity at the plasma membrane. *Mol Syst Biol* **7:** 540. doi:10.1038/msb.2011.72

Koizumi K, Sugiyama M, Fukuda H. 2000. A series of novel mutants of *Arabidopsis thaliana* that are defective in the formation of continuous vascular network: calling the auxin signal flow canalization hypothesis into question. *Development* **127:** 3197–3204.

Koizumi K, Naramoto S, Sawa S, Yahara N, Ueda T, Nakano A, Sugiyama M, Fukuda H. 2005. VAN3 ARF-GAP-mediated vesicle transport is involved in leaf vascular network formation. *Development* **132:** 1699–1711. doi:10.1242/dev.01716

Kramer EM. 2009. Auxin-regulated cell polarity: an inside job? *Trends Plant Sci* **14:** 242–247. doi:10.1016/j.tplants.2009.02.005

Kraus EJ, Brown NA, Hamner KC. 1936. Histological reactions of bean plants to indoleacetic acid. *Botanical Gazette* **98:** 370–420. doi:10.1086/334646

Krogan NT, Ckurshumova W, Marcos D, Caragea AE, Berleth T. 2012. Deletion of MP/ARF5 domains III and IV reveals a requirement for Aux/IAA regulation in *Arabidopsis* leaf vascular patterning. *New Phytol* **194:** 391–401. doi:10.1111/j.1469-8137.2012.04064.x

Łangowski L, Wabnik K, Li H, Vanneste S, Naramoto S, Tanaka H, Friml J. 2016. Cellular mechanisms for cargo delivery and polarity maintenance at different polar domains in plant cells. *Cell Discov* **2:** 16018. doi:10.1038/celldisc.2016.18

Lau S, De Smet I, Kolb M, Meinhardt H, Jürgens G. 2011. Auxin triggers a genetic switch. *Nat Cell Biol* **13:** 611–615. doi:10.1038/ncb2212

Lee BH, Johnston R, Yang Y, Gallavotti A, Kojima M, Travençolo BA, Costa LF, Sakakibara H, Jackson D. 2009. Studies of *aberrant phyllotaxy1* mutants of maize indicate complex interactions between auxin and cytokinin signaling in the shoot apical meristem. *Plant Physiol* **150:** 205–216. doi:10.1104/pp.109.137034

Lee SW, Feugier FG, Morishita Y. 2014. Canalization-based vein formation in a growing leaf. *J Theor Biol* **353:** 104–120. doi:10.1016/j.jtbi.2014.03.005

Lersten N. 1965. Histogenesis of leaf venation in *Trifolium wormskioldii* (Leguminosae). *Am J Bot* **52:** 767–774. doi:10.1002/j.1537-2197.1965.tb07247.x

Leyser O. 2018. Auxin signaling. *Plant Physiol* **176:** 465–479. doi:10.1104/pp.17.00765

Linh NM, Verna C, Scarpella E. 2018. Coordination of cell polarity and the patterning of leaf vein networks. *Curr Opin Plant Biol* **41:** 116–124. doi:10.1016/j.pbi.2017.09.009

Mähönen AP, Bishopp A, Higuchi M, Nieminen KM, Kinoshita K, Tormakangas K, Ikeda Y, Oka A, Kakimoto T, Helariutta Y. 2006. Cytokinin signaling and its inhibitor AHP6 regulate cell fate during vascular development. *Science* **311:** 94–98. doi:10.1126/science.1118875

Mansfield SG, Briarty LG. 1991. Early embryogenesis in *Arabidopsis thaliana*. II: The developing embryo. *Canadian Journal of Botany* **69:** 461–476. doi:10.1139/b91-063

Marcos D, Berleth T. 2014. Dynamic auxin transport patterns preceding vein formation revealed by live-imaging of *Arabidopsis* leaf primordia. *Front Plant Sci* **5:** 235. doi:10.3389/fpls.2014.00235

Marhava P, Bassukas AEL, Zourelidou M, Kolb M, Moret B, Fastner A, Schulze WX, Cattaneo P, Hammes UZ,

Schwechheimer C, et al. 2018. A molecular rheostat adjusts auxin flux to promote root protophloem differentiation. *Nature* **558:** 297–300. doi:10.1038/s41586-018-0186-z

Mattsson J, Sung ZR, Berleth T. 1999. Responses of plant vascular systems to auxin transport inhibition. *Development* **126:** 2979–2991.

Mattsson J, Ckurshumova W, Berleth T. 2003. Auxin signaling in *Arabidopsis* leaf vascular development. *Plant Physiol* **131:** 1327–1339. doi:10.1104/pp.013623

Mayer U, Buttner G, Jurgens G. 1993. Apical-basal pattern formation in the *Arabidopsis* embryo: studies on the role of the *gnom* gene. *Development* **117:** 149–162.

Mazur E, Benková E, Friml J. 2016. Vascular cambium regeneration and vessel formation in wounded inflorescence stems of *Arabidopsis*. *Sci Rep* **6:** 33754. doi:10.1038/srep33754

Mazur E, Kulik I, Hajný J, Friml J. 2020. Auxin canalization and vascular tissue formation by TIR1/AFB-mediated auxin signaling in *Arabidopsis*. *New Phytol* **226:** 1375–1383. doi:10.1111/nph.16446

Mitchison GJ. 1980a. Model for vein formation in higher-plants. *Proc Biol Sci* **207:** 79–109.

Mitchison GJ. 1980b. The dynamics of auxin transport. *Proc Biol Sci* **209:** 489–511.

Mitchison GJ. 1981. The polar transport of auxin and vein patterns in plants. *Philos Trans R Soc Lond B Biol Sci* **295:** 461–471. doi:10.1098/rstb.1981.0154

Naramoto S, Sawa S, Koizumi K, Uemura T, Ueda T, Friml J, Nakano A, Fukuda H. 2009. Phosphoinositide-dependent regulation of VAN3 ARF-GAP localization and activity essential for vascular tissue continuity in plants. *Development* **136:** 1529–1538. doi:10.1242/dev.030098

Naramoto S, Kleine-Vehn J, Robert S, Fujimoto M, Dainobu T, Paciorek T, Ueda T, Nakano A, Van Montagu MCE, Fukuda H, et al. 2010. ADP-ribosylation factor machinery mediates endocytosis in plant cells. *Proc Natl Acad Sci* **107:** 21890–21895. doi:10.1073/pnas.1016260107

Nelson T, Dengler N. 1997. Leaf vascular pattern formation. *Plant Cell* **9:** 1121–1135. doi:10.1105/tpc.9.7.1121

O'Connor DL, Runions A, Sluis A, Bragg J, Vogel JP, Prusinkiewicz P, Hake S. 2014. A division in PIN-mediated auxin patterning during organ initiation in grasses. *PLoS Comput Biol* **10:** e1003447. doi:10.1371/journal.pcbi.1003447

O'Connor DL, Elton S, Ticchiarelli F, Hsia MM, Vogel JP, Leyser O. 2017. Cross-species functional diversity within the PIN auxin efflux protein family. *eLife* **6:** e31804.

Okada K, Ueda J, Komaki MK, Bell CJ, Shimura Y. 1991. Requirement of the auxin polar transport system in early stages of *Arabidopsis* floral bud formation. *Plant Cell* **3:** 677–684. doi:10.2307/3869249

Paponov IA, Friz T, Budnyk V, Teale WD, Wüst F, Paponov M, Al-Babili S, Palme K. 2019. Natural auxin does not inhibit Brefeldin A induced PIN1 and PIN2 internalization in root cells. *Front Plant Sci* **10:** 574. doi:10.3389/fpls.2019.00574

Petrasek J, Mravec J, Bouchard R, Blakeslee JJ, Abas M, Seifertova D, Wisniewska J, Tadele Z, Kubes M, Covanova M, et al. 2006. PIN proteins perform a rate-limiting function in cellular auxin efflux. *Science* **312:** 914–918. doi:10.1126/science.1123542

Prabhakaran Mariyamma N, Clarke KJ, Yu H, Wilton EE, Van Dyk J, Hou H, Schultz EA. 2018. Members of the *Arabidopsis FORKED1-LIKE* gene family act to localize PIN1 in developing veins. *J Exp Bot* **69:** 4773–4790. doi:10.1093/jxb/ery248

Pray TR. 1955a. Foliar venation of angiosperms. II: Histogenesis of the venation of *Liriodendron*. *Am J Bot* **42:** 18–27. doi:10.1002/j.1537-2197.1955.tb11089.x

Pray TR. 1955b. Foliar venation of angiosperms. IV: Histogenesis of the venation of *Hosta*. *Am J Bot* **42:** 698–706. doi:10.1002/j.1537-2197.1955.tb10409.x

Przemeck GK, Mattsson J, Hardtke CS, Sung ZR, Berleth T. 1996. Studies on the role of the *Arabidopsis* gene *MONOPTEROS* in vascular development and plant cell axialization. *Planta* **200:** 229–237. doi:10.1007/BF00208313

Ramachandran P, Carlsbecker A, Etchells JP. 2017. Class III HD-ZIPs govern vascular cell fate: an HD view on patterning and differentiation. *J Exp Bot* **68:** 55–69. doi:10.1093/jxb/erw370

Raven JA. 1975. Transport of indole acetic acid in plant cells in relation to pH and electrical potential gradients, and its significance for polar IAA transport. *New Phytol* **74:** 163–172. doi:10.1111/j.1469-8137.1975.tb02602.x

Ravichandran SJ, Linh NM, Scarpella E. 2020. The canalization hypothesis—challenges and alternatives. *New Phytol* **227:** 1051–1059. doi:10.1111/nph.16605

Reinhardt D, Mandel T, Kuhlemeier C. 2000. Auxin regulates the initiation and radial position of plant lateral organs. *Plant Cell* **12:** 507–518. doi:10.1105/tpc.12.4.507

Reinhardt D, Pesce ER, Stieger P, Mandel T, Baltensperger K, Bennett M, Traas J, Friml J, Kuhlemeier C. 2003. Regulation of phyllotaxis by polar auxin transport. *Nature* **426:** 255–260. doi:10.1038/nature02081

Roberts LW. 1960. Experiments on xylem regeneration in stem wound responses in coleus. *Botanical Gazette* **121:** 201–208. doi:10.1086/336070

Rodriguez-Villalon A, Gujas B, Kang YH, Breda AS, Cattaneo P, Depuydt S, Hardtke CS. 2014. Molecular genetic framework for protophloem formation. *Proc Natl Acad Sci* **111:** 11551–11556. doi:10.1073/pnas.1407337111

Rodriguez-Villalon A, Gujas B, van Wijk R, Munnik T, Hardtke CS. 2015. Primary root protophloem differentiation requires balanced phosphatidylinositol-4,5-biphosphate levels and systemically affects root branching. *Development* **142:** 1437–1446. doi:10.1242/dev.118364

Rolland-Lagan AG, Prusinkiewicz P. 2005. Reviewing models of auxin canalization in the context of leaf vein pattern formation in *Arabidopsis*. *Plant J* **44:** 854–865. doi:10.1111/j.1365-313X.2005.02581.x

Roth-Nebelsick A, Uhl D, Mosbrugger V, Kerp H. 2001. Evolution and function of leaf venation architecture: a review. *Ann Bot* **87:** 553–566. doi:10.1006/anbo.2001.1391

Rubery PH, Sheldrake AR. 1974. Carrier-mediated auxin transport. *Planta* **118:** 101–121. doi:10.1007/BF00388387

Ruiz Sola MA, Coiro M, Crivelli S, Zeeman SC, Schmidt Kjølner Hansen S, Truernit E. 2017. *OCTOPUS-LIKE 2*, a novel player in *Arabidopsis* root and vascular development, reveals a key role for *OCTOPUS* family genes in root metaphloem sieve tube differentiation. *New Phytol* **216:** 1191–1204. doi:10.1111/nph.14751

Runions A, Fuhrer M, Lane B, Federl P, Rolland-Lagan AG, Prusinkiewicz P. 2005. Modeling and visualization of leaf venation patterns. *ACM Trans Graphics* **24:** 702–711. doi:10.1145/1073204.1073251

Runions A, Smith RS, Prusinkiewicz P. 2014. Computational models of auxin-driven development. In *Auxin and its role in plant development* (ed. E. Zažímalová et al.), pp. 315–357. Springer, Vienna.

Ruonala R, Ko D, Helariutta Y. 2017. Genetic networks in plant vascular development. *Annu Rev Genet* **51:** 335–359. doi:10.1146/annurev-genet-120116-024525

Sachs T. 1968. On the determination of the pattern of vascular tissues in peas. *Ann Bot* **32:** 781–790. doi:10.1093/oxfordjournals.aob.a084249

Sachs T. 1969. Polarity and the induction of organized vascular tissues. *Ann Bot* **33:** 263–275. doi:10.1093/oxfordjournals.aob.a084281

Sachs T. 1975. Control of the differentiation of vascular networks. *Ann Bot* **39:** 197–204. doi:10.1093/oxfordjournals.aob.a084930

Sachs T. 1981. The control of the patterned differentiation of vascular tissues. *Adv Bot Res* **9:** 151–262. doi:10.1016/S0065-2296(08)60351-1

Sachs T. 1989. The development of vascular networks during leaf development. *Curr Top Plant Biochem Physiol* **8:** 168–183.

Sachs T. 1991. Cell polarity and tissue patterning in plants. *Development* **13:** 83–93.

Sachs T. 2000. Integrating cellular and organismic aspects of vascular differentiation. *Plant Cell Physiol* **41:** 649–656. doi:10.1093/pcp/41.6.649

Sachs T. 2003. Collective specification of cellular development. *Bioessays* **25:** 897–903. doi:10.1002/bies.10328

Sauer M, Balla J, Luschnig C, Wisniewska J, Reinohl V, Friml J, Benkova E. 2006. Canalization of auxin flow by Aux/IAA-ARF-dependent feedback regulation of PIN polarity. *Genes Dev* **20:** 2902–2911. doi:10.1101/gad.390806

Sawa S, Koizumi K, Naramoto S, Demura T, Ueda T, Nakano A, Fukuda H. 2005. *DRP1A* is responsible for vascular continuity synergistically working with *VAN3* in *Arabidopsis*. *Plant Physiol* **138:** 819–826. doi:10.1104/pp.105.061689

Sawchuk MG, Head P, Donner TJ, Scarpella E. 2007. Time-lapse imaging of *Arabidopsis* leaf development shows dynamic patterns of procambium formation. *New Phytol* **176:** 560–571. doi:10.1111/j.1469-8137.2007.02193.x

Sawchuk MG, Edgar A, Scarpella E. 2013. Patterning of leaf vein networks by convergent auxin transport pathways. *PLoS Genet* **9:** e1003294. doi:10.1371/journal.pgen.1003294

Scacchi E, Salinas P, Gujas B, Santuari L, Krogan N, Ragni L, Berleth T, Hardtke CS. 2010. Spatio-temporal sequence of cross-regulatory events in root meristem growth. *Proc Natl Acad Sci* **107:** 22734–22739. doi:10.1073/pnas.1014716108

Scarpella E. 2017. The logic of plant vascular patterning. polarity, continuity and plasticity in the formation of the veins and of their networks. *Curr Opin Genet Dev* **45:** 34–43. doi:10.1016/j.gde.2017.02.009

Scarpella E, Francis P, Berleth T. 2004. Stage-specific markers define early steps of procambium development in *Arabidopsis* leaves and correlate termination of vein formation with mesophyll differentiation. *Development* **131:** 3445–3455. doi:10.1242/dev.01182

Scarpella E, Marcos D, Friml J, Berleth T. 2006. Control of leaf vascular patterning by polar auxin transport. *Genes Dev* **20:** 1015–1027. doi:10.1101/gad.1402406

Scheres B, Wolkenfelt H, Willemsen V, Terlouw M, Lawson E, Dean C, Weisbeek P. 1994. Embryonic origin of the *Arabidopsis* primary root and root-meristem initials. *Development* **120:** 2475–2487.

Schlereth A, Möller B, Liu W, Kientz M, Flipse J, Rademacher EH, Schmid M, Jürgens G, Weijers D. 2010. MONOPTEROS controls embryonic root initiation by regulating a mobile transcription factor. *Nature* **464:** 913–916. doi:10.1038/nature08836

Shevell DE, Leu WM, Gillmor CS, Xia G, Feldmann KA, Chua NH. 1994. EMB30 is essential for normal cell division, cell expansion, and cell adhesion in *Arabidopsis* and encodes a protein that has similarity to Sec7. *Cell* **77:** 1051–1062. doi:10.1016/0092-8674(94)90444-8

Sieburth LE. 1999. Auxin is required for leaf vein pattern in *Arabidopsis*. *Plant Physiol* **121:** 1179–1190. doi:10.1104/pp.121.4.1179

Sieburth LE, Muday GK, King EJ, Benton G, Kim S, Metcalf KE, Meyers L, Seamen E, Van Norman JM. 2006. *SCARFACE* encodes an ARF-GAP that is required for normal auxin efflux and vein patterning in *Arabidopsis*. *Plant Cell* **18:** 1396–1411. doi:10.1105/tpc.105.039008

Sinnott EW, Bloch R. 1944. Visible expression of cytoplasmic pattern in the differentiation of xylem strands. *Proc Natl Acad Sci* **30:** 388–392. doi:10.1073/pnas.30.12.388

Sinnott EW, Bloch R. 1945. The cytoplasmic basis of intercellular patterns in vascular differentiation. *Am J Bot* **32:** 151–156. doi:10.1002/j.1537-2197.1945.tb05100.x

Smit ME, Llavata-Peris CI, Roosjen M, van Beijnum H, Novikova D, Levitsky V, Sevilem I, Roszak P, Slane D, Jürgens G, et al. 2020. Specification and regulation of vascular tissue identity in the *Arabidopsis* embryo. *Development* **147:** dev186130.

Smith RS, Bayer EM. 2009. Auxin transport-feedback models of patterning in plants. *Plant Cell Environ* **32:** 1258–1271. doi:10.1111/j.1365-3040.2009.01997.x

Snow R. 1935. Activation of cambial growth by pure hormones. *New Phytol* **34:** 347–360. doi:10.1111/j.1469-8137.1935.tb06853.x

Steinmann T, Geldner N, Grebe M, Mangold S, Jackson CL, Paris S, Galweiler L, Palme K, Jurgens G. 1999. Coordinated polar localization of auxin efflux carrier PIN1 by GNOM ARF GEF. *Science* **286:** 316–318. doi:10.1126/science.286.5438.316

Stepanova AN, Robertson-Hoyt J, Yun J, Benavente LM, Xie DY, Doležal K, Schlereth A, Jürgens G, Alonso JM. 2008. TAA1-mediated auxin biosynthesis is essential for hormone crosstalk and plant development. *Cell* **133:** 177–191. doi:10.1016/j.cell.2008.01.047

Steynen QJ, Schultz EA. 2003. The *FORKED* genes are essential for distal vein meeting in *Arabidopsis*. *Development* **130:** 4695–4708. doi:10.1242/dev.00689

Stoma S, Lucas M, Chopard J, Schaedel M, Traas J, Godin C. 2008. Flux-based transport enhancement as a plausible unifying mechanism for auxin transport in meristem development. *PLoS Comput Biol* **4:** e1000207. doi:10.1371/journal.pcbi.1000207

Strader LC, Monroe-Augustus M, Bartel B. 2008. The IBR5 phosphatase promotes *Arabidopsis* auxin responses through a novel mechanism distinct from TIR1-mediated repressor degradation. *BMC Plant Biol* **8:** 41. doi:10.1186/1471-2229-8-41

Tan S, Zhang X, Kong W, Yang XL, Molnár G, Vondráková Z, Filepová R, Petrášek J, Friml J, Xue HW. 2020. The lipid code-dependent phosphoswitch PDK1-D6PK activates PIN-mediated auxin efflux in *Arabidopsis*. *Nat Plants* **6:** 556–569. doi:10.1038/s41477-020-0648-9

Thimann KV, Skoog F. 1934. On the inhibition of bud development and other functions of growth substance in *Vicia faba*. *Proc R Soc Lond B* **114:** 317–339. doi:10.1098/rspb.1934.0010

Thomas CL, Schmidt D, Bayer EM, Dreos R, Maule AJ. 2009. *Arabidopsis* plant homeodomain finger proteins operate downstream of auxin accumulation in specifying the vasculature and primary root meristem. *Plant J* **59:** 426–436. doi:10.1111/j.1365-313X.2009.03874.x

Thompson NP, Jacobs WP. 1966. Polarity of IAA effect on sieve-tube and xylem regeneration in *Coleus* and tomato stems. *Plant Physiol* **41:** 673–682. doi:10.1104/pp.41.4.673

Torres-Ruiz RA, Jurgens G. 1994. Mutations in the *FASS* gene uncouple pattern formation and morphogenesis in *Arabidopsis* development. *Development* **120:** 2967–2978.

Troll W. 1937. *Vergleichende morphologie der höhren pflanzen*. Gebrüder Borntraeger, Berlin.

Truernit E, Bauby H, Belcram K, Barthélémy J, Palauqui JC. 2012. OCTOPUS, a polarly localised membrane-associated protein, regulates phloem differentiation entry in *Arabidopsis thaliana*. *Development* **139:** 1306–1315. doi:10.1242/dev.072629

Ulmasov T, Hagen G, Guilfoyle TJ. 1997. ARF1, a transcription factor that binds to auxin response elements. *Science* **276:** 1865–1868. doi:10.1126/science.276.5320.1865

Ulmasov T, Hagen G, Guilfoyle TJ. 1999. Dimerization and DNA binding of auxin response factors. *Plant J* **19:** 309–319. doi:10.1046/j.1365-313X.1999.00538.x

van den Berg C, Willemsen V, Hendriks G, Weisbeek P, Scheres B. 1997. Short-range control of cell differentiation in the *Arabidopsis* root meristem. *Nature* **390:** 287–289. doi:10.1038/36856

Verna C, Sawchuk MG, Linh NM, Scarpella E. 2015. Control of vein network topology by auxin transport. *BMC Biol* **13:** 94. doi:10.1186/s12915-015-0208-3

Verna C, Ravichandran SJ, Sawchuk MG, Linh NM, Scarpella E. 2019. Coordination of tissue cell polarity by auxin transport and signaling. *eLife* **8:** e51061. doi:10.7554/eLife.51061

von Ettinghausen C. 1861. *Die Blatt-Skelette der Dikotyledonen*. Kais, Kön, Hof, und Staatsdr., Vienna.

Wabnik K, Kleine-Vehn J, Balla J, Sauer M, Naramoto S, Reinöhl V, Merks RM, Govaerts W, Friml J. 2010. Emergence of tissue polarization from synergy of intracellular and extracellular auxin signaling. *Mol Syst Biol* **6:** 447. doi:10.1038/msb.2010.103

Walker ML, Farcot E, Traas J, Godin C. 2013. The flux-based PIN allocation mechanism can generate either canalyzed or diffuse distribution patterns depending on geometry and boundary conditions. *PLoS ONE* **8:** e54802. doi:10.1371/journal.pone.0054802

Wangermann E. 1967. The effect of the leaf on differentiation of primary xylem in the internode of *Coleus blumei* Benth. *New Phytol* **66:** 747–754. doi:10.1111/j.1469-8137.1967.tb05442.x

Wangermann E. 1974. The pathway of transport of applied indolyl-acetic acid through internode segments. *New Phytol* **73:** 623–636. doi:10.1111/j.1469-8137.1974.tb01288.x

Weijers D, Benkova E, Jäger KE, Schlereth A, Hamann T, Kientz M, Wilmoth JC, Reed JW, Jürgens G. 2005. Developmental specificity of auxin response by pairs of ARF and Aux/IAA transcriptional regulators. *EMBO J* **24:** 1874–1885. doi:10.1038/sj.emboj.7600659

Weijers D, Schlereth A, Ehrismann JS, Schwank G, Kientz M, Jürgens G. 2006. Auxin triggers transient local signaling for cell specification in *Arabidopsis* embryogenesis. *Dev Cell* **10:** 265–270. doi:10.1016/j.devcel.2005.12.001

Went FW. 1928. Wuchsstoff und wachstum. *Rec Trav Bot Néerland* **25:** 11–16.

Wenzel CL, Schuetz M, Yu Q, Mattsson J. 2007. Dynamics of MONOPTEROS and PIN-FORMED1 expression during leaf vein pattern formation in *Arabidopsis thaliana*. *Plant J* **49:** 387–398. doi:10.1111/j.1365-313X.2006.02977.x

Willemsen V, Friml J, Grebe M, van den Toorn A, Palme K, Scheres B. 2003. Cell polarity and PIN protein positioning in *Arabidopsis* require *STEROL METHYLTRANSFERASE1* function. *Plant Cell* **15:** 612–625. doi:10.1105/tpc.008433

Wisniewska J, Xu J, Seifertova D, Brewer PB, Ruzicka K, Blilou I, Rouquie D, Benkova E, Scheres B, Friml J. 2006. Polar PIN localization directs auxin flow in plants. *Science* **312:** 883. doi:10.1126/science.1121356

Wolters H, Anders N, Geldner N, Gavidia R, Jürgens G. 2011. Coordination of apical and basal embryo development revealed by tissue-specific GNOM functions. *Development* **138:** 117–126. doi:10.1242/dev.059147

Xiao Y, Offringa R. 2020. PDK1 regulates auxin transport and *Arabidopsis* vascular development through AGC1 kinase PAX. *Nat Plants* **6:** 544–555. doi:10.1038/s41477-020-0650-2

Yoshida S, Barbier de Reuille P, Lane B, Bassel GW, Prusinkiewicz P, Smith RS, Weijers D. 2014. Genetic control of plant development by overriding a geometric division rule. *Dev Cell* **29:** 75–87. doi:10.1016/j.devcel.2014.02.002

An Essential Function for Auxin in Embryo Development

Subodh Verma, Venkata Pardha Saradhi Attuluri, and Hélène S. Robert

Mendel Centre for Genomics and Proteomics of Plants Systems, CEITEC MU - Central European Institute of Technology, Masaryk University, 625 00 Brno, Czech Republic

Correspondence: helene.robert.boisivon@ceitec.muni.cz

Embryogenesis in seed plants is the process during which a single cell develops into a mature multicellular embryo that encloses all the modules and primary patterns necessary to build the architecture of the new plant after germination. This process involves a series of cell divisions and coordinated cell fate determinations resulting in the formation of an embryonic pattern with a shoot-root axis and cotyledon(s). The phytohormone auxin profoundly controls pattern formation during embryogenesis. Auxin functions in the embryo through its maxima/minima distribution, which acts as an instructive signal for tissue specification and organ initiation. In this review, we describe how disruptions of auxin biosynthesis, transport, and response severely affect embryo development. Also, the mechanism of auxin action in the development of the shoot-root axis and the three-tissue system is discussed with recent findings. Biological tools that can be implemented to study the auxin function during embryo development are presented, as they may be of interest to the reader.

In seed plants, embryogenesis is the process of embryo development that begins with the fertilization of an ovule (de Vries and Weijers 2017). This process will determine the overall architecture of the new plant after germination. The coordinated growth of embryonic tissues governs the final shape and size of the seed that ultimately influence the overall quality and quantity of seed production. In recent years, multiomics approaches, including genetics, epigenetics, transcriptomics, and metabolomics are being used to advance our understanding on embryogenesis mostly in the model plant *Arabidopsis* (Robert et al. 2013; Palovaara et al. 2017; Gao et al. 2019; Han et al. 2019; Radoeva et al. 2019a; Zhou et al. 2019).

Like other developmental processes, embryo development involves the concerted action of several genetic and environmental factors, where the phytohormone auxin plays a prominent role (Mroue et al. 2018). Auxin allows proper polarity and patterning of the developing embryo through its distribution, which is maintained by a local auxin production together with well-coordinated transport (Friml et al. 2003; Cheng et al. 2007; Stepanova et al. 2008; Robert et al. 2013, 2015b; Locascio et al. 2014). The auxin signaling pathway interprets these auxin maxima into proper developmental programs (Hardtke and Berleth 1998; Hamann et al. 2002; Schlereth et al. 2010).

This review aims to offer a comprehensive overview of auxin actions during embryo devel-

Cite this article as *Cold Spring Harb Perspect Biol* doi: 10.1101/cshperspect.a039966

opment. We will highlight recent findings that together define the role of auxin in proper embryo patterning, including some of the latest tools to study this role.

PLANT EMBRYOGENESIS

During plant embryogenesis, the formation of an axial–radial pattern ultimately determines the developmental fates of the embryo (Capron et al. 2009). In *Arabidopsis*, it starts with the asymmetrical zygotic division that generates a small apical cell with dense cytoplasm and a large vacuolated basal cell (Fig. 1). After two rounds of longitudinal divisions and one round of transverse division, the apical cell converts into an octant-stage embryo (eight-celled). The basal cell divides anticlinally. It produces six to nine suspensor cells that help in connecting the embryo to the vascular system of the plant for nutritional purposes (Capron et al. 2009). The eight-celled embryo has two domains, an upper tier and a lower tier (Fig. 1). The upper tier domain of the embryo proper, corresponding to the apical domain, gives rise to the shoot apical meristem (SAM) and the cotyledons (embryonic leaves). The lower tier will generate the embryonic root and hypocotyl (protoderm, ground, and vascular tissues). At the globular stage, the upper suspensor cell, the hypophysis, divides asymmetrically to contribute to the formation of the root apical meristem (RAM).

Radial patterning begins when periclinal divisions give rise to the protoderm, an embryonic epidermis, resulting in a 16-celled embryo (Fig. 1). Divisions of the inner cells, aligned with the apicobasal axis, will form the vascular tissues surrounded by ground tissue cells that will generate endodermis and cortical cell types. The development of cotyledons marks the end of the globular stage, thus giving the embryo a bilateral symmetry and a heart-shaped appearance. Further, the development of cotyledons and cell elongation along the embryo apicobasal axis give rise to a torpedo-shaped embryo (Fig. 1). Accumulation of seed storage reserves such as proteins, fats, lipids, and carbohydrates marks the beginning of the seed-filling phase (Baud et al. 2008). Finally, in the maturation stage, the seed becomes tolerant to desiccation by losing water and enters into the state of dormancy.

PROPER EMBRYO DEVELOPMENT REQUIRES AUXIN BIOSYNTHETIC GENES

Auxin biosynthesis in plants is extremely complex, involving multiple pathways for de novo auxin production. Indole-3-acetic acid (IAA) is the predominant natural auxin in higher plants, which is mainly synthesized via tryptophan-dependent biosynthetic pathways, named after their first-formed intermediate metabolite (Mashiguchi et al. 2011; Ljung 2013; Zhao 2014). One of these pathways, the indole-3-pyruvate (IPyA) pathway, involves two steps, first the conversion of tryptophan to IPyA by the TRYPTOPHAN AMINOTRANSFERASE OF ARABIDOPSIS 1 (TAA1) family, followed by the oxidative decarboxylation of IPyA by the YUCCA (YUC) family to produce IAA. The importance of *TAA1* and *YUC* genes in the auxin biosynthesis pathway has been demonstrated by analyzing their mutant phenotypes. Loss-of-function mutants of *TAA1* and *TAA-RELATED 1* (*TAR1*) genes and multiple *yuc* mutants exhibit severe auxin-deficient phenotypes such as impaired embryogenesis, defects in vascular and floral development, and complete sterility, demonstrating the predominant developmental role of the TAA1-YUC pathway in auxin biosynthesis (Cheng et al. 2007; Stepanova et al. 2008; Robert et al. 2015a; Brumos et al. 2018).

In *Arabidopsis*, five *TAA* and 11 *YUC* genes have been identified (Cao et al. 2019). *YUC1*, *4*, *10*, and *11* display an overall similar expression pattern in developing embryos with concentrated expression in the apical region of the globular embryo and later in cotyledons (Cheng et al. 2007). The *yuc1 yuc4 yuc10 yuc11* quadruple mutants exhibited impaired embryo development, where the hypophysis failed to develop, and only one cotyledon was observed in most mutant embryos (Cheng et al. 2007). Similarly, GFP-TAA1 activity was observed in the apical parts of the embryo, and later in root meristematic regions. Subsequent analysis of their mutant phenotypes (*taa1 tar1 tar2*) corroborated their role in proper embryo development as

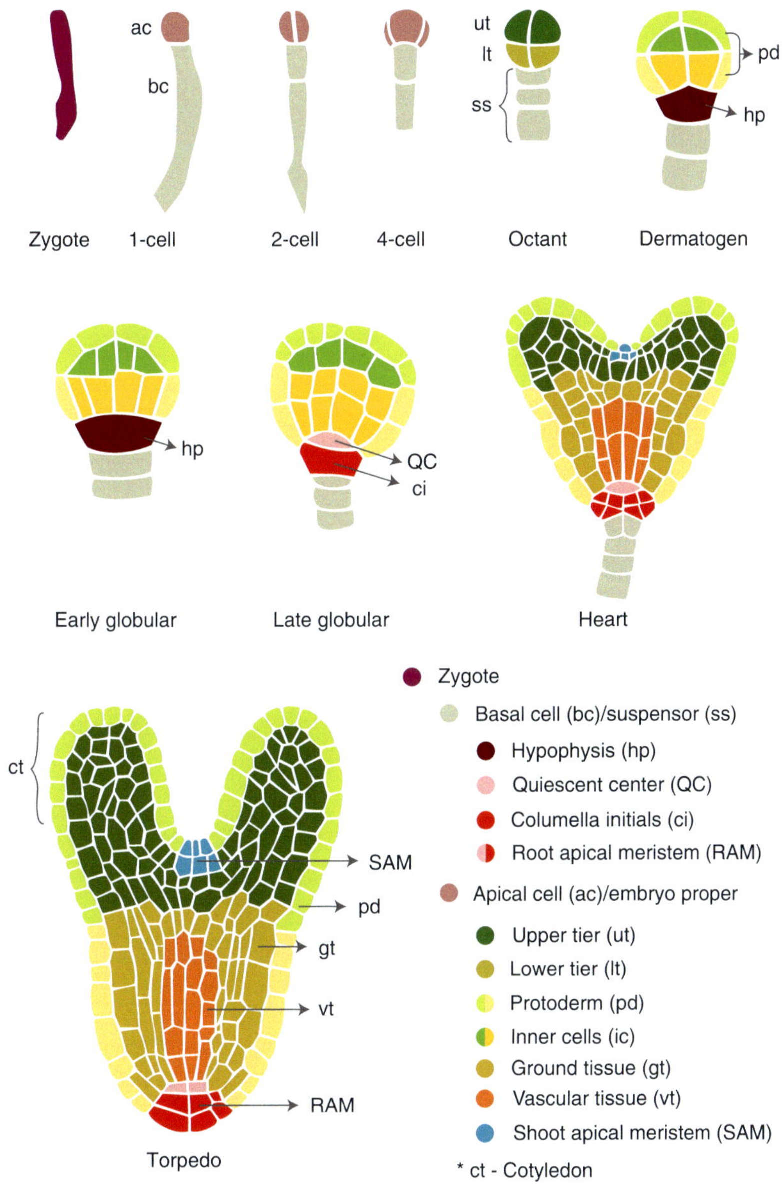

Figure 1. Schematic overview of *Arabidopsis* embryo morphogenesis. Embryo development is shown from the zygote to the torpedo stage. See the text for details.

the triple mutant exhibited developmental defects in the embryo similar to the quadruple *yuc* mutant phenotypes (Stepanova et al. 2008).

YUC3, *4*, and *9* are expressed in the suspensor of the (pre-)globular embryo, indicating their involvement in local auxin biosynthesis in the basal region (Fig. 2D,E; Robert et al. 2013). However, the respective mutants were phenotypically affected in the apical region, suggesting a non-cell-autonomous action of auxin biosynthesis in the development of the apical part of the embryo. Noteworthy, transport played an essential role as auxin is transported from the site of synthesis (suspensor) to the api-

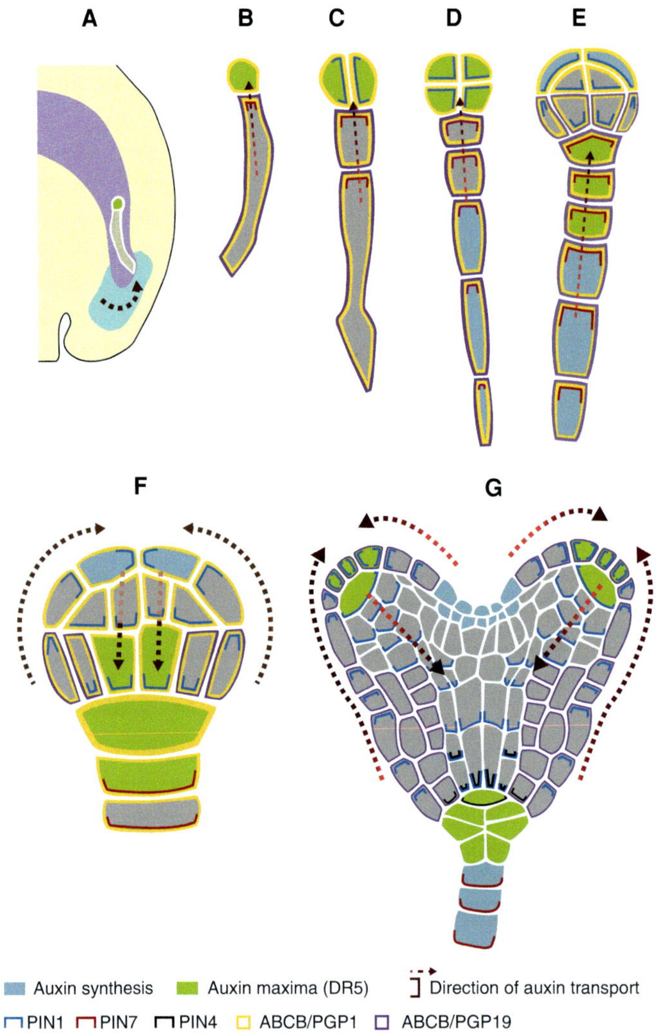

Figure 2. An illustration depicting auxin synthesis (in blue, contributed by TAA1 and YUC proteins), auxin maxima (in green), and transport (shown PIN1 [blue lines], PIN4 [black lines], PIN7 [red lines], ABCB/PGP1 [yellow lines], and ABCB/PGP19 [purple lines]) during *Arabidopsis* embryo development. (*A*) During the early development of an embryo, maternal tissue supplies auxin to the developing embryo, depicted by the arrow. (*B–G*) One-cell, two-cell, eight-cell, 16-cell, 32-cell, and heart-stage embryos, respectively. Dashed arrows depict the direction of auxin transported, extrapolated by the cellular localization of the auxin efflux carriers. Auxin influx carriers are not indicated.

cal embryonic region by PIN7 auxin carriers (Fig. 2; Friml et al. 2003; Robert et al. 2013). Likewise, the onset of auxin biosynthesis by *TAA1*, *YUC1*, and *YUC4* was observed in the apical region of the globular embryo (Robert et al. 2013). From there, auxin flow, mediated by PIN1 proteins, led to the formation of auxin maxima in the basal region, which is required for the specification of the future root pole (Fig. 2; Friml et al. 2003; Robert et al. 2013). In a recent study, it was shown that the auxin biosynthesis machinery involving *TAA1/TAR* and

YUC genes acts in the seed integuments to produce auxin. This maternally produced auxin contributes to early embryo development (Fig. 2; Robert et al. 2018).

AUXIN TRANSPORT IN GOVERNING EMBRYO DEVELOPMENT

Auxin is transported by translocation through the phloem, chemiosmotic gradient (passive membrane diffusion), and directional (polar) transport by influx and efflux carriers located in the plasma membrane. The current understanding of auxin transport during embryo development has been acquired by studying auxin reporter lines (such as DR5) and mutant phenotypes (Friml et al. 2003; Furutani et al. 2004; Blilou et al. 2005; Mravec et al. 2008; Ugartechea-Chirino et al. 2010; Robert et al. 2015b; Swarup and Bhosale. 2019).

Active auxin transport by cellular influx is mediated in the embryo by three importer proteins, AUX1 (AUXIN RESISTANT 1), LAX1 (LIKE AUX1), and LAX2, belonging to the AUX/LAX family, part of the amino acid permease superfamily (Mironova et al. 2017; Swarup and Bhosale 2019). *AUX1* expression begins from the 32-cell stage in the inner cells and later restricted to the provascular cells. *LAX2* follows a similar expression pattern, in addition to the uppermost suspensor cell and hypophysis. In contrast, *LAX1* is expressed in the apical cell, then restricted to the upper tier and later to the cotyledon tips (Robert et al. 2015b). These influx proteins were reported to be important for the formation and organization of shoot and root apex (Ugartechea-Chirino et al. 2010; Robert et al. 2015b).

Auxin efflux can only be achieved by active transport via efflux carriers. It is mediated by PIN (PIN-FORMED) and ABCB/PGP (B subfamily of ATP binding cassette transporters/P-glycoproteins) carriers (Fig. 2). Four PIN proteins (PIN1, 3, 4, and 7) are active during embryo development. Their redundant function is demonstrated by an absence or the presence of subtle embryonic phenotypes in any single *pin* mutants. However, a quadruple mutant (*pin1 pin3 pin4 pin7*) shows severe patterning defects stating the importance of PINs (Friml et al. 2003; Blilou et al. 2005). Nevertheless, the mutant embryos remain viable and can develop into seedlings (Verna et al. 2019).

PIN7 is restricted to the suspensor cells. Its expression is detectable after the zygotic division. It localizes to the apical membrane of the basal cell (Fig. 2B), driving the auxin from the integuments into the apical cell (Fig. 2A; Robert et al. 2018). Auxin accumulation in the apical cell is crucial for its specification into an embryo proper. The *pin7* mutant embryos are filamentous without proper apical cell establishment (Friml et al. 2003). At the 32-cell stage, PIN7 switches to the basal membranes of the suspensor cells, supposedly transporting auxin back to the seed integuments (Fig. 2F). Although PIN7 plays a critical role in transporting auxin between the seed tissues and the embryo, its regulation is poorly understood. Recently, Xiong et al. (2019) reported the importance of JANUS, a spliceosome subunit, in the expression regulation of *PIN7* in the basal cell. *PIN1* expresses in the apical cell lineage. Before the 32-cell stage, PIN1 shows no apparent polarity and is assumed to mediate a uniform auxin distribution within the cells of the embryo (Fig. 2B–E). However, coinciding with the PIN7 polarity switch at the 32-cell stage, PIN1 polarizes to the basal membranes of the inner cells (Fig. 2F). This polarity switch drives auxin into the upper suspensor cells, a critical step for the proper root pole formation. Also, PIN1 plays a key role in positioning and partitioning of cotyledon primordia and boundaries. *PIN3* and *PIN4* expression starts at the late heart stage in columella precursors and hypophysis, respectively (Fig. 2G; Friml et al. 2003).

ABCB/PGP1 and 19 contribute to the auxin efflux. ABCB1 localizes apolarly to the suspensor and all embryonic cells until the mid-globular stage, whereas *ABCB/PGP19* expression is restricted to suspensor lineage (Fig. 2B–F). Later, the *ABCB/PGP19* expression extends to the lower tier in the octant and 16-cell stages. However, at the late-globular and mid-heart stages, ABCB/PGP19 overlaps with PIN1 in protoderm cells and expresses in the ground tissue cells (Fig. 2G). In contrast to *pin* multiple mutants, *pgp1 pgp19* embryos are not morphogenetically defective. However, these transporters exert synergistic effects. Indeed, the *pin1 pgp1 pgp19* triple mutant displays stronger defects in the development of cotyledons than *pin1*, *pgp1*, and *pgp19* alone (Mravec et al. 2008).

AUXIN PERCEPTION AND RESPONSE IN MEDIATING EMBRYO DEVELOPMENT

The signaling mechanisms of auxin action are of central importance in modulating different developmental aspects and responses to environmental cues (Taylor-Teeples et al. 2016; Mroue et al. 2018). Auxin acts as a "molecular glue" that promotes interaction between an F-box protein as auxin receptor, TRANSPORT INHIBITOR RESPONSE1/AUXIN SIGNALING F-BOX (TIR1/AFB), and the members of Aux/IAA transcriptional repressors (Leyser 2018). This auxin-mediated binding between TIR1/AFB and Aux/IAA promotes the ubiquitination-dependent degradation of Aux/IAA proteins, which in turn releases the transcription factors (TFs) belonging to the AUXIN RESPONSE FACTOR (ARF) family (Lavy and Estelle 2016; Leyser 2018). These TFs bind to auxin response elements (AuxREs) present over the promoter sequences of different target genes and regulate their expression (Mironova et al. 2014).

The TIR1/AFB F-box proteins are the substrate recognition component of SCF (Skp1, Cullin, and F-box protein) ubiquitin-ligase complexes (Salehin et al. 2015; Leyser 2018). Six *AFB* genes, including *TIR1* and *AFB1-5*, have been identified in *Arabidopsis* (Dharmasiri et al. 2005a; Prigge et al. 2020). Genetic and molecular analysis revealed their involvement in mediating auxin response throughout plant development, including embryogenesis (Dharmasiri et al. 2005b; Parry et al. 2009; Prigge et al. 2016, 2020). Mutant phenotypes of all six *TIR1/AFB* genes were analyzed individually and in 63 mutant combinations (Dharmasiri et al. 2005b; Prigge et al. 2020). Embryogenesis was not affected in any of the single mutants, probably due to the functional redundancy. In contrast, the mutant combinations displayed a wide range of developmental defects during the root, shoot, gametophyte, and embryo development. Out of those mutants, the triple (*tir1 afb2 afb3*), quadruple (*tir1 afb2 afb3 afb5* and *tir1 afb1 afb2 afb3*), and sextuple (*tir1 afb1 afb2 afb3 afb4 afb5*) mutants exhibited severe embryonic phenotypes such as abnormal cell division in the upper and lower tiers of the embryo, lack of cotyledons, aberrant hypophysis development, and overproliferating suspensors (Dharmasiri et al. 2005b; Prigge et al. 2020). Also, one-quarter of the sextuple mutant embryos were either lethal or aborted. Besides, defects in radial patterning events were also observed in the *tir1 afb2 afb3 afb5* mutant. These studies demonstrate the roles of the TIR/AFB subunit of $SCF^{TIR1/AFB}$ complexes during embryo development.

In *Arabidopsis*, 15 out of 29 Aux/IAA short-lived repressors are expressed in embryo and suspensor (Weijers and Jürgens 2005; Ploense et al. 2009; Rademacher et al. 2012). Only a few have been functionally studied in combination with their ARF counterparts. ARFs, downstream components of the auxin response pathway, act either as activators or repressors to regulate the expression of their target genes. In *Arabidopsis*, out of 23 *ARF* genes, 12 *ARFs* exhibited diverse expression patterns throughout embryogenesis, suggesting their potential role in different developmental stages of *Arabidopsis* embryo (detailed in Rademacher et al. 2011). However, until now, strong embryonic phenotypes in *Arabidopsis* have only been observed for mutations in the *ARF5/MONOPTEROS* (*MP*) gene and its repressor *IAA12/BODENLOSS* (*BDL*) (Berleth and Jurgens 1993; Schlereth et al. 2010; Rademacher et al. 2012). Also, ARF9 and ARF13, and their common repressor IAA10, regulate hypophysis specification and suspensor cell identity in *Arabidopsis* embryo (Rademacher et al. 2012).

AUXIN IN SUSPENSOR-DERIVED EMBRYOGENESIS

Derived from the basal cell of the two-celled embryo, suspensor cells remain extraembryonic except for the uppermost suspensor cell. This uppermost suspensor cell becomes the hypophysis in dermatogen embryos and ultimately participates to the embryonic root development. The other suspensor cells become quiescent after the suspensor is fully developed by the early globular stage (Mansfield and Briarty 1991). Later, the suspensor degenerates via programmed cell death and, therefore, does not contribute to the mature seed (Peng and Sun 2018). In *Arabidopsis*, despite their quiescence, suspensor cells can develop secondary embryos after impairment of the proembryo through mutations, treatments that damage or kill embryonic cells (Schwartz et al. 1994; Vernon and Meinke 1994), cell ablation (Weijers et al. 2003), or laser disruption (Gooh et al. 2015; Liu et al. 2015). This suggests that suspensor cells have an innate capability to generate embryos, and this activity is suppressed by an embryonic signal. Rademacher et al. (2012) demonstrated that a cell-autonomous auxin response (mediated by IAA10 and ARF9) is required to maintain the extraembryonic identity of suspensor cells (Fig. 3B). When this local auxin response is inhibited, suspensor cells express embryonic marker genes and acquire embryonic fate, resulting in the formation of a second embryo and eventually twin-like seedlings. This auxin-mediated, suspensor-derived embryogenesis is a complex process, which involves several genes associated with auxin homeostasis and response (Radoeva et al. 2019b). Among them, the prominent ones are *bHLH* genes (*bHLH49*, *60*, *63*, and *100*) that mediate the auxin response and control suspensor proliferation and/or embryonic fate conversion. In particular, *bHLH49* is a direct mediator of the auxin response that controls the developmental potential of the suspensor (Radoeva et al. 2019b). Reprogramming of suspensor cells toward embryonic identity is triggered by defined regulators. This process is marked by the activation of cell divisions that generate cells in which reprogramming occurs, followed by loss of suspensor identity and activation of embryonic fate in newly formed cell clusters (Radoeva et al. 2020).

AUXIN-MEDIATED EMBRYO PATTERNING

In plant embryo, body poles are anatomically termed as apex and base (Friml et al. 2006). During apicobasal patterning, SAM and cotyledons are originated from the apex (upper tier), whereas the base (lower tier) produces hypocotyl, radicle, and RAM. In contrast, the development and arrangement of a three-tissue system (protodermal, vascular, and ground tissue) occur during radial patterning (Capron et al. 2009). In this part, we will review the role of auxin in pattern formation during embryogenesis.

Auxin Control of Cell Division

Embryo patterning relies on coordinated formative cell divisions that produce different cell layers by orienting the division plane (Yoshida et al. 2014). Besides, embryo development heavily depends on differential cell specification, which is predicted by geometric asymmetry during cell division. Auxin overrides the default geometric division rule and forces the cell to select another division wall, therefore, controlling pattern formation during embryo morphogenesis (Yoshida et al. 2014). Auxin affects division plane orientation by influencing the orientation of cortical microtubule (MT) arrays (CMAs) (Chakrabortty et al. 2018). A link between CMA orientation and division plane orientation has been established (Wasteneys 2002; Ehrhardt and Shaw 2006). Auxin-mediated MT stability, the influence of the cell shape, and reduced edge catastrophe are the factors governing CMA orientation that predicts division plane orientation, leading to oriented cell division in the early embryos (one-cell to 16-cell stages) (Chakrabortty et al. 2018). Cell-polarization relative to body axes is also crucial for patterning events. This polarity information influences division orientation during multicellular development, which is mediated by polar, edge-localized SOSEKI proteins (Yoshida et al. 2019). *SOSEKI1* (*SOK1*) has been identified in the transcriptome of the globular-stage embryo in which MP activity was locally inhibited (Möller et al. 2017). It was the second most down-regulated gene (Schlereth et al. 2010). It suggests that ARF5/MP transcriptionally regulates *SOK1*. *SOK1* expression is localized to apical edges of lower-tier inner cells at globular-stage embryo, and in the corners of the hypophysis and vascular cells at the transition to heart stage. Misexpression of *SOK1* leads to altered cell divi-

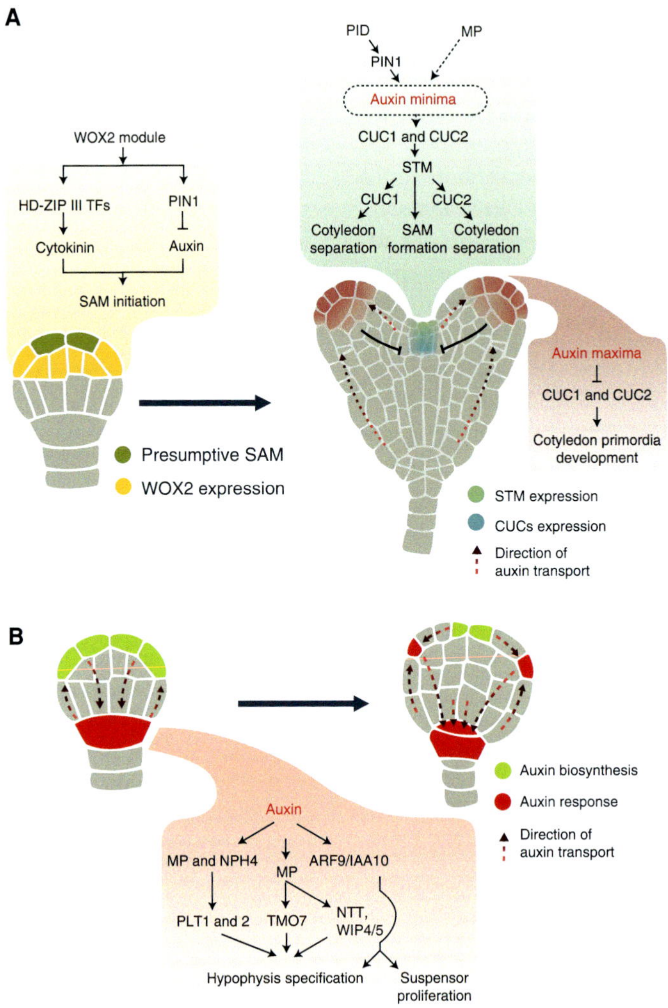

Figure 3. A model depicting initiation of shoot apical meristem (SAM) and root apical meristem (RAM) during embryogenesis. (*A*) At the globular stage, the WOX2 module acts in SAM initiation by promoting the expression of *HD-ZIP III* transcription factors (TFs) and *PIN1* in the apical region, which in turn enhances cytokinin response and represses auxin response, respectively. At the early heart stage, the expression of *CUC1* and *CUC2* is regulated by the activity of PIN1, PINOID (PID), and MP, most probably by creating an auxin gradient with a maximum in the cotyledon primordia and minimum in the center of the embryonic apex. *CUC* genes are preferentially expressed in the region with low auxin levels (i.e., between the cotyledon primordia where they promote cotyledon separation). CUC proteins are also required to activate *STM* expression. STM promotes SAM formation and also contributes to cotyledon separation by regulating *CUC1* and *CUC2* expression. PIN1 and PID-mediated auxin maxima suppress *CUC* genes expression in cotyledon primordia that eventually allows primordia to develop and also prevents (T-shaped arrow) the expression of *CUC* genes from expanding. (*B*) PIN-mediated basal auxin accumulation promotes MP and NPH4 activity, which is required for hypophysis specification. MP, through its targets (PLT1, PLT2, TMO7), triggers the asymmetric division of hypophysis. The ARF9/IAA10 module is required for hypophysis specification and to control suspensor cell proliferation. The dashed black circle and arrow in the green *inset* in *A* indicate signaling steps that need further experimental validation. The direction of auxin transport is represented by a dashed line.

sion orientations in all cell types in the embryo. However, the detailed cellular mechanism underlying SOK1 activity in division plane orientation is yet to be determined.

Initiation of Apicobasal Polarity: Emphasis on Auxin

Initial separation of the apical and basal cells after the first division of the zygote establishes a foundation for subsequent patterning events (Haecker et al. 2004; Laux et al. 2004; Capron et al. 2009). This crucial fate decision relies upon the differential expression of the *WUSCHEL-related homeobox* (*WOX*) TFs along the apicobasal axis (Haecker et al. 2004). *WOX2*, *WOX8/STPL* (*STIMPY-LIKE*; Wu et al. 2007), and *WOX9* (*STIMPY*; Wu et al. 2005) are expressed in different regions of the eight-celled embryo. *WOX2* is restricted to the apical region, whereas *WOX8* is expressed in the suspensor cells and hypophysis. Conversely, Lie et al. (2012) observed a weak expression of *WOX8* in the embryo proper. The expression of *WOX9* is confined to the upper suspensor cell at four-cell and eight-cell stages, and the lower tier domain of the eight-celled embryo. The shift of *WOX9* expression from the hypophysis to the lower tier is required for the specification of the lower tier domain identity (Haecker et al. 2004). These *WOX9* expression dynamics are absent in *arf5/mp* and *iaa12/bdl* mutants, suggesting that the MP/BDL-mediated auxin response is required for the changes of *WOX9* expression (Haecker et al. 2004).

WOX2 is the main regulator of embryonic shoot patterning in the cascade of *WOX* genes. The activity of *WOX1*, *WOX3/PRS* (*PRESSED FLOWER*), and *WOX5* is only relevant when WOX2 activity is absent. Conversely, whereas expressed in the basal domain, both the basal and apical embryonic lineages in the *wox8 wox9* double mutants are defective. Linking auxin to the WOX pathway, *wox8 wox9* double mutants showed reduced *PIN1* expression, suggesting that WOX8/WOX9 activity is required for normal PIN1 activity and localized auxin response maxima (Breuninger et al. 2008). Likewise, a genetic interaction between *WOX2*, *WOX8*, and *ARF5/MP* was observed during embryo patterning, with strongly enhanced defects in the *wox2 wox8 mp* triple mutants. WOX2, WOX8, and ARF5/MP act redundantly to regulate *PIN1* expression in the cotyledons (Breuninger et al. 2008). Taken together, these observations suggest that WOX genes control embryo patterning in concert with auxin-mediated responses and may involve other auxin-independent factors.

Role of Auxin in Initiating SAM

The SAM development begins with the specification of stem cells between the presumptive cotyledons in the globular embryo (Fig. 1; Kaplan and Cooke 1997; Mayer et al. 1998). During apical patterning, a transition from radial to bilateral symmetry takes place with the development of the cotyledon primordia. The development of cotyledons relies upon two coordinated processes: outgrowth of the cotyledon primordia and development of the cotyledon boundaries (Lie et al. 2012). The well-proportioned separation of cotyledon primordia is obligatory for the establishment of SAM (Aida et al. 1997, 1999).

SAM development and cotyledon separation are regulated by a network of genes involving *CUP-SHAPED COTYLEDON1* (*CUC1*), *CUC2*, *CUC3*, and *SHOOT MERISTEMLESS* (*STM*) (Barton and Poethig 1993; Aida et al. 1999; Takada et al. 2001; Vroemen et al. 2003). In agreement with their function, the three *CUC* genes have an overlapping expression domain in a central stripe of cells between the cotyledon primordia in early-heart-stage embryos (Fig. 3A). CUC1 and CUC2 act synergistically with STM to control SAM formation. CUC1 and CUC2 are required to activate STM expression, and STM in turn is necessary for CUC1 activity from early embryogenesis onward and spatial expression of CUC2 at the bending-cotyledon stage (Fig. 3A; Aida et al. 1999). Three essential factors of auxin transport and response pathway (i.e., PIN1, PINOID [PID]), and ARF5/MP, regulate *CUC1* and *CUC2* expression to confer cotyledon separation and establishment of the bilateral symmetry (Aida et al. 2002; Furutani et al. 2004). PID, a serine/threonine protein kinase, phosphorylates PIN1 proteins for their proper localization and activity (Huang et al. 2010;

Zhang et al. 2010). To establish the cotyledon primordia, PIN1 and PID synergistically promote auxin accumulation in cotyledon primordia, which in turn represses the expression of *CUC1* and *CUC2* therein. Indeed, *pin1* and *pid* mutant seedlings are both defective in the number of cotyledons (Furutani et al. 2004). Moreover, genetic and expression analysis indicated that PIN1 and ARF5/MP restrict *CUC1* expression to the central embryonic apex and activate *CUC2* expression in the cotyledon boundaries (Fig. 3A). It has been speculated that an auxin minimum in the presumptive SAM region due to the activity of PIN1 and MP may stimulate the expression of *CUC* genes between the cotyledon primordia (Fig. 3A). Furthermore, PIN1 promotes the establishment of bilateral symmetry in parallel to STM as *pin1 stm1* double mutants have completely fused cotyledons (Aida et al. 2002).

WOX2 and WOX8 redundantly promote the establishment of cotyledon boundaries by regulating the symmetric expression of the three *CUC* genes during apical patterning (Lie et al. 2012). WOX2, along with its redundant paralogs (i.e., WOX1, WOX3, and WOX5) (collectively termed as the WOX2 module), is required for the initiation of the embryonic apical meristem (Fig. 3A; Zhang et al. 2017). The WOX2 module promotes the development of a three-layered shoot meristem by regulating the cell division pattern in the apical domain, including the presumptive SAM region. By promoting *PIN1* expression, the WOX2 module prevents auxin accumulation in the SAM region. Furthermore, the WOX2 module is required to enhance the expression of HD-ZIP III genes (*PHABULOSA*, *PHAVULOTA*, and *REVOLUTA*) in the presumptive SAM region. These genes are known regulators of embryonic apical fate and are expressed in the central domain during shoot meristem formation (Smith and Long 2010). This suggests that HD-ZIP III TFs act downstream of the WOX2 module in the same genetic pathway to regulate stem cell initiation during embryonic apical patterning. In addition to the auxin pathway, the WOX2 module also balances the cytokinin pathway through the HD-ZIP III TFs, thus helping in the formation of embryonic shoot meristem (Fig. 3A; Zhang et al. 2017). Therefore, this module contributes to stem cell initiation and protects them from differentiation, thereby ensuring that the SAM region is correctly specified in the apical domain.

Role of Auxin in Embryonic Root Formation

The RAM harbors concentrically arrayed stem cells (initials of different tissue types) around a quiescent center (QC). The asymmetric division of the hypophysis marks the initiation of the RAM at the globular stage (Capron et al. 2009). This division generates an upper lens-shaped cell (progenitor of the QC) and a large basal cell (columella root cap progenitor). The QC functions in maintaining neighboring stem cells by keeping them in an undifferentiated state (van den Berg et al. 1997). The importance of the hypophysis specification in RAM initiation and the involvement of auxin signaling are reflected by the rootless phenotype of two well-studied mutants (i.e., *mp* and *bdl*) (Berleth and Jurgens 1993; Hamann et al. 1999, 2002; Weijers et al. 2006).

BDL and *MP* mRNAs are absent from the hypophysis until its division and confined to the central provascular cells (Hamann et al. 2002), suggesting that both proteins act non-cell-autonomously. Targets of the MP/BDL signaling pathway are AUX1, LAX2, and PIN1 auxin transporters (Schlereth et al. 2010; Robert et al. 2015b), acting in the embryonic inner cells to direct an auxin flow toward the hypophysis. The resulting accumulation of auxin in the hypophysis promotes the RAM specification. However, exogenous auxin treatment could not rescue the rootless phenotype of *mp* and *bdl* mutants, indicating the involvement of other MP/BDL-dependent signals in regulating hypophysis specification (Weijers et al. 2006).

Three direct targets of MP, named as TARGET OF MONOPTEROS (TMO) 3, TMO5, and TMO7, were identified in the embryo (Schlereth et al. 2010). TMO5 and TMO7 were found to be involved in MP-dependent embryonic root formation (Fig. 3B). TMO5 functions cell-autonomously as its expression was restricted to its transcriptional domain. In contrast,

TMO7 moves from its site of synthesis to the future hypophysis, indicating that TMO7 acts as an intercellular MP-dependent signal in the hypophysis for RAM specification. Other direct targets of MP, such as NO TRANSMITTING TRACT (NTT) and its paralogues WIP DOMAIN PROTEIN 4 (WIP4) and WIP5, are expressed in the hypophysis and redundantly promotes its asymmetric division, which marks the RAM initiation (Fig. 3B; Crawford et al. 2015).

The *PLETHORA* (*PLT*) genes are another indispensable gene network regulating the RAM establishment (Fig. 3B). The AP2 domain-containing TFs PLT1 and PLT2 are redundantly fundamental to the development and specification of the root stem cell niche in the embryo (Aida et al. 2004). The expression of both genes is confined to the basal region of the embryo, from where hypocotyl and embryonic root originate. Owing to the lack of PLT1 and PLT2 activity, hypophyseal derivatives divide aberrantly and fail to express QC markers. *PLT* genes act downstream of the auxin response, as suggested by their elevated expression after prolonged exogenous auxin treatment. Their basal embryonic expression depends on the redundant function of MP/ARF5 and its homolog NON-PHOTOTROPIC HYPOCOTYL4 (NPH4)/ARF7 (Hardtke et al. 2004). Moreover, the ectopic activity of PLT1 and PLT2 induces basal features (i.e., hypocotyl and root stem cell niche) in the embryonic shoot region, even in the absence of auxin. Furthermore, whereas a PIN1- and PIN4-dependent auxin maximum promotes root pole specification (Friml et al. 2003), these same PIN proteins are required to restrict the *PLT* expression domain to the basal region of the embryo to initiate the root primordia formation (Blilou et al. 2005). Altogether, these observations strongly support the notion that PIN-mediated accumulation of auxin in the basal domain is required to initiate and establish RAM through the transcriptional regulation of *PLT* genes.

It has been recently demonstrated that *PLT* genes are required for vascular regeneration in response to wounding, by directly regulating *CUC2* expression (Radhakrishnan et al. 2020). To promote vascular regeneration, both PLT and CUC2 modulate local auxin biosynthesis by regulating the *YUC4* expression in a coherent feedforward manner. Although this mechanism of the PLT-CUC2 module was observed in leaf, similar mechanisms may exist in the embryo as well.

Role of Auxin for the Formation and Patterning of the Vascular and Ground Tissues

Vascular tissues are essential for transport and structural stability/support. In a recent report, Smit et al. (2020) used a panel of vascular markers to understand the specification and ontogeny of the vascular tissue identity during embryogenesis. Most of these markers are active in the inner cells of a 16-cell embryo, indicating that these cells are the first to contribute to the vasculature (Smit et al. 2020). Vascular markers such as *SOK1* and *GATA20* lost their activity in *bdl* embryos, where the MP/ARF5 activity is blocked. The auxin-mediated signaling through MP/ARF5 appears to be required for vascular tissue initiation, specification, and patterning in the early embryo. Indeed, in *mp* embryos, abnormal periclinal vascular divisions are observed (De Rybel et al. 2013). TMO5, one direct target of MP, forms a heterodimer with LONESOME HIGHWAY (LHW) (Ohashi-Ito and Bergmann 2007) to mediate an auxin response in vascular tissue initiation. This TMO5-LHW dimer is required to regulate vascular indeterminacy and to promote periclinal cell divisions during vasculature formation. Consistent with their function, both proteins colocalize in vascular precursor cells at the mid-globular embryo. At the heart stage, TMO5 accumulation is restricted to xylem initials, whereas LHW accumulates over a wide area with maximum accumulation in the root pole and cotyledon primordia (De Rybel et al. 2013).

An auxin-cytokinin hormonal interaction is critical for vascular tissue development. The TMO5-LHW dimer promotes local cytokinin biosynthesis by regulating the expression of LONELY GUY4 (LOG4), an enzyme involved in the final step of cytokinin biosynthesis (Kuroha et al. 2009; De Rybel et al. 2014). This auxin-dependent cytokinin production is crucial for patterning of the vascular tissue and vascular growth. Mathematical modeling for the activity of the "auxin-MP-TMO5/LHW-LOG4-cytokinin" module in both embryonic and postembryonic roots indicates that this network is required to create two distinct domains during vascular tissue patterning: xylem axis with high auxin response and surrounding procambium cells with high cytokinin response. This network also involves the activity of PIN proteins whose localization dynamics are regulated by auxin-cytokinin intricacies (De Rybel et al. 2014).

While some of the molecular mechanisms for the vascular tissue initiation have been uncovered, the ones for the ground tissue initiation are still mysterious. Möller et al. (2017) observed aberrant division in the first ground tissue cells in the absence of ARF5/MP-mediated auxin response. Embryonic markers for ground tissue initiation were identified by transcriptome profiling of the early globular embryos, in which MP activity was blocked in the first ground tissue cells. Among them, known regulators of postembryonic ground tissue patterning and maintenance, such as SHORT-ROOT (SHR) and its direct target genes, were also present (Helariutta et al. 2000; Nakajima et al. 2001; Levesque et al. 2006). However, the SHR network does not affect division in the first ground tissue cells, suggesting the involvement of distinct regulators in establishing the ground tissue in the early embryo (Möller et al. 2017). Nevertheless, transcriptional regulation of SHR by MP activity implies, so far, an unpredicted role for SHR in initial ground tissue specification, which needs to be elucidated.

Overall, these observations suggest that the apicobasal embryonic polarity is largely regulated by PIN-mediated auxin transport, from a source (local production) to a sink (local signaling). In addition, the activity of ARF5/MP in auxin-dependent regulation of embryonic patterning is intricated within a multiple of signaling cascades.

TOOLS TO STUDY AUXIN FUNCTION IN EMBRYOS

In recent years, a few noteworthy tools joined the auxin toolbox. Among them, we will present two tools (hormone-activated Cas9-based repressors [HACRs], pronounced "hackers" and cellular markers for embryogenesis) that may be used for studying embryo development (Fig. 4).

HACRs

HACRs can be used as hormone reporters (auxin, gibberellins, and jasmonates) as well as degron-controlled genetic switches. HACRs are CRISPR-based genetic switches (Khakhar et al. 2018). HACRs are modular proteins designed to temper the expression of a target gene. It contains three domains (i.e., dCas9 [for the recognition of target sequence]), a degron, and a transcriptional regulator (to modulate the target gene expression). Khakhar et al. (2018) generated HACRs with degrons that respond to auxin, gibberellins, and jasmonates (Fig. 4A).

The auxin HACR functions as a radiometric auxin sensor similar to R2D2 (Liao et al. 2015). This HACR contains gRNAs targeting the *UBQ1* promoter, an auxin-sensitive degron (from IAA17), and a repressor domain from TOPLESS (TPL) (Fig. 4A). When introduced in plants expressing a radiometric readout (*pUBQ1:NLS-Venus; pUBQ1m:NLS-tdTomato*, in which *pUBQ1m* contains a mutated gRNA target site), this HACR displayed a similar expression pattern as in R2D2, indicative of tissular auxin distribution (Fig. 4A,B).

HACRs have a great potential for rewiring the signaling pathways. It includes establishing new feedback loops, targeting genes of one hormone pathway, and using degrons of another hormone. This level of granular control would be challenging to achieve with any other technologies. Applied to embryo development as a readout for patterning programs, HACRs would be a powerful tool to study cell-specific processes and the function of TFs and other effectors.

Arabidopsis Cellular Markers for Embryogenesis

To study the cellular organization and organellar dynamics of early embryo development, Liao and Weijers (2018) developed a series of fluorescent marker lines named "*Arabidopsis* cellu-

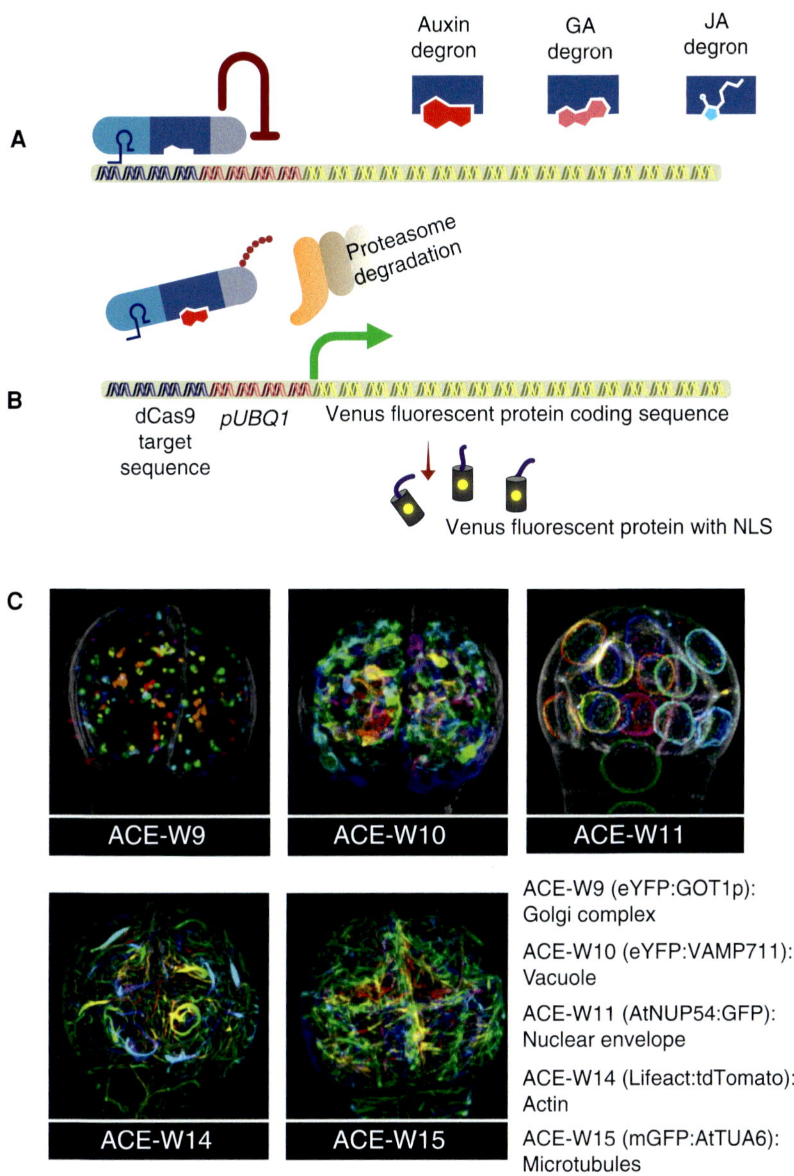

Figure 4. An illustration representing the mechanisms of HACRs (hormone-activated Cas9-based repressors) (*A*, *B*) and images representing a selection of the ACE-W lines (*C*). (*A*,*B*) The HACR system is designed to control the expression of NLS-Venus fluorescent proteins based on the concentration of auxin in the cell. In the absence of auxin (*A*), TOPLESS (TPL) represses the expression of NLS-Venus. The presence of auxin (*B*) in the cell targets the HACRs to proteasome-mediated degradation, thus derepressing the NLS-Venus expression. The HACR consists of gRNAs (light blue) targeting the *UBQ1* promoter, an auxin-sensitive degron (dark blue), and a repressor domain from TPL (gray). Degrons can be switched to monitor GA or JA levels. (*C*) Representative images of *Arabidopsis* 32-cell-stage embryos expressing ACE markers labeling Golgi complex (ACE-W9), vacuole (ACE-W10), nuclear envelope (ACE-W11), actin cytoskeleton (ACE-W14), and microtubules (ACE-W15). (Panel *C* from Liao and Weijers 2018; reprinted, with permission, from John Wiley © 2018.)

lar markers for embryogenesis" (ACE). The ACE expression is under the control of two different promoters, a meristematic *pRPS5A* (ACE-R) and an embryo-specific *pWOX2* (ACE-W). The ubiquitous expression of *pRPS5A* allows the imaging of the markers throughout embryogenesis. The *WOX2* promoter, on the other hand, has a limited window of expression until the early globular stage, with the advantage to be embryo-specific. These subcellular reporters label plasma membrane, connectivity between the cells (plasmodesmata), cytoskeleton (actin and MTs), various organelles (nuclear pore complex for nuclear labeling, vacuole), and components of endomembrane systems (Fig. 4C). Also, the series includes polarly localized proteins to mark apical, basal, central, and peripheral domains of the plasma membrane. These markers are useful to study the polarity of cells and the dynamics of organelles in wild-type and defective mutant embryos in various auxin-related processes.

CONCLUDING REMARKS

Our understanding of auxin function in embryo development has increased substantially in the past decade. The data available so far indicates that auxin acts through its instructive maxima/minima distribution and regulates cell specification and organ initiation during embryo patterning. Such auxin distribution is established and sustained by a structured interplay between local auxin biosynthesis and polar auxin transport. However, many questions remained to be answered regarding auxin synthesis and transport during embryo development. For example, we are far away from knowing the mechanisms behind the transcriptional regulation of auxin biosynthetic genes during embryo development. Yet, another challenge would be to identify the factors coordinating auxin biosynthesis and transport in this complex tissue. MP/ARF5 has been identified as a critical player of auxin-dependent responses in the embryo. However, many distinct combinations of AUX/IAA and ARFs may exist for individual cellular auxin responses. Systematic analysis of these components and identification of their target genes in the embryo will shed new light on the mechanisms by which auxin exerts its function during embryo development. Further, to reach a new level of understanding, biological tools to monitor auxin synthesis, signaling, and responses, notably in developing embryos, are newly available, among them, HACR. Combined with state-of-the-art imaging techniques, like light-sheet microscopy, such tools would enable breakthroughs in embryonic auxin biology, as we have recently seen for male germline differentiation (Valuchova et al. 2020).

ACKNOWLEDGMENTS

We apologize to those colleagues whose work was not included here due to space constraints. This work was supported by the European Regional Development Fund-Project "Centre for Experimental Plant Biology" (No. CZ.02.1.01/0.0/0.0/16_019/0000738). The authors declare no conflict of interest. The funders had no role in the writing of the manuscript or in the decision to publish the results.

REFERENCES

Aida M, Ishida T, Fukaki H, Fujisawa H, Tasaka M. 1997. Genes involved in organ separation in *Arabidopsis*: an analysis of the cup-shaped cotyledon mutant. *Plant Cell* **9:** 841–857. doi:10.1105/tpc.9.6.841

Aida M, Ishida T, Tasaka M. 1999. Shoot apical meristem and cotyledon formation during *Arabidopsis* embryogenesis: interaction among the CUP-SHAPED COTYLEDON and SHOOT MERISTEMLESS genes. *Development* **126:** 1563.

Aida M, Vernoux T, Furutani M, Traas J, Tasaka M. 2002. Roles of PIN-FORMED1 and MONOPTEROS in pattern formation of the apical region of the *Arabidopsis* embryo. *Development* **129:** 3965.

Aida M, Beis D, Heidstra R, Willemsen V, Blilou I, Galinha C, Nussaume L, Noh YS, Amasino R, Scheres B. 2004. The PLETHORA genes mediate patterning of the *Arabidopsis* root stem cell niche. *Cell* **119:** 109–120. doi:10.1016/j.cell.2004.09.018

Barton MK, Poethig RS. 1993. Formation of the shoot apical meristem in *Arabidopsis thaliana*: an analysis of development in the wild type and in the shoot meristemless mutant. *Development* **119:** 823.

Baud S, Dubreucq B, Miquel M, Rochat C, Lepiniec L. 2008. Storage reserve accumulation in *Arabidopsis*: metabolic and developmental control of seed filling. *Arabidopsis Book* **6:** e0113. doi:10.1199/tab.0113

Berleth T, Jurgens G. 1993. The role of the monopteros gene in organising the basal body region of the *Arabidopsis* embryo. *Development* **118:** 575.

Blilou I, Xu J, Wildwater M, Willemsen V, Paponov I, Friml J, Heidstra R, Aida M, Palme K, Scheres B. 2005. The PIN auxin efflux facilitator network controls growth and patterning in *Arabidopsis* roots. *Nature* **433:** 39–44. doi:10 .1038/nature03184

Breuninger H, Rikirsch E, Hermann M, Ueda M, Laux T. 2008. Differential expression of *WOX* genes mediates apical-basal axis formation in the *Arabidopsis* embryo. *Dev Cell* **14:** 867–876. doi:10.1016/j.devcel.2008.03.008

Brumos J, Robles LM, Yun J, Vu TC, Jackson S, Alonso JM, Stepanova AN. 2018. Local auxin biosynthesis is a key regulator of plant development. *Dev Cell* **47:** 306–318. e5. doi:10.1016/j.devcel.2018.09.022

Cao X, Yang H, Shang C, Ma S, Liu L, Cheng J. 2019. The roles of auxin biosynthesis YUCCA gene family in plants. *Int J Mol Sci* **20:** 6343.

Capron A, Chatfield S, Provart N, Berleth T. 2009. Embryogenesis: pattern formation from a single cell. *Arabidopsis Book* **7:** e0126. doi:10.1199/tab.0126

Chakrabortty B, Willemsen V, de Zeeuw T, Liao CY, Weijers D, Mulder B, Scheres B. 2018. A plausible microtubule-based mechanism for cell division orientation in plant embryogenesis. *Curr Biol* **28:** 3031–3043.e2. doi:10 .1016/j.cub.2018.07.025

Cheng Y, Dai X, Zhao Y. 2007. Auxin synthesized by the YUCCA flavin monooxygenases is essential for embryogenesis and leaf formation in *Arabidopsis. Plant Cell* **19:** 2430–2439. doi:10.1105/tpc.107.053009

Crawford BCW, Sewell J, Golembeski G, Roshan C, Long JA, Yanofsky MF. 2015. Genetic control of distal stem cell fate within root and embryonic meristems. *Science* **347:** 655–659. doi:10.1126/science.aaa0196

De Rybel B, Möller B, Yoshida S, Grabowicz I, Barbier de Reuille P, Boeren S, Smith RS, Borst JW, Weijers D. 2013. A bHLH complex controls embryonic vascular tissue establishment and indeterminate growth in *Arabidopsis. Dev Cell* **24:** 426–437. doi:10.1016/j.devcel.2012.12.013

De Rybel B, Adibi M, Breda AS, Wendrich JR, Smit ME, Novák O, Yamaguchi N, Yoshida S, Van Isterdael G, Palovaara J, et al. 2014. Integration of growth and patterning during vascular tissue formation in *Arabidopsis. Science* **345:** 1255215. doi:10.1126/science.1255215

de Vries SC, Weijers D. 2017. Plant embryogenesis. *Curr Biol* **27:** R870–R873. doi:10.1016/j.cub.2017.05.026

Dharmasiri N, Dharmasiri S, Estelle M. 2005a. The F-box protein TIR1 is an auxin receptor. *Nature* **435:** 441–445. doi:10.1038/nature03543

Dharmasiri N, Dharmasiri S, Weijers D, Lechner E, Yamada M, Hobbie L, Ehrismann JS, Jürgens G, Estelle M. 2005b. Plant development is regulated by a family of auxin receptor F box proteins. *Dev Cell* **9:** 109–119. doi:10.1016/j .devcel.2005.05.014

Ehrhardt DW, Shaw SL. 2006. Microtubule dynamics and organization in the plant cortical array. *Annu Rev Plant Biol* **57:** 859–875. doi:10.1146/annurev.arplant.57.032 905.105329

Friml J, Vieten A, Sauer M, Weijers D, Schwarz H, Hamann T, Offringa R, Jürgens G. 2003. Efflux-dependent auxin gradients establish the apical–basal axis of *Arabidopsis. Nature* **426:** 147–153. doi:10.1038/nature02085

Friml J, Benfey P, Benková E, Bennett M, Berleth T, Geldner N, Grebe M, Heisler M, Hejátko J, Jürgens G, et al. 2006. Apical-basal polarity: why plant cells don't stand on their heads. *Trends Plant Sci* **11:** 12–14. doi:10.1016/j.tplants .2005.11.010

Furutani M, Vernoux T, Traas J, Kato T, Tasaka M, Aida M. 2004. PIN-FORMED1 and PINOID regulate boundary formation and cotyledon development in *Arabidopsis* embryogenesis. *Development* **131:** 5021–5030. doi:10 .1242/dev.01388

Gao P, Xiang D, Quilichini TD, Venglat P, Pandey PK, Wang E, Gillmor CS, Datla R. 2019. Gene expression atlas of embryo development in *Arabidopsis. Plant Reprod* **32:** 93–104. doi:10.1007/s00497-019-00364-x

Gooh K, Ueda M, Aruga K, Park J, Arata H, Higashiyama T, Kurihara D. 2015. Live-cell imaging and optical manipulation of *Arabidopsis* early embryogenesis. *Dev Cell* **34:** 242–251. doi:10.1016/j.devcel.2015.06.008

Haecker A, Groß-Hardt R, Geiges B, Sarkar A, Breuninger H, Herrmann M, Laux T. 2004. Expression dynamics of WOX genes mark cell fate decisions during early embryonic patterning in *Arabidopsis thaliana. Development* **131:** 657–668. doi:10.1242/dev.00963

Hamann T, Mayer U, Jurgens G. 1999. The auxin-insensitive *bodenlos* mutation affects primary root formation and apical-basal patterning in the *Arabidopsis* embryo. *Development* **126:** 1387.

Hamann T, Benkova E, Bäurle I, Kientz M, Jürgens G. 2002. The *Arabidopsis* BODENLOS gene encodes an auxin response protein inhibiting MONOPTEROS-mediated embryo patterning. *Genes Dev* **16:** 1610–1615. doi:10.1101/ gad.229402

Han Q, Bartels A, Cheng X, Meyer A, An YC, Hsieh TF, Xiao W. 2019. Epigenetics regulates reproductive development in plants. *Plants (Basel)* **8:** 564. doi:10.3390/plants812 0564

Hardtke CS, Berleth T. 1998. The *Arabidopsis* gene MONOPTEROS encodes a transcription factor mediating embryo axis formation and vascular development. *EMBO J* **17:** 1405–1411. doi:10.1093/emboj/17.5.1405

Hardtke CS, Ckurshumova W, Vidaurre DP, Singh SA, Stamatiou G, Tiwari SB, Hagen G, Guilfoyle TJ, Berleth T. 2004. Overlapping and non-redundant functions of the *Arabidopsis* auxin response factors MONOPTEROS and NONPHOTOTROPIC HYPOCOTYL 4. *Development* **131:** 1089–1100. doi:10.1242/dev.00925

Helariutta Y, Fukaki H, Wysocka-Diller J, Nakajima K, Jung J, Sena G, Hauser MT, Benfey PN. 2000. The SHORT-ROOT gene controls radial patterning of the *Arabidopsis* root through radial signaling. *Cell* **101:** 555–567. doi:10 .1016/S0092-8674(00)80865-X

Huang F, Zago MK, Abas L, van Marion A, Galván-Ampudia CS, Offringa R. 2010. Phosphorylation of conserved PIN motifs directs *Arabidopsis* PIN1 polarity and auxin transport. *Plant Cell* **22:** 1129–1142. doi:10.1105/tpc.109 .072678

Kaplan DR, Cooke TJ. 1997. Fundamental concepts in the embryogenesis of dicotyledons: a morphological interpretation of embryo mutants. *Plant Cell* **9:** 1903. doi:10 .2307/3870553

Khakhar A, Leydon AR, Lemmex AC, Klavins E, Nemhauser JL. 2018. Synthetic hormone-responsive transcription factors can monitor and re-program plant development. *eLife* **7:** e34702. doi:10.7554/eLife.34702

Kuroha T, Tokunaga H, Kojima M, Ueda N, Ishida T, Nagawa S, Fukuda H, Sugimoto K, Sakakibara H. 2009. Functional analyses of *LONELY GUY* cytokinin-activating enzymes reveal the importance of the direct activation pathway in *Arabidopsis*. *Plant Cell* **21:** 3152–3169. doi:10.1105/tpc.109.068676

Laux T, Würschum T, Breuninger H. 2004. Genetic regulation of embryonic pattern formation. *Plant Cell* **16:** S190–S202. doi:10.1105/tpc.016014

Lavy M, Estelle M. 2016. Mechanisms of auxin signaling. *Development* **143:** 3226–3229. doi:10.1242/dev.131870

Levesque MP, Vernoux T, Busch W, Cui H, Wang JY, Blilou I, Hassan H, Nakajima K, Matsumoto N, Lohmann JU, et al. 2006. Whole-genome analysis of the SHORT-ROOT developmental pathway in *Arabidopsis*. *PLoS Biol* **4:** e143. doi:10.1371/journal.pbio.0040143

Leyser O. 2018. Auxin signaling. *Plant Physiol* **176:** 465–479. doi:10.1104/pp.17.00765

Liao CY, Weijers D. 2018. A toolkit for studying cellular reorganization during early embryogenesis in *Arabidopsis thaliana*. *Plant J* **93:** 963–976. doi:10.1111/tpj.13841

Liao CY, Smet W, Brunoud G, Yoshida S, Vernoux T, Weijers D. 2015. Reporters for sensitive and quantitative measurement of auxin response. *Nat Methods* **12:** 207–210. doi:10.1038/nmeth.3279

Lie C, Kelsom C, Wu X. 2012. *WOX2* and *STIMPY-LIKE/WOX8* promote cotyledon boundary formation in *Arabidopsis*. *Plant J* **72:** 674–682. doi:10.1111/j.1365-313X.2012.05113.x

Liu Y, Li X, Zhao J, Tang X, Tian S, Chen J, Shi C, Wang W, Zhang L, Feng X, et al. 2015. Direct evidence that suspensor cells have embryogenic potential that is suppressed by the embryo proper during normal embryogenesis. *Proc Natl Acad Sci* **112:** 12432–12437. doi:10.1073/pnas.1508651112

Ljung K. 2013. Auxin metabolism and homeostasis during plant development. *Development* **140:** 943–950. doi:10.1242/dev.086363

Locascio A, Roig-Villanova I, Bernardi J, Varotto S. 2014. Current perspectives on the hormonal control of seed development in *Arabidopsis* and maize: a focus on auxin. *Front Plant Sci* **5:** 412. doi:10.3389/fpls.2014.00412

Mansfield SG, Briarty LG. 1991. Early embryogenesis in *Arabidopsis thaliana*. II: The developing embryo. *Can J Bot* **69:** 461–476. doi:10.1139/b91-063

Mashiguchi K, Tanaka K, Sakai T, Sugawara S, Kawaide H, Natsume M, Hanada A, Yaeno T, Shirasu K, Yao H, et al. 2011. The main auxin biosynthesis pathway in *Arabidopsis*. *Proc Natl Acad Sci* **108:** 18512–18517. doi:10.1073/pnas.1108434108

Mayer KFX, Schoof H, Haecker A, Lenhard M, Jürgens G, Laux T. 1998. Role of WUSCHEL in regulating stem cell fate in the *Arabidopsis* shoot meristem. *Cell* **95:** 805–815. doi:10.1016/S0092-8674(00)81703-1

Mironova VV, Omelyanchuk NA, Wiebe DS, Levitsky VG. 2014. Computational analysis of auxin responsive elements in the *Arabidopsis thaliana* L. genome. *BMC Genomics* **15**(Suppl 12): S4. doi:10.1186/1471-2164-15-S12-S4

Mironova V, Teale W, Shahriari M, Dawson J, Palme K. 2017. The systems biology of auxin in developing embryos. *Trends Plant Sci* **22:** 225–235. doi:10.1016/j.tplants.2016.11.010

Möller BK, ten Hove CA, Xiang D, Williams N, López LG, Yoshida S, Smit M, Datla R, Weijers D. 2017. Auxin response cell-autonomously controls ground tissue initiation in the early *Arabidopsis* embryo. *Proc Natl Acad Sci* **114:** E2533–E2539. doi:10.1073/pnas.1616493114

Mravec J, Kubeš M, Bielach A, Gaykova V, Petrášek J, Skůpa P, Chand S, Benková E, Zažímalová E, Friml J. 2008. Interaction of PIN and PGP transport mechanisms in auxin distribution-dependent development. *Development* **135:** 3345–3354. doi:10.1242/dev.021071

Mroue S, Simeunovic A, Robert HS. 2018. Auxin production as an integrator of environmental cues for developmental growth regulation. *J Exp Bot* **69:** 201–212. doi:10.1093/jxb/erx259

Nakajima K, Sena G, Nawy T, Benfey PN. 2001. Intercellular movement of the putative transcription factor SHR in root patterning. *Nature* **413:** 307–311. doi:10.1038/35095061

Ohashi-Ito K, Bergmann DC. 2007. Regulation of the *Arabidopsis* root vascular initial population by LONESOME HIGHWAY. *Development* **134:** 2959–2968. doi:10.1242/dev.006296

Palovaara J, Saiga S, Wendrich JR, van 't Wout Hofland N, van Schayck JP, Hater F, Mutte S, Sjollema J, Boekschoten M, Hooiveld GJ, et al. 2017. Transcriptome dynamics revealed by a gene expression atlas of the early *Arabidopsis* embryo. *Nat Plants* **3:** 894–904. doi:10.1038/s41477-017-0035-3

Parry G, Calderon-Villalobos LI, Prigge M, Peret B, Dharmasiri S, Itoh H, Lechner E, Gray WM, Bennett M, Estelle M. 2009. Complex regulation of the TIR1/AFB family of auxin receptors. *Proc Natl Acad Sci* **106:** 22540–22545. doi:10.1073/pnas.0911967106

Peng X, Sun MX. 2018. The suspensor as a model system to study the mechanism of cell fate specification during early embryogenesis. *Plant Reprod* **31:** 59–65. doi:10.1007/s00497-018-0326-5

Ploense SE, Wu MF, Nagpal P, Reed JW. 2009. A gain-of-function mutation in *IAA18* alters *Arabidopsis* embryonic apical patterning. *Development* **136:** 1509–1517. doi:10.1242/dev.025932

Prigge MJ, Greenham K, Zhang Y, Santner A, Castillejo C, Mutka AM, O'Malley RC, Ecker JR, Kunkel BN, Estelle M. 2016. The *Arabidopsis* auxin receptor F-Box proteins AFB4 and AFB5 are required for response to the synthetic auxin picloram. *G3 (Bethesda)* **6:** 1383–1390. doi:10.1534/g3.115.025585

Prigge MJ, Platre M, Kadakia N, Zhang Y, Greenham K, Szutu W, Pandey BK, Bhosale RA, Bennett MJ, Busch W, et al. 2020. Genetic analysis of the *Arabidopsis* TIR1/AFB auxin receptors reveals both overlapping and specialized functions. *eLife* **9:** e54740. doi:10.7554/eLife.54740

Rademacher EH, Möller B, Lokerse AS, Llavata-Peris CI, van den Berg W, Weijers D. 2011. A cellular expression map of the *Arabidopsis* AUXIN RESPONSE FACTOR gene family. *Plant J* **68:** 597–606. doi:10.1111/j.1365-313X.2011.04710.x

Rademacher EH, Lokerse AS, Schlereth A, Llavata-Peris CI, Bayer M, Kientz M, Rios AF, Borst JW, Lukowitz W, Jürgens G, et al. 2012. Different auxin response machineries control distinct cell fates in the early plant embryo. *Dev Cell* **22:** 211–222. doi:10.1016/j.devcel.2011.10.026

Radhakrishnan D, Shanmukhan AP, Kareem A, Aiyaz M, Varapparambathu V, Toms A, Kerstens M, Valsakumar D, Landge AN, Shaji A, et al. 2020. A coherent feed-forward loop drives vascular regeneration in damaged aerial organs of plants growing in a normal developmental context. *Development* **147:** dev185710. doi:10.1242/dev.185710

Radoeva T, Vaddepalli P, Zhang Z, Weijers D. 2019a. Evolution, initiation, and diversity in early plant embryogenesis. *Dev Cell* **50:** 533–543. doi:10.1016/j.devcel.2019.07.011

Radoeva T, Lokerse AS, Llavata-Peris CI, Wendrich JR, Xiang D, Liao CY, Vlaar L, Boekschoten M, Hooiveld G, Datla R, et al. 2019b. A robust auxin response network controls embryo and suspensor development through a basic helix loop helix transcriptional module. *Plant Cell* **31:** 52–67. doi:10.1105/tpc.18.00518

Radoeva T, Albrecht C, Piepers M, De Vries SC, Weijers D. 2020. Suspensor-derived somatic embryogenesis in *Arabidopsis*. *Development* **147:** dev188912. doi:10.1242/dev.188912

Robert HS, Grones P, Stepanova AN, Robles LM, Lokerse AS, Alonso JM, Weijers D, Friml J. 2013. Local auxin sources orient the apical-basal axis in *Arabidopsis* embryos. *Curr Biol* **23:** 2506–2512. doi:10.1016/j.cub.2013.09.039

Robert HS, Crhak Khaitova L, Mroue S, Benková E. 2015a. The importance of localized auxin production for morphogenesis of reproductive organs and embryos in *Arabidopsis*. *J Exp Bot* **66:** 5029–5042. doi:10.1093/jxb/erv256

Robert HS, Grunewald W, Sauer M, Cannoot B, Soriano M, Swarup R, Weijers D, Bennett M, Boutilier K, Friml J. 2015b. Plant embryogenesis requires AUX/LAX-mediated auxin influx. *Development* **142:** 702–711. doi:10.1242/dev.115832

Robert HS, Park C, Gutièrrez CL, Wójcikowska B, Pěnčík A, Novák O, Chen J, Grunewald W, Dresselhaus T, Friml J, et al. 2018. Maternal auxin supply contributes to early embryo patterning in *Arabidopsis*. *Nat Plants* **4:** 548–553. doi:10.1038/s41477-018-0204-z

Salehin M, Bagchi R, Estelle M. 2015. $SCF^{TIR1/AFB}$-based auxin perception: mechanism and role in plant growth and development. *Plant Cell* **27:** 9–19. doi:10.1105/tpc.114.133744

Schlereth A, Möller B, Liu W, Kientz M, Flipse J, Rademacher EH, Schmid M, Jürgens G, Weijers D. 2010. MONOPTEROS controls embryonic root initiation by regulating a mobile transcription factor. *Nature* **464:** 913–916. doi:10.1038/nature08836

Schwartz BW, Yeung EC, Meinke DW. 1994. Disruption of morphogenesis and transformation of the suspensor in abnormal suspensor mutants of *Arabidopsis*. *Development* **120:** 3235–3245.

Smit ME, Llavata-Peris CI, Roosjen M, van Beijnum H, Novikova D, Levitsky V, Sevilem I, Roszak P, Slane D, Jürgens G, et al. 2020. Specification and regulation of vascular tissue identity in the *Arabidopsis* embryo. *Development* **147:** dev186130. doi:10.1242/dev.186130

Smith ZR, Long JA. 2010. Control of *Arabidopsis* apical–basal embryo polarity by antagonistic transcription factors. *Nature* **464:** 423–426. doi:10.1038/nature08843

Stepanova AN, Robertson-Hoyt J, Yun J, Benavente LM, Xie DY, Doležal K, Schlereth A, Jürgens G, Alonso JM. 2008. TAA1-mediated auxin biosynthesis is essential for hormone crosstalk and plant development. *Cell* **133:** 177–191. doi:10.1016/j.cell.2008.01.047

Swarup R, Bhosale R. 2019. Developmental roles of AUX1/LAX auxin influx carriers in plants. *Front Plant Sci* **10:** 1306. doi:10.3389/fpls.2019.01306

Takada S, Hibara K, Ishida T, Tasaka M. 2001. The CUP-SHAPED COTYLEDON1 gene of *Arabidopsis* regulates shoot apical meristem formation. *Development* **128:** 1127.

Taylor-Teeples M, Lanctot A, Nemhauser JL. 2016. As above, so below: auxin's role in lateral organ development. *Dev Biol* **419:** 156–164. doi:10.1016/j.ydbio.2016.03.020

Ugartechea-Chirino Y, Swarup R, Swarup K, Péret B, Whitworth M, Bennett M, Bougourd S. 2010. The AUX1 LAX family of auxin influx carriers is required for the establishment of embryonic root cell organization in *Arabidopsis thaliana*. *Ann Bot* **105:** 277–289. doi:10.1093/aob/mcp287

Valuchova S, Mikulkova P, Pecinkova J, Klimova J, Krumnikl M, Bainar P, Heckmann S, Tomancak P, Riha K. 2020. Imaging plant germline differentiation within *Arabidopsis* flowers by light sheet microscopy. *eLife* **9:** e52546. doi:10.7554/eLife.52546

van den Berg C, Willemsen V, Hendriks G, Weisbeek P, Scheres B. 1997. Short-range control of cell differentiation in the *Arabidopsis* root meristem. *Nature* **390:** 287–289. doi:10.1038/36856

Verna C, Ravichandran SJ, Sawchuk MG, Linh NM, Scarpella E. 2019. Coordination of tissue cell polarity by auxin transport and signaling. *eLife* **8:** e51061. doi:10.7554/eLife.51061

Vernon DM, Meinke DW. 1994. Embryogenic transformation of the suspensor in twin, a polyembryonic mutant of *Arabidopsis*. *Dev Biol* **165:** 566–573. doi:10.1006/dbio.1994.1276

Vroemen CW, Mordhorst AP, Albrecht C, Kwaaitaal MACJ, de Vries SC. 2003. The *CUP-SHAPED COTYLEDON3* gene is required for boundary and shoot meristem formation in *Arabidopsis*. *Plant Cell* **15:** 1563–1577. doi:10.1105/tpc.012203

Wasteneys GO. 2002. Microtubule organization in the green kingdom: chaos or self-order? *J Cell Sci* **115:** 1345–1354.

Weijers D, Jürgens G. 2005. Auxin and embryo axis formation: the ends in sight? *Curr Opin Plant Biol* **8:** 32–37. doi:10.1016/j.pbi.2004.11.001

Weijers D, Van Hamburg JP, Van Rijn E, Hooykaas PJ, Offringa R. 2003. Diphtheria toxin-mediated cell ablation reveals interregional communication during *Arabidopsis* seed development. *Plant Physiol* **133:** 1882–1892. doi:10.1104/pp.103.030692

Weijers D, Schlereth A, Ehrismann JS, Schwank G, Kientz M, Jürgens G. 2006. Auxin triggers transient local signaling for cell specification in *Arabidopsis* embryogenesis. *Dev Cell* **10:** 265–270. doi:10.1016/j.devcel.2005.12.001

Wu X, Dabi T, Weigel D. 2005. Requirement of homeobox gene STIMPY/WOX9 for *Arabidopsis* meristem growth and maintenance. *Curr Biol* **15:** 436–440. doi:10.1016/j.cub.2004.12.079

Wu X, Chory J, Weigel D. 2007. Combinations of WOX activities regulate tissue proliferation during *Arabidopsis* embryonic development. *Dev Biol* **309:** 306–316. doi:10.1016/j.ydbio.2007.07.019

Xiong F, Liu HH, Duan CY, Zhang BK, Wei G, Zhang Y, Li S. 2019. *Arabidopsis* JANUS regulates embryonic pattern formation through Pol II-mediated transcription of WOX2 and PIN7. *iScience* **19:** 1179–1188. doi:10.1016/j.isci.2019.09.004

Yoshida S, de Reuille PB, Lane B, Bassel GW, Prusinkiewicz P, Smith RS, Weijers D. 2014. Genetic control of plant development by overriding a geometric division rule. *Dev Cell* **29:** 75–87. doi:10.1016/j.devcel.2014.02.002

Yoshida S, Van Der Schuren A, Van Dop M, Van Galen L, Saiga S, Adibi M, Möller B, Colette A, Marhavy P, Smith R, et al. 2019. A SOSEKI-based coordinate system interprets global polarity cues in *Arabidopsis*. *Nat Plants* **5:** 160–166. doi:10.1038/s41477-019-0363-6

Zhang J, Nodzyński T, Pěnčík A, Rolčík J, Friml J. 2010. PIN phosphorylation is sufficient to mediate PIN polarity and direct auxin transport. *Proc Natl Acad Sci* **107:** 918–922. doi:10.1073/pnas.0909460107

Zhang Z, Tucker E, Hermann M, Laux T. 2017. A molecular framework for the embryonic initiation of shoot meristem stem cells. *Dev Cell* **40:** 264–277.e4. doi:10.1016/j.devcel.2017.01.002

Zhao Y. 2014. Auxin biosynthesis. *The Arabidopsis Book* **12:** e0173–e0173. doi:10.1199/tab.0173

Zhou X, Shi C, Zhao P, Sun M. 2019. Isolation of living apical and basal cell lineages of early proembryos for transcriptome analysis. *Plant Reprod* **32:** 105–111. doi:10.1007/s00497-018-00353-6

Auxin and Flower Development: A Blossoming Field

Mara Cucinotta,[1] Alex Cavalleri,[1] John William Chandler,[2] and Lucia Colombo[1]

[1]Dipartimento di Bioscienze, Università degli Studi di Milano, 20133 Milan, Italy
[2]Max Planck Institute for Plant Breeding Research, 50829 Cologne, Germany
Correspondence: lucia.colombo@unimi.it

The establishment of the species-specific floral organ body plan involves many coordinated spatiotemporal processes, which include the perception of positional information that specifies floral meristem and floral organ founder cells, coordinated organ outgrowth coupled with the generation and maintenance of inter-organ and inter-whorl boundaries, and the termination of meristem activity. Auxin is integrated within the gene regulatory networks that control these processes and plays instructive roles at the level of tissue-specific biosynthesis and polar transport to generate local maxima, perception, and signaling. Key features of auxin function in several floral contexts include cell nonautonomy, interaction with cytokinin gradients, and the central role of MONOPTEROS and ETTIN to regulate canonical and noncanonical auxin response pathways, respectively. *Arabidopsis* flowers are not representative of the enormous angiosperm floral diversity; therefore, comparative studies are required to understand how auxin underlies these developmental differences. It will be of great interest to compare the conservation of auxin pathways among flowering plants and to discuss the evolutionary role of auxin in floral development.

Angiosperm flowers display extraordinary diversity in their body plan and every aspect of morphology, including organ number and positioning, size, symmetry, color, and organogenesis. The study of select angiosperm species has elaborated the genetic basis of floral identity according to the coordinated expression of transcription factors. However, advances in auxin biology have demonstrated that auxin is a central developmental regulator of floral morphogenesis (Irish 2010; Wellmer et al. 2014).

Auxin is essential for the initiation of floral meristems (FMs) and organs and for terminating meristem growth; however, its role in determining positional information for floral organ initiation remains less mechanistically well understood than its function in organ outgrowth and development (Chandler 2011). Auxin can also function as a morphogen by generating response gradients, particularly in combination with those of cytokinins, which are instrumental in several aspects of floral organogenesis (Su et al. 2011).

This review summarizes how auxin coordinates diverse and complex regulatory gene networks in FM and organ initiation and development and focuses on the *Arabidopsis* model for which most is known. It also highlights the cen-

Cite this article as *Cold Spring Harb Perspect Biol* doi: 10.1101/cshperspect.a039974

tral role of canonical and noncanonical auxin response factors (ARFs) such as MONOPTEROS (MP) and ETTIN (ETT/ARF3), respectively.

AUXIN IS ESSENTIAL FOR FLORAL MERISTEM INITIATION

Floral organ primordia arise from FMs, which initiate as lateral organs at the inflorescence meristem (IM) periphery. The maintenance of the IM required a precise balance between the self-renewal of stem cells and cell differentiation. It has been proposed that the WUSCHEL (WUS) transcription factor achieves this balance by restricting auxin signaling in stem cells and simultaneously allowing low levels of auxin signaling output (Ma et al. 2019), although the precise mechanism of this rheostat function on select auxin signaling components remains to be elucidated.

Precise spatiotemporal local auxin biosynthesis within the IM and its polar transport and accumulation at specific sites from sources elsewhere in the plant are essential for FM initiation (Alvarez-Buylla et al. 2010; Yadav et al. 2019). This is evidenced by classical mutants, for example, in *PIN-FORMED1* (*PIN1*), which encodes a polar auxin efflux carrier (Gälweiler et al. 1998). Occasional flowers on *pin1* inflorescences lack sepals and have a variable number of abnormally shaped petals (Okada et al. 1991) and exogenous auxin application to *pin1-1* inflorescences restores flower formation at the point of application (Reinhardt et al. 2000). Mutation of the gene encoding the PINOID (PID) kinase that regulates PIN activity leads to flowers with supernumerary petals, few sepals and stamens, and mostly *pin*-like carpels (Bennett et al. 1995). Strong mutants of ARF5/MONOPTEROS (MP) produce naked inflorescences (Przemeck et al. 1996), although the hypomorphic *mpS319* allele occasionally produces rudimentary flowers (Alonso et al. 2003; Cole et al. 2009). Loss-of-function of *MACCHI-BOU4* (*MAB4*)/*ENHANCER OF PINOID* (*ENP*)/*NAKED PINS IN YUC MUTANTS 1* (*NPY*), hereafter *MAB4*, which encode NONPHOTOTROPIC HYPOCOTYL 3-like proteins that regulate PIN endocytosis, also causes *pin*-like inflorescences (Furutani et al. 2014), due to the abolition of auxin flow from external FM layers inward to create an auxin sink necessary for FM development after initiation.

Auxin response maxima prepattern sites of FM initiation at the IM periphery (Reinhardt et al. 2000), and in addition to the importance of the spatial distribution of auxin within the SAM as an instructive signal, recent high-resolution imaging of auxin response has revealed the importance of a dynamic temporal component to these auxin fluxes (Galvan-Ampudia et al. 2020). In this scenario, high-precision dynamic spatiotemporal auxin gradients within the IM periphery are coordinated with growth, to ensure that cells are exposed to a high level of auxin over time to activate organogenesis. The temporal aspect concerning the timing and duration of exposure of cells to high auxin is relevant because, according to phytomer theory, FMs initiate as lateral organs in the axils of bracts. *Arabidopsis* bracts are cryptic but are histologically and morphologically visible during early floral transition preceding FM initiation (Kwiatkowska 2008). Transcriptional reprogramming associated with bract initiation has been accessed in the *ap1 cal* mutant background by sorting cells that express the bract founder-cell marker *DORNROESCHEN-LIKE* (*DRNL*) (Frerichs et al. 2016). Minor transcriptional changes in genes in the auxin synthesis, transport, and response machinery suggest that its global activation might not be associated with bract initiation, but with FM initiation. This is consistent with spatially distinct domains of *DRNL* and *DR5* expression within the IM (Chandler et al. 2011) and that auxin response minima are required for the formation of axillary vegetative meristems in *Arabidopsis* and tomato (Wang et al. 2014a,b). Interacting concentration gradients of auxin and cytokinin establish the timing of FM outgrowth within the IM, which can be uncoupled from initiation (Besnard et al. 2014).

An early feature of FM initiation is inhibition of the pluripotency of stem cells that form FM primordia, by down-regulation of *SHOOT-MERISTEMLESS* (*STM*). Transcripts of *STM*

 Cite this article as *Cold Spring Harb Perspect Biol* doi: 10.1101/cshperspect.a039974

overaccumulate in the IM of *mpS319* and more so in an *mp arf3 arf4* triple mutant, resulting in complete loss of flower primordia. ETT directly binds to *STM* and silences its expression for FM fate acquisition. By contrast, MP does not bind *STM*, but directly activates expression of the repressor *FILAMENTOUS FLOWER* (*FIL*), which further contributes to *STM* silencing. ETT and FIL also directly bind and repress expression of *BREVIPEDICELLUS* (*BP*), which maintains meristematic identity. Although MP, ETTIN, and ARF4 repress *STM* and *BP* expression in parallel pathways, FIL physically interacts with ETTIN and ARF4 to create a complex that recruits chromatin modifiers such as histone deacetylase HDA19, which leads to target-gene silencing (Table 1; Fig. 1; Chung et al. 2019).

THE CENTRAL ROLE OF MP IN FM DEVELOPMENT

MONOPTEROS is the central regulator of FM development and translates a local auxin concentration maximum into a floral primordium by directly transcriptionally activating several transcription factors. These include the master regulator *LEAFY* (*LFY*), which, together with its target *APETALA1*, regulates the IM-to-FM identity transition (Weigel et al. 1992). *LFY* expression in incipient flower primordia is enhanced by high auxin levels, via the canonical MP/IAA12 (BODENLOS) module (Yamaguchi et al. 2013). *LFY* expression only partially rescues *mp-S319* flower defects and MP also transcriptionally activates *AINTEGUMENTA* (*ANT*) and *AINTEGUMENTA-LIKE6* (*AIL6*), which function in parallel and redundantly with LFY in FM initiation. Because the *lfy-6 ant-4 ail6-2* triple mutant initiates some flowers, additional auxin-regulated factors probably contribute to FM production. *ANT* and *AIL6* also function upstream of *LFY* and regulate its expression in response to auxin and in parallel to MP. Accordingly, they are expressed earlier than *LFY* (Yamaguchi et al. 2016) and delayed *LFY* expression in *ant ail6* delays FM production. However, expression of *ANT* and *AIL6* in *mp* is reduced but not abolished (Yamaguchi et al. 2013), suggesting that they are also regulated independently from MP, potentially via other ARFs.

In summary, MP triggers FM initiation and development by transcriptionally up-regulating *LFY*, *ANT*, and *AIL6* via two separate pathways in which *LFY* both targets ANT/AIL6 and acts in parallel with them. In addition, LFY positively feedbacks into the auxin pathway, to ensure maintenance of the auxin response network and to lock the auxin response in an "on" state that ensures commitment to flowering (Table 1; Fig. 1). Expression of *DR5er:GFP* positively correlates with LFY activity, and feedback regulation on auxin signaling by LFY occurs partly by up-regulation of *PID* expression (Li et al. 2013; Yamaguchi et al. 2013).

The reprogramming of MP targets in meristematic cells that develop into floral primordia involves chromatin remodeling via the SWI/SNF ATPases *BRAMA* (*BRM*) and *SPLAYED* (*SYD*), which increase chromatin accessibility in early floral primordia. Initiation of fewer flower primordia and *pin*-like inflorescences in *mpS319 syd* double mutants and *mpS319* plants expressing an artificial microRNA responsible for *BRAHMA* down-regulation demonstrate the functional relevance of *BRM* and *SYD* in FM initiation. Auxin enhances MP binding to BRM and SYD, which also bind the direct MP targets *FIL*, *LFY*, *ANT*, and *TMO3* (Wu et al. 2015). Therefore, a chromatin switch model explains how binding of AUX/IAA proteins to MP at low auxin concentrations prevents the recruitment of SWI/SNF factors by MP, which instead binds to the TOPLESS (TPL) corepressor and HISTONE DEACETYLASE19 (HDA19), which promote a repressive chromatin state. Increasing auxin concentrations degrade AUX/IAA proteins, enabling MP to recruit SWI/SNF-containing complexes, which promote a permissive chromatin state that up-regulates MP targets (Table 1; Fig. 1; Wu et al. 2015).

FLORAL ORGAN INITIATION

Floral morphogenesis involves the perception of positional information within the FM by floral organ founder cells, to become specified, acquire a particular fate and outgrow. These processes

Table 1. Summary of genes mentioned in the text

Gene	Protein function/relation with auxin	Floral developmental context	References
PIN-FORMED1 (PIN1)	Auxin efflux carrier	Flower initiation (FI), organ initiation (OI)	Okada et al. 1991
PINOID (PID)	Kinase that regulates PIN1 polarization	FI, OI	Bennett et al. 1995
ARF5/MONOPTEROS	Auxin response factor (ARF), transcription factor (TF)	FI, organ boundaries (OBs) OI, organ growth (OG)	Li et al. 2013; Yamaguchi et al. 2013
MACCHI-BOU4 (MAB4)	Regulates PIN1 polarization	FI	Furutani et al. 2014
DORNROESCHEN-LIKE (DRNL)	AP2/ETHYLENE RESPONSE FACTOR TF, a floral-organ founder-cell marker whose expression precedes DR5 and genetically interacts with *PID*	FI, petal initiation	Chandler et al. 2011
SHOOTMERISTEMLESS (STM)	KNOX TF, *STM* is repressed by auxin	FI	Chung et al. 2019
ARF3/ETTIN (ETT)	ARF, TF	FI, gynoecium growth, megaspore mother cell (MMC), floral meristem determinacy (FMD)	Sessions et al. 1997; Chung et al. 2019
ARF4	ARF, TF	FI	Chung et al. 2019
FILAMENTOUS FLOWER (FIL)	TF, *FIL* is activated by auxin	FI	Chung et al. 2019
BREVIPEDICELLUS (BP)	KNOX TF, *BP* is repressed by auxin	FI	Chung et al. 2019
HISTONE DEACETYLASE19 (HDA19)	Chromatin remodeling factor, interacts with ETT and ARF4	FI	Chung et al. 2019
LEAFY	Transcriptional master regulator connected with auxin maximum maintenance, *LFY* is activated by MP	FI	Weigel et al. 1992; Li et al. 2013; Yamaguchi et al. 2013
AINTEGUMENTA (ANT)	AP2-domain TF transcriptionally up-regulated by MP	FI, OI	Yamaguchi et al. 2013)
AINTEGUMENTA-LIKE6 (AIL6)	AP2-domain TF transcriptionally up-regulated by MP	FI	Yamaguchi et al. 2013
IAA12/BODENLOS	Aux/IAA protein that interacts with MP	FI	Yamaguchi et al. 2013
TOPLESS (TPL)	Corepressor of MP and ETT targets	FI	Wu et al. 2015
BRAMA (BRM)	SWI/SNF ATPs chromatin remodeling interactor of MP	FI	Wu et al. 2015

Continued

 Cite this article as *Cold Spring Harb Perspect Biol* doi: 10.1101/cshperspect.a039974

Table 1. *Continued*

Gene	Protein function/relation with auxin	Floral developmental context	References
SPLAYED (SYD)	SWI/SNF ATPs chromatin remodeling interactor of MP	FI	Wu et al. 2015
TARGET OF MONOPTEROS3 (TMO3)	AP2 TF transcriptionally up-regulated by MP	FI, ovules	Wu et al. 2015; Cucinotta et al. 2016
YUCCA (YUC)	Flavin-monooxygenases that catalyse IAA biosynthesis	FI, OI	Cheng et al. 2006
DORNROESCHEN (DRN)	AP2/ETHYLENE RESPONSE FACTOR that genetically interacts with *PIN*	FI, OI	Chandler and Werr 2014
PUCHI	AP2/ETHYLENE RESPONSE FACTOR that genetically interacts with *PID*	FI, OI	Chandler and Werr 2014
PERIANTHIA (PAN)	bZIP TF that regulates YUC1 and YUC4	OB, petal initiation	Maier et al. 2011
PRESSED FLOWER (PRS)	WUSCHEL-RELATED HOMEOBOX, *PRS* expression is induced by auxin	Petal initiation	Chandler and Werr 2014; Caggiano et al. 2017
ABCB1/19	Auxin efflux carrier	FI, anther growth, pollen	Zhao et al. 2013; Cecchetti et al. 2015
CUP-SHAPED COTYLEDON (CUC)	NAC-domain TFs regulated by auxin, *CUC* function is connected with polar auxin transport (PAT)	OB, ovules	Laufs et al. 2004; Mallory et al. 2005; Galbiati et al. 2013
RABBIT EARS (RBE)	C2H2 zinc finger protein connection with PAT	OB, petal growth	Lampugnani et al. 2013
HANABA TARANU (HAN)	GATA3-type transcriptional repressor with a putative role in PAT regulation	OB	Zhang et al. 2013
PETAL LOSS (PTL)	Trihelix TF connected with PAT	OB, petal growth	Lampugnani et al. 2013
AUX1	Auxin influx carrier	Petal growth	Lampugnani et al. 2013
SUPERMAN (SUP)	C2H2 zinc-finger protein that represses *YUC* expression, *SUP* is regulated by LFY	OB	Sakai et al. 1995; Xu et al. 2018
DEVELOPMENT-RELATED MYB-LIKE1 (DRMY1)	MYB-like TF that regulates auxin response	Sepal outgrowth	Zhu et al. 2020
JAGGED (JAG)	Zinc-finger protein that regulates auxin response	Petal growth	Dinneny et al. 2004; Sauret-Güeto et al. 2013
ARGOS	Auxin-inducible gene that regulates ANT	Petal growth	Hu et al. 2003

Continued

Table 1. *Continued*

Gene	Protein function/relation with auxin	Floral developmental context	References
BIGPETALp	bHLH TF that interacts with ARF8	Petal growth	Varaud et al. 2011
ARF8	ARF, TF	Petal growth, stamen growth	Nagpal et al. 2005; Varaud et al. 2011
ARF8.4	ARF, TF	Stamen growth	Ghelli et al. 2018
ARF6	ARF, TF	Stamen growth	Nagpal et al. 2005
SHORT INTERNODES (*SHI*)	Zinc-finger TFs that regulate auxin biosynthesis	Anther growth	Estornell et al. 2018
STYLISH1/2 (*STY1/2*)	Zinc-finger TFs that regulate auxin biosynthesis	Anther growth, gynoecium growth	Trigueros et al. 2009; Estornell et al. 2018
TIR1/AFB	Nuclear auxin receptor	Pollen, female gametophyte (FG)	Cecchetti et al. 2008, 2017
DEFECTIVE IN ANTHER DEHISCENCE1 (*DAD1*)	Enzyme for JA biosynthesis transcriptionally up-regulated by ARF6/ARF8	Anther dehiscence	Nagpal et al. 2005; Tabata et al. 2010
MYB26	TF, *MYB26* is induced by ARF6/ARF8	Anther dehiscence	Yang et al. 2017
ARF17	ARF, TF	Anther dehiscence	Xu et al. 2019
MYB108	TF, *MYB108* is transcriptionally up-regulated by ARF17	Anther dehiscence	Xu et al. 2019
IAA19	Aux/IAA protein	Anther dehiscence	Tashiro et al. 2009; Cecchetti et al. 2017
AGAMOUS (*AG*)	MADS-box TF, regulates *YUC4*	FMD	Liu et al. 2011, 2014; Yamaguchi et al. 2018
WUSCHEL (*WUS*)	Homeobox TF that controls the stem-cell pool, regulated by ETT	FMD	Sun et al. 2009; Liu et al. 2011; Huang et al. 2017
APETALA 2 (*AP2*)	AP2/EREBP TF, negatively regulates *ETT*	FMD	Liu et al. 2014
CRABS CLAW (*CBR*)	YABBY TF that transcriptionally up-regulates *TRN* and *YUC4*	FMD	Yamaguchi et al. 2018
TORNADO (*TRN2*)	Tetraspanin protein that regulates auxin homeostasis	FMD	Yamaguchi et al. 2018
CHROMATIN REMODELING 11/17 (*CHR11/17*)	Recruited by AG to activate *YUC*	FMD	Yamaguchi et al. 2018
KANADI1/4 (*KAN1/4*)	GARP TF that interacts with ETT	Gynoecium growth	Pekker et al. 2005
SPATULA (*SPT*)	bHLH TF that regulates PAT	Gynoecium growth	Moubayidin and Ostergaard 2014

Continued

Table 1. *Continued*

Gene	Protein function/relation with auxin	Floral developmental context	References
INDEHISCENT (IND)	bHLH TF that regulates PAT	Gynoecium growth	Moubayidin and Ostergaard 2014; Kuhn et al. 2019
PIN-FORMED3 (PIN3)	Auxin efflux carrier	Gynoecium growth	Gremski et al. 2007
NGATHA (NGA3)	RAV family TF that regulates *YUCs*	Gynoecium growth	Trigueros, et al. 2009
*TOPLESS-RELATED2 (*TPR2)*	Corepressor of ETT targets	Gynoecium growth	Benjamins et al. 2001; Gremski et al. 2007
HECATE (HEC)	bHLH TF that regulates PAT	Gynoecium growth	Gremski et al. 2007

The gene or protein function and relationship to auxin is described in the second column, and the third column shows the specific developmental context within flower development with which these genes are involved. Genes whose proteins have a direct molecular link with auxin are shown in bold.

are associated with the establishment and maintenance of organ boundaries.

Expression of the synthetic auxin response reporters DR5::GFP and DII (van Mourik et al. 2012; Goldental-Cohen et al. 2017) demonstrate that auxin response maxima colocalize with sites of organ initiation. However, it is unclear whether auxin specifies floral organ founder cells, or merely activates outgrowth. Modeling of auxin responses (van Mourik et al. 2012) and the ABCE model of floral organ identity acquisition propose that concentric organ whorls are specified centripetally from outer to inner (Theißen and Saedler 2001). However, *Arabidopsis* floral body plan establishment is more complex and the *DRNL* founder-cell marker reveals biphasic development in which sepals initiate unidirectionally from the center of the IM, petals and lateral stamens arise from adjacent domains, and medial stamens initiate from a ring of *DRNL* prepatterning (Chandler and Werr 2014). Expression of *DRNL* precedes auxin response maxima for some floral organs (Chandler et al. 2011) and the response of *DRNL* to auxin remains enigmatic (Comelli et al. 2020). However, DRNL and its close relatives DRN and PUCHI redundantly determine numbers of all floral organs at the founder-cell level, partly cell-nonautonomously (Table 1; Fig. 1; Chandler and Werr 2017). All three proteins integrate into auxin pathways: *DRN* genetically interacts with *pin1*, *DRNL* with *pid*, *yuc1* and *yuc4* (Chandler et al. 2011), and *PUCHI* with *pid* (Chandler and Werr 2014). PERIANTHIA (PAN) regulates the number of *Arabidopsis* perianth organs by targeting YUC-mediated auxin biosynthesis (Maier et al. 2011). The radially symmetrical arrangement of five perianth-whorl organs in *pan* (Running and Meyerowitz 1996) derives from bifurcations of sepal founder-cell populations at abaxial and adaxial positions and is partially dependent on *PRESSED FLOWER* (*PRS*) activity (Chandler and Werr 2014), which promotes cell proliferation in lateral floral primordium domains (Matsumoto 2001). In leaf primordia, auxin up-regulates *PRS* and its paralogue *WOX1* (Caggiano et al. 2017) via MP (Guan et al. 2017). Increased expression of DR5::GFP in the leaf base of *prs wox1* and the increased sensitivity of *prs* leaves to auxin transport inhibitors implies a role for PRS in auxin transport (Nakata et al. 2018). By analogy, positive feedback between auxin and *PRS* might exist during sepal initiation. Mutation of *ETT* also leads to one supernumerary petal and sepal (Sessions et al. 1997; Pekker et al. 2005).

FLORAL ORGAN BOUNDARY ESTABLISHMENT

Appropriate spatial positioning of floral organs requires the establishment and maintenance of

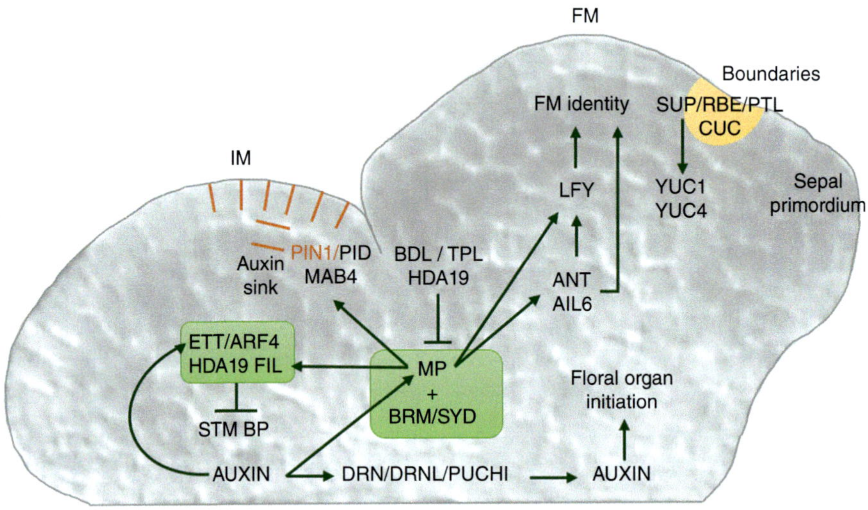

Figure 1. The gene regulatory networks involving auxin that regulate early floral meristem (FM) initiation and development. Schematic diagram of gene regulatory networks superimposed onto an image of an *Arabidopsis* FM emerging from the flank of the inflorescence meristem (IM), to illustrate how auxin coordinates FM initiation and identity and floral organ initiation (see text for details). (ANT) AINTEGUMENTA, (AIL6) AINTEGUMENTA-LIKE6, (ARF4) AUXIN RESPONSE FACTOR4, (BDL) BODENLOS, (BP) BREVIPEDICELLUS, (BRM) BRAHMA, (CUC) CUP-SHAPED COTYLEDON, (DRN) DORNRÖSCHEN, (DRNL) DORNRÖSCHEN-LIKE, (ETT) ETTIN, (FIL) FILAMENTOUS FLOWER, (HDA19) HISTONE DEACETYLASE19, (MAB4) MACCHI-BOU4, (PID) PINOID, (PIN1) PIN-FORMED1, (RBE) RABBIT EARS, (PTL) PETAL LOSS, (PYD) SPLAYED, (STM) SHOOTMERISTEMLESS, (SUP) SUPERMAN, (TPL) TOPLESS, (YUC1) YUCCA1, (YUC4) YUCCA4.

inter-organ boundary zones within floral whorls. In addition to separating organs, boundary domains act as organizing centers to developmentally pattern adjacent tissues. Many transcription factors have boundary functions that relate to auxin signaling (Yu and Huang 2016). A boundary between the IM and FM ensures that the FM buttress separates from the IM. It involves polar auxin transport by ATP-BINDING-CASSETTE B19 (ABCB19) and in *abcb19* mutants, auxin levels increase within the boundary (Zhao et al. 2013). Boundaries are zones of low auxin signaling (Wang et al. 2016), which repress cell division and expansion (Žádníková and Simon 2014; Wang et al. 2016).

The *CUP-SHAPED COTYLEDON* (*CUC*) genes *CUC1*, *CUC2*, and *CUC3* are expressed in different floral organ boundaries (Takada et al. 2001; Hibara et al. 2006; Xu et al. 2008). Their combinatorial mutation leads to fused floral organs (Laufs et al. 2004; Mallory et al. 2005) and *CUC1* and *CUC2* promote carpel margin meristem (CMM) initiation and positioning (Kamiuchi et al. 2014). Auxin represses *CUC2* expression (Furutani et al. 2004), consistent with boundary domains having low auxin levels, high *CUC* levels, and repressed cell division. Developmental contexts such as cotyledon initiation, leaf serration formation, and ovule primordium initiation have suggested how CUC2 activate floral organ initiation in adjacent whorls, by up-regulating PIN1 expression to promote an auxin convergence point, which creates an auxin maximum where organ founder cells are specified (Vernoux et al. 2000; Furutani et al. 2004; Heisler et al. 2005; Bilsborough et al. 2011; Galbiati et al. 2013). The auxin maximum then represses *CUC2* expression and restricts it to the boundary in a negative feedback loop (Vernoux et al. 2000; Bilsborough et al. 2011).

Posttranscriptional regulation of *CUC1* and *CUC2* by three *miRNA164* isoforms and mutation of *MIR164c* (also named *EARLY EXTRA PETALS1* [*EEP1*]), leads to supernumerary petals and is associated with additional boundary domains in the second whorl (Baker et al. 2005). After petal primordium formation, *MIR164c*/*EEP1* expression is repressed in part by RABBIT EARS (RBE) (Huang et al. 2012), which also regulates sepal boundaries cell-non-autonomously. *RBE* is also regulated by PETAL LOSS (PTL) in the intersepal zone, which establishes sepal boundaries concertedly with CUC1 and CUC2 (Lampugnani et al. 2013). *HANABA TARANU* (*HAN*) is expressed at organ boundaries and *han* mutants display fused sepals and fewer floral organs, which might be caused by impaired polar auxin transport by analogy to embryogenic HAN function, which regulates auxin flux to the hypophysis (Nawy et al. 2010; Zhang et al. 2013).

Auxin response within the intersepal boundary zone is controlled by PTL (Brewer 2004), which synergistically regulates petal initiation with RBE and PTL, illustrating that cell-non-autonomy is another important facet of auxin function (Griffith et al. 1999; Brewer 2004; Takeda et al. 2004; Krizek et al. 2006). In *ptl* mutants, petals arise sporadically and are absent in later-arising flowers (Griffith et al. 1999) and sites of petal initiation are restored by directing auxin synthesis within the intersepal zone (Lampugnani et al. 2013). Thus, auxin confers positional information for petal initiation by functioning as a mobile inter-whorl signal. RBE and PTL both function within the same pathway, with *RBE* downstream of *PTL*. The petal defects of *ptl* and *rbe* mutants are enhanced by mutation of the auxin influx carrier AUX1 (Lampugnani et al. 2013). Disruption of auxin efflux via PID and PIN1 in the petal initiation zone generally increases petal number, dependent on PTL function. *RBE* also defines the second-whorl domain independently of *PTL*, by regulating several boundary genes (Fig. 2).

The SUPERMAN (SUP) C2H2 transcription factor, which is closely related to *RBE* (Sakai et al. 1995), limits *APETALA3* expression in whorl 4, and *sup* mutants produce supernumerary stamens (Bowman et al. 1992). *SUP* expression spans the whorl 3/4 boundary (Prunet et al. 2017) and reduces local auxin biosynthesis by directly binding to *YUC1* and *YUC4* and recruiting Polycomb-repressive complex 2 to repress their expression (Table 1; Fig. 2; Xu et al. 2018). This up-regulation of auxin biosynthesis in *sup* both expands the stem-cell domain between whorls 3 and 4 cell-non-autonomously and prolongs its activity (Xu et al. 2018), leading to supernumerary stamens as the fate of cells in whorl 4 changes from female to male (Prunet et al. 2017). In conclusion, the fine-tuning of auxin synthesis or activity in inter-organ or inter-whorl boundaries is one mechanism for controlling cell divisions and stem-cell availability for organ differentiation.

FLORAL ORGAN OUTGROWTH

The concentration of free auxin, indicated by *DR5::GUS*, is high in the tips of floral organ primordia in all whorls (Aloni et al. 2006), which drives rapid organ outgrowth. After the specification of sepal founder cells in a stereotypic order (Chandler and Werr 2014), sepal outgrowth in *Arabidopsis* is coordinated by DEVELOPMENT-RELATED MYB-LIKE1 (DRMY1) (Zhu et al. 2020). Coordinated auxin and cytokinin signaling regulates sepal outgrowth, analogous to the temporal regulation of FM outgrowth (Besnard et al. 2014). *drmy1* mutants display variable sepal sizes due to delayed outgrowth, which results from diffuse auxin and cytokinin response domains in sepal primordia. Wild-type DRMY1 signaling involves initial auxin responses in the outer primordium cell layer that encroach into the underlying cytokinin signaling domain. These localized and focused auxin and cytokinin signaling domains define regions of competency for outgrowth, in which *DR5::GFP* accumulates at the sepal tips and *TCS* at the base. In tomato, the MADS-box gene *SlMBP21* negatively regulates sepal cell expansion, probably partly via auxin-related pathways.

Auxin contributes to patterning petal shape and size, which depend on cell division during early development and cell expansion during

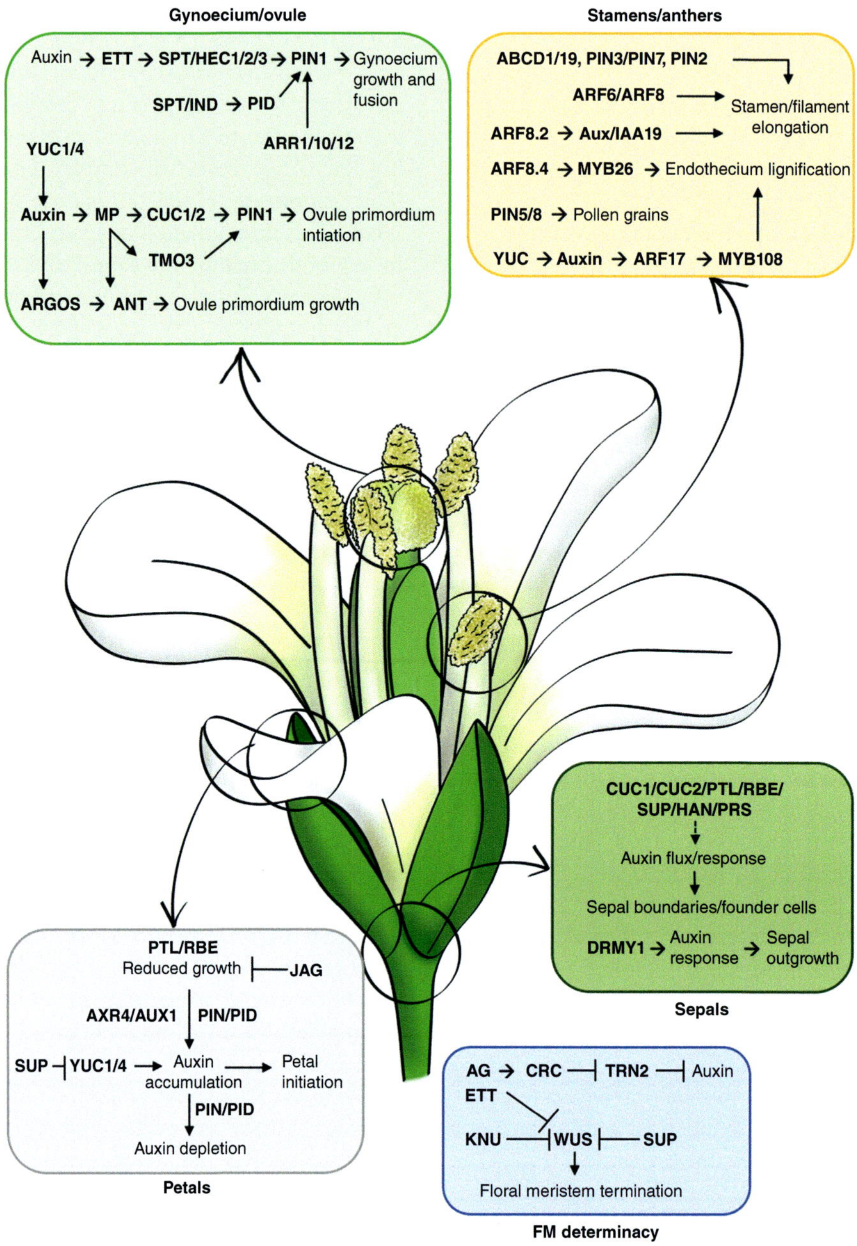

Figure 2. The gene regulatory networks involving auxin that orchestrate development of the different flower organs. A representation of an *Arabidopsis* flower linked to schematic diagrams to show how auxin integrates into the gene regulatory networks that coordinate gynoecium, stamen, petal, and sepal initiation and growth. An additional diagram shows the gene regulatory network involving auxin that relates to floral meristem (FM) determinacy and termination (see text for details). (AG) AGAMOUS, (ARF8/ARF17) AUXIN RESPONSE FACTOR7/17, (AUX1) AUXIN EFFLUX CARRIER1, (Aux/IAA19) AUX/IAA PROTEIN19, (AXR4) AUXIN-RESISTANT4, (CRC) CRABS CLAW, (DRMY1) DEVELOPMENT-RELATED MYB-LIKE1, (HAN) HANABA TARANU, (HEC1/HEC2/HEC3) HECATE1/2/3, (JAG) JAGGED, (KNU) KNUCKLES, (MYB26/MYB108) MYB-DOMAIN PROTEIN26/08, (SPT) SPATULA, (TMO3) TARGET OF MONOPTEROS3, (TRN) TORNADO, (WUS) WUSCHEL (see additional abbreviations in Fig. 1).

late development (Smyth et al. 1990). Many genes directly involved with auxin contribute to petal growth, including *JAGGED* (*JAG*), *ARGOS*, *ANT*, *BIGPETALp* (*BPEp*), and *ARF8*. JAG regulates growth by modulating the switch from cell proliferation to cell expansion (Sauret-Güeto et al. 2013). *jag* petals are narrow with a reduced distal region and a jagged or serrated margin (Dinneny et al. 2004). In *jag*, *DR5::GFP* expression is reduced, but is broader when JAG is ectopically expressed. Moreover, JAG directly represses *PTL* (Sauret-Güeto et al. 2013).

Overexpression of the auxin-inducible gene *ARGOS* increases petal size by increasing the number and size of petal epidermal cells. This can be blocked by loss of *ANT* function, suggesting that ARGOS positively regulates *ANT* during petal organogenesis (Hu et al. 2003). Conversely, BPEp and ARF8 together regulate petal size by negatively affecting mitotic growth and restricting cell expansion during early and late petal development, respectively (Varaud et al. 2011), and *arf8* and *bpe* mutants are associated with the up-regulation of early auxin-responsive genes. Notably, interaction between BPEp and ARF8 occurs via the BPEp carboxy-terminal domain, which shares sequence similarity with motif III of ARF and AUXIN/INDOLE-3-ACETIC ACID proteins (Varaud et al. 2011).

The stereotypic conical shape of cells in the *Arabidopsis* adaxial petal epidermis requires auxin signaling and apoplastic pH changes (Dang et al. 2020). Decreased auxin concentration, transport, or signaling abolishes cell-wall acidification and affects the height and shape of conical cells; notably, *arf6-2 arf8-3* displays severe defects in conical cell shape and outgrowth (Tabata et al. 2010).

In *Arabidopsis*, stamens develop asynchronously: primordia of the four long medial stamens arise contemporaneously with petals during stage 5 of flower development, whereas the two short lateral stamens develop slightly later (Bowman et al. 1989). Reduction in auxin biosynthesis, especially by *yuc2* and *yuc6* or polar transport (*pin1* and *pid*), reduces stamen number. The fewer-stamen phenotype of *ett* is enhanced by inhibiting auxin transport, leading to third-whorl organs that often consist of filamentous structures that lack anthers. Moreover, *ABCB*/*PGP* genes that encode auxin efflux carriers are expressed in young stamen primordia, as well as in tapetum and pollen mother cells (PMCs) entering meiosis, and *abcb1* and *abcb19* show unsynchronized and precocious meiosis (Cecchetti et al. 2015).

The role of auxin during premeiotic phases of stamen development is poorly understood, partly because *DR5* expression is only detected after microspore release (Cheng et al. 2006; Feng et al. 2006; Cecchetti et al. 2008, 2015). The R2D2 auxin reporter revealed that auxin-mediated DII depletion occurs in the locules early in anther development and overlaps with YUC1 and YUC4 activity (Cheng et al. 2006; Ståldal et al. 2012). Auxin biosynthesis is positively regulated by SHI/STY transcription factors, which are active throughout anther development and affect early anther patterning (Estornell et al. 2018).

Late stamen development involves the coordination of rapid filament elongation with anther dehiscence and pollen maturation and finally terminates with mature pollen-grain release. Anthers of *abcb1 abcb19* overproliferate tapetum cells and show enhanced septum lignification, suggesting that auxin transport regulates cell division and lignification (Cecchetti et al. 2015). Inhibition of auxin efflux (Cecchetti et al. 2017) suggests that auxin flows from the tapetum to the middle layer, where auxin response maxima are important for coordinated pollen maturation and anther dehiscence.

Auxin signaling also represses precocious entry into subsequent developmental steps in several anther cell layers, and anthers of mutants in genes encoding the four TIR/AFB nuclear auxin receptors show premature dehiscence and pollen release compared to wild-type (Cecchetti et al. 2008, 2017). Mutations in ARF1, ARF2 (Ellis et al. 2005), ETT (Sessions et al. 1997), ARF4 (Pekker et al. 2005), ARF5 (Przemeck et al. 1996), ARF17 (Xu et al. 2019), ARF6, and ARF8 (Nagpal et al. 2005) all cause defects in male reproductive tissues. Stamens of *arf6 arf8* are short with indehiscent anthers due to reduced jasmonic acid (JA) biosynthesis (Nagpal et al. 2005; Tabata et al. 2010). Furthermore, restriction of *ARF6* and *ARF8* domains by miR167 cleavage is required for anther development and fertility (Wu et al. 2006).

Several flower-specific, intron-retaining splice variants of *ARF8* affect stamen development in *Arabidopsis* (Ghelli et al. 2018). ARF8.4 transcriptionally activates *MYB26*, which regulates secondary thickening and anther dehiscence (Yang et al. 2007, 2017), and ARF8.2 regulates JA biosynthesis by activating DAD1 to promote the final steps of anther dehiscence. The ARF17–MYB108 complex also regulates anther dehiscence in parallel with the ARF8.4–MYB26 pathway (Xu et al. 2019). The stamen-specific AUX/IAA gene *IAA19* controls filament elongation (Tashiro et al. 2009) and is regulated by auxin (Cecchetti et al. 2017). *IAA19* is also regulated by ARF8.4, but is hardly affected by ARF8.2 and ARF8.1, highlighting the specificity of ARF8 isoforms during early stamen development (Table 1; Fig. 2; Ghelli et al. 2018).

FLORAL MERISTEM DETERMINACY AND GYNOECIUM DEVELOPMENT

The final floral development stage involves termination of FM growth following realization of the species-specific floral body plan. In *Arabidopsis*, the master regulator of FM determinacy, AGAMOUS (AG), represses the stem-cell marker *WUS* by recruiting Polycomb group (PcG) proteins to the *WUS* locus and in parallel, activating the expression of the *WUS* repressor *KNUCKLES* (*KNU*) (Sun et al. 2009; Liu et al. 2011). APETALA2 (AP2) antagonizes AG function by activating *WUS* expression (Huang et al. 2017). AG directly activates *CRABS CLAW* (*CRC*), which represses *TORNADO* (*TRN2*), a regulator of auxin homeostasis, leading to a shift from meristem maintenance to organ differentiation (Yamaguchi et al. 2017). AG and CRC both activate the *YUC4* promotor and AG recruits CHROMATIN REMODELING FACTOR11 and 17 (CHR11/17), which increases chromatin accessibility (Yamaguchi et al. 2018). Therefore, the regulation of auxin homeostasis by multiple pathways is an important component of FM determinacy (Table 1; Fig. 2).

The AG–AP2 FM determinacy pathway is also mediated by ETT. ETT and AG coordinately repress *WUS* expression in the FM and combined loss of ETT and AG strongly enhances defects in FM indeterminacy. Moreover, *ETT* expression is indirectly positively regulated by AG. Binding of ETT to *WUS* is enhanced by AG activity, but independently from the *WUS* repression pathway mediated by *KNU*. In addition, *ETT* function in FM determinacy is negatively regulated by AP2 (Liu et al. 2014).

The appropriate termination of FM growth relates to gynoecium initiation in the FM center, which depletes the meristematic cell pool. The absence of a gynoecium in *ag* mutants indicates that FM determinacy is essential for female organ differentiation (Marsch-Martínez and de Folter 2016; Yamaguchi et al. 2017).

Gynoecium development involves the establishment of apical–basal and adaxial–abaxial axes. Mutation of *ETT* perturbs gynoecium organogenesis along the apical–basal axis, leading to reduced valve development and the ectopic expansion of transmitting tract and stigma (Sessions and Zambryski 1995; Alvarez and Smyth 1998). A long-standing model for gynoecium development along the apical–basal axis derived from observations that inhibition of auxin transport and auxin biosynthesis or transport mutants all generate a valveless gynoecium topped with stigmatic tissue. This suggests that a high local auxin biosynthesis at the gynoecium tip establishes a basipetally decreasing auxin morphogenetic gradient that specifies style, ovary, and gynophore (Nemhauser et al. 2000). However, *DR5rev::GFP* reports no auxin gradient, only auxin response maxima in carpel founder cells at stage 7, the provasculature at stage 8, apical tissue at stage 9, and the vasculature and stigmatic papillae at stage 12 (Larsson et al. 2013). Additionally, specification of lateral and medial gynoecium domains, including valve outgrowth, depends on PIN1-mediated lateral auxin maxima (Larsson et al. 2014). A refined, so-called early action model (Hawkins and Liu 2014) based on early polarity establishment in lateral organs such as leaves, proposes that abaxial–adaxial polarity in early carpel primordia is critical for subsequent development. This is consistent with YUC1 and YUC4 activity at the start of organ initiation (Cheng et al. 2006), which potentially contributes to establishing an adaxial–abaxial boundary. Adaxial–abaxial axis formation also involves protein interactions between KANADI (KAN)1, KAN4, and ETT (Pekker et al. 2005). Therefore, gynoecium patterning involves two distinct phases: the early establishment of abaxial–adaxial identity within the mediolateral axis domains by coordinated auxin synthesis, transport, and signaling. Subsequently, this lateromedial domain is a prerequisite for establishing an apical–basal axis (Hawkins and Liu 2014).

During late gynoecium development, two additional foci of auxin response maxima initiate in the medial domain in addition to two in the lateral domain in the gynoecium apex. Immediately preceding style development, these four foci fuse into a ring due to SPATULA (SPT)/INDEHISCENT (IND)-dependent rearrangement of PIN proteins, which involves the direct repression of *PID* and leads to the acquisition of radial symmetry (Moubayidin and Østergaard 2014). During style and gynoecium development, auxin promotes the function of an ETT–IND complex that also leads to *PID* repression and finely regulates PIN localization and auxin flux (Moubayidin and Østergaard 2017; Kuhn et al. 2019). As well as interaction with IND, SPT associates with HECATE to regulate *PIN1* and *PIN3* expression and block cytokinin signaling (Gremski et al. 2007).

Because auxin biosynthesis mutants still differentiate stigmatic tissue, late gynoecium development is probably independent from the early action model and dependent on local auxin synthesis via *YUC* expression activated by STYLISH1/2 and NGA3 transcription factors in the gynoecium tip (Sohlberg et al. 2006; Trigueros et al. 2009; Eklund et al. 2011). SQUAMOSA PROMOTER-BINDING PROTEIN-LIKE transcription factors also contribute to style, septum, and transmitting tract development, probably upstream of YUC4 and by interactions with CRC, ETT, SPT, and STY1, which are all involved in auxin homeostasis or responsiveness (Xing et al. 2013).

The direct binding of auxin to the carboxy-terminal domain of ETT (Kuhn et al. 2020) reveals an additional layer of auxin regulation that triggers a rapid and reversible switch of chromatin states during gynoecium development. ETT cannot interact with Aux/IAA proteins or mediate canonical auxin signaling because it lacks a PB1 domain. In low-auxin conditions, the ETT EAR repressor motif physically interacts with members of the TPL and TPL-RELATED (TPR) family of corepressors, which recruit histone deacetylase HDA19 to maintain chromatin of ETT targets in the gynoecium, such as *PID* (Benjamins et al. 2001) and *HECATE1* (*HEC1*) (Gremski et al. 2007) in a repressed state. At higher auxin levels, binding of IAA to ETT causes its dissociation from TPL/TPR2, which prevents HDA19 from deacetylating histones at ETT target loci and leads to their transcriptional activation.

AUXIN AND OVULE DEVELOPMENT

Auxin maintains the meristematic activity of the placenta and determines the sites of ovule initiation and ovule identity. In most auxin-related mutants, defects in CMM formation cause the reduction or absence of placental tissue and ovules (Cucinotta et al. 2014; Reyes-Olalde and de Folter 2019). In particular, auxin concentration maxima generated by PIN1 in the placenta determine sites of ovule primordium initiation (Benková et al. 2003; Ceccato et al. 2013; Galbiati et al. 2013). Placental PIN1 expression is promoted by cytokinins via transcription of the direct MP targets *CYTOKININ RESPONSE FACTOR2*/*TARGET OF MONOPTEROS3* and *CUC1*/*CUC2* (Galbiati et al. 2013; Cucinotta et al. 2016, 2020), and a lower PIN1 level in *crf2* and *cuc2-1 CUC1_RNAi* mutants results in fewer ovule primordia. After ovule initiation, ANT promotes primordium cell proliferation and integument growth. Correct *ANT* expression is also controlled by auxin via MP and ARGOS (Hu et al. 2003; Galbiati et al. 2013).

Auxin is critical for the somatic-to-germline fate transition, and for downstream events of megasporogenesis and female gametophyte (FG) development. PIN1 also drives an auxin maximum in the nucellus at the ovule tip in the L1 surrounding the megaspore mother cell (MMC) (Benková et al. 2003; Ceccato et al.

2013). ETT is a negative regulator of MMC identity and is expressed in central ovule regions, where it restricts *WUS* expression, similar to its role in the FM. In the nucellus, *ETT* expression is tightly regulated by TAS3 tasiRNA, and deregulation of its expression causes supernumerary MMCs (Su et al. 2017).

After MMC enlargement, meiosis generates four haploid megaspores, but only the functional megaspore undergoes three mitoses to form an FG consisting of seven cells of four distinct cell types. During megagametogenesis, auxin response is localized to the chalazal pole in the sporophytic tissues surrounding the FG (Lituiev et al. 2013). This suggested a model in which the auxin cell-non-autonomously indirectly affects cell specification within the gametophyte by functioning in sporophytic tissues (Lituiev et al. 2013). Auxin helps to pattern the micropylar pole of the FG where egg and synergid cells are located (Skinner and Sundaresan 2018), and loss of synergid identity and the occasional acquisition of egg identity results when TIR1 and ARFs are down-regulated or inactivated in the young FG (Panoli et al. 2015; Liu et al. 2018). Conversely, ectopic *YUC* expression shifts micropylar cell fates toward the chalazal end of the gametophyte, conferring synergid and egg-cell marker expression to the central cell and antipodal cells (Panoli et al. 2015). It remains unknown whether auxin influences mitosis and cell identity within the TM, which could be addressed by live imaging.

CONCLUDING REMARKS

Despite detailed understanding of how auxin integrates into the gene networks that regulate floral development, and major advances in characterizing components of the auxin response machinery, higher-order layers of complexity involved in auxin response continue to emerge. However, understanding how auxin confers positional information for organ initiation requires knowledge concerning spatiotemporal auxin responses at single-cell resolution. Despite the central role of MP in coordinating auxin responses, the complete genome-wide repertoire of direct ARF targets has not been captured and paradigmatic canonical auxin response signaling might involve as-yet-unknown ARF-binding sequences.

Most importantly, although *Arabidopsis* is the best-studied plant development model, its flowers are hardly representative of angiosperm floral diversity and the molecular-genetic basis of many features of floral development that underlie differences in the floral body plan and morphogenesis in other species have not been considered; some examples include the unidirectional organ initiation, the existence of overlapping temporal organ initiation in different whorl, and common primordia in many legumes (Ferrandiz et al. 1999; Tucker 2003; Benlloch et al. 2015) or flowers that show centripetal, centrifugal (or basipetal), or more chaotic sequences of organ development within a single flower or whorl (Rudall 2010). One recurring theme in floral development is how auxin gradients orchestrate aspects of floral development in concert with those in cytokinin. Despite the characterization of auxin signaling components in comparative species, this needs to be combined with studies on comparative floral morphology at a high spatiotemporal resolution, to identify evolutionarily conserved aspects of the regulation of floral development by auxin to be identified, and to elucidate further levels of complexity not evident in *Arabidopsis*.

ACKNOWLEDGMENTS

A.C. and L.C. were supported by Ministero dell'Istruzione, dell'Università e della Ricerca (MIUR). M.C., A.C., and L.C. were funded by MIUR-PRIN 2012. We thank Sara Comai for drawing the flower in Figure 2.

REFERENCES

Aloni R, Aloni E, Langhans M, Ullrich CI. 2006. Role of auxin in regulating *Arabidopsis* flower development. *Planta* **223:** 315–328. doi:10.1007/s00425-005-0088-9

Alonso JM, Stepanova AN, Leisse TJ, Kim CJ, Chen H, Shinn P, Stevenson DK, Zimmerman J, Barajas P, Cheuk R, et al. 2003. Genome-wide insertional mutagenesis of *Arabidopsis thaliana*. *Science* **301:** 653–657. doi:10.1126/science.1086391

Alvarez J, Smyth DR. 1998. Genetic pathways controlling carpel development in *Arabidopsis thaliana*. *J Plant Res* **111:** 295–298. doi:10.1007/BF02512187

Alvarez-Buylla ER, Benítez M, Corvera-Poiré A, Cador C, De Folter S, De Buen AG, Garay-Arroyo A, García-Ponce B, Jaimes-Miranda F, Rigoberto V, et al. 2010. Flower development. *Arabidopsis Book* **8:** e0127. doi:10.1199/tab.0127

Baker CC, Sieber P, Wellmer F, Meyerowitz EM. 2005. The early extra petals1 mutant uncovers a role for microRNA miR164c in regulating petal number in *Arabidopsis*. *Curr Biol* **15:** 303–315. doi:10.1016/j.cub.2005.02.017

Benjamins R, Quint A, Weijer D, Hooykaas P, Offringa R. 2001. The PINOID protein kinase regulates organ development in *Arabidopsis* by enhancing polar auxin transport. *Development* **128:** 4057-4067.

Benková E, Michniewicz M, Sauer M, Teichmann T, Seifertová D, Jürgens G, Friml J. 2003. Local, efflux-dependent auxin gradients as a common module for plant organ formation. *Cell* **115:** 591–602. doi:10.1016/S0092-8674(03)00924-3

Benlloch R, Berbel A, Ali L, Gohari G, Millán T, Madueño F. 2015. Genetic control of inflorescence architecture in legumes. *Front Plant Sci* **6:** 543. doi:10.3389/fpls.2015.00543

Bennett SRM, Alvarez J, Bossinger G, Smyth DR. 1995. Morphogenesis in pinoid mutants of *Arabidopsis* thaliana. *Plant J* **8:** 505–520. doi:10.1046/j.1365-313X.1995.8040505.x

Besnard F, Refahi Y, Morin V, Marteaux B, Brunoud G, Chambrier P, Rozier F, Mirabet V, Legrand J, Lainé S, et al. 2014. Cytokinin signalling inhibitory fields provide robustness to phyllotaxis. *Nature* **505:** 417–421. doi:10.1038/nature12791

Bilsborough GD, Runions A, Barkoulas M, Jenkins HW, Hasson A, Galinha C, Laufs P, Hay A, Prusinkiewicz P, Tsiantis M. 2011. Model for the regulation of *Arabidopsis* thaliana leaf margin development. *Proc Natl Acad Sci* **108:** 3424–3429. doi:10.1073/pnas.1015162108

Bowman JL, Smyth DR, Meyerowitz EM. 1989. Genes directing flower development in *Arabidopsis*. *Plant Cell* **1:** 37–52.

Bowman JL, Sakai H, Jack T, Weigel D, Mayer U, Meyerowitz EM. 1992. SUPERMAN a regulator of floral homeotic genes in *Arabidopsis*. *Development* **114:** 599–615.

Brewer PB. 2004. PETAL LOSS, a trihelix transcription factor gene, regulates perianth architecture in the *Arabidopsis* flower. *Development* **131:** 4035–4045. doi:10.1242/dev.01279

Caggiano MP, Yu X, Bhatia N, Larsson A, Ram H, Ohno CK, Sappl P, Meyerowitz EM, Jönsson H, Heisler MG. 2017. Cell type boundaries organize plant development. *eLife* **6:** e27421. doi:10.7554/eLife.27421

Ceccato L, Masiero S, Sinha Roy D, Bencivenga S, Roig-Villanova I, Ditengou FA, Palme K, Simon R, Colombo L. 2013. Maternal control of PIN1 is required for female gametophyte development in *Arabidopsis*. *PLoS ONE* **8:** e66148. doi:10.1371/journal.pone.0066148

Cecchetti V, Altamura MM, Falasca G, Costantino P, Cardarelli M. 2008. Auxin regulates *Arabidopsis* anther dehiscence, pollen maturation, and filament elongation. *Plant Cell* **20:** 1760–1774. doi:10.1105/tpc.107.057570

Cecchetti V, Brunetti P, Napoli N, Fattorini L, Altamura MM, Costantino P, Cardarelli M. 2015. ABCB1 and ABCB19 auxin transporters have synergistic effects on early and late *Arabidopsis* anther development. *J Integr Plant Biol* **57:** 1089–1098. doi:10.1111/jipb.12332

Cecchetti V, Celebrin D, Napoli N, Ghelli R, Brunetti P, Costantino P, Cardarelli M. 2017. An auxin maximum in the middle layer controls stamen development and pollen maturation in *Arabidopsis*. *New Phytol* **213:** 1194–1207. doi:10.1111/nph.14207

Chandler JW. 2011. Founder cell specification. *Trends Plant Sci.* **16:** 607–613. doi:10.1016/j.tplants.2011.08.005

Chandler JW, Werr W. 2014. *Arabidopsis* floral phytomer development: auxin response relative to biphasic modes of organ initiation. *J Exp Bot* **65:** 3097–3110. doi:10.1093/jxb/eru153

Chandler J, Werr W. 2017. DORNRÖSCHEN, DORNRÖSCHEN-LIKE, and PUCHI redundantly control floral meristem identity and organ initiation in *Arabidopsis*. *J Exp Bot* **68:** 3457–3472. doi:10.1093/jxb/erx208

Chandler JW, Cole M, Jacobs B, Comelli P, Werr W. 2011. Genetic integration of DORNRÖSCHEN and DORNRÖSCHEN-LIKE reveals hierarchical interactions in auxin signalling and patterning of the *Arabidopsis* apical embryo. *Plant Mol Biol* **75:** 223–236. doi:10.1007/s11103-010-9721-5

Cheng Y, Dai X, Zhao Y. 2006. Auxin biosynthesis by the YUCCA flavin monooxygenases controls the formation of floral organs and vascular tissues in *Arabidopsis*. *Genes Dev* **20:** 1790–1799. doi:10.1101/gad.1415106

Chung Y, Zhu Y, Wu MF, Simonini S, Kuhn A, Armenta-Medina A, Jin R, Østergaard L, Gillmor CS, Wagner D. 2019. Auxin response factors promote organogenesis by chromatin-mediated repression of the pluripotency gene SHOOTMERISTEMLESS. *Nat Commun* **10:** 886.

Cole M, Chandler J, Weijers D, Jacobs B, Comelli P, Werr W. 2009. DORNROSCHEN is a direct target of the auxin response factor MONOPTEROS in the *Arabidopsis* embryo. *Development* **136:** 1643–1651. doi:10.1242/dev.032177

Comelli P, Glowa D, Frerichs A, Engelhorn J, Chandler JW, Werr W. 2020. Functional dissection of the DORNRÖSCHEN-LIKE enhancer 2 during embryonic and phyllotactic patterning. *Planta* **251:** 90. doi:10.1007/s00425-020-03381-7

Cucinotta M, Colombo L, Roig-villanova I. 2014. Ovule development, a new model for lateral organ formation. *Front Plant Sci* **5:** 1–12. doi:10.3389/fpls.2014.00117

Cucinotta M, Manrique S, Guazzotti A, Quadrelli NE, Mendes MA, Benkova E, Colombo L. 2016. Cytokinin response factors integrate auxin and cytokinin pathways for female reproductive organ development. *Development* **143:** 4419–4424. doi:10.1242/dev.143545

Cucinotta M, Di Marzo M, Guazzotti A, de Folter S, Kater MM, Colombo L. 2020. Gynoecium size and ovule number are interconnected traits that impact seed yield. *J Exp Bot* **71:** 2479–2489. doi:10.1093/jxb/eraa050

Dang X, Chen B, Liu F, Ren H, Liu X, Zhou J, Qin J, Lin D. 2020. Auxin signaling-mediated apoplastic pH modification functions in petal conical cell shaping. *Cell Rep* **30:** 3904–3916.e3. doi:10.1016/j.celrep.2020.02.087

Dinneny JR, Yadegari R, Fischer RL, Yanofsky MF, Weigel D. 2004. The role of JAGGED in shaping lateral organs. *Development* **131:** 1101–1110. doi:10.1242/dev.00949

Eklund DM, Cierlik I, Ståldal V, Claes AR, Vestman D, Chandler J, Sundberg E. 2011. Expression of *Arabidopsis SHORT INTERNODES/STYLISH* family genes in auxin biosynthesis zones of aerial organs is dependent on a GCC box-like regulatory element. *Plant Physiol* **157:** 2069–2080. doi:10.1104/pp.111.182253

Ellis CM, Nagpal P, Young JC, Hagen G, Guilfoyle TJ, Reed JW. 2005. AUXIN RESPONSE FACTOR1 and AUXIN RESPONSE FACTOR2 regulate senescence and floral organ abscission in *Arabidopsis thaliana. Development* **132:** 4563–4574. doi:10.1242/dev.02012

Estornell LH, Landberg K, Cierlik I, Sundberg E. 2018. SHI/STY genes affect pre- and post-meiotic anther processes in auxin sensing domains in *Arabidopsis. Front Plant Sci* **9:** 150. doi:10.3389/fpls.2018.00150

Feng XL, Ni WM, Elge S, Mueller-Roeber B, Xu ZH, Xue HW. 2006. Auxin flow in anther filaments is critical for pollen grain development through regulating pollen mitosis. *Plant Mol Biol* **61:** 215–226. doi:10.1007/s11103-006-0005-z

Ferrandiz C, Navarro C, Gomez MD, Canas LA, Beltran JP. 1999. Flower development in *Pisum sativum*: from the war of the whorls to the battle of the common primordia. *Dev Genet* **25:** 280–290.

Frerichs A, Thoma R, Abdallah AT, Frommolt P, Werr W, Chandler JW. 2016. The founder-cell transcriptome in the *Arabidopsis apetala1 cauliflower* inflorescence meristem. *BMC Genomics* **17:** 855. doi:10.1186/s12864-016-3189-x

Furutani M, Vernoux T, Traas J, Kato T, Tasaka M, Aida M. 2004. PIN-FORMED1 and PINOID regulate boundary formation and cotyledon development in *Arabidopsis* embryogenesis. *Development* **131:** 5021–5030. doi:10.1242/dev.01388

Furutani M, Nakano Y, Tasaka M. 2014. MAB4-induced auxin sink generates local auxin gradients in *Arabidopsis* organ formation. *Proc Natl Acad Sci* **111:** 1198–1203. doi:10.1073/pnas.1316109111

Galbiati F, Sinha Roy D, Simonini S, Cucinotta M, Ceccato L, Cuesta C, Simaskova M, Benková E, Kamiuchi Y, Aida M, et al. 2013. An integrative model of the control of ovule primordia formation. *Plant J* **76:** 446–455. doi:10.1111/tpj.12309

Galvan-Ampudia CS, Cerutti G, Legrand J, et al. 2020. Temporal integration of auxin information for the regulation of patterning. *eLife* **9:** e55832.

Gälweiler L, Guan C, Müller A, Wisman E, Mendgen K, Yephremov A, Palme K. 1998. Regulation of polar auxin transport by AtPIN1 in *Arabidopsis* vascular tissue. *Sci* **282:** 2226–2230. doi:10.1126/science.282.5397.2226

Ghelli R, Brunetti P, Napoli N, De Paolis A, Cecchetti V, Tsuge T, Serino G, Matsui M, Mele G, Rinaldi G, et al. 2018. A newly identified flower-specific splice variant of *AUXIN RESPONSE FACTOR8* regulates stamen elongation and endothecium lignification in *Arabidopsis. Plant Cell* **30:** 620–637. doi:10.1105/tpc.17.00840

Goldental-Cohen S, Israeli A, Ori N, Yasuor H. 2017. Auxin response dynamics during wild-type and entire flower development in tomato. *Plant Cell Physiol* **58:** 1661–1672. doi:10.1093/pcp/pcx102

Gremski K, Ditta G, Yanofsky MF. 2007. The HECATE genes regulate female reproductive tract development in *Arabidopsis thaliana. Development* **134:** 3593–3601. doi:10.1242/dev.011510

Griffith ME, da Silva Conceição A, Smyth DR. 1999. PETAL LOSS gene regulates initiation and orientation of second whorl organs in the *Arabidopsis* flower. *Development* **126:** 5635–5644.

Guan C, Wu B, Yu T, Wang Q, Krogan NT, Liu X, Jiao Y. 2017. Spatial auxin signaling controls leaf flattening in *Arabidopsis. Curr Biol* **27:** 2940–2950.e4. doi:10.1016/j.cub.2017.08.042

Hawkins C, Liu Z. 2014. A model for an early role of auxin in *Arabidopsis* gynoecium morphogenesis. *Front Plant Sci* **5:** 327.

Heisler MG, Ohno C, Das P, Sieber P, Reddy GV, Long JA, Meyerowitz EM. 2005. Patterns of auxin transport and gene expression during primordium development revealed by live imaging of the *Arabidopsis* inflorescence meristem. *Curr Biol* **15:** 1899–1911. doi:10.1016/j.cub.2005.09.052

Hibara K, Karim MR, Takada S, Taoka K, Furutani M, Aida M, Tasaka M. 2006. *Arabidopsis CUP-SHAPED COTYLEDON3* regulates postembryonic shoot meristem and organ boundary formation. *Plant Cell* **18:** 2946–2957. doi:10.1105/tpc.106.045716

Hu Y, Xie Q, Chua NH. 2003. The *Arabidopsis* auxin-inducible gene *ARGOS* controls lateral organ size. *Plant Cell* **15:** 1951–1961. doi:10.1105/tpc.013557

Huang T, López-Giráldez F, Townsend JP, Irish VF. 2012. RBE controls microRNA164 expression to effect floral organogenesis. *Dev* **139:** 2161–2169. doi:10.1242/dev.075069

Huang Z, Shi T, Zheng B, Yumul RE, Liu X, You C, Gao Z, Xiao L, Chen X. 2017. *APETALA2* antagonizes the transcriptional activity of *AGAMOUS* in regulating floral stem cells in *Arabidopsis thaliana. New Phytol* **215:** 1197–1209. doi:10.1111/nph.14151

Irish VF. 2010. The flowering of *Arabidopsis* flower development. *Plant J* **61:** 1014–1028. doi:10.1111/j.1365-313X.2009.04065.x

Kamiuchi Y, Yamamoto K, Furutani M, Tasaka M, Aida M. 2014. The CUC1 and CUC2 genes promote carpel margin meristem formation during *Arabidopsis* gynoecium development. *Front Plant Sci* **5:** 165. doi:10.3389/fpls.2014.00165

Krizek BA, Lewis MW, Fletcher JC. 2006. *RABBIT EARS* is a second-whorl repressor of *AGAMOUS* that maintains spatial boundaries in *Arabidopsis* flowers. *Plant J* **45:** 369–383. doi:10.1111/j.1365-313X.2005.02633.x

Kuhn A, Runciman B, Tasker-Brown W, Østergaard L. 2019. Two auxin response elements fine-tune *PINOID* expression during gynoecium development in *Arabidopsis thaliana. Biomolecules* **9:** 526. doi:10.3390/biom9100526

Kuhn A, Ramans Harborough S, McLaughlin HM, Natarajan B, Verstraeten I, Friml J, Kepinski S, Østergaard L. 2020. Direct ETTIN-auxin interaction controls chromatin states in gynoecium development. *eLife* **9:** e51787. doi:10.7554/eLife.51787

Kwiatkowska D. 2008. Flowering and apical meristem growth dynamics. *J Exp Bot* **59:** 187–201. doi:10.1093/jxb/erm290

Lampugnani ER, Kilinc A, Smyth DR. 2013. Auxin controls petal initiation in *Arabidopsis*. *Development* **140:** 185–194. doi:10.1242/dev.084582

Larsson E, Franks RG, Sundberg E. 2013. Auxin and the *Arabidopsis thaliana* gynoecium. *J Exp Bot* **64:** 2619–2627. doi:10.1093/jxb/ert099

Larsson E, Roberts CJ, Claes R, Franks RG, Sundberg E. 2014. Polar auxin transport is essential for medial versus lateral tissue specification and vascular-mediated valve outgrowth in *Arabidopsis* gynoecia. *Plant Physiol* **166:** 1998–2012. doi:10.1104/pp.114.245951

Laufs P, Peaucelle A, Morin H, Traas J. 2004. MicroRNA regulation of the CUC genes is required for boundary size control in *Arabidopsis* meristems. *Development* **131:** 4311–4322. doi:10.1242/dev.01320

Li W, Zhou Y, Liu X, Yu P, Cohen JD, Meyerowitz EM. 2013. LEAFY controls auxin response pathways in floral primordium formation. *Sci Signal* **6:** ra23.

Lituiev DS, Krohn NG, Muller B, Jackson D, Hellriegel B, Dresselhaus T, Grossniklaus U. 2013. Theoretical and experimental evidence indicates that there is no detectable auxin gradient in the angiosperm female gametophyte. *Development* **140:** 4544–4553. doi:10.1242/dev.098301

Liu X, Kim YJ, Müller R, Yumul RE, Liu C, Pan Y, Cao X, Goodrich J, Chen X. 2011. *AGAMOUS* terminates floral stem cell maintenance in *Arabidopsis* by directly repressing *WUSCHEL* through recruitment of Polycomb Group proteins. *Plant Cell* **23:** 3654–3670. doi:10.1105/tpc.111.091538

Liu X, Dinh TT, Li D, Shi B, Li Y, Cao X, Guo L, Pan Y, Jiao Y, Chen X. 2014. *AUXIN RESPONSE FACTOR 3* integrates the functions of *AGAMOUS* and *APETALA2* in floral meristem determinacy. *Plant J* **80:** 629–641. doi:10.1111/tpj.12658

Liu Z, Miao L, Huo R, Song X, Johnson C, Kong L, Sundaresan V, Yu X. 2018. ARF2–ARF4 and ARF5 are essential for female and male gametophyte development in *Arabidopsis*. *Plant Cell Physiol* **59:** 179–189. doi:10.1093/pcp/pcx174

Ma Y, Miotk A, Šutiković Z, Ermakova O, Wenzl C, Medzihradszky A, Gaillochet C, Forner J, Utan G, Brackmann K, et al. 2019. WUSCHEL acts as an auxin response rheostat to maintain apical stem cells in *Arabidopsis*. *Nat Commun* **10:** 5093.

Maier AT, Stehling-Sun S, Offenburger SL, Lohmann JU. 2011. The bZIP transcription factor PERIANTHIA: a multifunctional hub for meristem control. *Front Plant Sci* **2:** 79. doi:10.3389/fpls.2011.00079

Mallory AC, Bartel DP, Bartel B. 2005. MicroRNA-directed regulation of *Arabidopsis AUXIN RESPONSE FACTOR17* is essential for proper development and modulates expression of early auxin response genes. *Plant Cell* **17:** 1360–1375. doi:10.1105/tpc.105.031716

Marsch-Martínez N, de Folter S. 2016. Hormonal control of the development of the gynoecium. *Curr Opin Plant Biol* **29:** 104–114. doi:10.1016/j.pbi.2015.12.006

Matsumoto N. 2001. A homeobox gene, PRESSED FLOWER, regulates lateral axis-dependent development of *Arabidopsis* flowers. *Genes Dev* **15:** 3355–3364. doi:10.1101/gad.931001

Moubayidin L, Østergaard L. 2014. Dynamic control of auxin distribution imposes a bilateral-to-radial symmetry switch during gynoecium development. *Curr Biol* **24:** 2743–2748. doi:10.1016/j.cub.2014.09.080

Moubayidin L, Østergaard L. 2017. Gynoecium formation: an intimate and complicated relationship. *Curr Opin Genet Dev* **45:** 15–21. doi:10.1016/j.gde.2017.02.005

Nagpal P, Ellis CM, Weber H, Ploense SE, Barkawi LS, Guilfoyle TJ, Hagen G, Alonso JM, Cohen JD, Farmer EE, et al. 2005. Auxin response factors ARF6 and ARF8 promote jasmonic acid production and flower maturation. *Development* **132:** 4107–4118. doi:10.1242/dev.01955

Nakata MT, Tameshige T, Takahara M, Mitsuda N, Okada K. 2018. The functional balance between the *WUSCHEL-RELATED HOMEOBOX1* gene and the phytohormone auxin is a key factor for cell proliferation in *Arabidopsis* seedlings. *Plant Biotechnol* **35:** 141–154. doi:10.5511/plantbiotechnology.18.0427a

Nawy T, Bayer M, Mravec J, Friml J, Birnbaum KD, Lukowitz W. 2010. The GATA factor HANABA TARANU is required to position the proembryo boundary in the early *Arabidopsis* embryo. *Dev Cell* **19:** 103–113. doi:10.1016/j.devcel.2010.06.004

Nemhauser JL, Feldman LJ, Zambryski PC. 2000. Auxin and ETTIN in *Arabidopsis* gynoecium morphogenesis. *Development* **127:** 3877–3888.

Okada K, Ueda J, Komaki MK, Bell CJ, Shimura Y. 1991. Requirement of the auxin polar transport system in early stages of *Arabidopsis* floral bud formation. *Plant Cell* **3:** 677–684. doi:10.2307/3869249

Panoli A, Martin MV, Alandete-Saez M, Simon M, Neff C, Swarup R, Bellido A, Yuan L, Pagnussat GC, Sundaresan V. 2015. Auxin import and local auxin biosynthesis are required for mitotic divisions, cell expansion and cell specification during female gametophyte development in *Arabidopsis thaliana*. *PLoS ONE* **10:** e0126164. doi:10.1371/journal.pone.0126164

Pekker I, Alvarez JP, Eshed Y. 2005. Auxin response factors mediate *Arabidopsis* organ asymmetry via modulation of KANADI activity. *Plant Cell* **17:** 2899–2910. doi:10.1105/tpc.105.034876

Prunet N, Yang W, Das P, Meyerowitz EM, Jack TP. 2017. *SUPERMAN* prevents class B gene expression and promotes stem cell termination in the fourth whorl of *Arabidopsis thaliana* flowers. *Proc Natl Acad Sci* **114:** 7166–7171. doi:10.1073/pnas.1705977114

Przemeck GH, Mattsson J, Hardtke C, Sung ZR, Berleth T. 1996. Studies on the role of the *Arabidopsis* gene MONOPTEROS in vascular development and plant cell axialization. *Planta* **200:** 229–237. doi:10.1007/BF00208313

Reinhardt D, Mandel T, Kuhlemeier C. 2000. Auxin regulates the initiation and radial position of plant lateral organs. *Plant Cell* **12:** 507–518. doi:10.1105/tpc.12.4.507

Reyes-Olalde JI, de Folter S. 2019. Control of stem cell activity in the carpel margin meristem (CMM) in *Arabidopsis*. *Plant Reprod* **32:** 123–136. doi:10.1007/s00497-018-00359-0

Rudall PJ. 2010. All in a spin: centrifugal organ formation and floral patterning. *Curr Opin Plant Biol* **13:** 108–114. doi:10.1016/j.pbi.2009.09.019

Running MP, Meyerowitz EM. 1996. Mutations in the PERIANTHIA gene of *Arabidopsis* specifically alter floral organ number and initiation pattern. *Development* **122:** 1261–1269.

Sakai H, Medrano LJ, Meyerowitz EM. 1995. Role of SUPERMAN in maintaining *Arabidopsis* floral whorl boundaries. *Nature* **378:** 199–203. doi:10.1038/378199a0

Sauret-Güeto S, Schiessl K, Bangham A, Sablowski R, Coen E. 2013. JAGGED controls *Arabidopsis* petal growth and shape by interacting with a divergent polarity field. *PLoS Biol* **11:** e1001550. doi:10.1371/journal.pbio.1001550

Sessions RA, Zambryski PC. 1995. *Arabidopsis* gynoecium structure in the wild type and in *ettin* mutants. *Development* **121:** 1519–1532.

Sessions RA, Nemhauser JL, McColl A, Roe JL, Feldmann KA, Zambryski PC. 1997. ETTIN patterns the *Arabidopsis* floral meristem and reproductive organs. *Development* **124:** 4481–4491.

Skinner DJ, Sundaresan V. 2018. Recent advances in understanding female gametophyte development. *F1000Res* **7:** 804. doi:10.12688/f1000research.14508.1

Smyth DR, Bowman JL, Meyerowitz EM. 1990. Early flower development in *Arabidopsis*. *Plant Cell* **2:** 755–767.

Sohlberg JJ, Myrenås M, Kuusk S, Lagercrantz U, Kowalczyk M, Sandberg G, Sundberg E. 2006. *STY1* regulates auxin homeostasis and affects apical-basal patterning of the *Arabidopsis* gynoecium. *Plant J* **47:** 112–123. doi:10.1111/j.1365-313X.2006.02775.x

Ståldal V, Cierlik I, Chen S, Landberg K, Baylis T, Myrenås M, Sundström JF, Eklund DM, Ljung K, Sundberg E. 2012. The *Arabidopsis thaliana* transcriptional activator STYLISH1 regulates genes affecting stamen development, cell expansion and timing of flowering. *Plant Mol Biol* **78:** 545–559. doi:10.1007/s11103-012-9888-z

Su Y-H, Liu Y-B, Zhang X-S. 2011. Auxin–cytokinin interaction regulates meristem development. *Mol Plant* **4:** 616–625. doi:10.1093/mp/ssr007

Su Z, Zhao L, Zhao Y, Li S, Won S, Cai H, Wang L, Li Z, Chen P, Qin Y, et al. 2017. The THO complex non-cell-autonomously represses female germline specification through the TAS3-ARF3 module. *Curr Biol* **27:** 1597–1609.e2. doi:10.1016/j.cub.2017.05.021

Sun B, Xu Y, Ng KH, Ito T. 2009. A timing mechanism for stem cell maintenance and differentiation in the *Arabidopsis* floral meristem. *Genes Dev* **23:** 1791–1804. doi:10.1101/gad.1800409

Tabata R, Ikezaki M, Fujibe T, Aida M, Tian C, Ueno Y, Yamamoto KT, Machida Y, Nakamura K, Ishiguro S. 2010. *Arabidopsis* AUXIN RESPONSE FACTOR6 and 8 regulate jasmonic acid biosynthesis and floral organ development via repression of class 1 KNOX genes. *Plant Cell Physiol* **51:** 164–175. doi:10.1093/pcp/pcp176

Takada S, Hibara K, Ishida T, Tasaka M. 2001. The CUP-SHAPED COTYLEDON1 gene of *Arabidopsis* regulates shoot apical meristem formation. *Development* **128:** 1127–1135.

Takeda S, Matsumoto N, Okada K. 2004. RABBIT EARS, encoding a SUPERMAN-like zinc finger protein, regulates petal development in *Arabidopsis thaliana*. *Development* **131:** 425–434. doi:10.1242/dev.00938

Tashiro S, Tian CE, Watahiki MK, Yamamoto KT. 2009. Changes in growth kinetics of stamen filaments cause inefficient pollination in *massugu2*, an auxin insensitive, dominant mutant of *Arabidopsis thaliana*. *Physiol Plant* **137:** 175–187. doi:10.1111/j.1399-3054.2009.01271.x

Theißen G, Saedler H. 2001. Floral quartets. *Nature* **409:** 469–471. doi:10.1038/35054172

Trigueros M, Navarrete-Gómez M, Sato S, Christensen SK, Pelaz S, Weigel D, Yanofsky MF, Ferrándiz C. 2009. The *NGATHA* genes direct style development in the *Arabidopsis* gynoecium. *Plant Cell* **21:** 1394–1409. doi:10.1105/tpc.109.065508

Tucker SC. 2003. Floral development in legumes. *Plant Physiol* **131:** 911–926. doi:10.1104/pp.102.017459

Van Mourik S, Kaufmann K, van Dijk ADJ, Angenent GC, Merks RMH, Molenaar J. 2012. Simulation of organ patterning on the floral meristem using a polar auxin transport model. *PLoS ONE* **7:** e28762. doi:10.1371/journal.pone.0028762

Varaud E, Brioudes F, Szécsi J, Leroux J, Brown S, Perrot-Rechenmann C, Bendahmane M. 2011. AUXIN RESPONSE FACTOR8 regulates *Arabidopsis* petal growth by interacting with the bHLH transcription factor BIGPETALp. *Plant Cell* **23:** 973–983. doi:10.1105/tpc.110.081653

Vernoux T, Kronenberger J, Grandjean O, Laufs P, Traas J. 2000. PIN-FORMED 1 regulates cell fate at the periphery of the shoot apical meristem. **5165:** 5157–5165.

Wang Q, Kohlen W, Rossmann S, Vernoux T, Theres K. 2014a. Auxin depletion from the leaf axil conditions competence for axillary meristem formation in *Arabidopsis* and tomato. *Plant Cell* **26:** 2068–2079. doi:10.1105/tpc.114.123059

Wang Y, Wang J, Shi B, Yu T, Qi J, Meyerowitz EM, Jiao Y. 2014b. The stem cell niche in leaf axils is established by auxin and cytokinin in *Arabidopsis*. *Plant Cell* **26:** 2055–2067. doi:10.1105/tpc.114.123083

Wang Q, Hasson A, Rossmann S, Theres K. 2016. *Divide et impera*: boundaries shape the plant body and initiate new meristems. *New Phytol* **209:** 485–498. doi:10.1111/nph.13641

Weigel D, Alvarez J, Smyth DR, Yanofsky MF, Meyerowitz EM. 1992. LEAFY controls floral meristem identity in *Arabidopsis*. *Cell* **69:** 843–859. doi:10.1016/0092-8674(92)90295-N

Wellmer F, Bowman JL, Davies B, Ferrándiz C, Fletcher JC, Franks RG, Graciet E, Gregis V, Ito T, Jack TP, et al. 2014. Flower development: open questions and future directions. *Methods Mol Biol* **1110:** 103–124. doi:10.1007/978-1-4614-9408-9_5

Wu M-F, Tian Q, Reed JW. 2006. *Arabidopsis* microRNA167 controls patterns of ARF6 and ARF8 expression, and regulates both female and male reproduction. *Development* **133:** 4211–4218. doi:10.1242/dev.02602

Wu M, Yamaguchi N, Xiao J, Bargmann B, Estelle M, Sang Y, Wagner D. 2015. Auxin-regulated chromatin switch directs acquisition of flower primordium founder fate. *eLife* **4:** 1–19.

Xing S, Salinas M, Garcia-Molina A, Höhmann S, Berndtgen R, Huijser P. 2013. *SPL8* and miR156-targeted *SPL* genes redundantly regulate *Arabidopsis* gynoecium differential patterning. *Plant J* **75:** 566–577. doi:10.1111/tpj.12221

Xu B, Li Z, Zhu Y, Wang H, Ma H, Dong A, Huang H. 2008. *Arabidopsis* genes *AS1*, *AS2*, and *JAG* negatively regulate boundary-specifying genes to promote sepal and petal development. *Plant Physiol* **146:** 566–575. doi:10.1104/pp.107.113787

Xu Y, Prunet N, Gan E, Wang Y, Stewart D, Wellmer F, Huang J, Yamaguchi N, Tatsumi Y, Kojima M, et al. 2018. SUPERMAN regulates floral whorl boundaries through control of auxin biosynthesis. *EMBO J* **37:** e97499.

Xu XF, Wang B, Feng YF, Xue JS, Qian XX, Liu SQ, Zhou J, Yu YH, Yang NY, Xu P, et al. 2019. AUXIN RESPONSE FACTOR17 directly regulates *MYB108* for anther dehiscence. *Plant Physiol* **181:** 645–655. doi:10.1104/pp.19.00576

Yadav S, Kumar H, Yadav RK. 2019. Local auxin biosynthesis promotes stem cell differentiation and organogenesis in *Arabidopsis* shoot apex. bioRxiv doi:10.1101/819342

Yamaguchi N, Wu MF, Winter CM, Berns MC, Nole-Wilson S, Yamaguchi A, Coupland G, Krizek B, Wagner D. 2013. A molecular framework for auxin-mediated initiation of flower primordia. *Dev Cell* **24:** 271–282. doi:10.1016/j.devcel.2012.12.017

Yamaguchi N, Jeong CW, Nole-Wilson S, Krizek BA, Wagner D. 2016. AINTEGUMENTA and AINTEGUMENTA-LIKE6/PLETHORA3 induce *LEAFY* expression in response to auxin to promote the onset of flower formation in *Arabidopsis*. *Plant Physiol* **170:** 283–293. doi:10.1104/pp.15.00969

Yamaguchi N, Huang J, Xu Y, Tanoi K, Ito T. 2017. Fine-tuning of auxin homeostasis governs the transition from floral stem cell maintenance to gynoecium formation. *Nat Commun* **8:** 1125. doi:10.1038/s41467-017-01252-6

Yamaguchi N, Huang J, Tatsumi Y, Abe M, Sugano SS, Kojima M, Takebayashi Y, Kiba T, Yokoyama R, Nishitani K, et al. 2018. Chromatin-mediated feed-forward auxin biosynthesis in floral meristem determinacy. *Nat Commun* **9:** 5290. doi:10.1038/s41467-018-07763-0

Yang XY, Li JG, Pei M, Gu H, Chen ZL, Qu LJ. 2007. Overexpression of a flower-specific transcription factor gene AtMYB24 causes aberrant anther development. *Plant Cell Rep* **26:** 219–228. doi:10.1007/s00299-006-0229-z

Yang C, Song J, Ferguson AC, Klisch D, Simpson K, Mo R, Taylor B, Mitsuda N, Wilson ZA. 2017. Transcription factor MYB26 is key to spatial specificity in anther secondary thickening formation. *Plant Physiol* **175:** 333–350. doi:10.1104/pp.17.00719

Yu H, Huang T. 2016. Molecular mechanisms of floral boundary formation in *Arabidopsis*. *Int J Mol Sci* **17:** 317. doi:10.3390/ijms17030317

Žádníková P, Simon R. 2014. How boundaries control plant development. *Curr Opin Plant Biol* **17:** 116–125. doi:10.1016/j.pbi.2013.11.013

Zhang X, Zhou Y, Ding L, Wu Z, Liu R, Meyerowitz EM. 2013. Transcription repressor HANABA TARANU controls flower development by integrating the actions of multiple hormones, floral organ specification genes, and GATA3 family genes in *Arabidopsis*. *Plant Cell* **25:** 83–101. doi:10.1105/tpc.112.107854

Zhao H, Liu L, Mo H, Qian L, Cao Y, Cui S, Li X, Ma L. 2013. The ATP-binding cassette transporter ABCB19 regulates postembryonic organ separation in *Arabidopsis*. *PLoS ONE* **8:** e60809. doi:10.1371/journal.pone.0060809

Zhu M, Chen W, Mirabet V, Hong L, Bovio S, Strauss S, Schwarz EM, Tsugawa S, Wang Z, Smith RS, et al. 2020. Robust organ size requires robust timing of initiation orchestrated by focused auxin and cytokinin signalling. *Nat Plants* **6:** 686–698. doi:10.1038/s41477-020-0666-7

Auxin Interactions with Other Hormones in Plant Development

Serina M. Mazzoni-Putman, Javier Brumos, Chengsong Zhao, Jose M. Alonso, and Anna N. Stepanova

Department of Plant and Microbial Biology, North Carolina State University, Raleigh, North Carolina 27695, USA

Correspondence: atstepan@ncsu.edu

Auxin is a crucial growth regulator that governs plant development and responses to environmental perturbations. It functions at the heart of many developmental processes, from embryogenesis to organ senescence, and is key to plant interactions with the environment, including responses to biotic and abiotic stimuli. As remarkable as auxin is, it does not act alone, but rather solicits the help of, or is solicited by, other endogenous signals, including the plant hormones abscisic acid, brassinosteroids, cytokinins, ethylene, gibberellic acid, jasmonates, salicylic acid, and strigolactones. The interactions between auxin and other hormones occur at multiple levels: hormones regulate one another's synthesis, transport, and/or response; hormone-specific transcriptional regulators for different pathways physically interact and/or converge on common target genes; etc. However, our understanding of this crosstalk is still fragmentary, with only a few pieces of the gigantic puzzle firmly established. In this review, we provide a glimpse into the complexity of hormone interactions that involve auxin, underscoring how patchy our current understanding is.

The plant hormone auxin is an essential growth regulator central to a wide variety of developmental processes, environmental adaptation, and phenotypic plasticity (for review, see Enders and Strader 2015 and Lavy and Estelle 2016). The name auxin comes from the Greek word "auxein," meaning "to grow." The best-studied form of auxin, indole-3-acetic acid (IAA), is synthesized from the amino acid tryptophan (Trp) through a simple, two-step pathway. Trp aminotransferases of the Trp AMINOTRANSFERASE OF ARABIDOPSIS1 (TAA1)/TAA1-RELATED (TAR) family convert Trp to indole-3-pyruvic acid (IPyA), which is then metabolized to IAA by flavin-containing monooxygenases, YUCCAs (YUCs). The availability of biologically active IAA is controlled by auxin-catabolizing enzymes of the DIOXYGENASE OF AUXIN OXIDATION (DAO) family that oxidize IAA to 2-oxindole-3-acetic acid, auxin-conjugating enzymes such as the IAA amidosynthetases GRETCHEN HAGEN3 (GH3) and glucosyltransferases that inactivate auxin by linking it to amino acids or sugars, respectively, and by auxin transporters that move auxin in and out of the cell and between cells (for review, see Enders and Strader 2015). Free IAA can enter plant cells passively or be

Cite this article as *Cold Spring Harb Perspect Biol* doi: 10.1101/cshperspect.a039990

actively imported via AUXIN1 (AUX1)/LIKE AUX1 (LAX) influx carriers. IAA can also exit cells via efflux carriers of the PIN-FORMED (PIN) and P-GLYCOPROTEIN/ATP-BINDING CASETTE transporter families. Auxin perception takes place inside the cell, predominantly in the nucleus, where IAA binds to the TRANSPORT INHIBITOR-RESISTANT1 (TIR1)/AUXIN SIGNALING F-BOX (AFB) family of auxin receptors and promotes their interaction with the auxin coreceptors, Aux/IAAs. In the absence of the hormone, Aux/IAAs associate with AUXIN RESPONSE FACTOR (ARF) transcription factors (TFs) and block their transcriptional activity via the recruitment of TOPLESS (TPL) and chromatin-remodeling machinery. In the presence of IAA, Aux/IAAs undergo ubiquitin-mediated proteasomal degradation triggered by the SKP-CULLIN-F-BOX (SCF^{TIR}) E3-ligase complex, releasing ARFs and enabling ARF-mediated transcriptional regulation of auxin-response genes (for review, see Enders and Strader 2015 and Lavy and Estelle 2016).

While auxin governs multiple aspects of plant development, physiology, and environmental competence, it does not act in isolation. In every process where the contributions of auxin have been explored, it appears to enlist or be enlisted by other endogenous signals and external cues, enabling the plant to tailor its growth and development to the specific conditions it happens to be in. This article aims to present the current state of knowledge in the area of auxin interactions with other plant hormones, specifically abscisic acid (ABA), brassinosteroid (BR), cytokinin (CK), ethylene (ET), gibberellic acid (GA), jasmonate (JA), salicylic acid (SA), and strigolactone (SL). We chose to structure this manuscript by the pairs of plant hormones, with the caveat that in many processes multiple players are involved and we are only beginning to untangle the full complexity of signal crosstalk in plants. We have included graphical representations of key auxin interactions with other hormones during the processes of seed germination, root, shoot, and fruit development in Figures 1–4, respectively. Given the breadth of the auxin interaction network, we were unable to discuss all relevant studies and wish to apologize to those researchers whose work we could not describe in light of space limitations in this article.

AUXIN–ABA INTERACTIONS

ABA, originally named "abscisin II" for its role in the abscission of cotton fruits (Addicott et al. 1968), is a sequisterpene that belongs to the terpenoid class of metabolites. It is primarily synthesized in the vasculature and guard cells and transported by ABA transporters (for review, see Emenecker and Strader 2020). ABA is perceived by soluble PYRABACTIN RESISTANCE1 (PYR1)/PYR1-LIKE (PYL)/REGULATORY COMPONENT OF ABA RECEPTOR proteins, leading to the inhibition of PROTEIN PHOSPHATASE 2C (PP2C). Formation of the ABA-PYR-PP2C complex causes the accumulation of phosphorylated protein kinases in subclass III of the SNF1-RELATED PROTEIN KINASE2 (SnRK2) family, which phosphorylates various target proteins, including ABA-RESPONSIVE ELEMENT-BINDING FACTORs, to achieve the appropriate cellular response (for review, see Emenecker and Strader 2020).

Seed Dormancy and Germination

ABA is the major hormone involved in the establishment and maintenance of seed dormancy (Fig. 1; for review, see Bentsink and Koornneef 2008). Among ABA signaling components in *Arabidopsis*, TFs ABA-INSENSITIVE3 (ABI3), ABI4, and ABI5 were identified as positive regulators of ABA signaling and negative regulators of seed germination (for review, see Emenecker and Strader 2020).

The B3-domain TF ABI3 is transcriptionally induced by ABA, and disruption of this gene's function reduces seed dormancy and allows germination in the presence of exogenous ABA (Koornneef et al. 1984). Remarkably, *ABI3* transcription is also induced by exogenous IAA and positively regulated by ARF10 and ARF16 (Liu et al. 2013b), which are both believed to function as transcriptional repressors (Fig. 1; Wang et al. 2005b). Even though a potential auxin response element (AuxRE) is present in the *ABI3* promoter, neither ARF10 nor ARF16 directly bind

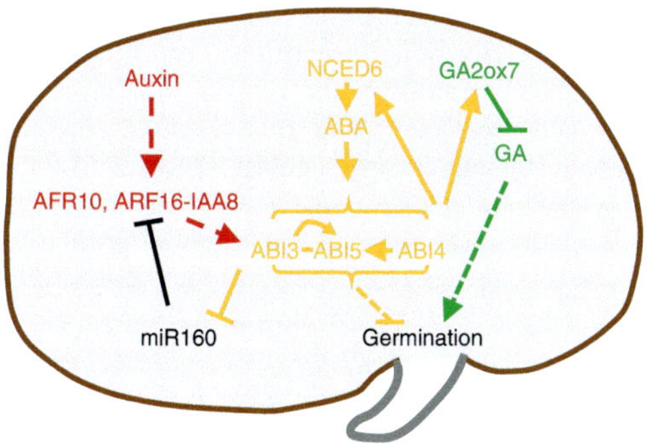

Figure 1. Auxin interactions with abscisic acid (ABA) in the control of seed germination. Seed dormancy is primarily controlled by the balance between ABA and gibberellic acid (GA), which oppose one another in the regulation of seed germination: ABA inhibits, whereas GA promotes seed sprouting. Auxin, acting through IAA8 and ARF10/ARF16, works to enhance seed dormancy by inducing *ABI3* expression. ABI3, in turn, down-regulates *miR160*, an miRNA that inhibits *ARF10/16*, enabling feedback regulation of ARF activity by ABI3. Transcription factors (TFs) ABI3, ABI4, and ABI5 all contribute to blocking seed germination. ABI4 regulates ABA production and GA metabolism by inducing an ABA biosynthesis gene, *NCED6*, and a GA catabolism gene, *GA2ox7*. Additional interactions that take place between ABA and GA, but do not involve auxin, are omitted in this schematic. Arrowheads represent positive regulation; blunt arrows represent negative regulation; solid and interrupted lines represent direct and indirect regulation, respectively; dashes between proteins represent direct interaction. Colors represent auxin (red), ABA (orange), and GA (green) pathway components. Black is used to depict genes/proteins that do not belong to a specific hormone pathway, as well as the developmental processes the network regulates.

the *ABI3* promoter, suggesting indirect action of these TFs by repressing a repressor of *ABI3* (Liu et al. 2013b). In addition, a loss-of-function mutant of *IAA8*, *iaa8-1*, shows delayed seed germination, and the IAA8 protein can associate with an AuxRE within the *ABI3* promoter, suggesting that IAA8 may bind to the *ABI3* promoter via yet unidentified ARFs to regulate seed germination (Hussain et al. 2020). Previous studies revealed the ability of ARFs, including ARF16, to interact with IAA8 (Piya et al. 2014), but it remains to be determined which specific ARF-Aux/IAA combinations bind the *ABI3* promoter to regulate its transcription. Interestingly, an older study discovered that ABI3 directly represses the transcription of *miR160B*, an miRNA that targets *ARF10* and *ARF16*, and thus up-regulates *ABI3* through a potentially complex feedback loop (Fig. 1; Tian et al. 2004).

ABI4, an APETALA2 (AP2)-domain TF, controls various developmental processes (for review, see Chandrasekaran et al. 2020). It regulates seed dormancy by directly activating an ABA biosynthetic gene, *NINE-CIS-EPOXYCAROTENOID DIOXYGENASE6* (*NCED6*), and a GA catabolic gene, *GA 2-BETA-DIOXYGENASE7* (*GA2ox7*) (Fig. 1; Shu et al. 2013, 2016). Furthermore, *abi4* mutant seeds are insensitive to auxin-mediated inhibition of seed germination and show reduced dormancy (Rohde et al. 2000). The prospective role of ABI4 in mediating auxin effects on germination is further supported by a recent study that showed that *YUC4*-overexpressing plants display enhanced sensitivity to ABA during seed germination, whereas the same construct in the *abi4* background leads to wild-type germination, indicating that *ABI4* is required for the ABA hypersensitivity of *YUC4*-overexpressing lines during germination (Munguia-Rodriguez et al. 2020). Thus, ABI4 is another convergence point between ABA and auxin during inhibition of germination (Fig. 1).

Finally, ABI5, a bZIP TF, functions downstream of ABI3 and is also a substrate of activated SnRK2s (Fig. 1; Yu et al. 2015). It binds ABA-responsive element (ABRE)-containing promoters (Finkelstein and Lynch 2000; Hossain et al. 2010; Zhou et al. 2013). While the *Arabidopsis abi5* mutant shows seed germination defects, rice plants with mutations in the *ABI5-LIKE1* (*OsABL1*) gene, which is induced by both ABA and auxin, show normal seed germination but suppressed ABA-triggered root growth inhibition and hypersensitivity to exogenous auxin (Yang et al. 2011). In addition, the OsABL1 protein can directly bind ABREs in vitro and the expression levels of several ABRE-containing genes potentially related to auxin metabolism or signaling are altered in *abl1* mutants, indicating that OsABL1 plays a role in the crosstalk between ABA and auxin (Yang et al. 2011).

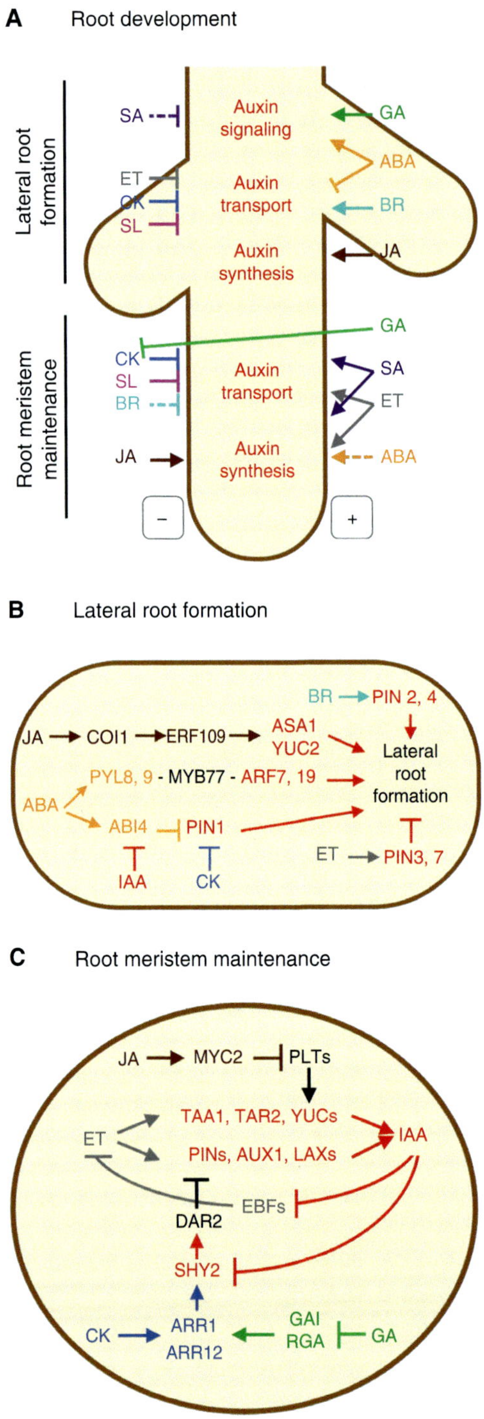

Figure 2. Auxin interactions with other hormones in roots. (*A*) Hormone interactions regulating auxin biosynthesis, transport, and signaling during lateral root (LR) formation and root meristem maintenance. The minus sign signifies repression of the root processes by the hormones listed on the *left* side of the root. The plus sign signifies promotion of the root processes by the hormones listed on the *right* side of the root. (*B*) Molecular network of the hormone interactions mediating LR formation. Abscisic acid (ABA) induces LR growth through PYL8. The interaction of PYL8, PYL9, with MYB77 promotes the crosstalk with the auxin signaling pathway via ARF7 and ARF19. ABA up-regulates the expression of *ABI4*, which represses the expression of *PIN1* to modify auxin transport. The negative effect of this network on LR formation can be reinforced by the cytokinin (CK)-mediated repression of *PIN1* or weakened by auxin-mediated inhibition of *ABI4*. Ethylene induces the expression of *PIN3* and *PIN7*, which reduce the local accumulation of auxin in the LR initiation sites, and thereby decrease LR formation. Jasmonate (JA) boosts the formation of LRs by inducing the expression of *ERF109*, which promotes auxin biosynthesis by up-regulating the expression of *ASA1* and *YUC2*. Brassinosteroid (BR) has a positive effect on the formation of LRs by inducing the expression of *PIN2* and *PIN4*. (*C*) Molecular network of hormone interactions in root meristem maintenance. Ethylene (ET) induces auxin transport and biosynthesis, promoting the accumulation of auxin and enabling root meristem maintenance. Auxin dampens the negative effect of the EBFs on ET, reinforcing meristem function. CK and gibberellic acid (GA) have opposite effects on meristem maintenance by increasing and decreasing the activity of ARR1/ARR12 CK signaling components, respectively, which in turn represses the expression of *PIN*s through SHY2 and DAR2. The JA-mediated repression of PLTs reduces the expression of the *YUC*s and lowers the production of auxin, thus negatively affecting root meristem maintenance. Arrowheads and blunt arrows represent positive and negative regulation, respectively. Solid and interrupted lines in panel *A* are used to depict direct and indirect effects, respectively. A dash between two proteins in panel *B* indicates direct physical interaction. Hormones, protein/genes in these hormonal pathways, and their actions are denoted by the following color pallet, ABA is orange, BR is turquoise, CK is blue, ET is gray, GA is green, IAA is red, JA is brown, salicylic acid (SA) is purple, and strigolactone (SL) is pink. Black is used to depict genes/proteins that do not belong to a specific hormone pathway, as well as the developmental processes the network regulates.

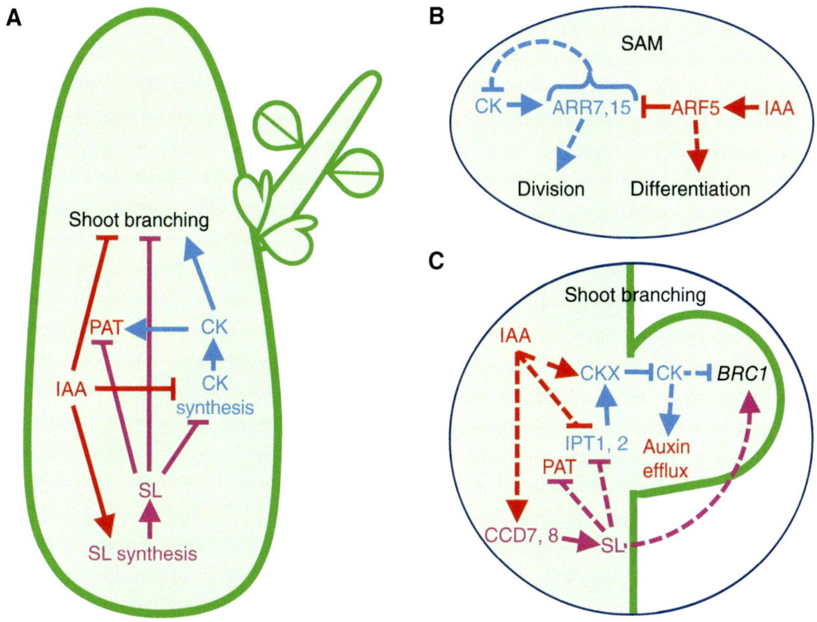

Figure 3. Auxin interactions with other hormones during shoot development. (*A*) Auxin and cytokinin (CK) oppose one another in the regulation of shoot branching. Strigolactone (SL) acts as a second messenger for auxin signaling: auxin from the shoot promotes SL production in the root and shoot, which then acts to inhibit branching. (*B*) Auxin acts to promote differentiation in the shoot apical meristem while CK promotes stem cell division. (*C*) During branching, shoot-derived auxin travels down the main stem where it promotes SL biosynthesis. This SL then travels up the shoot and inhibits CK production and polar auxin transport (PAT) in the shoot, thereby blocking branching. Auxin regulates local metabolism of CK, reducing CK levels, while CK promotes auxin efflux from the developing bud. CK signaling in the bud promotes bud release and branch outgrowth. CK and SL also oppose one another in regulating the expression of *BRC1*, a negative regulator of bud outgrowth. Red denotes auxin and its actions; blue denotes CK and its actions; magenta denotes SL and its actions. Arrowheads represent positive regulation; blunt arrows represent negative regulation; in panels *B* and *C*, dashed lines represent indirect regulation.

Root Development

Primary root (PR) growth regulated by auxin depends on Aux/IAA-ARF signaling modules. ARF2 directly binds to AuxREs in the promoter of the zinc finger homeodomain TF *HOMEOBOX PROTEIN33* (*HB33*) and negatively regulates its expression (Wang et al. 2011). Transgenic plants overexpressing *HB33* or RNAi lines with reduced *HB33* levels are more sensitive and more resistant to ABA respectively in the seed germination and PR growth assays, indicating that HB33 is a positive regulator in the ABA-mediated processes of seed germination and PR growth (Wang et al. 2011). Likewise, in wheat, TaARF4 targets *TaHB33* and two *TaGH3* genes to concomitantly regulate ABA sensitivity and IAA homeostasis to control root growth (Wang et al. 2019).

Auxin-ABA crosstalk is also involved in lateral root (LR) development (Fig. 2A,B). The ABA receptor mutant *pyl8* shows reduced LR growth in the presence of exogenous ABA, indicating that ABA signaling through PYL8 promotes LR growth (Fig. 2B). When *pyl8* seedlings are exposed to both ABA and IAA, LR growth is rescued, suggesting that *pyl8* seedlings may have an auxin deficiency or reduced auxin response (Zhao et al. 2014). A previous study showed that PYL8 interacts with MYB77 (*Arabidopsis* Interactome Mapping Consortium 2011) that can promote LR growth by interacting with ARF7

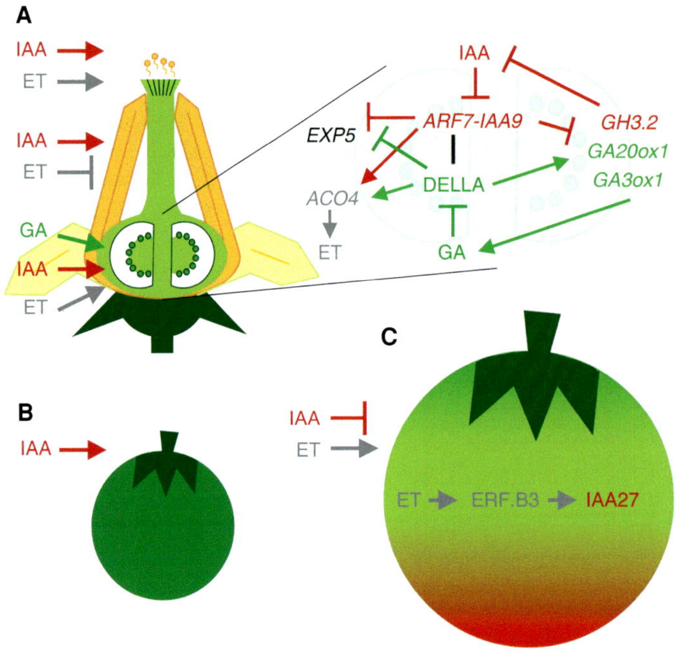

Figure 4. Auxin interactions with other hormones in fruit development. (*A*) Hormone interactions in flower development and fruit set. Auxin and ethylene (ET) promote pollen germination and pollen tube growth. Auxin induces and ET represses stamen development. ET positively regulates pistil and ovule development. Auxin and gibberellic acid (GA) promote fruit initiation. (*Inset*) Molecular network of the hormone crosstalk during tomato fruit initiation. Fertilization triggers auxin-mediated GA synthesis. Auxin inhibits the ARF7-IAA9 complex, releasing the repression of key GA biosynthetic genes. GA–auxin interaction promotes fruit growth by inducing *EXP5* and reduces the production of ET by repressing *ACO4*. (*B*) Auxin regulates early cell division and fruit development phases. (*C*) Fruit ripening is promoted by ethylene and repressed by auxin. During ripening, a key molecular interaction between ET and IAA is mediated by ERF.B3 and IAA27. Hormones and their actions are denoted by the following colors: ET is gray, GA is green, and IAA is red. Arrowheads and blunt arrows represent positive and negative regulation.

in the SOLITARY ROOT (SLR, IAA14)-ARF7/ARF19 module (Fukaki et al. 2005; Shin et al. 2007). Furthermore, PYL8 is functionally redundant with paralogous PYL9 (Xing et al. 2016) and the two proteins interact with several MYBs, including MYB77, in *Arabidopsis* to integrate ABA and auxin signals in the regulation of LR growth (Fig. 2B; Zhao et al. 2014).

In addition to controlling LR growth through the PYL–MYB77 interaction, ABA also regulates LRs via the core ABA-SnRK2 signaling pathway (Fig. 2B; for review, see Emenecker and Strader 2020). *abi4* mutants possess longer LRs, suggesting that ABI4 inhibits LR growth. *ABI4* expression in roots is induced by ABA but repressed by auxin (Shkolnik-Inbar and Bar-Zvi 2010). Furthermore, PIN1 levels are decreased in *ABI4*-overexpression lines but increased in *abi4* mutants. Thus, ABI4 likely mediates ABA-triggered inhibition of LR growth by suppressing PIN1 expression. Similarly, *ABI5* also regulates root growth by modulating the accumulation of PIN proteins (Yuan et al. 2014).

Under osmotic/salt stress conditions, another TF, WRKY46, can bind the promoters of the auxin-conjugating enzyme-encoding genes *UDP-GLYCOSYLTRANSFERASE 84B2*, *INDOLE-3-ACETATE BETA-GLUCOSYLTRANSFERASE*, and *GH3.1*, and an ABA signaling gene *ABI4*. Importantly, loss-of-function *wrky46* mutants and overexpression of *WRKY46* significantly reduce and increase LR development, respectively, suggesting that WRKY46 modulates LR development through the coregulation of ABA signaling and auxin homeostasis (Ding et al. 2015).

Pathogen Resistance

Black spot disease, a major disease in oilseed *Brassica* species, is caused by a group of pathogens including *Alternaria brassicae*, *Alternaria brassicicola*, and *Alternaria raphanin*. Numerous ABA and auxin mutants in *Arabidopsis* show altered susceptibility to *A. brassicicola*, suggesting that both ABA and auxin are involved in the response against this pathogen (Adie et al. 2007; Qi et al. 2012). A recent study showed that the auxin response factors *ARF10*, *ARF16*, and *ARF17* are up-regulated in the resistant species *Sinapis alba* upon challenge with *A. brassicicola*, but not in the susceptible species *Brassica juncea* (Mukherjee et al. 2019). Pathogen-induced expression of *Arabidopsis ARF10* in *B. juncea* enhances tolerance to *A. brassicicola,* and several ABA-responsive genes, including *ABI3*, *ABI4*, and *ABI5*, are up-regulated in the most tolerant transgenic lines. Furthermore, ARF10 interacts with the AuxREs in the *ABI5* promoter, suggesting that the binding of ARF10 to *ABI5* modulates auxin-ABA crosstalk to regulate resistance to *A. brassicicola* (Mukherjee et al. 2019).

AUXIN–BR INTERACTIONS

BRs, so named because the first example was identified in rapeseed (*Brassica napus*), are the only class of steroid hormones found in plants to date. BRs are a group of many compounds that act as extracellular ligands. While there are multiple routes involved in BR production, all of them proceed through triterpenoid pathways from campesterol (for review, see Chung and Choe 2013). BRs are believed to be synthesized in the endoplasmic reticulum and DWARF4 (DWF4) catalyzes a rate-limiting step of BR synthesis (for review, see Planas-Riverola et al. 2019 and Nolan et al. 2020). Unlike other hormones that can be transported throughout the plant, BRs typically act locally. BRs bind to the plasma membrane-localized receptor, BR-INSENSITIVE1 (BRI1), inducing a conformational change that allows interaction with coreceptors, such as BRI1-ASSOCIATED KINASE1 (BAK1). This interaction initiates a signaling cascade that leads to the activation of two TFs, BRASSINAZOLE-RESISTANT1 (BZR1) and BRI1 EMS SUPPRESSOR1 (BES1). In the presence of BR, BZR1 and BES1 are stabilized by a PP2A-dependent dephosphorylation and induce the transcriptional program of BR-dependent genes. In the absence of BR, signaling through the BRI1–BAK1 complex is blocked by autoinhibition and the association of inhibitory proteins, such as BRI1 KINASE INHIBITOR, which disrupts the BRI1–BAK1 interaction. BZR1 and BES1 are inactivated via phosphorylation by BR-INSENSITIVE2 (BIN2), leading to cytoplasmic retention and degradation of these TFs (for review, see Planas-Riverola et al. 2019 and Nolan et al. 2020).

BR-deficient plants have a dwarfed phenotype, consistent with the primary role of BRs in cell elongation. BR signaling has also been implicated in a broad range of plant biological processes, from control of cell division to biotic and abiotic stress responses (for review, see Lv and Li 2020 and Nolan et al. 2020) and, as outlined below, this phytohormone coordinates with auxin to regulate many aspects of plant development.

Root Development

BR signaling is critical for both cell elongation (Mouchel et al. 2004) and cell-cycle progression in the root (Gonzalez-Garcia et al. 2011; Hacham et al. 2011), with both BR-deficient and BR-activated mutants displaying smaller root meristems. In contrast to the synergy reported in shoot elongation, transcriptomic analysis of *Arabidopsis* root tips revealed that many auxin and BR coresponsive genes are regulated in opposing directions, suggesting these two hormones can act as checks on one another in the root to define different root cell types or functions (Chaiwanon and Wang 2015).

BR has a dose-dependent effect on PR elongation (Clouse et al. 1996; Müssig et al. 2002) and promotes the formation of LRs by regulating polar auxin transport (PAT) (Fig. 2A; Bao et al. 2004). BR treatment enhances shoot and root PAT in rapeseed (Li et al. 2005). *Arabidopsis* BR biosynthesis and signaling mutants (*diminuto1* and *bri1*) show decreased PAT, while *pin1* mutants display decreased sensitivity to BR-induced inhibition of root elongation, and *pin2* mutants are deficient in BR-induced LR formation. These results point to the modulation of PAT as a mechanism by which BR regulates root development (Li et al. 2005). Studies in *Arabidopsis* root meristems implicate transcriptional and posttranscriptional regulation of *PIN2* and *PIN4* as a potential mechanism for BR-controlled cell growth and proliferation in the root meristem (Hacham et al. 2012).

Another node of auxin–BR interaction during root development is the transcriptional coregulator BREVIS RADIX (BRX), which promotes the expression of a crucial BR biosynthesis gene, *CONSTITUTIVE PHOTOMORPHOGENESIS AND DWARFISM* (*CPD*). In *Arabidopsis*, *brx* mutants have altered root meristems, with increased root branching and shorter roots (Mouchel et al. 2004). These plants also display a BR-deficient phenotype and attenuated auxin transcriptional program (Mouchel et al. 2006). Exogenous BR treatment rescues the BR-deficient phenotype and restores auxin-responsive gene expression. Interestingly, auxin also induces the expression of *BRX*, thus promoting BR production (Mouchel et al. 2006). More recently, plasma-membrane-associated BRX has been shown to inhibit PIN protein activity in developing *Arabidopsis* root protophloem sieve elements by negatively regulating PROTEIN KINASE ASSOCIATED WITH BRX (PAX) (which stimulates PIN activity) (Marhava et al. 2018) and by promoting the endocytic removal of PIN1 proteins from the plasma membrane (Marhava et al. 2020). Through its actions on PAX and PIN1, BRX inhibits auxin efflux. In turn, auxin has been shown to promote BRX protein turnover (Scacchi et al. 2009), creating a check on BRX activity if local auxin concentrations become too high.

Shoot Development

BR and auxin act cooperatively to promote leaf lamina bending in rice, and cell elongation in numerous species, including bean, cucumber, maize, pea, rice, squash, and tomato (for review, see Mandava 1988, Clouse and Sasse 1998, and Park 1998). Further studies in *Arabidopsis* confirmed that these two hormones act synergistically to promote hypocotyl elongation under many conditions (Tanaka, 2003; Nemhauser et al. 2004). Accordingly, *Arabidopsis* Aux/IAA gain-of-function mutants display reduced BR sensitivity in hypocotyl elongation assays (Nakamura et al. 2006) and BR treatment enhances shoot PAT in rapeseed (Li et al. 2005). Coordinated BR signaling and PAT also control vascular patterning in *Arabidopsis* shoots, with auxin maxima defining the location and BR signaling determining the number of vascular bundles (Ibanes et al. 2009).

BR signaling is important for proper light responses, and *Arabidopsis* BR mutants, such as *de-etiolated2* and *cpd*, display a constitutive photomorphogenic phenotype (Li et al. 1996; Szekeres et al. 1996). Both auxin and BR signaling are required for shade avoidance (Keuskamp et al. 2011). In response to low blue light conditions, each hormone regulates a distinct set of genes, the full complement of which is needed for a proper shade avoidance response. While no direct link was established, the NAC TF ATAF2, which regulates auxin biosynthesis (Huh et al. 2012), was also found to regulate BR turnover in *Arabidopsis* by binding to the promoters and suppressing the expression of two BR catabolic enzyme genes, *PHYB-4 ACTIVATION-TAGGED SUPPRESSOR1* and *SUPPRESSOR OF PHYB-4 7* (Peng et al. 2015). In this way, ATAF2 promotes hypocotyl elongation by stimulating auxin biosynthesis and inhibiting BR degradation, whereas light and BR suppress *ATFAF2* expression, creating a feedback regulatory circuit (Peng et al. 2015).

Unequal auxin distribution leads to the curving of shoots seen in gravitropism, and BR application promotes shoot gravitropism in bean (Meudt 1987; Park 1998). Moreover, BR-mediated enhancement of root gravitropic curving in maize requires PAT (Kim et al. 2000). In *Arabidopsis*, BR treatment accelerates root and shoot gravitropism and auxin reporter activity following gravistimulation (Li et al. 2005). Fol-

 Cite this article as *Cold Spring Harb Perspect Biol* doi: 10.1101/cshperspect.a039990

lowing BR treatment, *PIN2* expression is enhanced and PIN2 protein localization mimics the protein distribution seen with gravistimulation. These findings suggest that the interaction between BR and auxin in gravitropism converges on PAT.

Another study found that in dark-grown *Arabidopsis* seedlings, BRs inhibit shoot gravitropism (Vandenbussche et al. 2011) and that the mechanism of this inhibition involves a complex interaction between BR, ET, and auxin signaling (Vandenbussche et al. 2013). In particular, several Aux/IAA proteins and ARFs, ARF7 and ARF19, are implicated in BR-regulated gravitropic responses. However, this inhibition is seen in seedlings grown in low sugar conditions, and the effect is lost when germinating the seedlings on vertical plates or supplementing the growth media with sugar (Vandenbussche et al. 2011). The authors speculate that the impact of BR signaling on gravitropism is due to a weakening of the hypocotyl cell wall. Indeed, further investigation found interactions between glucose, BRs, and potentially PAT during gravitropism and LR development (Singh et al. 2014; Gupta et al. 2015). Clearly, the role of auxin–BR interactions in shoot development is multifaceted, with tissue- and environment-specific effects.

Other Contexts

The first identified BR-responsive gene, *BUR1*, was found to be also induced by auxin treatment in soybean, albeit at a later time point (18 h vs. 2 h) (Zurek and Clouse 1994). This potential overlap in transcriptional control has been further investigated by many studies (Goda et al. 2002, 2004; Müssig et al. 2002; Yin et al. 2002; Nemhauser et al. 2004). Auxin and BR appear to cooperatively regulate some genes, while they are antagonistic in the control of others; this is fitting, considering the variable roles these two hormones play in different aspects of plant development. While these common transcriptional programs may be due to direct regulation of BR biosynthesis or PAT, one study suggests that these two hormones converge on common promoter motifs found in coregulated genes (Nemhauser et al. 2004) and ChIP data reveal that BZR1/BES1 bind to many auxin-responsive genes (Sun et al. 2010; Yu et al. 2011).

Inhibition of BR biosynthesis impairs the growth-promoting action of auxin (Vert et al. 2008), but together the two hormones enhance and potentiate the expression of common target genes. Closer examination revealed that the BIN2 kinase regulates auxin signaling by phosphorylating ARF2 and suppressing DNA binding by this negative regulator of auxin responses, thereby promoting transcription of its auxin target genes (Vert et al. 2008). BIN2, in turn, is negatively regulated by BR (Peng et al. 2008), supporting the hypothesis that BR blocks the BIN2-mediated activation of the auxin transcriptional program. Additionally, BR signaling is implicated in the control of *Aux/IAA* gene expression and activity in *Arabidopsis* (Nakamura et al. 2003, 2006; Cho et al. 2014). During *Arabidopsis* LR development, the ARF7 and ARF19 proteins become phosphorylated by BIN2, disrupting their interaction with Aux/IAAs and enhancing the auxin response (Cho et al. 2014). A screen for auxin transport mutants in *Arabidopsis* revealed that BR signaling promotes the nuclear accumulation of auxin by negatively regulating *PIN-LIKES* mRNA and protein expression (Sun et al. 2020). BR has also been shown to regulate auxin catabolism in barley (Sadura et al. 2019). Mutants in key BR biosynthetic (*HvDWARF*, *HvCPD*) and signaling (*HvBRI1*) genes had wild-type levels of total active auxins but altered levels of methylated and oxidized auxins.

Auxin signaling, in turn, increases BR sensitivity by enhancing expression of the BR receptor *BRI1* (Sakamoto et al. 2013). In rice, the *OsBRI1* promoter harbors an auxin-responsive element (AuxRE) that is bound by OsARF11 and is required for the up-regulation of *OsBRI1* and BR phenotypes, providing a direct connection between auxin signaling and BR perception (Sakamoto et al. 2013). DWF4, which catalyzes a rate-limiting step in BR synthesis, is transcriptionally induced by auxin in *Arabidopsis* and many auxin-inducible genes are not activated in the presence of a BR inhibitor (Chung et al. 2011), suggesting that the full auxin transcriptional response requires intact BR signaling.

AUXIN–CK INTERACTIONS

Auxin and CK work to balance one another in regulating plant developmental processes. In fact, CK was discovered based on its interaction with auxin to promote the growth and division of cultured plant cells (Miller et al. 1955, 1956). The two hormones are generally antagonistic of one another. For example, while both hormones are required for callus formation in cell culture, an excess of CK promotes shoot development, whereas excess auxin favors root development. These two hormones have a storied scientific past (for review, see Schaller et al. 2015 and Kieber and Schaller 2018). Here, we highlight the many levels of interaction between them. Most of the following studies take place in *Arabidopsis*, and while it is known that CK and auxin regulate one another in several plant species, the extent to which the observations made in *Arabidopsis* can be extrapolated to other species remains to be discovered.

CKs are adenine-derived hormones that are synthesized in a series of steps involving ISOPENTENYL TRANSFERASE (IPT), CYTOCHROME P450 FAMILY 735A, and LONELY GUY (LOG) enzymes. CKs are inactivated by conjugation to sugar molecules or by degradation at the hands of enzymes such as the CK OXIDASE/DEHYDROGENASE (CKXs). When CK binds to its receptors, ARABIDOPSIS HISTIDINE KINASEs (AHKs), the receptor autophosphorylates, initiating a phosphorylation cascade. The ARABIDOPSIS HISTIDINE PHOSPHOTRANSMITTER (AHP) proteins transfer phosphates from the AHKs to the type-B ARABIDOPSIS RESPONSE REGULATORs (ARRs), activating them. AHP1,2,3, and 5 are believed to positively regulate CK signaling, while AHP4 may be a negative regulator. AHP6, which lacks the histidine kinase activity of the other AHPs, also inhibits CK signaling through an unclear mechanism. AHP-mediated phosphorylation of the type-B ARRs induces the CK transcriptional response. One family of genes that is rapidly induced by CK are the type-A ARRs, a set of proteins that act in a feedback manner to repress CK signaling (Hutchison et al. 2006; for review, see Kieber and Schaller 2018).

CK induces auxin biosynthesis in several *Arabidopsis* tissues and this induction requires CK signaling through AHPs and type-A ARRs (Jones et al. 2010). These ARRs have also been shown to be a node of auxin-mediated control of CK signaling (Overvoorde et al. 2005; Müller and Sheen 2008; Lee et al. 2009; Zhao et al. 2010). Using several auxin and CK mutants, mutually inhibitory effects of the two hormones were found in a suite of developmental phenotypes (Kurepa et al. 2019). The authors proposed that the auxin signaling proteins ARF7 and IAA17 promote type-A *ARR* expression to inhibit CK signaling. However, while this route for auxin control of CK signaling has been previously investigated, the findings are not concordant; thus, the mechanism, and even direction, of type-A ARR regulation by auxin remains in question. It has also been widely demonstrated that CK regulates auxin signaling by altering the expression of PINs (Růžička et al. 2009; for review, see Schaller et al. 2015).

Likewise, auxin influences CK biosynthesis in *Arabidopsis* and other species via regulation of the *IPT* genes (Zhang et al. 1995; Miyawaki et al. 2004; Nordstrom et al. 2004; Tanaka et al. 2006; Cheng et al. 2013). As is often the case, the situation is probably far more complex, with auxin shown to also promote CK turnover by enhancing *CKX* expression (Palni et al. 1988; Werner et al. 2006).

Root Development

CK and auxin generally oppose one another in root development (Fig. 2; Kurepa et al. 2019) and the balance between cell division and differentiation relies on the interaction between auxin and CK (for review, see Jing and Strader 2019). CK promotes a smaller meristem via differentiation of cells in the transition zone, while auxin favors a larger meristem by promoting cell division (Blilou et al. 2005; Dello Ioio et al. 2007; Moubayidin et al. 2010). This effect of CK on meristem size is dependent upon auxin transport, as it is not observed in a *pin2/3/7* triple mutant (Dello Ioio et al. 2007). Additionally, *Arabidopsis CKX*-overexpressing lines have reduced *PIN2* and *PIN4* expression and abnormal roots with expanded auxin maxima and larger columella cells, suggesting that CK acts to regulate the expression of *PIN* genes and auxin transport, thereby balancing elongation and cell division (Pernisova et al. 2009).

The Aux/IAA SHORT HYPOCOTYL2 (SHY2, IAA3) has been implicated in the regulation of auxin transport by CK (Fig. 2C). *SHY2* transcription is induced by CK, presumably through the type-B ARRs, ARR1 and ARR12 (Dello Ioio et al. 2008; Moubayidin et al. 2010). SHY2 reduces the expression of *PIN1*, *PIN3*, and *PIN7* transcripts and protein, as assessed by the accumulation of translational fusion reporters (Dello Ioio et al. 2008). The SHY2-dependent decrease in PIN protein levels results in a local redistribution of auxin, cell differentiation, and a smaller meristem (Dello Ioio et al. 2008; Moubayidin et al. 2010). SHY2, in turn, inhibits the auxin-induced expression of an *IPT5* promoter reporter fusion construct, completing an auxin-CK regulatory circuit (Dello Ioio et al. 2008).

Auxin, on the other hand, promotes SHY2 degradation (Fig. 2C; Tian et al. 2002), restoring PIN expression, auxin localization, and cell division (Dello Ioio et al. 2008). SHY2 may target *PIN* genes directly or it could be an indirect interaction through repression of the ubiquitin-binding protein DA1-RELATED PROTEIN2 (DAR2) (Peng et al. 2013). The effects of auxin or CK on root meristem size are lost in *dar2* mutants and genetic analyses place *DAR2* downstream of CK signaling and *SHY2*. Further, *dar2* mutants have reduced auxin transport toward the root meristem, and reduced CK regulation of PIN3 and PIN7 translational fusions (Peng et al. 2013). Studies of the *Arabidopsis* root meristem and developing vasculature found that the CK-induced reduction of PIN proteins (via induction of *SHY2*) involved the transcriptional coregulator BRX (Scacchi et al. 2010). Whereas the mechanism is unclear, it likely involves reciprocal transcriptional repression between SHY2 and BRX and competition for MONOPTEROS (MP, ARF5) binding.

Arabidopsis ARR7 and *ARR15*, type-A feedback repressors of CK signaling, are up-regulated by auxin in the root stem cell niche (Müller and Sheen 2008). Type-A ARRs, in turn, regulate PIN proteins via an unknown mechanism. The *Arabidopsis arr3/4/5/6/7/8/9/15* mutant has a smaller meristem and reduced PIN1, PIN3, and PIN4 protein levels. PIN7 levels in the stele are decreased, but are higher in the root cap, suggesting that CK signaling acts on PIN proteins to alter auxin distribution and regulate meristem size and differentiation (Zhang et al. 2011). Furthermore, *Arabidopsis* type-B ARR1 and ARR12 inhibit the expression of the auxin influx transporters *LAX2* and *AUX1*, with ARR1 directly binding the promoter of *LAX2* (Zhang et al. 2013). ARR1 also promotes auxin biosynthesis, at least in part via the transcriptional activation of *ANTHRANILATE SYNTHASE BETA SUBUNIT1* (*ASB1*) that codes for a rate-limiting enzyme in the biosynthesis of the auxin precursor Trp (Moubayidin et al. 2013). *ARR1* suppression by SCARECROW (SCR), a critical regulator of root meristem activity, down-regulates auxin biosynthesis to maintain the root meristem (Moubayidin et al. 2013). Additionally, ARR12 works in concert with RETINOBLASTOMA-RELATED (RBR) protein to activate *ARF19* transcription in the root apical meristem, promoting cell differentiation (Rademacher et al. 2011; Perilli et al. 2013).

The role of auxin–CK interactions in LR development has been extensively studied (for review, see Jing and Strader 2019). In *Arabidopsis* LRs, CK regulates the expression level and pattern of *PIN* genes (Laplaze et al. 2007), and mutants for several CK signaling components display improper PIN localization (Marhavý et al. 2011; Chang et al. 2013; Moreira et al. 2013). CK was found to direct endocytic recycling of PIN1 toward degradation in LR primordia, thereby reducing the accumulation of PIN1 proteins on the plasma membrane and inhibiting LR initiation (Fig. 2A,B; Marhavý et al. 2011). During later stages of LR formation, CK was found to deplete anticlinal PIN1 protein accumulation, directing auxin flow to promote LR growth (Marhavý et al. 2014). *Ara-*

bidopsis ahp6 mutant plants display a mild defect in LR emergence and cell division in LR primordia, and have abnormal PIN1 protein localization, all reminiscent of CK treatment effects (Moreira et al. 2013). The *ahp6* mutant phenotype is lost in combination with *CKX*-overexpressing lines, suggesting that AHP6 may function to reduce CK signaling and maintain proper LR cell division. The *AHP6* gene is transcriptionally up-regulated by auxin in vascular tissues, pointing to auxin-triggered inhibition of CK signaling (via AHP6) during vascular patterning in *Arabidopsis* roots (Bishopp et al. 2011). If *AHP6* is also auxin regulated in LR primordia, it could serve as a general point of integration for auxin and CK signaling. More recently, CK was shown to exert its effects on LR root formation through TRANSPORTER OF IBA1 (TOB1) that blocks LR formation (Michniewicz et al. 2019). TOB1 is an indole-3-butyruc acid (IBA) transporter transcriptionally induced by CK signaling. IBA is an auxin precursor, suggesting that CK signaling may alter auxin distribution by promoting the relocalization of auxin intermediates.

Auxin plays a well-established role in gravitropism and regulates differential cell expansion to achieve root turning. In *Arabidopsis*, CK modifies the distribution of auxin by enhancing asymmetric localization of PIN1 proteins to the basal side of cells, redirecting the flow of auxin to promote root gravitropism (Marhavý et al. 2014). Enhancement of an AUX1 fluorescent reporter signal was also reported in *CKX2*- and *CKX3*-overexpression lines in *Arabidopsis*, suggesting that CK also regulates auxin influx (Pernisova et al. 2016).

During LR formation, emerging roots establish an angle of growth with a characteristic displacement from the gravity vector (the so-called gravitropic set point angle). This allows the root system to expand outward from the PR. CK signaling has also been shown to act in the gravitropic response by opposing bending toward the gravity vector (Waidmann et al. 2019). A genome-wide association study (GWAS) screen identified *CKX2* variants as a factor in determining LR angle. CK treatment increases the angle of LRs in *Arabidopsis* and rapeseed, whereas *Arabidopsis* CK receptor mutants show decreased LR angles. Enhanced CK signaling was observed on the topside of the emerging LR and inhibited root growth, complementing the auxin-mediated growth repression on the underside of the root. Whereas a direct interaction was not defined, treatment with the auxin transport inhibitor naphthylphthalamic acid (NPA) reduced CK reporter activity and blocked the formation of a CK signaling gradient. The balanced repression from auxin and CK promotes radial expansion of the root system by preventing the LR from completing a full 90° turn to align with gravity, demonstrating how these two hormones coordinate to fine-tune LR angles (Waidmann et al. 2019).

Similarly, this auxin-CK competition is believed to be at play in LR hydrotropism. The ER-associated MIZU-KUSSEI1 (MIZ1) protein is essential for a proper hydrotropic response (Kobayashi et al. 2007; Yamazaki et al. 2012). When overexpressed in *Arabidopsis*, MIZ1 increases CK sensitivity, reduces LR number, and lowers free auxin levels, but exogenous auxin supplementation rescued the LR phenotype (Moriwaki et al. 2011). Interestingly, CK promotes the localization of MIZ1 protein at root primordia, suggesting that MIZ1 serves as a node for auxin-CK crosstalk during LR formation.

Several other genes have been implicated as nodes of auxin-CK communication during root developmental processes. AUXIN UP-REGULATED F-BOX PROTEIN1 is transcriptionally induced by auxin and proposed to mediate CK-regulated cell expansion in the root by regulating *PIN* expression and, possibly, by targeting ARR1 for degradation (Zheng et al. 2011). The TF *TMO5* is transcriptionally up-regulated by auxin (Schlereth et al. 2010) and enhances the expression of the CK-biosynthesis gene, *LOG4*, helping determine vascular patterning in the *Arabidopsis* root (De Rybel et al. 2014). Of note, the *PLETHORA* (*PLT*) TFs, key regulators of root development that are induced by auxin (Aida et al. 2004; Blilou et al. 2005; Galinha et al. 2007) and regulate both auxin transport and synthesis (Pinon et al. 2013; Santuari et al. 2016), are repressed by CK in *Arabidopsis* roots (Dello Ioio et al. 2008). Finally, CK has been implicated in promoting auxin responses during nodulation in *Lotus japonicas* and *Medicago truncatula* (for review, see Kohlen et al. 2018), with nodule formation defects observed in loss-of-function mutants for CK signaling and auxin biosynthesis genes. On the other hand, CK treatment, CK signaling gain-of-function mutants, and chemical inhibition of PAT lead to the formation of nodule-like structures in the absence of rhizobia in *L. japonicas* and alfalfa (*Medicago sativa*) (for review, see Kohlen et al. 2018).

Shoot Development

In contrast to their roles in root meristem development, auxin acts to promote shoot apical meristem differentiation while CK promotes division of undifferentiated cells (for review, see Azizi et al. 2015). There is a general theme of high auxin levels reducing CK biosynthesis or promoting CK turnover, while rising CK levels disrupt auxin distribution by modulating PIN proteins (Fig. 3A). In *Arabidopsis* hypocotyl explants, CK up-regulates *PIN3* and *PIN6*, but reduces the expression of *PIN2*, shaping auxin distribution, and thus altering the cell division versus differentiation balance (Pernisova et al. 2009). On the other hand, during shoot induction from callus, auxin production increases, activating ARF3, which in turn suppresses *IPT5* (Cheng et al. 2013). In this way, auxin acts to restrict CK activity to the site of future meristems. Accordingly, decapitation of pea plants, thereby removing young expanding leaves, which are a major auxin source in the stem, induces *PsIPT1* and *PsIPT2*, while exogenous auxin inhibits them (Tanaka et al. 2006). In *Cremastra appendiculata* pseudobulbs, decapitation results in a turnover of auxin and induction of CK biosynthesis (Lv et al. 2018). Treatment with the auxin transport inhibitors NPA and 2,3,5-triiodobenzoic acid up-regulated *CaIPT*, and reduced the expression of *CaCKX*, promoting an accumulation of CK and pseudobulb branching.

Auxin and CK are spatially regulated and dependent upon one another to organize the formation of axillary meristems in multiple species (Wang et al. 2014; Dierck et al. 2016; Qiu et al. 2019). To confer apical dominance, auxin blocks axillary growth and CK promotes it. Auxin originating from the shoot apex down-regulates CK biosynthesis in stems, reducing CK levels in axillary buds, and CK regulates auxin efflux from branches (Fig. 3A; for review, see Müller and Leyser 2011). For example, in pea, CK promotes polarization of PIN1 protein in axillary buds (Kalousek et al. 2010) and increases levels of PIN3, 4, and 7 fluorescent fusion proteins in the xylem parenchyma of *Arabidopsis* main stems (Waldie and Leyser 2018). Correspondingly, CK was found to regulate both auxin transport and bud activation during branching in pea (Kotov and Kotova 2018).

The type-A ARRs *ARR7* and *ARR15* are induced by CK in the *Arabidopsis* shoot apical meristem but repressed by ARF5 (Fig. 3B; Zhao et al. 2010). As type-A ARRs inhibit CK signaling, this interaction reveals an important point of coordination for these two hormones in maintenance of the shoot stem cell niche. The roles of CK and auxin in regulating axillary branching are not yet fully elucidated and some studies of type-A and type-B ARRs seem to present paradoxical findings (Müller et al. 2015; Xu et al. 2015; Waldie and Leyser 2018; Zha et al. 2019). How these two hormones coordinate branching is surely complex, involving signals from other hormones and/or nutrients.

CK and auxin signaling also interact during secondary shoot vasculature development in *Populus*, where CK was found to regulate the auxin gradients that induce cambium formation, but the underlying molecular mechanism was not determined. Overproduction of CK promoted the formation of cambium through increased meristem cell divisions and IAA accumulation (Immanen et al. 2016), while CK treatment of wild-type plants reduced IAA accumulation and favored phloem development following injury (Chen et al. 2019).

The maize *aberrant phyllotaxy1* (*abph1*) mutant displays altered phyllotaxis (Giulini et al. 2004) and reduced auxin levels (Lee et al. 2009). *ZmABPH1* encodes a type-A ARR that is up-regulated by auxin and whose action depends on auxin transport (Lee et al. 2009). Both CK and ABPH1 induce *ZmPIN1* transcription, and ZmPIN1 localizes to sites of incipient leaf primordium formation, highlighting a link between auxin, CK, and the control of leaf primordia formation (Lee et al. 2009). Leaf shape is also modulated by auxin–CK interaction. Tomato *Sliaa9* antisense plants exhibit a range of phenotypes with simpler, rather than compound, leaf morphology and less lobed, even entire, leaf margins (Wang et al. 2014). Further studies revealed that compound leaf development in tomato is regulated by CK signaling, and this effect of CK is dependent upon proper SlPIN localization (Shani et al. 2010).

Regulation of auxin signaling by CK is also critical during flower development. Treatment of *Arabidopsis* plants with CK or NPA results in similar gynoecium phenotypes with disrupted apicobasal patterns (Zuniga-Mayo et al. 2014). A PIN1 translational GREEN FLUORESCENT PROTEIN (GFP) fusion, which normally displays restricted expression, was detected throughout the developing inflorescence after CK treatment, suggesting that CK mediates its effect on *Arabidopsis* gynoecium patterning at least in part by modulating PIN1 expression and localization (Zuniga-Mayo et al. 2014). Indeed, *Arabidopsis* CK receptor mutants show that CK perception is necessary for proper *PIN1* expression during ovule development (Bencivenga et al. 2012). This effect on *PIN1* is dependent upon the TFs SPOROCYTELESS and BELL1, which are transcriptionally up-regulated and down-regulated by CK, respectively (Bencivenga et al. 2012). More recently, both CK and the TF SPATULA (SPT) were found to induce the expression of the auxin genes *TAA1* and *PIN3* during *Arabidopsis gynoecium* development (Reyes-Olalde et al. 2017). The effects of CK were found to be mediated by SPT, which up-regulates the expression of type-B *ARR1* and *ARR12*. These ARRs, in turn, induce the transcription of *TAA1* and *PIN3*, although it is unclear whether they bind alone or cooperatively with SPT.

On the other hand, auxin was found to promote the expression of *AHP6*, which inhibits CK signaling, to dictate the patterning of developing inflorescences in *Arabidopsis* (Besnard et al. 2014). And in floral meristems, auxin acts through ARF3 to inhibit CK signaling by repressing *IPT*, *LOG*, and *AHK4* (Zhang et al. 2018). During sepal formation, DEVELOPMENT RELATED MYB-LIKE1 (DRMY1) helps regulate auxin and CK signaling to define sepal initiation and size (Zhu et al. 2020). *Arabidopsis drmy1-2* plants display weaker and more diffuse auxin reporter activity, but stronger and more diffuse CK reporter activity. This lack of proper spatiotemporal signaling results in delayed and smaller sepals (Zhu et al. 2020). These findings paint a complex picture of how auxin and CK regulate one another to create the patterns of hormone signaling required for proper flower development.

Other Contexts

Auxin-CK crosstalk has been implicated in myriad additional processes, including coregulation of common target genes (Hurny et al. 2020) and responses to various biotic (Boivin et al. 2016; Hurny et al. 2020) and abiotic (Wang et al. 2006; Tognetti et al. 2017; Bielach et al. 2017; Yang et al. 2017) factors. Whether these interactions are coincidental or true crosstalk requires further exploration. A greater understanding of the interplay between these two hormones could be informative for adapting crops to unfavorable and changing growth conditions.

AUXIN–ET INTERACTIONS

ET is a gaseous plant hormone first discovered for its effects on leaf abscission and ripening (for review, see Abeles et al. 1973). The crosstalk between ET and auxin is key for proper plant development and manifests itself in a wide range of processes. Plants produce ET from the amino acid L-methionine (Met) that is converted into *S*-adenosyl-L-methionine (AdoMet) by AdoMet synthetase (Giovanelli et al. 1985). The first committed and rate-limiting step in the ET biosynthesis pathway is the conversion of AdoMet into 1-aminocyclopropane-1-carboxylate (ACC) by ACC SYNTHASE (ACS). In a second step, ACC OXIDASE (ACO) turns ACC into ET (for review, see Yang and Hoffman 1984). Once synthesized, ET moves from cell to cell by diffusion without requiring specific transporters.

ET is sensed by ER- and Golgi-localized receptors of the ET RECEPTOR/ET RESPONSE

SENSOR/ET-INSENSITIVE4 (EIN4) transmembrane protein family (for review, see Binder 2020). In the absence of ET, the receptors continuously activate CONSTITUTIVE TRIPLE RESPONSE1 (CTR1), a serine/threonine protein kinase that phosphorylates and inactivates the transmembrane ER-localized protein EIN2 (for review, see Binder 2020). Phosphorylated EIN2 is targeted for degradation by two F-Box EIN2-TARGETING PROTEINs and the master TFs of the ET response, EIN3/EIN3-LIKE1 (EIL1), are targeted for degradation by the F-Box EIN3-BINDING F-BOX PROTEINs (EBFs) (for review, see Ju and Chang 2015 and Binder 2020). In the presence of ET, the receptors are turned off, inactivating CTR1, reducing the phosphorylation of EIN2, and triggering the cleavage of the carboxy-terminal end of EIN2. The EIN2 C-end blocks translation of EBF1 and EBF2 in the cytoplasm and promotes the activity of EIN3/EIL1 in the nucleus (Merchante et al. 2015; for review, see Binder 2020), eliciting the transcriptional regulation of thousands of downstream genes that mediate ET responses (Binder et al. 2004).

The first clue to the importance of auxin-ET crosstalk came from an observation that a remarkable number of auxin mutants are ET-insensitive, including auxin transport mutants such as *aux1* and *pin2* (*eir*1); mutants with impaired auxin biosynthesis, such as *weak ET-insensitive2* (*wei2*), *wei7*, and *wei8*; auxin perception mutants, such as *tir1*; and mutants in components of auxin signal transduction, including *axr2* (*iaa7*) and *axr3* (*iaa17*) (for review, see Merchante and Stepanova 2017). These findings suggest that proper levels of auxin biosynthesis, transport, signaling, and response are required for the normal response of plants to ET.

Root Development

ET promotes local auxin biosynthesis in roots, locally inducing *TAA1* and *TAR2* and several *YUC* genes, which contributes to ET-triggered PR shortening (Fig. 2A,C; Růžička et al. 2007; Stepanova et al. 2007; Swarup et al. 2007). The identification of the small molecule L-kynurenine (Kyn) (He et al. 2011) as a potent inhibitor of auxin biosynthesis provides an illustrative example of the intricate crosstalk between auxin and ET. Kyn blocks the conversion of L-Trp into IPyA, catalyzed by TAA1, inhibiting ET-induced auxin production. Kyn suppresses the short-root phenotype of ET-treated wild-type plants and of untreated *ctr1* mutants that display constitutive ET responses, supporting the notion that the root growth inhibition triggered by ET is mediated by TAA1-dependent auxin biosynthesis. Elevated levels of auxin, on the other hand, promote the stabilization of EIN3 in the nucleus by suppressing its EBF-dependent degradation. Thus, blocking auxin accumulation with Kyn inhibits EIN3 nuclear accumulation and represses root responses to ET (He et al. 2011).

The epidermis of the root elongation zone is a key site for ET-induced root growth inhibition (Vaseva et al. 2018). ET promotes the transition from the mitotic cycle to endoreduplication, reducing cell division and, therefore, the size of the meristem and root growth (Street et al. 2015). The increased activity of the auxin reporter DR5 in the root elongation zone in ET-treated plants is linked to the ability of ET to arrest cell elongation and, thus, PR growth (Růžička et al. 2007; Stepanova et al. 2007; Swarup et al. 2007). ET inhibits PR growth by transcriptionally up-regulating genes involved in auxin biosynthesis and auxin transport, thereby stimulating auxin translocation from the meristem to the elongation zone and increasing local auxin levels above the physiological threshold required for the cells to become fully sensitized to ET (Růžička et al. 2007; Stepanova et al. 2007; Swarup et al. 2007).

Auxin plays a pivotal role in stimulating LR formation by priming pericycle cells, inducing cell division, and promoting root emergence and elongation (for review, see Lavenus et al. 2013). ET reduces local accumulation of auxin required for LR formation by increasing the expression levels of *PIN3* and *PIN7* and, thus, shifting auxin away from the initiation sites (Fig. 2A,B). Correspondingly, ET causes a prominent decrease of DR5 reporter levels in the regions of LR emergence, leading to a reduction in LR emergence (Lewis et al. 2011).

Shoot Development

Shortly after germination, seedlings form the apical hook to protect the shoot apical meristem while emerging through the soil. ET up-regulates the expression of auxin biosynthetic genes such as *TAA1*, *TAR2*, and several *YUC* genes in the hypocotyl. ET also modulates auxin transport, inducing *AUX1*, *LAX1*, and *PIN3* and inhibiting *PIN1* and *PIN4* expression, to achieve preferential auxin localization on the inner side of the hook. This ET-mediated auxin gradient initiates and maintains the temporary hook structure (Vandenbussche et al. 2010; Zádníková et al. 2010). Consistent with the role of auxin–ET interactions in apical hook formation, the ET-insensitive mutant *ein2* displays a defective hook that can be corrected with exogenous auxin application, consistent with the notion that ET induces the boost in auxin production required for proper hook development (Vandenbussche et al. 2010). Moreover, ET stimulates the transcription of an *N*-acetyltransferase-like gene *HOOKLESS1* (*HLS1*), which down-regulates the expression of *ARF2*, a repressor of the auxin response (Li et al. 2004). This repression of a repressor results in the general induction of auxin responses. Accordingly, *hls1* mutants do not form a hook, highlighting another point of integration of the auxin and ET cues during apical hook development (Lehman et al. 1996).

Fruit Development

Auxin-ET crosstalk is fundamental for the development of the male and female reproductive organs (An et al. 2012). Auxin induces, whereas ET represses, stamen development, and both promote pollen germination and the growth of the pollen tube (Fig. 4A; for review, see An et al. 2020). Genes involved in ET and auxin biosynthesis and signaling are highly expressed during pistil development and specifically in ovules at anthesis, suggesting that these two hormones coregulate the process of fruit set, but act at different time points (for review, see An et al. 2020). In tomato, auxin governs the initial phases of fruit development and ET controls the ripening of the fruit by promoting the degradation of chlorophylls, conversion of xanthophylls to carotenes, initiation of the climacteric ET production, etc. (Fig. 4B,C; Fraser et al. 1994). Fruits treated with auxin exhibit a delay in the climacteric transition to the ET-mediated ripening processes, preservation of high levels of xanthophylls and chlorophyll, and the inhibition of genes involved in carotenoid biosynthesis (Su et al. 2015).

Tomato transgenic lines with reduced expression of *SlIAA3* exhibit auxin and ET phenotypes, with delayed ripening and reduced apical dominance, auxin sensitivity, and petiole epinasty, suggesting that *SlIAA3* is a link between the auxin and ET response pathways (Chaabouni et al. 2009). Furthermore, both ET and auxin induce the expression of the TF *ET RESPONSE FACTOR.B3* (*SlERF.B3*), a major player in the regulation of ET responses and fruit ripening (Liu et al. 2013a). *SlERF.B3* integrates ET and auxin signals by binding to the promoter and inducing the expression of *SlIAA27* (Fig. 4C; Liu et al. 2018).

Other Contexts

As mentioned above, in the IPyA pathway of auxin biosynthesis, TAA1 and TARs catalyze the conversion of Trp into IPyA that is subsequently used by the YUCs to produce auxin (Mashiguchi et al. 2011; Stepanova et al. 2011; Won et al. 2011). The VAS1 enzyme directs IPyA away from the YUCs, using it and the ET biosynthetic precursor, Met, to produce Trp and 2-oxo-4-methylthiobutyric acid, reducing IPyA availability and, in turn, auxin production. Correspondingly, *vas1* mutants exhibit elevated levels of both IAA and the ET precursor ACC. VAS1 therefore represents a point of interaction at the metabolic level between auxin and ET biosynthesis (Zheng et al. 2013).

AUXIN–GA INTERACTIONS

GAs take their name from the growth-modifying fungus *Gibberella fujikuroi* that triggers a “foolish seedling disease” in rice. Since GA was first identified in the 1930s, more than 130 GAs have been discovered in plants, fungi, and bacteria. However, only a handful, GA_1, GA_3, GA_4, and GA_7, are

 Cite this article as *Cold Spring Harb Perspect Biol* doi: 10.1101/cshperspect.a039990

biologically active, and most nonbioactive GAs are precursors or deactivated catabolites of the bioactive forms. GAs are derived from *trans*-geranylgeranyl diphosphate (GGPP), a common C20 precursor for diterpenoids. First, GGPP is converted to the tetracyclic hydrocarbon intermediate, *ent*-kaurene, by the diterprene cyclases ENT-COPALYL DIPHOSPHATE SYNTHASE and ENT-KAURENE SYNTHASE. Then, *ent*-kaurene is converted to GA_{12} by a plastid membrane-bound ENT-KAURENE OXIDASE and an endoplasmic reticulum-bound ENT-KAURENOIC ACID OXIDASE. In the third step, the conversion of GA_{12} to various intermediates and bioactive GAs is mediated in the cytosol by GA20-OXIDASE and GA3-OXIDASE through two parallel pathways. Bioactive GAs can be deactivated by GA2-OXIDASE (GA2ox) (for review, see Binenbaum et al. 2018).

The soluble receptor GIBBERELLIN-INSENSITIVE DWARF1 (GID1) binds to GA in the nucleus triggering a conformational change that promotes association with the transcriptional regulators DELLAs. The formation of the GID1-GA-DELLA complex enhances DELLA binding to GID2/SLY1 F-box proteins, which triggers the degradation of DELLA via the 26S proteasome pathway, activating the GA response. In the absence of GA, DELLA proteins bind to TFs, repressing the GA response (for review, see Hernández-García et al. 2020). As the sites of GA biosynthesis are not always colocalized with the expression domains of GA perception genes, GA movement is thought to be essential. Although GA efflux transporters have not been identified yet, the NITRATE TRANSPORTER1/PEPTIDE TRANSPORTER family of proteins in *Arabidopsis* have been described as bona fide influx carriers (for review, see Binenbaum et al. 2018). In addition, two members of the SWEET transporter family (SWEET13 and SWEET14) are also capable of transporting GA (for review, see Binenbaum et al. 2018).

While auxin plays essential roles in almost all developmental processes (for review, see Gallei et al. 2020), GAs play important roles in both cell division and cell elongation, such as seed germination, stem/shoot elongation, and floral organ development (Ubeda-Tomás et al. 2012). It is generally considered that auxin acts upstream of GA by activating GA biosynthesis (Hu et al. 2018a), but there are many levels of interaction between these two hormones, as reviewed below.

Root Development

As described above, the auxin-CK regulatory circuit ARR1-SHY2-PIN controls root meristem size by balancing cell differentiation with cell division (Fig. 2C; Moubayidin et al. 2010). ARR1 physically interacts with DELLA proteins GA-INSENSITIVE (GAI) and REPRESSOR OF GA (RGA) and this interaction enhances its transactivation activity (Rosa et al. 2015). Thus, during the meristem growth phase, a high level of GA represses *ARR1* expression by promoting degradation of the DELLA proteins, which results in a low level of SHY2, thus promoting cell division (Fig. 2C; Rosa et al. 2015).

Shoot Development

In the absence of light, seedlings undergo skotomorphogenesis/etiolation, resulting in an elongated hypocotyl, presence of an apical hook and small and closed cotyledons (Von Arnim and Deng 1996). Several PHYTOCHROME-INTERACTING FACTORs (PIFs) including PIF1, PIF3, PIF4, and PIF6, play a critical role in etiolation (Huq and Quail 2002; Kim et al. 2003; Huq et al. 2004; Monte et al. 2004). DELLAs physically interact with PIFs, preventing PIF binding to their targets, which results in inhibition of hypocotyl growth (de Lucas et al. 2008; Feng et al. 2008). On the other hand, ARF6 and ARF8 also regulate hypocotyl elongation (Tian et al. 2004; Nagpal et al. 2005). Genome-wide analyses indicate that ARF6 shares a large number of target genes with the BR signaling TF BZR1 and with the light/temperature-regulated TF PIF4 (Oh et al. 2014), two components of the PIF4-BZR1-DELLA module that integrates signals from light, BR, and GA (Bai et al. 2012a,b), suggesting that ARF6 may interact with DELLA. ARF6, ARF7, and ARF8 were confirmed to physically associate with the DELLA protein RGA. Through this interaction, RGA blocks ARF6 binding to the promoters of its target genes, suggesting that GA promotes cell elongation in the *Arabidopsis* hypocotyl by enhancing auxin/ARF-mediated responses (Oh et al. 2014).

Downstream from DELLA and PIF signaling, the two *Arabidopsis* paralogous GATA TFs, GATA NITRATE-INDUCIBLE CARBON-METABOLISM INVOLVED (GNC) and GNC-LIKE (GNL), were identified as direct transcriptional targets of PIFs (Richter et al. 2010). Single and double *gnc* and *gnl* mutant seeds germinate faster than wild-type, with the double mutant germinating even on plates containing the GA biosynthesis inhibitor paclobutrazol (Richter et al. 2010). Furthermore, *gnc gnl* partially suppresses the GA biosynthesis mutant *ga1*, suggesting that GNC and GNL are repressors of GA signaling (Richter et al. 2010). *gnc gnl* also suppresses *arf2* phenotypes, suggesting that, genetically, *GNC* and *GNL* act downstream of *ARF2* (Richter et al. 2013). Consistent with this idea, ARF2 and ARF7 can directly bind to the promoters of *GNC* and *GNL* and inhibit their expression. Thus, *GNC* and *GNL* represent another point of convergence for the crosstalk between auxin and GA (Richter et al. 2013).

Another potential node of auxin–GA interaction is *miR319*, an important player in shoot organ morphogenesis (Curaba et al. 2014) that acts as a positive regulator of auxin signaling by indirectly repressing *SHY2* and *SMALL AUXIN UP RNA39* in leaf morphogenesis (Tian et al. 2002; Palatnik et al. 2003; Kant et al. 2009). Interestingly, *miR319* can affect leaf cell differentiation by targeting *LANCEOLATE*, which indirectly inhibits GA biosynthesis (Ori et al. 2007; Yanai et al. 2011), thus implicating *miR319* in the auxin-GA crosstalk controlling leaf organogenesis (Curaba et al. 2014).

Similar to primary stem growth, cambial growth in secondary stems is also regulated by auxin and GA. Consistent with previous studies in *Arabidopsis* and in pea (Ross et al. 2000; Frigerio et al. 2006), poplar *PttGA20ox1* and *PttGA20ox4* transcript levels are decreased in the stem of decapitated trees and induced by IAA, indicating that auxin stimulates the expression of GA biosynthesis genes in cambial growth. Furthermore, GA-only treatment increases cell division in the cambial zone of decapitated poplar trees (i.e., under auxin depletion) (Björklund et al. 2007). GA, in turn, increases local auxin levels by promoting expression of a putative auxin efflux transporter, *PttPIN1* (Björklund et al. 2007). A recent study demonstrated that GA can redirect protein trafficking to the plasma membrane, thus coregulating multiple processes, including PIN-dependent auxin fluxes (Salanenka et al. 2018).

GAs not only promote vegetative growth, but also induce developmental phase transitions. In *Arabidopsis*, the GA pathway plays a major role in flowering time under short-day conditions, promoting flowering through the activation of floral integrator genes such as *SUPPRESSION OF OVEREXPRESSION OF CONSTANS1* and *LEAFY* (*LFY*) (Blázquez et al. 1998; Bonhomme et al. 2000; Moon et al. 2007). On the other hand, MP (ARF5) plays a critical role in flower primordium initiation (Przemeck et al. 1996) and directly induces *LFY* (Yamaguchi et al. 2013; Wu et al. 2015). Thus, LFY is yet another point of convergence for the crosstalk between GA and auxin. In addition, MP physically interacts with BRAHMA (BRM) and SPLAYED, two related *Arabidopsis* SWI/SNF-subgroup ATPases (Wu et al. 2015), and BRM binds the promoters of GA biosynthetic genes such as *GA3ox1* as an activator (Archacki et al. 2013). Furthermore, several DELLA proteins physically interact with an SWI/SNF subunit SWI3C (Sarnowska et al. 2013). Thus, crosstalk between GA and auxin during flowering may also be dependent on MP- and DELLA-mediated interactions with chromatin-remodeling complexes.

Fruit Development

Exogenous application of diverse plant growth substances, mainly auxins and GAs, can induce parthenocarpic fruit set and development (Gorguet et al. 2005; Srivastava and Handa 2005). Consistent with these observations, mutations affecting auxin signaling or GA biosynthesis genes (such as *AtARF8*, *SlARF7*, *SlIAA9*, and *SlDELLA*) can also induce parthenocarpic fruits in *Arabidopsis* and tomato (Wang et al. 2005a, 2009, 2011; Goetz et al. 2007; de Jong et al. 2009; Mounet et al. 2012). It was proposed that fertilization triggers auxin-mediated GA synthesis (Dorcey et al. 2009). A recent study in tomato showed that SlARF7 and five other activator SlARFs interact with SlDELLA and SlIAA9, and that SlARF7 and SlDELLA directly associate with the promoters of GA biosynthesis (*GA20ox1* and *GA3ox1*) and auxin catabolism (*GH3.2*) genes (Fig. 4A; Hu et al. 2018a). The direct interaction between the activator SlARF7 and the repressor SlIAA9 may turn the SlARF7/SlIAA9 complex into a transcriptional repressor, whereas SlDELLA blocks SlARF7 binding to the promoters of its target genes, thus antagonizing the repression by SlARF7/SlIAA9 of GA- and auxin-related genes (Hu et al. 2018a). In contrast, the effect of SlARF7/SlIAA9 and SlDELLA on downstream growth-related genes, such as *EXPANSIN5* (*EXP5*) and *ACC OXIDASE4* (*ACO4*), is additive (Fig. 4A). Taken together, these findings reveal that direct crosstalk between SlDELLA-mediated GA and SlARF7/SlIAA9-mediated auxin signaling events coregulates fruit initiation in tomato (Hu et al. 2018a).

Interestingly, this DELLA/ARF-mediated regulation central to fruit initiation in tomato appears to be conserved in grape (Zhang et al. 2019b), also a true botanical fruit derived from ovaries, and in strawberry (Zhou et al. 2020), which makes accessory fruits derived from receptacles. FveARF6 and FveARF8 interact with the DELLA repressor FveRGA1. Like SlARF7 (Hu et al. 2018a; Zhou et al. 2020), FveARF8 interacts with FveRGA1 and FveIAA4 through distinct protein domains, suggesting that FveARF8 may be simultaneously repressed by FveRGA1 and FveIAA4 prior to fertilization (Zhou et al. 2020). Furthermore, FveARF8 can directly bind to the promoter and repress the expression of *FveGID1c*, suggesting that the auxin-GA crosstalk in strawberry fruits is multifaceted (Zhou et al. 2020).

AUXIN–JA INTERACTIONS

JA is a fatty-acid-derived hormone that takes its name from jasmine oil of *Jasminum grandiflorum* (Demole et al. 1962). JA regulates not only plant stress responses but also plant growth and development, in part through its tissue-specific interactions with auxin. JA biosynthesis has been extensively reviewed (for review, see Fonseca et al. 2009 and Ruan et al. 2019). Briefly, JAs are lipid-derived molecules produced via two main pathways. The octadecane pathway uses linolenic acid (18:3) as a precursor to produce 12-oxo-phytodienoic acid (OPDA) and the hexadecane pathway uses hexadecatrienoic acid (16:3) yielding dinor-oxo-phytodienoic acid (dn-OPDA) (Chini et al. 2018). These first steps of both pathways take place in the chloroplasts. Further reactions lead to JA production in the peroxisomes and its modification to methyl-JA (MeJA) or conjugation to isoleucine (Ile) to make JA-Ile in the cytoplasm. The bioactive form of JA is JA-Ile, which mediates plant responses to environmental and developmental cues. JA-Ile can promote resistance to a broad range of plant pathogenic bacteria, fungi, and herbivores (Campos et al. 2014; Machado et al. 2016, 2017).

JA biosynthesis and accumulation of JA-Ile are promoted in response to specific endogenous and environmental cues. JA-Ile binds to the CORONATINE-INSENSITIVE1 (COI1) receptor, inducing the degradation of the JASMONATE ZIM-DOMAIN (JAZ) proteins and releasing the MYC2 TFs from the JAZ-MYC2 complex. The MYC2 TFs then induce the expression of JA-responsive genes (Chini et al. 2018).

The JA signaling pathway presents certain similarities to the auxin signaling pathway. COI1 encodes an F-box protein, related to TIR1/AFB family of F-box proteins that bind auxin. Once JA-Ile is bound, COI1 functions in E3-ubiquitin ligase-mediated proteolysis of the targeted JAZ proteins. The JAZ repressors play a negative role in the JA signaling pathway similar to the function of Aux/IAAs in auxin signaling. Likewise, the MYC TFs up-regulate the expression of downstream genes as activator ARFs do in the auxin signaling pathway (Campos et al. 2016). Generally, while auxin is considered to be the growth-promoting hormone, JA is known to repress plant growth.

Root Development

Auxin promotes PR meristem activity and cell division but can also inhibit PR growth by reducing cell elongation in the elongation zone (for review, see Overvoorde et al. 2010 and Vaseva et al. 2018). In contrast, JA treatments inhibit PR meristem activity, decreasing cell number in the root meristem, as well as reduce cell size in both the meristem and the elongation zone, thus leading to the reduction of the overall root growth (Fig. 2A; for review, see Wasternack and Hause 2013). Auxin distribution defines the levels and expression patterns of *PLETHORAs* (*PLTs*), AP2-domain TFs that mediate the establishment and maintenance of the root stem cell niche and cell proliferation (Mähönen et al. 2014). ARF2 positively regulates *PLT1* and *PLT2* expression (Promchuea et al. 2017). In turn, PTLs stimulate the production of auxin by inducing the expression of the auxin biosynthesis genes *ANTHRANILATE SYNTHASE ALPHA SUBUNIT1* (*ASA1*), *YUC1*, and *YUC4* (Pinon et al. 2013). On the other hand, PLTs are also regulated by JA (Chen et al. 2011). Upon JA perception, *PLT* genes are repressed in the root stem cell niche (Fig. 2C). In JA signaling mutants, such as *coi1* or *myc2*, JA-mediated regulation of *PLT* is abolished, whereas in a Trp- and auxin-deficient mutant *asa1* and in an auxin overproducing mutant *yuc1D*, *PLT*s display normal responses to JA (Chen et al. 2011). Furthermore, *asa1* and *yuc1D* mutants are fully sensitive to JA in the PR growth assay. These results argue that the JA-triggered inhibition of the *PLTs* and the effect of JA on root growth are independent of the effects of *PLTs* on auxin biosynthesis, and that auxin and JA coregulate root growth via *PLT*s independently of one another (Chen et al. 2011).

In LR development, there is evidence for direct crosstalk between JA and auxin (for review, see Wasternack and Hause 2013). Exogenous application of JA or an increase in endogenous JA levels caused by the induction of JA biosynthetic genes result in enhanced auxin production and signaling (Fig. 2A; Cai et al. 2015). The *coi1-1* JA receptor mutant exhibits an uneven distribution of LRs and is unable to promote the formation of additional LRs in response to JA. In wild-type plants, JA treatment boosts the formation of LRs by inducing the expression of the *ERF109* TF in the LR primordium, and in the tip and base of LRs (Cai et al. 2014). ERF109 binds to the promoters of auxin biosynthesis genes such as *YUC2* (Cheng et al. 2006) and *ASA1* (Sun et al. 2009), increasing auxin levels and promoting the emergence of LRs (Fig. 2B). Consistent with the idea of auxin acting downstream of JA in this process, mutants with compromised auxin signaling, such as *solitary root* (*iaa14*) and the double mutant *arf7 arf19*, fail to increase LR formation in response to JA (Raya-González et al. 2012). These observations support the idea that JA effects in LR development are mediated by auxin.

The formation of adventitious roots (ARs) is another developmental process affected by auxin–JA interactions (Gutierrez et al. 2012). Auxin controls the levels of active JA-Ile by regulating the expression of several *GH3* genes, *GH3.3*, *GH3.5*, and *GH3.6*, whose protein products are believed to conjugate JA to amino acids, modulate the levels of both free JA and JA-Ile and, thus, fine-tune AR formation (Gutierrez et al. 2012). Auxin effects are mediated by several ARFs, with ARF6 and ARF8 up-regulating and ARF17 repressing these *GH3*s (Gutierrez et al. 2012). The formation of JA conjugates by the GH3s reduces the level of free, active JA-Ile and promotes the formation of ARs. Furthermore, JA induces *ERF115*, a TF that activates CK signaling by up-regulating *ARR5* and *ARR7* and CK biosynthesis by inducing the expression of *IPT3*, encoding one of the rate-limiting enzymes in CK production. Thus, JA represses AR formation by modifying CK homeostasis through *ERF115* activity (Lakehal et al. 2020).

Upon root damage, the synergistic interaction between JA and auxin signaling pathways favors the activation of the root stem cell division and tissue regeneration (Zhou et al. 2019). In intact roots, ERF109 and ERF115 TFs keep the quiescent center (QC) of the root undifferentiated (Zhou et al. 2019). These TFs work in conjunction with auxin and *CYCLIN D6;1* (*CYCD6;1*) to prevent the QC from dividing, while promoting cell division in other regions of the root meristem. Root wounding triggers the rapid systemic production of JA and the local accumulation of auxin at the sites of injury due to impaired PAT upon tissue damage. JA perception

 Cite this article as *Cold Spring Harb Perspect Biol* doi: 10.1101/cshperspect.a039990

activates MYC2, which directly induces *ERF109*, which in turn targets and up-regulates *CYCD6;1* and promotes auxin biosynthesis by inducing *ASA1* (Zhang et al. 2019a) and *YUC2* (Cai et al. 2014). JA and auxin cooperatively induce the expression of *ERF115*, a key TF in tissue regeneration that promotes division of the cells directly surrounding the wounded site to replenish the damaged cells (Ye et al. 2020).

As both IAA and JA-Ile are sensed by SCF E3-ligase complexes, SCF^{TIR} and SCF^{COI}, respectively, that share multiple components, disruptions of these complexes (e.g., defects in the generic SCF subunits or in upstream players) result in impaired responses to both hormones (Dharmasiri et al. 2007; Moon et al. 2007). For example, single *auxin resistant1* (*axr1*) and double *axr1 axr1-like* mutants defective in the RUB1-activating enzyme E1 display PR insensitivity to synthetic auxin 2,4-D and to methyl-JA (Dharmasiri et al. 2007). RUB1 modification of the Cullin SCF subunit promotes SCF activity. Thus, the ability to assemble functional SCF complexes is apparently necessary for both JA and auxin signaling.

Shoot Development

Crosstalk between the auxin signaling pathway and JA production regulates the development of floral organs according to external cues and flower phenology (Reeves et al. 2012). The auxin response factors, ARF6 and ARF8, govern the late stages of flower development leading to anthesis. The development of *arf6 arf8* mutant flowers is arrested at stage 12, resulting in flowers with short petals and stamen filaments, and immature anthers and gynoecium. The *arf6 arf8* flowers never undergo anthesis and are largely male and female sterile. ARF6 and ARF8 regulate flower development, in part through the TEOSINTE BRANCHED/CYCLOIDEA/PCF4-mediated induction of JA synthesis, which in turn up-regulates the expression of the JA-responsive TFs *MYB21* and *MYB24* that control petal, stamen, and gynoecium development (Reeves et al. 2012). Accordingly, mutants in the JA biosynthesis or signaling pathways often exhibit delayed anther dehiscence, low pollen viability, and compromised filament elongation (for review, see Song et al. 2013). During flower development, the *arf6 arf8* mutant produces low levels of JA (Tabata et al. 2010). Consequently, application of exogenous JA can rescue the petal elongation and anther dehiscence defects, but not the stamen filaments or gynoecium developmental deficiencies of the *arf6 arf8* double mutant (Nagpal et al. 2005), clearly indicating that JA acts downstream of the auxin signaling and response pathways with respect to petal and stamen development.

Exogenous applications of JA or auxin induce and repress leaf senescence, respectively. JA promotes leaf senescence through the COI1 signaling pathway. The MYC TFs, including MYC2, induce the expression of genes involved in senescence and chlorophyll degradation (Qi et al. 2015; Zhu et al. 2015). The WRKY57 TF plays a major role repressing the expression of senescence-associated genes (Jiang et al. 2014). Upon JA application, *WRKY57* transcription is induced, whereas WRKY57 protein is rapidly turned over by the 26S proteasome pathway preventing its accumulation. In contrast, auxin treatment promotes both WRKY57 transcript and protein accumulation. Furthermore, WRKY57 physically interacts with JAZ4 and JAZ8, as well as with IAA29, negative regulators of the JA and auxin signaling pathways, respectively. This competition for WRKY57 between JAZs and IAA29 is thought to also contribute to the antagonistic role of JA and auxin in the regulation of leaf senescence (Jiang et al. 2014).

AUXIN–SA INTERACTIONS

SA is a phenolic compound produced by plants in response to pathogen exposure and, along with JA, plays a central role in plant defenses to biotic stress (for review, see Lefevere et al. 2020). Remarkably, auxin and SA share a common precursor, chorismate, which is produced via the shikimate pathway in the chloroplast (Pérez-Llorca et al. 2019). Chorismate can be converted into Trp to synthesize auxin or into isochorismate to produce SA. The SA biosynthesis pathway is thought to occur through two independent routes, one relying on PHENYLALANINE AMMONIA-LYASE (PAL) and the other on ISOCHORISMATE SYNTHASE (ICS), both residing in the cytosol (Dempsey and Klessig 2017). In the first reaction of the PAL pathway, PAL catalyzes the conversion of phenylalanine into *trans*-cinnamic acid, which is later metabolized to either ortho-coumaric acid or benzoic acid intermediates to finally produce SA. In the ICS pathway, ICS is responsible for the conversion of chorismate into isochorismate. Then, avrPphB SUSCEPTIBLE3 (PBS3) conjugates isochorismate with glutamate to produce isochorismate-9-glutamate, which can be converted by the acyltransferase ENHANCED PSEUDOMONAS SUSCEPTIBILITY 1 (EPS1) or spontaneously catabolized into SA (for review, see Lefevere et al. 2020).

Members of the *NON-EXPRESSOR OF PATHOGENESIS-RELATED* (*NPR*) gene family are postulated to be SA receptors. In the absence of SA, NPR1 is located in the cytoplasm in an oligomeric state. Upon SA binding, NPR1 experiences a conformational change that causes the complexes to dissociate into monomers that migrate to the nucleus where NPR1 interacts with the TGACG-BINDING FACTOR (TGA) family of TFs to induce the expression of *PATHOGENESIS-RELATED* genes (Wu et al. 2012). NPR3 and NPR4 bind SA and act as adaptor proteins regulating the activity of the CUL3 E3 LIGASE that degrades NPR1 depending on the SA concentration in the cell. In healthy plants, the low levels of SA trigger the degradation of NPR1 by NPR3, NPR4, and the proteasome, thus blocking the induction of defense genes. As SA levels increase upon pathogen attack, SA-bound NPR3 and NPR4 lose their ability to promote the degradation of NPR1, enabling stabilized NPR1 to induce *PATHOGENESIS-RELATED* genes (Fu et al. 2012).

Besides the NPRs, additional SA-binding proteins may be involved in specific NPR1-independent SA immune responses and in SA-mediated regulation of growth. SA produced by plants is involved not only in plant defenses but also in plant growth and development (Wang et al. 2007). Auxin and SA interact to modulate these processes.

Root Development

Treatments with exogenous SA shape the development of the root, at least in part by affecting auxin production and distribution. PIN efflux transporters are stabilized at the plasma membrane by SA (Du et al. 2013). Low concentrations of SA (below 50 µM) induce the formation of ARs but reduce the size of the PR meristem, whereas high concentrations of SA (above 50 µM) hamper all root developmental processes (Pasternak et al. 2019). SA has an inhibitory effect on root elongation, in part due to the activation of auxin synthesis via *TAA1* induction in the epidermis of the elongation zone (Fig. 2A; Pasternak et al. 2019), similar to the expression boost triggered by ET (Stepanova et al. 2008).

The mechanisms governing AR formation are defined by the crosstalk between SA, auxin, ET, and JA (Pasternak et al. 2019). The amido-synthases encoded by the *GH3* gene family are able to conjugate not only IAA, but also JA and SA, to amino acids (for review, see Woodward and Bartel 2005 and Zhang et al. 2007). The activity of GH3 enzymes controls the endogenous levels of active hormones, ultimately influencing the number of ARs and other morphometric traits (Staswick et al. 2005; Sorin et al. 2006; Gutierrez et al. 2012). For example, transgenic lines overexpressing *GH3.5* exhibit reduced levels of free auxins and SA, but elevated levels of their aspartate conjugated forms, and show dramatic morphological defects including severe dwarfism (Westfall et al. 2016). Furthermore, by regulating the levels of the IAA and SA, GH3s affect plant responses to pathogens (Zhang et al. 2007) and to a wide range of abiotic stresses including drought, freezing, and salinity, consistent with the central role of the GH3 substrates in these processes (Park et al. 2007). In summary, the crosstalk between SA and auxin affects the architecture of the root and its interaction with environmental cues.

Shoot Development

During apical hook formation, ET and GA enhance the uneven distribution of auxin, promoting the formation of an exaggerated apical hook, whereas JA and SA disrupt the differential distribution of auxin and, thus, the formation of the apical hook (Wang and Guo 2019). In the presence of SA, NPR1 monomers migrate to the nucleus and interact with EIN3 and EIL1, blocking the expression of EIN3- and EIL1-target genes, including those involved in auxin biosynthesis and transport and, thus, the formation of the apical hook (Huang et al. 2020). This SA-mediated effect on the formation of the apical hook is NPR1-dependent, with formation of the apical hook being impaired in *NPR1* overexpression lines and enhanced in *npr1* loss-of-function mutants (Huang et al. 2020).

Pathogen Response

Auxin and SA exhibit antagonistic functions during plant defense (Wang et al. 2007). A number of pathogens are capable of either producing auxin or inducing auxin biosynthesis in the host to modify the plant developmental programs to their own benefit (Chen et al. 2007; for review, see Robert-Seilaniantz et al. 2007). The extra auxin loosens the cell walls and promotes cell elongation, favoring pathogen attack and enabling the development of symptoms.

To counteract pathogens, plants have evolved mechanisms to dampen the effects of the excess auxin produced during an attack (Spaepen et al. 2007). One primary method is to accumulate high levels of SA that inhibit the response to auxin at multiple levels. SA down-regulates the expression of several *PIN* genes, reducing auxin transport (Armengot et al. 2014). Furthermore, SA dampens auxin signaling by inhibiting the expression of *TIR1* and *AFBs*, stabilizing the Aux/IAA repressors and, thus, blocking the global response to auxin (Wang et al. 2007). In addition, mutants that overaccumulate SA have lower IAA levels relative to wild-type, suggesting that the inhibitory effect of SA on auxin is in part due to a reduction in auxin signaling and response (Wang et al. 2007). SA impairs the production of auxin by inhibiting *CATALASE2* (*CAT2*) activity. *CAT2* is down-regulated in response to SA, leading to an increase in hydrogen peroxide that triggers the sulfenylation of TRP SYNTHETASE B SUBUNIT1, reducing the production of Trp and, thus, hampering the biosynthesis of auxin (Yuan et al. 2017). CAT2 coordinates the action of SA not only on auxin but also on JA production, thus affecting plant susceptibility to pathogens (Yuan et al. 2017).

AUXIN–SL INTERACTIONS

SLs are the most recent addition to the phytohormone family. They were discovered based on their involvement in promoting the germination of parasitic plants of the *Striga* genus (Cook et al. 1966). SLs are produced from carotenoids. While the complete picture of SL synthesis is still unfolding, in general, SL synthesis starts with the conversion of β-carotene to carlactone. This process is catalyzed by three enzymes: DWARF27 (D27), CAROTENOID CLEAVAGE7 (CCD7), and CCD8 (for review, see Omoarelojie et al. 2019). Most SL production takes place in the root, but the shoots also make SL. SL can be transported to the shoot to mediate shoot developmental processes, such as branching (Gomez-Roldan et al. 2008; Umehara et al. 2008), or secreted into the rhizosphere to mediate symbiotic and parasitic relationships (Cook et al. 1966; Akiyama et al. 2005).

The current model for SL signaling is that the SL molecule binds to its receptor, α/β-hydrolase D14, initiating a signaling cascade. In the presence of SL, D14 forms a complex with an F-box protein D3 (in *Arabidopsis*, MORE AXILLARY BRANCHES, MAX2). This complex directs protein targets, such as rice D53 (in *Arabidopsis*, three redundant members of the SUPPRESSOR OF MAX2 1-LIKE (SMXL) family) for ubiquitin-mediated degradation. The loss of these target proteins relieves the transcriptional repression of SL-regulated genes, triggering some SL-induced responses (for review, see Omoarelojie et al. 2019 and Bürger and Chory 2020). In the final step, D14 hydrolyzes and inactivates the SL molecule (Seto et al. 2019). There is also evidence for feedback regulation of SL signaling. In pea, the SL pathway mutants *ramosus* (*rms3*, *rms4*, and *rms5*, defective in the orthologs of rice *OsD14*, *OsD3*, and *OsCCD7*, respectively) show increased expression of *PsRMS1* (*OsCCD8*), suggestive of feedback regulation, but plants harboring the *rms2* mutation have reduced expression of *PsRMS1* (Foo et al. 2005). *PsRMS2* was found to encode an ortholog of the TIR1 auxin receptor that up-regulates SL biosynthesis, providing further support for the connection between auxin response and SL biosynthesis in pea (Ligerot et al. 2017).

Root Development

SL treatment of rapeseed seedlings promotes both shoot and root growth and rapidly reduces endogenous auxin levels, suggesting that auxin–SL interactions are at play both above and below ground (Ma et al. 2020). The *Arabidopsis* SL mutants *max1*, *max2*, and *max4* (an ortholog of *OsCCD8*) have a shorter PR and fewer meristematic cells relative to wild-type, as well as expanded auxin reporter activity (Ruyter-Spira et al. 2011). Application of a synthetic SL to the roots of plants increases the size of the root meristem and transition zone, blocks LR formation and reduces the auxin content of leaves. SL treatment also inhibits the expression of PIN 1, 3, and 7 translational fusions with GFP in the root tip, suggesting that an interaction between SL and auxin transport is critical for balancing root growth (Ruyter-Spira et al. 2011). Similarly, SL promotes *PIN2* expression and PIN2 polar localization at the plasma membrane in *Arabidopsis* root tips, implicating SL in the regulation of PIN-mediated auxin transport in roots (Fig. 2A; Pandya-Kumar et al. 2014).

Arabidopsis max2, *max3* (orthologous to *OsCCD7*), and *max4* mutants all display an increased number of LRs, and SL treatment reduces LR number and increases root hair length in *max3* and *max4* (Kapulnik et al. 2011). SL regulates LR development in rice via inhibition of auxin transport (Fig. 2A; Sun et al. 2019). SL signaling (*d3*) and biosynthesis (*d10*) mutants display an increased number of LRs, which is further enhanced by auxin treatment, but attenuated by NPA. The effect of auxin is abrogated by application of SL in *d10*, but not in *d3*, providing evidence that SL acts to block auxin-mediated LR growth.

SHY2, involved in auxin and CK signaling, promotes auxin accumulation to inhibit LR development (Goh et al. 2012). The SL signaling mutant *max2* displays a similar LR phenotype as a *shy2* loss-of-function mutant, and both are insensitive to SL treatment (Koren et al. 2013). Conversely, wild-type plants and a *shy2* gain-of-function mutant are both sensitive to exogenous SL. The authors propose that SHY2 acts to integrate signals from multiple hormone pathways, including SL and auxin (Koren et al. 2013).

SL rice mutants have a decreased number of ARs per tiller, but higher levels of endogenous auxin and auxin reporter activity (Sun et al. 2015). Treatment of the *d10* mutant with exogenous SL increases the number of ARs, decreases the expression of auxin transport genes, and inhibits auxin reporter activity. Moreover, while NPA treatment does not alter AR number in the *d3* or *d10* mutants, it does reduce AR number in wild-type plants to a level akin to SL mutants. Together, these data show that, in rice, SL promotes AR formation, which may be mediated by its actions on PAT. However, studies in *Arabidopsis* and pea suggest that, unlike in rice, SL and auxin might be largely independent of one another in their control of AR development (Rasmussen et al. 2012).

In tomato, SL treatment reduces the abundance and length of root hairs, and blocks auxin-mediated inhibition of root elongation (Koltai et al. 2010). Exogenous auxin is unable to rescue the SL phenotype, but treatment with 2,4-D, a synthetic auxin analog that is a poor substrate for auxin efflux carriers, rescues the root hair and root elongation phenotypes, suggesting that, in tomato, SL modulates root development via effects on auxin transport.

SLs are exuded from plant roots as a signal to stimulate arbuscular mycorrhizal (AM) colonization (Akiyama et al. 2005). In tomato roots, AM fungi induce *SlIAA27* expression (Bassa et al. 2013) and silencing of *SlIAA27* reduces AM colonization (Guillotin et al. 2017). Interestingly, silencing of *SlIAA27* also reduces the expression of SL biosynthesis genes *SlNSP1*, *SlD27*, and *SlMAX1*, and treatment with exogenous SL restores the AM phenotype of *SlIAA27*-silenced plants (Guillotin et al. 2017). Further evidence

 Cite this article as *Cold Spring Harb Perspect Biol* doi: 10.1101/cshperspect.a039990

in pea supports the idea that auxin and SL interact to promote AM symbiosis. An auxin-deficient mutant, *bushy* (*bsh*), exhibits lower *PsCCD8* expression, reduced SL in root exudates, and decreased mycorrhizal colonization (Foo 2013). In wild-type plants, blocking endogenous auxin transport to roots by stem girdling also reduces SL levels in root exudates, and exogenous SL treatment restores colonization in the *bsh* mutant, suggesting that shoot-derived auxin acts through SL to promote AM colonization.

Shoot Development

It has been proposed that SL inhibits branching by acting like a second messenger for auxin: an auxin signal from the young expanding leaves in the primary shoot promotes SL production in the main stem and root, and then SL travels up into axillary meristems to influence branching (Fig. 3A; Brewer et al. 2009). Auxin-mediated apical dominance requires intact SL signaling in pea (Beveridge et al. 2000) and *Arabidopsis* (Sorefan et al. 2003), and SL treatment prevents decapitation-induced branching in pea (Brewer et al. 2009). Blocking PAT in the bud with NPA takes days to have an effect, whereas SL treatment rapidly inhibits bud outgrowth, suggesting that SL, and not the establishment of auxin export, regulates bud release (Brewer et al. 2009). Auxin has been shown to promote SL production during shoot branching by increasing expression of the genes encoding CCD7 and CCD8 in rice, pea, and *Arabidopsis* (Fig. 3C; Sorefan et al. 2003; Bainbridge et al. 2005; Foo et al. 2005; Johnson et al. 2006; Zou et al. 2006; Arite et al. 2007; Hayward et al. 2009). In *Arabidopsis*, auxin-triggered induction of SL biosynthesis is lost in the auxin-insensitive mutant *axr1* (Hayward et al. 2009), but synthetic SL can still inhibit *axr1* shoot branching (Brewer et al. 2009), supporting the idea that auxin acts through AXR1 to promote SL biosynthesis and, thus, inhibit shoot development.

SLs have been shown to alter PAT by decreasing the expression of *PIN* genes and/or the membrane abundance and localization of PIN proteins, leading to the inhibition of bud development in *Arabidopsis* (Fig. 3A,C; Bennett et al. 2006; Crawford et al. 2010; Shinohara et al. 2013). *Arabidopsis* SL mutants have increased auxin transport, up-regulated PIN protein expression, and enhanced branching (which is lost upon NPA treatment or when an SL mutant is crossed to the *pin1* mutant), indicating that SL inhibits branching via regulation of auxin transport (Bennett et al. 2006; Lazar and Goodman 2006). When the *pin3/4/7* triple mutant was crossed to highly branched SL mutants (*max2* and *max4*), the quadruple mutants had reduced branching and less auxin transport in the stem than the SL single mutants, bringing their phenotypes closer to wild-type (van Rongen et al. 2019). Likewise, the auxin transport mutant *abcb19*, when crossed to SL mutants, displayed reduced branching. These findings suggest that several auxin transporters interact with SL to regulate auxin transport and branching. Similar to *Arabidopsis*, rice SL mutants exhibit increased auxin levels (Arite et al. 2007). However, conflicting experiments in pea found that SL mutants do not have altered auxin transport and SL did not require PAT to inhibit branching (Brewer et al. 2009, 2015). This variation could be due to differences in species, the type of mutants analyzed, or experimental design.

Auxin–SL interactions are at play in several other plant species as well. In tall fescue roots and leaves, expression of the SL signaling component genes *FaD3* and *FaD14* is up-regulated by exogenous auxin, but down-regulated by NPA, providing evidence for auxin-mediated regulation of SL perception and signal transduction (Hu et al. 2018b, 2019). In peach, both auxin and SL inhibit axillary bud development through actions involving CK (Li et al. 2018). Auxin-treated plants show decreased expression of *PpIPTs* after decapitation, and SL reduces auxin transport by decreasing *PpPIN1* expression (Fig. 3C). In turn, decreased auxin transport inhibits CK biosynthesis and bud outgrowth. Likewise, in apple, SL treatment and NPA treatment have similar phenotypic effects, with both treatments decreasing the expression of *MsPIN1*, in contrast to CK treatment that enhances *MsPIN1* transcript levels (Tan et al. 2019).

Many of the seemingly irreconcilable results may be due to the effects of a third hormone critical to branching and apical dominance: CK. CK and SL are both regulated by auxin, and converge on the TF BRANCHED1 (BRC1) (Dun et al. 2012), known to play a critical role in the SL-mediated control of branching in multiple species (Brewer et al. 2009; Minakuchi et al. 2010; Braun et al. 2012; Dun et al. 2012). CK and SL oppose one another in regulating the expression of *PsBRC1* and branching (Fig. 3C; Dun et al. 2012), adding another layer of complexity to the interpretations of the auxin-SL relationship. The interactions between auxin and CK are discussed in more detail in the Auxin–CK section above.

Other Contexts

In *Arabidopsis*, low inorganic phosphate (Pi) conditions increase root hair density (Ma et al. 2001) and LR density, which coincides with up-regulation of the auxin receptor *TIR1* (Mayzlish-Gati et al. 2012). Induction of *TIR1* by low Pi is lost in a *max2* mutant, but IAA treatment rescues the mutant root hair phenotype. While this effect is specific to a certain developmental time window, it suggests that SL can regulate root hair formation in response to low Pi conditions by up-regulating *TIR1* and enhancing auxin signaling. Another study in *Arabidopsis* found that SL mediates the low Pi response by altering PIN2 polarity (Kumar et al. 2015), suggesting that SL regulates not only auxin perception, but also auxin transport. Similarly, low Pi or low N conditions in rice induce SL biosynthesis (Sun et al. 2014), and mutations in SL signaling and biosynthesis genes (*OsD3*; *OsD10* and *OsD27*, respectively) lead to reduced root responses to Pi or N deficiency. Low Pi, low N, or SL treatment inhibit IAA transport from shoot to root and reduce the activity of an auxin reporter construct, suggesting that auxin–SL interactions also mediate response to nutrient availability in rice.

In tall fescue under heat stress, auxin inhibits root elongation, but SL promotes it (Hu et al. 2018b). SL and/or NPA treatment result in similar root phenotypes and SL treatment inhibits the expression of auxin transport genes (*FaTIR1*, *FaPIN1*, *FaPIN2*, and *FaPIN5*) in root tips, especially under heat stress conditions. Nearly identical results were found in tall fescue leaves (Hu et al. 2019), suggesting that SL and auxin coordinate in a general heat stress response.

CONCLUDING REMARKS

The complexity of the hormone crosstalk and the fragmentary view that we have been able to generate up to this point highlight the critical need to balance the efforts of dissecting the function of individual hormones with investigating the molecular mechanisms of hormone interactions. Although studying auxin and other growth regulators individually has proven fruitful, a more holistic outlook on signal interaction is required to better understand the true complexity of the role of these hormones in plant growth and development. What has been holding plant biologists back from generating a more comprehensive view of the hormone interaction network and from uncovering the full complexity of the signal crosstalk? Perhaps, one of the existing limitations is the lack of adequate tools to monitor multiple hormones in parallel with a cellular resolution, which is a prerequisite for identifying and implicating additional players in the process of interest or explaining a pleiotropic phenotype of a mutant. To this end, the development of multihormone reporters or biosensors that enable simultaneous detection of several hormones is clearly a pressing need. The second major roadblock in studying hormone crosstalk is the time and effort required to generate desired multigene mutant combinations to resolve gene functional redundancy and to simultaneously deregulate several interacting pathways. Fortunately, higher-order mutant generation has now been significantly accelerated with the implementation of genome-editing technologies in plants, paving the way to rapidly make mutant combinations that were previously difficult, or even impossible, to generate via traditional crosses (e.g., due to linkage). We hope that the continuous optimization and wide adoption of the latest molecular genetic tools in plant biology, such as base editing, biosensors, and single cell sequencing, will make the study of hormone interactions less daunting, will attract new talent, and shed much-needed light on the web of signal crosstalk in plants.

ACKNOWLEDGMENTS

The work in the Alonso-Stepanova laboratory is supported by the National Science Foundation Grants 1444561 and 1940829 to J.M.A. and A.N.S., and 1750006 to A.N.S.

REFERENCES

Abeles FB, Morgan PW, Saltveit ME Jr. 1973. *Ethylene in plant biology*. Academic, New York.

Addicott FT, Lyon JL, Ohkuma K, Thiessen WE, Carns HR, Smith OE, Cornforth JW, Milborrow BV, Ryback G, Wareing PF. 1968. Abscisic acid: a new name for abscisin II (dormin). *Science* **159:** 1493. doi:10.1126/science.159.3822.1493

Adie BAT, Pérez-Pérez J, Pérez-Pérez MM, Godoy M, Sánchez-Serrano J, Schmelz EA, Solano R. 2007. ABA is an essential signal for plant resistance to pathogens affecting JA biosynthesis and the activation of defenses in *Arabidopsis*. *Plant Cell* **19:** 1665–1681. doi:10.1105/tpc.106.048041

Aida M, Beis D, Heidstra R, Willemsen V, Blilou I, Galinha C, Nussaume L, Noh YS, Amasino R, Scheres B. 2004. The PLETHORA genes mediate patterning of the *Arabidopsis* root stem cell niche. *Cell* **119:** 109–120. doi:10.1016/j.cell.2004.09.018

Akiyama K, Matsuzaki K, Hayashi H. 2005. Plant sesquiterpenes induce hyphal branching in arbuscular mycorrhizal fungi. *Nature* **435:** 824–827. doi:10.1038/nature03608

An F, Zhang X, Zhu Z, Ji Y, He W, Jiang Z, Li M, Guo H. 2012. Coordinated regulation of apical hook development by gibberellins and ethylene in etiolated *Arabidopsis* seedlings. *Cell Res* **22:** 915–927. doi:10.1038/cr.2012.29

An J, Althiab Almasaud R, Bouzayen M, Zouine M, Chervin C. 2020. Auxin and ethylene regulation of fruit set. *Plant Sci* **292:** 110381. doi:10.1016/j.plantsci.2019.110381

Arabidopsis Interactome Mapping Consortium. 2011. Evidence for network evolution in an *Arabidopsis* interactome map. *Science* **333:** 601–607. doi:10.1126/science.1203877

Archacki R, Buszewicz D, Sarnowski TJ, Sarnowska E, Rolicka AT, Tohge T, Fernie AR, Jikumaru Y, Kotlinski M, Iwanicka-Nowicka R, et al. 2013. BRAHMA ATPase of the SWI/SNF chromatin remodeling complex acts as a positive regulator of gibberellin-mediated responses in *Arabidopsis*. *PLoS ONE* **8:** e58588. doi:10.1371/journal.pone.0058588

Arite T, Iwata H, Ohshima K, Maekawa M, Nakajima M, Kojima M, Sakakibara H, Kyozuka J. 2007. DWARF10, an RMS1/MAX4/DAD1 ortholog, controls lateral bud outgrowth in rice. *Plant J* **51:** 1019–1029. doi:10.1111/j.1365-313X.2007.03210.x

Armengot L, Marquès-Bueno MM, Soria-Garcia A, Müller M, Munné-Bosch S, Martínez MC. 2014. Functional interplay between protein kinase CK2 and salicylic acid sustains *PIN* transcriptional expression and root development. *Plant J* **78:** 411–423. doi:10.1111/tpj.12481

Azizi P, Rafii MY, Maziah M, Abdullah SN, Hanafi MM, Latif MA, Rashid AA, Sahebi M. 2015. Understanding the shoot apical meristem regulation: a study of the phytohormones, auxin and cytokinin, in rice. *Mech Dev* **135:** 1–15. doi:10.1016/j.mod.2014.11.001

Bai M, Shang J, Oh E, Fan M, Bai Y, Zentella R, Sun T, Wang Z. 2012a. Brassinosteroid, gibberellin, and phytochrome impinge on a common transcription module in *Arabidopsis*. *Nat Cell Biol* **14:** 810–817. doi:10.1038/ncb2546

Bai M, Fan M, Oh E, Wang Z. 2012b. A triple helix-loop-helix/basic helix-loop-helix cascade controls cell elongation downstream of multiple hormonal and environmental signaling pathways in *Arabidopsis*. *Plant Cell* **24:** 4917–4929. doi:10.1105/tpc.112.105163

Bainbridge K, Sorefan K, Ward S, Leyser O. 2005. Hormonally controlled expression of the *Arabidopsis* MAX4 shoot branching regulatory gene. *Plant J* **44:** 569–580. doi:10.1111/j.1365-313X.2005.02548.x

Bao F, Shen J, Brady SR, Muday GK, Asami T, Yang Z. 2004. Brassinosteroids interact with auxin to promote lateral root development in *Arabidopsis*. *Plant Physiol* **134:** 1624–1631. doi:10.1104/pp.103.036897

Bassa C, Etemadi M, Combier JP, Bouzayen M, Audran-Delalande C. 2013. *Sl-IAA27* gene expression is induced during arbuscular mycorrhizal symbiosis in tomato and in *Medicago truncatula*. *Plant Signal Behav* **8:** e25637. doi:10.4161/psb.25637

Bencivenga S, Simonini S, Benková E, Colombo L. 2012. The transcription factors BEL1 and SPL are required for cytokinin and auxin signaling during ovule development in *Arabidopsis*. *Plant Cell* **24:** 2886–2897. doi:10.1105/tpc.112.100164

Bennett T, Sieberer T, Willett B, Booker J, Luschnig C, Leyser O. 2006. The *Arabidopsis* MAX pathway controls shoot branching by regulating auxin transport. *Curr Biol* **16:** 553–563. doi:10.1016/j.cub.2006.01.058

Bentsink L, Koornneef M. 2008. Seed dormancy and germination. *The Arabidopsis Book* **6:** e0119. doi:10.1199/tab.0119

Besnard F, Refahi Y, Morin V, Marteaux B, Brunoud G, Chambrier P, Rozier F, Mirabet V, Legrand J, Lainé S, et al. 2014. Cytokinin signalling inhibitory fields provide robustness to phyllotaxis. *Nature* **505:** 417–421. doi:10.1038/nature12791

Beveridge CA, Symons GM, Turnbull CG. 2000. Auxin inhibition of decapitation-induced branching is dependent on graft-transmissible signals regulated by genes *Rms1* and *Rms2*. *Plant Physiol* **123:** 689–698. doi:10.1104/pp.123.2.689

Bielach A, Hrtyan M, Tognetti VB. 2017. Plants under stress: involvement of auxin and cytokinin. *Int J Mol Sci* **18:** 1427. 10.3390/ijms18071427.

Binder BM. 2020. Ethylene signaling in plants. *J Biol Chem* **295:** 7710–7725. doi:10.1074/jbc.REV120.010854

Binder BM, Mortimore LA, Stepanova AN, Ecker JR, Bleecker AB. 2004. Short-term growth responses to ethylene in *Arabidopsis* seedlings are EIN3/EIL1 independent. *Plant Physiol* **136:** 2921–2927. doi:10.1104/pp.104.050393

Binenbaum J, Weinstain R, Shani E. 2018. Gibberellin localization and transport in plants. *Trends Plant Sci* **23:** 410–421. doi:10.1016/j.tplants.2018.02.005

Bishopp A, Help H, El-Showk S, Weijers D, Scheres B, Friml J, Benková E, Mähönen AP, Helariutta Y. 2011. A mutually inhibitory interaction between auxin and cytokinin specifies vascular pattern in roots. *Curr Biol* **21:** 917–926. doi:10.1016/j.cub.2011.04.017

Björklund S, Antti H, Uddestrand I, Moritz T, Sundberg B. 2007. Cross-talk between gibberellin and auxin in development of *Populus* wood: gibberellin stimulates polar auxin transport and has a common transcriptome with auxin. *Plant J* **52:** 499–511. doi:10.1111/j.1365-313X.2007.03250.x

Blázquez MA, Green R, Nilsson O, Sussman MR, Weigel D. 1998. Gibberellins promote flowering of *Arabidopsis* by activating the *LEAFY* promoter. *Plant Cell* **10:** 791–800. doi:10.1105/tpc.10.5.791

Blilou I, Xu J, Wildwater M, Willemsen V, Paponov I, Friml J, Heidstra R, Aida M, Palme K, Scheres B. 2005. The PIN auxin efflux facilitator network controls growth and patterning in *Arabidopsis* roots. *Nature* **433:** 39–44. doi:10.1038/nature03184

Boivin S, Fonouni-Farde C, Frugier F. 2016. How auxin and cytokinin phytohormones modulate root microbe interactions. *Front Plant Sci* **7:** 1240. doi:10.3389/fpls.2016.01240

Bonhomme F, Kurz B, Melzer S, Bernier G, Jacqmard A. 2000. Cytokinin and gibberellin activate *SaMADS A*, a gene apparently involved in regulation of the floral transition in *Sinapis alba*. *Plant J* **24:** 103–111. doi:10.1046/j.1365-313x.2000.00859.x

Braun N, de Saint Germain A, Pillot JP, Boutet-Mercey S, Dalmais M, Antoniadi I, Li X, Maia-Grondard A, Le Signor C, Bouteiller N, et al. 2012. The pea TCP transcription factor PsBRC1 acts downstream of strigolactones to control shoot branching. *Plant Physiol* **158:** 225–238. doi:10.1104/pp.111.182725

Brewer PB, Dun EA, Ferguson BJ, Rameau C, Beveridge CA. 2009. Strigolactone acts downstream of auxin to regulate bud outgrowth in pea and *Arabidopsis*. *Plant Physiol* **150:** 482–493. doi:10.1104/pp.108.134783

Brewer PB, Dun EA, Gui R, Mason MG, Beveridge CA. 2015. Strigolactone inhibition of branching independent of polar auxin transport. *Plant Physiol* **168:** 1820–1829. doi:10.1104/pp.15.00014

Bürger M, Chory J. 2020. The many models of strigolactone signaling. *Trends Plant Sci* **25:** 395–405. doi:10.1016/j.tplants.2019.12.009

Cai X, Xu P, Zhao P, Liu R, Yu L, Xiang C. 2014. *Arabidopsis* ERF109 mediates cross-talk between jasmonic acid and auxin biosynthesis during lateral root formation. *Nat Commun* **5:** 5833. doi:10.1038/ncomms6833

Cai X, Xu P, Wang Y, Xiang C. 2015. Activated expression of *AtEDT1/HDG11* promotes lateral root formation in *Arabidopsis* mutant *edt1* by upregulating jasmonate biosynthesis. *J Integr Plant Biol* **57:** 1017–1030. doi:10.1111/jipb.12347

Campos ML, Kang J, Howe GA. 2014. Jasmonate-triggered plant immunity. *J Chem Ecol* **40:** 657–675. doi:10.1007/s10886-014-0468-3

Campos ML, Yoshida Y, Major IT, de Oliveira Ferreira D, Weraduwage SM, Froehlich JE, Johnson BF, Kramer DM, Jander G, Sharkey TD, et al. 2016. Rewiring of jasmonate and phytochrome B signalling uncouples plant growth-defense tradeoffs. *Nat Commun* **7:** 12570. doi:10.1038/ncomms12570

Chaabouni S, Jones B, Delalande C, Wang H, Li Z, Mila I, Frasse P, Latché A, Pech J, Bouzayen M. 2009. Sl-IAA3, a tomato Aux/IAA at the crossroads of auxin and ethylene signalling involved in differential growth. *J Exp Bot* **60:** 1349–1362. doi:10.1093/jxb/erp009

Chaiwanon J, Wang ZY. 2015. Spatiotemporal brassinosteroid signaling and antagonism with auxin pattern stem cell dynamics in *Arabidopsis* roots. *Curr Biol* **25:** 1031–1042. doi:10.1016/j.cub.2015.02.046

Chandrasekaran U, Luo X, Zhou W, Shu K. 2020. Multifaceted signaling networks mediated by abscisic acid insensitive 4. *Plant Commun* **1:** 100040. doi:10.1016/j.xplc.2020.100040

Chang L, Ramireddy E, Schmülling T. 2013. Lateral root formation and growth of *Arabidopsis* is redundantly regulated by cytokinin metabolism and signalling genes. *J Exp Bot* **64:** 5021–5032. doi:10.1093/jxb/ert291

Chen Z, Agnew JL, Cohen JD, He P, Shan L, Sheen J, Kunkel BN. 2007. Pseudomonas syringae type III effector AvrRpt2 alters *Arabidopsis thaliana* auxin physiology. *Proc Natl Acad Sci* **104:** 20131–20136. doi:10.1073/pnas.0704901104

Chen Q, Sun J, Zhai Q, Zhou W, Qi L, Xu L, Wang B, Chen R, Jiang H, Qi J, et al. 2011. The basic helix-loop-helix transcription factor MYC2 directly represses *PLETHORA* expression during jasmonate-mediated modulation of the root stem cell niche in *Arabidopsis*. *Plant Cell* **23:** 3335–3352. doi:10.1105/tpc.111.089870

Chen JJ, Wang LY, Immanen J, Nieminen K, Spicer R, Helariutta Y, Zhang J, He XQ. 2019. Differential regulation of auxin and cytokinin during the secondary vascular tissue regeneration in *Populus* trees. *New Phytol* **224:** 188–201. doi:10.1111/nph.16019

Cheng Y, Dai X, Zhao Y. 2006. Auxin biosynthesis by the YUCCA flavin monooxygenases controls the formation of floral organs and vascular tissues in *Arabidopsis*. *Genes Dev* **20:** 1790–1799. doi:10.1101/gad.1415106

Cheng ZJ, Wang L, Sun W, Zhang Y, Zhou C, Su YH, Li W, Sun TT, Zhao XY, Li XG, et al. 2013. Pattern of auxin and cytokinin responses for shoot meristem induction results from the regulation of cytokinin biosynthesis by AUXIN RESPONSE FACTOR3. *Plant Physiol* **161:** 240–251. doi:10.1104/pp.112.203166

Chini A, Monte I, Zamarreño AM, Hamberg M, Lassueur S, Reymond P, Weiss S, Stintzi A, Schaller A, Porzel A, et al. 2018. An OPR3-independent pathway uses 4,5-didehydrojasmonate for jasmonate synthesis. *Nat Chem Biol* **14:** 171–178. doi:10.1038/nchembio.2540

Cho H, Ryu H, Rho S, Hill K, Smith S, Audenaert D, Park J, Han S, Beeckman T, Bennett MJ, et al. 2014. A secreted peptide acts on BIN2-mediated phosphorylation of ARFs to potentiate auxin response during lateral root development. *Nat Cell Biol* **16:** 66–76. doi:10.1038/ncb2893

Chung Y, Choe S. 2013. The regulation of brassinosteroid biosynthesis in *Arabidopsis*. *Crit Rev Plant Sci* **32:** 396–410. doi:10.1080/07352689.2013.797856

Chung Y, Maharjan PM, Lee O, Fujioka S, Jang S, Kim B, Takatsuto S, Tsujimoto M, Kim H, Cho S, et al. 2011. Auxin stimulates DWARF4 expression and brassinosteroid biosynthesis in *Arabidopsis*. *Plant J* **66:** 564–578. doi:10.1111/j.1365-313X.2011.04513.x

Clouse SD, Sasse JM. 1998. BRASSINOSTEROIDS: essential regulators of plant growth and development. *Annu Rev Plant Physiol Plant Mol Biol* **49:** 427–451. doi:10.1146/annurev.arplant.49.1.427

Clouse SD, Langford M, McMorris TC. 1996. A brassinosteroid-insensitive mutant in *Arabidopsis thaliana* exhibits multiple defects in growth and development. *Plant Physiol* **111:** 671–678. doi:10.1104/pp.111.3.671

Cook CE, Whichard LP, Turner B, Wall ME, Egley GH. 1966. Germination of Witchweed (*Striga lutea* Lour.): isolation and properties of a potent stimulant. *Science* **154:** 1189–1190. doi:10.1126/science.154.3753.1189

Crawford S, Shinohara N, Sieberer T, Williamson L, George G, Hepworth J, Muller D, Domagalska MA, Leyser O. 2010. Strigolactones enhance competition between shoot branches by dampening auxin transport. *Development* **137:** 2905–2913. doi:10.1242/dev.051987

Curaba J, Singh MB, Bhalla PL. 2014. miRNAs in the cross-talk between phytohormone signalling pathways. *J Exp Bot* **65:** 1425–1438. doi:10.1093/jxb/eru002

de Jong M, Wolters-Arts M, Feron R, Mariani C, Vriezen WH. 2009. The *Solanum lycopersicum* auxin response factor 7 (*Sl* ARF7) regulates auxin signaling during tomato fruit set and development. *Plant J* **57:** 160–170. doi:10.1111/j.1365-313X.2008.03671.x

Dello Ioio R, Linhares FS, Scacchi E, Casamitjana-Martinez E, Heidstra R, Costantino P, Sabatini S. 2007. Cytokinins determine *Arabidopsis* root-meristem size by controlling cell differentiation. *Curr Biol* **17:** 678–682. doi:10.1016/j.cub.2007.02.047

Dello Ioio R, Nakamura K, Moubayidin L, Perilli S, Taniguchi M, Morita MT, Aoyama T, Costantino P, Sabatini S. 2008. A genetic framework for the control of cell division and differentiation in the root meristem. *Science* **322:** 1380–1384. doi:10.1126/science.1164147

de Lucas M, Davière J, Rodríguez-Falcón M, Pontin M, Iglesias-Pedraz JM, Lorrain S, Fankhauser C, Blázquez MA, Titarenko E, Prat S. 2008. A molecular framework for light and gibberellin control of cell elongation. *Nature (London)* **451:** 480–484. doi:10.1038/nature06520

Demole E, Lederer E, Mercier D. 1962. Isolement et détermination de la structure du jasmonate de méthyle, constituant odorant caractéristique de l'essence de jasmin [Isolation and determination of the structure of methyl jasmonate, a characteristic fragrant constituent of jasmine essence]. *Helv Chim Acta* **45:** 675–685. doi:10.1002/hlca.19620450233

Dempsey DA, Klessig DF. 2017. How does the multifaceted plant hormone salicylic acid combat disease in plants and are similar mechanisms utilized in humans? *BMC Biol* **15:** 23. doi:10.1186/s12915-017-0364-8

De Rybel B, Adibi M, Breda AS, Wendrich JR, Smit ME, Novák O, Yamaguchi N, Yoshida S, Van Isterdael G, Palovaara J, et al. 2014. Plant development. Integration of growth and patterning during vascular tissue formation in *Arabidopsis*. *Science* **345:** 1255215. doi:10.1126/science.1255215

Dharmasiri N, Dharmasiri S, Weijers D, Karunarathna N, Jurgens G, Estelle M. 2007. AXL and AXR1 have redundant functions in RUB conjugation and growth and development in *Arabidopsis*. *Plant J* **52:** 114–123. doi:10.1111/j.1365-313X.2007.03211.x

Dierck R, De Keyser E, De Riek J, Dhooghe E, Van Huylenbroeck J, Prinsen E, Van Der Straeten D. 2016. Change in auxin and cytokinin levels coincides with altered expression of branching genes during axillary bud outgrowth in chrysanthemum. *PLoS ONE* **11:** e0161732. doi:10.1371/journal.pone.0161732

Ding ZJ, Yan JY, Li CX, Li GX, Wu YR, Zheng SJ. 2015. Transcription factor WRKY46 modulates the development of *Arabidopsis* lateral roots in osmotic/salt stress conditions via regulation of ABA signaling and auxin homeostasis. *Plant J* **84:** 56–69. doi:10.1111/tpj.12958

Dorcey E, Urbez C, Blázquez MA, Carbonell J, Perez-Amador M. 2009. Fertilization-dependent auxin response in ovules triggers fruit development through the modulation of gibberellin metabolism in *Arabidopsis*. *Plant J* **58:** 318–332. doi:10.1111/j.1365-313X.2008.03781.x

Du Y, Tejos R, Beck M, Himschoot E, Li H, Robatzek S, Vanneste S, Friml J. 2013. Salicylic acid interferes with clathrin-mediated endocytic protein trafficking. *Proc Natl Acad Sci* **110:** 7946–7951. doi:10.1073/pnas.1220205110

Dun EA, de Saint Germain A, Rameau C, Beveridge CA. 2012. Antagonistic action of strigolactone and cytokinin in bud outgrowth control. *Plant Physiol* **158:** 487–498. doi:10.1104/pp.111.186783

Emenecker RJ, Strader LC. 2020. Auxin-abscisic acid interactions in plant growth and development. *Biomolecules* **10:** 281. doi:10.3390/biom10020281

Enders TA, Strader LC. 2015. Auxin activity: past, present, and future. *Am J Bot* **102:** 180–196. doi:10.3732/ajb.1400285

Feng S, Martinez C, Gusmaroli G, Wang Y, Zhou J, Wang F, Chen L, Yu L, Iglesias-Pedraz JM, Kircher S, et al. 2008. Coordinated regulation of *Arabidopsis thaliana* development by light and gibberellins. *Nature* **451:** 475–479. doi:10.1038/nature06448

Finkelstein RR, Lynch TJ. 2000. The *Arabidopsis* abscisic acid response gene *ABI5* encodes a basic leucine zipper transcription factor. *Plant Cell* **12:** 599–609. doi:10.1105/tpc.12.4.599

Fonseca S, Chico JM, Solano R. 2009. The jasmonate pathway: the ligand, the receptor and the core signalling module. *Curr Opin Plant Biol* **12:** 539–547. doi:10.1016/j.pbi.2009.07.013

Foo E. 2013. Auxin influences strigolactones in pea mycorrhizal symbiosis. *J Plant Physiol* **170:** 523–528. doi:10.1016/j.jplph.2012.11.002

Foo E, Bullier E, Goussot M, Foucher F, Rameau C, Beveridge CA. 2005. The branching gene *RAMOSUS1* mediates interactions among two novel signals and auxin in pea. *Plant Cell* **17:** 464–474. doi:10.1105/tpc.104.026716

Fraser PD, Truesdale MR, Bird CR, Schuch W, Bramley PM. 1994. Carotenoid biosynthesis during tomato fruit development (evidence for tissue-specific gene expression). *Plant Physiol* **105:** 405–413. doi:10.1104/pp.105.1.405

Frigerio M, Alabadí D, Pérez-Gómez J, García-Cárcel L, Phillips AL, Hedden P, Blázquez MA. 2006. Transcriptional regulation of gibberellin metabolism genes by auxin signaling in *Arabidopsis*. *Plant Physiol* **142:** 553–563. doi:10.1104/pp.106.084871

Fu ZQ, Yan S, Saleh A, Wang W, Ruble J, Oka N, Mohan R, Spoel SH, Tada Y, Zheng N, et al. 2012. NPR3 and NPR4 are receptors for the immune signal salicylic acid in plants. *Nature* **486:** 228–232. doi:10.1038/nature11162

Fukaki H, Nakao Y, Okushima Y, Theologis A, Tasaka M. 2005. Tissue-specific expression of stabilized SOLITARY-ROOT/IAA14 alters lateral root development in *Arabidopsis*. *Plant J* **44:** 382–395. doi:10.1111/j.1365-313X.2005.02537.x

Galinha C, Hofhuis H, Luijten M, Willemsen V, Blilou I, Heidstra R, Scheres B. 2007. PLETHORA proteins as dose-dependent master regulators of *Arabidopsis* root development. *Nature* **449:** 1053–1057. doi:10.1038/nature06206

Gallei M, Luschnig C, Friml J. 2020. Auxin signalling in growth: Schrödinger's cat out of the bag. *Curr Opin Plant Biol* **53:** 43–49. doi:10.1016/j.pbi.2019.10.003

Giovanelli J, Mudd SH, Datko AH. 1985. Quantitative analysis of pathways of methionine metabolism and their regulation in *Lemna*. *Plant Physiol* **78:** 555–560. doi:10.1104/pp.78.3.555

Giulini A, Wang J, Jackson D. 2004. Control of phyllotaxy by the cytokinin-inducible response regulator homologue ABPHYL1. *Nature* **430:** 1031–1034. doi:10.1038/nature02778

Goda H, Shimada Y, Asami T, Fujioka S, Yoshida S. 2002. Microarray analysis of brassinosteroid-regulated genes in *Arabidopsis*. *Plant Physiol* **130:** 1319–1334. doi:10.1104/pp.011254

Goda H, Sawa S, Asami T, Fujioka S, Shimada Y, Yoshida S. 2004. Comprehensive comparison of auxin-regulated and brassinosteroid-regulated genes in *Arabidopsis*. *Plant Physiol* **134:** 1555–1573. doi:10.1104/pp.103.034736

Goetz M, Hooper LC, Johnson SD, Rodrigues JCM, Vivian-Smith A, Koltunow AM. 2007. Expression of aberrant forms of *AUXIN RESPONSE FACTOR8* stimulates parthenocarpy in *Arabidopsis* and tomato. *Plant Physiol (Bethesda)* **145:** 351–366. doi:10.1104/pp.107.104174

Goh T, Kasahara H, Mimura T, Kamiya Y, Fukaki H. 2012. Multiple AUX/IAA-ARF modules regulate lateral root formation: the role of *Arabidopsis* SHY2/IAA3-mediated auxin signalling. *Philos Trans R Soc Lond B Biol Sci* **367:** 1461–1468. doi:10.1098/rstb.2011.0232

Gomez-Roldan V, Fermas S, Brewer PB, Puech-Pagès V, Dun EA, Pillot JP, Letisse F, Matusova R, Danoun S, Portais JC, et al. 2008. Strigolactone inhibition of shoot branching. *Nature* **455:** 189–194. doi:10.1038/nature07271

Gonzalez-Garcia MP, Vilarrasa-Blasi J, Zhiponova M, Divol F, Mora-Garcia S, Russinova E, Cano-Delgado AI. 2011. Brassinosteroids control meristem size by promoting cell cycle progression in *Arabidopsis* roots. *Development* **138:** 849–859. doi:10.1242/dev.057331

Gorguet B, Heusden AW, Lindhout P. 2005. Parthenocarpic fruit development in tomato. *Plant Biol (Stuttgart, Germany)* **7:** 131–139. doi:10.1055/s-2005-837494

Guillotin B, Etemadi M, Audran C, Bouzayen M, Bécard G, Combier JP. 2017. *Sl-IAA27* regulates strigolactone biosynthesis and mycorrhization in tomato (var. *MicroTom*). *New Phytol* **213:** 1124–1132. doi:10.1111/nph.14246

Gupta A, Singh M, Laxmi A. 2015. Interaction between glucose and brassinosteroid during the regulation of lateral root development in *Arabidopsis*. *Plant Physiol* **168:** 307–320. doi:10.1104/pp.114.256313

Gutierrez L, Mongelard G, Floková K, Păcurar DI, Novák O, Staswick P, Kowalczyk M, Păcurar M, Demailly H, Geiss G, et al. 2012. Auxin controls *Arabidopsis* adventitious root initiation by regulating jasmonic acid homeostasis. *Plant Cell* **24:** 2515–2527. doi:10.1105/tpc.112.099119

Hacham Y, Holland N, Butterfield C, Ubeda-Tomas S, Bennett MJ, Chory J, Savaldi-Goldstein S. 2011. Brassinosteroid perception in the epidermis controls root meristem size. *Development* **138:** 839–848. doi:10.1242/dev.061804

Hacham Y, Sela A, Friedlander L, Savaldi-Goldstein S. 2012. BRI1 activity in the root meristem involves post-transcriptional regulation of PIN auxin efflux carriers. *Plant Signal Behav* **7:** 68–70. doi:10.4161/psb.7.1.18657

Hayward A, Stirnberg P, Beveridge C, Leyser O. 2009. Interactions between auxin and strigolactone in shoot branching control. *Plant Physiol* **151:** 400–412. doi:10.1104/pp.109.137646

He W, Brumos J, Li H, Ji Y, Ke M, Gong X, Zeng Q, Li W, Zhang X, An F, et al. 2011. A small-molecule screen identifies L-kynurenine as a competitive inhibitor of TAA1/TAR activity in ethylene-directed auxin biosynthesis and root growth in *Arabidopsis*. *Plant Cell* **23:** 3944–3960. doi:10.1105/tpc.111.089029

Hernández-García J, Briones-Moreno A, Blázquez MA. 2020. Origin and evolution of gibberellin signaling and metabolism in plants. *Semin Cell Dev Biol* **109:** 46–54. doi:10.1016/j.semcdb.2020.04.009

Hossain MA, Cho J, Han M, Ahn C, Jeon J, An G, Park PB. 2010. The ABRE-binding bZIP transcription factor OsABF2 is a positive regulator of abiotic stress and ABA signaling in rice. *J Plant Physiol* **167:** 1512–1520. doi:10.1016/j.jplph.2010.05.008

Hu J, Israeli A, Ori N, Sun T. 2018a. The interaction between DELLA and ARF/IAA mediates crosstalk between gibberellin and auxin signaling to control fruit initiation in tomato. *Plant Cell* **30:** 1710–1728. doi:10.1105/tpc.18.00363

Hu Q, Zhang S, Huang B. 2018b. Strigolactones and interaction with auxin regulating root elongation in tall fescue under different temperature regimes. *Plant Sci* **271:** 34–39. doi:10.1016/j.plantsci.2018.03.008

Hu Q, Zhang S, Huang B. 2019. Strigolactones promote leaf elongation in tall fescue through upregulation of cell cycle genes and downregulation of auxin transport genes in tall fescue under different temperature regimes. *Int J Mol Sci* **20:** 1836. 10.3390/ijms20081836.

Huang P, Dong Z, Guo P, Zhang X, Qiu Y, Li B, Wang Y, Guo H. 2020. Salicylic acid suppresses apical hook formation via NPR1-mediated repression of EIN3 and EIL1 in *Arabidopsis*. *Plant Cell* **32:** 612–629. doi:10.1105/tpc.19.00658

Huh SU, Lee SB, Kim HH, Paek KH. 2012. ATAF2, a NAC transcription factor, binds to the promoter and regulates NIT2 gene expression involved in auxin biosynthesis. *Mol Cells* **34:** 305–313. doi:10.1007/s10059-012-0122-2

Huq E, Quail PH. 2002. PIF4, a phytochrome-interacting bHLH factor, functions as a negative regulator of phytochrome B signaling in *Arabidopsis*. *EMBO J* **21:** 2441–2450. doi:10.1093/emboj/21.10.2441

Huq E, Al-Sady B, Hudson M, Kim C, Apel K, Quail PH. 2004. Phytochrome-interacting factor 1 is a critical bHLH regulator of chlorophyll biosynthesis. *Science* **305:** 1937–1941. doi:10.1126/science.1099728

Hurný A, Cuesta C, Cavallari N, Ötvös K, Duclercq J, Dokládal L, Montesinos JC, Gallemí M, Semerádová H, Rauter T, et al. 2020. SYNERGISTIC ON AUXIN AND CYTOKININ 1 positively regulates growth and attenuates soil pathogen resistance. *Nat Commun* **11:** 2170–2175. doi:10.1038/s41467-020-15895-5

Hussain S, Kim SH, Bahk S, Ali A, Nguyen XC, Yun D, Chung WS. 2020. The auxin signaling repressor IAA8 promotes seed germination through down-regulation of ABI3 transcription in *Arabidopsis*. *Front Plant Sci* **11:** 111. doi:10.3389/fpls.2020.00111

Hutchison CE, Li J, Argueso C, Gonzalez M, Lee E, Lewis MW, Maxwell BB, Perdue TD, Schaller GE, Alonso JM, et al. 2006. The *Arabidopsis* histidine phosphotransfer proteins are redundant positive regulators of cytokinin signaling. *Plant Cell* **18:** 3073–3087. doi:10.1105/tpc.106.045674

Ibanes M, Fabregas N, Chory J, Cano-Delgado AI. 2009. Brassinosteroid signaling and auxin transport are required to establish the periodic pattern of *Arabidopsis* shoot vascular bundles. *Proc Natl Acad Sci* **106:** 13630–13635. doi:10.1073/pnas.0906416106

Immanen J, Nieminen K, Smolander OP, Kojima M, Alonso Serra J, Koskinen P, Zhang J, Elo A, Mähönen AP, Street N, et al. 2016. Cytokinin and auxin display distinct but interconnected distribution and signaling profiles to stimulate cambial activity. *Curr Biol* **26:** 1990–1997. doi:10.1016/j.cub.2016.05.053

Jiang Y, Liang G, Yang S, Yu D. 2014. *Arabidopsis* WRKY57 functions as a node of convergence for jasmonic acid- and auxin-mediated signaling in jasmonic acid-induced leaf senescence. *Plant Cell* **26:** 230–245. doi:10.1105/tpc.113.117838

Jing H, Strader LC. 2019. Interplay of auxin and cytokinin in lateral root development. *Int J Mol Sci* **20:** 486. doi:10.3390/ijms20030486

Johnson X, Brcich T, Dun EA, Goussot M, Haurogné K, Beveridge CA, Rameau C. 2006. Branching genes are conserved across species. Genes controlling a novel signal in pea are coregulated by other long-distance signals. *Plant Physiol* **142:** 1014–1026. doi:10.1104/pp.106.087676

Jones B, Gunnerås SA, Petersson SV, Tarkowski P, Graham N, May S, Dolezal K, Sandberg G, Ljung K. 2010. Cytokinin regulation of auxin synthesis in *Arabidopsis* involves a homeostatic feedback loop regulated via auxin and cytokinin signal transduction. *Plant Cell* **22:** 2956–2969. doi:10.1105/tpc.110.074856

Ju C, Chang C. 2015. Mechanistic insights in ethylene perception and signal transduction. *Plant Physiol* **169:** 85–95. doi:10.1104/pp.15.00845

Kalousek P, Buchtová D, Balla J, Reinöhl V, Procházka S. 2010. Cytokinins and polar transport of auxin in axillary pea buds. *Plant J* **65:** 571–577.

Kant S, Bi Y, Zhu T, Rothstein SJ. 2009. *SAUR39*, a small auxin-up RNA gene, acts as a negative regulator of auxin synthesis and transport in rice. *Plant Physiol (Bethesda)* **151:** 691–701. doi:10.1104/pp.109.143875

Kapulnik Y, Resnick N, Mayzlish-Gati E, Kaplan Y, Wininger S, Hershenhorn J, Koltai H. 2011. Strigolactones interact with ethylene and auxin in regulating root-hair elongation in *Arabidopsis*. *J Exp Bot* **62:** 2915–2924. doi:10.1093/jxb/erq464

Keuskamp DH, Sasidharan R, Vos I, Peeters AJ, Voesenek LA, Pierik R. 2011. Blue-light-mediated shade avoidance requires combined auxin and brassinosteroid action in *Arabidopsis* seedlings. *Plant J* **67:** 208–217. doi:10.1111/j.1365-313X.2011.04597.x

Kieber JJ, Schaller GE. 2018. Cytokinin signaling in plant development. *Development* **145:** dev149344. 10.1242/dev.149344

Kim SK, Chang SC, Lee EJ, Chung WS, Kim YS, Hwang S, Lee JS. 2000. Involvement of brassinosteroids in the gravitropic response of primary root of maize. *Plant Physiol* **123:** 997–1004. doi:10.1104/pp.123.3.997

Kim J, Yi H, Choi G, Shin B, Song P, Choi G. 2003. Functional characterization of phytochrome interacting factor 3 in phytochrome-mediated light signal transduction. *Plant Cell* **15:** 2399–2407. doi:10.1105/tpc.014498

Kobayashi A, Takahashi A, Kakimoto Y, Miyazawa Y, Fujii N, Higashitani A, Takahashi H. 2007. A gene essential for hydrotropism in roots. *Proc Natl Acad Sci* **104:** 4724–4729. doi:10.1073/pnas.0609929104

Kohlen W, Ng JLP, Deinum EE, Mathesius U. 2018. Auxin transport, metabolism, and signalling during nodule initiation: indeterminate and determinate nodules. *J Exp Bot* **69:** 229–244. doi:10.1093/jxb/erx308

Koltai H, Dor E, Hershenhorn J, Joel DM, Weininger S, Lekalla S, Shealtiel H, Bhattacharya C, Eliahu E, Resnick N, et al. 2010. Strigolactones' effect on root growth and root-hair elongation may be mediated by auxin-efflux carriers. *J Plant Growth Regul* **29:** 129–136. doi:10.1007/s00344-009-9122-7

Koornneef M, Reuling G, Karssen CM. 1984. The isolation and characterization of abscisic acid-insensitive mutants of *Arabidopsis thaliana*. *Physiol Plantarum* **61:** 377–383. doi:10.1111/j.1399-3054.1984.tb06343.x

Koren D, Resnick N, Mayzlish Gati E, Belausov E, Weininger S, Kapulnik Y, Koltai H. 2013. Strigolactone signaling in the endodermis is sufficient to restore root responses and involves SHORT HYPOCOTYL 2 (SHY2) activity. *New Phytol* **198:** 866–874. doi:10.1111/nph.12189

Kotov AA, Kotova LM. 2018. Auxin-cytokinin interactions in the regulation of correlative inhibition in two-branched pea seedlings. *J Exp Bot* **69:** 2967–2978. doi:10.1093/jxb/ery117

Kumar M, Pandya-Kumar N, Dam A, Haor H, Mayzlish-Gati E, Belausov E, Wininger S, Abu-Abied M, McErlean CS, Bromhead LJ, et al. 2015. *Arabidopsis* response to low-phosphate conditions includes active changes in actin filaments and PIN2 polarization and is dependent on strigolactone signalling. *J Exp Bot* **66:** 1499–1510. doi:10.1093/jxb/eru513

Kurepa J, Shull TE, Smalle JA. 2019. Antagonistic activity of auxin and cytokinin in shoot and root organs. *Plant Direct* **3:** e00121. doi:10.1002/pld3.121

Lakehal A, Dob A, Rahneshan Z, Novak O, Escamez S, Alallaq S, Strnad M, Tuominen H, Bellini C. 2020. ETHYLENE RESPONSE FACTOR 115 integrates jasmonate and cytokinin signaling machineries to repress adventitious rooting in *Arabidopsis*. *New Phytol* **228:** 1611–1626. doi:10.1111/nph.16794

Laplaze L, Benkova E, Casimiro I, Maes L, Vanneste S, Swarup R, Weijers D, Calvo V, Parizot B, Herrera-Rodriguez M, et al. 2007. Cytokinins act directly on lateral root founder cells to inhibit root initiation. *Plant Cell* **19:** 3889–3900. doi:10.1105/tpc.107.055863

Lavenus J, Goh T, Roberts I, Guyomarc'h S, Lucas M, De Smet I, Fukaki H, Beeckman T, Bennett M, Laplaze L. 2013. Lateral root development in *Arabidopsis*: fifty shades of auxin. *Trends Plant Sci* **18:** 450–458. doi:10.1016/j.tplants.2013.04.006

Lavy M, Estelle M. 2016. Mechanisms of auxin signaling. *Development* **143:** 3226–3229. doi:10.1242/dev.131870

Lazar G, Goodman HM. 2006. MAX1, a regulator of the flavonoid pathway, controls vegetative axillary bud outgrowth in *Arabidopsis*. *Proc Natl Acad Sci* **103:** 472–476. doi:10.1073/pnas.0509463102

Lee BH, Johnston R, Yang Y, Gallavotti A, Kojima M, Travençolo BA, Costa Lda F, Sakakibara H, Jackson D. 2009. Studies of *aberrant phyllotaxy1* mutants of maize indicate complex interactions between auxin and cytokinin signaling in the shoot apical meristem. *Plant Physiol* **150:** 205–216. doi:10.1104/pp.109.137034

Lefevere H, Bauters L, Gheysen G. 2020. Salicylic acid biosynthesis in plants. *Front Plant Sci* doi:10.3389/fpls.2020.00338

Lehman A, Black R, Ecker JR. 1996. HOOKLESS1, an ethylene response gene, is required for differential cell elongation in the *Arabidopsis* hypocotyl. *Cell* **85:** 183–194. doi:10.1016/S0092-8674(00)81095-8

Lewis DR, Negi S, Sukumar P, Muday GK. 2011. Ethylene inhibits lateral root development, increases IAA transport and expression of PIN3 and PIN7 auxin efflux carriers. *Development* **138:** 3485–3495. doi:10.1242/dev.065102

Li J, Nagpal P, Vitart V, McMorris TC, Chory J. 1996. A role for brassinosteroids in light-dependent development of *Arabidopsis*. *Science* **272:** 398–401. doi:10.1126/science.272.5260.398

Li H, Johnson P, Stepanova A, Alonso JM, Ecker JR. 2004. Convergence of signaling pathways in the control of differential cell growth in *Arabidopsis*. *Dev Cell* **7:** 193–204. doi:10.1016/j.devcel.2004.07.002

Li L, Xu J, Xu ZH, Xue HW. 2005. Brassinosteroids stimulate plant tropisms through modulation of polar auxin transport in *Brassica* and *Arabidopsis*. *Plant Cell* **17:** 2738–2753. doi:10.1105/tpc.105.034397

Li M, Wei Q, Xiao Y, Peng F. 2018. The effect of auxin and strigolactone on ATP/ADP isopentenyltransferase expression and the regulation of apical dominance in peach. *Plant Cell Rep* **37:** 1693–1705. doi:10.1007/s00299-018-2343-0

Ligerot Y, de Saint Germain A, Waldie T, Troadec C, Citerne S, Kadakia N, Pillot JP, Prigge M, Aubert G, Bendahmane A, et al. 2017. The pea branching RMS2 gene encodes the PsAFB4/5 auxin receptor and is involved in an auxin-strigolactone regulation loop. *PLoS Genet* **13:** e1007089. doi:10.1371/journal.pgen.1007089

Liu M, Pirrello J, Kesari R, Mila I, Roustan J, Li Z, Latché A, Pech J, Bouzayen M, Regad F. 2013a. A dominant repressor version of the tomato *Sl-ERF.B3* gene confers ethylene hypersensitivity via feedback regulation of ethylene signaling and response components. *Plant J* **76:** 406–419. doi:10.1111/tpj.12305

Liu X, Zhang H, Zhao Y, Feng Z, Li Q, Yang H, Luan S, Li J, He Z. 2013b. Auxin controls seed dormancy through stimulation of abscisic acid signaling by inducing ARF-mediated *ABI3* activation in *Arabidopsis*. *Proc Natl Acad Sci* **110:** 15485–15490. doi:10.1073/pnas.1304651110

Liu M, Chen Y, Chen Y, Shin J, Mila I, Audran C, Zouine M, Pirrello J, Bouzayen M. 2018. The tomato Ethylene Response Factor Sl-ERF.B3 integrates ethylene and auxin signaling via direct regulation of *Sl-Aux/IAA27*. *New Phytol* **219:** 631–640. doi:10.1111/nph.15165

Lv M, Li J. 2020. Molecular mechanisms of brassinosteroid-mediated responses to changing environments in *Arabidopsis*. *Int J Mol Sci* **21:** 2737. 10.3390/ijms21082737

Lv X, Zhang M, Li X, Ye R, Wang X. 2018. Transcriptome profiles reveal the crucial roles of auxin and cytokinin in the "shoot branching" of *Cremastra appendiculata*. *Int J Mol Sci* **19:** 3354. 10.3390/ijms19113354

Ma Z, Bielenberg DG, Brown KM, Lynch JP. 2001. Regulation of root hair density by phosphorus availability in *Arabidopsis thaliana*. *Plant Cell Environ* **24:** 459–467. doi:10.1046/j.1365-3040.2001.00695.x

Ma N, Wan L, Zhao W, Liu H, Li J, Zhang C. 2020. Exogenous strigolactones promote lateral root growth by reducing the endogenous auxin level in rapeseed. *J Integrative Agriculture* **19:** 465–482. doi:10.1016/S2095-3119(19)62810-8

Machado RAR, McClure M, Hervé MR, Baldwin IT, Erb M. 2016. Benefits of jasmonate-dependent defenses against vertebrate herbivores in nature. *eLife* **5:** e13720.

Machado RAR, Baldwin IT, Erb M. 2017. Herbivory-induced jasmonates constrain plant sugar accumulation and growth by antagonizing gibberellin signaling and not by promoting secondary metabolite production. *New Phytol* **215:** 803–812. doi:10.1111/nph.14597

Mähönen AP, Ten Tusscher K, Siligato R, Smetana O, Díaz-Triviño S, Salojärvi J, Wachsman G, Prasad K, Heidstra R, Scheres B. 2014. PLETHORA gradient formation mechanism separates auxin responses. *Nature* **515:** 125–129. doi:10.1038/nature13663

Mandava NB. 1988. Plant growth-promoting brassinosteroids. *Annu Rev Plant Physiol Plant Mol Biol* **39:** 23–52. doi:10.1146/annurev.pp.39.060188.000323

Marhava P, Bassukas AEL, Zourelidou M, Kolb M, Moret B, Fastner A, Schulze WX, Cattaneo P, Hammes UZ, Schwechheimer C, et al. 2018. A molecular rheostat adjusts auxin flux to promote root protophloem differentiation. *Nature* **558:** 297–300. doi:10.1038/s41586-018-0186-z

Marhava P, Aliaga Fandino AC, Koh SWH, Jelínková A, Kolb M, Janacek DP, Breda AS, Cattaneo P, Hammes UZ, Petrášek J, et al. 2020. Plasma membrane domain patterning and self-reinforcing polarity in *Arabidopsis*. *Dev Cell* **52:** 223–235.e5. doi:10.1016/j.devcel.2019.11.015

Marhavý P, Bielach A, Abas L, Abuzeineh A, Duclercq J, Tanaka H, Pařezová M, Petrášek J, Friml J, Kleine-Vehn J, et al. 2011. Cytokinin modulates endocytic trafficking of PIN1 auxin efflux carrier to control plant organogenesis. *Dev Cell* **21:** 796–804. doi:10.1016/j.devcel.2011.08.014

Marhavý P, Duclercq J, Weller B, Feraru E, Bielach A, Offringa R, Friml J, Schwechheimer C, Murphy A, Benková E. 2014. Cytokinin controls polarity of PIN1-dependent auxin transport during lateral root organogenesis. *Curr Biol* **24:** 1031–1037. doi:10.1016/j.cub.2014.04.002

Mashiguchi K, Tanaka K, Sakai T, Sugawara S, Kawaide H, Natsume M, Hanada A, Yaeno T, Shirasu K, Yao H, et al. 2011. The main auxin biosynthesis pathway in *Arabidopsis*. *Proc Natl Acad Sci* **108:** 18512–18517. doi:10.1073/pnas.1108434108

Mayzlish-Gati E, De-Cuyper C, Goormachtig S, Beeckman T, Vuylsteke M, Brewer PB, Beveridge CA, Yermiyahu U, Kaplan Y, Enzer Y, et al. 2012. Strigolactones are involved in root response to low phosphate conditions in *Arabidopsis*. *Plant Physiol* **160:** 1329–1341. doi:10.1104/pp.112.202358

Merchante C, Stepanova AN. 2017. The triple response assay and its use to characterize ethylene mutants in *Arabidopsis*. *Methods Mol Biol* **1573:** 163–209. doi:10.1007/978-1-4939-6854-1_13

Merchante C, Brumos J, Yun J, Hu Q, Spencer KR, Enríquez P, Binder BM, Heber S, Stepanova AN, Alonso JM. 2015. Gene-specific translation regulation mediated by the hormone-signaling molecule EIN2. *Cell* **163:** 684–697. doi:10.1016/j.cell.2015.09.036

Meudt WJ. 1987. Investigations on the mechanism of the brassinosteroid response. VI: Effect of brassinolide on gravitropism of bean hypocotyls. *Plant Physiol* **83:** 195–198. doi:10.1104/pp.83.1.195

Michniewicz M, Ho CH, Enders TA, Floro E, Damodaran S, Gunther LK, Powers SK, Frick EM, Topp CN, Frommer WB, et al. 2019. TRANSPORTER OF IBA1 links auxin and cytokinin to influence root architecture. *Dev Cell* **50:** 599–609.e4. doi:10.1016/j.devcel.2019.06.010

Miller CO, Skoog F, Von Saltza MH, Strong FM. 1955. Kinetin, a cell division factor from DEOXYRIBONUCLEIC ACID. *J Am Chem Soc* **77:** 1392. doi:10.1021/ja01610a105

Miller CO, Skoog F, Okumura FS, Von Saltza MH, Strong FM. 1956. Isolation, structure and synthesis of kinetin, a substance promoting cell division. *J Am Chem Soc* **78:** 1375–1380. doi:10.1021/ja01588a032

Minakuchi K, Kameoka H, Yasuno N, Umehara M, Luo L, Kobayashi K, Hanada A, Ueno K, Asami T, Yamaguchi S, et al. 2010. FINE CULM1 (FC1) works downstream of strigolactones to inhibit the outgrowth of axillary buds in rice. *Plant Cell Physiol* **51:** 1127–1135. doi:10.1093/pcp/pcq083

Miyawaki K, Matsumoto-Kitano M, Kakimoto T. 2004. Expression of cytokinin biosynthetic isopentenyltransferase genes in *Arabidopsis*: tissue specificity and regulation by auxin, cytokinin, and nitrate. *Plant J* **37:** 128–138. doi:10.1046/j.1365-313X.2003.01945.x

Monte E, Tepperman JM, Al-Sady B, Kaczorowski KA, Alonso JM, Ecker JR, Li X, Zhang Y, Quail PH. 2004. The phytochrome-interacting transcription factor, PIF3, acts early, selectively, and positively in light-induced chloroplast development. *Proc Natl Acad Sci* **101:** 16091–16098. doi:10.1073/pnas.0407107101

Moon J, Zhao Y, Dai X, Zhang W, Gray WM, Huq E, Estelle M. 2007. A new CULLIN 1 mutant has altered responses to hormones and light in *Arabidopsis*. *Plant Physiol* **143:** 684–696. doi:10.1104/pp.106.091439

Moreira S, Bishopp A, Carvalho H, Campilho A. 2013. AHP6 inhibits cytokinin signaling to regulate the orientation of pericycle cell division during lateral root initiation. *PLoS ONE* **8:** e56370. doi:10.1371/journal.pone.0056370

Moriwaki T, Miyazawa Y, Kobayashi A, Uchida M, Watanabe C, Fujii N, Takahashi H. 2011. Hormonal regulation of lateral root development in *Arabidopsis* modulated by *MIZ1* and requirement of GNOM activity for *MIZ1* function. *Plant Physiol* **157:** 1209–1220. doi:10.1104/pp.111.186270

Moubayidin L, Perilli S, Dello Ioio R, Di Mambro R, Costantino P, Sabatini S. 2010. The rate of cell differentiation controls the *Arabidopsis* root meristem growth phase. *Curr Biol* **20:** 1138–1143. doi:10.1016/j.cub.2010.05.035

Moubayidin L, Di Mambro R, Sozzani R, Pacifici E, Salvi E, Terpstra I, Bao D, van Dijken A, Dello Ioio R, Perilli S, et al. 2013. Spatial coordination between stem cell activity and cell differentiation in the root meristem. *Dev Cell* **26:** 405–415. doi:10.1016/j.devcel.2013.06.025

Mouchel CF, Briggs GC, Hardtke CS. 2004. Natural genetic variation in *Arabidopsis* identifies BREVIS RADIX, a novel regulator of cell proliferation and elongation in the root. *Genes Dev* **18:** 700–714. doi:10.1101/gad.1187704

Mouchel CF, Osmont KS, Hardtke CS. 2006. BRX mediates feedback between brassinosteroid levels and auxin signalling in root growth. *Nature* **443:** 458–461. doi:10.1038/nature05130

Mounet F, Moing A, Kowalczyk M, Rohrmann J, Petit J, Garcia V, Maucourt M, Yano K, Deborde C, Aoki K, et al. 2012. Down-regulation of a single auxin efflux transport protein in tomato induces precocious fruit development. *J Exp Bot* **63:** 4901–4917. doi:10.1093/jxb/ers167

Mukherjee A, Mazumder M, Jana J, Archana KS, Mondal B, De A, Ghosh S, Saha U, Bose R, Chatterjee S, et al. 2019. Enhancement of ABA sensitivity through conditional expression of the *ARF10* gene in *Brassica juncea* reveals fertile plants with tolerance against *Alternaria brassicicola*. *MPMI* **32:** 1429–1447. doi:10.1094/MPMI-05-19-0132-R

Müller D, Leyser O. 2011. Auxin, cytokinin and the control of shoot branching. *Ann Bot* **107:** 1203–1212. doi:10.1093/aob/mcr069

Müller B, Sheen J. 2008. Cytokinin and auxin interaction in root stem-cell specification during early embryogenesis. *Nature* **453:** 1094–1097. doi:10.1038/nature06943

Müller D, Waldie T, Miyawaki K, To JP, Melnyk CW, Kieber JJ, Kakimoto T, Leyser O. 2015. Cytokinin is required for escape but not release from auxin mediated apical dominance. *Plant J* **82:** 874–886. doi:10.1111/tpj.12862

Munguía-Rodríguez AG, López-Bucio JS, Ruiz-Herrera LF, Ortiz-Castro R, Guevara-García AA, Marsch-Martínez N, Carreón-Abud Y, López-Bucio J, Martínez-Trujillo M. 2020. YUCCA4 overexpression modulates auxin biosynthesis and transport and influences plant growth and development via crosstalk with abscisic acid in *Arabidopsis thaliana*. *Genet Mol Biol* **43:** e20190221. doi:10.1590/1678-4685-gmb-2019-0221

Müssig C, Fischer S, Altmann T. 2002. Brassinosteroid-regulated gene expression. *Plant Physiol* **129:** 1241–1251. doi:10.1104/pp.011003

Nagpal P, Ellis CM, Weber H, Ploense SE, Barkawi LS, Guilfoyle TJ, Hagen G, Alonso JM, Cohen JD, Farmer EE, et al. 2005. Auxin response factors ARF6 and ARF8 promote jasmonic acid production and flower maturation. *Development* **132:** 4107–4118. doi:10.1242/dev.01955

Nakamura A, Higuchi K, Goda H, Fujiwara MT, Sawa S, Koshiba T, Shimada Y, Yoshida S. 2003. Brassinolide induces *IAA5*, *IAA19*, and DR5, a synthetic auxin response element in *Arabidopsis*, implying a cross talk point of brassinosteroid and auxin signaling. *Plant Physiol* **133:** 1843–1853. doi:10.1104/pp.103.030031

Nakamura A, Nakajima N, Goda H, Shimada Y, Hayashi K, Nozaki H, Asami T, Yoshida S, Fujioka S. 2006. *Arabidopsis Aux/IAA* genes are involved in brassinosteroid-mediated growth responses in a manner dependent on organ type. *Plant J* **45:** 193–205. doi:10.1111/j.1365-313X.2005.02582.x

Nemhauser JL, Mockler TC, Chory J. 2004. Interdependency of brassinosteroid and auxin signaling in *Arabidopsis*. *PLoS Biol* **2:** E258. doi:10.1371/journal.pbio.0020258

Nolan TM, Vukašinović N, Liu D, Russinova E, Yin Y. 2020. Brassinosteroids: multidimensional regulators of plant growth, development, and stress responses. *Plant Cell* **32:** 295–318. doi:10.1105/tpc.19.00335

Nordstrom A, Tarkowski P, Tarkowska D, Norbaek R, Astot C, Dolezal K, Sandberg G. 2004. Auxin regulation of cytokinin biosynthesis in *Arabidopsis thaliana*: a factor of potential importance for auxin-cytokinin-regulated development. *Proc Natl Acad Sci* **101:** 8039–8044. doi:10.1073/pnas.0402504101

Oh E, Zhu J, Ryu H, Hwang I, Wang Z. 2014. TOPLESS mediates brassinosteroid-induced transcriptional repression through interaction with BZR1. *Nat Commun* **5:** 4140. doi:10.1038/ncomms5140

Omoarelojie LO, Kulkarni MG, Finnie JF, Van Staden J. 2019. Strigolactones and their crosstalk with other phytohormones. *Ann Bot* **124:** 749–767. doi:10.1093/aob/mcz100

Ori N, Cohen AR, Etzioni A, Brand A, Yanai O, Shleizer S, Menda N, Amsellem Z, Efroni I, Pekker I, et al. 2007. Regulation of LANCEOLATE by miR319 is required for compound-leaf development in tomato. *Nat Genet* **39:** 787–791. doi:10.1038/ng2036

Overvoorde PJ, Okushima Y, Alonso JM, Chan A, Chang C, Ecker JR, Hughes B, Liu A, Onodera C, Quach H, et al. 2005. Functional genomic analysis of the *AUXIN/INDOLE-3-ACETIC ACID* gene family members in *Arabidopsis thaliana*. *Plant Cell* **17:** 3282–3300. doi:10.1105/tpc.105.036723

Overvoorde P, Fukaki H, Beeckman T. 2010. Auxin control of root development. *Cold Spring Harb Perspect Biol* **2:** a001537. doi:10.1101/cshperspect.a001537

Palatnik JF, Allen E, Wu X, Schommer C, Schwab R, Carrington JC, Weigel D. 2003. Control of leaf morphogenesis by microRNAs. *Nature* **425:** 257–263. doi:10.1038/nature01958

Palni LM, Burch L, Horgan R. 1988. The effect of auxin concentration on cytokinin stability and metabolism. *Planta* **174:** 231–234. doi:10.1007/BF00394775

Pandya-Kumar N, Shema R, Kumar M, Mayzlish-Gati E, Levy D, Zemach H, Belausov E, Wininger S, Abu-Abied M, Kapulnik Y, et al. 2014. Strigolactone analog GR24 triggers changes in PIN2 polarity, vesicle trafficking and actin filament architecture. *New Phytol* **202:** 1184–1196. doi:10.1111/nph.12744

Park WJ. 1998. Effect of epibrassinolide on hypocotyl growth of the tomato mutant diageotropica. *Planta* **207:** 120–124. doi:10.1007/s004250050463

Park J, Park J, Kim Y, Staswick PE, Jeon J, Yun J, Kim S, Kim J, Lee Y, Park C. 2007. GH3-mediated auxin homeostasis links growth regulation with stress adaptation response in *Arabidopsis*. *J Biol Chem* **282:** 10036–10046. doi:10.1074/jbc.M610524200

Pasternak T, Groot EP, Kazantsev FV, Teale W, Omelyanchuk N, Kovrizhnykh V, Palme K, Mironova VV. 2019. Salicylic acid affects root meristem patterning via auxin distribution in a concentration-dependent manner. *Plant Physiol* **180:** 1725–1739. doi:10.1104/pp.19.00130

Peng P, Yan Z, Zhu Y, Li J. 2008. Regulation of the *Arabidopsis* GSK3-like kinase BRASSINOSTEROID-INSENSITIVE 2 through proteasome-mediated protein degradation. *Mol Plant* **1:** 338–346. doi:10.1093/mp/ssn001

Peng Y, Ma W, Chen L, Yang L, Li S, Zhao H, Zhao Y, Jin W, Li N, Bevan MW, et al. 2013. Control of root meristem size by DA1-RELATED PROTEIN2 in *Arabidopsis*. *Plant Physiol* **161:** 1542–1556. doi:10.1104/pp.112.210237

Peng H, Zhao J, Neff MM. 2015. ATAF2 integrates *Arabidopsis* brassinosteroid inactivation and seedling photomorphogenesis. *Development* **142:** 4129–4138. doi:10.1242/dev.124347

Pérez-Llorca M, Muñoz P, Müller M, Munné-Bosch S. 2019. Biosynthesis, metabolism and function of auxin, salicylic acid and melatonin in climacteric and non-climacteric fruits. *Front Plant Sci* **10:** 136. doi:10.3389/fpls.2019.00136

Perilli S, Perez-Perez J, Di Mambro R, Peris CL, Diaz-Trivino S, Del Bianco M, Pierdonati E, Moubayidin L, Cruz-Ramirez A, Costantino P, et al. 2013. RETINOBLASTOMA-RELATED protein stimulates cell differentiation in the *Arabidopsis* root meristem by interacting with cytokinin signaling. *Plant Cell* **25:** 4469–4478. doi:10.1105/tpc.113.116632

Pernisova M, Klima P, Horak J, Valkova M, Malbeck J, Soucek P, Reichman P, Hoyerova K, Dubova J, Friml J, et al. 2009. Cytokinins modulate auxin-induced organogenesis in plants via regulation of the auxin efflux. *Proc Natl Acad Sci* **106:** 3609–3614. doi:10.1073/pnas.0811539106

Pernisova M, Prat T, Grones P, Harustiakova D, Matonohova M, Spichal L, Nodzynski T, Friml J, Hejatko J. 2016. Cytokinins influence root gravitropism via differential regulation of auxin transporter expression and localization in *Arabidopsis*. *New Phytol* **212:** 497–509. doi:10.1111/nph.14049

Pinon V, Prasad K, Grigg SP, Sanchez-Perez GF, Scheres B. 2013. Local auxin biosynthesis regulation by PLETHORA transcription factors controls phyllotaxis in *Arabidopsis*. *Proc Natl Acad Sci* **110:** 1107–1112. doi:10.1073/pnas.1213497110

Piya S, Shrestha SK, Binder B, Stewart CN, Hewezi T. 2014. Protein-protein interaction and gene co-expression maps of ARFs and Aux/IAAs in *Arabidopsis*. *Front Plant Sci* **5:** 744. doi:10.3389/fpls.2014.00744

Planas-Riverola A, Gupta A, Betegón-Putze I, Bosch N, Ibañes M, Caño-Delgado AI. 2019. Brassinosteroid signaling in plant development and adaptation to stress. *Development* **146:** dev151894. 10.1242/dev.151894

Promchuea S, Zhu Y, Chen Z, Zhang J, Gong Z. 2017. ARF2 coordinates with PLETHORAs and PINs to orchestrate ABA-mediated root meristem activity in *Arabidopsis*. *J Integr Plant Biol* **59:** 30–43. doi:10.1111/jipb.12506

Przemeck GH, Mattsson J, Hardtke C, Sung ZR, Berleth T. 1996. Studies on the role of the *Arabidopsis* gene MONOPTEROS in vascular development and plant cell axialization. *Planta* **200:** 229–237. doi:10.1007/BF00208313

Qi L, Yan J, Li Y, Jiang H, Sun J, Chen Q, Li H, Chu J, Yan C, Sun X, et al. 2012. *Arabidopsis thaliana* plants differentially modulate auxin biosynthesis and transport during defense responses to the necrotrophic pathogen *Alternaria brassicicola*. *New Phytol* **195:** 872–882. doi:10.1111/j.1469-8137.2012.04208.x

Qi T, Wang J, Huang H, Liu B, Gao H, Liu Y, Song S, Xie D. 2015. Regulation of jasmonate-induced leaf senescence by antagonism between bHLH subgroup IIIe and IIId factors in *Arabidopsis*. *Plant Cell* **27:** 1634–1649. doi:10.1105/tpc.15.00110

Qiu Y, Guan SC, Wen C, Li P, Gao Z, Chen X. 2019. Auxin and cytokinin coordinate the dormancy and outgrowth of axillary bud in strawberry runner. *BMC Plant Biol* **19:** 528. doi:10.1186/s12870-019-2151-x

Rademacher EH, Möller B, Lokerse AS, Llavata-Peris CI, van den Berg W, Weijers D. 2011. A cellular expression map of the *Arabidopsis* AUXIN RESPONSE FACTOR gene family. *Plant J* **68:** 597–606. doi:10.1111/j.1365-313X.2011.04710.x

Rasmussen A, Mason MG, De Cuyper C, Brewer PB, Herold S, Agusti J, Geelen D, Greb T, Goormachtig S, Beeckman T, et al. 2012. Strigolactones suppress adventitious rooting in *Arabidopsis* and pea. *Plant Physiol* **158:** 1976–1987. doi:10.1104/pp.111.187104

Raya-González J, Pelagio-Flores R, López-Bucio J. 2012. The jasmonate receptor COI1 plays a role in jasmonate-induced lateral root formation and lateral root positioning in *Arabidopsis thaliana*. *J Plant Physiol* **169:** 1348–1358. doi:10.1016/j.jplph.2012.05.002

Reeves PH, Ellis CM, Ploense SE, Wu M, Yadav V, Tholl D, Chételat A, Haupt I, Kennerley BJ, Hodgens C, et al. 2012. A regulatory network for coordinated flower maturation. *PLoS Genet* **8:** e1002506. doi:10.1371/journal.pgen.1002506

Reyes-Olalde JI, Zúñiga-Mayo VM, Serwatowska J, Chavez Montes RA, Lozano-Sotomayor P, Herrera-Ubaldo H, Gonzalez-Aguilera KL, Ballester P, Ripoll JJ, Ezquer I, et al. 2017. The bHLH transcription factor SPATULA enables cytokinin signaling, and both activate auxin biosynthesis and transport genes at the medial domain of the gynoecium. *PLoS Genet* **13:** e1006726. doi:10.1371/journal.pgen.1006726

Richter R, Behringer C, Müller IK, Schwechheimer C. 2010. The GATA-type transcription factors GNC and GNL/CGA1 repress gibberellin signaling downstream from DELLA proteins and PHYTOCHROME-INTERACTING FACTORS. *Genes Dev* **24:** 2093–2104. doi:10.1101/gad.594910

Richter R, Behringer C, Zourelidou M, Schwechheimer C. 2013. Convergence of auxin and gibberellin signaling on the regulation of the GATA transcription factors GNC and GNL in *Arabidopsis thaliana*. *Proc Natl Acad Sci* **110:** 13192–13197. doi:10.1073/pnas.1304250110

Robert-Seilaniantz A, Navarro L, Bari R, Jones JDG. 2007. Pathological hormone imbalances. *Curr Opin Plant Biol* **10:** 372–379. doi:10.1016/j.pbi.2007.06.003

Rohde A, De Rycke R, Beeckman T, Engler G, Van Montagu M, Boerjan W. 2000. ABI3 affects plastid differentiation in dark-grown *Arabidopsis* seedlings. *Plant Cell* **12:** 35–52. doi:10.1105/tpc.12.1.35

Rosa NM, Pfeiffer A, Hill K, Locascio A, Bhalerao RP, Miskolczi P, Grønlund AL, Wanchoo-Kohli A, Thomas SG, Bennett MJ, et al. 2015. Genome wide binding site analysis reveals transcriptional coactivation of cytokinin-responsive genes by DELLA proteins. *PLoS Genet* **11:** e1005337.

Ross JJ, O'neill DP, Smith JJ, Kerckhoffs LH, Elliott RC. 2000. Evidence that auxin promotes gibberellin A_1 biosynthesis in pea. *Plant J* **21:** 547–552. doi:10.1046/j.1365-313x.2000.00702.x

Ruan J, Zhou Y, Zhou M, Yan J, Khurshid M, Weng W, Cheng J, Zhang K. 2019. Jasmonic acid signaling pathway in plants. *Int J Mol Sci* **20:** 2479.

Ruyter-Spira C, Kohlen W, Charnikhova T, van Zeijl A, van Bezouwen L, de Ruijter N, Cardoso C, Lopez-Raez JA, Matusova R, Bours R, et al. 2011. Physiological effects of the synthetic strigolactone analog GR24 on root system architecture in *Arabidopsis*: another belowground role for strigolactones? *Plant Physiol* **155:** 721–734. doi:10.1104/pp.110.166645

Růžička K, Ljung K, Vanneste S, Podhorská R, Beeckman T, Friml J, Benková E. 2007. Ethylene regulates root growth through effects on auxin biosynthesis and transport-dependent auxin distribution. *Plant Cell* **19:** 2197–2212. doi:10.1105/tpc.107.052126

Růžička K, Šimášková M, Duclercq J, Petrášek J, Zažímalová E, Simon S, Friml J, Van Montagu, Marc CE, Benková E. 2009. Cytokinin regulates root meristem activity via modulation of the polar auxin transport. *Proc Natl Acad Sci* **106:** 4284–4289. doi:10.1073/pnas.0900060106

Sadura I, Pociecha E, Dziurka M, Oklestkova J, Novak O, Gruszka D, Janeczko A. 2019. Mutations in the HvDWARF, HvCPD and HvBRI1 genes-involved in brassinosteroid biosynthesis/signalling: altered photosynthetic efficiency, hormonal homeostasis and tolerance to high/low temperatures in barley. *J Plant Growth Regul* **38:** 1062–1081. doi:10.1007/s00344-019-09914-z

Sakamoto T, Morinaka Y, Inukai Y, Kitano H, Fujioka S. 2013. Auxin signal transcription factor regulates expression of the brassinosteroid receptor gene in rice. *Plant J* **73:** 676–688. doi:10.1111/tpj.12071

Salanenka Y, Verstraeten I, Löfke C, Tabata K, Naramoto S, Glanc M, Friml J. 2018. Gibberellin DELLA signaling targets the retromer complex to redirect protein trafficking to the plasma membrane. *Proc Natl Acad Sci* **115:** 3716–3721. doi:10.1073/pnas.1721760115

Santuari L, Sanchez-Perez GF, Luijten M, Rutjens B, Terpstra I, Berke L, Gorte M, Prasad K, Bao D, Timmermans-Hereijgers JL, et al. 2016. The PLETHORA gene regulatory network guides growth and cell differentiation in *Arabidopsis* roots. *Plant Cell* **28:** 2937–2951. doi:10.1105/tpc.16.00656

Sarnowska EA, Rolicka AT, Bucior E, Cwiek P, Tohge T, Fernie AR, Jikumaru Y, Kamiya Y, Franzen R, Schmelzer E, et al. 2013. DELLA-interacting SWI3C core subunit of switch/sucrose nonfermenting chromatin remodeling complex modulates gibberellin responses and hormonal cross talk in *Arabidopsis*. *Plant Physiol* **163:** 305–317. doi:10.1104/pp.113.223933

Scacchi E, Osmont KS, Beuchat J, Salinas P, Navarrete-Gomez M, Trigueros M, Ferrandiz C, Hardtke CS. 2009. Dynamic, auxin-responsive plasma membrane-to-nucleus movement of *Arabidopsis* BRX. *Development* **136:** 2059–2067. doi:10.1242/dev.035444

Scacchi E, Salinas P, Gujas B, Santuari L, Krogan N, Ragni L, Berleth T, Hardtke CS. 2010. Spatio-temporal sequence of cross-regulatory events in root meristem growth. *Proc Natl Acad Sci* **107:** 22734–22739. doi:10.1073/pnas.1014716108

Schaller GE, Bishopp A, Kieber JJ. 2015. The yin-yang of hormones: cytokinin and auxin interactions in plant development. *Plant Cell* **27:** 44–63. doi:10.1105/tpc.114.133595

Schlereth A, Möller B, Liu W, Kientz M, Flipse J, Rademacher EH, Schmid M, Jürgens G, Weijers D. 2010. MONOPTEROS controls embryonic root initiation by regulating a mobile transcription factor. *Nature* **464:** 913–916. doi:10.1038/nature08836

Seto Y, Yasui R, Kameoka H, Tamiru M, Cao M, Terauchi R, Sakurada A, Hirano R, Kisugi T, Hanada A, et al. 2019. Strigolactone perception and deactivation by a hydrolase receptor DWARF14. *Nat Commun* **10:** 191–197. doi:10.1038/s41467-018-08124-7

Shani E, Ben-Gera H, Shleizer-Burko S, Burko Y, Weiss D, Ori N. 2010. Cytokinin regulates compound leaf development in tomato. *Plant Cell* **22:** 3206–3217. doi:10.1105/tpc.110.078253

Shin R, Burch AY, Huppert KA, Tiwari SB, Murphy AS, Guilfoyle TJ, Schachtman DP. 2007. The *Arabidopsis* transcription factor MYB77 modulates auxin signal transduction. *Plant Cell* **19:** 2440–2453. doi:10.1105/tpc.107.050963

Shinohara N, Taylor C, Leyser O. 2013. Strigolactone can promote or inhibit shoot branching by triggering rapid depletion of the auxin efflux protein PIN1 from the plasma membrane. *PLoS Biol* **11:** e1001474. doi:10.1371/journal.pbio.1001474

Shkolnik-Inbar D, Bar-Zvi D. 2010. *ABI4* mediates abscisic acid and cytokinin inhibition of lateral root formation by reducing polar auxin transport in *Arabidopsis*. *Plant Cell* **22:** 3560–3573. doi:10.1105/tpc.110.074641

Shu K, Zhang H, Wang S, Chen M, Wu Y, Tang S, Liu C, Feng Y, Cao X, Xie Q. 2013. ABI4 regulates primary seed dormancy by regulating the biogenesis of abscisic acid and gibberellins in *Arabidopsis*. *PLoS Genet* **9:** e1003577. doi:10.1371/journal.pgen.1003577

Shu K, Chen Q, Wu Y, Liu R, Zhang H, Wang P, Li Y, Wang S, Tang S, Liu C, et al. 2016. ABI4 mediates antagonistic effects of abscisic acid and gibberellins at transcript and protein levels. *Plant J* **85:** 348–361. doi:10.1111/tpj.13109

Singh M, Gupta A, Laxmi A. 2014. Glucose control of root growth direction in *Arabidopsis thaliana*. *J Exp Bot* **65:** 2981–2993. doi:10.1093/jxb/eru146

Song S, Qi T, Huang H, Xie D. 2013. Regulation of stamen development by coordinated actions of jasmonate, auxin, and gibberellin in *Arabidopsis*. *Mol Plant* **6:** 1065–1073. doi:10.1093/mp/sst054

Sorefan K, Booker J, Haurogne K, Goussot M, Bainbridge K, Foo E, Chatfield S, Ward S, Beveridge C, Rameau C, et al. 2003. MAX4 and RMS1 are orthologous dioxygenase-like genes that regulate shoot branching in *Arabidopsis* and pea. *Genes Dev* **17:** 1469–1474. doi:10.1101/gad.256603

Sorin C, Negroni L, Balliau T, Corti H, Jacquemot M, Davanture M, Sandberg G, Zivy M, Bellini C. 2006. Proteomic analysis of different mutant genotypes of *Arabidopsis* led to the identification of 11 proteins correlating with adventitious root development. *Plant Physiol* **140:** 349–364. doi:10.1104/pp.105.067868

Spaepen S, Vanderleyden J, Remans R. 2007. Indole-3-acetic acid in microbial and microorganism-plant signaling. *FEMS Microbiol Rev* **31:** 425–448. doi:10.1111/j.1574-6976.2007.00072.x

Srivastava A, Handa AK. 2005. Hormonal regulation of tomato fruit development: a molecular perspective. *J Plant Growth Regul* **24:** 67–82. doi:10.1007/s00344-005-0015-0

Staswick PE, Serban B, Rowe M, Tiryaki I, Maldonado MT, Maldonado MC, Suza W. 2005. Characterization of an *Arabidopsis* enzyme family that conjugates amino acids to indole-3-acetic acid. *Plant Cell* **17:** 616–627. doi:10.1105/tpc.104.026690

Stepanova AN, Yun J, Likhacheva AV, Alonso JM. 2007. Multilevel interactions between ethylene and auxin in *Arabidopsis* roots. *Plant Cell* **19:** 2169–2185. doi:10.1105/tpc.107.052068

Stepanova AN, Robertson-Hoyt J, Yun J, Benavente LM, Xie D, Doležal K, Schlereth A, Jürgens G, Alonso JM. 2008. TAA1-mediated auxin biosynthesis is essential for hormone crosstalk and plant development. *Cell* **133:** 177–191. doi:10.1016/j.cell.2008.01.047

Stepanova AN, Yun J, Robles LM, Novak O, He W, Guo H, Ljung K, Alonso JM. 2011. The *Arabidopsis* YUCCA1 flavin monooxygenase functions in the indole-3-pyruvic acid branch of auxin biosynthesis. *Plant Cell* **23:** 3961–3973. doi:10.1105/tpc.111.088047

Street IH, Aman S, Zubo Y, Ramzan A, Wang X, Shakeel SN, Kieber JJ, Schaller GE. 2015. Ethylene inhibits cell proliferation of the *Arabidopsis* root meristem. *Plant Physiol* **169:** 338–350. doi:10.1104/pp.15.00415

Su L, Diretto G, Purgatto E, Danoun S, Zouine M, Li Z, Roustan J, Bouzayen M, Giuliano G, Chervin C. 2015. Carotenoid accumulation during tomato fruit ripening is modulated by the auxin-ethylene balance. *BMC Plant Biol* **15:** 114.

Sun J, Xu Y, Ye S, Jiang H, Chen Q, Liu F, Zhou W, Chen R, Li X, Tietz O, et al. 2009. *Arabidopsis ASA1* is important for jasmonate-mediated regulation of auxin biosynthesis and transport during lateral root formation. *Plant Cell* **21:** 1495–1511. doi:10.1105/tpc.108.064303

Sun Y, Fan XY, Cao DM, Tang W, He K, Zhu JY, He JX, Bai MY, Zhu S, Oh E, et al. 2010. Integration of brassinosteroid signal transduction with the transcription network for plant growth regulation in *Arabidopsis*. *Dev Cell* **19:** 765–777. doi:10.1016/j.devcel.2010.10.010

Sun H, Tao J, Liu S, Huang S, Chen S, Xie X, Yoneyama K, Zhang Y, Xu G. 2014. Strigolactones are involved in phosphate- and nitrate-deficiency-induced root development and auxin transport in rice. *J Exp Bot* **65:** 6735–6746. doi:10.1093/jxb/eru029

Sun H, Tao J, Hou M, Huang S, Chen S, Liang Z, Xie T, Wei Y, Xie X, Yoneyama K, et al. 2015. A strigolactone signal is required for adventitious root formation in rice. *Ann Bot* **115:** 1155–1162. doi:10.1093/aob/mcv052

Sun H, Xu F, Guo X, Wu D, Zhang X, Lou M, Luo F, Zhao Q, Xu G, Zhang Y. 2019. A Strigolactone signal inhibits secondary lateral root development in rice. *Front Plant Sci* **10:** 1527. doi:10.3389/fpls.2019.01527

Sun L, Feraru E, Feraru MI, Waidmann S, Wang W, Passaia G, Wang ZY, Wabnik K, Kleine-Vehn J. 2020. PIN-LIKES coordinate brassinosteroid signaling with nuclear auxin input in *Arabidopsis thaliana*. *Curr Biol* **30:** 1579–1588. e6. doi:10.1016/j.cub.2020.02.002

Swarup R, Perry P, Hagenbeek D, Van Der Straeten D, Beemster GTS, Sandberg G, Bhalerao R, Ljung K, Bennett MJ. 2007. Ethylene upregulates auxin biosynthesis in *Arabidopsis* seedlings to enhance inhibition of root cell elongation. *Plant Cell* **19:** 2186–2196. doi:10.1105/tpc.107.052100

Szekeres M, Németh K, Koncz-Kálmán Z, Mathur J, Kauschmann A, Altmann T, Rédei GP, Nagy F, Schell J, Koncz C. 1996. Brassinosteroids rescue the deficiency of CYP90, a cytochrome P450, controlling cell elongation and de-etiolation in *Arabidopsis*. *Cell* **85:** 171–182. doi:10.1016/S0092-8674(00)81094-6

Tabata R, Ikezaki M, Fujibe T, Aida M, Tian C, Ueno Y, Yamamoto KT, Machida Y, Nakamura K, Ishiguro S. 2010. *Arabidopsis* auxin response factor6 and 8 regulate jasmonic acid biosynthesis and floral organ development via repression of class 1 KNOX genes. *Plant Cell Physiol* **51:** 164–175. doi:10.1093/pcp/pcp176

Tan M, Li G, Chen X, Xing L, Ma J, Zhang D, Ge H, Han M, Sha G, An N. 2019. Role of cytokinin, strigolactone, and auxin export on outgrowth of axillary buds in apple. *Front Plant Sci* **10:** 616. doi:10.3389/fpls.2019.00616

Tanaka K. 2003. Physiological roles of brassinosteroids in early growth of *Arabidopsis*: brassinosteroids have a synergistic relationship with gibberellin as well as auxin in light-grown hypocotyl elongation. *J Plant Growth Regul* **22:** 259–271. doi:10.1007/s00344-003-0119-3

Tanaka M, Takei K, Kojima M, Sakakibara H, Mori H. 2006. Auxin controls local cytokinin biosynthesis in the nodal stem in apical dominance. *Plant J* **45:** 1028–1036. doi:10.1111/j.1365-313X.2006.02656.x

Tian Q, Uhlir NJ, Reed JW. 2002. *Arabidopsis* SHY2/IAA3 inhibits auxin-regulated gene expression. *Plant Cell* **14:** 301–319. doi:10.1105/tpc.010283

Tian C, Muto H, Higuchi K, Matamura T, Tatematsu K, Koshiba T, Yamamoto KT. 2004. Disruption and overexpression of *auxin response factor 8* gene of *Arabidopsis* affect hypocotyl elongation and root growth habit, indicating its possible involvement in auxin homeostasis in light condition. *Plant J* **40:** 333–343. doi:10.1111/j.1365-313X.2004.02220.x

Tognetti VB, Bielach A, Hrtyan M. 2017. Redox regulation at the site of primary growth: auxin, cytokinin and ROS crosstalk. *Plant Cell Environ* **40:** 2586–2605. doi:10.1111/pce.13021

Ubeda-Tomás S, Beemster GTS, Bennett MJ. 2012. Hormonal regulation of root growth: integrating local activities into global behaviour. *Trends Plant Sci* **17:** 326–331. doi:10.1016/j.tplants.2012.02.002

Umehara M, Hanada A, Yoshida S, Akiyama K, Arite T, Takeda-Kamiya N, Magome H, Kamiya Y, Shirasu K, Yoneyama K, et al. 2008. Inhibition of shoot branching by new terpenoid plant hormones. *Nature* **455:** 195–200. doi:10.1038/nature07272

Vandenbussche F, Petrásek J, Zádníková P, Hoyerová K, Pesek B, Raz V, Swarup R, Bennett M, Zazímalová E, Benková E, et al. 2010. The auxin influx carriers AUX1 and LAX3 are involved in auxin-ethylene interactions during apical hook development in *Arabidopsis thaliana* seedlings. *Development* **137:** 597–606. doi:10.1242/dev.040790

Vandenbussche F, Suslov D, De Grauwe L, Leroux O, Vissenberg K, Van der Straeten D. 2011. The role of brassinosteroids in shoot gravitropism. *Plant Physiol* **156:** 1331–1336. doi:10.1104/pp.111.177873

Vandenbussche F, Callebert P, Zadnikova P, Benkova E, Van Der Straeten D. 2013. Brassinosteroid control of shoot gravitropism interacts with ethylene and depends on auxin signaling components. *Am J Bot* **100:** 215–225. doi:10.3732/ajb.1200264

van Rongen M, Bennett T, Ticchiarelli F, Leyser O. 2019. Connective auxin transport contributes to strigolactone-mediated shoot branching control independent of the transcription factor BRC1. *PLoS Genet* **15:** e1008023. doi:10.1371/journal.pgen.1008023

Vaseva II, Qudeimat E, Potuschak T, Du Y, Genschik P, Vandenbussche F, Van Der Straeten D. 2018. The plant hormone ethylene restricts *Arabidopsis* growth via the epidermis. *Proc Natl Acad Sci* **115:** E4130–E4139. doi:10.1073/pnas.1717649115

Vert G, Walcher CL, Chory J, Nemhauser JL. 2008. Integration of auxin and brassinosteroid pathways by Auxin Response Factor 2. *Proc Natl Acad Sci* **105:** 9829–9834. doi:10.1073/pnas.0803996105

Von Arnim A, Deng X. 1996. Light control of seedling development. *Annu Rev Plant Physiol Plant Mol Biol* **47:** 215–243. doi:10.1146/annurev.arplant.47.1.215

Waidmann S, Ruiz Rosquete M, Schöller M, Sarkel E, Lindner H, LaRue T, Petřík I, Dünser K, Martopawiro S, Sasidharan R, et al. 2019. Cytokinin functions as an asymmetric and anti-gravitropic signal in lateral roots. *Nat Commun* **10:** 3540–3544. doi:10.1038/s41467-019-11483-4

Waldie T, Leyser O. 2018. Cytokinin targets auxin transport to promote shoot branching. *Plant Physiol* **177:** 803–818. doi:10.1104/pp.17.01691

Wang Y, Guo H. 2019. On hormonal regulation of the dynamic apical hook development. *New Phytol* **222:** 1230–1234. doi:10.1111/nph.15626

Wang H, Jones B, Li Z, Frasse P, Delalande C, Regad F, Chaabouni S, Latché A, Pech JC, Bouzayen M. 2005a. The tomato *Aux/IAA* transcription factor *IAA9* is involved in fruit development and leaf morphogenesis. *Plant Cell* **17:** 2676–2692. doi:10.1105/tpc.105.033415

Wang J, Wang L, Mao Y, Cai W, Xue H, Chen X. 2005b. Control of root cap formation by microRNA-targeted auxin response factors in *Arabidopsis*. *Plant Cell* **17:** 2204–2216. doi:10.1105/tpc.105.033076

Wang G, Römheld V, Li C, Bangerth F. 2006. Involvement of auxin and CKs in boron deficiency induced changes in apical dominance of pea plants (*Pisum sativum* L.). *J Plant Physiol* **163:** 591–600. doi:10.1016/j.jplph.2005.09.014

Wang D, Pajerowska-Mukhtar K, Culler AH, Dong X. 2007. Salicylic acid inhibits pathogen growth in plants through repression of the auxin signaling pathway. *Curr Biol* **17:** 1784–1790. doi:10.1016/j.cub.2007.09.025

Wang H, Schauer N, Usadel B, Frasse P, Zouine M, Hernould M, Latché A, Pech J, Fernie AR, Bouzayen M. 2009. Regulatory features underlying pollination-dependent and -independent tomato fruit set revealed by transcript and primary metabolite profiling. *Plant Cell* **21:** 1428–1452. doi:10.1105/tpc.108.060830

Wang L, Hua D, He J, Duan Y, Chen Z, Hong X, Gong Z. 2011. Auxin response factor2 (ARF2) and its regulated homeodomain gene HB33 mediate abscisic acid response in *Arabidopsis*. *PLoS Genet* **7:** e1002172. doi:10.1371/journal.pgen.1002172

Wang Y, Wang J, Shi B, Yu T, Qi J, Meyerowitz EM, Jiao Y. 2014. The stem cell niche in leaf axils is established by auxin and cytokinin in *Arabidopsis*. *Plant Cell* **26:** 2055–2067. doi:10.1105/tpc.114.123083

Wang J, Wang R, Mao X, Li L, Chang X, Zhang X, Jing R. 2019. TaARF4 genes are linked to root growth and plant height in wheat. *Ann Bot* **124:** 903–915. doi:10.1093/aob/mcy218

Wasternack C, Hause B. 2013. Jasmonates: biosynthesis, perception, signal transduction and action in plant stress response, growth and development. An update to the 2007 review in *Annals of Botany*. *Ann Bot* **111:** 1021–1058. doi:10.1093/aob/mct067

Werner T, Köllmer I, Bartrina I, Holst K, Schmülling T. 2006. New insights into the biology of cytokinin degradation. *Plant Biol (Stuttg)* **8:** 371–381. doi:10.1055/s-2006-923928

Westfall CS, Sherp AM, Zubieta C, Alvarez S, Schraft E, Marcellin R, Ramirez L, Jez JM. 2016. *Arabidopsis thaliana* GH3.5 acyl acid amido synthetase mediates metabolic crosstalk in auxin and salicylic acid homeostasis. *Proc Natl Acad Sci* **113:** 13917–13922. doi:10.1073/pnas.1612635113

Won C, Shen X, Mashiguchi K, Zheng Z, Dai X, Cheng Y, Kasahara H, Kamiya Y, Chory J, Zhao Y. 2011. Conversion of tryptophan to indole-3-acetic acid by TRYPTOPHAN AMINOTRANSFERASES OF *ARABIDOPSIS* and YUCCAs in *Arabidopsis*. *Proc Natl Acad Sci* **108:** 18518–18523. doi:10.1073/pnas.1108436108

Woodward AW, Bartel B. 2005. Auxin: regulation, action, and interaction. *Ann Bot* **95:** 707–735. doi:10.1093/aob/mci083

Wu Y, Zhang D, Chu JY, Boyle P, Wang Y, Brindle ID, De Luca V, Després C. 2012. The *Arabidopsis* NPR1 protein is a receptor for the plant defense hormone salicylic acid. *Cell Rep* **1:** 639–647. doi:10.1016/j.celrep.2012.05.008

Wu M, Yamaguchi N, Xiao J, Bargmann B, Estelle M, Sang Y, Wagner D. 2015. Auxin-regulated chromatin switch directs acquisition of flower primordium founder fate. *eLife* **4:** e09269. doi:10.7554/eLife.09269

Xing L, Zhao Y, Gao J, Xiang C, Zhu JK. 2016. The ABA receptor PYL9 together with PYL8 plays an important role in regulating lateral root growth. *Sci Rep* **6:** 27177. doi:10.1038/srep27177

Xu J, Zha M, Li Y, Ding Y, Chen L, Ding C, Wang S. 2015. The interaction between nitrogen availability and auxin, cytokinin, and strigolactone in the control of shoot branching in rice (*Oryza sativa* L.). *Plant Cell Rep* **34:** 1647–1662. doi:10.1007/s00299-015-1815-8

Yamaguchi N, Wu M, Winter CM, Berns MC, Nole-Wilson S, Yamaguchi A, Coupland G, Krizek BA, Wagner D. 2013. A molecular framework for auxin-mediated initiation of flower primordia. *Dev Cell* **24:** 271–282. doi:10.1016/j.devcel.2012.12.017

Yamazaki T, Miyazawa Y, Kobayashi A, Moriwaki T, Fujii N, Takahashi H. 2012. MIZ1, an essential protein for root hydrotropism, is associated with the cytoplasmic face of the endoplasmic reticulum membrane in *Arabidopsis* root cells. *FEBS Lett* **586:** 398–402. doi:10.1016/j.febslet.2012.01.008

Yanai O, Shani E, Russ D, Ori N. 2011. Gibberellin partly mediates LANCEOLATE activity in tomato: gibberellin partly mediates LANCEOLATE activity. *Plant J* **68:** 571–582. doi:10.1111/j.1365-313X.2011.04716.x

Yang SF, Hoffman NE. 1984. Ethylene biosynthesis and its regulation in higher plants. *Ann Rev Plant Physiol* **35:** 155–189. doi:10.1146/annurev.pp.35.060184.001103

Yang X, Yang Y, Xue L, Zou M, Liu J, Chen F, Xue H. 2011. Rice ABI5-Like1 regulates abscisic acid and auxin responses by affecting the expression of ABRE-containing genes. *Plant Physiol* **156:** 1397–1409. doi:10.1104/pp.111.173427

Yang ZB, Liu G, Liu J, Zhang B, Meng W, Müller B, Hayashi KI, Zhang X, Zhao Z, De Smet I, et al. 2017. Synergistic action of auxin and cytokinin mediates aluminum-induced root growth inhibition in *Arabidopsis*. *EMBO Rep* **18:** 1213–1230. doi:10.15252/embr.201643806

Ye B, Shang G, Pan Y, Xu Z, Zhou C, Mao Y, Bao N, Sun L, Xu T, Wang J. 2020. AP2/ERF transcription factors integrate age and wound signals for root regeneration. *Plant Cell* **32:** 226–241. doi:10.1105/tpc.19.00378

Yin Y, Wang ZY, Mora-Garcia S, Li J, Yoshida S, Asami T, Chory J. 2002. BES1 accumulates in the nucleus in response to brassinosteroids to regulate gene expression and promote stem elongation. *Cell* **109:** 181–191. doi:10.1016/S0092-8674(02)00721-3

Yu X, Li L, Zola J, Aluru M, Ye H, Foudree A, Guo H, Anderson S, Aluru S, Liu P, et al. 2011. A brassinosteroid transcriptional network revealed by genome-wide identification of BESI target genes in *Arabidopsis thaliana*. *Plant J* **65:** 634–646. doi:10.1111/j.1365-313X.2010.04449.x

Yu F, Wu Y, Xie Q. 2015. Precise protein post-translational modifications modulate ABI5 activity. *Trends Plant Sci* **20:** 569–575. doi:10.1016/j.tplants.2015.05.004

Yuan TT, Xu HH, Zhang KX, Guo TT, Lu YT. 2014. Glucose inhibits root meristem growth via *ABA INSENSITIVE 5*, which represses PIN1 accumulation and auxin activity in

Arabidopsis. *Plant Cell Environ* **37:** 1338–1350. doi:10.1111/pce.12233

Yuan H, Liu W, Lu Y. 2017. CATALASE2 coordinates SA-mediated repression of both auxin accumulation and JA biosynthesis in plant defenses. *Cell Host Microbe* **21:** 143–155. doi:10.1016/j.chom.2017.01.007

Zádníková P, Petrásek J, Marhavy P, Raz V, Vandenbussche F, Ding Z, Schwarzerová K, Morita MT, Tasaka M, Hejátko J, et al. 2010. Role of PIN-mediated auxin efflux in apical hook development of *Arabidopsis thaliana*. *Development* **137:** 607–617. doi:10.1242/dev.041277

Zha M, Imran M, Wang Y, Xu J, Ding Y, Wang S. 2019. Transcriptome analysis revealed the interaction among strigolactones, auxin, and cytokinin in controlling the shoot branching of rice. *Plant Cell Rep* **38:** 279–293. doi:10.1007/s00299-018-2361-y

Zhang R, Zhang X, Wang J, Letham DS, McKinney SA, Higgins TJV. 1995. The effect of auxin on cytokinin levels and metabolism in transgenic tobacco tissue expressing an *ipt* gene. *Planta* **196:** 84–94. doi:10.1007/BF00193221

Zhang Z, Li Q, Li Z, Staswick PE, Wang M, Zhu Y, He Z. 2007. Dual regulation role of *GH3.5* in salicylic acid and auxin signaling during *Arabidopsis*-*Pseudomonas syringae* interaction. *Plant Physiol* **145:** 450–464. doi:10.1104/pp.107.106021

Zhang W, To JP, Cheng CY, Schaller GE, Kieber JJ. 2011. Type-A response regulators are required for proper root apical meristem function through post-transcriptional regulation of PIN auxin efflux carriers. *Plant J* **68:** 1–10. doi:10.1111/j.1365-313X.2011.04668.x

Zhang W, Swarup R, Bennett M, Schaller GE, Kieber JJ. 2013. Cytokinin induces cell division in the quiescent center of the *Arabidopsis* root apical meristem. *Curr Biol* **23:** 1979–1989. doi:10.1016/j.cub.2013.08.008

Zhang K, Wang R, Zi H, Li Y, Cao X, Li D, Guo L, Tong J, Pan Y, Jiao Y, et al. 2018. AUXIN RESPONSE FACTOR3 regulates floral meristem determinacy by repressing cytokinin biosynthesis and signaling. *Plant Cell* **30:** 324–346. doi:10.1105/tpc.17.00705

Zhang G, Zhao F, Chen L, Pan Y, Sun L, Bao N, Zhang T, Cui C, Qiu Z, Zhang Y, et al. 2019a. Jasmonate-mediated wound signalling promotes plant regeneration. *Nat Plants* **5:** 491–497. doi:10.1038/s41477-019-0408-x

Zhang W, Abdelrahman M, Jiu S, Guan L, Han J, Zheng T, Jia H, Song C, Fang J, Wang C. 2019b. VvmiR160s/VvARFs interaction and their spatio-temporal expression/cleavage products during GA-induced grape parthenocarpy. *BMC Plant Biol* **19:** 111. doi:10.1186/s12870-019-1719-9

Zhao Z, Andersen SU, Ljung K, Dolezal K, Miotk A, Schultheiss SJ, Lohmann JU. 2010. Hormonal control of the shoot stem-cell niche. *Nature* **465:** 1089–1092. doi:10.1038/nature09126

Zhao Y, Xing L, Wang X, Hou Y, Gao J, Wang P, Duan C, Zhu X, Zhu J. 2014. The ABA receptor PYL8 promotes lateral root growth by enhancing MYB77-dependent transcription of auxin-responsive genes. *Sci Signal* **7:** ra53. doi:10.1126/scisignal.2005051

Zheng X, Miller ND, Lewis DR, Christians MJ, Lee KH, Muday GK, Spalding EP, Vierstra RD. 2011. AUXIN UP-REGULATED F-BOX PROTEIN1 regulates the cross talk between auxin transport and cytokinin signaling during plant root growth. *Plant Physiol* **156:** 1878–1893. doi:10.1104/pp.111.179812

Zheng Z, Guo Y, Novák O, Dai X, Zhao Y, Ljung K, Noel JP, Chory J. 2013. Coordination of auxin and ethylene biosynthesis by the aminotransferase VAS1. *Nat Chem Biol* **9:** 244–246. doi:10.1038/nchembio.1178

Zhou X, Yuan F, Wang M, Guo A, Zhang Y, Xie CG. 2013. Molecular characterization of an ABA insensitive 5 orthologue in *Brassica oleracea*. *Biochem Biophys Res Commun* **430:** 1140–1146. doi:10.1016/j.bbrc.2012.12.023

Zhou W, Lozano-Torres JL, Blilou I, Zhang X, Zhai Q, Smant G, Li C, Scheres B. 2019. A jasmonate signaling network activates root stem cells and promotes regeneration. *Cell* **177:** 942–956.e14. doi:10.1016/j.cell.2019.03.006

Zhou J, Sittmann J, Guo L, Huang X, Pulapaka A, Liu Z. 2020. Gibberellin and auxin signaling genes *RGA1* and *ARF8* repress accessory fruit initiation in diploid strawberry. *Plant Physiol* doi:10.1093/plphys/kiaa087

Zhu X, Chen J, Xie Z, Gao J, Ren G, Gao S, Zhou X, Kuai B. 2015. Jasmonic acid promotes degreening via MYC2/3/4- and ANAC019/055/072-mediated regulation of major chlorophyll catabolic genes. *Plant J* **84:** 597–610. doi:10.1111/tpj.13030

Zhu M, Chen W, Mirabet V, Hong L, Bovio S, Strauss S, Schwarz EM, Tsugawa S, Wang Z, Smith RS, et al. 2020. Robust organ size requires robust timing of initiation orchestrated by focused auxin and cytokinin signalling. *Nat Plants* **6:** 686–698. doi:10.1038/s41477-020-0666-7

Zou J, Zhang S, Zhang W, Li G, Chen Z, Zhai W, Zhao X, Pan X, Xie Q, Zhu L. 2006. The rice *HIGH-TILLERING DWARF1* encoding an ortholog of *Arabidopsis* MAX3 is required for negative regulation of the outgrowth of axillary buds. *Plant J* **48:** 687–698. doi:10.1111/j.1365-313X.2006.02916.x

Zuniga-Mayo VM, Reyes-Olalde JI, Marsch-Martinez N, de Folter S. 2014. Cytokinin treatments affect the apical-basal patterning of the *Arabidopsis* gynoecium and resemble the effects of polar auxin transport inhibition. *Front Plant Sci* **5:** 191.

Zurek DM, Clouse SD. 1994. Molecular cloning and characterization of a brassinosteroid-regulated gene from elongating soybean (*Glycine max* L.) epicotyls. *Plant Physiol* **104:** 161–170. doi:10.1104/pp.104.1.161

Casting the Net—Connecting Auxin Signaling to the Plant Genome

Yanfei Ma,[1] Sebastian Wolf,[2] and Jan U. Lohmann[1]

[1]Department of Stem Cell Biology, [2]Cell Wall Signalling Group, Centre for Organismal Studies, Heidelberg University, D-69120 Heidelberg, Germany

Correspondence: jan.lohmann@cos.uni-heidelberg.de

Auxin represents one of the most potent and most versatile hormonal signals in the plant kingdom. Built on a simple core of only a few dedicated components, the auxin signaling system plays important roles for diverse aspects of plant development, physiology, and defense. Key to the diversity of context-dependent functional outputs generated by cells in response to this small molecule are gene duplication events and sub-functionalization of signaling components on the one hand, and a deep embedding of the auxin signaling system into complex regulatory networks on the other hand. Together, these evolutionary innovations provide the mechanisms to allow each cell to display a highly specific auxin response that suits its individual requirements. In this review, we discuss the regulatory networks connecting auxin with a large number of diverse pathways at all relevant levels of the signaling system ranging from biosynthesis to transcriptional response.

As sessile organisms, plants are exposed to a continuously changing environment, which is shaped by a plethora of abiotic and biotic parameters. To cope with these challenges, plants have evolved regulatory networks that allow them to constantly adapt their growth and developmental program to the current conditions. Given the decentralized nature of cellular decision making in plants, these networks need to fulfill a surprisingly large number of sometimes conflicting functions. First, they need to be able to integrate a substantial amount of information into a stable cellular program. For example, a plant may encounter a windy and cold, but very bright, day in spring with the soil moist and nutrients available in abundance. In this hypothetical environment, a plant finds excellent conditions for growth on the one hand with water and nutrient availability near the optimum, as well as enough light to drive a maximum of photosynthesis. On the other hand, UV irradiation, low temperatures, and wind impose substantial stress on the individual. To complicate matters for finding the optimal developmental solution, the opposing inputs mostly affect different parts of the plant, requiring tight communication between roots, leaves, and stems. Here, the plant encounters a second challenge, namely, the system-wide coupling of cell behavior. Whereas in the above example uni-

Cite this article as *Cold Spring Harb Perspect Biol* doi: 10.1101/cshperspect.a040006

form cell behavior across the entire plant likely is beneficial, there are cases, such as localized attack by herbivores or partial shading, where strictly local responses are required. A third challenge for the discussed regulatory networks is that they need to be deeply embedded into the developmental and metabolic programs, both for environmental sensing and for modulating growth and physiology. Last, the decision making has to be aware of temporal components to allow the plant to prepare for certain conditions expected in the future, based on previous exposures.

In this review, we will focus on regulatory networks that have evolved around the auxin signaling system, which plays key roles in diverse aspects of plant development, physiology, and defense. The most abundant and best-characterized auxin, indole acetic acid (IAA), is a small metabolite with similarity to tryptophan and able to move throughout the plant body in the phloem or by active and highly polar cell–cell transport (Okada et al. 1991; Gälweiler et al. 1998; Müller et al. 1998; Palme and Gälweiler 1999; Reinhardt et al. 2003; Petrášek et al. 2006; Tanaka et al. 2006; Adamowski and Friml 2015). The core auxin perception and signaling system is relatively simple, consisting of only three dedicated intracellular components: receptor (Dharmasiri et al. 2005; Kepinski and Leyser 2005), transcriptional executor (Ulmasov et al. 1997a; Hardtke and Berleth 1998), and inhibitor protein (Ulmasov et al. 1997b; Rouse et al. 1998; Hamann et al. 2002), which are coupled to the proteasome system (Gray et al. 2001). However, the pathway has evolved by gene duplication, and in many plants all components are represented by multiple copies with diverse expression patterns and biochemical properties (Vernoux et al. 2011; Weijers and Wagner 2016; Kato et al. 2020), giving rise to an enormous range in direct cellular responses to auxin through combinatorial logic. Importantly, the medium- and long-term auxin response is further shaped by intricate connections of this core auxin-sensing pathway with a dazzling array of signaling systems through transcriptional cross regulation, protein modification, or chromatin remodeling.

CONTROL OF AUXIN HOMEOSTASIS AND TRANSPORT

Many of the biological effects of auxin appear highly localized, such as organ initiation or tropic responses, and it follows that the hormone must be able to act with high spatiotemporal specificity. Indeed, it has turned out that local accumulation of active auxin is tightly regulated at multiple levels including biosynthesis, transport, as well as conjugation and degradation of this small metabolite. Since genetic analysis had suggested that auxin transport would suffice to create local effects, localized auxin biosynthesis, inactivation, or sensing was thought to play only a minor role until recently (Cheng et al. 2006, 2007; Stepanova et al. 2008; Tao et al. 2008; Brumos et al. 2018). However, with the elucidation of auxin anabolic and catabolic pathways and the expression patterns of the underlying genes, it became clear that this level of control is more important than previously recognized. Not only do many auxin metabolic genes exhibit highly specific spatiotemporal expression patterns, they also respond dynamically to a variety of stimuli, lending strong support to the importance of complex gene regulatory networks controlling auxin homeostasis.

Auxin (IAA) biosynthesis can follow both tryptophan (Trp)-dependent and -independent pathways, but currently only the Trp-dependent TAA-YUCCA pathway is well understood at the genetic and mechanistic level (Zhao 2018). To produce IAA from Trp, two sequential steps need to be carried out. The first step is catalyzed by enzymes encoded by members of the TRYPTOPHAN AMINOTRANSFERASE OF ARABIDOPSIS (TAA) gene family (Tao et al. 2008), whereas the second and rate-limiting step relies on flavin-containing monooxygenases encoded by the *YUCCA* (*YUC*) gene family (Zhao et al. 2001; Won et al. 2011). *YUC* genes show highly specific expression patterns and higher order *yucca* mutants show severe development defects (Cheng et al. 2006), lending support to the idea that levels of local auxin synthesis via the transcriptional regulation of *YUC*s is essential for the correct execution of developmental programs (Eklund et al. 2010; Cui et al. 2013;

Martínez-Fernández et al. 2014; Tang et al. 2017; Zhou et al. 2018; Blakeslee et al. 2019; Cao et al. 2019). An interesting example is the induction of *YUC* expression by the STYLISH (STY) transcription factor (TF) during fruit development (Eklund et al. 2010). *STY* expression is dependent on the activity of the AP2 TF DORNROESCHEN-LIKE (Eklund et al. 2011), which itself has been shown to act downstream of auxin signals (Cole et al. 2009), hinting at intricate feedforward connections. Similarly, in the root, auxin biosynthesis is controlled by the brassinosteroid (BR) signaling pathway via *TAA* and *YUC* genes in a cell-type-specific fashion, which in turn controls meristem development (Vragović et al. 2015). A subset of *YUC* genes have also been shown to be regulated at the epigenetic level. TFL2/LHP1 (Rizzardi et al. 2011), the HP1 homolog in plants, localizes to promoters of *YUC* genes and is required for their transcriptional activation. This interaction is functionally relevant, since in *tfl2/lhp1* mutants, the activity of the auxin output reporter DR5 is substantially reduced. In addition to these developmental inputs, altered *YUC* gene expression is also involved in the response to external stimuli, such as light (Li et al. 2012; Zhou et al. 2018), temperature (Sun et al. 2012), and biotic stress (Liu et al. 2016). For example, the growth responses downstream of shade perception and increase in ambient temperature are tightly connected and require direct binding of PHYTOCHROME INTERACTING FACTORS (PIFs) such as PIF4 and PIF7 to the promoters of *YUC* genes to increase local auxin biosynthesis (Hornitschek et al. 2012; Li et al. 2012; Hersch et al. 2014).

While the TAA-dependent step in the pathway does not appear to be rate limiting, *taa* mutants still exhibit substantial reduction in the levels of free IAA, along with obvious phenotypes (Stepanova et al. 2008). In line with the hypothesis that regulation of *TAA* genes is the primary mode of action to modulate auxin levels in response to environmental stimuli, *TAA* transcription has been shown to be under control of the phytohormones ethylene (Stepanova et al. 2008) and cytokinin (CK) (Yan et al. 2017), as well as light quality (Tao et al. 2008), high temperature (Franklin et al. 2011), or gravity (Cui et al. 2013). In addition to the transcriptional level, local auxin biosynthesis via TAA1 can also be regulated at the protein level by the switch of its enzymatic activity via T101 phosphorylation (Wang et al. 2020).

Acting downstream of auxin biosynthesis, IAA conjugation and degradation are also embedded into the regulatory networks underlying dynamic local auxin homeostasis. IAA conjugation to glutamic acid (IAA-Glu) or other amino acids, is the first step toward auxin degradation. The conjugating process is catalyzed by the Gretchen Hagen 3 (GH3) enzyme family, which are IAA-amido synthetases and first identified as auxin rapid-induced transcripts from soybean (Hagen and Guilfoyle 1985; Wright et al. 1987; Hagen et al. 1991). The feedback loop between auxin-induced increase in *GH3* mRNA accumulation and the conversion of IAA to inactive IAA-Glu by GH3 enzymatic activity provides a reset of local auxin levels. Importantly, the activation of *GH3* expression is also employed by other pathways, such as by CK signaling. In the root, the CK output TF ARABIDOPSIS RESPONSE REGULATOR 1 (ARR1) stimulates GH3.17, which transforms IAA into IAA-Glu in the lateral root cap, creating an auxin minimum defining the transition zone between cell division and cell differentiation in root meristem (Di Mambro et al. 2017, 2019). Unlike IAA-Asp or IAA-Glu, other IAA-amino acid conjugates can serve as a storage of IAA, such as IAA-Ala, and IAA-Leu, which can be hydrolyzed back to free IAA (Woodward and Bartel 2005). However, the oxidization of IAA into 2-oxoindole-3-acetic acid (OxIAA) is chemically irreversible, leading to the inactivation of IAA (Peer et al. 2013). The regulation of IAA oxidization was first revealed in a *dao* mutant (dioxygenase for auxin oxidation) in rice, which encodes a 2-oxoglutarate-dependent-Fe (II) dioxygenase, oxidizing free IAA into OxIAA (Zhao et al. 2013). Main defects of *dao* mutants in flowers and seeds highlight importance of auxin concentration in local contexts. Both *GH3* and *DAO* genes are induced by auxin, and inactivate auxin in a concentration-dependent manner, which serves as a dynamic feedback to maintain optimal auxin levels (Mellor et al. 2016). In addition to regulatory conjugation of IAA, the abundance of its precursor IPA is also controlled by the formation of IPA-glucose (IPA-Glc), a reaction mediated by an UDP-glycosyltransferase (UGT76F1). This mechanism contributes significantly to auxin homeostasis via competing with YUC enzymes for the shared substrate IPA (Chen et al. 2020). Again, these steps are tightly integrated into the regulatory landscape and it has been shown that the environmental integrator PIF4 can bind to the promoter of both *YUC* and *UGT76F1* (Chen et al. 2020), suggesting a switch-like activity to tune growth to current conditions.

Whereas synthesis and breakdown only recently have emerged as important players in setting up local gradients of auxin, transport processes have attracted attention for many decades. Directed cell-to-cell transport of auxin has been demonstrated to lie at the heart of a plethora of developmental and adaptive processes, including tissue patterning, body polarity, organ specification, and tropisms. In line with the massive influence on many aspects of plant life, auxin transport is dynamically regulated both at the organismal, as well as the tissue level. One fundamental property influencing this process is the acidic environment found in the cell wall, the extracellular matrix of plants (Wolf et al. 2012), which dictates that auxin is partially protonated. This part of the auxin pool can passively diffuse across the plasma membrane (PM), while the rest is actively transported into cells by the influx carrier AUX1/LAX (Vanneste and Friml 2009; Swarup and Péret 2012) and the nitrate transporter NRT1.1 (Krouk et al. 2010).

The molecules responsible for the proton gradient between apoplast and cytoplasm are PM-localized H^+-ATPase proton pumps. Interestingly, the transport capacity of the pumps can be regulated by auxin itself and environmental cues, such as light (Kinoshita and Shimazaki 1999; Takahashi et al. 2012) and the mechanism for modulating H^+-ATPase activity is through protein modification. Both auxin and blue light activate pump activity via SMALL AUXIN UP RNA (SAUR) proteins (McClure and Guilfoyle 1987; Hagen and Guilfoyle 2002) that in turn inhibit type 2C protein phosphatase D (PP2C-D) enzymes, which usually inhibit H^+-ATPases (Leyser et al. 1996; Kinoshita and Shimazaki 1999; Takahashi et al. 2012; Spartz et al. 2014; Fendrych et al. 2016; Ren et al. 2018). The transcriptional regulation of SAUR proteins indeed represents a key component of the rapid auxin-induced changes in elongation growth that initiated auxin research more than 150 years ago (Darwin 1880). The SAUR H^+-ATPase module also has been shown to couple apoplast pH regulation with other phytohormones or environmental cues. For instance, BR signaling is able to induce phosphorylation of H^+-ATPase in hypocotyl elongation either through transcriptional activation of SAURs (Minami et al. 2019) or through a fast BR-regulated PM protein–protein interaction between the BR signaling receptor BRI1 and the H^+-ATPase (Fig. 1; Caesar et al. 2011; Miao et al. 2018). In addition, an alternative pathway based on a ligand-receptor system, involving RAPID ALKALINIZATION FACTOR (RALF) and the receptor-like kinase FERONIA, is able to elicit a transient apoplast alkalinization via inhibitory phosphorylation of H^+-ATPases (Haruta et al. 2014).

Because deprotonated IAA is essentially incapable of crossing the PM, auxin influx carriers AUXIN1/LIKE-AUX1 (AUX1/LAX) are required for its influx in cells (Swarup and Péret 2012). AUX1, the founder of the AUX1/LAX family, is an amino acid permease-like protein (Bennett et al. 1996) that functions as a H^+/IAA^- symporter (Yang et al. 2006). AUX/LAX family members in *Arabidopsis* are all proven functional auxin transporters (Péret et al. 2012) and are involved in a large variety of auxin-mediated developmental processes, such as AUX1 in root gravitropism, root hair development, and phyllotaxy (Bennett et al. 1996; Marchant et al. 1999, 2002); LAX2 in vascular development and quiescent center (QC) maintenance (Péret et al. 2012; Zhang et al. 2013); AUX1 and LAX3 redundant in lateral root initiation and apical hook formation (Hobbie and Estelle 1995; Vandenbussche et al. 2010; Péret et al. 2012); and AUX1, LAX1, and LAX2 redundant in leaf patterning and embryogenesis (Bainbridge et al. 2008; Robert et al. 2015). Specific expression patterns of those family members suggest that these

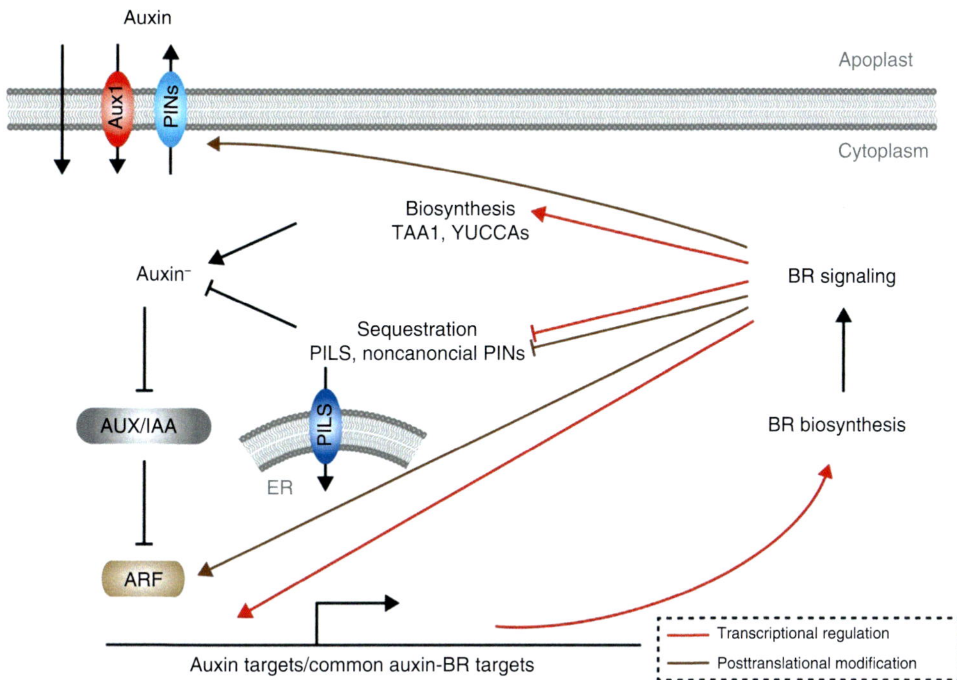

Figure 1. Regulatory network connecting auxin and brassinosteroid signaling. (ER) Endoplasmic reticulum, (BR) brassinosteroid.

genes are part of an intricate regulatory network and linked to developmental programs; however, most of the underlying mechanistic details still remain to be elucidated. Two examples of transcriptional control of *AUX1*/*LAX* genes show the prominent connection to hormone signaling: CK treatment promotes the binding of B-type ARR 1 and 12 TFs to the promoter of *LAX2* and causes QC cell division (Zhang et al. 2013). Once again highlighting the significance of feedback regulation in the auxin network, *LAX3* expression has shown to be activated by direct promoter binding of the nuclear auxin signaling module ARF7-LBD29 (Porco et al. 2016).

Once inside the cell, auxin is largely present as ionic IAA^- and therefore cannot exit the cell by diffusion. Hence, directional transport of auxin depends on two classes of auxin efflux carrier proteins, namely, the ATP-binding cassette of the B subfamily (ABCB) (Geisler et al. 2017) and PIN-FORMED (PIN) families (Okada et al. 1991; Gälweiler et al. 1998; Wisniewska et al. 2006; Adamowski and Friml 2015). A third class of transporters, PIN-like proteins (PILS) (Barbez et al. 2012), together with noncanonical PIN proteins (Bennett et al. 2014), regulate the intracellular auxin homeostasis by importing the hormone to the endoplasmic reticulum (ER). Since auxin mostly acts in the nucleus, this mechanism helps to create local response minima by exploiting the ER as auxin sink (Mravec et al. 2009; Barbez et al. 2012; Béziat et al. 2017; Feraru et al. 2019; Sun et al. 2020).

As the dominant directional auxin efflux carriers, PIN protein families are polarly localized at the PM and PIN localization predicts auxin flux direction (Petrášek et al. 2006; Vanneste and Friml 2009). Analysis of expression patterns and mutant phenotypes of *PIN* genes demonstrated that PINs are the major determinants of local auxin accumulation (Vieten et al. 2007). Polar localization of PINs on the PM is dependent on controlled exo- and endocytosis (Geldner et al. 2001; Paciorek et al. 2005; Kleine-Vehn et al. 2011) and posttranslational modification represents a major regulatory step in this context (Friml et al. 2004; Abas et al. 2006; Michniewicz et al. 2007; Kleine-Vehn et al. 2009; Leitner et al. 2012; Marhavý et al. 2014; Zourelidou et al. 2014; Weller et al. 2017; Grones et al. 2018; Marhava et al. 2018; Hajný et al. 2020).

Directional auxin transport represents an important module for the spatiotemporal regulation of cell behavior. Using this module, auxin feedbacks are integrated with developmental and environmental signals. For example, *PIN* gene expression and subsequent subcellular transport protein polarization is enhanced by auxin, which generates a powerful positive feedback loop that plays important roles in many developmental processes, such as phyllotaxis (Bhatia et al. 2016), leaf serrations (Bilsborough et al. 2011), canalization (Sauer et al. 2006; Hajný et al. 2020), lateral root initiation (Chen et al. 2015; Wang et al. 2015), and meristem maintenance (Krogan et al. 2016). This mechanism is based on the transcriptional activation of auxin pathway genes, including *PINs* (Paponov et al. 2008), which is dependent on auxin nuclear signaling and auxin concentration (Vieten et al. 2005). At the molecular level, ARFs can bind to the promoters of *PIN* genes in a variety of tissues (Chen et al. 2015; Krogan et al. 2016; Simonini et al. 2017) and are responsible for the reinforcement of auxin signaling and auxin transport (Sauer et al. 2006; Bilsborough et al. 2011; Chen et al. 2015; Bhatia et al. 2016; Krogan et al. 2016; Weijers 2016). For example, a small number of MP/ARF5-positive cells at the periphery of the shoot apical meristem (SAM) promote polarization of PIN1 protein in neighboring cells toward the MP/ARF5-expressing cells, thereby further enhancing auxin influx into the founder cells, which eventually will give rise to a new organ primordium (Bhatia et al. 2016). Similarly, during the early steps of early lateral root formation, ARF7 activates *PIN3* expression via direct binding to its promoter. Interestingly, ARF7 also activates expression of the MYB TF FOUR LIPS (FLP), which further enhances *PIN3* expression. This feedforward regulation helps sustain *PIN3* expression and likely makes the system more robust against initial auxin fluctuations (Chen et al. 2015). Another prominent example is the *Arabidopsis* fruit, where highly specific auxin accumulation patterns are achieved by intricate spatiotemporal control of *PIN* gene activity and PIN protein localization. First, auxin signaling output is switched from a bilateral to a radial pattern by the direct promotion of *PIN1* and *PIN3* expression through bHLH TFs of the HECATE clade, which likely act together with the SPATULA (SPT) bHLH factor (Schuster et al. 2015). Importantly, at the fruit tip, auxin transporters need to switch to an apolar localization to allow for accumulation of auxin in these cells and this transition is dependent on transcriptional repression of *PINOID* (*PID*) by the INDIHISCENT (IND)/SPT module (Moubayidin and Østergaard 2014). A similar process is required to pattern the valve margins to support seed dispersal. Here, *IND* directly represses *PID* and activates the related *WAG2* kinase to promote PIN3 delocalization, leading to a local minimum of auxin accumulation, which in turn is the signal for valve margin development (Sorefan et al. 2009). Another important example is the differential deployment of the auxin-PIN1-CUC2 module during leaf development in *Arabidopsis thaliana* and *Cardamine hirsuta*, which provides the substrate for leaf shape variations during evolution (Kierzkowski et al. 2019).

Several other hormone pathways are integrated with auxin signaling at the level of the auxin transport machinery. For example, numerous studies have shown that CK signaling can interfere with polar auxin transport through direct transcriptional control of *PINs* (Pernisová et al. 2009; Ruzicka et al. 2009; Bishopp et al. 2011a,b; Xie et al. 2018), which has in some cases been shown to be mediated directly through B-ARR TFs. In addition, CK has been shown to interfere with auxin signaling–mediated reinforcement of *PIN* expression through modulating the expression of *SHY2*/*IAA3* (Dello Ioio et al. 2008). CK signaling can also influence PIN activity through posttranslational regulation (Marhavý et al. 2011, 2014; Zhang et al. 2011; Waldie and Leyser 2018), for example, by triggering depletion of PINs from specific polar membrane domains, thus enhancing polarity (Marhavý et al. 2011, 2014). A particularly intriguing example is on display during lateral root formation. Auxin signaling is involved in all stages of lateral root development, which is initiated by a local auxin response maximum in xylem pole pericycle cells near the root tip (Fukaki et al. 2002, 2005; Wilmoth et al. 2005; De Smet et al. 2007, 2010; Dubrovsky et al. 2008; Moreno-Risueno et al. 2010; Xuan et al. 2016). During further steps, auxin flux continues to be regulated in a complex and dynamic manner. For example, auxin itself induces expression of the auxin influx carriers *AUX1* and *LAX3* in the cells overlaying lateral root primordia. This, in turn, leads to auxin-mediated expression of cell wall remodeling enzymes, which helps to separate cells in the tissues that have to be crossed by the lateral root before emergence (Swarup et al. 2008). CK, on the other hand, negatively affects lateral root initiation and regulates expression and polarity of PIN proteins (Laplaze et al. 2007; Ruzicka et al. 2009; Marhavý et al. 2011, 2014; Bielach et al. 2012). Interestingly, CK produced in and around lateral root primordia forms an inhibitory field preventing the formation of further primordia and thus is important for the regulation of lateral root spacing (Chang et al. 2015). Besides CK, BRs (Retzer et al. 2019), strigolactones (Crawford et al. 2010; Shinohara et al. 2013), gibberellic acid (Willige et al. 2011; Löfke et al. 2013), and other phytohormones significantly influence polar auxin transport (reviewed in Semeradova et al. 2020).

CONTROL OF AUXIN SIGNALING

While the core auxin signaling system is simple and only consists of three basic modules, namely, receptor, inhibitor, and transcriptional output regulator, these elements have been multiplied by gene duplication events and co-opted in a multitude of regulatory programs. In *Arabidopsis*, there are six TIR1/AFB-like receptors, 29 Aux/IAA signaling inhibitors, and 23 ARF transcriptional regulators providing an extensive network for functional specification. Supporting the idea that diverse combinations of signaling components will result in context-dependent outputs, all members of the system display highly specific spatiotemporal expression patterns (Rademacher et al. 2011; Vernoux et al. 2011; Weijers and Wagner 2016; Kato et al. 2020). Recently, it was shown that activator *ARF* expression patterns are driven by a combination of constitutively open chromatin at their loci and the interaction of specific transcriptional repressors with the regulatory regions of individual ARF-encoding genes (Truskina et al. 2021). Importantly, because many WRKY regulators were found among the repressors, these findings suggest strong connectivity of activator ARFs to transcriptional inputs relaying environmental information. In addition, key components of developmental circuits, such as WUS, CUC2, and KNAT1, were found to bind to *ARF* promoters, supporting the view that ARFs may serve as central nodes integrating exogenous and endogenous signals.

Another important aspect is that many components, most prominently the Aux/IAA auxin-response inhibitors, are transcriptionally activated by auxin (Abel et al. 1995), providing a potent feedback regulation in the network. Importantly, these interactions are mediated by ARF binding to promoters of *Aux/IAA* genes, which therefore have served as models to elucidate auxin output regulation (Krogan et al. 2014). One prominent example for a developmentally controlled input is the posttranscriptional control of *ARF3* mRNA accumulation by *trans*-acting siRNAs (ta-siRNA). Here, several gene-regulatory layers are superimposed to result in spatiotemporal refinement of *ARF3* and *ARF4* activity. First, the *TAS3* mRNA is transcribed, which is then targeted by *miR390* for phased processing into ta-siRNAs, which in turn target *ARF3* and *ARF4* transcripts for degradation. This intricate regulatory regime is required for a number of developmental contexts, including leaf development, lateral root growth, or developmental timing (Allen et al. 2005; Adenot et al. 2006; Fahlgren et al. 2006; Garcia et al. 2006; Marin et al. 2010).

In addition to their specific expression patterns, auxin signaling components exhibit divergent biochemical functions, which greatly contribute to the complexity of the resulting regulatory network. Most notably, ARF TFs, responsible for the transcriptional output of the system, show considerable sequence divergence and can functionally be grouped in those that have the potential to activate transcription (activator ARFs) and those that do not (repressor ARFs) (Mutte et al. 2018). The sequence divergence has also brought about specific DNA-binding behaviors and preferences for orientation and spacing of DNA-binding motifs, which together specify ARF-dependent gene regulatory output (Boer et al. 2014; Freire-Rios et al. 2020; Lanctot and Nemhauser 2020).

In addition to the transcriptional and biochemical level, specificity is also provided by the protein–protein interaction preferences between ARFs and Aux/IAAs, which in turn dictate the transcriptional output of the pathway (Vernoux et al. 2011; Weijers and Wagner 2016). As a rule of thumb, activator ARFs tend to show interactions with a large number of Aux/IAAs, whereas repressor ARFs do not (Vernoux et al. 2011). In addition to the complex dimerization patterns observed within the ARF-Aux/IAA universe, ARFs also directly interact with a number of other transcriptional regulators from diverse families, such as MYB, MADS-box, KANADI, RGA (GRAS), and bHLH (Shin et al. 2007; Varaud et al. 2011; Kelley et al. 2012; Smaczniak et al. 2012; Oh et al. 2014; José Ripoll et al. 2015). Importantly, MP interacts with the SWI/SNF chromatin-remodeling factors BRAHMA (BRM) and SPLAYED (SYD), which act to open chromatin at target loci. These positive regulators of gene expression compete with the binding of Aux/IAA proteins, which in turn interact with proteins of the TOPLESS family of transcriptional corepressors to repress target expression via histone deacetylation (Long et al. 2006; Szemenyei et al. 2008; Wu et al. 2015). Because auxin sensing triggers the degradation of Aux/IAAs, this represents a major switch for downstream gene activity. Another example for an inhibitory ARF interactor is the bHLH TF HECATE1, the activity of which seems to counteract the capacity of MP to drive auxin-dependent transcription in the periphery of the SAM (Gaillochet et al. 2017).

Recent studies have revealed that ARF6 forms a trimeric complex with PIF4 and the BR responsive TF BZR1 to integrate multiple cues into regulation of cell elongation. DNA binding of the three TFs to their common target loci is inhibited by DELLA proteins, thus GA signaling is required to release inhibition (Oh et al. 2014). BR and auxin signaling are connected at multiple levels in addition to convergence on target genes (Nemhauser et al. 2004). For example, auxin transcriptionally regulates BR biosynthesis and signaling (Chung et al. 2011; Yoshimitsu et al. 2011; Sakamoto et al. 2013; Zhang et al. 2015), whereas a BR input is wired into auxin signaling through interaction of the GSK3-like kinase BIN2, a BR signaling component, and ARFs (Fig. 1). BIN2 phosphorylates the (repressive) B-class ARF2, which decreases DNA binding of the TF, and thus enhances auxin response (Vert et al. 2008). BIN2 can also phosphorylate ARF7/19 to promote DNA binding by disturbing interaction with IAAs, thus enhancing auxin response. However, in this case, BIN2 does not seem to provide BR input but acts downstream of the signaling peptide TDIF (CLE41/44) and its receptor PXY (Cho et al. 2014).

It has recently emerged that posttranslational modification of ARFs seems to be a more general mechanism to mediate integration of inputs form other pathways. For example, ARF2 can also be phosphorylated in response to potassium stress (Zhao et al. 2016), whereas root branching toward water involves ARF7 SUMOylation. Here, SUMO negatively affects DNA binding activity and enhances interaction with IAA3, thus attenuating auxin signaling (Orosa-Puente et al. 2018).

CONTROL OF AUXIN OUTPUT

For auxin to be able to play so many divergent roles, the output of the signaling pathway needs to be tightly integrated with other regulatory programs. Importantly, since many activities are highly local, this also requires that the competence for specific auxin responses is limited to few cells by independent repression of downstream genes. Plant stem cell systems, or meristems, represent prominent examples for this scenario, since auxin plays essential, but highly specific roles in subsets of meristematic cells. In the root meristem, auxin accumulates maximally in niche and stem cells and is required for the maintenance of both populations (Sabatini et al. 1999; Friml et al. 2002; Vanneste and Friml 2009); however, adjacent cells that are also exposed to high levels of auxin do not exhibit stem cell fate. In the shoot meristem, high auxin signaling output is excluded from niche and stem cells, but required for inducing differentiation in adjacent cells (Vernoux et al. 2011; Luo et al. 2018; Ma et al. 2019). And finally, auxin provides a locally restricted signal for the development of wood from the cambium, despite a substantially broader accumulation in this tissue (Brackmann et al. 2018). All of these cases have in common that auxin signaling interacts with regulatory modules controlled by WUSCHEL-RELATED HOMEOBOX (WOX) TFs. WUSCHEL (WUS) is a master regulator in shoot apical stem cells (Mayer et al. 1998; Leibfried et al. 2005; Busch et al. 2010; Yadav et al. 2013; Ma et al. 2019), whose family members play conserved functions in stem cell induction, organization, and maintenance, such as WOX5 in root QC cells (Sarkar et al. 2007), WOX4 in cambium stem cells (Brackmann et al. 2018), and root cambium organizer cells (Smetana et al. 2019), as well as other WOXs in embryonic stem cells (WOX2) (Zhang et al. 2017), leaf mediolateral axis cells (WOX1 and WOX3/PRS) (Nakata et al. 2012; Guan et al. 2017; Zhang et al. 2020), and de novo organogenesis (WOX11/12) (Liu et al. 2014). In the stem cambium, WOX4 promotes stem cell fate and is activated by ARF3/4, while it is repressed by ARF5/MP (Brackmann et al. 2018). These results showed that auxin not only is required to position the stem cells, but at the same time to define the fate of one of the derived cell lineages. In the SAM, auxin acts mainly as a local signal to trigger differentiation in the periphery; however, SAM treatments even with high doses and in the absence of auxin transport do not lead to stem cell termination (Reinhardt et al. 2003; Guenot et al. 2012). This robustness is conferred by the local activity of WUS in niche and stem cells by acting as a global repressor of auxin-dependent transcriptional activation (Ma et al. 2019). Interestingly, even in the shoot, stem cells require a basal level of auxin signaling, and it turns out that WUS acts as an auxin response rheostat via the promotion of histone deacetylation (Ma et al. 2019). The dependence of stem cells on low auxin signaling may be connected to the cross talk of auxin with CK (Fig. 2). In the SAM, auxin acts to repress the expression of type-A ARRs, negative feedback regulators of CK via direct binding of MP to ARR promoters, thereby synergizing with the activity of WUS (Leibfried et al. 2005; Zhao et al. 2010). Breaking this interaction may lead to reduced local CK signaling and hence a termination of the stem cell pool. Similarly, during embryonic stem cell initiation, WOX2 represses auxin signaling and in parallel activates CK, which shields the stem cells from differentiation (Zhang et al. 2017). Later, during the initiation of the root meristem, auxin induced expression of type-A ARRs, which in turn limits CK output in the hypophysis to the cells that will give rise to the QC (Müller and Sheen 2008). Along these lines, auxin signaling is required to restrict WOX5 expression to the QC of the mature root meristem, probably through microRNA *MIR169*-targeted ARF10 and ARF16 auxin-response factors (Ding and Friml 2010). Importantly, WOX5 expression and QC function depend on PLETHORA (PLT) proteins, AP2 TFs that play important roles in setting the balance between cell proliferation and differentiation in roots and shoots (Aida et al. 2004; Galinha et al. 2007; Cruz-Ramírez et al. 2013; Mähönen et al. 2014; Shimotohno et al. 2018). *PLT* expression, in turn, depends on auxin, and PLT proteins promote auxin biosynthesis and signaling, forming a positive feedback loop (Santuari et al. 2016). PLT proteins form a relatively stable and inert gradient that is generally aligned with the auxin-response gradient in the root meristem (Sabatini et al. 2003; Billou et al. 2005; Grieneisen et al. 2007), but uncouples developmental patterning from rapid auxin fluctuations (Mähönen et al. 2014), for example, those observed during growth adjustments and tropic responses (Shih et al. 2015; Fendrych et al. 2018; Harmer and Brooks 2018). This auxin-PLT module is antagonized by CK signaling to accomplish robust zonation of the longitudinal root meristem (Di Mambro et al. 2017; Salvi et al. 2020).

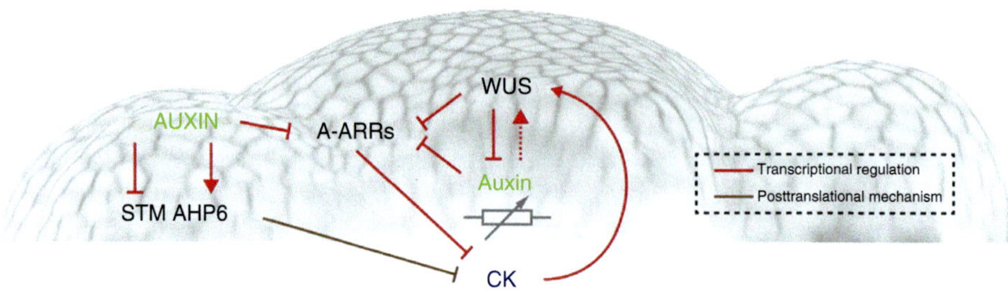

Figure 2. Regulatory network connecting auxin and cytokinin signaling in the shoot apical meristem (SAM). (CK) Cytokinin.

Several other regulators interact with auxin to coordinate developmental decisions. At the shoot apex, differentiation is repressed by the class I KNOX homeobox TF SHOOTMERISTEMLESS (STM), whose combined activity with WUS is required for stem cell fate (Lenhard et al. 2002). To allow differentiation at the periphery, STM expression is turned off in response to local PIN1-mediated auxin accumulation (Heisler et al. 2005) via histone deacetylation-mediated transcriptional silencing (Chung et al. 2019). This epigenetic reprogramming of the *STM* locus is achieved directly by ARF3/ARF4 and indirectly by MP through the intermediary regulator FILAMENTOUS FLOWER (FIL) (Chung et al. 2019). Interestingly, *STM* at the same time receives positive regulatory inputs by mechanical signals (Landrein et al. 2015), providing an important integration point, which likely serves to properly position organ primordia at the shoot apex. In parallel to this system, auxin signaling at the SAM interacts with an elaborate prepattern setup by the polarity factors KANADI (KAN) and REVOLUTA (REV), one of the HD-ZIP III TFs. While *REV* is expressed throughout a wide central domain, but excluded from the periphery due to the activity of the mobile miR165/166, KAN accumulation is restricted to the very periphery. Interestingly, both factors repress organ formation, directing maximal *PIN1* expression and auxin response to the boundary of their expression domains (Caggiano et al. 2017). This not only defines a competence zone for differentiation, but also prepatterns emerging organ primordia, because the cells in the specification field inherit the *REV*/*KAN* expression domains, which in turn will instruct dorsoventral patterning of leaves.

The developing vascular system of the root provides another great example for how auxin signaling is embedded into developmentally regulated transcriptional networks. In a first step, a local auxin maximum established by polar auxin transport confers provascular initial identity to some of the inner cells of the embryo at the early globular stage. MP plays a crucial role in the formation of vascular tissue and activates a suite of downstream TFs termed TARGET OF MONOPTEROS (TMO) (Hardtke and Berleth 1998; Schlereth et al. 2010). One of these, namely, the bHLH regulator TMO7, moves from the basal cell of the embryo proper to the topmost cell of the suspensor to trigger hypophysis development. In contrast, TMO5, another MP target in the embryo, acts cell-autonomously by dimerizing with LONESOME HIGHWAY, but holds the key to setting up the vascular CK signaling domain to promote periclinal cell divisions in both the embryo and the primary root (De Rybel et al. 2013, 2014; Ohashi-Ito et al. 2014). In xylem precursor cells, the TMO5/LHW complex promotes the expression of *LONELY GUY4* (*LOG4*) encoding for a rate-limiting enzyme in CK biosynthesis (De Rybel et al. 2014). Concomitantly, xylem precursor cells are desensitized against the CK they produce through auxin-induced expression of the CK signaling inhibitor AHP6 (Mähönen et al. 2006; Bishopp et al. 2011a; Ohashi-Ito et al. 2014). However, within the xylem axis, *AHP6* is only expressed in protoxylem precursor cells, due to the presence of auxin-induced HD-ZIP

III TFs in the central xylem cells (Carlsbecker et al. 2010; Bishopp et al. 2011a; Ursache et al. 2014). As these central cells are also largely insensitive toward CK, the existence of another, yet unidentified inhibitory factor besides AHP6 has been postulated. HD-ZIP III TFs also feed back on the auxin response as PHABULSOSA promotes the expression of both MP and the non-auxin-responsive *IAA20*, indicating an elaborate regulation of auxin signaling strength (Müller et al. 2016). Taken together, high auxin response in xylem precursor cells promotes the biosynthesis of CK and simultaneously cell autonomously interferes with CK signaling. Xylem precursor cell-derived CK thus acts non–cell autonomously in the adjacent procambial domain, where CK signaling promotes formative (periclinal) cell divisions (Mähönen et al. 2000, 2006). In addition, CK in the procambium, which is also derived from phloem transport cells, is required for the expression and subcellular localization of PIN proteins that direct auxin flow toward the xylem axis (Bishopp et al. 2011a,b). This intricate network is further stabilized by a group of DOF TFs called PEARs, which are induced in protophloem sieve-element precursor cells by CK. These factors move to neighboring procambial cells and promote cell division and in addition enhance HD-ZIP III expression, which in turn antagonize PEARs (Miyashima et al. 2019). Interestingly, the bilateral symmetry of the root vasculature, established through the mutually inhibitory interdigitation of auxin and CK signaling, also impacts on tissues outside of the stele (Andersen et al. 2018).

Later in development, during secondary root growth, MP and HD-ZIP III activities synergize to activate a canonical WOX organizer module, in this case built around WOX4. Importantly, vascular stem cell fate is induced non–cell autonomously in the procambial cell adjacent to the WOX4 expressing organizer cell, leading to a direct and indirect induction of xylem and phloem identity, respectively, via a common stem cell (Smetana et al. 2019). This elegant coupling of highly local auxin signaling, CK production, and mobile factors allow for a stepwise elaboration of a complex vascular system from a single inductive cell in the embryo.

Taken together, research in past decades has uncovered that the auxin signaling system is deeply embedded into a multitude of regulatory networks at all levels from biosynthesis to transcriptional output and from the organismal to the subcellular scale. This very embedding appears to be the mechanism allowing one small molecule to play so many and so different roles during plant life.

REFERENCES

Abas L, Benjamins R, Malenica N, Paciorek TT, Wiřniewska J, Moulinier-Anzola JC, Sieberer T, Friml J, Luschnig C. 2006. Intracellular trafficking and proteolysis of the *Arabidopsis* auxin-efflux facilitator PIN2 are involved in root gravitropism. *Nat Cell Biol* **8:** 249–256. doi:10.1038/ncb1369

Abel S, Nguyen MD, Theologis A. 1995. The PS-IAA4/5-like family of early auxin-inducible mRNAs in *Arabidopsis thaliana*. *J Mol Biol* **251:** 533–549. doi:10.1006/jmbi.1995.0454

Adamowski M, Friml J. 2015. PIN-dependent auxin transport: action, regulation, and evolution. *Plant Cell* **27:** 20–32. doi:10.1105/tpc.114.134874

Adenot X, Elmayan T, Lauressergues D, Boutet S, Bouché N, Gasciolli V, Vaucheret H. 2006. DRB4-dependent TAS3 *trans*-acting siRNAs control leaf morphology through AGO7. *Curr Biol* **16:** 927–932. doi:10.1016/j.cub.2006.03.035

Aida M, Beis D, Heidstra R, Willemsen V, Blilou I, Galinha C, Nussaume L, Noh YS, Amasino R, Scheres B. 2004. The PLETHORA genes mediate patterning of the *Arabidopsis* root stem cell niche. *Cell* **119:** 109–120. doi:10.1016/j.cell.2004.09.018

Allen E, Xie Z, Gustafson AM, Carrington JC. 2005. MicroRNA-directed phasing during *trans*-acting siRNA biogenesis in plants. *Cell* **121:** 207–221. doi:10.1016/j.cell.2005.04.004

Andersen TG, Naseer S, Ursache R, Wybouw B, Smet W, De Rybel B, Vermeer JEM, Geldner N. 2018. Diffusible repression of cytokinin signalling produces endodermal symmetry and passage cells. *Nature* **555:** 529–533. doi:10.1038/nature25976

Bainbridge K, Guyomarc'h S, Bayer E, Swarup R, Bennett M, Mandel T, Kuhlemeier C. 2008. Auxin influx carriers stabilize phyllotactic patterning. *Genes Dev* **22:** 810–823. doi:10.1101/gad.462608

Barbez E, Kubeš M, Rolčík J, Béziat C, Pěnčík A, Wang B, Rosquete MR, Zhu J, Dobrev PI, Lee Y, et al. 2012. A novel putative auxin carrier family regulates intracellular auxin homeostasis in plants. *Nature* **485:** 119–122. doi:10.1038/nature11001

Bennett MJ, Marchant A, Green HG, May ST, Ward SP, Millner PA, Walker AR, Schulz B, Feldmann KA. 1996. *Arabidopsis* AUX1 gene: a permease-like regulator of root gravitropism. *Science* **273:** 948–950. doi:10.1126/science.273.5277.948

Bennett T, Brockington SF, Rothfels C, Graham SW, Stevenson D, Kutchan T, Rolf M, Thomas P, Wong GKS, Leyser O, et al. 2014. Paralogous radiations of PIN proteins with multiple origins of noncanonical PIN structure. *Mol Biol Evol* **31:** 2042–2060. doi:10.1093/molbev/msu147

Béziat C, Barbez E, Feraru MI, Lucyshyn D, Kleine-Vehn J. 2017. Light triggers PILS-dependent reduction in nuclear auxin signalling for growth transition. *Nat Plants* **3:** 17105. doi:10.1038/nplants.2017.105

Bhatia N, Bozorg B, Larsson A, Ohno C, Jönsson H, Heisler MG. 2016. Auxin acts through MONOPTEROS to regulate plant cell polarity and pattern phyllotaxis. *Curr Biol* **26:** 3202–3208. doi:10.1016/j.cub.2016.09.044

Bielach A, Podlešáková K, Marhavý P, Duclercq J, Cuesta C, Müller B, Grunewald W, Tarkowski P, Benkováa E. 2012. Spatiotemporal regulation of lateral root organogenesis in *Arabidopsis* by cytokinin. *Plant Cell* **24:** 3967–3981. doi:10.1105/tpc.112.103044

Billou I, Xu J, Wildwater M, Willemsen V, Paponov I, Frimi J, Heldstra R, Aida M, Palme K, Scheres B. 2005. The PIN auxin efflux facilitator network controls growth and patterning in *Arabidopsis* roots. *Nature* **433:** 39–44. doi:10.1038/nature03184

Bilsborough GD, Runions A, Barkoulas M, Jenkins HW, Hasson A, Galinha C, Laufs P, Hay A, Prusinkiewicz P, Tsiantis M. 2011. Model for the regulation of *Arabidopsis thaliana* leaf margin development. *Proc Natl Acad Sci* **108:** 3424–3429. doi:10.1073/pnas.1015162108

Bishopp A, Help H, El-Showk S, Weijers D, Scheres B, Friml J, Benková E, Mähönen AP, Helariutta Y. 2011a. A mutually inhibitory interaction between auxin and cytokinin specifies vascular pattern in roots. *Curr Biol* **21:** 917–926. doi:10.1016/j.cub.2011.04.017

Bishopp A, Lehesranta S, Vatén A, Help H, El-Showk S, Scheres B, Helariutta K, Mähönen AP, Sakakibara H, Helariutta Y. 2011b. Phloem-transported cytokinin regulates polar auxin transport and maintains vascular pattern in the root meristem. *Curr Biol* **21:** 927–932. doi:10.1016/j.cub.2011.04.049

Blakeslee JJ, Spatola Rossi T, Kriechbaumer V. 2019. Auxin biosynthesis: spatial regulation and adaptation to stress. *J Exp Bot* **70:** 5041–5049. doi:10.1093/jxb/erz283

Boer DR, Freire-Rios A, Van Den Berg WAM, Saaki T, Manfield IW, Kepinski S, López-Vidrieo I, Franco-Zorrilla JM, De Vries SC, Solano R, et al. 2014. Structural basis for DNA binding specificity by the auxin-dependent ARF transcription factors. *Cell* **156:** 577–589. doi:10.1016/j.cell.2013.12.027

Brackmann K, Qi J, Gebert M, Jouannet V, Schlamp T, Grünwald K, Wallner ES, Novikova DD, Levitsky VG, Agustí J, et al. 2018. Spatial specificity of auxin responses coordinates wood formation. *Nat Commun* **9:** 875. doi:10.1038/s41467-018-03256-2

Brumos J, Robles LM, Yun J, Vu TC, Jackson S, Alonso JM, Stepanova AN. 2018. Local auxin biosynthesis is a key regulator of plant development. *Dev Cell* **47:** 306–318.e5. doi:10.1016/j.devcel.2018.09.022

Busch W, Miotk A, Ariel FD, Zhao Z, Forner J, Daum G, Suzaki T, Schuster C, Schultheiss SJ, Leibfried A, et al. 2010. Transcriptional control of a plant stem cell niche. *Dev Cell* **18:** 841–853. doi:10.1016/j.devcel.2010.03.012

Caesar K, Elgass K, Chen Z, Huppenberger P, Witthöft J, Schleifenbaum F, Blatt MR, Oecking C, Harter K. 2011. A fast brassinolide-regulated response pathway in the plasma membrane of *Arabidopsis thaliana. Plant J* **66:** 528–540. doi:10.1111/j.1365-313X.2011.04510.x

Caggiano MP, Yu X, Bhatia N, Larsson A, Ram H, Ohno CK, Sappl P, Meyerowitz EM, Jönsson H, Heisler MG. 2017. Cell type boundaries organize plant development. *eLife* **6:** e27421. doi:10.7554/eLife.27421

Cao X, Yang H, Shang C, Ma S, Liu L, Cheng J. 2019. The roles of auxin biosynthesis YUCCA gene family in plants. *Int J Mol Sci* **20:** 8–10.

Carlsbecker A, Lee JY, Roberts CJ, Dettmer J, Lehesranta S, Zhou J, Lindgren O, Moreno-Risueno MA, Vatén A, Thitamadee S, et al. 2010. Cell signalling by microRNA165/6 directs gene dose-dependent root cell fate. *Nature* **465:** 316–321. doi:10.1038/nature08977

Chang L, Ramireddy E, Schmülling T. 2015. Cytokinin as a positional cue regulating lateral root spacing in *Arabidopsis. J Exp Bot* **66:** 4759–4768. doi:10.1093/jxb/erv252

Chen Q, Liu Y, Maere S, Lee E, Van Isterdael G, Xie Z, Xuan W, Lucas J, Vassileva V, Kitakura S, et al. 2015. A coherent transcriptional feed-forward motif model for mediating auxin-sensitive PIN3 expression during lateral root development. *Nat Commun* **6:** 8821. doi:10.1038/ncomms9821

Chen L, Huang XX, Zhao SM, Xiao DW, Xiao LT, Tong JH, Wang WS, Li YJ, Ding Z, Hou BK. 2020. IPyA glucosylation mediates light and temperature signaling to regulate auxin-dependent hypocotyl elongation in *Arabidopsis. Proc Natl Acad Sci* **117:** 6910–6917. doi:10.1073/pnas.2000172117

Cheng Y, Dai X, Zhao Y. 2006. Auxin biosynthesis by the YUCCA flavin monooxygenases controls the formation of floral organs and vascular tissues in *Arabidopsis. Genes Dev* **20:** 1790–1799. doi:10.1101/gad.1415106

Cheng Y, Dai X, Zhao Y. 2007. Auxin synthesized by the YUCCA flavin monooxygenases is essential for embryogenesis and leaf formation in *Arabidopsis. Plant Cell* **19:** 2430–2439. doi:10.1105/tpc.107.053009

Cho H, Ryu H, Rho S, Hill K, Smith S, Audenaert D, Park J, Han S, Beeckman T, Bennett MJ, et al. 2014. A secreted peptide acts on BIN2-mediated phosphorylation of ARFs to potentiate auxin response during lateral root development. *Nat Cell Biol* **16:** 66–76. doi:10.1038/ncb2893

Chung Y, Maharjan PM, Lee O, Fujioka S, Jang S, Kim B, Takatsuto S, Tsujimoto M, Kim H, Cho S, et al. 2011. Auxin stimulates DWARF4 expression and brassinosteroid biosynthesis in *Arabidopsis. Plant J* **66:** 564–578. doi:10.1111/j.1365-313X.2011.04513.x

Chung Y, Zhu Y, Wu MF, Simonini S, Kuhn A, Armenta-Medina A, Jin R, Østergaard L, Gillmor CS, Wagner D. 2019. Auxin response factors promote organogenesis by chromatin-mediated repression of the pluripotency gene SHOOTMERISTEMLESS. *Nat Commun* **10:** 886. doi:10.1038/s41467-019-08861-3

Cole M, Chandler J, Weijers D, Jacobs B, Comelli P, Werr W. 2009. DORNRÖSCHEN is a direct target of the auxin response factor MONOPTEROS in the *Arabidopsis* embryo. *Development* **136:** 1643–1651. doi:10.1242/dev.032177

Crawford S, Shinohara N, Sieberer T, Williamson L, George G, Hepworth J, Müller D, Domagalska MA, Leyser O. 2010. Strigolactones enhance competition between shoot branches by dampening auxin transport. *Development* **137:** 2905–2913. doi:10.1242/dev.051987

Cruz-Ramírez A, Díaz-Triviño S, Wachsman G, Du Y, Arteága-Vázquez M, Zhang H, Benjamins R, Blilou I, Neef AB, Chandler V, et al. 2013. A SCARECROW-RETINOBLASTOMA protein network controls protective quiescence in the *Arabidopsis* root stem cell organizer. *PLoS Biol* **11:** e1001724. doi:10.1371/journal.pbio.1001724

Cui D, Zhao J, Jing Y, Fan M, Liu J, Wang Z, Xin W, Hu Y. 2013. The *Arabidopsis* IDD14, IDD15, and IDD16 cooperatively regulate lateral organ morphogenesis and gravitropism by promoting auxin biosynthesis and transport. *PLoS Genet* **9:** e1003759. doi:10.1371/journal.pgen.1003759

Darwin C. 1880. *The power of movements in plants.* John Murray, London.

Dello Ioio R, Nakamura K, Moubayidin L, Perilli S, Taniguchi M, Morita MT, Aoyama T, Costantino P, Sabatini S. 2008. A genetic framework for the control of cell division and differentiation in the root meristem. *Science* **322:** 1380–1384. doi:10.1126/science.1164147

De Rybel B, Möller B, Yoshida S, Grabowicz I, Barbier de Reuille P, Boeren S, Smith RS, Borst JW, Weijers D. 2013. A bHLH complex controls embryonic vascular tissue establishment and indeterminate growth in *Arabidopsis. Dev Cell* **24:** 426–437. doi:10.1016/j.devcel.2012.12.013

De Rybel B, Adibi M, Breda AS, Wendrich JR, Smit ME, Novák O, Yamaguchi N, Yoshida S, Van Isterdael G, Palovaara J, et al. 2014. Integration of growth and patterning during vascular tissue formation in *Arabidopsis. Science* **345:** 1255215. doi:10.1126/science.1255215

De Smet I, Tetsumura T, De Rybel B, dit Frey NF, Laplaze L, Casimiro I, Swarup R, Naudts M, Vanneste S, Audenaert D, et al. 2007. Auxin-dependent regulation of lateral root positioning in the basal meristem of *Arabidopsis. Development* **134:** 681–690. doi:10.1242/dev.02753

De Smet I, Lau S, Voß U, Vanneste S, Benjamins R, Rademacher EH, Schlereth A, De Rybel B, Vassileva V, Grunewald W, et al. 2010. Bimodular auxin response controls organogenesis in *Arabidopsis. Proc Natl Acad Sci* **107:** 2705–2710. doi:10.1073/pnas.0915001107

Dharmasiri N, Dharmasiri S, Estelle M. 2005. The F-box protein TIR1 is an auxin receptor. *Nature* **435:** 441–445. doi:10.1038/nature03543

Di Mambro R, De Ruvo M, Pacifici E, Salvi E, Sozzani R, Benfey PN, Busch W, Novak O, Ljung K, Di Paola L, et al. 2017. Auxin minimum triggers the developmental switch from cell division to cell differentiation in the *Arabidopsis* root. *Proc Natl Acad Sci* **114:** E7641–E7649. doi:10.1073/pnas.1705833114

Di Mambro R, Svolacchia N, Dello Ioio R, Pierdonati E, Salvi E, Pedrazzini E, Vitale A, Perilli S, Sozzani R, Benfey PN, et al. 2019. The lateral root cap acts as an auxin sink that controls meristem size. *Curr Biol* **29:** 1199–1205.e4. doi:10.1016/j.cub.2019.02.022

Ding Z, Friml J. 2010. Auxin regulates distal stem cell differentiation in *Arabidopsis* roots. *Proc Natl Acad Sci* **107:** 12046–12051. doi:10.1073/pnas.1000672107

Dubrovsky JG, Sauer M, Napsucialy-Mendivil S, Ivanchenko MG, Friml J, Shishkova S, Celenza J, Benkova E, Benková E. 2008. Auxin acts as a local morphogenetic trigger to specify lateral root founder cells. *Proc Natl Acad Sci* **105:** 8790–8794. doi:10.1073/pnas.0712307105

Eklund D, Ståldal V, Valsecchi I, Cierlik I, Eriksson C, Hiratsu K, Ohme-Takagi M, Sundström JF, Thelander M, Ezcurra I, et al. 2010. The *Arabidopsis thaliana* STYLISH1 protein acts as a transcriptional activator regulating auxin biosynthesis. *Plant Cell* **22:** 349–363. doi:10.1105/tpc.108.064816

Eklund D, Cierlik I, Ståldal V, Claes AR, Vestman D, Chandler J, Sundberg E. 2011. Expression of *Arabidopsis SHORT INTERNODES/STYLISH* family genes in auxin biosynthesis zones of aerial organs is dependent on a GCC box-like regulatory element. *Plant Physiol* **157:** 2069–2080. doi:10.1104/pp.111.182253

Fahlgren N, Montgomery TA, Howell MD, Allen E, Dvorak SK, Alexander AL, Carrington JC. 2006. Regulation of AUXIN RESPONSE FACTOR3 by TAS3 ta-siRNA affects developmental timing and patterning in *Arabidopsis. Curr Biol* **16:** 939–944. doi:10.1016/j.cub.2006.03.065

Fendrych M, Leung J, Friml J. 2016. Tir1/AFB-Aux/IAA auxin perception mediates rapid cell wall acidification and growth of *Arabidopsis* hypocotyls. *eLife* **5:** e19048. doi:10.7554/eLife.19048

Fendrych M, Akhmanova M, Merrin J, Glanc M, Hagihara S, Takahashi K, Uchida N, Torii KU, Friml J. 2018. Rapid and reversible root growth inhibition by TIR1 auxin signalling. *Nat Plants* **4:** 453–459. doi:10.1038/s41477-018-0190-1

Feraru E, Feraru MI, Barbez E, Waidmann S, Sun L, Gaidora A, Kleine-Vehn J. 2019. PILS6 is a temperature-sensitive regulator of nuclear auxin input and organ growth in *Arabidopsis thaliana. Proc Natl Acad Sci* **116:** 3893–3898. doi:10.1073/pnas.1814015116

Franklin KA, Lee SH, Patel D, Kumar SV, Spartz AK, Gu C, Ye S, Yu P, Breen G, Cohen JD, et al. 2011. PHYTOCHROME-INTERACTING FACTOR 4 (PIF4) regulates auxin biosynthesis at high temperature. *Proc Natl Acad Sci* **108:** 20231–20235. doi:10.1073/pnas.1110682108

Freire-Rios A, Tanaka K, Crespo I, Van der Wijk E, Sizentsova Y, Levitsky V, Lindhoud S, Fontana M, Hohlbein J, Roeland Boer D, et al. 2020. Architecture of DNA elements mediating ARF transcription factor binding and auxin-responsive gene expression in *Arabidopsis. Proc Natl Acad Sci* **117:** 24557–24566. doi:10.1073/pnas.2009554117

Friml J, Benková E, Blilou I, Wisniewska J, Hamann T, Ljung K, Woody S, Sandberg G, Scheres B, Jürgens G, et al. 2002. AtPIN4 mediates sink-driven auxin gradients and root patterning in *Arabidopsis. Cell* **108:** 661–673. doi:10.1016/S0092-8674(02)00656-6

Friml J, Yang X, Michniewicz M, Weijers D, Quint A, Tietz O, Benjamins R, Ouwerkerk PBF, Ljung K, Sandberg G, et al. 2004. A PINOID-dependent binary switch in apical-basal PIN polar targeting directs auxin efflux. *Science* **306:** 862–865. doi:10.1126/science.1100618

Fukaki H, Tameda S, Masuda H, Tasaka M. 2002. Lateral root formation is blocked by a gain-of-function mutation in the *solitary-root/IAA14* gene of *Arabidopsis. Plant J* **29:** 153–168. doi:10.1046/j.0960-7412.2001.01201.x

Fukaki H, Nakao Y, Okushima Y, Theologis A, Tasaka M. 2005. Tissue-specific expression of stabilized SOLITARY-ROOT/IAA14 alters lateral root development in *Arabidopsis*. *Plant J* **44:** 382–395. doi:10.1111/j.1365-313X.2005.02537.x

Gaillochet C, Stiehl T, Wenzl C, Ripoll JJ, Bailey-Steinitz LJ, Li L, Pfeiffer A, Miotk A, Hakenjos JP, Forner J, et al. 2017. Control of plant cell fate transitions by transcriptional and hormonal signals. *eLife* **6:** e30135. doi:10.7554/eLife.30135

Galinha C, Hofhuis H, Luijten M, Willemsen V, Blilou I, Heidstra R, Scheres B. 2007. PLETHORA proteins as dose-dependent master regulators of *Arabidopsis* root development. *Nature* **449:** 1053–1057. doi:10.1038/nature06206

Gälweiler L, Guan C, Müller A, Wisman E, Mendgen K, Yephremov A, Palme K. 1998. Regulation of polar auxin transport by AtPIN1 in *Arabidopsis* vascular tissue. *Science* **282:** 2226–2230. doi:10.1126/science.282.5397.2226

Garcia D, Collier SA, Byrne ME, Martienssen RA. 2006. Specification of leaf polarity in *Arabidopsis* via the *trans*-acting siRNA pathway. *Curr Biol* **16:** 933–938. doi:10.1016/j.cub.2006.03.064

Geisler M, Aryal B, Di Donato M, Hao P. 2017. A critical view on ABC transporters and their interacting partners in auxin transport. *Plant Cell Physiol* **58:** 1601–1614. doi:10.1093/pcp/pcx104

Geldner N, Firml J, Stierhof YD, Jürgens G, Palme K. 2001. Auxin transport inhibitors block PIN1 and vesicle trafficking. *Nature* **413:** 425–428. doi:10.1038/35096571

Gray WM, Kepinski S, Rouse D, Leyser O, Estelle M. 2001. Auxin regulates SCFTIR1-dependent degradation of AUX/IAA proteins. *Nature* **414:** 271–276. doi:10.1038/35104500

Grieneisen VA, Xu J, Marée AFM, Hogeweg P, Scheres B. 2007. Auxin transport is sufficient to generate a maximum and gradient guiding root growth. *Nature* **449:** 1008–1013. doi:10.1038/nature06215

Grones P, Abas M, Hajný J, Jones A, Waidmann S, Kleine-Vehn J, Friml J. 2018. PID/WAG-mediated phosphorylation of the *Arabidopsis* PIN3 auxin transporter mediates polarity switches during gravitropism. *Sci Rep* **8:** 10279. doi:10.1038/s41598-018-28188-1

Guan C, Wu B, Yu T, Wang Q, Krogan NT, Liu X, Jiao Y. 2017. Spatial auxin signaling controls leaf flattening in *Arabidopsis*. *Curr Biol* **27:** 2940–2950. doi:10.1016/j.cub.2017.08.042

Guenot B, Bayer E, Kierzkowski D, Smith RS, Mandel T, Zádníková P, Benková E, Kuhlemeier C. 2012. Pin1-independent leaf initiation in *Arabidopsis*. *Plant Physiol* **159:** 1501–1510. doi:10.1104/pp.112.200402

Hagen G, Guilfoyle TJ. 1985. Rapid induction of selective transcription by auxins. *Mol Cell Biol* **5:** 1197–1203. doi:10.1128/MCB.5.6.1197

Hagen G, Guilfoyle T. 2002. Auxin-responsive gene expression: genes, promoters and regulatory factors. *Plant Mol Biol* **49:** 373–385. doi:10.1023/A:1015207114117

Hagen G, Martin G, Li Y, Guilfoyle TJ. 1991. Auxin-induced expression of the soybean GH3 promoter in transgenic tobacco plants. *Plant Mol Biol* **17:** 567–579. doi:10.1007/BF00040658

Hajný J, Prát T, Rydza N, Rodriguez L, Tan S, Verstraeten I, Domjan D, Mazur E, Smakowska-Luzan E, Smet W, et al. 2020. Receptor kinase module targets PIN-dependent auxin transport during canalization. *Science* **370:** 550–557. doi:10.1126/science.aba3178

Hamann T, Benkova E, Bäurle I, Kientz M, Jürgens G. 2002. The *Arabidopsis* BODENLOS gene encodes an auxin response protein inhibiting MONOPTEROS-mediated embryo patterning. *Genes Dev* **16:** 1610–1615. doi:10.1101/gad.229402

Hardtke CS, Berleth T. 1998. The *Arabidopsis* gene MONOPTEROS encodes a transcription factor mediating embryo axis formation and vascular development. *EMBO J* **17:** 1405–1411. doi:10.1093/emboj/17.5.1405

Harmer SL, Brooks CJ. 2018. Growth-mediated plant movements: hidden in plain sight. *Curr Opin Plant Biol* **41:** 89–94. doi:10.1016/j.pbi.2017.10.003

Haruta M, Sabat G, Stecker K, Minkoff BB, Sussman MR. 2014. A peptide hormone and its receptor protein kinase regulate plant cell expansion. *Science* **343:** 408–411. doi:10.1126/science.1244454

Heisler MG, Ohno C, Das P, Sieber P, Reddy GV, Long JA, Meyerowitz EM. 2005. Patterns of auxin transport and gene expression during primordium development revealed by live imaging of the *Arabidopsis* inflorescence meristem. *Curr Biol* **15:** 1899–1911. doi:10.1016/j.cub.2005.09.052

Hersch M, Lorrain S, De Wit M, Trevisan M, Ljung K, Bergmann S, Fankhauser C. 2014. Light intensity modulates the regulatory network of the shade avoidance response in *Arabidopsis*. *Proc Natl Acad Sci* **111:** 6515–6520. doi:10.1073/pnas.1320355111

Hobbie L, Estelle M. 1995. The axr4 auxin-resistant mutants of *Arabidopsis thaliana* define a gene important for root gravitropism and lateral root initiation. *Plant J* **7:** 211–220. doi:10.1046/j.1365-313X.1995.7020211.x

Hornitschek P, Kohnen MV, Lorrain S, Rougemont J, Ljung K, López-Vidriero I, Franco-Zorrilla JM, Solano R, Trevisan M, Pradervand S, et al. 2012. Phytochrome interacting factors 4 and 5 control seedling growth in changing light conditions by directly controlling auxin signaling. *Plant J* **71:** 699–711. doi:10.1111/j.1365-313X.2012.05033.x

José Ripoll J, Bailey LJ, Mai QA, Wu SL, Hon CT, Chapman EJ, Ditta GS, Estelle M, Yanofsky MF. 2015. MicroRNA regulation of fruit growth. *Nat Plants* **1:** 1–9. doi:10.1038/nplants.2015.36

Kato H, Mutte SK, Suzuki H, Crespo I, Das S, Radoeva T, Fontana M, Yoshitake Y, Hainiwa E, van den Berg W, et al. 2020. Design principles of a minimal auxin response system. *Nat Plants* **6:** 473–482. doi:10.1038/s41477-020-0662-y

Kelley DR, Arreola A, Gallagher TL, Gasser CS. 2012. ETTIN (ARF3) physically interacts with KANADI proteins to form a functional complex essential for integument development and polarity determination in *Arabidopsis*. *Development* **139:** 1105–1109. doi:10.1242/dev.067918

Kepinski S, Leyser O. 2005. The *Arabidopsis* F-box protein TIR1 is an auxin receptor. *Nature* **435:** 446–451. doi:10.1038/nature03542

Kierzkowski D, Runions A, Vuolo F, Strauss S, Lymbouridou R, Routier-Kierzkowska AL, Wilson-Sánchez D, Jenke H,

Galinha C, Mosca G, et al. 2019. A growth-based framework for leaf shape development and diversity. *Cell* **177:** 1405–1418.e17. doi:10.1016/j.cell.2019.05.011

Kinoshita T, Shimazaki KI. 1999. Blue light activates the plasma membrane H^+-ATPase by phosphorylation of the C-terminus in stomatal guard cells. *EMBO J* **18:** 5548–5558. doi:10.1093/emboj/18.20.5548

Kleine-Vehn J, Huang F, Naramoto S, Zhang J, Michniewicz M, Offringa R, Friml J. 2009. PIN auxin efflux carrier polarity is regulated by PINOID kinase-mediated recruitment into GNOM-independent trafficking in *Arabidopsis*. *Plant Cell* **21:** 3839–3849. doi:10.1105/tpc.109.071639

Kleine-Vehn J, Wabnik K, Martinière A, Łangowski Ł, Willig K, Naramoto S, Leitner J, Tanaka H, Jakobs S, Robert S, et al. 2011. Recycling, clustering, and endocytosis jointly maintain PIN auxin carrier polarity at the plasma membrane. *Mol Syst Biol* **7:** 540. doi:10.1038/msb.2011.72

Krogan NT, Yin X, Ckurshumova W, Berleth T. 2014. Distinct subclades of *Aux/IAA* genes are direct targets of ARF5/MP transcriptional regulation. *New Phytol* **204:** 474–483. doi:10.1111/nph.12994

Krogan NT, Marcos D, Weiner AI, Berleth T. 2016. The auxin response factor MONOPTEROS controls meristem function and organogenesis in both the shoot and root through the direct regulation of *PIN* genes. *New Phytol* **212:** 42–50. doi:10.1111/nph.14107

Krouk G, Lacombe B, Bielach A, Perrine-Walker F, Malinska K, Mounier E, Hoyerova K, Tillard P, Leon S, Ljung K, et al. 2010. Nitrate-regulated auxin transport by NRT1.1 defines a mechanism for nutrient sensing in plants. *Dev Cell* **18:** 927–937. doi:10.1016/j.devcel.2010.05.008

Lanctot A, Nemhauser JL. 2020. It's Morphin' time: how multiple signals converge on ARF transcription factors to direct development. *Curr Opin Plant Biol* **57:** 1–7. doi:10.1016/j.pbi.2020.04.008

Landrein B, Kiss A, Sassi M, Chauvet A, Das P, Cortizo M, Laufs P, Takeda S, Aida M, Traas J, et al. 2015. Mechanical stress contributes to the expression of the STM homeobox gene in *Arabidopsis* shoot meristems. *eLife* **4:** e07811. doi:10.7554/eLife.07811

Laplaze L, Benkova E, Casimiro I, Maes L, Vanneste S, Swarup R, Weijers D, Calvo V, Parizot B, Herrera-Rodriguez MB, et al. 2007. Cytokinins act directly on lateral root founder cells to inhibit root initiation. *Plant Cell* **19:** 3889–3900. doi:10.1105/tpc.107.055863

Leibfried A, To JPCC, Busch W, Stehling S, Kehle A, Demar M, Kieber JJ, Lohmann JU. 2005. WUSCHEL controls meristem function by direct regulation of cytokinin-inducible response regulators. *Nature* **438:** 1172–1175. doi:10.1038/nature04270

Leitner J, Petrášek J, Tomanov K, Retzer K, Pařezová M, Korbei B, Bachmair A, Zažímalová E, Luschnig C. 2012. Lysine63-linked ubiquitylation of PIN2 auxin carrier protein governs hormonally controlled adaptation of *Arabidopsis* root growth. *Proc Natl Acad Sci* **109:** 8322–8327. doi:10.1073/pnas.1200824109

Lenhard M, Jürgens G, Laux T. 2002. The WUSCHEL and SHOOTMERISTEMLESS genes fulfil complementary roles in *Arabidopsis* shoot meristem regulation. *Development* **129:** 3195–3206.

Leyser HMO, Pickett FB, Dharmasiri S, Estelle M. 1996. Mutations in the AXR3 gene of *Arabidopsis* result in altered auxin response including ectopic expression from the SAUR-AC1 promoter. *Plant J* **10:** 403–413. doi:10.1046/j.1365-313x.1996.10030403.x

Li L, Ljung K, Breton G, Schmitz RJ, Pruneda-Paz J, Cowing-Zitron C, Cole BJ, Ivans LJ, Pedmale UV, Jung HS, et al. 2012. Linking photoreceptor excitation to changes in plant architecture. *Genes Dev* **26:** 785–790. doi:10.1101/gad.187849.112

Liu J, Sheng L, Xu Y, Li J, Yang Z, Huang H, Xu L. 2014. *WOX11* and *12* are involved in the first-step cell fate transition during de novo root organogenesis in *Arabidopsis*. *Plant Cell* **26:** 1081–1093. doi:10.1105/tpc.114.122887

Liu G, Gao S, Tian H, Wu W, Robert HS, Ding Z. 2016. Local transcriptional control of YUCCA regulates auxin promoted root-growth inhibition in response to aluminium stress in *Arabidopsis*. *PLoS Genet* **12:** 1–25.

Löfke C, Zwiewka M, Heilmann I, Van Montagu MCE, Teichmann T, Friml J. 2013. Asymmetric gibberellin signaling regulates vacuolar trafficking of PIN auxin transporters during root gravitropism. *Proc Natl Acad Sci* **110:** 3627–3632. doi:10.1073/pnas.1300107110

Long JA, Ohno C, Smith ZR, Meyerowitz EM. 2006. TOPLESS regulates apical embryonic fate in *Arabidopsis*. *Science* **312:** 1520–1523. doi:10.1126/science.1123841

Luo L, Zeng J, Wu H, Tian Z, Zhao Z. 2018. A molecular framework for auxin-controlled homeostasis of shoot stem cells in *Arabidopsis*. *Mol Plant* **11:** 899–913. doi:10.1016/j.molp.2018.04.006

Ma Y, Miotk A, Šutiković Z, Ermakova O, Wenzl C, Medzihradszky A, Gaillochet C, Forner J, Utan G, Brackmann K, et al. 2019. WUSCHEL acts as an auxin response rheostat to maintain apical stem cells in *Arabidopsis*. *Nat Commun* **10:** 5093. doi:10.1038/s41467-019-13074-9

Mähönen AP, Bonke M, Kauppinen L, Riikonen M, Benfey PN, Helariutta Y. 2000. A novel two-component hybrid molecule regulates vascular morphogenesis of the *Arabidopsis* root. *Genes Dev* **14:** 2938–2943. doi:10.1101/gad.189200

Mähönen AP, Bishopp A, Higuchi M, Nieminen KM, Kinoshita K, Törmäkangas K, Ikeda Y, Oka A, Kakimoto T, Helariutta Y. 2006. Cytokinin signaling and its inhibitor AHP6 regulate cell fate during vascular development. *Science* **311:** 94–98. doi:10.1126/science.1118875

Mähönen AP, Ten Tusscher K, Siligato R, Smetana O, Díaz-Triviño S, Salojärvi J, Wachsman G, Prasad K, Heidstra R, Scheres B. 2014. PLETHORA gradient formation mechanism separates auxin responses. *Nature* **515:** 125–129. doi:10.1038/nature13663

Marchant A, Kargul J, May ST, Muller P, Delbarre A, Perrot-Rechenmann C, Bennett MJ, Perrot-Rechenmann C, Bennett MJ. 1999. AUX1 regulates root gravitropism in *Arabidopsis* by facilitating auxin uptake within root apical tissues. *EMBO J* **18:** 2066–2073. doi:10.1093/emboj/18.8.2066

Marchant A, Bhalerao R, Casimiro I, Eklöf J, Casero PJ, Bennett M, Sandberg G. 2002. AUX1 promotes lateral root formation by facilitating indole-3-acetic acid distribution between sink and source tissues in the *Arabidopsis* seedling. *Plant Cell* **14:** 589–597. doi:10.1105/tpc.010354

Marhava P, Bassukas AEL, Zourelidou M, Kolb M, Moret B, Fastner A, Schulze WX, Cattaneo P, Hammes UZ,

Schwechheimer C, et al. 2018. A molecular rheostat adjusts auxin flux to promote root protophloem differentiation. *Nature* **558:** 297–300. doi:10.1038/s41586-018-0186-z

Marhavý P, Bielach A, Abas L, Abuzeineh A, Duclercq J, Tanaka H, Pařezová M, Petrášek J, Friml J, Kleine-Vehn J, et al. 2011. Cytokinin modulates endocytic trafficking of PIN1 auxin efflux carrier to control plant organogenesis. *Dev Cell* **21:** 796–804. doi:10.1016/j.devcel.2011.08.014

Marhavý P, Duclercq J, Weller B, Feraru E, Bielach A, Offringa R, Friml J, Schwechheimer C, Murphy A, Benková E. 2014. Cytokinin controls polarity of PIN1-dependent auxin transport during lateral root organogenesis. *Curr Biol* **24:** 1031–1037. doi:10.1016/j.cub.2014.04.002

Marin E, Jouannet V, Herz A, Lokerse AS, Weijers D, Vaucheret H, Nussaume L, Crespi MD, Maizel A. 2010. mir390, *Arabidopsis* TAS3 tasiRNAs, and their *AUXIN RESPONSE FACTOR* targets define an autoregulatory network quantitatively regulating lateral root growth. *Plant Cell* **22:** 1104–1117. doi:10.1105/tpc.109.072553

Martínez-Fernández I, Sanchís S, Marini N, Balanzá V, Ballester P, Navarrete-Gómez M, Oliveira AC, Colombo L, Ferrándiz C. 2014. The effect of NGATHA altered activity on auxin signaling pathways within the *Arabidopsis* gynoecium. *Front Plant Sci* **5:** 210.

Mayer KFX, Schoof H, Haecker A, Lenhard M, Jürgens G, Laux T. 1998. Role of WUSCHEL in regulating stem cell fate in the *Arabidopsis* shoot meristem. *Cell* **95:** 805–815. doi:10.1016/S0092-8674(00)81703-1

McClure BA, Guilfoyle T. 1987. Characterization of a class of small auxin-inducible soybean polyadenylated RNAs. *Plant Mol Biol* **9:** 611–623. doi:10.1007/BF00020537

Mellor N, Band LR, Pěňík A, Novák O, Rashed A, Holman T, Wilson MH, Vo U, Bishopp A, King JR, et al. 2016. Dynamic regulation of auxin oxidase and conjugating enzymes *AtDAO1* and *GH3* modulates auxin homeostasis. *Proc Natl Acad Sci* **113:** 11022–11027. doi:10.1073/pnas.1604458113

Miao R, Wang M, Yuan W, Ren Y, Li Y, Zhang N, Zhang J, Kronzucker HJ, Xu W. 2018. Comparative analysis of *Arabidopsis* ecotypes reveals a role for brassinosteroids in root hydrotropism. *Plant Physiol* **176:** 2720–2736. doi:10.1104/pp.17.01563

Michniewicz M, Zago MK, Abas L, Weijers D, Schweighofer A, Meskiene I, Heisler MG, Ohno C, Zhang J, Huang F, et al. 2007. Antagonistic regulation of PIN phosphorylation by PP2A and PINOID directs auxin flux. *Cell* **130:** 1044–1056. doi:10.1016/j.cell.2007.07.033

Minami A, Takahashi K, Inoue SI, Tada Y, Kinoshita T. 2019. Brassinosteroid induces phosphorylation of the plasma membrane H^+-ATPase during hypocotyl elongation in *Arabidopsis thaliana. Plant Cell Physiol* **60:** 935–944. doi:10.1093/pcp/pcz005

Miyashima S, Roszak P, Sevilem I, Toyokura K, Blob B, Heo JO, Mellor N, Help-Rinta-Rahko H, Otero S, Smet W, et al. 2019. Mobile PEAR transcription factors integrate positional cues to prime cambial growth. *Nature* **565:** 490–494. doi:10.1038/s41586-018-0839-y

Moreno-Risueno MA, Van Norman JM, Moreno A, Zhang J, Ahnert SE, Benfey PN. 2010. Oscillating gene expression determines competence for periodic *Arabidopsis* root branching. *Science* **329:** 1306–1311. doi:10.1126/science.1191937

Moubayidin L, Østergaard L. 2014. Dynamic control of auxin distribution imposes a bilateral-to-radial symmetry switch during gynoecium development. *Curr Biol* **24:** 2743–2748. doi:10.1016/j.cub.2014.09.080

Mravec J, Skůpa P, Bailly A, Hoyerová K, Křeček P, Bielach A, Petrášek J, Zhang J, Gaykova V, Stierhof YD, et al. 2009. Subcellular homeostasis of phytohormone auxin is mediated by the ER-localized PIN5 transporter. *Nature* **459:** 1136–1140. doi:10.1038/nature08066

Müller B, Sheen J. 2008. Cytokinin and auxin interaction in root stem-cell specification during early embryogenesis. *Nature* **453:** 1094–1097. doi:10.1038/nature06943

Müller A, Guan C, Gälweiler L, Tänzler P, Huijser P, Marchant A, Parry G, Bennett M, Wisman E, Palme K. 1998. AtPIN2 defines a locus of *Arabidopsis* for root gravitropism control. *EMBO J* **17:** 6903–6911. doi:10.1093/emboj/17.23.6903

Müller CJ, Valdés AE, Wang G, Ramachandran P, Beste L, Uddenberg D, Carlsbecker A. 2016. PHABULOSA mediates an auxin signaling loop to regulate vascular patterning in *Arabidopsis. Plant Physiol* **170:** 956–970. doi:10.1104/pp.15.01204

Mutte SK, Kato H, Rothfels C, Melkonian M, Wong GKS, Weijers D. 2018. Origin and evolution of the nuclear auxin response system. *eLife* **7:** e33399. doi:10.7554/eLife.33399

Nakata M, Matsumoto N, Tsugeki R, Rikirsch E, Laux T, Okada K. 2012. Roles of the middle domain-specific *WUSCHEL-RELATED HOMEOBOX* genes in early development of leaves in *Arabidopsis. Plant Cell* **24:** 519–535. doi:10.1105/tpc.111.092858

Nemhauser JL, Mockler TC, Chory J. 2004. Interdependency of brassinosteroid and auxin signaling in *Arabidopsis. PLoS Biol* **2:** e258. doi:10.1371/journal.pbio.0020258

Oh E, Zhu JY, Bai MY, Arenhart RA, Sun Y, Wang ZY. 2014. Cell elongation is regulated through a central circuit of interacting transcription factors in the *Arabidopsis* hypocotyl. *eLife* **3:** e03031. doi:10.7554/eLife.03031

Ohashi-Ito K, Saegusa M, Iwamoto K, Oda Y, Katayama H, Kojima M, Sakakibara H, Fukuda H. 2014. A bHLH complex activates vascular cell division via cytokinin action in root apical meristem. *Curr Biol* **24:** 2053–2058. doi:10.1016/j.cub.2014.07.050

Okada K, Ueda J, Komaki MK, Bell CJ, Shimura Y. 1991. Requirement of the auxin polar transport system in early stages of *Arabidopsis* floral bud formation. *Plant Cell* **3:** 677–684. doi:10.2307/3869249

Orosa-Puente B, Leftley N, von Wangenheim D, Banda J, Srivastava AK, Hill K, Truskina J, Bhosale R, Morris E, Srivastava M, et al. 2018. Root branching toward water involves posttranslational modification of transcription factor ARF7. *Science* **362:** 1407–1410. doi:10.1126/science.aau3956

Paciorek T, Zažímalová E, Ruthardt N, Petrášek J, Stierhof YD, Kleine-Vehn J, Morris DA, Emans N, Jürgens G, Geldner N, et al. 2005. Auxin inhibits endocytosis and promotes its own efflux from cells. *Nature* **435:** 1251–1256. doi:10.1038/nature03633

Palme K, Gälweiler L. 1999. PIN-pointing the molecular basis of auxin transport. *Curr Opin Plant Biol* **2:** 375–381. doi:10.1016/S1369-5266(99)00008-4

Paponov IA, Paponov M, Teale W, Menges M, Chakrabortee S, Murray JAH, Palme K. 2008. Comprehensive transcriptome analysis of auxin responses in *Arabidopsis*. *Mol Plant* **1:** 321–337. doi:10.1093/mp/ssm021

Peer WA, Cheng Y, Murphy AS. 2013. Evidence of oxidative attenuation of auxin signalling. *J Exp Bot* **64:** 2629–2639. doi:10.1093/jxb/ert152

Péret B, Swarup K, Ferguson A, Seth M, Yang Y, Dhondt S, James N, Casimiro I, Perry P, Syed A, et al. 2012. *AUX/LAX* genes encode a family of auxin influx transporters that perform distinct functions during *Arabidopsis* development. *Plant Cell* **24:** 2874–2885. doi:10.1105/tpc.112.097766

Pernisová M, Klíma P, Horák J, Válková M, Malbeck J, Souček P, Reichman P, Hoyerová K, Dubová J, Friml J, et al. 2009. Cytokinins modulate auxin-induced organogenesis in plants via regulation of the auxin efflux. *Proc Natl Acad Sci* **106:** 3609–3614. doi:10.1073/pnas.0811539106

Petrášek J, Mravec J, Bouchard R, Blakeslee JJ, Abas M, Seifertová D, Wiśniewska J, Tadele Z, Kubeš M, Čovanová M, et al. 2006. PIN proteins perform a rate-limiting function in cellular auxin efflux. *Science* **312:** 914–918. doi:10.1126/science.1123542

Porco S, Larrieu A, Du Y, Gaudinier A, Goh T, Swarup K, Swarup R, Kuempers B, Bishopp A, Lavenus J, et al. 2016. Lateral root emergence in *Arabidopsis* is dependent on transcription factor LBD29 regulation of auxin influx carrier *LAX3*. *Development* **143:** 3340–3349. doi:10.1242/dev.136283

Rademacher EH, Möller B, Lokerse AS, Llavata-Peris CI, Van Den Berg W, Weijers D. 2011. A cellular expression map of the *Arabidopsis* AUXIN RESPONSE FACTOR gene family. *Plant J* **68:** 597–606. doi:10.1111/j.1365-313X.2011.04710.x

Reinhardt D, Pesce ER, Stieger P, Mandel T, Baltensperger K, Bennett M, Traas J, Friml J, Kuhlemeier C. 2003. Regulation of phyllotaxis by polar auxin transport. *Nature* **426:** 255–260. doi:10.1038/nature02081

Ren H, Park MY, Spartz AK, Wong JH, Gray WM. 2018. A subset of plasma membrane-localized PP2C.D phosphatases negatively regulate SAUR-mediated cell expansion in *Arabidopsis*. *PLoS Genet* **14:** e1007455. doi:10.1371/journal.pgen.1007455

Retzer K, Akhmanova M, Konstantinova N, Malínská K, Leitner J, Petrášek J, Luschnig C. 2019. Brassinosteroid signaling delimits root gravitropism via sorting of the *Arabidopsis* PIN2 auxin transporter. *Nat Commun* **10:** 5516. doi:10.1038/s41467-019-13543-1

Rizzardi K, Landberg K, Nilsson L, Ljung K, Sundås-Larsson A. 2011. TFL2/LHP1 is involved in auxin biosynthesis through positive regulation of YUCCA genes. *Plant J* **65:** 897–906. doi:10.1111/j.1365-313X.2010.04470.x

Robert HS, Grunewald W, Sauer M, Cannoot B, Soriano M, Swarup R, Weijers D, Bennett M, Boutilier K, Friml J. 2015. Plant embryogenesis requires aux/lax-mediated auxin influx. *Development* **142:** 702–711. doi:10.1242/dev.115832

Rouse D, Mackay P, Stirnberg P, Estelle M, Leyser O. 1998. Changes in auxin response from mutations in an AUX/IAA gene. *Science* **279:** 1371–1373. doi:10.1126/science.279.5355.1371

Ruzicka K, Šimášková M, Duclercq J, Petrášek J, Zažímalová E, Simon S, Friml J, Van Montagu MCE, Benková E. 2009. Cytokinin regulates root meristem activity via modulation of the polar auxin transport. *Proc Natl Acad Sci* **106:** 4284–4289. doi:10.1073/pnas.0900060106

Sabatini S, Beis D, Wolkenfelt H, Murfett J, Guilfoyle T, Malamy J, Benfey P, Leyser O, Bechtold N, Weisbeek P, et al. 1999. An auxin-dependent distal organizer of pattern and polarity in the *Arabidopsis* root. *Cell* **99:** 463–472. doi:10.1016/S0092-8674(00)81535-4

Sabatini S, Heidstra R, Wildwater M, Scheres B. 2003. SCARECROW is involved in positioning the stem cell niche in the *Arabidopsis* root meristem. *Genes Dev* **17:** 354–358. doi:10.1101/gad.252503

Sakamoto T, Morinaka Y, Inukai Y, Kitano H, Fujioka S. 2013. Auxin signal transcription factor regulates expression of the brassinosteroid receptor gene in rice. *Plant J* **73:** 676–688. doi:10.1111/tpj.12071

Salvi E, Rutten JP, Di Mambro R, Polverari L, Licursi V, Negri R, Dello Ioio R, Sabatini S, Ten Tusscher K. 2020. A self-organized PLT/auxin/ARR-B network controls the dynamics of root zonation development in *Arabidopsis thaliana*. *Dev Cell* **53:** 431–443.e23. doi:10.1016/j.devcel.2020.04.004

Santuari L, Sanchez-Perez GF, Luijten M, Rutjens B, Terpstra I, Berke L, Gorte M, Prasad K, Bao D, Timmermans-Hereijgers JLPM, et al. 2016. The PLETHORA gene regulatory network guides growth and cell differentiation in *Arabidopsis* roots. *Plant Cell* **28:** 2937–2951. doi:10.1105/tpc.16.00656

Sarkar AK, Luijten M, Miyashima S, Lenhard M, Hashimoto T, Nakajima K, Scheres B, Heidstra R, Laux T. 2007. Conserved factors regulate signalling in *Arabidopsis thaliana* shoot and root stem cell organizers. *Nature* **446:** 811–814. doi:10.1038/nature05703

Sauer M, Balla J, Luschnig C, Wiśniewska J, Reinöhl V, Friml J, Benková E. 2006. Canalization of auxin flow by Aux/IAA-ARF-dependent feedback regulation of PIN polarity. *Genes Dev* **20:** 2902–2911. doi:10.1101/gad.390806

Schlereth A, Möller B, Liu W, Kientz M, Flipse J, Rademacher EH, Schmid M, Jürgens G, Weijers D. 2010. MONOPTEROS controls embryonic root initiation by regulating a mobile transcription factor. *Nature* **464:** 913–916. doi:10.1038/nature08836

Schuster C, Gaillochet C, Lohmann JU. 2015. *Arabidopsis HECATE* genes function in phytohormone control during gynoecium development. *Development* **142:** 3343–3350. doi:10.1242/dev.120444

Semeradova H, Montesinos JC, Benkova E. 2020. All roads lead to auxin: post-translational regulation of auxin transport by multiple hormonal pathways. *Plant Commun* **1:** 100048. doi:10.1016/j.xplc.2020.100048

Shih HW, Depew CL, Miller ND, Monshausen GB. 2015. The cyclic nucleotide-gated channel CNGC14 regulates root gravitropism in *Arabidopsis thaliana*. *Curr Biol* **25:** 3119–3125. doi:10.1016/j.cub.2015.10.025

Shimotohno A, Heidstra R, Blilou I, Scheres B. 2018. Root stem cell niche organizer specification by molecular convergence of PLETHORA and SCARECROW transcription factor modules. *Genes Dev* **32:** 1085–1100. doi:10.1101/gad.314096.118

Shin R, Burch AY, Huppert KA, Tiwari SB, Murphy AS, Guilfoyle TJ, Schachtman DP. 2007. The *Arabidopsis* transcription factor MYB77 modulates auxin signal transduction. *Plant Cell* **19:** 2440–2453. doi:10.1105/tpc.107.050963

Shinohara N, Taylor C, Leyser O. 2013. Strigolactone can promote or inhibit shoot branching by triggering rapid depletion of the auxin efflux protein PIN1 from the plasma membrane. *PLoS Biol* **11:** e1001474. doi:10.1371/journal.pbio.1001474

Simonini S, Bencivenga S, Trick M, Østergaard L. 2017. Auxin-induced modulation of ETTIN activity orchestrates gene expression in *Arabidopsis*. *Plant Cell* **29:** 1864–1882. doi:10.1105/tpc.17.00389

Smaczniak C, Immink RGH, Muiño JM, Blanvillain R, Busscher M, Busscher-Lange J, Dinh QD, Liu S, Westphal AH, Boeren S, et al. 2012. Characterization of MADS-domain transcription factor complexes in *Arabidopsis* flower development. *Proc Natl Acad Sci* **109:** 1560–1565. doi:10.1073/pnas.1112871109

Smetana O, Mäkilä R, Lyu M, Amiryousefi A, Sánchez Rodríguez F, Wu MF, Solé-Gil A, Leal Gavarrón M, Siligato R, Miyashima S, et al. 2019. High levels of auxin signalling define the stem-cell organizer of the vascular cambium. *Nature* **565:** 485–489. doi:10.1038/s41586-018-0837-0

Sorefan K, Girin T, Liljegren SJ, Ljung K, Robles P, Galván-Ampudia CS, Offringa R, Friml J, Yanofsky MF, Østergaard L. 2009. A regulated auxin minimum is required for seed dispersal in *Arabidopsis*. *Nature* **459:** 583–586. doi:10.1038/nature07875

Spartz AK, Ren H, Park MY, Grandt KN, Lee SH, Murphy AS, Sussman MR, Overvoorde PJ, Gray WM. 2014. SAUR inhibition of PP2C-D phosphatases activates plasma membrane H^+-ATPases to promote cell expansion in *Arabidopsis*. *Plant Cell* **26:** 2129–2142. doi:10.1105/tpc.114.126037

Stepanova AN, Robertson-Hoyt J, Yun J, Benavente LM, Xie DY, Doležal K, Schlereth A, Jürgens G, Alonso JM. 2008. TAA1-mediated auxin biosynthesis is essential for hormone crosstalk and plant development. *Cell* **133:** 177–191. doi:10.1016/j.cell.2008.01.047

Sun J, Qi L, Li Y, Chu J, Li C. 2012. PIF4-mediated activation of YUCCA8 expression integrates temperature into the auxin pathway in regulating *Arabidopsis* hypocotyl growth. *PLoS Genet* **8:** e1002594. doi:10.1371/journal.pgen.1002594

Sun L, Feraru E, Feraru MI, Waidmann S, Wang W, Passaia G, Wang ZY, Wabnik K, Kleine-Vehn J. 2020. PIN-LIKES coordinate brassinosteroid signaling with nuclear auxin input in *Arabidopsis thaliana*. *Curr Biol* **30:** 1579–1588.e6. doi:10.1016/j.cub.2020.02.002

Swarup R, Péret B. 2012. AUX/LAX family of auxin influx carriers—an overview. *Front Plant Sci* **3:** 225. doi:10.3389/fpls.2012.00225

Swarup K, Benková E, Swarup R, Casimiro I, Péret B, Yang Y, Parry G, Nielsen E, De Smet I, Vanneste S, et al. 2008. The auxin influx carrier LAX3 promotes lateral root emergence. *Nat Cell Biol* **10:** 946–954. doi:10.1038/ncb1754

Szemenyei H, Hannon M, Long JA. 2008. TOPLESS mediates auxin-dependent transcriptional repression during *Arabidopsis* embryogenesis. *Science* **319:** 1384–1386. doi:10.1126/science.1151461

Takahashi K, Hayashi KI, Kinoshita T. 2012. Auxin activates the plasma membrane H^+-ATPase by phosphorylation during hypocotyl elongation in *Arabidopsis*. *Plant Physiol* **159:** 632–641. doi:10.1104/pp.112.196428

Tanaka H, Dhonukshe P, Brewer PB, Friml J. 2006. Spatiotemporal asymmetric auxin distribution: a means to coordinate plant development. *Cell Mol Life Sci* **63:** 2738–2754. doi:10.1007/s00018-006-6116-5

Tang LP, Zhou C, Wang SS, Yuan J, Zhang XS, Su YH. 2017. FUSCA3 interacting with LEAFY COTYLEDON2 controls lateral root formation through regulating *YUCCA4* gene expression in *Arabidopsis thaliana*. *New Phytol* **213:** 1740–1754. doi:10.1111/nph.14313

Tao Y, Ferrer JL, Ljung K, Pojer F, Hong F, Long JA, Li L, Moreno JE, Bowman ME, Ivans LJ, et al. 2008. Rapid synthesis of auxin via a new tryptophan-dependent pathway is required for shade avoidance in plants. *Cell* **133:** 164–176. doi:10.1016/j.cell.2008.01.049

Truskina J, Han J, Chrysanthou E, Galvan-Ampudia CS, Lainé S, Brunoud G, Macé J, Bellows S, Legrand J, Bågman AM, et al. 2021. A network of transcriptional repressors modulates auxin responses. *Nature* **589:** 116–119. doi:10.1038/s41586-020-2940-2

Ulmasov T, Hagen G, Guilfoyle TJ. 1997a. ARF1, a transcription factor that binds to auxin response elements. *Science* **276:** 1865–1868. doi:10.1126/science.276.5320.1865

Ulmasov T, Murfett J, Hagen G, Guilfoyle TJ. 1997b. Aux/IAA proteins repress expression of reporter genes containing natural and highly active synthetic auxin response elements. *Plant Cell* **9:** 1963–1971.

Ursache R, Miyashima S, Chen Q, Vatén A, Nakajima K, Carlsbecker A, Zhao Y, Helariutta Y, Dettmer J. 2014. Tryptophan-dependent auxin biosynthesis is required for HD-ZIP III-mediated xylem patterning. *Development* **141:** 1250–1259. doi:10.1242/dev.103473

Vandenbussche F, Petrášek J, Žádníková P, Hoyerová K, Pešek B, Raz V, Swarup R, Bennett M, Zažímolavá E, Benkova E, et al. 2010. The auxin influx carriers AUX1 and LAX3 are involved in auxin–ethylene interactions during apical hook development in *Arabidopsis thaliana* seedlings. *Development* **137:** 597–606. doi:10.1242/dev.040790

Vanneste S, Friml J. 2009. Auxin: a trigger for change in plant development. *Cell* **136:** 1005–1016. doi:10.1016/j.cell.2009.03.001

Varaud E, Brioudes F, Szécsi J, Leroux J, Brown S, Perrot-Rechenmann C, Bendahmane M. 2011. AUXIN RESPONSE FACTOR8 regulates *Arabidopsis* petal growth by interacting with the bHLH transcription factor BIGPETALp. *Plant Cell* **23:** 973–983. doi:10.1105/tpc.110.081653

Vernoux T, Brunoud G, Farcot E, Morin V, Van Den Daele H, Legrand J, Oliva M, Das P, Larrieu A, Wells D, et al. 2011. The auxin signalling network translates dynamic input into robust patterning at the shoot apex. *Mol Syst Biol* **7:** 508. doi:10.1038/msb.2011.39

Vert G, Walcher CL, Chory J, Nemhauser JL. 2008. Integration of auxin and brassinosteroid pathways by auxin response factor 2. *Proc Natl Acad Sci* **105:** 9829–9834. doi:10.1073/pnas.0803996105

Vieten A, Vanneste S, Wiśniewska J, Benková E, Benjamins R, Beeckman T, Luschnig C, Friml J. 2005. Functional redundancy of PIN proteins is accompanied by auxin-dependent cross-regulation of PIN expression. *Development* **132:** 4521–4531. doi:10.1242/dev.02027

Vieten A, Sauer M, Brewer PB, Friml J. 2007. Molecular and cellular aspects of auxin-transport-mediated development. *Trends Plant Sci* **12:** 160–168. doi:10.1016/j.tplants.2007.03.006

Vragović K, Sela A, Friedlander-Shani L, Fridman Y, Hacham Y, Holland N, Bartom E, Mockler TC, Savaldi-Goldstein S, Selaa A, et al. 2015. Translatome analyses capture of opposing tissues pecific brassinosteroid signals orchestrating root meristem differentiation. *Proc Natl Acad Sci* **112:** 923–928. doi:10.1073/pnas.1417947112

Waldie T, Leyser O. 2018. Cytokinin targets auxin transport to promote shoot branching. *Plant Physiol* **177:** 803–818. doi:10.1104/pp.17.01691

Wang HZ, Yang KZ, Zou JJ, Zhu LL, Xie ZD, Morita MT, Tasaka M, Friml J, Grotewold E, Beeckman T, et al. 2015. Transcriptional regulation of PIN genes by FOUR LIPS and MYB88 during *Arabidopsis* root gravitropism. *Nat Commun* **6:** 8822. doi:10.1038/ncomms9822

Wang Q, Qin G, Cao M, Chen R, He Y, Yang L, Zeng Z, Yu Y, Gu Y, Xing W, et al. 2020. A phosphorylation-based switch controls TAA1-mediated auxin biosynthesis in plants. *Nat Commun* **11:** 679. doi:10.1038/s41467-020-14395-w

Weijers D. 2016. Phyllotaxis: a Matthew effect in auxin action. *Curr Biol* **26:** R1233–R1235. doi:10.1016/j.cub.2016.10.019

Weijers D, Wagner D. 2016. Transcriptional responses to the auxin hormone. *Annu Rev Plant Biol* **67:** 539–574. doi:10.1146/annurev-arplant-043015-112122

Weller B, Zourelidou M, Frank L, Barbosa ICR, Fastner A, Richter S, Jürgens G, Hammes UZ, Schwechheimer C. 2017. Dynamic PIN-FORMED auxin efflux carrier phosphorylation at the plasma membrane controls auxin efflux-dependent growth. *Proc Natl Acad Sci* **114:** E887–E896. doi:10.1073/pnas.1614380114

Willige BC, Isono E, Richter R, Zourelidou M, Schwechheimer C. 2011. Gibberellin regulates PIN-FORMED abundance and is required for auxin transport-dependent growth and development in *Arabidopsis thaliana*. *Plant Cell* **23:** 2184–2195. doi:10.1105/tpc.111.086355

Wilmoth JC, Wang S, Tiwari SB, Joshi AD, Hagen G, Guilfoyle TJ, Alonso JM, Ecker JR, Reed JW. 2005. NPH4/ARF7 and ARF19 promote leaf expansion and auxin-induced lateral root formation. *Plant J* **43:** 118–130. doi:10.1111/j.1365-313X.2005.02432.x

Wisniewska J, Xu J, Seifartová D, Brewer PB, Růžička K, Blilou L, Rouquié D, Benková E, Scheres B, Friml J. 2006. Polar PIN localization directs auxin flow in plants. *Science* **312:** 883. doi:10.1126/science.1121356

Wolf S, Hématy K, Höfte H. 2012. Growth control and cell wall signaling in plants. *Annu Rev Plant Biol* **63:** 381–407. doi:10.1146/annurev-arplant-042811-105449

Won C, Shen X, Mashiguchi K, Zheng Z, Dai X, Cheng Y, Kasahara H, Kamiya Y, Chory J, Zhao Y. 2011. Conversion of tryptophan to indole-3-acetic acid by TRYPTOPHAN AMINOTRANSFERASES OF ARABIDOPSIS and YUCCAs in *Arabidopsis*. *Proc Natl Acad Sci* **108:** 18518–18523. doi:10.1073/pnas.1108436108

Woodward AW, Bartel B. 2005. Auxin: regulation, action, and interaction. *Ann Bot* **95:** 707–735. doi:10.1093/aob/mci083

Wright RM, Hagen G, Guilfoyle T. 1987. An auxin-induced polypeptide in dicotyledonous plants. *Plant Mol Biol* **9:** 625–634. doi:10.1007/BF00020538

Wu MF, Yamaguchi N, Xiao J, Bargmann B, Estelle M, Sang Y, Wagner D. 2015. Auxin-regulated chromatin switch directs acquisition of flower primordium founder fate. *eLife* **4:** e09269. doi:10.7554/eLife.09269

Xie M, Chen H, Huang L, O'Neil RC, Shokhirev MN, Ecker JR. 2018. A B-ARR-mediated cytokinin transcriptional network directs hormone cross-regulation and shoot development. *Nat Commun* **9:** 1604. doi:10.1038/s41467-018-03921-6

Xuan W, Band LR, Kumpf RP, Van Damme D, Parizot B, De Rop G, Opdenacker D, Möller BK, Skorzinski N, Njo MF, et al. 2016. Cyclic programmed cell death stimulates hormone signaling and root development in *Arabidopsis*. *Science* **351:** 384–387. doi:10.1126/science.aad2776

Yadav RK, Perales M, Gruel J, Ohno C, Heisler M, Girke T, Jönsson H, Reddy GV, Venugopala Reddy G. 2013. Plant stem cell maintenance involves direct transcriptional repression of differentiation program. *Mol Syst Biol* **9:** 654. doi:10.1038/msb.2013.8

Yan Z, Liu X, Ljung K, Li S, Zhao W, Yang F, Wang M, Tao Y. 2017. Type B response regulators act as central integrators in transcriptional control of the auxin biosynthesis enzyme TAA1. *Plant Physiol* **175:** 1438–1454. doi:10.1104/pp.17.00878

Yang Y, Hammes UZ, Taylor CG, Schachtman DP, Nielsen E. 2006. High-affinity auxin transport by the AUX1 influx carrier protein. *Curr Biol* **16:** 1123–1127. doi:10.1016/j.cub.2006.04.029

Yoshimitsu Y, Tanaka K, Fukuda W, Asami T, Yoshida S, Hayashi KI, Kamiya Y, Jikumaru Y, Shigeta T, Nakamura Y, et al. 2011. Transcription of DWARF4 plays a crucial role in auxin-regulated root elongation in addition to brassinosteroid homeostasis in *Arabidopsis thaliana*. *PLoS ONE* **6:** e23851. doi:10.1371/journal.pone.0023851

Zhang W, To JPC, Cheng CY, Eric Schaller G, Kieber JJ. 2011. Type-A response regulators are required for proper root apical meristem function through post-transcriptional regulation of PIN auxin efflux carriers. *Plant J* **68:** 1–10. doi:10.1111/j.1365-313X.2011.04668.x

Zhang W, Swarup R, Bennett M, Schaller GE, Kieber JJ. 2013. Cytokinin induces cell division in the quiescent center of the *Arabidopsis* root apical meristem. *Curr Biol* **23:** 1979–1989. doi:10.1016/j.cub.2013.08.008

Zhang S, Wang S, Xu Y, Yu C, Shen C, Qian Q, Geisler M, Jiang DA, Qi Y. 2015. The auxin response factor, OsARF19, controls rice leaf angles through positively regulating OsGH3-5 and OsBRI1. *Plant Cell Environ* **38:** 638–654. doi:10.1111/pce.12397

Zhang Z, Tucker E, Hermann M, Laux T. 2017. A molecular framework for the embryonic initiation of shoot meristem stem cells. *Dev Cell* **40:** 264–277.e4. doi:10.1016/j.devcel.2017.01.002

Zhang Z, Runions A, Mentink RA, Kierzkowski D, Karady M, Hashemi B, Huijser P, Strauss S, Gan X, Ljung K, et al. 2020. A WOX/auxin biosynthesis module controls growth to shape leaf form. *Curr Biol* **30:** 1–12.

Zhao Y. 2018. Essential roles of local auxin biosynthesis in plant development and in adaptation to environmental changes. *Annu Rev Plant Biol* **69:** 417–435. doi:10.1146/annurev-arplant-042817-040226

Zhao Y, Christensen SK, Fankhauser C, Cashman JR, Cohen JD, Weigel D, Chory J. 2001. A role for flavin monooxygenase-like enzymes in auxin biosynthesis. *Science* **291:** 306–309. doi:10.1126/science.291.5502.306

Zhao Z, Andersen SU, Ljung K, Dolezal K, Miotk A, Schultheiss SJ, Lohmann JU. 2010. Hormonal control of the shoot stem-cell niche. *Nature* **465:** 1089–1092. doi:10.1038/nature09126

Zhao Z, Zhang Y, Liu X, Zhang X, Liu S, Yu X, Ren Y, Zheng X, Zhou K, Jiang L, et al. 2013. A role for a dioxygenase in auxin metabolism and reproductive development in rice. *Dev Cell* **27:** 113–122. doi:10.1016/j.devcel.2013.09.005

Zhao S, Zhang ML, Ma TL, Wang Y. 2016. Phosphorylation of ARF2 relieves its repression of transcription of the K^+ transporter gene *HAK5* in response to low potassium stress. *Plant Cell* **28:** 3005–3019. doi:10.1105/tpc.16.00684

Zhou Y, Zhang D, An J, Yin H, Fang S, Chu J, Zhao Y, Li J. 2018. TCP transcription factors regulate shade avoidance via directly mediating the expression of both *PHYTOCHROME INTERACTING FACTOR*s and auxin biosynthetic genes. *Plant Physiol* **176:** 1850–1861. doi:10.1104/pp.17.01566

Zourelidou M, Absmanner B, Weller B, Barbosa ICR, Willige BC, Fastner A, Streit V, Port SA, Colcombet J, van Bentem SDLF, et al. 2014. Auxin efflux by PIN-FORMED proteins is activated by two different protein kinases, D6 PROTEIN KINASE and PINOID. *eLife* **3:** e02860. doi:10.7554/eLife.02860

Uncovering How Auxin Optimizes Root Systems Architecture in Response to Environmental Stresses

Nicola Leftley, Jason Banda, Bipin Pandey, Malcolm Bennett, and Ute Voß

Plant and Crop Sciences, School of Biosciences, Sutton Bonington Campus, The University of Nottingham, Loughborough LE12 5RD, United Kingdom

Correspondence: ute.voss@nottingham.ac.uk

Since colonizing land, plants have developed mechanisms to tolerate a broad range of abiotic stresses that include flooding, drought, high salinity, and nutrient limitation. Roots play a key role acclimating plants to these as their developmental plasticity enables them to grow toward more favorable conditions and away from limiting or harmful stresses. The phytohormone auxin plays a key role translating these environmental signals into developmental outputs. This is achieved by modulating auxin levels and/or signaling, often through cross talk with other hormone signals like abscisic acid (ABA) or ethylene. In our review, we discuss how auxin controls root responses to water, osmotic and nutrient-related stresses, and describe how the synthesis, degradation, transport, and response of this key signaling hormone helps optimize root architecture to maximize resource acquisition while limiting the impact of abiotic stresses.

Plant roots explore soil to acquire often scarce resources such as water and nutrients (Kiba and Krapp 2016). Soil constitutes a highly heterogenous environment (Morris et al. 2017) where resources can be locally limiting or replete and therefore act as stresses that restrict plant growth. Plants have developed myriad mechanisms to locally sense resources and translate them into optimizing root architecture, allowing growth toward more favorable conditions (Koevoets et al. 2016). Plants also rapidly inhibit their growth in response to stress, termed an "acute response," followed by recovery and acclimation to the new conditions on a physiological and morphological level (Skirycz and Inzé 2010).

The phytohormone auxin is an important regulator of almost all aspects of plant growth and development (Salehin et al. 2015) and also plays a key role controlling acute growth arrest and mediating long-term morphological changes (Skirycz and Inzé 2010). These plant responses are initiated by modifying auxin synthesis, transport, signaling, and/or degradation (Teale et al. 2006). Modifying these regulatory pathways results in differences in auxin distribution at the organ and cellular level, which in turn alter growth rates, growth directions (tropisms), and organ initiation (Vanneste and Friml 2009). Several articles have reviewed auxin regulation of plant growth (Paque and Weijers

Cite this article as *Cold Spring Harb Perspect Biol* doi: 10.1101/cshperspect.a040014

2016), tropisms (Muday 2001), and organ patterning (Banda et al. 2019). In this review, we discuss current understanding of how abiotic stresses modulate root growth and development by targeting the auxin synthesis, degradation, transport, and response machinery.

AUXIN-REGULATED ROOT BRANCHING IS CONTROLLED BY SOIL WATER AVAILABILITY

The highly heterogeneous nature of soil has a marked impact on root systems architecture (Morris et al. 2017). The ability of a plant to efficiently absorb water and nutrients from the complex soil environment is dependent on its root being able to sense these resources, and then optimizing its architecture to the spatial availability of these resources (Osmont et al. 2007; Morris et al. 2017; Banda et al. 2019). For example, roots have developed the ability to distinguish wet and dry microenvironments in the soil and modify the positioning of lateral roots (LRs) accordingly. This adaptive mechanism, termed LR hydropatterning, positions newly emerged lateral branches toward the soil water source (Bao et al. 2014). Noninvasive X-ray microCT imaging revealed that when maize roots grew down a large soil macropore, LRs preferentially emerged where roots were in direct contact with soil (Fig. 1A). This response can also be observed when growing seedlings on vertical agar plates, where LRs preferentially grow into or parallel to, rather away from, the agar surface (Bao et al. 2014; von Wangenheim et al. 2020). The importance of this response is highlighted by its conservation in both monocot and eudicot species (Bao et al. 2014).

LR hydropatterning depends on the perception of moisture across the primary root (PR) circumference. A root's ability to sense external water distribution appears to be linked to its uptake during cell expansion in the root elongation zone (Robbins and Dinneny 2018). This hydrological information influences the specification of xylem pole pericycle (XPP) cells in the basal meristem/elongation zone to become new LR primordia (von Wangenheim et al. 2020). The selection of LR founder cells based on external water availability was first reported to be regulated via an auxin response gradient by Bao et al. (2014). When seedling roots were grown along vertical agar plates, auxin biosynthesis (*wei8-1*), and transport (*pin2/3/7*) mutants were observed to exhibit a higher proportion of LRs emerging toward the air than wild-type (WT). This suggests the auxin synthesis and transport machinery partially control the sites of hormone accumulation during an LR hydropatterning response (Bao et al. 2014).

LR hydropatterning is also dependent on the auxin response factor 7 (ARF7) since roots of *arf7* mutant seedlings no longer show a bias for their LRs to emerge toward the agar surface (Orosa-Puente et al. 2018). *ARF7* is an important regulatory gene during every phase of LR development (Okushima et al. 2005; Lavenus et al. 2013, 2015). The ARF7 transcription factor induces the asymmetric expression of downstream LR regulatory genes, including *LBD16* in XPP cells in the elongation zone (Orosa-Puente et al. 2018). Asymmetric *LBD16* expression (which is required for LR hydropatterning) is regulated by the posttranslational modification (PTM) of ARF7 by the small ubiquitin-like modifier (SUMO) machinery. Mutating all four ARF7 SUMOylation sites disrupts the transcription factor's ability to rescue the *arf7* mutant's LR hydropatterning defect and restore *LBD16* asymmetric expression (Orosa-Puente et al. 2018). ARF7 SUMOylation disrupts its DNA-binding activity and promotes recruitment of a subset of Aux/IAA repressor proteins including IAA3/SHY2 (Fig. 1B). ARF7 and IAA3 interact via their SUMO PTM and SUMO interaction motif (SIM), respectively. Mutating ARF7 SUMO and IAA3 SIM sites disrupts their ability to interact and, in the case of the dominant IAA3 mutant form *shy2-2*, blocks its ability to disrupt LR development (Orosa-Puente et al. 2018). A gradient of ARF7 SUMOylation is hypothesized to form in response to a hydropatterning stimulus, causing LRs to preferentially form and emerge on the side of the root exposed to water (Fig. 1B). However, it remains unclear how the ARF7 SUMOylation gradient is regulated by external water availability. Components of the SUMO machinery represent promising regulators for monitoring soil

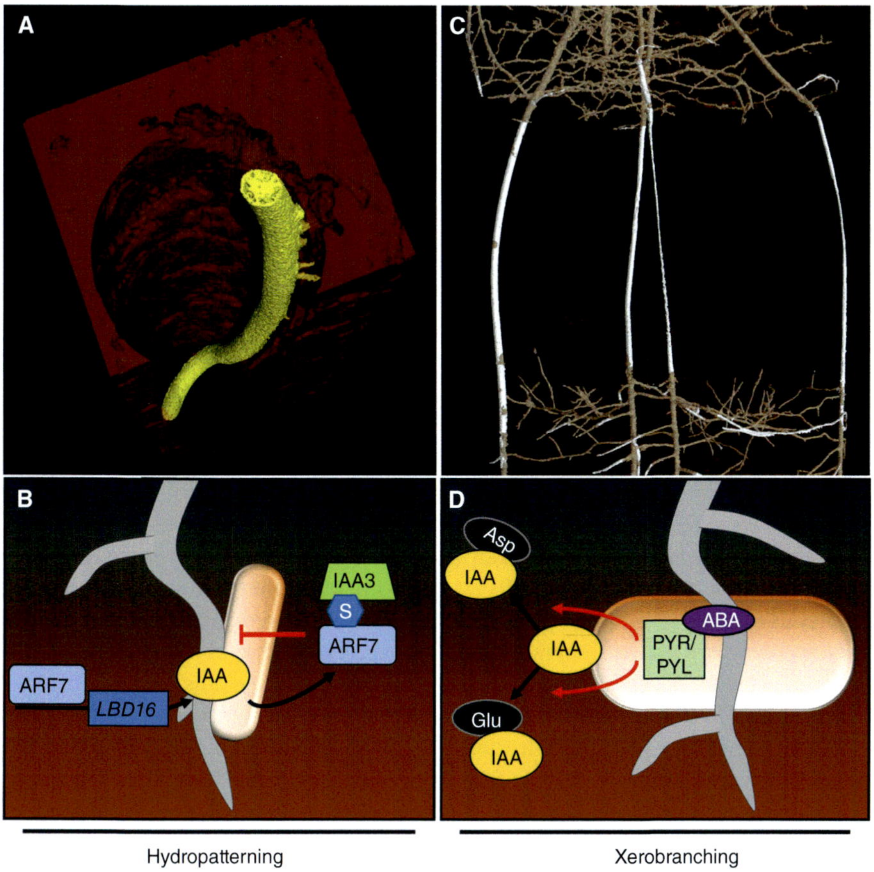

Figure 1. Auxin-regulated root-adaptive responses to water availability. (*A*) X-ray microCT image of a maize root growing down a macropore reveals lateral branches only form on the side in contact with soil employing the hydropatterning response. (*B*) Hydropatterning is regulated via the posttranslational modification of ARF7 by SUMO (blue hexagon "S") in cells on the side of the root exposed to the air-filled macropore (denoted by bubble). SUMOylated ARF7 recruits the repressor Aux/IAA3, blocking auxin-dependent transcription and lateral root development. On the wet side of the root, ARF7 is not SUMOylated and therefore can activate LBD16 gene expression, leading to lateral root initiation. (*C*) X-ray microCT image of a barley root growing through an air-filled space, causing branching to cease until later reentering soil due to the xerobranching response. (*D*) A transient reduction in water uptake causes abscisic acid (ABA) to accumulate in root tip tissues during growth through the air space (denoted by bubble). ABA triggers its receptor PYR/PYL, which results in an increase of indole-3-acetic acid (IAA) conjugation to aspartate (Asp) and glutamate (Glu) by GH3 enzymes. This leads to a drop in free IAA levels, negatively affecting lateral root initiation.

water availability that merit further investigation to address this important question.

In contrast to hydropatterning, when a root grows through an air space it temporarily loses contact with soil and is therefore deprived of an external water source. When barley and maize roots were grown through an air-filled space in a soil column, their roots transiently repressed root branching until roots grew into the next soil profile (Orman-Ligeza et al. 2018). MicroCT imaging noninvasively visualized this adaptive response in soil, highlighting the plasticity of root branching (Fig. 1C; Orman-Ligeza et al. 2018). Imaging also revealed the close relationship between moisture and LR formation, since growing roots through a water (rather than air)-filled space resulted in branching along the entire root radius (von Wangenheim et al. 2020).

An earlier study, using aeroponically grown barley roots transiently deprived of a water source, observed LR development was blocked at the very first stage of organ initiation (Babé et al. 2012). This novel water-related root-adaptive response has been termed xerobranching (Orman-Ligeza et al. 2018).

The xerobranching response takes place in the elongation zone (Orman-Ligeza et al. 2018). The elongation zone is vital for both LR hydropatterning and xerobranching responses (Bao et al. 2014; Orman-Ligeza et al. 2018). Whereas both water-related responses superficially appear similar, mutant experiments have revealed distinct signals regulate each process. LR hydropatterning is regulated by auxin and not the water-stress-related signal abscisic acid (ABA), as mutants disrupting its synthesis exhibit a normal hydropatterning response (Bao et al. 2014). In contrast, levels of ABA were shown to rapidly increase in root tips exposed to a xerobranching stimulus (Orman-Ligeza et al. 2018). It is well known that ABA (and its signaling pathway) is induced if water availability is limited in roots (Babé et al. 2012; Vishwakarma et al. 2017; Dietrich 2018). Moreover, exogenous addition of ABA in the aeroponic spray could mimic water deficiency and cause a significant reduction in LR formation (Orman-Ligeza et al. 2018). Finally, the ABA-insensitive receptor mutant *pyr/pyl 112458* did not form an LR repression zone, suggesting a key role of ABA signaling in xerobranching. Intriguingly, *pyr/pyl 112458* mutants exhibited no LR hydropatterning defect, suggesting these two water-related adaptive responses employ distinct hormone signals and response pathways.

Although ABA appears to regulate xerobranching, also auxin plays a significant role. Repression of LR branching could be partially blocked by exogenous addition of the auxin analog indole-3-butyric acid (IBA) pre-drought stress (Babé et al. 2012). Interestingly, the plasma-membrane permeable auxin NAA was able to fully restore organ initiation in the LR repression zone, while treatment with actively transported IAA could only partially restore LR number (Orman-Ligeza et al. 2018). Additionally, profiling of auxin precursors and conjugates in excised root tips revealed a decrease in free IAA levels and an increase in auxin conjugates. These results imply a role for active auxin conjugation that impair auxin accumulation and create the LR repression zone (Orman-Ligeza et al. 2018). Collectively, these results suggest that, in contrast to LR hydropatterning, cross talk exists between ABA signaling and auxin accumulation necessary for repression of LR initiation during a xerobranching response (Fig. 1D).

FLOODING-INDUCED ADVENTITIOUS ROOT FORMATION RELIES ON AUXIN AND ETHYLENE

Water is crucial for root growth and development; however, an excess of water due to flooding, can severely limit this. Flooding can be subdivided into two main conditions: fully submerged or waterlogged (whereby only the root is submerged). Both conditions cause a drop in the rate of oxygen diffusion, about 10,000 times compared to that in air (Armstrong 1980), which can severely affect plant performance (Voesenek and Bailey-Serres 2015). Plant roots have evolved numerous adaptive responses to survive flooding such as formation of shallower adventitious roots (ARs) and air-filled gaps between cells termed aerenchyma, together with induction of shoot hyponasty and elongation (Visser et al. 2003). The gaseous hormone ethylene quickly builds up when plants are flooded, and this signal is thought to be the main regulator of this adaptive response (Jackson 1985). Auxin has also been shown to perform a crucial function through cross talk with ethylene to control multiple flooding-related developmental adaptations in roots (Fig. 2A; Visser and Voesenek 2005).

During prolonged floods, several plant species including rice (*Oryza sativa*), maize (*Zea mays*), sunflower (*Helianthus annuus*), barley, wheat, and *Arabidopsis* exchange their dying PR system for ARs. ARs contain more aerenchyma than PRs, thereby greatly increasing gas exchange capability (Visser and Voesenek 2005). ARs emerge from the stem or hypocotyl and although they do not necessarily originate from the pericycle, their development follows a

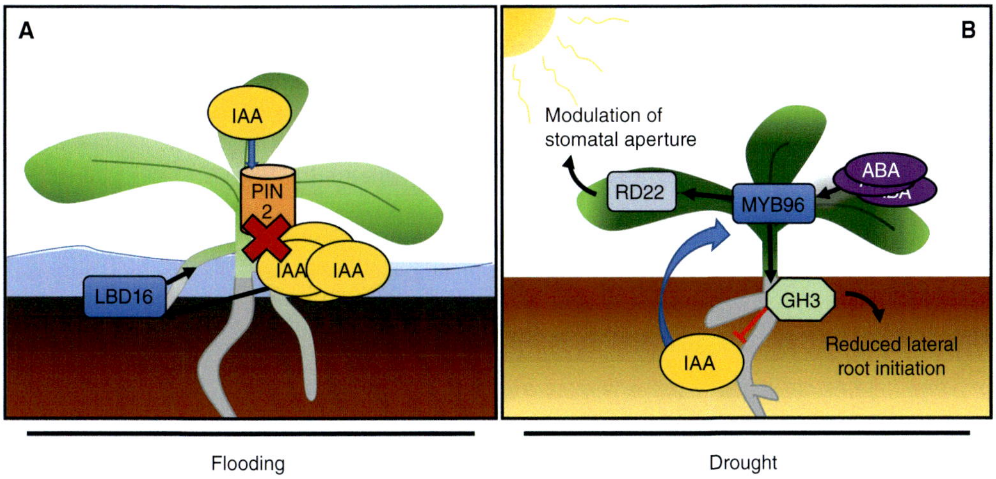

Figure 2. Systematic adaptive responses caused by flooding and drought stress. (*A*) During prolonged drought, exposure plants increase the initiation of adventitious roots as indicated in this simplified model. This initiation is induced by the inhibition of PIN2, decreasing auxin transport from shoot to root. The accumulation of indole-3-acetic acid (IAA) in the stem causes the induction of LBD16 (through auxin response factors [ARFs]) and the subsequent initiation of adventitious roots. (*B*) During long drought periods, plants accumulate abscisic acid (ABA) in leaf tissue. This accumulation leads to regulation of multiple genes of which one is the transcription factor MYB96. Through MYB96 induction, ABA regulates gene expression of RD22 and GH3. RD22 is important for drought responses in the shoot by modulating stomatal aperture. GH3 promotes conjugation of IAA in the root, thereby limiting auxin levels and restricting the initiation and development of lateral roots.

similar auxin-controlled pathway to LR formation (Bellini et al. 2014). Polar auxin transport is crucial for flooding-induced AR formation, as the auxin transport inhibitor *N*-1-naphthylphalamic acid (NPA) inhibits ARs formation upon flooding (Visser et al. 1996). Additionally, mutants that affect expression and polar localization of the auxin efflux carrier PIN-formed 2 (PIN2) exhibit delays in AR emergence (Fig. 2B; Vidoz et al. 2010; Ivanchenko et al. 2015; Dawood et al. 2016).

In addition to polar transport, auxin signaling is also a target for regulation during flooding in multiple plant species. In *Rumex palustris,* while auxin levels were unaltered during flooding, auxin sensitivity increased due to ethylene accumulation, triggering increased AR formation (Visser et al. 1996). This observation reveals cross talk between ethylene and auxin signaling components. In the tomato sp. *Solanum dulcamara*, flooding induced expression of *LBD16*, *LBD18*, *LBD29*, and *PUCHI* orthologs in AR primordia (Dawood et al. 2016), revealing a very similar auxin-regulated response pathway to de novo LR formation is induced under flooding during AR development (Goh et al. 2019). Recent work has revealed a role for auxin coreceptor TIR1/AFB F-box protein AFB2 in AR formation, suggesting auxin signaling might play a more prominent role than previously thought (da Costa et al. 2020).

A role for auxin signaling during flooding-induced aerenchyma formation is less well understood. Aerenchyma, which are longitudinal interconnected air spaces formed through programmed cell death, become increasingly important under long-term flooding, contributing to gaseous exchange between aerial and submerged root tissues (Jackson and Armstrong 1999). Oxygen-deficient conditions, simulating water logging in rice, increases the level of induced lysigenous aerenchyma created by programmed cell death (Colmer and Voesenek 2009). Treatment with auxin transport inhibitor NPA strongly reduced lysigenous aerenchyma formation (Yamauchi et al. 2019). Furthermore, the dominant-negative *iaa13* mutant, which strongly reduced auxin signaling, showed a steep decline in the number of aerenchyma, which could partially be restored by ethylene treatment (Yamauchi et al. 2020). Further studies are necessary to dissect cross talk between auxin and ethylene pathways in flooding adaptive mechanisms.

DROUGHT RESPONSES RELY ON CROSS TALK BETWEEN AUXIN AND OTHER HORMONES

Under water-deficient conditions, roots play a crucial adaptive role by perceiving differences in water availability in the first instance and subsequently modifying root growth rate, direction, and branching toward available water while avoiding dry areas (Seo et al. 2009; Robbins and Dinneny 2015). Drought stress also activates several signaling pathways involving ABA and ethylene (Tiwari et al. 2017). Cross talk between auxin and these pathways mediates a plethora of stress-adapted responses to regulate plant growth (Seo et al. 2009; Verma et al. 2016).

Auxin plays an important part in the plant's morphological response to drought. Levels of auxin decrease significantly under water deficit, although, at a higher auxin concentration, drought tolerance can be improved (Du et al. 2012). Hormonal cross talk between ABA and auxin under drought stress conditions optimizes root growth and branching toward moist soil, while limiting growth in dry soil (Fig. 2B). In *Arabidopsis*, ABA is able to induce the MYB96 transcription factor that in turn induces *RD22* (*RESPONSIVE TO DEHYDRATION 22*) in the shoot (Seo et al. 2009, 2011). RD22 then modulates stomatal aperture during drought stress (Seo et al. 2011). Auxin can also induce MYB96, which then regulates a subset of the auxin-responsive GH3 (GRETCHEN HAGEN 3) genes under drought conditions in the root (Seo et al. 2009). The GH3 IAA-amido synthetases conjugate auxin to amino acids, thereby removing IAA from the active pool (Casanova-Sáez and Voß 2019). The addition of exogenous IAA leads to an accumulation of MYB96 transcript in the pericycle (Seo et al. 2009), indicating MYB96 regulates LR meristem activation. MYB96 OX mutants exhibit a reduction in LR number that was also seen in GH3 OX mutant lines (Seo et al. 2009). Essentially, MYB96 has been shown to mediate ABA-auxin cross talk that modulates auxin homeostasis during LR meristem activation with the aid of GH3 (Fig. 2B).

In *O. sativa*, OsGH3.13 is induced under drought stress (Du et al. 2012; Tiwari et al. 2017). The activation of OsGH3.13 leads to a reduction in endogenous free IAA but an increase in ABA levels, resulting in marked changes in plant architecture. This also enhanced expression of LEA (late embryogenesis abundant) genes leading to increased drought tolerance (Zhang et al. 2009; Du et al. 2012). Overexpressing OsGH3-2 was found to reduce IAA and ABA levels but led to increased water loss and thereby drought sensitivity, suggesting only precise tissue-specific GH3 induction increases drought tolerance (Zhang et al. 2009; Du et al. 2012). OsGH3-2 overexpressing lines display a clear reduction in LR number that is also observed in other auxin-deficient mutants, suggesting a key role in regulating LR development (Kong et al. 2019). In addition, active auxin levels can be further increased by modulating its synthesis in response to drought stress. Auxin is a tryptophan (Trp)-derived molecule synthesized by parallel routes, the most abundant being the IPyA pathway employing TAA1/TAR and YUCCA (YUC) enzymes (Casanova-Sáez and Voß 2019). It was determined in *Arabidopsis* and previously in potato that overexpressing YUC6 results in improved drought tolerance (Kim et al. 2013; Park et al. 2013; Cha et al. 2015). Collectively, these results suggest that changes in auxin homeostasis can influence synthesis of other signals such as ABA and thereby confer drought tolerance.

OSMOTIC STRESS RESPONSES RELY ON TISSUE-SPECIFIC CHANGES IN AUXIN METABOLISM

Osmotic stresses, including high salinity, leads to reduced water availability through changes in osmotic pressure and ionic toxicity (Chaves and Oliveira 2004; Munns and Tester 2008). Soil salinization is having an increasingly negative impact on global agriculture. Six percent of irrigated and 20% of the world's total cultivable land is affected by high salinity (FAO 2009). Salt stress tolerance varies significantly between crops. Barley (*Hordeum vulgare*), the most salt-tolerant cereal, can endure up to 250 mM NaCl (equivalent to 50% seawater), beyond which survival rates drop drastically. Other economically important cereals, such as durum wheat (*Triticum turgidum* ssp.), maize (*Z. mays*), and sorghum (*Sorghum bicolor*), are less tolerant to salinity (Maas and Hoffman 1977). Generally, leaf and shoot development are more sensitive to salinity than root growth. High salinity-specific growth adaptations in the root include inhibition of PR growth, halotropism, increased LR density, and reduced root hair length (West et al. 2004; Wang et al. 2009; Galvan-Ampudia and Testerink 2011). Salt-induced root growth inhibition is correlated with reduced auxin response (Wang et al. 2009; Zolla et al. 2010; Liu et al. 2015). Measurements of endogenous IAA in leaves and roots revealed IAA levels are significantly lower in plants grown in high salt conditions (Du et al. 2013; Liu et al. 2015). Generally, lower auxin levels and auxin signaling enhances salt/osmotic stress tolerance (Wang et al. 2009; Zolla et al. 2010; Liu et al. 2015).

Modulating auxin synthesis, storage, transport, degradation, and signaling all appear to be involved in translating osmotic and salt stress signals into developmental adaptations. Consistently, auxin inactivation genes in *Arabidopsis* (Park et al. 2007), cotton (Kirungu et al. 2019), or rice (Du et al. 2012) are generally up-regulated in response to osmotic stress. Auxin can be inactivated either by DAO-encoded IAA oxidases and conjugation to amino acids by GH3s or to sugars by UGT enzymes (Casanova-Sáez and Voß 2019). A remarkable proportion of *Arabidopsis* root-specific auxin inactivation genes are up-regulated upon salt treatment resulting in higher auxin turnover rates and lower root auxin levels (Korver et al. 2018). This suggests that auxin homeostasis plays an important role in regulating root growth in response to stress and that auxin-mediated growth mechanisms are important for adapting to challenging environments. For example, high salinity induces overexpression of the IPyA auxin synthesis pathway, leading to increased IAA synthesis and salinity tolerance in several species including *Arabidopsis* (Lee et al. 2012; Shi et al. 2014; Cha et al. 2015), poplar (Ke et al. 2015), cucumber (Yan et al. 2016), and potato (Kim et al. 2013). Furthermore, tissue-specific microarray data during salt stress from *Arabidopsis* roots indicate a tissue-specific shift of YUCCA expression rather than a general change in expression rates. For example, YUCCAs are strongly expressed in the columella under control conditions but are strongly expressed in the epidermis and cortex during salt stress. Similarly, the IAOx IAA synthesis pathway regulates salt stress-induced root branching. IAOx pathway genes CYP79B2/3 are mainly expressed in the QC, the differentiation zone (Korver et al. 2018), and specifically in cells underlying newly developing LRs and LR primordia (Ljung et al. 2005). *cyp79b2;cyp79b3* double-knockout mutants exhibit decreased LR growth during salt stress (Julkowska et al. 2017). Salt-regulated transcriptional changes of the IAOx auxin synthesis machinery likely lead to a reduction in root tip auxin levels and therefore a reduction in root meristem growth, while higher auxin levels toward the differentiation zone promote root branching (Kilian et al. 2007; Dinneny et al. 2008; Korver et al. 2018).

Osmotic stress also impacts auxin distribution through differential regulation of PIN protein levels and localization during halotropism, for example. This adaptive growth response allows directional root growth away from high salt concentrations by reducing the gravitropic response (Dinneny et al. 2008; Sun et al. 2008; Rosquete and Kleine-Vehn 2013). PIN2, which is a key regulator of root gravitropism, plays a critical role during halotropism (Abas et al. 2006). PIN2 is internalized on the side of the root exposed to high salt, leading to reduced shootward auxin transport. Therefore, shootward auxin transport and cellular auxin concentrations are higher on the side exposed to low salt, leading to reduced cell elongation. Consequently, the root bends away from the local high salt environment. Interestingly this adaptive response does not occur when exposed to mannitol gradients (a nonanionic osmotic stress) but only when exposed to anionic osmotic gradients (Sun et al. 2008). In contrast, osmotic treatments with NaCl or mannitol lead to a quick internalization of PIN1 and PIN2 in root meristems (Nakayama et al. 2012; Galvan-Ampudia et al. 2013; Zwiewka et al. 2015). This raises the interesting question why some effects on PIN localization are salt specific, such as PIN2 internalization during halotropism, while others are generic to osmotic stress.

ROOT RESPONSES TO LOW EXTERNAL PHOSPHATE LEVELS ARE REGULATED BY AUXIN

Phosphate (P) deficiency is considered a major abiotic stress significantly reducing global crop yields. P deficiency is exacerbated by its unequal distribution in top (versus sub)-soil and its tendency to form organic and inorganic complexes that limit its mobility in soil and availability to plants (Shen et al. 2011; Pandey et al. 2013). To tackle this, plants modify their root system architecture to enhance foraging for P in top soils (Williamson et al. 2001; Band et al. 2012) and reshuffle metabolic programs to reduce internal P shortages (Esfahani et al. 2016). For example, shallower root angles maximize topsoil foraging under P-deficient conditions (Lynch and Brown 2001; Ramaekers et al. 2010; Giri et al. 2018). Gravitropic sensing in columella cells at the root tip is regulated by the low P-responsive gene *RICE MORPHOLOGY DETERMINANT 1* (*RMD1*) in rice. RMD1 is localized on the statolith surface in columella cells where it regulates the binding of actin filament and statoliths, thus controlling statolith sedimentation in response to gravitropic stimuli in the root tip (Huang et al. 2018). Under low P conditions, RMD levels are elevated, slowing gravity-sensing statolith sedimentation, which causes rice crown roots to grow at a shallower angle to improve P foraging in topsoil (Huang et al. 2018). Increased root hair elongation and LR branching are also considered a hallmark of P deficiency in rice, *Arabidopsis*, wheat, barley, and other crops (Lynch 2011; Péret et al. 2014). These root responses will serve to increase P uptake capacity by increasing root surface area to forage for this immobile soil nutrient. Root hair elongation under low P stress is regulated by increased auxin synthesis, transport, and response in *Arabidopsis* and rice (Bhosale et al. 2018; Giri et al. 2018).

P deficiency appears to orchestrate root system architecture modifications in large part by regulation of the auxin response machinery. Levels of auxin in P-starved root tips do not change drastically, yet their auxin responses are hyperactivated (Bhosale et al. 2018). Auxin-inducible expression of *ARF19* by ARF7 increases under low P, which then activates RSL2 and RSL4 TFs to enhance root hair elongation in *Arabidopsis* (Fig. 3A,B; Bhosale et al. 2018). Similarly, in rice, Os*ARF16* is induced under low P conditions and *osarf16* mutants severely inhibit root hair elongation and LR branching capacity (Shen et al. 2013). On the other hand, *OsARF12* and *OsARF25* act as negative regulators of P homeostasis and regulate PR elongation and branching capacity (Wang et al. 2014). Moreover, *Arabidopsis* plants overexpressing the auxin signaling receptor *TIR1* showed a hypersensitive response to low P conditions (Pérez-Torres et al. 2008). P deficiency thus promotes the auxin response pathway to trigger changes in root architecture. In contrast, auxin biosynthesis genes such as *TAA1* exhibit limited transcriptional changes under low P conditions yet, rather puzzlingly, *taa1* mutants disrupt the low P root hair elongation response (Bhosale et al. 2018). This may suggest that TAA1 activity is regulated by low P at the posttranscriptional level. Fascinatedly, Wang et al. (2020) recently reported the TAA1 activity is indeed controlled at the posttranscriptional level via phosphorylation. The authors also report TAA1 activity is controlled by an auxin feedback loop, in part, via the receptor TMK4. However, it is currently unclear whether external signals like low P also control TAA1 activity via this phosphorylation-based mechanism. Intriguingly, this TAA1-based mechanism appears to be highly conserved across plant evolution (Wang et al. 2020), revealing a fascinating new tier of regulation for auxin synthesis.

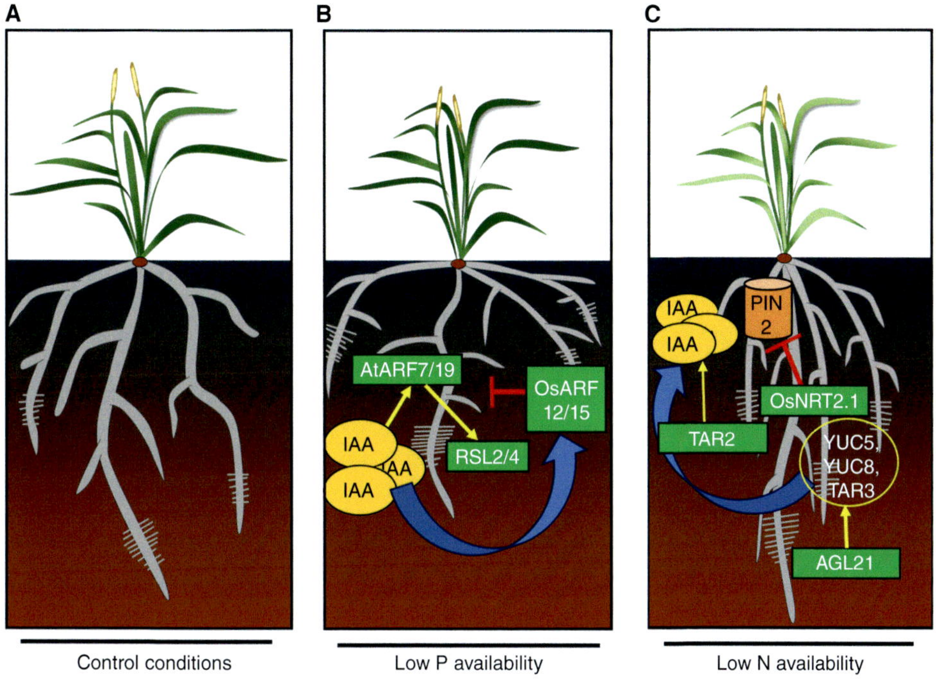

Figure 3. Phosphorus (P) and nitrogen (N) availability steer root system architecture. (*A*) Representative cartoon of a cereal root architecture. (*B*) Auxin response and levels are enhanced in P-deficient roots. The elevated auxin response causes the activation of ARF7 and ARF19 regulating the expression of *RSL2* and *RSL4* transcription factors, ultimately regulating the root hair elongation. OsARF12 and OsARF15 act as negative regulators of P homeostasis and regulate the branching density and root hair elongation in rice. (*C*) Root architecture of a low N (nitrogen) stressed plants showing enhanced forging capacity by elongating root hairs, longer branching, and steep root angle. Low N stress increases the IAA level via up-regulating expression of auxin biosynthetic genes. For example, TAR2 expression is increased under low N and results in higher IAA levels, which in turn reshapes the root architecture to favor maximum N uptake. AGL21 also positively regulates the expression *YUC5*, *YUC8*, and *TAR3*, which increases IAA synthesis rate under low N conditions. OsNRT2.1, which acts as an auxin influx facilitator, regulates lateral branching in rice by controlling *PIN2* expression.

AUXIN DYNAMICS SHAPE ROOT ARCHITECTURE UNDER LOW N REGIMES

Like P deficiency, N (nitrogen) deficiency in agricultural soils is a pervasive global issue and its deficiency causes significant drops in crop yields (Billen et al. 2013). To counter the impact of N deficiency in soils and to boost crop yield, farmers apply an excess of N fertilizers, which can be just as harmful as excess P for aquatic systems. Unlike the immobile nature of P, which accumulates in topsoils, highly soluble N forms like nitrate are prone to leach down into deeper soil layers with water (Meisinger and Delgado 2002). To access N from deeper soil layers, crops of several species increase their rooting depth to maximize N foraging, including rice, barley, and maize (Bonser et al. 1996; Trachsel et al. 2013; Ogawa et al. 2014; Yu et al. 2015b). Increased root branching length and steepness of growth angle are considered useful traits under N-limited conditions (Kiba and Krapp 2016), providing greater exposure to mobile N sources. Therefore, selection of long and steep branching genotypes with high yielding capacity could be useful to ameliorate N deficiency.

Each step of LR development: initiation, patterning, emergence, elongation, and maturation is tightly regulated by auxin (Casimiro et al. 2001; De Smet et al. 2007; Lavenus et al. 2013). Therefore, it comes as no surprise that auxin plays a pivotal role in inducing branching under

N-limited conditions. NRT1.1, a nitrate transporter, has been elegantly shown to also function as an auxin influx facilitator during nitrate (NO_3)-limiting conditions (Krouk et al. 2010). In addition, NRT1.1 regulates *AUXIN SIGNALING F-BOX 3 (AFB3)* expression in a NO_3-dependent manner, directly connecting downstream NRT1.1 signaling with auxin signaling (Vidal et al. 2014). Its homolog in rice, *OsNRT2.1*, positively regulates LR development under low NO_3- conditions. However, unlike *Arabidopsis*, *osnrt2.1* suppresses *OsPIN2* expression and forms fewer LRs (Huang et al. 2015).

To uncover the relationship between the auxin machinery and low N signaling, several genes have been identified. For example, the auxin synthesis gene *TAR2* is up-regulated under low NO_3- conditions, causing enhanced IAA levels in LR primordia. Additionally, *tar2* mutants produced very few LRs, revealing a need for enhanced auxin biosynthesis under low NO_3- conditions (Ma et al. 2014). *AGL21*, a MADS box gene induced under low N conditions, is also an auxin-responsive gene. Overexpressing of *AGL21* exhibited longer lateral branches and, in contrast to overexpressing lines, *agl21* mutant roots showed significantly fewer, smaller lateral branches under low NO_3- conditions. Interestingly, auxin biosynthetic genes (*YUC5*, *YUC8*, and *TAR3*) were up-regulated in *AGL21* overexpressing transgenics, explaining how higher IAA levels can help plants to elongate lateral branches in overexpressing roots especially under low NO_3- conditions (Fig. 3C; Yu et al. 2014).

NUTRIENT AND HEAVY METAL STRESS AFFECT AUXIN HOMEOSTASIS AND ROOT ARCHITECTURE

Other nutrients and metals in soil have also been shown to signal via the root auxin machinery. For example, potassium (K) deficiency impacts root growth and development via inhibition of PR growth. Recently, AKT1, a low K sensor, was found to induce degradation of the PIN1 protein under low K conditions. *akt1* mutants did not exhibit PR growth inhibition under low K conditions as compared to WT, suggesting AKT1 is a crucial component for root responses to K deficiency (Li et al. 2017).

Cadmium (Cd) is a toxic heavy metal and environmental pollutant. Cd is primarily stored in root xylem and transported to aerial tissues. However, excess Cd inhibits root growth and lateral branching. Cd causes increased expression of auxin reporters in poplar roots primarily in vascular tissues (Elobeid et al. 2012). This suggests that auxin-mediated root growth inhibition protects plants from the toxic effects of Cd. External application of Cd can perturb auxin biosynthesis and transport and has been shown to suppress LR formation in rice (Ronzan et al. 2018). Interestingly, the rice *aux1* mutant is highly sensitive to external Cd stress and shows reduced LR branching, root hair elongation, and PR growth (Cha et al. 2015; Yu et al. 2015a).

Similar to Cd stress, arsenate V (As) is a toxic metalloid, which affects root systems architecture. As it is transported through Pi transporters and in contaminated soils, its accumulation in plant organs negatively affects both plant and human health (Meharg and Macnair 1992). External application of As to *Arabidopsis* roots exhibited a reduction in root branching and a slight increase in PR length (Piacentini et al. 2020). Interestingly, external application of As affected *aux1*, *pin1*, and *pin2* mutants more than WT plants, revealing that auxin transport machinery plays an important role mitigating As stress (Krishnamurthy and Rathinasabapathi 2013).

CONCLUDING REMARKS

Our review highlights that auxin is a critical regulator of root responses to environmental signals in soil, enabling plants to maximize resource acquisition. Auxin synthesis, inactivation, transport, and signaling are modified by stress-related signaling pathways such as ABA or ethylene, allowing integration of local environmental conditions into growth outputs. Whereas our understanding of how stress signaling is transmitted to transcriptome changes has grown substantially over the past decade, we are only just starting to understand the role of PTMs like SUMO on auxin responses to abiotic

 Cite this article as *Cold Spring Harb Perspect Biol* doi: 10.1101/cshperspect.a040014

stresses (Orosa-Puente et al. 2018). In addition, our knowledge on how changes at transcriptional and posttranslational levels are differentially regulated in individual cells, tissues, and organs in response to stress is very limited (Wang et al. 2020). Individual cells and tissues differentially regulate their auxin levels and subsequent growth responses. For example, salt stress reduces PR tip auxin levels and response but increases LR branching (high auxin). This suggests a need for spatiotemporally fine-tuning auxin metabolism, transport, and signaling to ensure optimal adaptation of growth responses to external conditions. In addition, where and how stress sensing is translated into local versus systemic changes in auxin distribution and signaling has not been addressed to date but would significantly add to understanding how auxin contributes to plant stress responses. Future research will shed light on how micro-environmental conditions shape specific plant stress responses and how these are regulated on gene and protein levels. As the need for new crop varieties adapted to specific environmental conditions will increase with global climate change, this knowledge will inform breeding programs and allow the development of new varieties with increased fitness for future adverse conditions.

ACKNOWLEDGMENTS

We thank the University of Nottingham (Nottingham Research Fellowship to U.V.); BBSRC Divining Root Grant (BB/T001437/1) to N.L. and M.B.; BBSRC Doctoral Training Partnering award to J.B. and M.B.; and Future Food Beacon PhD+ Fellowship and Royal Society Challenge Grant (CHG\R1\170040) to B.P. and M.B. We apologize to colleagues whose work could not be cited owing to space limitations.

REFERENCES

Abas L, Benjamins R, Malenica N, Paciorek T, Wišniewska J, Moulinier-Anzola JC, Sieberer T, Friml J, Luschnig C. 2006. Intracellular trafficking and proteolysis of the *Arabidopsis* auxin-efflux facilitator PIN2 are involved in root gravitropism. *Nat Cell Biol* **8:** 249–256. doi:10.1038/ncb1369

Armstrong W. 1980. Aeration in higher plants. *Adv Bot Res* **7:** 225–332. doi:10.1016/S0065-2296(08)60089-0

Babé A, Lavigne T, Séverin JP, Nagel KA, Walter A, Chaumont F, Batoko H, Beeckman T, Draye X. 2012. Repression of early lateral root initiation events by transient water deficit in barley and maize. *Philos Trans R Soc B Biol Sci* **367:** 1534–1541. doi:10.1098/rstb.2011.0240

Band LR, Wells DM, Larrieu A, Sun J, Middleton AM, French AP, Brunoud G, Sato EM, Wilson MH, Péret B, et al. 2012. Root gravitropism is regulated by a transient lateral auxin gradient controlled by a tipping-point mechanism. *Proc Natl Acad Sci* **109:** 4668–4673. doi:10.1073/pnas.1201498109

Banda J, Bellande K, von Wangenheim D, Goh T, Guyomarc'h S, Laplaze L, Bennett MJ. 2019. Lateral root formation in *Arabidopsis*: a well-ordered LRexit. *Trends Plant Sci* **24:** 826–839. doi:10.1016/j.tplants.2019.06.015

Bao Y, Aggarwal P, Robbins NE, Sturrock CJ, Thompson MC, Tan HQ, Tham C, Duan L, Rodriguez PL, Vernoux T, et al. 2014. Plant roots use a patterning mechanism to position lateral root branches toward available water. *Proc Natl Acad Sci* **111:** 9319–9324. doi:10.1073/pnas.1400966111

Bellini C, Pacurar DI, Perrone I. 2014. Adventitious roots and lateral roots: similarities and differences. *Annu Rev Plant Biol* **65:** 639–666. doi:10.1146/annurev-arplant-050213-035645

Bhosale R, Giri J, Pandey BK, Giehl RFH, Hartmann A, Traini R, Truskina J, Leftley N, Hanlon M, Swarup K, et al. 2018. A mechanistic framework for auxin dependent *Arabidopsis* root hair elongation to low external phosphate. *Nat Commun* **9:** 1409. doi:10.1038/s41467-018-03851-3

Billen G, Garnier J, Lassaletta L. 2013. The nitrogen cascade from agricultural soils to the sea: modelling nitrogen transfers at regional watershed and global scales. *Philos Trans R Soc B Biol Sci* **368:** 20130123. doi:10.1098/rstb.2013.0123

Bonser AM, Lynch J, Snapp S. 1996. Effect of phosphorus deficiency on growth angle of basal roots in *Phaseolus vulgaris*. *New Phytol* **132:** 281–288. doi:10.1111/j.1469-8137.1996.tb01847.x

Casanova-Sáez R, Voß U. 2019. Auxin metabolism controls developmental decisions in land plants. *Trends Plant Sci* **24:** 741–754. doi:10.1016/j.tplants.2019.05.006

Casimiro I, Marchant A, Bhalerao RP, Beeckman T, Dhooge S, Swarup R, Graham N, Inzé D, Sandberg G, Casero PJ, et al. 2001. Auxin transport promotes *Arabidopsis* lateral root initiation. *Plant Cell* **13:** 843–852. doi:10.1105/tpc.13.4.843

Cha J-Y, Kim W-Y, Kang SB, Kim JI, Baek D, Jung IJ, Kim MR, Li N, Kim H-J, Nakajima M, et al. 2015. A novel thiol-reductase activity of *Arabidopsis* YUC6 confers drought tolerance independently of auxin biosynthesis. *Nat Commun* **6:** 8041. doi:10.1038/ncomms9041

Chaves MM, Oliveira MM. 2004. Mechanisms underlying plant resilience to water deficits: prospects for water-saving agriculture. *J Exp Bot* **55:** 2365–2384. doi:10.1093/jxb/erh269

Colmer TD, Voesenek LACJ. 2009. Flooding tolerance: suites of plant traits in variable environments. *Funct Plant Biol* **36:** 665–681. doi:10.1071/FP09144 da Costa CT, Offringa R, Fett-Neto AG. 2020. The role of auxin transporters and receptors in adventitious rooting of *Arabidopsis thaliana* pre-etiolated flooded seedlings. *Plant Sci* **290:** 110294. doi:10.1016/j.plantsci.2019.110294

Dawood T, Yang X, Visser EJW, Te Beek TAH, Kensche PR, Cristescu SM, Lee S, Floková K, Nguyen D, Mariani C, et al. 2016. A co-opted hormonal cascade activates dormant adventitious root primordia upon flooding in *Solanum dulcamara*. *Plant Physiol* **170:** 2351–2364. doi:10.1104/pp.15.00773

De Smet I, Tetsumura T, De Rybel B, dit Frey NF, Laplaze L, Casimiro I, Swarup R, Naudts M, Vanneste S, Audenaert D, et al. 2007. Auxin-dependent regulation of lateral root positioning in the basal meristem of *Arabidopsis*. *Development* **134:** 681–690. doi:10.1242/dev.02753

Dietrich D. 2018. Hydrotropism: how roots search for water. *J Exp Bot* **69:** 2759–2771. doi:10.1093/jxb/ery034

Dinneny JR, Long TA, Wang JY, Jung JW, Mace D, Pointer S, Barron C, Brady SM, Schiefelbein J, Benfey PN. 2008. Cell identity mediates the response of *Arabidopsis* roots to abiotic stress. *Science* **320:** 942–945. doi:10.1126/science.1153795

Du H, Wu N, Fu J, Wang S, Li X, Xiao J, Xiong L. 2012. A GH3 family member, OsGH3-2, modulates auxin and abscisic acid levels and differentially affects drought and cold tolerance in rice. *J Exp Bot* **63:** 6467–6480. doi:10.1093/jxb/ers300

Du H, Liu H, Xiong L. 2013. Endogenous auxin and jasmonic acid levels are differentially modulated by abiotic stresses in rice. *Front Plant Sci* **4:** 397.

Elobeid M, Göbel C, Feussner I, Polle A. 2012. Cadmium interferes with auxin physiology and lignification in poplar. *J Exp Bot* **63:** 1413–1421. doi:10.1093/jxb/err384

Esfahani MN, Kusano M, Nguyen KH, Watanabe Y, Van Ha C, Saito K, Sulieman S, Herrera-Estrella L, Tran LSP. 2016. Adaptation of the symbiotic *Mesorhizobium*-chickpea relationship to phosphate deficiency relies on reprogramming of whole-plant metabolism. *Proc Natl Acad Sci* **113:** E4610–E4619. doi:10.1073/pnas.1609440113

Food and Agricultural Organization of the United Nations (FAO). 2009. High level expert forum—how to feed the world in 2050. Economic and Social Development, Rome.

Galvan-Ampudia CS, Testerink C. 2011. Salt stress signals shape the plant root. *Curr Opin Plant Biol* **14:** 296–302. doi:10.1016/j.pbi.2011.03.019

Galvan-Ampudia CS, Julkowska MM, Darwish E, Gandullo J, Korver RA, Brunoud G, Haring MA, Munnik T, Vernoux T, Testerink C. 2013. Halotropism is a response of plant roots to avoid a saline environment. *Curr Biol* **23:** 2044–2050. doi:10.1016/j.cub.2013.08.042

Giri J, Bhosale R, Huang G, Pandey BK, Parker H, Zappala S, Yang J, Dievart A, Bureau C, Ljung K, et al. 2018. Rice auxin influx carrier OsAUX1 facilitates root hair elongation in response to low external phosphate. *Nat Commun* **9:** 1810. doi:10.1038/s41467-018-04280-y

Goh T, Toyokura K, Yamaguchi N, Okamoto Y, Uehara T, Kaneko S, Takebayashi Y, Kasahara H, Ikeyama Y, Okushima Y, et al. 2019. Lateral root initiation requires the sequential induction of transcription factors LBD16 and PUCHI in *Arabidopsis thaliana*. *New Phytol* **224:** 749–760. doi:10.1111/nph.16065

Huang S, Chen S, Liang Z, Zhang C, Yan M, Chen J, Xu G, Fan X, Zhang Y. 2015. Knockdown of the partner protein OsNAR2.1 for high-affinity nitrate transport represses lateral root formation in a nitrate-dependent manner. *Sci Rep* **5:** 18192. doi:10.1038/srep18192

Huang G, Liang W, Sturrock CJ, Pandey BK, Giri J, Mairhofer S, Wang D, Muller L, Tan H, York LM, et al. 2018. Rice actin binding protein RMD controls crown root angle in response to external phosphate. *Nat Commun* **9:** 1–9.

Ivanchenko MG, Zhu J, Wang B, Medvecká E, Du Y, Azzarello E, Mancuso S, Megraw M, Filichkin S, Dubrovsky JG, et al. 2015. The cyclophilin a DIAGEOTROPICA gene affects auxin transport in both root and shoot to control lateral root formation. *Development* **142:** 712–721. doi:10.1242/dev.113225

Jackson MB. 1985. Ethylene and responses of plants to soil waterlogging and submergence. *Annu Rev Phytopathol* **36:** 145–174.

Jackson MB, Armstrong W. 1999. Formation of aerenchyma and the processes of plant ventilation in relation to soil flooding and submergence. *Plant Biol* **1:** 274–287. doi:10.1111/j.1438-8677.1999.tb00253.x

Julkowska MM, Koevoets IT, Mol S, Hoefsloot H, Feron R, Tester MA, Keurentjes JJB, Korte A, Haring MA, de Boer GJ, et al. 2017. Genetic components of root architecture remodeling in response to salt stress. *Plant Cell* **29:** 3198–3213. doi:10.1105/tpc.16.00680

Ke Q, Wang Z, Ji CY, Jeong JC, Lee HS, Li H, Xu B, Deng X, Kwak SS. 2015. Transgenic poplar expressing *Arabidopsis* YUCCA6 exhibits auxin-overproduction phenotypes and increased tolerance to abiotic stress. *Plant Physiol Biochem* **94:** 19–27. doi:10.1016/j.plaphy.2015.05.003

Kiba T, Krapp A. 2016. Plant nitrogen acquisition under low availability: regulation of uptake and root architecture. *Plant Cell Physiol* **57:** 707–714. doi:10.1093/pcp/pcw052

Kilian J, Whitehead D, Horak J, Wanke D, Weinl S, Batistic O, D'Angelo C, Bornberg-Bauer E, Kudla J, Harter K. 2007. The AtGenExpress global stress expression data set: protocols, evaluation and model data analysis of UV-B light, drought and cold stress responses. *Plant J* **50:** 347–363. doi:10.1111/j.1365-313X.2007.03052.x

Kim JI, Baek D, Park HC, Chun HJ, Oh DH, Lee MK, Cha JY, Kim WY, Kim MC, Chung WS, et al. 2013. Overexpression of *Arabidopsis* YUCCA6 in potato results in high-auxin developmental phenotypes and enhanced resistance to water deficit. *Mol Plant* **6:** 337–349. doi:10.1093/mp/sss100

Kirungu JN, Magwanga RO, Lu P, Cai X, Zhou Z, Wang X, Peng R, Wang K, Liu F. 2019. Functional characterization of Gh_A08G1120 (GH3.5) gene reveal their significant role in enhancing drought and salt stress tolerance in cotton. *BMC Genet* **20:** 62. doi:10.1186/s12863-019-0756-6

Koevoets IT, Venema JH, Elzenga JTM, Testerink C. 2016. Roots withstanding their environment: exploiting root system architecture responses to abiotic stress to improve crop tolerance. *Front Plant Sci* **7:** 1335. doi:10.3389/fpls.2016.01335

Kong W, Zhong H, Deng X, Gautam M, Gong Z, Zhang Y, Zhao G, Liu C, Li Y. 2019. Evolutionary analysis of GH3 genes in six *Oryza* species/subspecies and their expression under salinity stress in *Oryza sativa* ssp. Japonica. *Plants (Basel)* **8:** 30. doi:10.3390/plants8020030

Korver RA, Koevoets IT, Testerink C. 2018. Out of shape during stress: a key role for auxin. *Trends Plant Sci* **23:** 783–793. doi:10.1016/j.tplants.2018.05.011

Krishnamurthy A, Rathinasabapathi B. 2013. Auxin and its transport play a role in plant tolerance to arsenite-induced oxidative stress in *Arabidopsis thaliana*. *Plant, Cell Environ* **36:** 1838–1849. doi:10.1111/pce.12093

Krouk G, Lacombe B, Bielach A, Perrine-Walker F, Malinska K, Mounier E, Hoyerova K, Tillard P, Leon S, Ljung K, et al. 2010. Nitrate-regulated auxin transport by NRT1.1 defines a mechanism for nutrient sensing in plants. *Dev Cell* **18:** 927–937. doi:10.1016/j.devcel.2010.05.008

Lavenus J, Goh T, Roberts I, Guyomarc'h S, Lucas M, De Smet I, Fukaki H, Beeckman T, Bennett M, Laplaze L. 2013. Lateral root development in *Arabidopsis*: fifty shades of auxin. *Trends Plant Sci* **18:** 450–458. doi:10.1016/j.tplants.2013.04.006

Lavenus J, Goh T, Guyomarc'H S, Hill K, Lucas M, Voß U, Kenobi K, Wilson MH, Farcot E, Hagen G, et al. 2015. Inference of the *Arabidopsis* lateral root gene regulatory network suggests a bifurcation mechanism that defines primordia flanking and central zones. *Plant Cell* **27:** 1368–1388. doi:10.1105/tpc.114.132993

Lee M, Jung JH, Han DY, Seo PJ, Park WJ, Park CM. 2012. Activation of a flavin monooxygenase gene YUCCA7 enhances drought resistance in *Arabidopsis*. *Planta* **235:** 923–938. doi:10.1007/s00425-011-1552-3

Li J, Wu WH, Wang Y. 2017. Potassium channel AKT1 is involved in the auxin-mediated root growth inhibition in *Arabidopsis* response to low K^+ stress. *J Integr Plant Biol* **59:** 895–909. doi:10.1111/jipb.12575

Liu W, Li RJ, Han TT, Cai W, Fu ZW, Lu YT. 2015. Salt stress reduces root meristem size by nitric oxide-mediated modulation of auxin accumulation and signaling in *Arabidopsis*. *Plant Physiol* **168:** 343–356. doi:10.1104/pp.15.00030

Ljung K, Hull AK, Celenza J, Yamada M, Estelle M, Normanly J, Sandberg G. 2005. Sites and regulation of auxin biosynthesis in *Arabidopsis* roots. *Plant Cell* **17:** 1090–1104. doi:10.1105/tpc.104.029272

Lynch JP. 2011. Root phenes for enhanced soil exploration and phosphorus acquisition: tools for future crops. *Plant Physiol* **156:** 1041–1049. doi:10.1104/pp.111.175414

Lynch JP, Brown KM. 2001. Topsoil foraging—an architectural adaptation of plants to low phosphorus availability. *Plant Soil* **237:** 225–237. doi:10.1023/A:1013324727040

Ma W, Li J, Qu B, He X, Zhao X, Li B, Fu X, Tong Y. 2014. Auxin biosynthetic gene TAR2 is involved in low nitrogen-mediated reprogramming of root architecture in *Arabidopsis*. *Plant J* **78:** 70–79. doi:10.1111/tpj.12448

Maas EV, Hoffman GJ. 1977. Crop salt tolerance—current assessment. *J Irrig Drain Div* **103:** 115–134. doi:10.1061/JRCEA4.0001137

Meharg AA, Macnair MR. 1992. Genetic correlation between arsenate tolerance and the rate of influx of arsenate and phosphate in *Holcus lanatus* L. *Heredity (Edinb)* **69:** 336–341. doi:10.1038/hdy.1992.133

Meisinger JJ, Delgado JA. 2002. Principles for managing nitrogen leaching. *J Soil Water Conservation* **57:** 485–498.

Morris EC, Griffiths M, Golebiowska A, Mairhofer S, Burr-Hersey J, Goh T, von Wangenheim D, Atkinson B, Sturrock CJ, Lynch JP, et al. 2017. Shaping 3D root system architecture. *Curr Biol* **27:** R919–R930. doi:10.1016/j.cub.2017.06.043

Muday GK. 2001. Auxins and tropisms. *J Plant Growth Regul* **20:** 226–243. doi:10.1007/s003440010027

Munns R, Tester M. 2008. Mechanisms of salinity tolerance. *Annu Rev Plant Biol* **59:** 651–681. doi:10.1146/annurev.arplant.59.032607.092911

Nakayama N, Smith RS, Mandel T, Robinson S, Kimura S, Boudaoud A, Kuhlemeier C. 2012. Mechanical regulation of auxin-mediated growth. *Curr Biol* **22:** 1468–1476. doi:10.1016/j.cub.2012.06.050

Ogawa S, Valencia MO, Ishitani M, Selvaraj MG. 2014. Root system architecture variation in response to different $NH4^+$ concentrations and its association with nitrogen-deficient tolerance traits in rice. *Acta Physiol Plant* **36:** 2361–2372. doi:10.1007/s11738-014-1609-6

Okushima Y, Overvoorde PJ, Arima K, Alonso JM, Chan A, Chang C, Ecker JR, Hughes B, Lui A, Nguyen D, et al. 2005. Functional genomic analysis of the AUXIN RESPONSE FACTOR gene family members in *Arabidopsis thaliana*: unique and overlapping functions of ARF7 and ARF19. *Plant Cell* **17:** 444–463. doi:10.1105/tpc.104.028316

Orman-Ligeza B, Morris EC, Parizot B, Lavigne T, Babé A, Ligeza A, Klein S, Sturrock C, Xuan W, Novák O, et al. 2018. The xerobranching response represses lateral root formation when roots are not in contact with water. *Curr Biol* **28:** 3165–3173.e5. doi:10.1016/j.cub.2018.07.074

Orosa-Puente B, Leftley N, von Wangenheim D, Banda J, Srivastava AK, Hill K, Truskina J, Bhosale R, Morris E, Srivastava M, et al. 2018. Root branching toward water involves posttranslational modification of transcription factor ARF7. *Science* **362:** 1407–1410. doi:10.1126/science.aau3956

Osmont KS, Sibout R, Hardtke CS. 2007. Hidden branches: developments in root system architecture. *Annu Rev Plant Biol* **58:** 93–113. doi:10.1146/annurev.arplant.58.032806.104006

Pandey J, Singh AV, Singh A, Singh R. 2013. Impacts of changing atmospheric deposition chemistry on nitrogen and phosphorus loading to Ganga River (India). *Bull Environ Contam Toxicol* **91:** 184–190. doi:10.1007/s00128-013-1016-5

Paque S, Weijers D. 2016. Q&A: auxin: the plant molecule that influences almost anything. *BMC Biol* **14:** 67. doi:10.1186/s12915-016-0291-0

Park JE, Park JY, Kim YS, Staswick PE, Jeon J, Yun J, Kim SY, Kim J, Lee YH, Park CM. 2007. GH3-mediated auxin homeostasis links growth regulation with stress adaptation response in *Arabidopsis*. *J Biol Chem* **282:** 10036–10046. doi:10.1074/jbc.M610524200

Park HC, Cha JY, Yun DJ. 2013. Roles of YUCCAs in auxin biosynthesis and drought stress responses in plants. *Plant Signal Behav* **8:** e24495. doi:10.4161/psb.24495

Péret B, Desnos T, Jost R, Kanno S, Berkowitz O, Nussaume L. 2014. Root architecture responses: in search of phosphate. *Plant Physiol* **166:** 1713–1723. doi:10.1104/pp.114.244541

Pérez-Torres CA, López-Bucio J, Cruz-Ramírez A, Ibarra-Laclette E, Dharmasiri S, Estelle M, Herrera-Estrella L. 2008. Phosphate availability alters lateral root development in *Arabidopsis* by modulating auxin sensitivity via a mechanism involving the TIR1 auxin receptor. *Plant Cell* **20:** 3258–3272. doi:10.1105/tpc.108.058719

Piacentini D, Corpas FJ, D'Angeli S, Altamura MM, Falasca G. 2020. Cadmium and arsenic-induced-stress differentially modulates *Arabidopsis* root architecture, peroxisome distribution, enzymatic activities and their nitric oxide content. *Plant Physiol Biochem* **148:** 312–323. doi:10.1016/j.plaphy.2020.01.026

Ramaekers L, Remans R, Rao IM, Blair MW, Vanderleyden J. 2010. Strategies for improving phosphorus acquisition efficiency of crop plants. *Field Crop Res* **117:** 169–176. doi:10.1016/j.fcr.2010.03.001

Robbins NE, Dinneny JR. 2015. The divining root: moisture-driven responses of roots at the micro- and macro-scale. *J Exp Bot* **66:** 2145–2154. doi:10.1093/jxb/eru496

Robbins NE, Dinneny JR. 2018. Growth is required for perception of water availability to pattern root branches in plants. *Proc Natl Acad Sci* **115:** E822–E831. doi:10.1073/pnas.1710709115

Ronzan M, Piacentini D, Fattorini L, Della Rovere F, Eiche E, Riemann M, Altamura MM, Falasca G. 2018. Cadmium and arsenic affect root development in *Oryza sativa* L. negatively interacting with auxin. *Environ Exp Bot* **151:** 64–75. doi:10.1016/j.envexpbot.2018.04.008

Rosquete MR, Kleine-Vehn J. 2013. Halotropism: turning down the salty date. *Curr Biol* **23:** R927–R929. doi:10.1016/j.cub.2013.08.020

Salehin M, Bagchi R, Estelle M. 2015. ScfTIR1/AFB-based auxin perception: mechanism and role in plant growth and development. *Plant Cell* **27:** 9–19. doi:10.1105/tpc.114.133744

Seo PJ, Xiang F, Qiao M, Park JY, Lee YN, Kim SG, Lee YH, Park WJ, Park CM. 2009. The MYB96 transcription factor mediates abscisic acid signaling during drought stress response in *Arabidopsis. Plant Physiol* **151:** 275–289. doi:10.1104/pp.109.144220

Seo PJ, Lee SB, Suh MC, Park MJ, Park CM. 2011. The MYB96 transcription factor regulates cuticular wax biosynthesis under drought conditions in *Arabidopsis. Plant Cell* **23:** 1138–1152. doi:10.1105/tpc.111.083485

Shen J, Yuan L, Zhang J, Li H, Bai Z, Chen X, Zhang W, Zhang F. 2011. Phosphorus dynamics: from soil to plant. *Plant Physiol* **156:** 997–1005. doi:10.1104/pp.111.175232

Shen C, Wang S, Zhang S, Xu Y, Qian Q, Qi Y, Jiang DA. 2013. OsARF16, a transcription factor, is required for auxin and phosphate starvation response in rice (*Oryza sativa* L.). *Plant Cell Environ* **36:** 607–620. doi:10.1111/pce.12001

Shi H, Chen L, Ye T, Liu X, Ding K, Chan Z. 2014. Modulation of auxin content in *Arabidopsis* confers improved drought stress resistance. *Plant Physiol Biochem PPB* **82:** 209–217. doi:10.1016/j.plaphy.2014.06.008

Skirycz A, Inzé D. 2010. More from less: plant growth under limited water. *Curr Opin Biotechnol* **21:** 197–203. doi:10.1016/j.copbio.2010.03.002

Sun F, Zhang W, Hu H, Li B, Wang Y, Zhao Y, Li K, Liu M, Li X. 2008. Salt modulates gravity signaling pathway to regulate growth direction of primary roots in *Arabidopsis. Plant Physiol* **146:** 178–188. doi:10.1104/pp.107.109413

Teale WD, Paponov IA, Palme K. 2006. Auxin in action: signalling, transport and the control of plant growth and development. *Nat Rev Mol Cell Biol* **7:** 847–859. doi:10.1038/nrm2020

Tiwari S, Lata C, Chauhan PS, Prasad V, Prasad M. 2017. A functional genomic perspective on drought signalling and its crosstalk with phytohormone-mediated signalling pathways in plants. *Curr Genomics* **18:** 469–482. doi:10.2174/1389202918666170605083319

Trachsel S, Kaeppler SM, Brown KM, Lynch JP. 2013. Maize root growth angles become steeper under low N conditions. *Field Crop Res* **140:** 18–31. doi:10.1016/j.fcr.2012.09.010

Vanneste S, Friml J. 2009. Auxin: a trigger for change in plant development. *Cell* **136:** 1005–1016. doi:10.1016/j.cell.2009.03.001

Verma V, Ravindran P, Kumar PP. 2016. Plant hormone-mediated regulation of stress responses. *BMC Plant Biol* **16:** 86. doi:10.1186/s12870-016-0771-y

Vidal EA, Álvarez JM, Gutiérrez RA. 2014. Nitrate regulation of AFB3 and NAC4 gene expression in *Arabidopsis* roots depends on NRT1.1 nitrate transport function. *Plant Signal Behav* **9:** e28501. doi:10.4161/psb.28501

Vidoz ML, Loreti E, Mensuali A, Alpi A, Perata P. 2010. Hormonal interplay during adventitious root formation in flooded tomato plants. *Plant J* **63:** 551–562. doi:10.1111/j.1365-313X.2010.04262.x

Vishwakarma K, Upadhyay N, Kumar N, Yadav G, Singh J, Mishra RK, Kumar V, Verma R, Upadhyay RG, Pandey M, et al. 2017. Abscisic acid signaling and abiotic stress tolerance in plants: a review on current knowledge and future prospects. *Front Plant Sci* **8:** 161. doi:10.3389/fpls.2017.00161

Visser EJW, Voesenek LACJ. 2005. Acclimation to soil flooding-sensing and signal-transduction. *Plant Soil* **274:** 197–214. doi:10.1007/s11104-004-1650-0

Visser EJW, Cohen JD, Barendse GWM, Blom CWPM, Voesenek LACJ. 1996. An ethylene-mediated increase in sensitivity to auxin induces adventitious root formation in flooded *Rumex palustris* Sm. *Plant Physiol* **112:** 1687–1692. doi:10.1104/pp.112.4.1687

Visser EJW, Voesenek LACJ, Vartapetian BB, Jackson MB. 2003. Flooding and plant growth. *Ann Bot* **91:** 107–109. doi:10.1093/aob/mcg014

Voesenek LACJ, Bailey-Serres J. 2015. Flood adaptive traits and processes: an overview. *New Phytol* **206:** 57–73. doi:10.1111/nph.13209

von Wangenheim D, Banda J, Schmitz A, Boland J, Bishopp A, Maizel A, Stelzer EHK, Bennett M. 2020. Early developmental plasticity of lateral roots in response to asymmetric water availability. *Nat Plants* **6:** 73–77. doi:10.1038/s41477-019-0580-z

Wang Y, Li K, Li X. 2009. Auxin redistribution modulates plastic development of root system architecture under salt stress in *Arabidopsis thaliana. J Plant Physiol* **166:** 1637–1645. doi:10.1016/j.jplph.2009.04.009

Wang S, Zhang S, Sun C, Xu Y, Chen Y, Yu C, Qian Q, Jiang DA, Qi Y. 2014. Auxin response factor (OsARF12), a novel regulator for phosphate homeostasis in rice (*Oryza sativa*). *New Phytol* **201:** 91–103. doi:10.1111/nph.12499

Wang Q, Qin G, Cao M, Chen R, He Y, Yang L, Zeng Z, Yu Y, Gu Y, Xing W, et al. 2020. A phosphorylation-based switch controls TAA1-mediated auxin biosynthesis in plants. *Nat Commun* **11:** 679. doi:10.1038/s41467-020-14395-w

West G, Inzé D, Beemster GTS. 2004. Cell cycle modulation in the response of the primary root of *Arabidopsis* to salt stress. *Plant Physiol* **135:** 1050–1058. doi:10.1104/pp.104.040022

Williamson LC, Ribrioux SPCP, Fitter AH, Ottoline Leyser HM. 2001. Phosphate availability regulates root system architecture in *Arabidopsis*. *Plant Physiol* **126:** 875–882. doi:10.1104/pp.126.2.875

Yamauchi T, Tanaka A, Inahashi H, Nishizawa NK, Tsutsumi N, Inukai Y, Nakazono M. 2019. Fine control of aerenchyma and lateral root development through AUX/IAA- and ARF-dependent auxin signaling. *Proc Natl Acad Sci* **116:** 20770–20775. doi:10.1073/pnas.1907181116

Yamauchi T, Tanaka A, Tsutsumi N, Inukai Y, Nakazono M. 2020. A role for auxin in ethylene-dependent inducible aerenchyma formation in rice roots. *Plants (Basel)* **9:** 610. doi:10.3390/plants9050610

Yan S, Che G, Ding L, Chen Z, Liu X, Wang H, Zhao W, Ning K, Zhao J, Tesfamichael K, et al. 2016. Different cucumber CsYUC genes regulate response to abiotic stresses and flower development. *Sci Rep* **6:** 20760. doi:10.1038/srep20760

Yu LH, Miao ZQ, Qi GF, Wu J, Cai XT, Mao JL, Xiang CB. 2014. MADS-Box transcription factor AGL21 regulates lateral root development and responds to multiple external and physiological signals. *Mol Plant* **7:** 1653–1669. doi:10.1093/mp/ssu088

Yu C, Sun C, Shen C, Wang S, Liu F, Liu Y, Chen Y, Li C, Qian Q, Aryal B, et al. 2015a. The auxin transporter, OsAUX1, is involved in primary root and root hair elongation and in Cd stress responses in rice (*Oryza sativa* L.). *Plant J* **83:** 818–820. doi:10.1111/tpj.12929

Yu P, Li X, White PJ, Li C. 2015b. A large and deep root system underlies high nitrogen-use efficiency in maize production. *PLoS ONE* **10:** e0126293. doi:10.1371/journal.pone.0126293

Zhang SW, Li CH, Cao J, Zhang YC, Zhang SQ, Xia YF, Sun DY, Sun Y. 2009. Altered architecture and enhanced drought tolerance in rice via the down-regulation of indole-3-acetic acid by TLD1/OsGH3.13 activation. *Plant Physiol* **151:** 1889–1901. doi:10.1104/pp.109.146803

Zolla G, Heimer YM, Barak S. 2010. Mild salinity stimulates a stress-induced morphogenic response in *Arabidopsis thaliana* roots. *J Exp Bot* **61:** 211–224. doi:10.1093/jxb/erp290

Zwiewka M, Nodzyński T, Robert S, Vanneste S, Friml J. 2015. Osmotic stress modulates the balance between exocytosis and clathrin-mediated endocytosis in *Arabidopsis thaliana*. *Mol Plant* **8:** 1175–1187. doi:10.1016/j.molp.2015.03.007

Auxin Plays Multiple Roles during Plant–Pathogen Interactions

Barbara N. Kunkel and Joshua M.B. Johnson

Department of Biology, Washington University in St. Louis, St. Louis, Missouri 63130, USA
Correspondence: kunkel@wustl.edu

The plant hormone auxin governs many aspects of normal plant growth and development. Auxin also plays an important role in plant–microbe interactions, including interactions between plant hosts and pathogenic microorganisms that cause disease. It is now well established that indole-3-acetic acid (IAA), the most well-studied form of auxin, promotes disease in many plant–pathogen interactions. Recent studies have shown that IAA can act both as a plant hormone that modulates host signaling and physiology to increase host susceptibility and as a microbial signal that directly impacts the pathogen to promote virulence, but large gaps in our understanding remain. In this article, we review recent studies on the roles that auxin plays during plant–pathogen interactions and discuss the virulence mechanisms that many plant pathogens have evolved to manipulate host auxin signaling and promote pathogenesis.

To colonize their plant hosts, pathogenic microbes must enter plant tissue, evade or suppress basal defense, obtain nutrients and water, and grow to high levels. In many cases, this ultimately results in the development of disease symptoms, such as abnormal plant growth or development and/or generation of chlorotic or necrotic lesions. Plant pathogens have evolved a variety of virulence strategies that allow them to achieve these ends. One common theme that has emerged from studies of plant–pathogen interactions is that pathogens often manipulate plant hormone signaling to promote disease, especially to suppress host defenses or to alter plant development.

The roles of plant hormone signaling in plant–pathogen interactions have been discussed in several recent reviews (Robert-Seilaniantz et al. 2011; Kazan and Lyons 2014; Ma and Ma 2016; Bürger and Chory 2019; Han and Kahmann 2019), so here we focus on summarizing studies that provide new insight into the roles that auxin and host auxin responses play during plant–pathogen interactions. One discovery that we find especially exciting is that, in addition to functioning as a plant hormone to regulate host physiology and defense, auxin can also act as a microbial signaling molecule that modulates pathogen gene expression.

Because of space limitations, we are not able to discuss the demonstrated and potential roles of auxins in beneficial interactions with mycorrhizal fungi, nodulating bacteria, or plant growth–promoting rhizobacteria (PGPR), nor in the establishment and function of the plant microbiome. We refer readers interested in these areas to reviews by others (Grunewald et al. 2009; Singh et al. 2019; Legein et al. 2020).

Cite this article as *Cold Spring Harb Perspect Biol* doi: 10.1101/cshperspect.a040022

OVERVIEW OF PLANT–PATHOGEN INTERACTIONS

Many different microorganisms, including viruses, bacteria, fungi, oomycetes, and nematodes, can infect and cause disease on plants. Collectively, pathogenic organisms can colonize all plant tissues (e.g., leaves, fruits, seeds, vascular tissue, and roots) and give rise to a variety of different disease symptoms, such as chlorotic or necrotic lesions on aerial portions of the plant, wilting, altered growth habits (e.g., stunting or loss of apical dominance), and formation of tumor-like growths (e.g., galls, knots, and cankers) (Agrios 1997). The majority of pathogens, with the exception of viruses and vascular pathogens, colonize the intercellular space between plant cells. This space, known as the apoplast, is believed to be a relatively inhospitable environment for microbes, as it is relatively poor in water and nutrients (Beattie 2011; Xin et al. 2018). The apoplast also contains antimicrobial compounds that are either constitutively present or that are secreted into the extracellular space as part of an induced defense response (Heath 2000; Wang et al. 2007). Thus, to successfully colonize and grow to high levels within the apoplast, pathogens must be able to tolerate or detoxify these antimicrobial compounds, evade or suppress any additional host defenses, and further modify host tissue to render it a suitable place for pathogen growth.

Plant Defense Responses

Plants constantly come into contact with many different microorganisms in their environment, many of them potential pathogens. Despite this fact, it is important to note that plant disease is rare. This is due in part to the ability of plants to detect microbes in their proximity and induce basal defense responses that prevent most microbes from colonizing plant tissue. Recognition of potential pathogens is dependent upon plasma membrane–localized receptors, known as pattern recognition receptors (PRRs). These receptors have evolved to recognize conserved molecules commonly associated with microbes (microbial-associated molecular patterns [MAMPs]), such as flagellin peptide or extracellular polysaccharides (EPS), or molecules associated with tissue damage caused by microbial attack (DAMPs). Recognition of one or more of these molecules results in activation of defense responses, including production of reactive oxygen species (ROS), activation of mitogen-activated protein (MAP) kinase cascades, expression of defense-related genes, synthesis of antimicrobial compounds, and the accumulation of one or more defense hormones (Boller and Felix 2009; Bigeard et al. 2015; Boutrot and Zipfel 2017; Saijo et al. 2018).

The plant defense hormones salicylic acid (SA), jasmonates (JAs), and ethylene play central roles in protection against microbial attack. SA is important for defense against biotrophic and hemibiotrophic pathogens that infect living tissue. In contrast, JAs and ethylene activate defense responses that provide protection against necrotrophic organisms that rapidly kill plant cells and grow on dying or dead tissue. The regulation of plant defenses by these three hormones is not as simple as the previous statement may imply. In fact, the regulation of plant defense hormone signaling is complicated and involves intricate regulatory interactions in a complex signaling network (Katagiri and Tsuda 2010). Further, additional plant hormones including auxin, abscisic acid (ABA), gibberellins (GAs), cytokinins, and brassinolides (BRs) also play roles in plant–pathogen interactions, for example, by modulating the SA/JA/ethylene signaling network (Robert-Seilaniantz et al. 2011; Kazan and Lyons 2014; Kunkel and Harper 2018; Bürger and Chory 2019). A fascinating area of research is the investigation of mechanisms pathogens have evolved to take advantage of this intrinsic hormone-signaling network to promote disease. A major focus of this review is how pathogens manipulate auxin signaling.

Pathogen Virulence Strategies

Given that plants are very effective at protecting themselves against generic environmental microbes, pathogenic organisms have had to evolve strategies and virulence factors to enable them to evade or suppress basal host defenses to colonize plant tissue. Once that is accomplished, they further modify the physiology of their hosts to render the plant tissue amenable for pathogen growth, such as stimulating water and nutrient release into the apoplastic space at the infection site. In the case of gall-forming pathogens such as *Agrobacterium tumefaciens* or cyst nematodes, this includes coercing the plant to form novel structures to house and feed the pathogen.

As mentioned above, many plant pathogens are extracellular and colonize the apoplast. Thus, to successfully modify the biology of plant cells at the infection site, these pathogens rely on secreted virulence factors. These secreted factors either remain in the apoplast and impact plant cells from outside or enter the plant cell cytosol to directly target host cell processes. Examples of apoplastic factors include cell wall degrading enzymes and EPS. Virulence factors that function inside plant cells include plant hormones and hormone analogs and fungal and oomycete effector proteins that can be translocated into plant cells, as well as bacterial effector proteins that are secreted directly into the host cell cytosol via a dedicated type III secretion system (T3SS) (Melotto and Kunkel 2013; Kunkel and Harper 2018; Han and Kahmann 2019). Pathogen effector proteins have been the focus of much exciting research in recent years, and several have been shown to target plant defense hormone signaling, and to promote disease susceptibility and symptom development. Several recent reviews provide more comprehensive summaries of type III secreted and RXLR effectors and their functions (Büttner 2016; Toruño et al. 2016; Kunkel and Harper 2018; Xin et al. 2018; Han and Kahmann 2019; Boevink et al. 2020).

As will be discussed in more detail below, many pathogens manipulate host auxin signaling to promote disease. Strategies to accomplish this include production of auxins, including indole-3-acetic acid (IAA) and phenyl acetic acid (PAA), inhibition of auxin signaling in interactions in which auxin contributes to resistance, and enhancing auxin signaling to promote pathogenesis in interactions in which auxin increases susceptibility (Ma and Ma 2016; Han and Kahmann 2019).

A BRIEF INTRODUCTION TO AUXIN SIGNALING

As the role of host auxin responses during plant–pathogen interactions is a main focus of this review, we provide a brief introduction to plant auxin signaling pathways. We refer the reader to the literature cited within for more details on this fascinating area of plant biology.

Auxin induces transcriptional changes in plant cells by promoting ubiquitin-mediated degradation of a family of AUX/IAA transcriptional repressor proteins. In the absence of auxin, the AUX/IAA proteins bind to and inhibit the function of transcription factors called auxin response factors (ARFs) to repress expression of auxin-responsive genes. When auxin levels increase in the cell, auxin binds to auxin sensor proteins (also called auxin coreceptors) belonging to the TRANSPORT INHIBITOR RESPONSE1/AUXIN F-BOX (TIR1/AFB) family of F-box proteins and promotes the interaction between TIR1/AFB proteins and AUX/IAA proteins. This results in the recruitment of the AUX/IAA proteins to the SCF^{Tir} ubiquitin protein ligase, ubiquitination, and degradation of the transcriptional repressors, activation of the ARFs, and expression of auxin-responsive genes (Fig. 1; Mockaitis and Estelle 2008; Lavy and Estelle 2016; Dubey et al. 2021).

Investigation of the roles of auxin signaling during plant growth, development, and responses to environmental stimuli has been greatly facilitated by the availability of a variety of *Arabidopsis thaliana* mutants and transgenic lines that are impaired in various aspects of auxin signaling. For example, plants carrying the dominant *axr2-1* allele, which encodes a mutant form of an AUX/IAA protein that is not ubiquitinated and degraded upon auxin treatment, can be used to examine the impact of impaired auxin responses on the process of interest (Timpte et al. 1994; Djami-Tchatchou et al. 2020). Likewise, a variety of higher-order mutants disrupted for various combinations of the TIR1/AFB auxin coreceptors can be used to examine the effect of disrupted auxin perception (Djami-Tchatchou et al. 2020). Several studies mentioned in this review take advantage of these types of genetic

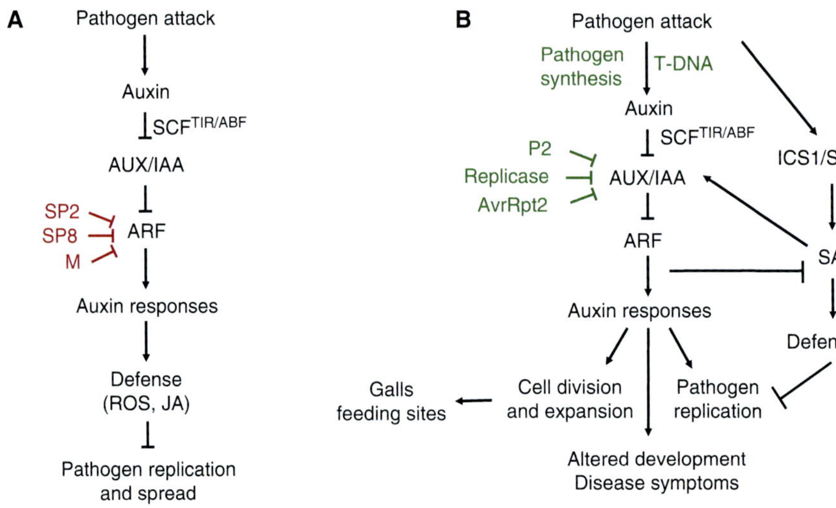

Figure 1. Host auxin signaling during plant–pathogen interactions. (*A*) Infection by some necrotrophic pathogens, such as *Alternaria* and *Rhizoctonia*, and some plant viruses stimulates auxin signaling, resulting in activation of defenses that inhibit pathogen replication and spread. The SP2 protein from rice stripe virus, the SP8 protein from Southern rice black-streaked dwarf virus, and the M protein from rice stripe mosaic virus interact with different regions of OsARF17 and modulate its function, thereby inhibiting activation of auxin-dependent defenses (Zhang et al. 2019, 2020). (*B*) Biotrophic pathogens, such as *Pseudomonas* spp., *Agrobacterium tumefaciens*, and *Pantoea agglomerans*, stimulate auxin signaling to promote disease. Stimulation of auxin signaling can occur at many points in the signaling process including auxin synthesis by the pathogen (Kunkel and Harper 2018). Enhanced auxin signaling can promote production of disease symptoms, development of galls and feeding sites, and/or suppress salicylic acid (SA)-mediated defenses. SA-mediated defense signaling and auxin signaling are mutually antagonistic (Wang et al. 2007; Cui et al. 2013). Several pathogens have evolved virulence factors to promote auxin signaling. The *Pseudomonas syringae* T3S-effector protein AvrRpt2 enhances auxin sensitivity by promoting degradation of AUX/IAA proteins (Cui et al. 2013). The 126/183 kDa replicase protein from tobacco mosaic virus inhibits PAP1/IAA16 function (Padmanabhan et al. 2006), and the P2 protein from rice dwarf virus targets OsIAA10 and prevents its degradation (Jin et al. 2016). Virulence factors shown in red inhibit auxin signaling. Virulence factors shown in green stimulate auxin responses. (SCF) Skp, Cullin, F-box ubiquitin protein ligase complex, (TIR/AFB) TRANSPORT INHIBITOR RESPONSE1/AUXIN F-BOX, F-box proteins (auxin coreceptor), (AUX/IAA) AUXIN/INDOLE ACETIC ACID (transcriptional repressors), (ARF) AUXIN RESPONSE FACTOR (transcription factor), (ROS) reactive oxygen species, (JAs) jasmonates, (ICS1/SID2) isochorismate synthase (SA synthesis), (SA) salicylic acid.

resources to investigate the roles of auxin signaling during plant–pathogen interactions.

THE ROLES OF AUXIN AS A PLANT-SIGNALING MOLECULE IN PLANT–PATHOGEN INTERACTIONS

The role of host auxin signaling during pathogen infection, and whether it promotes disease or defense, depends on the pathogen and the nature of the interaction. For example, activation of auxin signaling enhances resistance against some necrotrophic pathogens and plant viruses (Fig. 1A). In contrast, auxin signaling tends to promote disease in plants challenged with biotrophic or hemibiotrophic pathogens (Fig. 1B; Kazan and Manners 2009; Robert-Seilaniantz et al. 2011; Kazan and Lyons 2014; Ma and Ma 2016; Kunkel and Harper 2018; Bürger and Chory 2019).

Auxin Signaling and Promotion of Plant Defense

As mentioned above, the primary defense hormones are SA, JAs, and ethylene (ET). Thus, there are only a few reports in the literature of auxin and auxin signaling playing a role in activation of plant defense responses. For example, in interactions involving necrotrophic fungi (e.g., *Botrytis*, *Rhizoctonia*, and *Alternaria*), infection can result in elevated IAA levels, and application of exogenous auxin results in reduced disease susceptibility (Llorente et al. 2008; Mah et al. 2012; Qi et al. 2012; Qiao et al. 2020). Consistent with a role for auxin in defense in these interactions, disruption of auxin signaling due to mutations that stabilize AUX/IAA transcriptional repressors (e.g., *axr* mutants) or overexpression of small RNAs that negatively regulate expression of TIR1/AFB family auxin receptors, results in enhanced disease susceptibility (Llorente et al. 2008; Qi et al. 2012; Qiao et al. 2020). In contrast, overexpression of the TIR1/AFB family auxin coreceptor FBL55 in rice resulted in enhanced resistance to *Rhizoctonia solani* (Qiao et al. 2020). In most of the above studies, the mechanism through which auxin signaling results in increased resistance was not investigated. It seems reasonable to speculate that in many cases auxin signaling leads to activation of JA-mediated defenses, which are protective against necrotrophs (Fu and Wang 2011; Qi et al. 2012; Zhang et al. 2017). The observation that auxin signaling activates defense in these interactions raises the question of whether necrotrophic pathogens have evolved strategies to perturb auxin signaling to promote susceptibility, similar to what has been reported for their ability to interfere with JA signaling (Kazan and Lyons 2014; Zhang et al. 2017).

We could find evidence of only a few interactions in which resistance to infection by plant viruses is associated with activation of auxin-mediated defense signaling. In two recent studies, Zhang et al. (2019, 2020) discovered that several viral proteins can interfere with host auxin signaling by interacting with plant ARFs (Fig. 1A). The SP8 protein from Southern rice black-streaked dwarf virus (SRBSDV) is a transcriptional repressor that can interact with and inhibit the carboxy-terminal dimerization domain (CTD) of OsARF17, an activator ARF. Similarly, the M protein from rice stripe mosaic virus (RSMV) and the SP2 protein from rice stripe virus (RSV) interact with and inhibit OsARF17. Like SP8, the RSMV M protein also interacts with the CTD domain. However, the RSV SP2 protein interacts with the DNA-binding domain of OsARF17 and interferes with its ability to bind DNA (Zhang et al. 2020). Thus, three different viruses have evolved strategies to interfere with host auxin signaling, presumably to promote virulence, by targeting different functions of the same host-signaling component. It is interesting to contemplate what might make OsARF17 such a popular target. A question for the future is whether this particular activator ARF has a critical function in auxin-mediated defenses against viruses and possibly also against other pathogens.

Consistent with the hypothesis that auxin signaling promotes antiviral defense, studies involving both overexpression lines and CRISPR/cas9 mutants demonstrated that auxin signaling plays a critical role in defense against both SRBSDV and the related RBSDV (Zhang et al. 2019, 2020). These studies also showed that increased susceptibility in transgenic rice lines with reduced auxin signaling was correlated with a reduction in JA signaling and ROS in infected plants, suggesting that auxin signaling defends against viruses through activation of JA-dependent defenses and accumulation of ROS (Zhang et al. 2019).

Auxin Signaling and Promotion of Disease Susceptibility

There are many examples of host auxin signaling enhancing disease susceptibility. For example, application of exogenous auxins promote susceptibility to a number of biotrophic and hemibiotrophic pathogens, including the bacterial pathogens *Pseudomonas syringae* and *Xanthomonas oryzae* (Navarro et al. 2006; Chen et al. 2007; Wang et al. 2007; Fu et al. 2011), the fungal pathogen *Magnaporthe grisea* (Fu et al. 2011), and the oomycete *Phytophthora parasitica* (Evangelisti et al. 2013). Additionally, *A. thaliana* lines with elevated levels of free IAA exhibit increased susceptibility to *P. syringae* (Mutka et al. 2013; Djami-Tchatchou et al. 2020) and plant lines with impaired auxin perception or signaling (e.g., *axr2/3* or *tir1/afb* mutants) exhibit reduced susceptibility to several biotrophs and hemibiotrophs (Wang et al. 2007; Kidd et al. 2011; Fousia et al. 2018; French et al. 2018; Djami-Tchatchou et al. 2020).

Perhaps the most compelling evidence that auxin signaling promotes disease is the growing number of reports of pathogens that stimulate host auxin signaling to promote pathogenesis and disease development. These microbes use a variety of strategies, including synthesis of auxin (Patten et al. 2013; Duca et al. 2014; Kunkel and Harper 2018) and deployment of pathogen virulence proteins that act within host cells to enhance host auxin signaling (Chen et al. 2007; Kazan and Lyons 2014; Ludwig-Müller 2015; Ma and Ma 2016).

There are several mechanisms by which host auxin signaling promotes disease, and they fall into two general categories: (1) manipulation of host auxin biology (e.g., signaling, homeostasis, and/or transport) to directly promote disease development, and (2) suppression of host defenses by taking advantage of regulatory "cross talk" inherent in hormone signaling networks (Fig. 1B; Robert-Seilaniantz et al. 2011; Kazan and Lyons 2014; Ma and Ma 2016; Kunkel and Harper 2018). Some examples from the first category have been observed in viral pathogens (Fig. 1B). Infection by plant viruses often results in changes in hormone homeostasis that results in disease symptoms, including severe stunting, altered shoot architecture, and tissue chlorosis. These symptoms are commonly reflected in the names of the viruses, such as rice dwarf virus (RDV) and tobacco mosaic virus (TMV). Several viral pathogens have evolved mechanisms to perturb or reprogram auxin signaling as a strategy to promote disease development (Islam et al. 2019). For example, the TMV 126/183 kDa replicase protein interacts with several AUX/IAA proteins to prevent their localization to the nucleus (Padmanabhan et al. 2006), leading to changes in auxin-responsive gene expression and development of disease symptoms (Padmanabhan et al. 2005). Likewise, RDV modifies host auxin signaling through the action of its P2 protein, which interacts with the region II degron domain of the rice OsIAA10 transcriptional repressor. This interferes with the normal interaction between OsIAA10 and OsTIR1 and prevents degradation of OsIAA10 upon auxin binding. Consistent with a role for OsIAA10 in RDV infection, transgenic rice plants expressing a stabilized OsIAA10 variant exhibited phenotypes reminiscent of RDV-infected plants, even in the absence of virus, and plants expressing reduced levels of OsIAA10 were less susceptible to RDV infection (Jin et al. 2016).

The virulence strategies of gall-forming pathogens such as *A. tumefaciens*, *Pseudomonas savastanoi*, *Pantoea agglomerans* (formerly *Erwinia herbicola*), and parasitic nematodes involve manipulation of host auxin physiology to directly stimulate abnormal plant cell division and expansion, to form tumor-like structures to shelter and feed the pathogen (Fig. 1B). In these interactions, the pathogens either make auxin themselves (*P. savastanoi*, *P. agglomerans*) (Patten et al. 2013; Duca et al. 2014), stimulate plant cells at the site of infection to synthesize auxin (*A. tumefaciens*) (Thomashow et al. 1986; Mashiguchi et al. 2019), or redirect auxin flow within the plant so that auxin accumulates at the feeding site and stimulates cell division or syncytium development (nematodes) (Grunewald et al. 2009; Kyndt et al. 2016). Presumably, nutrient flow into the infected host tissue is also altered to support the infection site, but this is not well understood.

The discovery that auxin can also promote disease caused by biotrophic pathogens such as *P. syringae*, that do not result in dramatic changes in plant growth or development, was surprising at first. Insight into the mode of auxin action in these interactions was provided by the observation that auxin signaling and SA signaling can be mutually antagonistic (Fig. 1B; Wang et al. 2007). Additional evidence for cross talk between auxin and SA signaling has come from more recent studies demonstrating that the virulence-promoting effect of enhanced auxin signaling involves down-regulation of SA-mediated defenses (McClerklin et al. 2018; Djami-Tchatchou et al. 2020). Further, the observation that *A. thaliana* auxin signaling mutants express elevated SA-mediated defenses and that the reduced disease susceptibility phenotype of these mutant plants can be genetically rescued by

 Cite this article as *Cold Spring Harb Perspect Biol* doi: 10.1101/cshperspect.a040022

alterations in SA synthesis or accumulation further support the hypothesis that auxin promotes disease by suppressing SA-mediated defenses (Wang et al. 2007; Fousia et al. 2018; Djami-Tchatchou et al. 2020). However, as will be discussed below, there is growing evidence that auxin can also promote disease susceptibility via one or more additional SA-independent mechanism (Chen et al. 2004; Kidd et al. 2011; Mutka et al. 2013; Djami-Tchatchou et al. 2020).

AUXIN AS A MICROBIAL SIGNAL

In addition to promoting pathogenesis by manipulating host auxin signaling, auxin may act directly to promote virulence of several pathogens. There is growing evidence that IAA acts as a molecular signal, both in plant-associated and non-plant-associated microbes, to regulate gene expression (Spaepen and Vanderleyden 2011; Kunkel and Harper 2018). However, the majority of what is known to date about the mechanisms used by microorganisms to sense and respond to IAA comes from studies of non-plant-associated bacteria.

IAA Regulates Genes Involved in Stress Tolerance, Metabolism, and Antibiotic Production

There are several examples in which IAA dramatically alters gene expression in non-plant-associated bacteria. For instance, in *Escherichia coli*, IAA regulates genes involved in central metabolism, which is believed to be why IAA-treated *E. coli* cells exhibited increased tolerance to several environmental stresses, including osmotic, heat and cold shock, and oxidative stress (Bianco et al. 2006a,b). Likewise, Donati et al. observed induction of genes involved in stress responses in *Bradyrhizobium japonicum* in response to IAA (Donati et al. 2013).

Several bacteria also have the ability to use IAA as a carbon source. In many of these organisms, IAA acts as a signal to up-regulate IAA catabolism genes, which are encoded in an IAA catabolic (*iac* or *iaa*) operon responsible for aerobic breakdown of IAA (Laird et al. 2020). This operon has been identified in several diverse bacterial species, including *Pseudomonas putida*, *Enterobacter soli*, *Acinetobacter baumannii*, *Paraburkholderia phytofirmans*, *Caballeronia glathei*, *Aromatoleum evansii*, and *Aromatoleum aromaticum* (Laird et al. 2020).

In *A. baumannii*, regulation of the *iac* operon by IAA is conferred by IacR, a MarR-type transcriptional regulator that negatively regulates the *iac* operon. In the absence of IAA, IacR binds the promoter of the *iac* operon. IAA is hypothesized to bind to IacR and release it from the *iac* promoter, thereby allowing RNA polymerase access to the promoter to initiate transcription (Shu et al. 2015). Similarly, in *E. soli*, IAA derepresses the *iac* operon via a MarR-type transcriptional repressor (Greenhut et al. 2018). Donoso et al. also identified an *iac* gene cluster in the PGPR *P. phytofirmans* strain *PsJN*. The *iac* operon in this strain has a different gene organization and lacks the *iacR* MarR family repressor. Instead, the *PsJN* genome carries other genes adjacent to the *iac* operon, including genes encoding a LysR-type regulator and a putative two-component signal transduction system, composed of *iacS* and *iacR1*. Expression of these regulatory genes is induced by IAA and is required for IAA-induced transcription from the *iac* promoter (Donoso et al. 2017). This is the only sensory system that has been identified in bacteria in which IAA does not directly target a transcription factor and instead may initiate a large-scale physiological response downstream of a two-component system. It will be interesting to learn whether other microbes use a similar system to sense and respond to IAA.

Adapting to a New Environment

An additional example of auxin-regulated gene expression has been observed in the soil bacterium *Serratia plymuthica* (Matilla et al. 2018), which produces a wide range of antibiotics including andrimid. Production of andrimid is regulated by AdmX, a transcriptional activator with an amino-terminal DNA-binding domain and a LysR-type ligand-binding domain. AdmX has been shown to bind IAA, which may change its conformation. The proposed change in conformation upon IAA binding is thought to reduce AdmX's ability to bind DNA, thereby disabling its transcriptional activation function. Thus, IAA lowers andrimid production in this strain. The biological relevance of reducing andrimid production is unclear, but one possibility is that this results in a physiological or metabolic change that enables *S. plymuthica* to acclimate to a new environment.

In several plant-associated bacteria, IAA may act as a signal of a new environment. Two examples of bacteria that show dramatic changes in gene expression in response to IAA are *Azospirillum brasilense* (Vande Broek et al. 2005; Van Puyvelde et al. 2011) and *Rhizobium etli* (Spaepen et al. 2009). In *A. brasilense*, IAA activates a positive feedback loop in which IAA biosynthesis genes are up-regulated (Vande Broek et al. 2005). Van Puyvelde et al. (2011) then showed extensive transcriptional responses to IAA by analyzing gene-expression patterns in auxin biosynthesis mutants. Interestingly, in wild-type *A. brasilense*, exposure to IAA altered expression of genes encoding transport proteins, cell surface proteins, and proteins involved in type VI secretion. These observations support the hypothesis that IAA may stimulate bacteria to alter gene expression to increase their competitiveness in the rhizosphere. Spaepen et al. observed that in *R. etli*, genes involved in motility and attachment are differentially regulated by IAA. Because these genes are believed to be important for symbiosis with plants, the authors hypothesized that IAA plays a role in promoting *R. etli* plant interactions (Spaepen et al. 2009).

Bacterial plant pathogens have also evolved the ability to use IAA as a signal to modulate gene expression and virulence mechanisms. One well-known pathogen that relies on its own ability to alter host auxin biology is *A. tumefaciens*, the causal agent of crown gall disease. This pathogen expresses *virulence* (*vir*) genes during pathogenesis that are ultimately responsible for genetic transformation of plant cells so that they will abnormally produce auxins and cytokinins and stimulate the formation of galls at the site of infection (Thomashow et al. 1986; Mashiguchi et al. 2019). After sufficient accumulation of IAA in the infected tissue and thus pathogen *vir* genes are presumably no longer required, the high level of IAA inhibits their expression (Liu and Nester 2006). A large number of additional non-*vir* genes are also regulated by IAA, suggesting IAA may also regulate processes important for bacterial persistence after gall formation (Yuan et al. 2008).

Plant pathogens in the *Pseudomonas* genus also respond to IAA by altering gene expression. Experiments to investigate the possibility that IAA regulates virulence gene expression in *P. syringae* strain PtoDC3000 were prompted by studies suggesting that auxin promotes disease development via a mechanism independent of suppressing SA-mediated host defenses (Chen et al. 2004; Mutka et al. 2013). In a recent study, Djami-Tchatchou et al. (2020) demonstrated that IAA regulates expression of known virulence genes, including genes involved in T3S, in PtoDC3000 growing in culture. However, rather than promoting expression of T3SS-related genes, as was initially expected given that IAA promotes pathogenesis, they observed that T3SS-related genes were down-regulated in response to IAA. However, other virulence genes, such as *tvrR*, which encodes a TetR family transcription factor, were up-regulated in response to IAA. Similar effects of IAA on expression of these virulence genes were also observed *in planta*. These findings led to the hypothesis that IAA acts as a signaling molecule that coordinates expression of virulence genes required during different stages of pathogenesis (Djami-Tchatchou et al. 2020). IAA has also been shown to down-regulate expression of T3SS genes in *P. savastanoi* (Aragon et al. 2014; Ryu 2015). Currently, we do not know how *P. syrinage* and other plant pathogens sense and respond to auxin. Elucidating the mechanisms involved will greatly enhance our understanding of the roles of IAA as a pathogen signal during infection.

CONCLUDING REMARKS

It is evident from the examples discussed above that auxin plays several different roles during plant–pathogen interactions. It should come as no surprise that one of its primary roles is as a plant hormone, and that alterations in host auxin signaling and physiology can either result in enhanced defense or increased disease, depending on the pathogen. Perhaps more surprising are the more recent discoveries that auxin also acts as a microbial signaling molecule that can directly impact pathogen biology, for example, by modulating virulence gene expression. These findings have opened up an exciting new area of research focused on elucidating the mechanisms by which pathogens sense and respond to IAA.

REFERENCES

*Reference is also in this collection.

Agrios GN. 1997. *Plant pathology*, 4th ed. Academic, San Diego.

Aragon IM, Perez-Martinez I, Moreno-Perez A, Cerezo M, Ramos C. 2014. New insights into the role of indole-3-acetic acid in the virulence of *Pseudomonas savastanoi* pv. *savastanoi*. *FEMS Microbiol Lett* **356:** 184–192. doi:10.1111/1574-6968.12413

Beattie GA. 2011. Water relations in the interaction of foliar bacterial pathogens with plants. *Annu Rev Phytopathol* **49:** 533–555. doi:10.1146/annurev-phyto-073009-114436

Bianco C, Imperlini E, Calogero R, Senatore B, Amoresano A, Carpentieri A, Pucci P, Defez R. 2006a. Indole-3-acetic acid improves *Escherichia coli*'s defences to stress. *Arch Microbiol* **185:** 373–382. doi:10.1007/s00203-006-0103-y

Bianco C, Imperlini E, Calogero R, Senatore B, Pucci P, Defez R. 2006b. Indole-3-acetic acid regulates the central metabolic pathways in *Escherichia coli*. *Microbiolog (Reading)* **152:** 2421–2431. doi:10.1099/mic.0.28765-0

Bigeard J, Colcombet J, Hirt H. 2015. Signaling mechanisms in pattern-triggered immunity (PTI). *Mol Plant* **8:** 521–539. doi:10.1016/j.molp.2014.12.022

Boevink PC, Birch PRJ, Turnbull D, Whisson SC. 2020. Devastating intimacy: the cell biology of plant–*Phytophthora* interactions. *New Phytol* **228:** 445–458. doi:10.1111/nph.16650

Boller T, Felix G. 2009. A renaissance of elicitors: perception of microbe-associated molecular patterns and danger signals by pattern-recognition receptors. *Annu Rev Plant Biol* **60:** 379–406. doi:10.1146/annurev.arplant.57.032905.105346

Boutrot F, Zipfel C. 2017. Function, discovery, and exploitation of plant pattern recognition receptors for broad-spectrum disease resistance. *Annu Rev Phytopathol* **55:** 257–286. doi:10.1146/annurev-phyto-080614-120106

Bürger M, Chory J. 2019. Stressed out about hormones: how plants orchestrate immunity. *Cell Host Microbe* **26:** 163–172. doi:10.1016/j.chom.2019.07.006

Büttner D. 2016. Behind the lines-actions of bacterial type III effector proteins in plant cells. *FEMS Microbiol Rev* **40:** 894–937. doi:10.1093/femsre/fuw026

Chen Z, Kloek AP, Cuzick A, Moeder W, Tang D, Innes RW, Klessig DF, McDowell JM, Kunkel BN. 2004. The *Pseudomonas syringae* type III effector AvrRpt2 functions downstream or independently of SA to promote virulence on *Arabidopsis thaliana*. *Plant J* **37:** 494–504. doi:10.1111/j.1365-313X.2003.01984.x

Chen Z, Agnew JL, Cohen JD, He P, Shan L, Sheen J, Kunkel BN. 2007. *Pseudomonas syringae* type III effector AvrRpt2 alters *Arabidopsis thaliana* auxin physiology. *Proc Natl Acad Sci* **104:** 20131–20136. doi:10.1073/pnas.0704901104

Cui F, Wu S, Sun W, Coaker G, Kunkel B, He P, Shan L. 2013. The *Pseudomonas syringae* type III effector AvrRpt2 promotes pathogen virulence via stimulating *Arabidopsis* auxin/indole acetic acid protein turnover. *Plant Physiol* **162:** 1018–1029. doi:10.1104/pp.113.219659

Djami-Tchatchou AT, Harrison GA, Harper CP, Wang R, Prigge MJ, Estelle M, Kunkel BN. 2020. Dual role of auxin in regulating plant defense and bacterial virulence gene expression during *Pseudomonas syringae* PtoDC3000 pathogenesis. *Mol Plant Microbe Interact* **33:** 1059–1071. doi:10.1094/MPMI-02-20-0047-R

Donati AJ, Lee HI, Leveau JH, Chang WS. 2013. Effects of indole-3-acetic acid on the transcriptional activities and stress tolerance of *Bradyrhizobium japonicum*. *PLoS ONE* **8:** e76559. doi:10.1371/journal.pone.0076559

Donoso R, Leiva-Novoa P, Zúñiga A, Timmermann T, Recabarren-Gajardo G, González B. 2017. Biochemical and genetic bases of indole-3-acetic acid (auxin phytohormone) degradation by the plant-growth-promoting rhizobacterium *Paraburkholderia phytofirmans* PsJN. *Appl Environ Microbiol* **83:** e01991-16. doi:10.1128/AEM.01991-16

* Dubey SM, Calixte Serre NB, Oulehlová D, Vittal P, Fendrych M. 2021. No time for transcription—rapid auxin responses in plants. *Cold Spring Harb Perspect Biol* doi:10.1101/cshperspect.a039891

Duca D, Lorv J, Patten CL, Rose D, Glick BR. 2014. Indole-3-acetic acid in plant-microbe interactions. *Antonie van Leeuwenhoek* **106:** 85–125. doi:10.1007/s10482-013-0095-y

Evangelisti E, Govetto B, Minet-Kebdani N, Kuhn ML, Attard A, Ponchet M, Panabières F, Gourgues M. 2013. The *Phytophthora parasitica* RXLR effector penetration-specific effector 1 favours *Arabidopsis thaliana* infection by interfering with auxin physiology. *New Phytol* **199:** 476–489. doi:10.1111/nph.12270

Fousia S, Tsafouros A, Roussos PA, Tjamos SE. 2018. Increased resistance to *Verticillium dahliae* in *Arabidopsis* plants defective in auxin signalling. *Plant Pathol* **67:** 1749–1757. doi:10.1111/ppa.12881

French E, Kim BS, Rivera-Zuluaga K, Iyer-Pascuzzi AS. 2018. Whole root transcriptomic analysis suggests a role for auxin pathways in resistance to *Ralstonia solanacearum* in tomato. *Mol Plant Microbe Interact* **31:** 432–444. doi:10.1094/MPMI-08-17-0209-R

Fu J, Wang S. 2011. Insights into auxin signaling in plant–pathogen interactions. *Front Plant Sci* **2:** 74.

Fu J, Liu H, Li Y, Yu H, Li X, Xiao J, Wang S. 2011. Manipulating broad-spectrum disease resistance by suppressing pathogen-induced auxin accumulation in rice. *Plant Physiol* **155:** 589–602. doi:10.1104/pp.110.163774

Greenhut IV, Slezak BL, Leveau JHJ. 2018. *Iac* gene expression in the indole-3-acetic acid-degrading soil bacterium

Enterobacter soli LF7. *Appl Environ Microbiol* **84:** e01057-18. doi:10.1128/AEM.01057-18

Grunewald W, van Noorden G, van Isterdael G, Beeckman T, Gheysen G, Mathesius U. 2009. Manipulation of auxin transport in plant roots during *Rhizobium* symbiosis and nematode parasitism. *Plant Cell* **21:** 2553–2562. doi:10.1105/tpc.109.069617

Han X, Kahmann R. 2019. Manipulation of phytohormone pathways by effectors of filamentous plant pathogens. *Front Plant Sci* **10:** 822. doi:10.3389/fpls.2019.00822

Heath M. 2000. Nonhost resistance and nonspecific plant defenses. *Curr Opin Plant Biol* **3:** 315–319. doi:10.1016/S1369-5266(00)00087-X

Islam W, Naveed H, Zaynab M, Huang ZQ, Chen HYH. 2019. Plant defense against virus diseases; growth hormones in highlights. *Plant Signal Behav* **14:** 1596719. doi:10.1080/15592324.2019.1596719

Jin L, Qin QQ, Wang Y, Pu YY, Liu LF, Wen X, Ji SY, Wu JG, Wei CH, Ding B, et al. 2016. Rice dwarf virus P2 protein hijacks auxin signaling by directly targeting the rice OsIAA10 protein, enhancing viral infection and disease development. *Plos Pathog* **12:** e1005847. doi:10.1371/journal.ppat.1005847

Katagiri F, Tsuda K. 2010. Understanding the plant immune system. *Mol Plant Microbe Interact* **23:** 1531–1536. doi:10.1094/MPMI-04-10-0099

Kazan K, Lyons R. 2014. Intervention of phytohormone pathways by pathogen effectors. *Plant Cell* **26:** 2285–2309. doi:10.1105/tpc.114.125419

Kazan K, Manners JM. 2009. Linking development to defense: auxin in plant–pathogen interactions. *Trends Plant Sci* **14:** 373–382. doi:10.1016/j.tplants.2009.04.005

Kidd BN, Kadoo NY, Dombrecht B, Tekeoglu M, Gardiner DM, Thatcher LF, Aitken EA, Schenk PM, Manners JM, Kazan K. 2011. Auxin signaling and transport promote susceptibility to the root-infecting fungal pathogen *Fusarium oxysporum* in *Arabidopsis*. *Mol Plant Microbe Interact* **24:** 733–748. doi:10.1094/MPMI-08-10-0194

Kunkel BN, Harper CP. 2018. The roles of auxin during interactions between bacterial plant pathogens and their hosts. *J Exp Bot* **69:** 245–254. doi:10.1093/jxb/erx447

Kyndt T, Goverse A, Haegeman A, Warmerdam S, Wanjau C, Jahani M, Engler G, de Almeida Engler J, Gheysen G. 2016. Redirection of auxin flow in *Arabidopsis thaliana* roots after infection by root-knot nematodes. *J Exp Bot* **67:** 4559–4570. doi:10.1093/jxb/erw230

Laird TS, Flores N, Leveau JHJ. 2020. Bacterial catabolism of indole-3-acetic acid. *Appl Microbiol Biotechnol* **104:** 9535–9550. doi:10.1007/s00253-020-10938-9

Lavy M, Estelle M. 2016. Mechanisms of auxin signaling. *Development* **143:** 3226–3229. doi:10.1242/dev.131870

Legein M, Smets W, Vandenheuvel D, Eilers T, Muyshondt B, Prinsen E, Samson R, Lebeer S. 2020. Modes of action of microbial biocontrol in the phyllosphere. *Front Microbiol* **11:** 1619. doi:10.3389/fmicb.2020.01619

Liu P, Nester EW. 2006. Indoleacetic acid, a product of transferred DNA, inhibits *vir* gene expression and growth of *Agrobacterium tumefaciens* C58. *Proc Natl Acad Sci* **103:** 4658–4662. doi:10.1073/pnas.0600366103

Llorente F, Muskett P, Sánchez-Vallet A, López G, Ramos B, Sánchez-Rodráguez C, Jordá L, Parker J, Molina A. 2008. Repression of the auxin response pathway increases *Arabidopsis* susceptibility to necrotrophic fungi. *Mol Plant* **1:** 496–509. doi:10.1093/mp/ssn025

Ludwig-Müller J. 2015. Bacteria and fungi controlling plant growth by manipulating auxin: balance between development and defense. *J Plant Physiol* **172:** 4–12. doi:10.1016/j.jplph.2014.01.002

Ma KW, Ma WB. 2016. Phytohormone pathways as targets of pathogens to facilitate infection. *Plant Mol Biol* **91:** 713–725. doi:10.1007/s11103-016-0452-0

Mah KM, Uppalapati SR, Tang YH, Allen S, Shuai B. 2012. Gene expression profiling of *Macrophomina phaseolina* infected *Medicago truncatula* roots reveals a role for auxin in plant tolerance against the charcoal rot pathogen. *Physiol Mol Plant Pathol* **79:** 21–30.

Mashiguchi K, Hisano H, Takeda-Kamiya N, Takebayashi Y, Ariizumi T, Gao YB, Ezura H, Sato K, Zhao Y, Hayashi K, et al. 2019. *Agrobacterium tumefaciens* enhances biosynthesis of two distinct auxins in the formation of crown galls. *Plant Cell Physiol* **60:** 29–37.

Matilla MA, Daddaoua A, Chini A, Morel B, Krell T. 2018. An auxin controls bacterial antibiotics production. *Nucleic Acids Res* **46:** 11229–11238.

McClerklin SA, Lee SG, Harper CP, Nwumeh R, Jez JM, Kunkel BN. 2018. Indole-3-acetaldehyde dehydrogenase-dependent auxin synthesis contributes to virulence of *Pseudomonas syringae* strain DC3000. *PLoS Pathog* **14:** e1006811.

Melotto M, Kunkel BN. 2013. Virulence strategies of plant pathogenic bacteria. In *The prokaryotes* (ed. Rosenberg E, DeLong EF, Lory S, Stackebrandt E, Thompson F). Springer, Berlin, Heidelberg. doi:10.1007/978-3-642-30141-4_62

Mockaitis K, Estelle M. 2008. Auxin receptors and plant development: a new signaling paradigm. *Ann Rev Cell Dev Biol* **24:** 55–80.

Mutka AM, Fawley S, Tsao T, Kunkel BN. 2013. Auxin promotes susceptibility to *Pseudomonas syringae* via a mechanism independent of suppression of salicylic acid-mediated defenses. *Plant J* **74:** 746–754. doi:10.1111/tpj.12157

Navarro L, Dunoyer P, Jay F, Arnold B, Dharmasiri N, Estelle M, Voinnet O, Jones JD. 2006. A plant miRNA contributes to antibacterial resistance by repressing auxin signaling. *Science* **312:** 436–439. doi:10.1126/science.1126088

Padmanabhan MS, Goregaoker SP, Golem S, Shiferaw H, Culver JN. 2005. Interaction of the tobacco mosaic virus replicase protein with the Aux/IAA protein PAP1/IAA26 is associated with disease development. *J Virol* **79:** 2549–2558. doi:10.1128/JVI.79.4.2549-2558.2005

Padmanabhan MS, Shiferaw H, Culver JN. 2006. The *Tobacco mosaic virus* replicase protein disrupts the localization and function of interacting Aux/IAA proteins. *Mol Plant Microbe Interact* **19:** 864–873. doi:10.1094/MPMI-19-0864

Patten CL, Blakney AJ, Coulson TJ. 2013. Activity, distribution, and function of indole-3-acetic acid biosynthetic pathways in bacteria. *Crit Rev Microbiol* **39:** 395–415. doi:10.3109/1040841X.2012.716819

Qi LL, Yan J, Li YN, Jiang HL, Sun JQ, Chen Q, Li HX, Chu JF, Yan CY, Sun XH, et al. 2012. *Arabidopsis thaliana* plants differentially modulate auxin biosynthesis and transport during defense responses to the necrotrophic pathogen *Alternaria brassicicola*. *New Phytol* **195:** 872–882. doi:10.1111/j.1469-8137.2012.04208.x

Qiao LL, Zheng LY, Sheng C, Zhao HW, Jin HL, Niu DD. 2020. Rice siR109944 suppresses plant immunity to sheath blight and impacts multiple agronomic traits by affecting auxin homeostasis. *Plant J* **102:** 948–964. doi:10.1111/tpj.14677

Robert-Seilaniantz A, Grant M, Jones JD. 2011. Hormone crosstalk in plant disease and defense: more than just jasmonate–salicylate antagonism. *Annu Rev Phytopathol* **49:** 317–343. doi:10.1146/annurev-phyto-073009-114447

Ryu CM. 2015. Against friend and foe: type 6 effectors in plant-associated bacteria. *J Microbiol* **53:** 201–208. doi:10.1007/s12275-015-5055-y

Saijo Y, Loo EP, Yasuda S. 2018. Pattern recognition receptors and signaling in plant–microbe interactions. *Plant J* **93:** 592–613. doi:10.1111/tpj.13808

Shu HY, Lin LC, Lin TK, Chen HP, Yang HH, Peng KC, Lin GH. 2015. Transcriptional regulation of the *iac* locus from *Acinetobacter baumannii* by the phytohormone indole-3-acetic acid. *Antonie van Leeuwenhoek* **107:** 1237–1247. doi:10.1007/s10482-015-0417-3

Singh D, Raina TK, Kumar A, Singh J, Prasad R. 2019. Plant microbiome: a reservoir of novel genes and metabolites. *Plant Gene* **18:** 100177. doi:10.1016/j.plgene.2019.100177

Spaepen S, Vanderleyden J. 2011. Auxin and plant–microbe interactions. *Cold Spring Harb Perspect Biol* **3:** a001438. doi:10.1101/cshperspect.a001438

Spaepen S, Das F, Luyten E, Michiels J, Vanderleyden J. 2009. Indole-3-acetic acid-regulated genes in *Rhizobium etli* CNPAF512. *FEMS Microbiol Lett* **291:** 195–200. doi:10.1111/j.1574-6968.2008.01453.x

Thomashow MF, Hugly S, Buchholz WG, Thomashow LS. 1986. Molecular basis for the auxin-independent phenotype of crown gall tumor tissues. *Science* **231:** 616–618. doi:10.1126/science.3511528

Timpte C, Wilson AK, Estelle M. 1994. The *axr2-1* mutation of *Arabidopsis thaliana* is a gain-of-function mutation that disrupts an early step in auxin response. *Genetics* **138:** 1239–1249. doi:10.1093/genetics/138.4.1239

Toruño TY, Stergiopoulos I, Coaker G. 2016. Plant–pathogen effectors: cellular probes interfering with plant defenses in spatial and temporal manners. *Ann Rev Phytopathol* **54:** 419–441. doi:10.1146/annurev-phyto-080615-100204

Vande Broek A, Gysegom P, Ona O, Hendrickx N, Prinsen E, Van Impe J, Vanderleyden J. 2005. Transcriptional analysis of the *Azospirillum brasilense* indole-3-pyruvate decarboxylase gene and identification of a *cis*-acting sequence involved in auxin responsive expression. *Mol Plant Microbe Interact* **18:** 311–323. doi:10.1094/MPMI-18-0311

Van Puyvelde S, Cloots L, Engelen K, Das F, Marchal K, Vanderleyden J, Spaepen S. 2011. Transcriptome analysis of the rhizosphere bacterium *Azospirillum brasilense* reveals an extensive auxin response. *Microb Ecol* **61:** 723–728. doi:10.1007/s00248-011-9819-6

Wang D, Pajerowska-Mukhtar K, Culler AH, Dong X. 2007. Salicylic acid inhibits pathogen growth in plants through repression of the auxin signaling pathway. *Curr Biol* **17:** 1784–1790. doi:10.1016/j.cub.2007.09.025

Xin XF, Kvitko B, He SY. 2018. *Pseudomonas syringae*: what it takes to be a pathogen. *Nat Rev Microbiol* **16:** 316–328. doi:10.1038/nrmicro.2018.17

Yuan ZC, Haudecoeur E, Faure D, Kerr KF, Nester EW. 2008. Comparative transcriptome analysis of *Agrobacterium tumefaciens* in response to plant signal salicylic acid, indole-3-acetic acid and γ-amino butyric acid reveals signalling cross-talk and *Agrobacterium*–plant co-evolution. *Cell Microbiol* **10:** 2339–2354. doi:10.1111/j.1462-5822.2008.01215.x

Zhang L, Zhang F, Melotto M, Yao J, He SY. 2017. Jasmonate signaling and manipulation by pathogens and insects. *J Exp Bot* **68:** 1371–1385.

Zhang HH, Tan XX, Li LL, He YQ, Hong GJ, Li JM, Lin L, Cheng Y, Yan F, Chen JP, et al. 2019. Suppression of auxin signalling promotes rice susceptibility to *Rice black streaked dwarf virus* infection. *Mol Plant Pathol* **20:** 1093–1104. doi:10.1111/mpp.12814

Zhang HH, Li LL, He YQ, Qin QQ, Chen CH, Wei ZY, Tan XX, Xie KL, Zhang RF, Hong GJ, et al. 2020. Distinct modes of manipulation of rice auxin response factor OsARF17 by different plant RNA viruses for infection. *Proc Natl Acad Sci* **117:** 9112–9121. doi:10.1073/pnas.1918254117

Auxin–Environment Integration in Growth Responses to Forage for Resources

Jorge J. Casal[1,2] and José M. Estevez[2,3]

[1]Universidad de Buenos Aires, Consejo Nacional de Investigaciones Científicas y Técnicas (CONICET), Instituto de Investigaciones Fisiológicas y Ecológicas Vinculadas a la Agricultura (IFEVA), Facultad de Agronomía, Buenos Aires 1417, Argentina

[2]Fundación Instituto Leloir and IIBBA-CONICET, Buenos Aires C1405BWE, Argentina

[3]Centro de Biotecnología Vegetal, Facultad de Ciencias de la Vida, Universidad Andres Bello and Millennium Institute for Integrative Biology (iBio), Santiago 8370146, Chile

Correspondence: casal@ifeva.edu.ar; jestevez@leloir.org.ar

Plant fitness depends on the adequate morphological adjustment to the prevailing conditions of the environment. Therefore, plants sense environmental cues through their life cycle, including the presence of full darkness, light, or shade, the range of ambient temperatures, the direction of light and gravity vectors, and the presence of water and mineral nutrients (such as nitrate and phosphate) in the soil. The environmental information impinges on different aspects of the auxin system such as auxin synthesis, degradation, transport, perception, and downstream transcriptional regulation to modulate organ growth. Although a single environmental cue can affect several of these points, the relative impacts differ significantly among the various growth processes and cues. While stability in the generation of precise auxin gradients serves to guide the basic developmental pattern, dynamic changes in the auxin system fine-tune body shape to optimize the capture of environmental resources.

Given the role of auxin in the control of complex developmental programs that require stability, such as the organization of the shoot meristem (Vernoux and Pernisová 2020), we might expect homeostasis of auxin signaling status in response to environmental fluctuations. However, in this review we show that a wide array of environmental cues target auxin metabolism, perception, and/or transcriptional regulation to modify body shape in *Arabidopsis thaliana*. The cellular abundance of auxin depends on its metabolism (Casanova-Sáez et al. 2020) and transport (Schwechheimer et al. 2020) and is sensed by the coreceptors of the TRANSPORT INHIBITOR RESPONSE1/AUXIN SIGNALING F-BOX (TIR1/AFB) and the AUXIN RESISTANT/INDOLE-3 ACETIC ACID INDUCIBLE (Aux/IAA) families (Strader and Morffy 2020). The AUXIN RESPONSE FACTOR (ARF) family involves transcription factors that directly bind the promoters of auxin-responsive genes, such as *SMALL AUXIN-UP RNA* (*SAUR*) and *Aux/IAA* genes, to positively regulate their expression (except for negative ARFs). AUX/IAA proteins

Cite this article as *Cold Spring Harb Perspect Biol* doi: 10.1101/cshperspect.a040030

recruit transcriptional repressors to the regulatory region of auxin-responsive genes by binding to ARFs. Auxin binding to its coreceptors induces the proteasome-dependent degradation of Aux/IAA, releasing the activity of ARF transcription factors to induce the auxin-dependent expression profile.

Plants show remarkable plasticity (Casal et al. 2004). They rely on cues from the above- and below-ground environments to adjust their growth and developmental patterns to capture the available resources and use them efficiently. For instance, plants require light for photosynthesis. The occurrence of full darkness in the soil indicates that this resource is not available yet. For aerial tissues, the changes in spectral distribution and irradiance with respect to incident sunlight indicate the presence, nearness, and size of competitors for light, and ambient temperature weights the negative consequences of being shaded. Meanwhile, the direction of incident light indicates the location of impediments to reach light. In all these cases, the plant rapidly projects the foliage toward the places where the light resource is predicted to be more available. Plants also require water and mineral nutrients, and while gravity guides root growth toward the direction where these resources are more likely to be available, the occurrence of localized sources within the soil triggers changes in the root growth pattern that optimize their capture. Therefore, plants forage for light, water, and nutrients. In all these cases, auxin is between environmental sensing and the growth response.

FORAGING FOR LIGHT

Buried in the Darkness of the Soil

Seeds can germinate buried at certain deepness, where there is more water available than in the upper layers of the soil. There, the seedlings experience full darkness because light typically penetrates only a few millimeters into the soil. In response to this environment, the hypocotyl (*A. thaliana* seedlings) grows at a fast rate to minimize the time between germination and the emergence of the shoot to receive light for photosynthesis (i.e., the duration of a period when growth relies on seed reserves [heterotrophy]). Cotyledon expansion is rudimentary, which helps to save energy and avoids excessive attrition caused by soil roughness (Fig. 1). The upper section of the hypocotyl forms the plumular hook, which reduces the risk of apical meristem damage on its way through the soil. This growth pattern is called skotomorphogenesis (Arsovski et al. 2012) and the seedlings are described as etiolated. Auxin plays a fundamental role in skotomorphogenesis, and gain-of-function mutants producing stable Aux/IAA proteins that repress auxin-induced gene expression show a partially de-etiolated phenotype (Nagpal et al. 2000; Tian et al. 2002).

Near the soil surface, the perception of UV-A/blue-light gradients by the plasma-membrane-associated kinases, phototropins (phot1, phot2; we follow the nomenclature established for photoreceptors by Quail et al. [1994], where the holoprotein containing the chromophore attached to the apoprotein is in lower case), induces hypocotyl bending toward the light (Liscum et al. 2020). This phototropic response guides the shoot to avoid physical barriers that might prevent successful emergence from the soil. Hypocotyl bending involves redistribution of growth-promoting auxin away from the illuminated side of the hypocotyl, and the formation of a gradient resulting from the regulation of auxin carriers (Fankhauser and Christie 2015). In darkness, phot1 interacts with PHYTOCHROME KINASE SUBSTRATE 4 (PKS4) and NON-PHOTOTROPIC HYPOCOTYL 3 (NPH3) at the plasma membrane. Light activation of phot1 leads to its dimerization and autophosphorylation, direct phosphorylation of PKS4, and indirect dephosphorylation of NPH3, which are considered important early events in downstream signaling (Liscum et al. 2020). phot1 also directly phosphorylates the multidrug resistance transporter ATP-BINDING-CASSETTE B19 (ABCB19), apparently reducing its polar auxin transport capacity to favor the formation of the horizontal gradient (Christie et al. 2011). The action of PIN-FORMED (PIN) efflux carriers (Schwechheimer et al. 2020) is important and PIN3 is laterally distributed in response to unilateral light (Ding et al. 2011). However, the exact

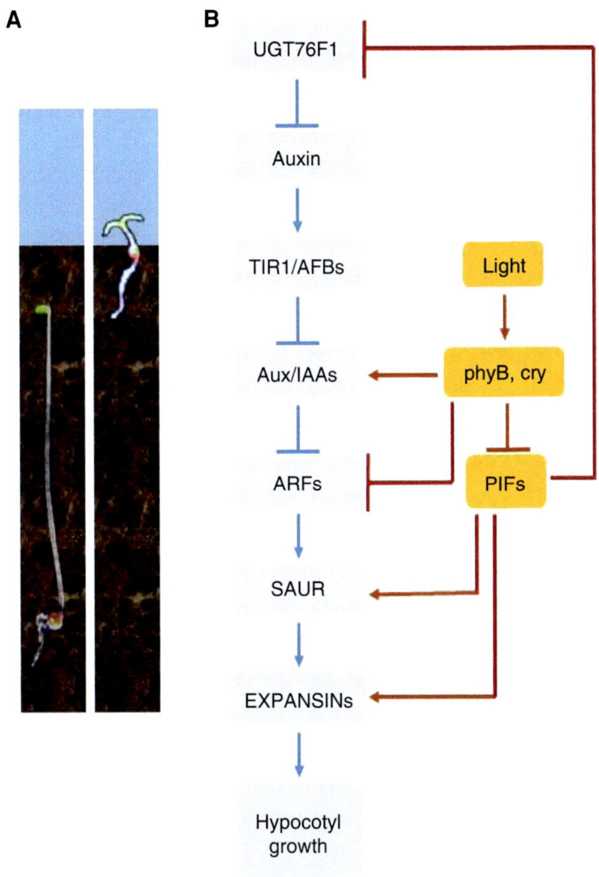

Figure 1. The transition between skoto- and photomorphogenesis involves changes in the auxin system. (*A*) Etiolated (dark grown, *left*) and de-etiolated (light grown, *right*) seedling of *Arabidopsis thaliana*. (*B*) Light perceived by phytochromes (phy) and cryptochromes (cry) represses auxin levels and auxin signaling by direct interaction of these photosensory receptors with Aux/IAA and ARF proteins and by repressing the activity of PHYTOCHROME INTERACTING FACTORS (PIFs), which enhance auxin abundance and the expression of auxin targets.

mechanisms by which the phot1-PKS4-NPH3 complex establishes the auxin gradient remain to be elucidated (Fankhauser and Christie 2015).

As the uppermost regions of the shoot emerge from the soil, light activates photosensory receptors such as phytochromes (including phyA and phyB) (Burgie and Vierstra 2014) and cryptochromes (cry1 and cry2) (Wang and Lin 2020). Red light activates phyB, far-red light (proportionally more abundant in shaded environments) is the most effective wave band to activate phyA (Casal et al. 2014) and blue light activates cry1 and cry2. If the seedlings are in darkness, after synthesis, phytochromes remain in the cytosol in the inactive form. Light induces the migration of phyA and phyB to the nucleus (Kircher et al. 1999; Yamaguchi et al. 1999), where, together with cry1 and cry2, they negatively regulate auxin signaling (Fig. 1).

One mechanism of regulation involves direct physical interaction of photosensory receptors with components of the auxin system. phyB and cry1 bind to Aux/IAA proteins favoring their dissociation from the TIR1 coreceptor and increasing their stability to negatively regulate auxin-induced gene expression (Tian et al.

2003; Xu et al. 2018). phyA might have a similar action (Colón-Carmona et al. 2000). Furthermore, the physical interaction of active phyB or cry1 with ARFs reduce the binding of these transcription factors to their target promoters (Fig. 1; Mao et al. 2020).

In darkness, the basic helix-loop-helix transcription factors PHYTOCHROME INTERACTING FACTOR (PIF) (Lee and Choi 2017; Pham et al. 2018) such as PIF3 and PIF4 bind the promoters of *SAUR* genes, promoting their expression in the hypocotyl and having the opposite effect in the cotyledons (Sun et al. 2016). SAUR proteins inhibit the activity of type 2C protein phosphatases belonging to the D clade (PP2C.D) that in turn inhibit plasma membrane proton-ATPases, favoring apoplast acidification (Spartz et al. 2014). Apoplast acidification enhances the activity of the wall-loosening protein expansins that promote growth (Cosgrove 2005). In response to light, active phyA and/or phyB bind PIFs and induce their phosphorylation and degradation in the proteasome (Lee and Choi 2017; Pham et al. 2018), whereas active cry1 and cry2 reduce the transcriptional activity of bound PIFs (Ma et al. 2016; Pedmale et al. 2016). Consequently, several *SAUR* genes are light-induced in cotyledons and/or repressed in hypocotyls, at least partially accounting for the differential growth responses of these organs to light (Sun et al. 2016). Compared to full darkness, light causes some reduction in auxin levels in the hypocotyl and the opposite response in the cotyledons (Sun et al. 2016). Indole-3-pyruvic acid can be converted to auxin by YUCCA (YUC) or glucosylated by the UDP-glycosyltransferase UGT76F1, which therefore reduces auxin levels (Casanova-Sáez et al. 2020; Chen et al. 2020). PIF4 binds to the *UGT76F1* gene promoter to repress its activity and light enhances *UGT76F1* to promote de-etiolation (Chen et al. 2020).

In summary, auxin signaling is elevated in darkness and, during de-etiolation, light reduces auxin levels and auxin sensitivity via direct physical interaction between the photosensory receptors and the core transcriptional machinery (Aux/IAAs and ARFs) and via the photosensory receptors–PIFs pathway directly affecting auxin-controlled or auxin metabolic genes.

Under Shade

"Sun plants" initiate shade-avoidance responses when exposed to the shade of neighboring vegetation or even to the threat of that possibility, anticipated by light reflected by potential competitors that still do not shade (Casal 2013; Ballaré and Pierik 2017). Shade-avoidance responses typically include the promotion of stem growth and petiole growth and the increased angle of insertion of the leaves to place the foliage at higher and better light strata of the canopy. Warm temperatures induce similar morphological responses (Quint et al. 2016; Casal and Balasubramanian 2019) that help cool down the tissues (Crawford et al. 2012) and enhance shade avoidance (Romero-Montepaone et al. 2020). The combination of shade and warm temperatures is predicted to be a complex scenario in terms of carbon balance due to low photosynthesis and high respiration.

Under sunlight conditions, the activity of the photosensory receptor phyB represses the action of the transcription factor PIF7 by cytoplasmic retention (Huang et al. 2018b) and that of PIF4 and PIF5 by facilitating their phosphorylation and degradation in the proteasome (Lorrain et al. 2008). The changes in spectral distribution (lowering of the red to far-red ratio) and overall irradiance caused by neighbor plants reduce the activity of phyB (Sellaro et al. 2019). Then, PIF7, PIF4, and PIF5 bind to the promoters of *YUC* genes to enhance auxin synthesis (Hornitschek et al. 2012; Li et al. 2012). The reduced blue light levels of shade lower cry1 and cry2 activity, enhancing these responses (de Wit et al. 2016; Pedmale et al. 2016). Warm temperatures enhance auxin levels by a mechanism that shows significant overlap with that of shade avoidance. Warm temperatures decrease the activity of phyB by accelerating the reversion of its active form to its inactive form (Jung et al. 2016; Legris et al. 2016), a mechanism that is different from the photochemical reaction that reduces phyB activity under shade. The latter in itself increases the activity of PIFs but temperature has additional points of action. At high temperature, an RNA hairpin within the 5′ untranslated region of *PIF7* transcripts leads to enhanced translation, in-

Cite this article as *Cold Spring Harb Perspect Biol* doi: 10.1101/cshperspect.a040030

creasing PIF7 protein abundance (Chung et al. 2020). In addition, the EVENING COMPLEX, which contains ELF3, represses the expression of *PIF4* and warm temperatures reduce the association of the transcription factor ELF3 to the promoter of *PIF4* (Box et al. 2015). Warm temperature reduces the binding of the EVENING COMPLEX to DNA in vitro, suggesting that the complex acts as a direct thermosensor (Silva et al. 2020). Therefore, warm temperatures increase the expression and protein abundance of PIF4. In turn, PIF4 and PIF7 bind the promoters of *YUC8*, *TRYPTOPHAN AMINOTRANSFERASE OF ARABIDOPSIS* (*TAA1*), and other genes to increase auxin levels (Franklin et al. 2011; Sun et al. 2012; Chung et al. 2020; Fiorucci et al. 2020).

Both neighbor cues and warm temperatures increase the levels of auxin in the cotyledons to promote the growth of the hypocotyl (Procko et al. 2014; Bellstaedt et al. 2019), and this activation appears to occur in the epidermis rather than in vascular tissues of the cotyledons (Kim et al. 2020). The sole induction of *YUC3* expression in cotyledons is enough to trigger hypocotyl elongation in seedlings grown in the absence of neighbor signals or warm temperatures (Kohnen et al. 2016). Auxin travels via PINs and ABCBs from the cotyledons, where neighbor cues or warmth increase its levels, to the growing hypocotyl. Neighbor cues reorient PIN3 from the basal to the lateral side of the cell membrane to drive auxin flux to the epidermis (Keuskamp et al. 2010), which is the growth-limiting tissue of the organ. Neighbor cues reaching the hypocotyl help to conserve the additional auxin received from the cotyledons by lowering the expression of *GH3.17*, which is involved in a type of auxin conjugation that is followed by degradation (Procko et al. 2016).

The promotion of hypocotyl growth is not only about extra auxin synthesized in the cotyledons. Neighbor cues also affect the expression of genes involved in auxin conjugation, transport, perception, and signaling, many of which are targets of PIFs (Iglesias et al. 2018). Persistent shade reinforces the growth response but 1 day after the beginning of the environmental cue, auxin shows the same levels observed before the stimulation (Bou-Torrent et al. 2014; de Wit et al. 2015; Pucciariello et al. 2018). The hypocotyl growth response still requires auxin, but now a higher sensitivity to auxin mediates the shade effect (Pucciariello et al. 2018). Persistent shade reduces the expression of *MIR393* and consequently increases the abundance of its targets, the auxin receptors (Pucciariello et al. 2018). In addition, persistent shade increases PIF4 abundance in the vascular tissues of the hypocotyl (Pucciariello et al. 2018) and there is evidence in favor of enhanced sensitivity to auxin mediated by PIF4 and PIF5 under shade (Hersch et al. 2014). Enhanced PIF4 abundance promotes the expression of its target genes *IAA19* and *IAA29*, which in turn reduces the expression of the *IAA17* gene to further accelerate growth (Pucciariello et al. 2018). Therefore, the shade-avoidance response initially involves increased auxin levels and subsequently a rewiring of auxin signaling that increases the sensitivity to the hormone. The promotion of *YUC8* expression by warm temperature peaks at 4 hours and then decreases sharply (Sun et al. 2012; Huai et al. 2018). Then, enhanced sensitivity to auxin also plays a role in the response to warm temperatures, which increase HSP90 protein levels to stabilize the auxin coreceptor TIR1 (Wang et al. 2016).

Shade favors hypocotyl phototropic bending toward the site from which the seedlings receive more blue light perceived by phot1 and phot2, indicative of the position of a gap within the canopy (Goyal et al. 2016; Boccaccini et al. 2020). The reduced activities of phyB and cry1 under shade converge to enhance PIF4 abundance, respectively, by enhancing protein stability and gene expression. This establishes a strong auxin signaling status in the hypocotyl and a more pronounced auxin gradient in response to unilateral blue light to enhance light foraging (Goyal et al. 2016; Boccaccini et al. 2020).

Shade and warmth also cause leaf hyponasty. During this response, the lower face of the petiole grows faster than its upper face and the leaf adopts a more upright position. Adding a drop of auxin at the tip of the leaf lamina is enough to induce the hyponastic response. The neighbor cue perceived at the tip of the leaf lamina increases the activity of PIFs, locally enhancing the expression of *YUC* genes and the synthesis of auxin (Michaud et al. 2017; Pantazopoulou et al. 2017). In response to neighbors or warm temperatures, auxin produced in the lamina is transported to the petiole by the vasculature and distributed by PIN3 preferentially to the abaxial (fast growing) face of the petiole (Michaud et al. 2017; Pantazopoulou et al. 2017; Park et al. 2019). The epigenetic repressor ASYMMETRIC LEAVES 1 (AS1) provides leaf polarity information by repressing PIF4 binding to the *PINOID* (*PID*) gene promoter and hence *PID* expression more strongly in the adaxial than the abaxial face of the petiole (Park et al. 2019). PID is a protein kinase that phosphorylates the auxin efflux transporter PIN3 (Schwechheimer et al. 2020) and regulates its polarization to the outer membrane of abaxial petiole cells by vesicle-to-membrane trafficking (Park et al. 2019).

Shade and warmth also promote the elongation of the petiole. At least in the case of the shade-avoidance response, processes in the petiole itself perceive the neighbor cues, and auxin added to the lamina is not effective (Michaud et al. 2017; Pantazopoulou et al. 2017). In addition to effects on auxin levels, neighbor cues upregulate the expression of *TIR1/AFBs* genes and enhance the accumulation of AFB1 protein in petioles with potential effects on auxin sensitivity (de Wit et al. 2015).

The environment is essentially dynamic across the diurnal cycle. Diurnal sensitivity to shade events depends on auxin signaling status and correlates with sensitivity to exogenously applied auxin (Sellaro et al. 2012). Furthermore, even under dense vegetation shade is often interrupted by light penetrating through gaps in the canopy. The occurrence of these interruptions depends on the interaction between the geometry of the canopy and solar angle, and therefore these events tend to repeat daily at a certain time in the photoperiod. phyB/phyA and the UV-B receptor UV-B RESISTANT 8 (UVR8), respectively, dominate the perception of afternoon and midday daily interruptions of shade (Sellaro et al. 2011; Moriconi et al. 2018). Light activation of phyB and phyA enhances the expression of *ELONGATED HYPOCOTYL 5* (HY5) and the HY5 transcription factor represses the expression of auxin-induced genes (Sellaro et al. 2011). UV-B perceived by UVR8 reduces the abundance of PIF4 and PIF5 and the expression of auxin synthesis genes (Hayes et al. 2014). The sensitivity to light interruptions of shade also depends on auxin signaling status (Sellaro et al. 2011).

FORAGING FOR WATER AND NUTRIENTS

General Orientation to Explore the Soil

The direction of the gravity vector provides information that roots use to orient their growth in soil. Gravity sensing occurs in root columella cells, where the position of starch-accumulating amyloplasts depends on the direction of gravity (Nakamura et al. 2019). When the root grows in the direction of gravity, PIN3 evenly distributes auxin toward expanding cells in the root elongation zone. Disruption of root position causes PIN3 phosphorylation and relocation, increasing the flux of auxin toward the epidermal cells of the lower face of the root (Nakamura et al. 2019). In response to auxin, these cells reduce their rate of expansion causing root bending to reestablish the match with the direction of the gravity vector. This response is very rapid as it initiates in <30 sec after auxin reaches the growing cells (Fendrych et al. 2018). The action of auxin requires intracellular perception and the assembly of the TIR1/AFB–Aux/IAA coreceptor complex but it does not involve changes in transcription. Therefore, the canonical auxin receptor system appears to have a nontranscriptional output that is important for root positive gravitropism (Fendrych et al. 2018).

Taking Advantage of Water and Nitrate Patches

Although gravitropism provides a general sense of direction for the root, the soil is a heterogeneous environment. Therefore, the root can find previously unexplored areas where water and nutrients are more readily available. The root tip senses water gradients, biasing the radial position of the initiation of lateral roots when this critical resource for plant growth and development is available at soil–water microsites, an adaptive response termed hydropatterning (Bennett 2020). Lateral root development depends on auxin and ARF7, among other ARFs (Cavallari et al. 2020). Hydropatterning involves a change in the activity of ARF7 (Orosa-Puente et al. 2018). In fact, ARF7 is differentially SUMOylated at several lysine residues in the root area exposed to air, but not in the water exposed root section, by SMALL UBIQUITIN MODIFIER (SUMO) protein(s) that remain unidentified. Once SUMOylated, ARF7-SUMO is recruited and inhibited by IAA3, blocking the expression of LATERAL ORGAN BOUNDARIES 16 (LBD16) and other downstream genes required for lateral root development (Orosa-Puente et al. 2018).

Nitrogen is an essential macronutrient, vital for plant growth and productivity. In the soil, nitrogen is present in inorganic (e.g., nitrate, ammonium) or organic forms (e.g., amino acids, urea, etc.). In many species, including *A. thaliana*, nitrate is the main source of nitrogen for the plant (Crawford and Forde 2002). In soil with relatively low levels of nitrogen, the root takes advantage of rich spots by generating new branches and increasing the density of root hairs in response to the local nitrate stimulus. The plasma membrane protein NITRATE TRANSPORTER1.1 (NRT1.1) plays a key role in this branching response. NRT1.1 transports nitrate (with dual-affinity), senses nitrate (defined as a nitrate transceptor after these two functions), and facilitates auxin uptake (Krouk et al. 2010). Under low nitrate conditions, phosphorylated NRT1.1 transports auxin out of the lateral root apex preventing the accumulation of auxin required for the growth and consequent emergence of the lateral root (Bouguyon et al. 2015). When the root reaches a region containing high nitrate levels, auxin transport by NRT1.1 is inhibited, auxin accumulates, and the lateral root grows, exploring the place where nitrate is available (Bouguyon et al. 2015).

Nitrate can reduce primary root growth corrected by shoot growth (Linkohr et al. 2002). While ammonium is retained by the soil matrix, after rain, nitrate moves freely in solution toward lower strata. Compared to ammonium, nitrate triggers a rapid enhancement of root growth by increasing the number of cells that enter the root elongation zone (Ötvös et al. 2020), which would help foraging for this nutrient in deep soil strata. This response depends on the precise modulation of auxin transport routes between cortex and epidermis. The root adjusts to nitrogen sources by using a sensitive fine-tuning mechanism for auxin transport, mediated by PIN2 phosphorylation, trafficking, and localization (Ötvös et al. 2020).

Staying Superficial When There Is Not Enough Phosphate

Phosphorus is an essential element for plant growth and development. Plants uptake phosphorus in the form of phosphate, which has low mobility in soil, is extremely reactive and usually precipitates with aluminum, iron, or calcium, making phosphorous one of the most frequently limiting nutrients. In response to low availability of phosphate, the growth of the primary root is restricted, root branching is more profuse, and the root hairs are longer (Fig. 2A). This pattern helps to explore the upper layers of the soils where plant detritus can release phosphorus.

Auxin plays a key role in the response of root system architecture to phosphate availability (Nacry et al. 2005). PHOSPHATE RESPONSE 1 (PHR1) and related MYB transcription factors are crucial for adaptive responses to low phosphate. In the presence of phosphate, nuclear proteins that bear the SPX domain, also present in yeast phosphate sensors, reduce the interaction between PHR1 and its target promoters by direct physical interaction (Puga et al. 2014). Phosphate availability increases the concentration of inositol polyphosphate signaling molecules, which are the physiological ligands of the SPX-domain proteins, favoring their interaction with PHR1 and preventing adaptive responses to low phosphate (Wild et al. 2016). The specific connections between phosphate sensing and changes in auxin signaling remain to be fully elucidated, but genes involved in auxin synthesis, sensing, and downstream control of transcription are possible PHR1 targets involved root growth responses to phosphate starvation (Crombez et al. 2019). Furthermore, *PHR1* is a

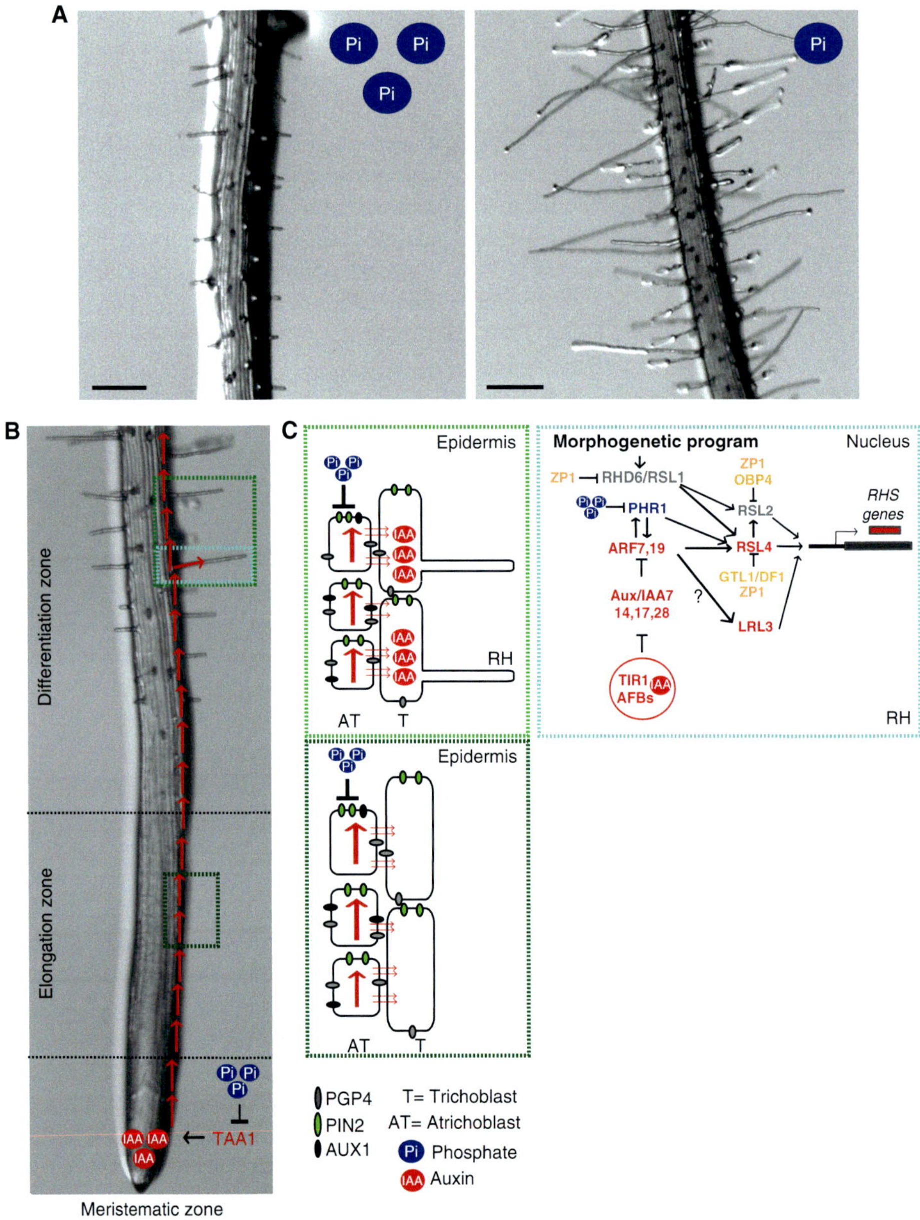

Figure 2. Low phosphate promotes hair root growth. (*A*) Detail of the root hairs of *Arabidopsis thaliana* seedlings grown at two different levels of phosphate (low and high). (*B*) Low phosphate increases auxin levels at the meristematic zone of the primary root and auxin travels from the meristematic zone through the lateral root cap and the epidermis to the differentiation zone. (*C*) This movement of auxin takes place via auxin carriers mostly in atrichoblasts to reach the root hairs (note negative regulation by phosphate). In the root hairs (RHs), auxin arriving from the meristematic zone is perceived to promote polar growth. *Right-hand* box shows the detail of nuclear events. Auxin signaling components (note negative regulation by phosphate) and auxin-responsive genes are in red, positive regulators of hair root growth are in gray, and negative regulators of hair root growth are in orange. (ZP1) ZINC FINGER PROTEIN 1 (Han et al. 2020), (GTL1/DF1) TRIHELIX TRANSCRIPTION FACTOR GT-2-LIKE 1 and its homolog (Shibata et al. 2018), (OBP4) OBF-BINDING PROTEIN 4 (Rymen et al. 2017), (*RHS*) *ROOT HAIR SPECIFIC* genes.

 Cite this article as *Cold Spring Harb Perspect Biol* doi: 10.1101/cshperspect.a040030

direct target of ARF7 and ARF19 in the root, a regulation that is important for the formation of lateral root (Huang et al. 2018a).

Low levels of phosphate enhance the expression of the *TAA1* gene at the primary root tip (Bhosale et al. 2018), auxin levels at the root apex and root hairs, and the expression of the *DR5::VENUS* reporter of auxin-induced transcriptional activity in the columella cells, quiescent center, stele, lateral root cap, and root hair zone (Fig. 2B; Nacry et al. 2005; Bhosale et al. 2018). The additional auxin synthesized at the root tip must reach the root hairs to promote their growth and the *taa1* mutant displays a poor hair-root response to low phosphate (Bhosale et al. 2018).

Basipetal (shoot-ward) auxin movement via the lateral root cap and epidermal cells relies on the auxin efflux carrier PIN2, and the auxin influx carrier AUXIN RESISTANT 1 (AUX1) at the plasma membrane (Fig. 2C; Cho et al. 2007a; Ganguly et al. 2010; Bhosale et al. 2018). PIN2 shows higher abundance in nonhair cells (atrichoblast) than in hair cells (trichoblasts) of the root epidermis (Löfke et al. 2015) and AUX1 is only expressed in atrichoblast (Jones et al. 2009), suggesting that basipetal auxin transport though nonhair cells sustains root hair development (Lee and Cho 2006; Jones et al. 2009). When phosphate levels are normal, PIN2 accumulates in the vacuole in a process mediated by SORTING NEXIN (SNX)-containing endosomes (Lin et al. 2020). There, PIN2 is degraded, reducing auxin transport of auxin to the hair roots. Under phosphate starvation, PHR1 induces substantial remodeling of phospholipids, releasing phosphorus for other uses in the cell (Pant et al. 2015). The phosphatidic acid resulting from these lipid changes, binds to SNX1, preventing PIN2 accumulation and hence degradation in the vacuole, to favor root hair growth (Lin et al. 2020). Auxin can reach the trichoblast by diffusion from the atrichoblasts and this process could be accelerated by efflux carriers in atrichoblast plasma membranes adjacent to the trichoblast. The potential role of a symplastic pathway via plasmodesmata (Mellor et al. 2020) remains to be elucidated. The auxin efflux carrier ABCB4/P-GLYCOPROTEIN 1 (PGP4) can reduce the auxin stimulus via a local control (Santelia et al. 2005; Cho et al. 2007b).

Once in the trichoblasts, auxin sensed in situ by the TIR1/AFBs and Aux/IAAs coreceptors promotes root hair growth. Seedlings with root hair–specific TIR1 overexpression (Ganguly et al. 2010) have long root hairs, whereas *tir1 afb1 afb3* mutants and *axr1* mutants (affecting another of the SCF^{TIR1} complex components) have short root hairs (Lincoln et al. 1990; Pitts et al. 1998; Dharmasiri et al. 2005), restored by suppressors of the *axr1* mutation (Cernac et al. 1997). Gain-of-function mutants of Aux/IAAs, such as IAA7 (*axr2*) (Lee and Cho 2006), IAA17 (*axr3*) (Knox et al. 2003), *iaa28* (Rogg et al. 2001), and IAA14 (*srl1*) (Fukaki et al. 2002) all show reduced root hair growth (Fig. 2C). ARFs control the expansion of root hairs (Mangano et al. 2017; Bhosale et al. 2018; Schoenaers et al. 2018). ARF7 and ARF19 are the most abundantly expressed ARFs in root hair cells (Bargmann et al. 2013) and IAA14 is a repressor of both (Fukaki et al. 2002). ARF7 and ARF19 are direct targets of PHR1 (Crombez et al. 2019) and ARF7 and ARF19 enhance *PHR1* expression (Huang et al. 2018a).

ARFs promote root hair growth by positively regulating the expression of several transcription factors such as ROOT HAIR DEFECTIVE-SIX-LIKE 2 (RSL2), RSL4 (Yi et al. 2010; Pires et al. 2013; Mangano et al. 2017), and LOTUS JAPONICUS RHL1-LIKE 3 (LRL3) (Karas et al. 2009; Tam et al. 2015), as well as other key components of polar growth such as the receptor-like kinase ERULUS (Schoenaers et al. 2018). RSL4 levels quantitatively relate to final root hair size (Datta et al. 2015). RSL1 and ROOT HAIR DEFECTIVE 6 (RHD6) are basic helix-loop-helix transcription factors (Menand et al. 2007; Pires et al. 2013) that act as master regulators, inducing the expression of their downstream basic helix-loop-helix transcription factors RSL2, RSL4, and LRL3 (Fig. 2C; Karas et al. 2009; Yi et al. 2010; Bruex et al. 2012).

The higher levels of auxin under low phosphate may mediate the early inhibition of primary root growth (Nacry et al. 2005), when the growth of epidermal cells is reduced (Sánchez-Calderón et al. 2005). However, low phosphate

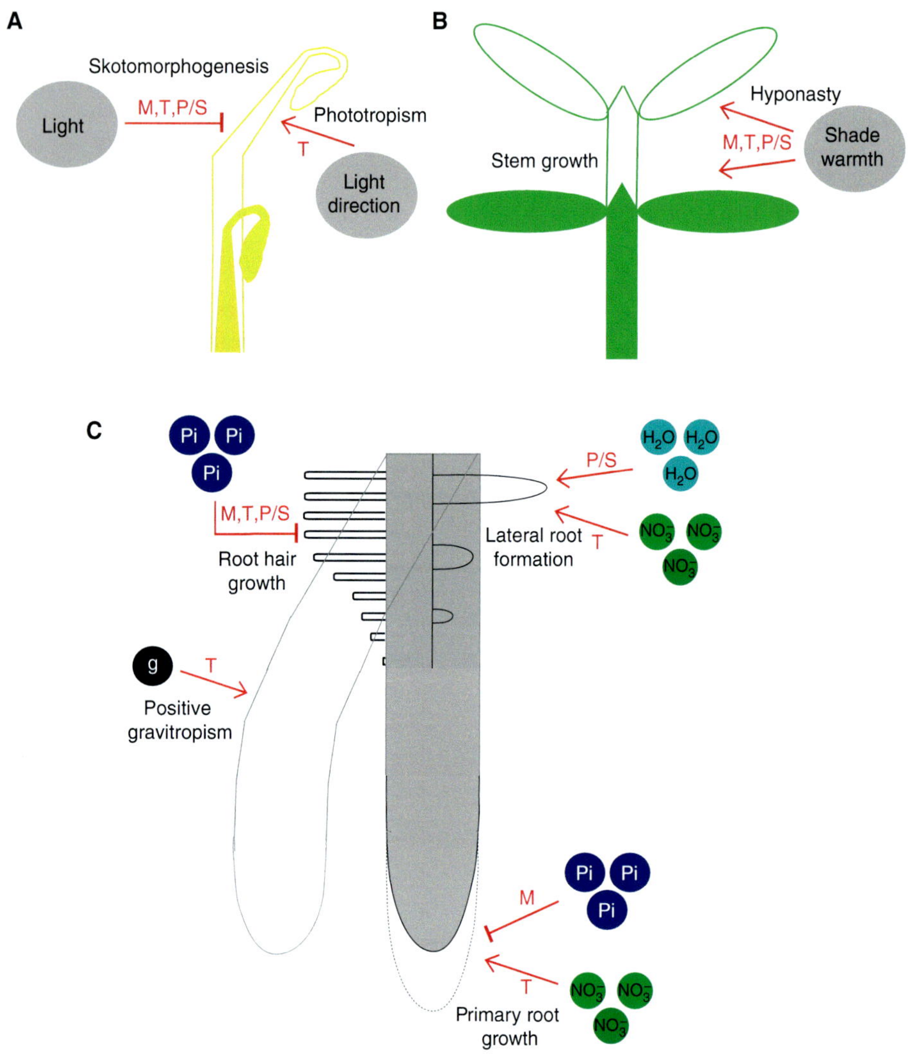

Figure 3. Environmental cues affect plant growth by modifying auxin metabolisms (M), transport (T), perception (P), and/or signaling (S). (*A*) Etiolated shoot responses. (*B*) De-etiolated shoot responses. (*C*) Root responses.

has a stronger effect on primary root growth at a later stage (Nacry et al. 2005), which might relate to a shift from indeterminate to determinate growth of the primary root under severely low phosphate (Sánchez-Calderón et al. 2005). In this case, there is a reduction in the number of cells in the meristem because of a lower rate of cell division plus accelerated cell differentiation, which could be caused by alterations in the activity and/or maintenance of the quiescent center (Sánchez-Calderón et al. 2005). These alterations in the quiescent center correlate with a reduction in the intensity of the *DR5:GUS* signal in the columella initials, the quiescent center, and the mature columella root cap (Sánchez-Calderón et al. 2005).

CONCLUDING REMARKS

Auxin provides endogenous cues that are crucial for the organization of the plant body. Here we describe how external cues convey to the auxin system information about the environment that plants have to face. Environmental signaling converges on different aspects of the auxin system to reshape the plant body in a way that reduces the chances of stress (Fig. 3). For instance, light and temperature conditions affect the metabolic processes that control auxin levels in the shoot (Franklin et al. 2011; Hornitschek et al. 2012; Li et al. 2012; Sun et al. 2012; Procko et al. 2016; Chen et al. 2020; Chung et al. 2020; Fiorucci et al. 2020), and phosphate controls auxin levels in the root (Bhosale et al. 2018). The direction of light and gravity and the availability of nitrate or phosphate affect auxin transport and modify its distribution within the organ (Bouguyon et al. 2015; Fankhauser and Christie 2015; Nakamura et al. 2019; Lin et al. 2020; Ötvös et al. 2020). Light and temperature conditions can also affect the levels of TIR1/AFBs or Aux/IAA coreceptors (Tian et al. 2003; de Wit et al. 2015; Wang et al. 2016; Pucciariello et al. 2018; Xu et al. 2018). Light, water, and phosphate cues can affect the abundance and/or intrinsic activity of the ARF transcription factors (Orosa-Puente et al. 2018; Mao et al. 2020).

Environmental cues introduce these changes by transcriptional and/or posttranscriptional regulation. Gene ontology terms related to auxin are often overrepresented in transcriptome responses to environmental cues. Shade (Kohnen et al. 2016) and warmth (Bellstaedt et al. 2019) provide clear examples, but we only partially understand the functional significance of these changes (Iglesias et al. 2018). Posttranscriptional regulation includes phosphorylation and relocalization (Nakamura et al. 2019; Ötvös et al. 2020), effects on turnover rates (Lin et al. 2020), and modified affinity (Bouguyon et al. 2015) of auxin carriers. Posttranscriptional regulation also includes the physical interaction of proteins carrying environmental information either with the Aux/IAAs to influence their interaction with the TIR1/ABFs coreceptors (Tian et al. 2003; Xu et al. 2018), with TIR1 to modify its stability (Wang et al. 2016) or with ARFs to prevent their interaction with the target promoters (Mao et al. 2020), and the SUMOylation of ARF transcription factors (Orosa-Puente et al. 2018). The list of growth processes and molecular mechanisms presented here does not pretend to be exhaustive. It simply illustrates that all aspects of the auxin system are under environmental control to substantially modify the patterns of plant growth and enhance fitness.

ACKNOWLEDGMENTS

Work at the Casal lab is supported by grants from Agencia Nacional de Promoción Científica y Tecnológica (ANPCYT, PICT-2018-01695), Universidad de Buenos Aires (20020170100505BA), and Consejo Nacional de Investigaciones Científicas y Técnicas (SNSF, SUIZ/17/04). Work at the Estevez lab is supported by grants from ICGEB (CRP/ARG16-03), ANPCYT (PICT2016-0132 and PICT2017-0066), and Fondo Nacional de Desarrollo Científico y Tecnológico (1200010) and Instituto Milenio iBio – Iniciativa Científica Milenio, MINECON.

REFERENCES

**Reference is also in this collection.*

Arsovski AA, Galstyan A, Guseman JM, Nemhauser JL. 2012. Photomorphogenesis. *Arab B* **10:** e0147. doi:10.1199/tab.0147

Ballaré CL, Pierik R. 2017. The shade-avoidance syndrome: multiple signals and ecological consequences. *Plant Cell Environ* **40:** 2530–2543. doi:10.1111/pce.12914

Bargmann BOR, Vanneste S, Krouk G, Nawy T, Efroni I, Shani E, Choe G, Friml J, Bergmann DC, Estelle M, et al. 2013. A map of cell type-specific auxin responses. *Mol Syst Biol* **9:** 688. doi:10.1038/msb.2013.40

Bellstaedt J, Trenner J, Lippmann R, Poeschl Y, Zhang X, Friml J, Quint M, Delker C. 2019. A mobile auxin signal connects temperature sensing in cotyledons with growth responses in hypocotyls. *Plant Physiol* **180:** 757–766. doi:10.1104/pp.18.01377

* Bennett M. 2020. Abiotic stress networks. *Cold Spring Harb Perspect Biol* doi:10.1101/cshperspect.a040014

Bhosale R, Giri J, Pandey BK, Giehl RFH, Hartmann A, Traini R, Truskina J, Leftley N, Hanlon M, Swarup K, et al. 2018. A mechanistic framework for auxin dependent *Arabidopsis* root hair elongation to low external phosphate. *Nat Commun* **9:** 1409. doi:10.1038/s41467-018-03851-3

Boccaccini A, Legris M, Krahmer J, Allenbach Petrolati L, Goyal A, Ampudia CG, Vernoux T, Karayekov E, Casal JJ, Fankhauser C. 2020. Low blue light enhances phototropism by releasing cryptochrome 1-mediated inhibition of *PIF4* expression. *Plant Physiol* **183:** 1780–1793. doi:10.1104/pp.20.00243

Bouguyon E, Brun F, Meynard D, Kubeš M, Pervent M, Leran S, Lacombe B, Krouk G, Guiderdoni E, Zažímalová E, et al. 2015. Multiple mechanisms of nitrate sensing by

Arabidopsis nitrate transceptor NRT1.1. *Nat Plants* **1:** 15015. doi:10.1038/nplants.2015.15

Bou-Torrent J, Galstyan A, Gallemí M, Cifuentes-Esquivel N, Molina-Contreras MJ, Salla-Martret M, Jikumaru Y, Yamaguchi S, Kamiya Y, Martínez-García JF. 2014. Plant proximity perception dynamically modulates hormone levels and sensitivity in *Arabidopsis*. *J Exp Bot* **65:** 2937–2947. doi:10.1093/jxb/eru083

Box MS, Huang BE, Domijan M, Jaeger KE, Khattak AK, Yoo SJ, Sedivy EL, Jones DM, Hearn TJ, Webb AAR, et al. 2015. ELF3 controls thermoresponsive growth in *Arabidopsis*. *Curr Biol* **25:** 194–199. doi:10.1016/j.cub.2014.10.076

Bruex A, Kainkaryam RM, Wieckowski Y, Kang YH, Bernhardt C, Xia Y, Zheng X, Wang JY, Lee MM, Benfey P, et al. 2012. A gene regulatory network for root epidermis cell differentiation in *Arabidopsis*. *PLoS Genet* **8:** e1002446. doi:10.1371/journal.pgen.1002446

Burgie ES, Vierstra RD. 2014. Phytochromes: an atomic perspective on photoactivation and signaling. *Plant Cell* **26:** 4568–4583. doi:10.1105/tpc.114.131623

Casal JJ. 2013. Photoreceptor signaling networks in plant responses to shade. *Annu Rev Plant Biol* **64:** 403–427. doi:10.1146/annurev-arplant-050312-120221

Casal JJ, Balasubramanian S. 2019. Thermomorphogenesis. *Annu Rev Plant Biol* **70:** 321–346. doi:10.1146/annurev-arplant-050718-095919

Casal JJ, Fankhauser C, Coupland G, Blázquez MA. 2004. Signalling for developmental plasticity. *Trends Plant Sci* **9:** 309–314. doi:10.1016/j.tplants.2004.04.007

Casal JJ, Candia AN, Sellaro R. 2014. Light perception and signalling by phytochrome A. *J Exp Bot* **65:** 2835–2845. doi:10.1093/jxb/ert379

* Casanova-Sáez R, Mateo-Bonmatí E, Ljung K. 2020. Auxin metabolism in plants. *Cold Spring Harb Perspect Biol* doi:10.1101/cshperspect.a039867

* Cavallari N, Artner C, Benkova E. 2020. Auxin-regulated lateral root organogenesis. *Cold Spring Harb Perspect Biol* doi:10.1101/cshperspect.a039941

Cernac A, Lincoln C, Lammer D, Estelle M, Schneider K, Lawson E, Dean C, Poethig S, Roberts K. 1997. The SAR1 gene of *Arabidopsis* acts downstream of the AXR1 gene in auxin response. *Development* **124:** 1583–1591.

Chen L, Huang XX, Zhao SM, Xiao DW, Xiao LT, Tong JH, Wang WS, Li YJ, Ding Z, Hou BK. 2020. IPyA glucosylation mediates light and temperature signaling to regulate auxin-dependent hypocotyl elongation in *Arabidopsis*. *Proc Natl Acad Sci* **117:** 6910–6917. doi:10.1073/pnas.2000172117

Cho M, Lee OR, Ganguly A, Cho HT. 2007a. Auxin-signaling: short and long. *J Plant Biol* **50:** 79–89. doi:10.1007/BF03030615

Cho M, Lee SH, Cho HT. 2007b. P-glycoprotein4 displays auxin efflux transporter-like action in *Arabidopsis* root hair cells and tobacco cells. *Plant Cell* **19:** 3930–3943. doi:10.1105/tpc.107.054288

Christie JM, Yang H, Richter GL, Sullivan S, Thomson CE, Lin J, Titapiwatanakun B, Ennis M, Kaiserli E, Lee OR, et al. 2011. Phot1 inhibition of ABCB19 primes lateral auxin fluxes in the shoot apex required for phototropism. *PLoS Biol* **9:** e1001076. doi:10.1371/journal.pbio.1001076

Chung BYW, Balcerowicz M, Di Antonio M, Jaeger KE, Geng F, Franaszek K, Marriott P, Brierley I, Firth AE, Wigge PA. 2020. An RNA thermoswitch regulates daytime growth in *Arabidopsis*. *Nat Plants* **6:** 522–532. doi:10.1038/s41477-020-0633-3

Colón-Carmona A, Chen DL, Yeh KC, Abel S. 2000. Aux/IAA proteins are phosphorylated by phytochrome in vitro. *Plant Physiol* **124:** 1728–1738. doi:10.1104/pp.124.4.1728

Cosgrove DJ. 2005. Growth of the plant cell wall. *Nat Rev Mol Cell Biol* **6:** 850–861. doi:10.1038/nrm1746

Crawford NM, Forde BG. 2002. Molecular and developmental biology of inorganic nitrogen nutrition. *Arab B* **1:** e0011. doi:10.1199/tab.0011

Crawford AJ, McLachlan DH, Hetherington AM, Franklin KA. 2012. High temperature exposure increases plant cooling capacity. *Curr Biol* **22:** R396–R397. doi:10.1016/j.cub.2012.03.044

Crombez H, Motte H, Beeckman T. 2019. Tackling plant phosphate starvation by the roots. *Dev Cell* **48:** 599–615. doi:10.1016/j.devcel.2019.01.002

Datta S, Prescott H, Dolan L. 2015. Intensity of a pulse of RSL4 transcription factor synthesis determines *Arabidopsis* root hair cell size. *Nat Plants* **1:** 15138. doi:10.1038/nplants.2015.138

de Wit M, Ljung K, Fankhauser C. 2015. Contrasting growth responses in lamina and petiole during neighbor detection depend on differential auxin responsiveness rather than different auxin levels. *New Phytol* **3:** 198–209.

de Wit M, Keuskamp DH, Bongers FJ, Martínez-Cerón C, Fankhauser C, Pierik R, Hornitschek P, Gommers CMM, Reinen E. 2016. Integration of phytochrome and cryptochrome signals determines plant growth during competition for light. *Curr Biol* **26:** 3320–3326. doi:10.1016/j.cub.2016.10.031

Dharmasiri N, Dharmasiri S, Estelle M. 2005. The F-box protein TIR1 is an auxin receptor. *Nature* **435:** 441–445. doi:10.1038/nature03543

Ding Z, Galván-Ampudia CS, Demarsy E, Łangowski Ł, Kleine-Vehn J, Fan Y, Morita MT, Tasaka M, Fankhauser C, Offringa R, et al. 2011. Light-mediated polarization of the PIN3 auxin transporter for the phototropic response in *Arabidopsis*. *Nat Cell Biol* **13:** 447–452. doi:10.1038/ncb2208

Fankhauser C, Christie JM. 2015. Plant phototropic growth. *Curr Biol* **25:** R384–R389. doi:10.1016/j.cub.2015.03.020

Fendrych M, Akhmanova M, Merrin J, Glanc M, Hagihara S, Takahashi K, Uchida N, Torii KU, Friml J. 2018. Rapid and reversible root growth inhibition by TIR1 auxin signalling. *Nat Plants* **4:** 453–459. doi:10.1038/s41477-018-0190-1

Fiorucci AS, Galvão VC, Ince YÇ, Boccaccini A, Goyal A, Allenbach Petrolati L, Trevisan M, Fankhauser C. 2020. PHYTOCHROME INTERACTING FACTOR 7 is important for early responses to elevated temperature in *Arabidopsis* seedlings. *New Phytol* **226:** 50–58. doi:10.1111/nph.16316

Franklin KA, Lee SH, Patel D, Kumar SV, Spartz AK, Gu C, Ye S, Yu P, Breen G, Cohen JD, et al. 2011. PHYTOCHROME-INTERACTING FACTOR 4 (PIF4) regulates auxin biosynthesis at high temperature. *Proc Natl Acad Sci* **108:** 20231–20235. doi:10.1073/pnas.1110682108

Fukaki H, Tameda S, Masuda H, Tasaka M. 2002. Lateral root formation is blocked by a gain-of-function mutation in the *SOLITARY-ROOT/IAA14* gene of *Arabidopsis*. *Plant J* **29:** 153–168. doi:10.1046/j.0960-7412.2001.01201.x

Ganguly A, Lee SH, Cho M, Lee OR, Yoo H, Cho HT. 2010. Differential auxin-transporting activities of PIN-FORMED proteins in *Arabidopsis* root hair cells. *Plant Physiol* **153:** 1046–1061. doi:10.1104/pp.110.156505

Goyal A, Karayekov E, Galvão VC, Ren H, Casal JJ, Fankhauser C. 2016. Shade promotes phototropism through phytochrome B-controlled auxin production. *Curr Biol* **26:** 3280–3287. doi:10.1016/j.cub.2016.10.001

Han G, Wei X, Dong X, Wang C, Sui N, Guo J, Yuan F, Gong Z, Li X, Zhang Y, et al. 2020. *Arabidopsis* ZINC FINGER PROTEIN1 acts downstream of GL2 to repress root hair initiation and elongation by directly suppressing bHLH genes. *Plant Cell* **32:** 206–225. doi:10.1105/tpc.19.00226

Hayes S, Velanis CN, Jenkins GI, Franklin KA. 2014. UV-B detected by the UVR8 photoreceptor antagonizes auxin signaling and plant shade avoidance. *Proc Natl Acad Sci* **111:** 11894–11899. doi:10.1073/pnas.1403052111

Hersch M, Lorrain S, de Wit M, Trevisan M, Ljung K, Bergmann S, Fankhauser C. 2014. Light intensity modulates the regulatory network of the shade avoidance response in *Arabidopsis*. *Proc Natl Acad Sci* **111:** 6515–6520. doi:10.1073/pnas.1320355111

Hornitschek P, Kohnen MV, Lorrain S, Rougemont J, Ljung K, López-Vidriero I, Franco-Zorrilla JM, Solano R, Trevisan M, Pradervand S, et al. 2012. Phytochrome interacting factors 4 and 5 control seedling growth in changing light conditions by directly controlling auxin signaling. *Plant J* **71:** 699–711. doi:10.1111/j.1365-313X.2012.05033.x

Huai J, Zhang X, Li J, Ma T, Zha P, Jing Y, Lin R. 2018. SEUSS and PIF4 coordinately regulate light and temperature signaling pathways to control plant growth. *Mol Plant* **11:** 928–942. doi:10.1016/j.molp.2018.04.005

Huang KL, Ma GJ, Zhang ML, Xiong H, Wu H, Zhao CZ, Liu CS, Jia HX, Chen L, Kjorven JO, et al. 2018a. The ARF7 and ARF19 transcription factors positively regulate *PHOSPHATE STARVATION RESPONSE1* in *Arabidopsis* roots. *Plant Physiol* **178:** 413–427. doi:10.1104/pp.17.01713

Huang X, Zhang Q, Jiang Y, Yang C, Wang Q, Li L. 2018b. Shade-induced nuclear localization of PIF7 is regulated by phosphorylation and 14-3-3 proteins in *Arabidopsis*. *eLife* **7:** e31636. doi:10.7554/eLife.31636

Iglesias MJ, Sellaro R, Zurbriggen MD, Casal JJ. 2018. Multiple links between shade avoidance and auxin networks. *J Exp Bot* **69:** 213–228. doi:10.1093/jxb/erx295

Jones AR, Kramer EM, Knox K, Swarup R, Bennett MJ, Lazarus CM, Leyser HMO, Grierson CS. 2009. Auxin transport through non-hair cells sustains root-hair development. *Nat Cell Biol* **11:** 78–84. doi:10.1038/ncb1815

Jung JH, Domijan M, Klose C, Biswas S, Ezer D, Gao M, Box MS, Charoensawan V, Cortijo S, Locke JC, et al. 2016. Phytochromes function as thermosensors in *Arabidopsis*. *Science* **354:** 886–889. doi:10.1126/science.aaf6005

Karas B, Amyot L, Johansen C, Sato S, Tabata S, Kawaguchi M, Szczyglowski K. 2009. Conservation of *lotus* and *Arabidopsis* basic helix-loop-helix proteins reveals new players in root hair development. *Plant Physiol* **151:** 1175–1185. doi:10.1104/pp.109.143867

Keuskamp DH, Pollmann S, Voesenek LACJ, Peeters AJM, Pierik R. 2010. Auxin transport through PIN-FORMED 3 (PIN3) controls shade avoidance and fitness during competition. *Proc Natl Acad Sci* **107:** 22740–22744. doi:10.1073/pnas.1013457108

Kim S, Hwang G, Kim S, Thi TN, Kim H, Jeong J, Kim J, Kim J, Choi G, Oh E. 2020. The epidermis coordinates thermoresponsive growth through the phyB-PIF4-auxin pathway. *Nat Commun* **11:** 1053. doi:10.1038/s41467-020-14905-w

Kircher S, Kozma-Bognar L, Kim L, Ádám E, Harter K, Schäfer E, Nagy F. 1999. Light-quality dependent nuclear import of the plant photoreceptors phytochrome A and B. *Plant Cell* **11:** 1445–1456.

Knox K, Grierson CS, Leyser O. 2003. AXR3 and SHY2 interact to regulate root hair development. *Development* **130:** 5769–5777. doi:10.1242/dev.00659

Kohnen MV, Schmid-Siegert E, Trevisan M, Petrolati LA, Sénéchal F, Müller-Moulé P, Maloof J, Xenarios I, Fankhauser C. 2016. Neighbor detection induces organ-specific transcriptomes, revealing patterns underlying hypocotyl-specific growth. *Plant Cell* **28:** 2889–2904. doi:10.1105/tpc.16.00463

Krouk G, Lacombe B, Bielach A, Perrine-Walker F, Malinska K, Mounier E, Hoyerova K, Tillard P, Leon S, Ljung K, et al. 2010. Nitrate-regulated auxin transport by NRT1.1 defines a mechanism for nutrient sensing in plants. *Dev Cell* **18:** 927–937. doi:10.1016/j.devcel.2010.05.008

Lee SH, Cho HT. 2006. PINOID positively regulates auxin efflux in *Arabidopsis* root hair cells and tobacco cells. *Plant Cell* **18:** 1604–1616. doi:10.1105/tpc.105.035972

Lee N, Choi G. 2017. Phytochrome-interacting factor from *Arabidopsis* to liverwort. *Curr Opin Plant Biol* **35:** 54–60. doi:10.1016/j.pbi.2016.11.004

Legris M, Klose C, Burgie E, Costigliolo Rojas C, Neme M, Hiltbrunner A, Wigge PA, Schäfer E, Vierstra RD, Casal JJ. 2016. Phytochrome B integrates light and temperature signals in *Arabidopsis*. *Science* **354:** 897–900. doi:10.1126/science.aaf5656

Li L, Ljung K, Breton G, Schmitz RJ, Pruneda-Paz J, Cowing-Zitron C, Cole BJ, Ivans LJ, Pedmale UV, Jung H-S, et al. 2012. Linking photoreceptor excitation to changes in plant architecture. *Genes Dev* **26:** 785–790. doi:10.1101/gad.187849.112

Lin DL, Yao HY, Jia LH, Tan JF, Xu ZH, Zheng WM, Xue HW. 2020. Phospholipase D-derived phosphatidic acid promotes root hair development under phosphorus deficiency by suppressing vacuolar degradation of PIN-FORMED2. *New Phytol* **226:** 142–155. doi:10.1111/nph.16330

Lincoln C, Britton JH, Estelle M. 1990. Growth and development of the axr1 mutants of *Arabidopsis*. *Plant Cell* **2:** 1071–1080.

Linkohr BI, Williamson LC, Fitter AH, Leyser HMO. 2002. Nitrate and phosphate availability and distribution have different effects on root system architecture of *Arabidopsis*. *Plant J* **29:** 751–760. doi:10.1046/j.1365-313X.2002.01251.x

Liscum E, Nittler P, Koskie K. 2020. The continuing arc toward phototropic enlightenment. *J Exp Bot* **71:** 1652–1658. doi:10.1093/jxb/eraa005

Löfke C, Dünser K, Scheuring D, Kleine-Vehn J. 2015. Auxin regulates SNARE-dependent vacuolar morphology restricting cell size. *eLife* **4:** e05868. doi:10.7554/eLife.05868

Lorrain S, Allen T, Duek PD, Whitelam GC, Fankhauser C. 2008. Phytochrome-mediated inhibition of shade avoidance involves degradation of growth-promoting bHLH transcription factors. *Plant J* **53:** 312–323. doi:10.1111/j.1365-313X.2007.03341.x

Ma D, Li X, Guo Y, Chu J, Fang S, Yan C, Noel JP, Liu H. 2016. Cryptochrome 1 interacts with PIF4 to regulate high temperature-mediated hypocotyl elongation in response to blue light. *Proc Natl Acad Sci* **113:** 224–229. doi:10.1073/pnas.1511437113

Mangano S, Denita-Juarez SP, Choi HS, Marzol E, Hwang Y, Ranocha P, Velasquez SM, Borassi C, Barberini ML, Aptekmann AA, et al. 2017. Molecular link between auxin and ROS-mediated polar growth. *Proc Natl Acad Sci* **114:** 5289–5294. doi:10.1073/pnas.1701536114

Mao Z, He S, Xu F, Wei X, Jiang L, Liu Y, Wang W, Li T, Xu P, Du S, et al. 2020. Photoexcited CRY1 and phyB interact directly with ARF6 and ARF8 to regulate their DNA-binding activity and auxin-induced hypocotyl elongation in *Arabidopsis*. *New Phytol* **225:** 848–865. doi:10.1111/nph.16194

Mellor NL, Voß U, Janes G, Bennett MJ, Wells DM, Band LR. 2020. Auxin fluxes through plasmodesmata modify root-tip auxin distribution. *Development* **147:** dev181669. doi:10.1242/dev.181669

Menand B, Yi K, Jouannic S, Hoffmann L, Ryan E, Linstead P, Schaefer DG, Dolan L. 2007. An ancient mechanism controls the development of cells with a rooting function in land plants. *Science* **316:** 1477–1480. doi:10.1126/science.1142618

Michaud O, Fiorucci A, Xenarios I, Fankhauser C. 2017. Local auxin production underlies a spatially restricted neighbor-detection response in *Arabidopsis*. *Proc Natl Acad Sci* **114:** 7444–7449. doi:10.1073/pnas.1702276114

Moriconi V, Binkert M, Costigliolo C, Sellaro R, Ulm R, Casal JJ. 2018. Perception of sunflecks by the UV-B photoreceptor UV RESISTANCE LOCUS8. *Plant Physiol* **177:** 75–81. doi:10.1104/pp.18.00048

Nacry P, Canivenc G, Muller B, Azmi A, Van Onckelen H, Rossignol M, Doumas P. 2005. A role for auxin redistribution in the responses of the root system architecture to phosphate starvation in *Arabidopsis*. *Plant Physiol* **138:** 2061–2074. doi:10.1104/pp.105.060061

Nagpal P, Walker LM, Young JC, Sonawala A, Timpte C, Estelle M, Reed JW. 2000. *AXR2* encodes a member of the Aux/IAA protein family. *Plant Physiol* **123:** 563–574. doi:10.1104/pp.123.2.563

Nakamura M, Nishimura T, Morita MT. 2019. Bridging the gap between amyloplasts and directional auxin transport in plant gravitropism. *Curr Opin Plant Biol* **52:** 54–60. doi:10.1016/j.pbi.2019.07.005

Orosa-Puente B, Leftley N, von Wangenheim D, Banda J, Srivastava AK, Hill K, Truskina J, Bhosale R, Morris E, Srivastava M, et al. 2018. Root branching toward water involves posttranslational modification of transcription factor ARF7. *Science* **362:** 1407–1410. doi:10.1126/science.aau3956

Ötvös K, Marconi M, Vega A, Brien JO, Johnson A, Abualia R, Antonielli L, Montesinos JC, Zhang Y, Tan S, et al. 2020. Modulation of root growth by nutrient-defined fine-tuning of polar auxin transport. bioRxiv 2020.06.19.160994

Pant BD, Burgos A, Pant P, Cuadros-Inostroza A, Willmitzer L, Scheible W. 2015. The transcription factor PHR1 regulates lipid remodeling and triacylglycerol accumulation in *Arabidopsis thaliana* during phosphorus starvation. *J Exp Bot* **66:** 1907–1918. doi:10.1093/jxb/eru535

Pantazopoulou CK, Bongers FJ, Küpers JJ, Reinen E, Das D, Evers JB, Anten NPR, Pierik R. 2017. Neighbor detection at the leaf tip adaptively regulates upward leaf movement through spatial auxin dynamics. *Proc Natl Acad Sci* **114:** 7450–7455. doi:10.1073/pnas.1702275114

Park YJ, Lee HJ, Gil KE, Kim JY, Lee JH, Lee H, Cho HT, Vu LD, De Smet I, Park CM. 2019. Developmental programming of thermonastic leaf movement. *Plant Physiol* **180:** 1185–1197. doi:10.1104/pp.19.00139

Pedmale UV, Huang SC, Zander M, Cole BJ, Hetzel J, Ljung K, Reis PAB, Sridevi P, Nito K, Nery JR, et al. 2016. Cryptochromes interact directly with PIFs to control plant growth in limiting blue light. *Cell* **164:** 233–245. doi:10.1016/j.cell.2015.12.018

Pham VN, Kathare PK, Huq E. 2018. Phytochromes and phytochrome interacting factors. *Plant Physiol* **176:** 1025–1038. doi:10.1104/pp.17.01384

Pires ND, Yi K, Breuninger H, Catarino B, Menand B, Dolan L. 2013. Recruitment and remodeling of an ancient gene regulatory network during land plant evolution. *Proc Natl Acad Sci* **110:** 9571–9576. doi:10.1073/pnas.1305457110

Pitts RJ, Cernac A, Estelle M. 1998. Auxin and ethylene promote root hair elongation in *Arabidopsis*. *Plant J* **16:** 553–560. doi:10.1046/j.1365-313x.1998.00321.x

Procko C, Crenshaw CM, Ljung K, Noel JP, Chory J. 2014. Cotyledon-generated auxin is required for shade-induced hypocotyl growth in *Brassica rapa*. *Plant Physiol* **165:** 1285–1301. doi:10.1104/pp.114.241844

Procko C, Burko Y, Jaillais Y, Ljung K, Long JA, Chory J. 2016. The epidermis coordinates auxin-induced stem growth in response to shade. *Genes Dev* **30:** 1529–1541. doi:10.1101/gad.283234.116

Pucciariello O, Legris M, Rojas CC, Iglesias MJ, Hernando CE, Dezar C, Vazquez M, Yanovsky MJ, Finlayson SA, Prat S, et al. 2018. Rewiring of auxin signaling under persistent shade. *Proc Natl Acad Sci* **115:** 5612–5617. doi:10.1073/pnas.1721110115

Puga MI, Mateos I, Charukesi R, Wang Z, Franco-Zorrilla JM, De Lorenzo L, Irigoyen ML, Masiero S, Bustos R, Rodríguez J, et al. 2014. SPX1 is a phosphate-dependent inhibitor of phosphate starvation response 1 in *Arabidopsis*. *Proc Natl Acad Sci* **111:** 14947–14952. doi:10.1073/pnas.1404654111

Quail PH, Briggs WR, Chory J, Hangarter RP, Harberd NP, Kendrick RE, Koornneef M, Parks B, Sharrock RA, Schäfer E, et al. 1994. Spotlight on phytochrome nomenclature. *Plant Cell* **6:** 468–471. doi:10.2307/3869926

Quint M, Delker C, Franklin KA, Wigge PA, Halliday KJ, van Zanten M. 2016. Molecular and genetic control of plant thermomorphogenesis. *Nat Plants* **2:** 15190. doi:10.1038/nplants.2015.190

Rogg LE, Lasswell J, Bartel B. 2001. A gain-of-function mutation in *IAA28* suppresses lateral root development. *Plant Cell* **13:** 465–480. doi:10.1105/tpc.13.3.465

Romero-Montepaone S, Poodts S, Fischbach P, Sellaro R, Zurbriggen MD, Casal JJ. 2020. Shade-avoidance responses become more aggressive in warm environments. *Plant Cell Environ* **43:** 1625–1636. doi:10.1111/pce.13720

Rymen B, Kawamura A, Schäfer S, Breuer C, Iwase A, Shibata M, Ikeda M, Mitsuda N, Koncz C, Ohme-Takagi M, et al. 2017. ABA suppresses root hair growth via the OBP4 transcriptional regulator. *Plant Physiol* **173:** 1750–1762. doi:10.1104/pp.16.01945

Sánchez-Calderón L, López-Bucio J, Chacón-López A, Cruz-Ramírez A, Nieto-Jacobo F, Dubrovsky JG, Herrera-Estrella L. 2005. Phosphate starvation induces a determinate developmental program in the roots of *Arabidopsis thaliana*. *Plant Cell Physiol* **46:** 174–184. doi:10.1093/pcp/pci011

Santelia D, Vincenzetti V, Azzarello E, Bovet L, Fukao Y, Düchtig P, Mancuso S, Martinoia E, Geisler M. 2005. MDR-like ABC transporter AtPGP4 is involved in auxin-mediated lateral root and root hair development. *FEBS Lett* **579:** 5399–5406. doi:10.1016/j.febslet.2005.08.061

Schoenaers S, Balcerowicz D, Breen G, Hill K, Zdanio M, Mouille G, Holman TJ, Oh J, Wilson MH, Nikonorova N, et al. 2018. The auxin-regulated CrRLK1L kinase ERULUS controls cell wall composition during root hair tip growth. *Curr Biol* **28:** 722–732.e6. doi:10.1016/j.cub.2018.01.050

* Schwechheimer C, Hammes U, Murphy A. 2020. Transport. *Cold Spring Harb Perspect Biol* doi:10.1101/cshperspect.a039875

Sellaro R, Yanovsky MJ, Casal JJ. 2011. Repression of shade-avoidance reactions by sunfleck induction of HY5 expression in *Arabidopsis*. *Plant J* **68:** 919–928. doi:10.1111/j.1365-313X.2011.04745.x

Sellaro R, Pacín M, Casal JJ. 2012. Diurnal dependence of growth responses to shade in *Arabidopsis*: role of hormone, clock, and light signaling. *Mol Plant* **5:** 619–628. doi:10.1093/mp/ssr122

Sellaro R, Smith RW, Legris M, Fleck C, Casal JJ. 2019. Phytochrome B dynamics departs from photoequilibrium in the field. *Plant Cell Environ* **42:** 606–617. doi:10.1111/pce.13445

Shibata M, Breuer C, Kawamura A, Clark NM, Rymen B, Braidwood L, Morohashi K, Busch W, Benfey PN, Sozzani R, et al. 2018. GTL1 and DF1 regulate root hair growth through transcriptional repression of *ROOT HAIR DEFECTIVE 6-LIKE 4* in *Arabidopsis*. *Dev* **145:** dev159707. doi:10.1242/dev.159707

Silva CS, Nayak A, Lai X, Hutin S, Hugouvieux V, Jung JH, López-Vidriero I, Franco-Zorrilla JM, Panigrahi KCS, Nanao MH, et al. 2020. Molecular mechanisms of evening complex activity in *Arabidopsis*. *Proc Natl Acad Sci* **117:** 6901–6909. doi:10.1073/pnas.1920972117

Spartz AK, Ren H, Park MY, Grandt KN, Lee SH, Murphy AS, Sussman MR, Overvoorde PJ, Gray WM. 2014. SAUR inhibition of PP2C-D phosphatases activates plasma membrane H^+-ATPases to promote cell expansion in *Arabidopsis*. *Plant Cell* **26:** 2129–2142. doi:10.1105/tpc.114.126037

* Strader L, Morffy N. 2020. Response mechanism and structural aspects. *Cold Spring Harb Perspect Biol* doi:10.1101/cshperspect.a039883

Sun J, Qi L, Li Y, Chu J, Li C. 2012. PIF4-mediated activation of YUCCA8 expression integrates temperature into the auxin pathway in regulating *Arabidopsis* hypocotyl growth. *PLoS Genet* **8:** e1002594. doi:10.1371/journal.pgen.1002594

Sun N, Wang J, Gao Z, Dong J, He H, Terzaghi W, Wei N, Deng XW, Chen H. 2016. *Arabidopsis* SAURs are critical for differential light regulation of the development of various organs. *Proc Natl Acad Sci* **113:** 6071–6076. doi:10.1073/pnas.1604782113

Tam THY, Catarino B, Dolan L. 2015. Conserved regulatory mechanism controls the development of cells with rooting functions in land plants. *Proc Natl Acad Sci* **112:** E3959–E3968. doi:10.1073/pnas.1416324112

Tian Q, Uhlir NJ, Reed JW. 2002. *Arabidopsis* SHY2/IAA3 inhibits auxin-regulated gene expression. *Plant Cell* **14:** 301–319. doi:10.1105/tpc.010283

Tian Q, Nagpal P, Reed JW. 2003. Regulation of *Arabidopsis* SHY2/IAA3 protein turnover. *Plant J* **36:** 643–651. doi:10.1046/j.1365-313X.2003.01909.x

* Vernoux T, Pernisová M. 2020. Auxin in the shoot meristem. *Cold Spring Harb Perspect Biol* doi:10.1101/cshperspect.a039925

Wang Q, Lin C. 2020. Mechanisms of cryptochrome-mediated photoresponses in plants. *Annu Rev Plant Biol* **71:** 103–129. doi:10.1146/annurev-arplant-050718-100300

Wang R, Zhang Y, Kieffer M, Yu H, Kepinski S, Estelle M. 2016. HSP90 regulates temperature-dependent seedling growth in *Arabidopsis* by stabilizing the auxin co-receptor F-box protein TIR1. *Nat Commun* **7:** 10269. doi:10.1038/ncomms10269

Wild R, Gerasimaite R, Jung JY, Truffault V, Pavlovic I, Schmidt A, Saiardi A, Jacob Jessen H, Poirier Y, Hothorn M, et al. 2016. Control of eukaryotic phosphate homeostasis by inositol polyphosphate sensor domains. *Science* **352:** 986–990. doi:10.1126/science.aad9858

Xu F, He S, Zhang J, Mao Z, Wang W, Li T, Hua J, Du S, Xu P, Li L, et al. 2018. Photoactivated CRY1 and phyB interact directly with AUX/IAA proteins to inhibit auxin signaling in *Arabidopsis*. *Mol Plant* **11:** 523–541. doi:10.1016/j.molp.2017.12.003

Yamaguchi R, Nakamura M, Mochizuki N, Kay SA, Nagatani A. 1999. Light-dependent translocation of a phytochrome B-GFP fusion protein to the nucleus in transgenic *Arabidopsis*. *J Cell Biol* **145:** 437–445. doi:10.1083/jcb.145.3.437

Yi K, Menand B, Bell E, Dolan L. 2010. A basic helix-loop-helix transcription factor controls cell growth and size in root hairs. *Nat Genet* **42:** 264–267. doi:10.1038/ng.529

On the Evolutionary Origins of Land Plant Auxin Biology

John L. Bowman,[1,4] Eduardo Flores Sandoval,[2,4] and Hirotaka Kato[3,4]

[1]School of Biological Science, Monash University, Melbourne, Victoria 3800, Australia
[2]Graduate School of Biostudies, Kyoto University, Kyoto 606-8502, Japan
[3]Graduate School of Science, Kobe University, Kobe 657-8501, Japan
Correspondence: john.bowman@monash.edu

Indole-3-acetic acid, that is, auxin, is a molecule found in a broad phylogenetic distribution of organisms, from bacteria to eukaryotes. In the ancestral land plant auxin was co-opted to be the paramount phytohormone mediating tropic responses and acting as a facilitator of developmental decisions throughout the life cycle. The evolutionary origins of land plant auxin biology genes can now be traced with reasonable clarity. Genes encoding the two enzymes of the land plant auxin biosynthetic pathway arose in the ancestral land plant by a combination of horizontal gene transfer from bacteria and possible neofunctionalization following gene duplication. Components of the auxin transcriptional signaling network have their origins in ancestral alga genes, with gene duplication and neofunctionalization of key domains allowing integration of a portion of the preexisting transcriptional network with auxin. Knowledge of the roles of orthologous genes in extant charophycean algae is lacking, but could illuminate the ancestral functions of both auxin and the co-opted transcriptional network.

In the late 19th century during the birth of modern plant physiology, effort was directed toward understanding the physiochemical nature of "correlations," that is, how different parts of plants interact with one another and with the environment to produce characteristic phenotypes. Julius Sachs proposed that "correlations" were caused by specific chemicals, not just nutritive substances, acting at low concentrations, which we would now recognize as plant hormones (Sachs 1880, 1882). In a number of simple, yet elegant experiments over the next 40 years, it was established that the effects of light and gravity were perceived at the shoot and root tips, respectively, with the stimulus subsequently communicated to distant regions where growth changes occurred (Darwin 1880), that the stimulus could pass through a wound gap filled by gelatin (Boysen Jensen 1910), and that asymmetric application to a cut shoot tip could induce the same growth changes (Paál 1919). Further, this substance could be infused from the tip into agar, which could then provide the growth substance to a decapitated shoot, a key finding that enabled biochemical determination of the growth substance, indole-3-acetic

[4]These authors contributed equally to this work.

Cite this article as *Cold Spring Harb Perspect Biol* doi: 10.1101/cshperspect.a040048

acid (IAA), that we now know as auxin (Went 1926, 1928). While most early experiments focused on angiosperms, in subsequent decades auxin was found to influence bryophyte development in a similar manner suggesting a conserved role throughout land plants (Fitting 1939; Heitz 1940; von Witsch 1940; Rousseau 1950; Halbsguth and Kohlenbach 1953; LaRue and Narayanaswami 1957). An adaptation of Went's experiment whereby chemicals infused into agar blocks applied asymmetrically to decapitated *Avena* plants were used to test whether various sources contained auxin activity. This test suggested that auxin, or related molecules, could be found in a remarkably wide variety of organisms including fungi, bacteria, and even mammalian (human) urine (Kögl and Kostermans 1934; Went and Thimann 1937).

During the early decades of the twentieth century, it became apparent that auxin elicited a wide variety of effects in different tissues and that auxin itself did not have specificity of action. This was first explicitly recognized by Cholodny, who noted that geotropic sensitivity to decapitated roots was restored by placing either another root tip or, alternatively, a shoot tip on the decapitated stump, thus demonstrating that auxins produced in the shoot and root tips are equivalent and that auxin acts in a context-dependent manner (Cholodny 1924). This notion was elaborated upon in subsequent years with Went and Thimann who hypothesized that "… auxins bring about some master-reaction within the cell, the results of which will be determined by the presence and amount of other factors, and by conditions of the experiment" (Went and Thimann 1937). This idea was reiterated in the previous version of this collection where auxin was described as a cellular "currency" that interacts with genetic programs that vary between cell types (Stewart and Nemhauser 2010). Likewise, auxin has been described as an "impetus" absent of specificity (Bennett and Leyser 2014). The discovery of a minimal auxin response system in the liverwort *Marchantia polymorpha* and the myriad of effects of auxin on its growth and development confirmed that there exists no tissue or cell-type specificity of action with respect to auxin signaling components, but rather since the origin of land plants, auxin has acted as a facilitator rather than a determiner of cell fate (Flores-Sandoval et al. 2015; Kato et al. 2015).

Recently, it has become apparent that auxin transcriptional signaling, as defined in land plants, evolved in the ancestral land plant, but other aspects of auxin biology, such as transport, are more ancient. The scope of this review will include (1) the origin and evolution of the molecular components of land plant auxin biology (Fig. 1), (2) the predicted auxin biology of the ancestral land plant, (3) how auxin was integrated into a preexisting transcriptional network, and (4) possible antecedent functions of auxin biology in algal relatives. In short, how did a ubiquitous molecule found in all domains of life become the key developmental hormone mediating tropisms and differentiation in land plants?

Figure 1. The predicted evolutionary origin of auxin biology genes based on genomic and transcriptomic analyses. (*A*) Origin of auxin biology genes are mapped onto the current consensus of Viridiplantae relationships based on phylogenomic studies, with representative genera depicted above. The phylogenetic delineations of Viridiplantae, Streptophytes, and charophytes are indicated. (*B*) Distribution of auxin biology gene families across the major land plant lineages highlights lineage-specific losses and species-specific gains (*). (*C*) Genetic redundancy within auxin biology gene families. Heat map of gene counts for fully sequenced genomes of species. Gene counts are indicated and color coding follows a log10 transformation. Families not previously accounted for in published genome studies were cross-corroborated using tblastx and a reciprocal blast against TAIR10 (www.arabidopsis.org). E-value thresholds for both BLASTS were $<1 \times 10^{-20}$. Sequences were derived from GenBank (www.ncbi.nlm.nih.gov), Phytozome (phytozome.jgi.doe.gov/pz/portal.html), MarpolBase (marchantia.info), and the *Klebsormidium nitens* (http://www.plantmorphogenesis.bio.titech.ac.jp/~algae_genome_project/klebsormidium/) genome website. Gene families are color coded according to their proposed function in auxin biology.

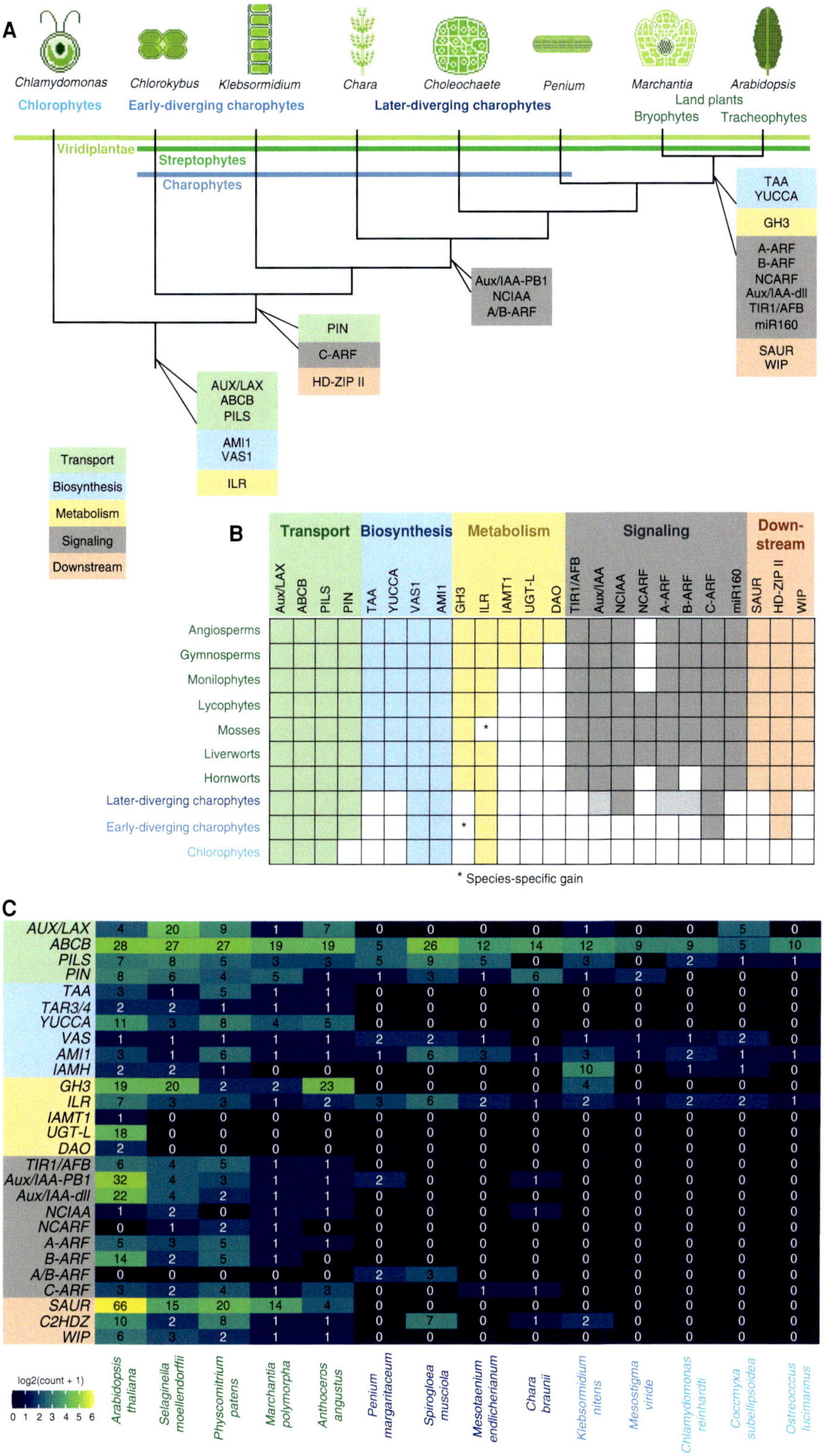

	Arabidopsis thaliana	*Selaginella moellendorffii*	*Physcomitrium patens*	*Marchantia polymorpha*	*Anthoceros angustus*	*Penium margaritaceum*	*Spirogloea musciola*	*Mesotaenium endlichenianum*	*Chara braunii*	*Klebsormidium nitens*	*Mesostigma viride*	*Chlamydomonas reinhardtii*	*Coccomyxa subellipsoidea*	*Ostreococcus lucimarinus*
AUX/LAX	4	20	9	1	7	0	0	0	0	1	0	0	5	0
ABCB	28	27	27	19	19	5	26	12	14	12	9	9	5	10
PILS	7	8	5	3	3	5	9	5	0	3	0	2	1	1
PIN	8	6	4	5	1	1	3	1	6	1	2	0	0	0
TAA	3	1	5	1	1	0	0	0	0	0	0	0	0	0
TAR3/4	2	2	1	1	1	0	0	0	0	0	0	0	0	0
YUCCA	11	3	8	4	5	0	0	0	0	0	0	0	0	0
VAS	1	1	1	1	1	2	2	1	0	1	1	1	2	0
AMI1	3	1	6	1	1	1	6	3	1	3	1	2	1	1
IAMH	2	2	1	0	0	0	0	0	0	10	0	1	1	0
GH3	19	20	2	2	23	0	0	0	0	4	0	0	0	0
ILR	7	3	3	1	2	3	6	2	1	2	1	2	2	1
IAMT1	1	0	0	0	0	0	0	0	0	0	0	0	0	0
UGT-L	18	0	0	0	0	0	0	0	0	0	0	0	0	0
DAO	2	0	0	0	0	0	0	0	0	0	0	0	0	0
TIR1/AFB	6	4	5	1	1	0	0	0	0	0	0	0	0	0
Aux/IAA-PB1	32	4	3	1	1	2	0	0	1	0	0	0	0	0
Aux/IAA-dII	22	4	2	1	1	0	0	0	0	0	0	0	0	0
NCIAA	1	2	0	1	1	0	0	0	1	0	0	0	0	0
NCARF	0	1	2	1	0	0	0	0	0	0	0	0	0	0
A-ARF	5	3	5	1	1	0	0	0	0	0	0	0	0	0
B-ARF	14	2	5	1	0	0	0	0	0	0	0	0	0	0
A/B-ARF	0	0	0	0	0	2	3	0	0	0	0	0	0	0
C-ARF	3	2	4	1	3	0	0	1	1	0	0	0	0	0
SAUR	66	15	20	14	4	0	0	0	0	0	0	0	0	0
C2HDZ	10	2	8	1	1	0	7	0	1	2	0	0	0	0
WIP	6	3	2	1	1	0	0	0	0	0	0	0	0	0

Figure 1. (*See facing page for legend.*)

ORIGIN AND EVOLUTION OF THE COMPONENTS

Auxin Biosynthesis

IAA is found widely throughout both bacteria and eukaryotes (Žižková et al. 2017; Morffy and Strader 2020). In bacteria, auxin is synthesized by plant pathogens or symbionts via independently evolved enzymes (e.g., iaaM and iaaH of *Agrobacterium*) that convert tryptophan (trp) into auxin via a variety of intermediaries (Kunkel and Harper 2018). In eukaryotes, biologically active auxins can be found in fungi (Chanclud and Morel 2016), brown algae (Basu et al. 2002; Sun et al. 2004; Le Bail et al. 2010), red algae (Ashen et al. 1999), and in Viridiplantae lineages, including Chlorophytes (Stirk et al. 2013; Žižková et al. 2017), charophycean algae (Sztein et al. 2000; Žižková et al. 2017), and embryophytes (Cooke et al. 2002). Whereas there is no evidence supporting an ancestral eukaryotic auxin biosynthetic pathway, that auxin triggers physiological or morphological changes in these eukaryotic lineages suggests independent recruitment of auxin for diverse biological functions (Wood and Berliner 1979; Vance 1987; Yokoya and Handro 1996; Basu et al. 2002; Jin et al. 2008; Le Bail et al. 2010; Ohtaka et al. 2017; Bogaert et al. 2019).

The genetic basis of auxin biosynthesis has been elucidated in land plants (Zhao et al. 2001; Stepanova et al. 2008; Tao et al. 2008), with the primary pathway being tryptophan-dependent: TRYPTOPHAN AMINOTRANSFERASEs (TAAs) convert trp to indole-3-pyruvic acid (IPyA) and subsequently IPyA is converted into IAA by YUCCA flavin monooxygenases (Mashiguchi et al. 2011; Stepanova et al. 2011). The origins of these two enzymes, via gene duplication of an ancestral algal gene and horizontal gene transfer from bacteria, respectively, provides compelling evidence that this auxin biosynthetic pathway is a land plant innovation (Fig. 1). The predicted control of cellular auxin homeostasis of an ancestral charophyte and the ancestral land plant are presented in Figure 2 and will be referred to in the following discussion.

Comparative genomics indicates the presence of a TAA/YUCCA biosynthetic pathway in the land plant ancestor (Bowman et al. 2017; Poulet and Kriechbaumer 2017; Li et al. 2018) with both TAA and YUCCA genes required for auxin production in bryophytes (Eklund et al. 2015) and Tracheophyta (Zhao et al. 2001; Stepanova et al. 2008; Tao et al. 2008). Notably, loss of this pathway in *M. polymorpha* results in complete loss of developmental patterning (Eklund et al. 2015) and the same is likely the case in the flowering plant *Arabidopsis* (Cheng et al. 2007), indicating the assembly of this pathway was critical for the evolution of land plants. Whether IPyA-synthesizing TAA genes exist in charophytes is equivocal as phylogenetic analyses place charophycean TAA-related genes as sister to two monophyletic land plant lineages (Wang et al. 2014, 2016; Romani 2017). One lineage contains embryophyte TAA orthologs encoding enzymes catalyzing the first step of auxin biosynthesis. The second lineage contains orthologs of *Arabidopsis TAR3* and *TAR4*, whose function(s) are as-yet uncharacterized. These genes encode a conserved EGF-Alliinase protein motif that has been repeatedly lost in the TAA lineage genes (Turnaev et al. 2015; Wang et al. 2016; Romani 2017). Charophytes experienced multiple lineage-specific losses of TAA-related (TAR) genes, except in *Klebsormidium* (Hori et al. 2014; Nishiyama et al. 2018; Cheng et al. 2019; Jiao et al. 2020), suggesting that this gene family is not bound by strong selective pressures in charophytes. Functional characterization of charophyte TAR enzymes and cross-lineage complementation experiments could clarify the origin of the embryophyte TAA genes. Pertinent questions are whether charophyte *TAR* mutants are auxin-deficient and whether algal *TAR* genes complement embryophyte mutant TAA alleles.

Although YUCCA flavin monooxygenases were proposed to exist in Chlorophytes (Khasin et al. 2018), charophytes (Wang et al. 2014), and brown algae (Bogaert et al. 2019), exhaustive phylogenetic (Yue et al. 2012, 2014; Bowman et al. 2017; Turnaev et al. 2020), and sequence similarity network analysis (Dai et al. 2013) indicate YUCCA flavin monooxygenases are a land plant innovation acquired by horizontal

Cite this article as *Cold Spring Harb Perspect Biol* doi: 10.1101/cshperspect.a040048

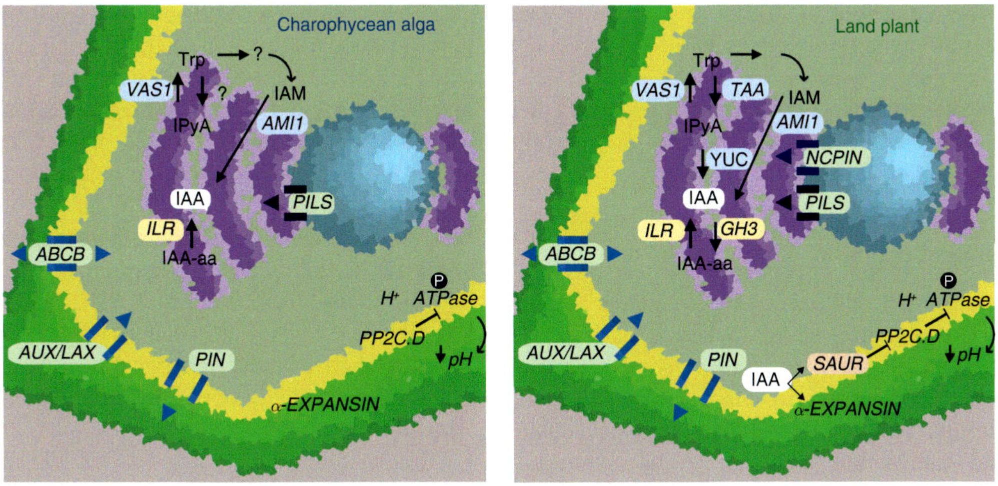

Figure 2. Control of auxin homeostasis in charophycean algae and land plants. Biosynthesis and metabolism: In land plants, tryptophan (trp) is the substrate for TRYPTOPHAN AMINOTRANSFERASE (TAA), which produces indole-3-pyruvic acid (IPyA) that is subsequently converted to indole-3-acetic acid (IAA) by YUCCA enzymatic activity. The conversion of trp to IPyA could be mediated by a TAA/TAR34 ortholog in charophycean algae but this needs experimental corroboration. VAS1 enzymes mediating the conversion of IPyA into Trp are conserved across Streptophytes. In land plants, GH3 enzymes conjugate free IAA into amino acid conjugates (IAA-aa) and ILRs conversely hydrolyze IAA-aa back into free IAA. Whereas GH3 genes have not been found in charophytes, ILR enzymes are conserved across Streptophytes, suggesting hydrolysis of IAA-amino acid conjugates into auxin may occur in charophycean algae. Amidases that convert IAM into IAA are conserved across Streptophytes, suggestive of a possible older auxin metabolic pathway. Transport: Auxin transporters of the PIN, PILS, ABCB, and AUX/LAX protein families were present in the common ancestor of Streptophytes, implying intracellular (PILS) and intercellular (AUX/LAX, ABCB, PIN) active transport was present in the charophycean algae ancestor. In land plants, NCPINs channel auxin into the endoplasmic reticulum (ER), a role perhaps filled by PILS proteins in algae. Cell expansion: Acid growth modulated by H^+ ATPases, PP2C-D phosphatases and EXPANSINs is present in charophycean algal lineages (e.g., *Nitella*). Given the absence of SAUR genes in charophycean algae, acid growth may not have been dependent on auxin levels in the Streptophyte ancestor. In land plants, acid growth is dependent on auxin, as IAA induces both SAUR and α-EXPANSIN transcription. SAURs inhibit PP2C-D phosphatases, allowing phosphorylation of H^+ ATPases and acidification of the apoplastic environment. Acid growth in turn is dependent on the cell wall loosening activity of α-EXPANSINs. In cell diagrams, the cell wall (green), cell membrane (yellow), ER (purple), and nucleus (blue) are color coded.

gene transfer from bacteria. Thus, if charophyte TAR genes can produce IPyA, IPyA may have had an IAA-independent function in the algal ancestor. VAS1 enzymes, which convert IPyA into Trp (Zheng et al. 2013; Matsuo et al. 2020) are conserved across Streptophytes, suggesting IPyA homeostatic levels were actively regulated in the Streptophyte ancestor (Fig. 2).

In at least some embryophytes, auxin is produced from other trp-dependent derivatives (for review, see Tivendale et al. 2014; Morffy and Strader 2020) such as indole-3-acetaldoxime (IAOx), indole-3-acetonitrile (IAN), indole-3-acetamide (IAM), and tryptamine (TAM). AMIDASEs (AMI) and IAM hydrolases (IAMH) that convert IAM to auxin, are found across Streptophytes, hinting at the presence of an ancestral biosynthetic pathway (Fig. 1). This is further supported by the presence of IAM in chlorophytes and charophycean algae (Stirk et al. 2013; Žižková et al. 2017), although enzymes converting Trp to IAM and whether reaction intermediaries exist are unknown (Fig. 2). Finally, auxin may be produced in Streptophytic algae via trp-independent derivatives (Wright et al. 1991; Normanly et al. 1993) such as indole, indole-3-glycerol phosphate (Ouyang et al. 2000; Ljung 2013) or indole-3-butyric acid (Strader et al. 2011). However, these pathways contribute little to the developmental and tropic roles of auxin in embryophytes and their evolutionary origins are largely unexplored.

Auxin Metabolism

In embryophytes, auxin exists in a "free" active form or can be converted, reversibly or irreversibly, into inactive conjugates. In angiosperms, the major natural auxin, IAA, is inactivated by conjugation with amino acids or sugars, oxidation, and methylation (Fig. 3A). Amide conjugation with various amino acids (IAA-aa) is catalyzed by members of the Streptophyte-specific GRETCHEN HAGEN3 (GH3) protein family, which is divided into three subgroups (Staswick et al. 2005). In angiosperms, group II members play a major role in auxin conjugation while those of groups I and III utilize jasmonic acid (JA) or benzoate as substrates, respectively (Staswick and Tiryaki 2004; Staswick et al. 2005; Okrent et al. 2009; Chiu et al. 2018). IAA-aa are also present in various bryophytes, although the amounts and conversion rates are lower than in angiosperms (Sztein et al. 1999; Záveská Drábková et al. 2015; Kaneko et al. 2020). The moss *Physcomitrium patens* has two group I GH3 genes (Bierfreund et al. 2004) whose encoded proteins can conjugate various amino acids to both IAA and JA in vitro (Ludwig-Müller et al. 2009). Double mutants exhibit increased auxin sensitivity with higher free-IAA and reduced IAA-aa levels (Ludwig-Müller et al. 2009). In algal species, the amount of IAA-Asp detected was almost at the detection limit (Žižková et al. 2017) and conversion rates are even slower than in bryophytes (Sztein et al. 2000). Among charophytes, only *Klebsormidium nitens* has known GH3 homologs; however, phylogenetic analysis suggests the *K. nitens* ancestor independently obtained *GH3* genes by horizontal gene transfer from bacteria (Fig. 3B).

Among amino acid conjugates, IAA-Ala and IAA-Leu inhibit root elongation similar to IAA and are thus hydrolyzable in *Arabidopsis* (Bartel and Fink 1995; LeClere et al. 2002; Rampey et al. 2004), suggesting their contribution as storage forms. Hydrolysis of IAA-aa is mediated by M20D peptidase family members IAA–LEUCINE RESISTANT1 (ILR1) and its homologs, ILR-LIKEs, and IAA-ALANINE RESISTANT3 (Bartel and Fink 1995; Davies et al. 1999; LeClere et al. 2002). ILR1 homologs are found broadly (Fig. 1), but not ubiquitously, throughout Streptophytes and even cyanobacteria (Campanella et al. 2014, 2018, 2019). Three *P. patens* genes have activity against a variety of auxin-aa conjugates (Campanella et al. 2019), but their higher sequence similarity and codon usage to bacterial

Figure 3. Indole-3-acetic acid (IAA) is modified into several inactive forms in land plants. (*A*) IAA is inactivated by conjugation with amino acids (IAA-aa: blue) or sugars (IAA-glc: green), oxidation (oxIAA: red), and methylation (IAA-Me: magenta), and reactivated by enzymes indicated. The enzyme box color indicates conservation among charophycean algae + land plants (blue), land plants (green), seed plants (orange), or angiosperms (red). (*B*) Phylogenetic reconstruction of the GH3 protein family. The tree highlights two factors to consider when tracing the evolutionary history of genes. First, similar sequences may not be orthologous, but rather derived ultimately from independent sources (e.g., the *Klebsormidium* and land plant *GH3* genes). Second, classifications based on analyses of sequences from only derived taxa (e.g., GH3 classes based on angiosperm sequences may fail to recognize phylogenetic diversity found outside the derived taxa). Protein sequences were collected from GenBank, Phytozome, and the *Klebsormidium nitens* genome website using blastp search with threshold E-value = 1.0×10^{-20}, and aligned using the MUSCLE program (Edgar 2004) implemented in MEGA X (Kumar et al. 2018). Alignment positions with more than 50% gaps were removed using the Phyutility program (version 2.2.6; code.google.com/p/phyutility/downloads/list) and then sequences with more than 30% gap in the alignment were removed. The phylogenetic tree was constructed in MEGA X using an LG + G model with bootstrap replicates = 500. Orange circles indicate bootstrap values more than 0.75. Black: Tracheophytes (*Arabidopsis thaliana* [Arth], *Amborella trichopoda* [Amtr], *Selaginella moerrendorffii* [Semo]), green: bryophytes (*Anthoceros angustus* [Anan], *Marchantia polymorpha* [Mapo], *Physcomitrium patens* [Phpa], *Sphagnum fallax* [Spfa]), blue: *K. nitens* (Klni), purple: bacteria. Groups I to III are based on data in Staswick et al. (2005).

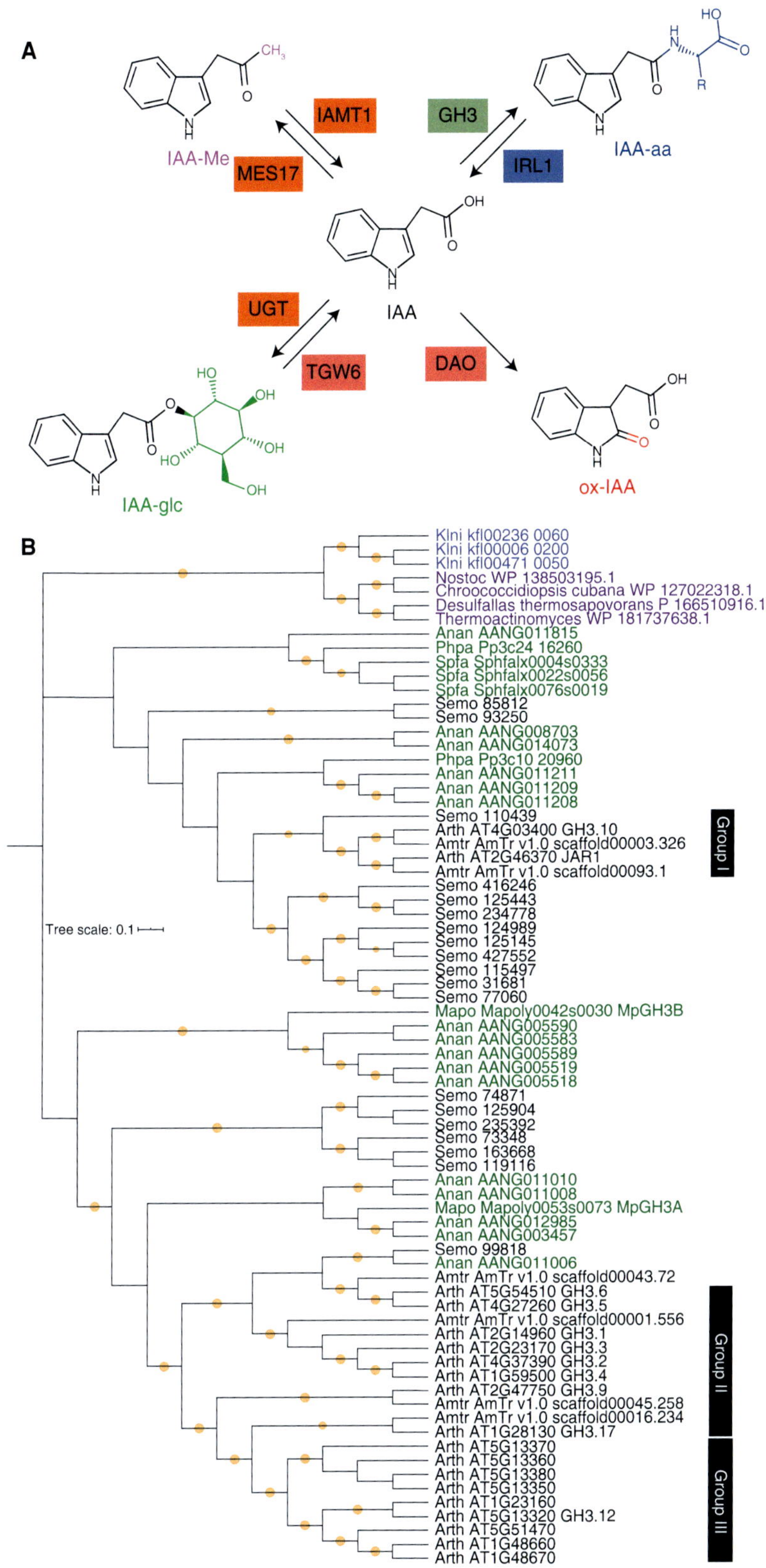

Figure 3. (*See facing page for legend.*)

genes rather than Streptophyte genes, and absence of homologs in 42 other mosses suggest a loss of the plant *ILR1* gene in the ancestral moss and independent acquisition in *P. patens* by horizontal gene transfer. *M. polymorpha* has a single plant-like ILR1 homolog (Mp*ILR1*), but MpILR1 has weak hydrolase activity in comparison with seed plant ILRs (Campanella et al. 2018). Notably, MpILR1 does not hydrolyze either of the two *M. polymorpha* auxin conjugates found under normal conditions, IAA-Val and IAA-Gly (Sztein et al. 1999; Campanella et al. 2018), suggesting MpILR1 may be disconnected with active regulation of free IAA. Thus, while the ancestral land plant likely possessed GH3-mediated amino acid conjugation (Fig. 2), its role might have been primarily inactivation rather than maintaining an accessible auxin pool. This scenario is consistent with earlier ideas of a progressively increasing complexity in the control of auxin homeostasis within land plants (Sztein et al. 2000; Cooke et al. 2002).

Other IAA conjugates have been characterized in angiosperms, and while such conjugates are also found in bryophytes and algae, genes encoding the relevant angiosperm enzymes are not conserved among earlier diverging lineages (Fig. 3). For example, IAA inactivation to 2-oxindole-3-acetic acid (oxIAA) is a major IAA oxidation pathway in seed plants (Östin et al. 1998) and is mediated in angiosperms by DIOXYGENASE FOR AUXIN OXIDATION (DAO) (Zhao et al. 2013; Porco et al. 2016; Zhang et al. 2016). Whereas various bryophytes accumulate oxIAA, and oxIAA has been detected in various algal lineages (Záveská Drábková et al. 2015; Žižková et al. 2017), phylogenetic analysis indicates DAO homologs are limited to angiosperms (Takehara et al. 2020). Likewise, ester-linked IAA–sugar conjugates (e.g., IAA-glucose) have been identified in angiosperms and bryophytes, although amounts in bryophytes are limited (Korasick et al. 2013; Záveská Drábková et al. 2015). Production of IAA-glucose in angiosperms is mediated by group L UDP glucosyltransferases (UGTs) (Jackson et al. 2001; Li et al. 2001; Jin et al. 2013), which are found only in seed plants (Wilson and Tian 2019). Enzymes hydrolyzing IAA-glucose have been identified in rice and wheat, but related genes appear monocot specific. Finally, methylated IAA (MeIAA), a presumed storage form, is detected in angiosperms (Korasick et al. 2013). In *Arabidopsis*, MeIAA is produced by the SABATH family enzyme IAA CARBOXYL-METHYLTRANSFERASE 1 (IAMT1) (Zubieta et al. 2003; Qin et al. 2005), with IAMT1 orthologs limited to seed plants. As most studies have approached the question from an angiosperm perspective, investigations in bryophytes are required to ascertain the involvement of these and other possible conjugates in auxin homeostasis.

Auxin Transport

Polar auxin transport (PAT)—the unidirectional intercellular flow of auxin between cells—has been reported in Tracheophytes (Okada et al. 1991), mosses (Poli et al. 2003), liverworts (Maravolo 1976), and *Chara* (Klämbt et al. 1992; Boot et al. 2012). Auxin concentrations resulting from PAT is contingent on the activity of influx, efflux, and intracellular carriers, with PAT-generated auxin gradients translated into transcriptional changes that determine many land plant morphogenetic processes.

The AUX/LAX protein family constitutes the only dedicated auxin influx carrier yet characterized in plants and was originally identified by auxin resistance of *aux* mutants (Maher and Martindale 1980). AUX/LAX proteins are plasma membrane (PM) localized (Swarup et al. 2004) and resemble amino acid permeases (Bennett et al. 1996). AUX/LAX homologs are present throughout the Viridiplantae (Figs. 1 and 2). Pending functional analyses, this suggests that the unicellular Viridiplantae ancestor had the potential to actively import environmental auxin and that multicellular charophytes and embryophytes retained and co-opted AUX/LAX proteins to regulate auxin flow within the plant body.

Streptophyte-specific PIN proteins act as auxin efflux carriers and have been classified into either canonical or noncanonical by virtue of their central hydrophilic domain or loop length, although this is a functional not a phylogenetic distinction (Petrásek et al. 2006;

Bennett et al. 2014; Vosolsobe et al. 2020). Canonical PINs, with long loops, are PM localized, while noncanonical PINs, with short loops, are localized to the endoplasmic reticulum (ER) membrane (Ding et al. 2012). Canonical PIN proteins may be polarized to reside on only specific cell faces, providing the basis for directional auxin transport (Naramoto 2017; Béziat and Kleine-Vehn 2018). Embryophyte noncanonical PINs evolved multiple times from an ancestral canonical PIN, and given their capacity to transport auxin analogs or conjugates, they are proposed to act as homeostatic modulators of auxin metabolites (Ding et al. 2012; Bennett et al. 2014). Whereas canonical PINs have been considered the ancestral type of land plant PIN (Bennett et al. 2014), phylogenetic analyses with charophyte sequences as outgroups are equivocal. Charophyte PIN proteins are not simply classified as canonical or noncanonical, but a *Klebsormidium* PIN can act as an auxin efflux carrier when heterologously expressed in angiosperms, suggesting that PIN-mediated PAT occurred in a Streptophyte ancestor (Skokan et al. 2019). An independent expansion of *PIN* gene family members in *Chara braunii* (Nishiyama et al. 2018) is correlated with the presence of PAT in other *Chara* species (Klämbt et al. 1992; Boot et al. 2012).

The ancient PIN-LIKES (PILS) protein family, likely present in the unicellular Viridiplantae ancestor (Fig. 1), is distantly related to the PIN family, sharing a transmembrane domain topology but with low amino acid sequence conservation (Barbez et al. 2012; Vosolsobe et al. 2020). PILS are proposed to canalize auxin into the ER (Fig. 2), thus dampening nuclear responses (Béziat et al. 2017; Feraru et al. 2019).

The ATP-binding cassette subfamily-B/multidrug resistant/P-glycoprotein (ABCB/MDR/PGP) subfamily B (ABCB) proteins also act as auxin efflux carriers in angiosperms (Noh et al. 2001; Blakeslee et al. 2007; Wu et al. 2010), with some also showing facultative auxin influx capabilities (Kamimoto et al. 2012). The ABCB subfamily is ancient, with members throughout Viridiplantae, with ABCB proteins having diverse functions, and thus functional studies in algae are required to authenticate any auxin-transport capabilities.

In summary, protein families with the capacity to import (AUX/LAX), export (ABCB), and sequester (PILS) auxin were present in the Viridiplantae ancestor, suggesting that ancestral unicellular green algae may have had the capacity to respond to fluctuating concentrations of auxin or auxin-like compounds in their environment. The evolution of these three protein families could have been an adaptation to excess auxin in the environment, if auxin was used as a nutrient or a quorum-sensing signal in unicellular algae (Vosolsobe et al. 2020). In contrast, PIN-mediated auxin efflux originated in the freshwater Streptophyte ancestor, with PAT evolving before the evolution of the canonical land plant auxin biosynthesis and signaling pathways, perhaps presaging the evolution of morphological complexity in this lineage.

Auxin Transcriptional Signaling

The current land plant auxin signaling model (Fig. 4) stems from four decades of experimental work in flowering plants, although it is an ancestral land plant innovation derived from both ancient and neofunctionalized genes. The conservation of the auxin signaling gene families across land plants has been corroborated by multiple genomic studies (Rensing et al. 2008; Banks et al. 2011; De Smet et al. 2011; Finet et al. 2013; Bowman et al. 2017; Mutte et al. 2018; Li et al. 2020; Zhang et al. 2020) and experimental studies in *P. patens* (Prigge et al. 2010; Lavy et al. 2016) and *M. polymorpha* (Flores-Sandoval et al. 2015; Kato et al. 2015, 2017). The *M. polymorpha* genome harbors single copies of each of the auxin signaling genes implying that this is the likely ancestral condition in land plants (Flores-Sandoval et al. 2015; Kato et al. 2015; Bowman et al. 2017). Most auxin signaling components have undergone diversification in both Tracheophytes and mosses, but have remained mostly single copy in liverworts and hornworts (Bowman et al. 2017; Li et al. 2020; Zhang et al. 2020).

Ultimately, changes in transcriptional patterns depend upon AUXIN RESPONSE FAC-

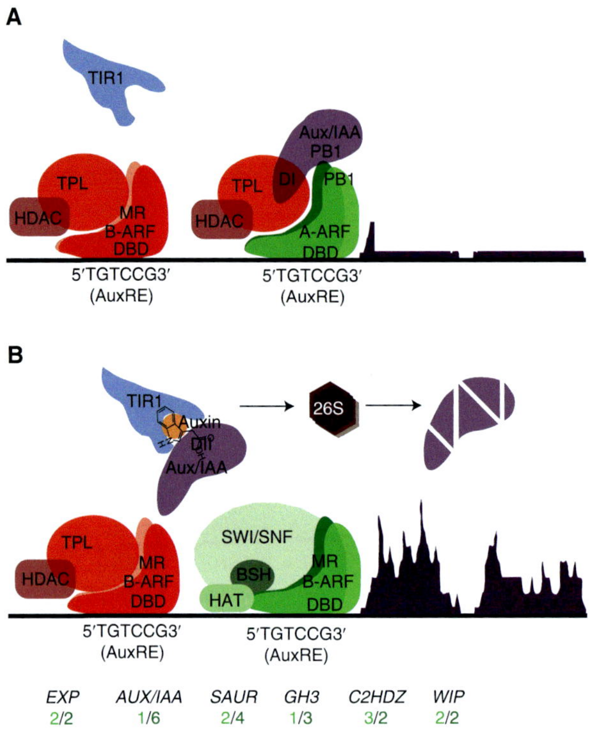

Figure 4. Nuclear auxin signaling in land plants. (*A*) In the absence of nuclear auxin, AUXIN/INDOLE-3-ACETIC ACID (AUX/IAA) repressor proteins interact with A-ARF proteins via a shared carboxy-terminal protein–protein interaction domain called Phox/Bem1 (PB1) (Kim et al. 1997; Korasick et al. 2014; Nanao et al. 2014) and recruit members of the TOPLESS (TPL) family of corepressors via an EAR-like domain I (DI). TPL proteins, in turn, recruit histone deacetylases forming a repressive complex (Tiwari et al. 2001, 2004; Szemenyei et al. 2008). B-ARF proteins also interact with TOPLESS corepressors and can compete with A-ARFs for Auxin Response Elements (AuxREs) in the promoters of auxin-inducible genes (Ulmasov et al. 1997; Kato et al. 2020). (*B*) Auxin is perceived in the nucleus by TRANSPORT INHIBITOR RESPONSE1/AUXIN SIGNALING F-BOX (TIR1/AFB) proteins. TIR1/AFBs are ubiquitin ligases by virtue of their amino-terminal F-box domain and recognize auxin in the pockets formed in their leucine-rich repeats (LRRs). Auxin "glues" TIR1 to the degron domain II (DII) of AUX/IAA proteins and ubiquitinates AUX/IAAs for degradation in the 26S proteasome, releasing A-ARF proteins from the TPL complex (Gray et al. 2001; Dharmasiri et al. 2005; Kepinski and Leyser 2005; Tan et al. 2007). This allows activation of downstream targets via recruitment of SWI/SNF ATPases, which in turn recruit BUSHY (BSH) proteins allowing HISTONE ACETYL TRANSFERASES (HATs) to promote an open chromatin state and further recruitment of activating transcription factors (Gray et al. 2001; Dharmasiri et al. 2005; Kepinski and Leyser 2005; Wu et al. 2015). In this scenario, A-ARF targets can still be inhibited by B-ARFs when both are coexpressed. Putative ancestral land plant targets of the auxin signaling pathway are shown. The number of species of bryophytes (light green) and Tracheophytes (dark green) where at least one member of each gene family is auxin-inducible is indicated (Abel and Theologis 1996; Catalá et al. 2000; Esmon et al. 2006; Lavy et al. 2016; Kato et al. 2017; Mutte et al. 2018).

TORs (ARFs), only some of which are integrated into the auxin signaling system (Fig. 5). The ancestral land plant possessed three *ARF* genes, A, B, and C (Finet et al. 2013), and a noncanonical ARF (NCARF) derived via gene duplication of the A-ARF and lacking a DNA-binding domain (DBD) (Flores-Sandoval et al. 2018; Mutte et al. 2018). The ancestral land plant harbored two evolutionarily related genes encoding the Aux/IAA repressor of A-ARFs and a noncanonical Aux/IAA (NCIAA), both lacking a DBD (Flores-Sandoval et al. 2018; Mutte et al.

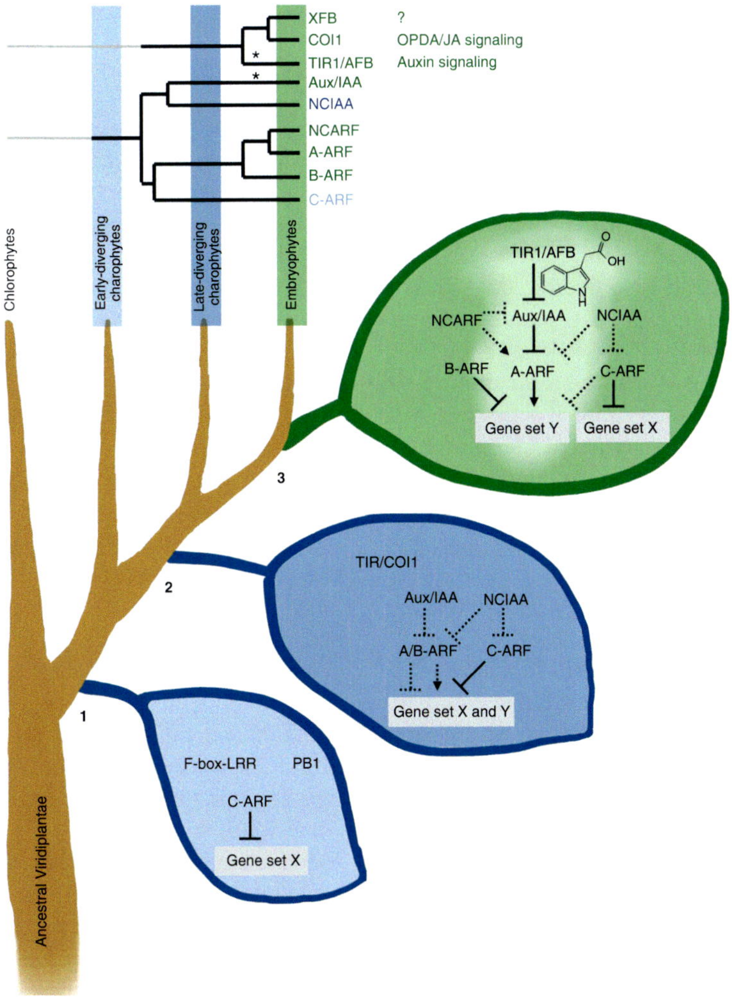

Figure 5. Evolutionary scenario of auxin signaling assembly. At the advent of Streptophyte evolution (**1**) C-type ARF transcription factors were established by combining preexisting small protein domains in ancestral Streptophyte. Prior to this time, proteins containing an F-box with a leucine-rich repeat (LRR) and others with PB1 domains were present but not functionally connected with each other. In the common ancestor of later-diverging charophytes and embryophytes (**2**; see Fig. 1 for phylogenetic delineations), an A/B-type ARF evolved, via gene duplication of the C-type ARF followed by neofunctionalization. The A/B-ARF and C-ARF transcription factors might have shared the target genes. In addition, two types of Aux/IAA-related PB1-containing proteins lacking the B3 DNA-binding domain (DBD) evolved, also likely via a gene duplication of the ancestral C-ARF followed by DBD loss and a further gene duplication. These proteins could act as negative regulators of the C-ARF and A/B-ARF proteins via their ability to heterodimerize but not bind to DNA. Thus, a complex network of related transcription factors with multiple levels of cross-regulation had evolved in the common ancestor of later-diverging charophytes and embryophytes. However, while the ancestral protein that would give rise to TIR1/AFB and COI1 was present, it is likely that it did not perceive either auxin or OPDA/JA, and its relationship, if any, to the ARF transcriptional network is unknown. After the divergence of the ancestral land plant from the most closely related extant charophycean algal lineage (**3**), the *A/B-ARF* gene gave rise to activator (A-ARF) and repressor (B-ARF) genes via gene duplication. In addition, a gene duplication of the A-ARF produced a paralog, NCARF, that subsequently lost its DBD. Gene duplication also generated dedicated receptors for auxin (TIR1/AFB) and OPDA/JA (COI1), along with a third as-yet-uncharacterized F-box protein (XFB). The TIR1/AFB protein became integrated with the Aux/IAA-ARF transcriptional pathway. This was accomplished by the acquisition of dII in an Aux/IAA protein and amino acid changes in the TIR1/AFB protein to form a binding pocket between the two proteins for auxin (noted by *). In addition, distinct target specification differences between A/B-ARF and C-ARF proteins progressed, perhaps driven by the auxin-dependent versus auxin-independent action of the two classes of ARF. Thus, in the ancestral land plant, the ARF transcriptional network was expanded with the activity of a core part of the network now integrated with auxin, while a portion of the network, including the ancestral C-ARF, remained auxin independent.

2018). ARFs possess an amino-terminal B3-type DBD (Ulmasov et al. 1997) that mediates dimerization and binding to two palindromically positioned AuxREs (Boer et al. 2014) and a carboxy-terminal Phox/Bem1 (PB1) protein–protein interaction domain (Kim et al. 1997; Korasick et al. 2014; Nanao et al. 2014). In between, the middle region (MR) confers activator or repressor identity (Ulmasov et al. 1999; Kato et al. 2015, 2020). A-ARFs have a glutamine-rich MR, function as transcriptional activators, and are generally subject to Aux/IAA regulation (Fig. 4). B- and C-ARFs are transcriptional repressors (Ulmasov et al. 1999; Kato et al. 2015, 2020) able to form complexes with TPL corepressors (Martin-Arevalillo et al. 2017, 2019; Choi et al. 2018) and act auxin independently (Piya et al. 2014; Flores-Sandoval et al. 2018; Mutte et al. 2018; Kato et al. 2020). B-ARFs can inhibit A-ARF activity via target competition, as their B3 domains are interchangeable (Kato et al. 2020). Hornworts sequenced to date lack B-ARFs (Li et al. 2020; Zhang et al. 2020), suggesting unknown mechanisms are responsible for auxin-independent A-ARF target repression in this lineage. NCARFs function as positive regulators of auxin responses in *M. polymorpha* but these genes have been lost in hornworts and Euphyllophytes (Mutte et al. 2018). NCIAAs are found across land plants, preferentially interacting with C-ARFs (Piya et al. 2014) with protein accumulation in response to auxin mediated via phosphorylation by MITOGEN-ACTIVATED PROTEIN KINASE 14 (MPK14) (Lv et al. 2020). Thus, the ancestral land plant possessed a complex network of ARF and ARF-related proteins (Fig. 5), with only the Aux/IAA and A-ARF proteins integrated with auxin signaling, with the remainder being auxin-independent (C-ARF) or indirectly associated (NCARF, B-ARF).

Across Streptophytes, conservation of auxin signaling genes is fragmented (Hori et al. 2014; Wang et al. 2020) and comparative genomic studies suggest that the complete canonical pathway only exists in land plants (Bowman et al. 2017; Mutte et al. 2018; Nishiyama et al. 2018; Cheng et al. 2019; Jiao et al. 2020). Whereas the Aux/IAA and NCIAA lineages have not been found in the genomes of early diverging charophytes (e.g., *Mesostigma*, *Chlorokybus*, and *Klebsormidium*) (Hori et al. 2014; Wang et al. 2020), ARFs have been found in *Chlorokybus* and *Mesostigma*, suggesting that the Aux/IAA and NCIAA lineages could have evolved from ancestral ARFs via loss of the B3-DBD domain and it might be presumed that they act as negative regulators in the charophyte transcriptional network (Fig. 5). Despite the early origin of Aux/IAA proteins, the acquisition of functional domains I and II in the ancestral land plant was an important neofunctionalization allowing part of the network to become auxin modulated (De Smet et al. 2011; Flores-Sandoval et al. 2018; Mutte et al. 2018; Nishiyama et al. 2018; Jiao et al. 2020). A further key innovation occurred in the ancestral *TIR1* gene. Charophyte genomes have F-box orthologs equally related to TIR1 and the COI1 OPDA/JA receptor proteins, and although these orthologs are predicted to form three-dimensional structures similar to either TIR1 or COI1, they lack key residues that allow binding to either auxin or OPDA/JA (Bowman et al. 2017; Monte et al. 2018; Mutte et al. 2018). Thus, coevolution of Aux/IAA domain II and specific amino acids in TIR1 was the key innovation in the evolution of auxin as a molecular glue between the two proteins and the integration of auxin into a preexisting transcriptional network.

Although ARFs have undergone multiple lineage-specific losses in charophytes (Hori et al. 2014), two types of ARFs (A/B and C-ARFs) were present in the common ancestor of later diverging charophytes and embryophytes (Flores-Sandoval et al. 2018; Mutte et al. 2018), while the Streptophyte ancestor only had C-ARFs (Martin-Arevalillo et al. 2019; Wang et al. 2020), suggesting C-ARFs may be ancestral. A *Chlorokybus* C-ARF can bind land plant AuxREs and interact with TPL in vitro, suggesting ARFs originated as transcriptional repressors (Martin-Arevalillo et al. 2019). Charophyte C-ARFs lack miR160-binding sites, suggesting the miR160/C-ARF regulatory loop is a land plant innovation (Mallory et al. 2005; Flores-Sandoval et al. 2018). Known charophyte A/B ARFs are sister to land plant A- and B-ARFs, whose divergence can be dated to the ancestral land plant, and their identity as activators or repressors is uncertain.

The evolution of the auxin signaling pathway is the product of the rewiring of a preexisting transcriptional network to respond to a novel ligand, providing an outstanding example of exaptation within the green lineage of life. Understanding how this novel pathway was used to pattern key features of the ancestral land plant still requires further functional studies in charophytes as well in Tracheophytes with free-living haploid and diploid generations. To date, tropisms, cell differentiation, and cell expansion (Fig. 2) represent likely biological processes controlled by auxin in the common ancestor of all land plants. While not explored here, the parallel evolution of the auxin and OPDA/JA signaling pathways from a common ancestral F-box protein, and potentially using common ancestral GH3-conjugating enzymes, deserves further investigation.

Downstream of Auxin Signaling

Whereas auxin transcriptional signaling is conserved throughout land plants, the question of whether there are universal, and hence ancestral targets, has only been addressed recently. The land plant auxin signaling pathway attains a steady state through multiple negative feedback loops involving A-ARFs inactivation via auxin-dependent AUX/IAA transcription (Abel et al. 1995; Abel and Theologis 1996) and the reduction of free IAA levels via down-regulation of YUCCA biosynthetic enzymes (Mutte et al. 2018) and up-regulation of GH3-conjugating enzymes (Hagen et al. 1991; Abel and Theologis 1996). While YUCCA down-regulation is consistent across embryophytes, auxin-dependent AUX/IAA transcription and GH3 activation have not been detected across all bryophyte lineages (Mutte et al. 2018).

Genes encoding two transcriptional repressors (Steindler et al. 1999; Jones and Dolan 2017; Ren et al. 2018), the CLASS-II HOMEODOMAIN LEUCINE ZIPPER (C2HDZ) and the WIP-DOMAIN C2H2 zinc fingers, are consistently up-regulated by auxin across land plants (Sawa et al. 2002; Crawford et al. 2015; Mutte et al. 2018). C2HDZ genes as targets evolved in the ancestral land plant as charophyte orthologs (Romani et al. 2018) are not auxin up-regulated (Mutte et al. 2018). It is unknown whether these transcription factors regulate auxin signaling through feedback effects or, alternatively, control conserved developmental or physiological responses.

SAUR (SMALL AUXIN UP-REGULATED RNA) genes are a distinct auxin-induced gene class (McClure et al. 1989) whose products do not influence feedback regulation, but rather promote cell expansion via acid growth (Chae et al. 2012; Spartz et al. 2017). Auxin-mediated SAUR gene up-regulation is observed widely in embryophytes, although it may have been lost in liverworts (Mutte et al. 2018). SAUR proteins promote acidification of the apoplast via inhibition of D-type PPC (PP2C.D) phosphatases, which in turn modulate H^+ ATPases (Cosgrove 1998; Spartz et al. 2014; Stortenbeker and Bemer 2019; Wong et al. 2019). α-EXPANSINs are another auxin-inducible protein family (Lavy et al. 2016; Kato et al. 2017; Mutte et al. 2018) potentially acting downstream of the SAUR/PP2C.D/AHA module. Whereas acid growth exists in *Nitella* (Métraux and Taiz 1977; Métraux et al. 1980), SAUR protein-mediated acid growth is a land plant innovation (Fig. 2).

AUXIN BIOLOGY OF THE ANCESTRAL LAND PLANT AND ANTECEDENT FUNCTIONS IN THE ALGAL ANCESTOR

Knowledge of the phylogenetic distribution of genes encoding the critical components of auxin biology allows the prediction of the auxin biology network in the ancestral plants and, further, functional analyses spanning the phylogenetic spectrum of extant land plants has provided insight into the functionality of the ancestral network. Using these criteria, the ancestral land plant is predicted to have possessed a system similar to that existing in *M. polymorpha*, with solitary paralogs of most components (Flores-Sandoval et al. 2015; Kato et al. 2015; Bowman et al. 2017; Mutte et al. 2018). Two most critical innovations shaped changes in auxin biology during the transition from an ancestral alga to land plants and form the basis of how auxin became a key land plant hormone. First, the acquisition via HGT from a bacterium of a *YUCCA* gene enabled assembly of a new auxin biosynthetic pathway (Yue et al. 2014; Bowman et al. 2017). Second, the integration of auxin into a subsection of a preexisting transcriptional network, which was also expanded in the ancestral land plant, established the response output of the pathway to be modulated by the presence of auxin (Bowman et al. 2017; Flores-Sandoval et al. 2018; Mutte et al. 2018). Two auxin-mediated responses may be common to all land plants. First, tropic growth responses are intimately connected with auxin biology throughout land plants and presumably this was the case in the ancestral land plants. Second, auxin biology is also critical for almost every characterized developmental process to proceed appropriately. In *M. polymorpha*, loss of the TAA/YUCCA auxin biosynthetic pathway results in a complete loss of differentiation with plants being composed of an amorphous mass of dividing cells (Eklund et al. 2015). Whereas complete loss of the TAA/YUCCA pathway has not been achieved in angiosperms due to a high level of genetic redundancy, various multiple mutant combinations result in a spectrum of developmental defects (for review, see Zhao 2018). With respect to development, one plausible hypothesis is that the ancestral algal transcriptional network controlled the balance between meristematic (i.e., cell division) and differentiated states and that this came under the influence of auxin in the ancestral land plant (Flores-Sandoval et al. 2018). In this scenario, the role of auxin as a facilitator rather than a specifier follows naturally, with specific tissue-/cell-type specification dependent upon the milieu of other transcription factors present in the cells transitioning from cell division to differentiation.

While there are credible (and testable) hypotheses concerning the role(s) of auxin in the ancestral land plants, the situation is enigmatic in the algal ancestor and in extant charophytes. There are two components to consider: (1) What was the role of the ancestral transcriptional network? One hypothesis was presented above. (2) What was the function of auxin in these algae? Development of charophyte models amenable to genome editing would facilitate addressing these questions, albeit the latter question may require specific knowledge of any auxin biosynthetic pathways in these organisms. Answers to these two questions could illuminate why auxin was co-opted from an as-yet-unknown ancestral role into a molecule critical for most aspects of land plant development.

ACKNOWLEDGMENTS

The authors acknowledge funding from The Australian Research Council (DP160100892 and DP170100049 to J.L.B.), the Japan Society for the Promotion of Science (standard postdoctoral fellowship P19381 to E.F.S. and KAKENHI through Grant No. 19K23751 to H.K.).

REFERENCES

Abel S, Theologis A. 1996. Early genes and auxin action. *Plant Physiol* **111:** 9–17. doi:10.1104/pp.111.1.9

Abel S, Nguyen MD, Theologis A. 1995. The PS-IAA4/5-like family of early auxin-inducible mRNAs in *Arabidopsis thaliana*. *J Mol Biol* **251:** 533–549. doi:10.1006/jmbi.1995.0454

Ashen JB, Cohen JD, Goff LJ. 1999. GC-SIM-MS detection and quantification of free indole-3-acetic acid in bacterial galls on the marine alga *Prionitis lanceolata* (rhodophyta). *J Phycology* **35:** 493–500. doi:10.1046/j.1529-8817.1999.3530493.x

Banks JA, Nishiyama T, Hasebe M, Bowman JL, Gribskov M, dePamphilis C, Albert VA, Aono N, Aoyama T, Ambrose BA, et al. 2011. The *Selaginella* genome identifies genetic changes associated with the evolution of vascular plants. *Science* **332:** 960–963. doi:10.1126/science.1203810

Barbez E, Kubeš M, Rolčík J, Béziat C, Pěnčík A, Wang B, Rosquete MR, Zhu J, Dobrev PI, Lee Y, et al. 2012. A novel putative auxin carrier family regulates intracellular auxin homeostasis in plants. *Nature* **485:** 119–122. doi:10.1038/nature11001

Bartel B, Fink GR. 1995. ILR1, an amidohydrolase that releases active indole-3-acetic acid from conjugates. *Science* **268:** 1745–1748. doi:10.1126/science.7792599

Basu S, Sun H, Brian L, Quatrano RL, Muday GK. 2002. Early embryo development in *Fucus distichus* is auxin sensitive. *Plant Physiol* **130:** 292–302. doi:10.1104/pp.004747

Bennett T, Leyser O. 2014. The auxin question: a philosophical overview. In *Auxin and its role in plant development* (ed. Zažímalová E, Petrášek J, Benková E), pp. 3–19. Springer, Vienna.

Bennett MJ, Marchant A, Green HG, May ST, Ward SP, Millner PA, Walker AR, Schulz B, Feldmann KA. 1996. *Arabidopsis AUX1* gene: a permease-like regulator of root gravitropism. *Science* **273:** 948–950. doi:10.1126/science.273.5277.948

Bennett T, Brockington SF, Rothfels C, Graham SW, Stevenson D, Kutchan T, Rolf M, Thomas P, Wong GK, Leyser O, et al. 2014. Paralogous radiations of PIN proteins with multiple origins of noncanonical PIN structure. *Mol Biol Evol* **31:** 2042–2060. doi:10.1093/molbev/msu147

Béziat C, Kleine-Vehn J. 2018. The road to auxin-dependent growth repression and promotion in apical hooks. *Curr Biol* **28:** R519–R525. doi:10.1016/j.cub.2018.01.069

Béziat C, Barbez E, Feraru MI, Lucyshyn D, Kleine-Vehn J. 2017. Light triggers PILS-dependent reduction in nuclear auxin signalling for growth transition. *Nat Plants* **3:** 17105. doi:10.1038/nplants.2017.105

Bierfreund NM, Tintelnot S, Reski R, Decker EL. 2004. Loss of GH3 function does not affect phytochrome-mediated development in a moss, *Physcomitrella patens*. *J Plant Physiol* **161:** 823–835. doi:10.1016/j.jplph.2003.12.010

Blakeslee JJ, Bandyopadhyay A, Lee OR, Mravec J, Titapiwatanakun B, Sauer M, Makam SN, Cheng Y, Bouchard R, Adamec J, et al. 2007. Interactions among PIN-FORMED and P-glycoprotein auxin transporters in *Arabidopsis*. *Plant Cell* **19:** 131–147. doi:10.1105/tpc.106.040782

Boer DR, Freire-Rios A, van den Berg WA, Saaki T, Manfield IW, Kepinski S, Lopez-Vidrieo I, Franco-Zorrilla JM, de Vries SC, Solano R, et al. 2014. Structural basis for DNA binding specificity by the auxin-dependent ARF transcription factors. *Cell* **156:** 577–589. doi:10.1016/j.cell.2013.12.027

Bogaert KA, Blommaert L, Ljung K, Beeckman T, De Clerck O. 2019. Auxin function in the brown alga *Dictyota dichotoma*. *Plant Physiol* **179:** 280–299. doi:10.1104/pp.18.01041

Boot KJ, Libbenga KR, Hille SC, Offringa R, van Duijn B. 2012. Polar auxin transport: an early invention. *J Exp Bot* **63:** 4213–4218. doi:10.1093/jxb/ers106

Bowman JL, Kohchi T, Yamato KT, Jenkins J, Shu S, Ishizaki K, Yamaoka S, Nishihama R, Nakamura Y, Berger F, et al. 2017. Insights into land plant evolution garnered from the *Marchantia polymorpha* genome. *Cell* **171:** 287–304.e15. doi:10.1016/j.cell.2017.09.030

Boysen Jensen P. 1910. Über die Leitung des phototropischen Reizes in Avena-keimpflanzen [About the conduction of the phototropic stimulus in Avena seedlings]. *Ber Dtsch Bot Ges* **28:** 118–120.

Campanella JJ, Zaben N, Enriquez D, Smalley JV, Ludwig-Müller J. 2014. *An enzymatic analysis of loblolly pine and sitka spruce auxin conjugate hydrolases and evolutionary implications*, pp. 79–88. International Society for Horticultural Science (ISHS), Leuven, Belgium.

Campanella JJ, Kurdach S, Bochis J, Smalley JV. 2018. Evidence for exaptation of the *Marchantia polymorpha* M20D peptidase MpILR1 into the tracheophyte auxin regulatory pathway. *Plant Physiol* **177:** 1595–1604. doi:10.1104/pp.18.00543

Campanella JJ, Kurdach S, Skibitski R, Smalley JV, Desind S, Ludwig-Müller J. 2019. Evidence for the early evolutionary loss of the M20D auxin amidohydrolase family from mosses and horizontal gene transfer from soil bacteria of cryptic hydrolase orthologues to *Physcomitrella patens*. *J Plant Growth Regulation* **38:** 1428–1438. doi:10.1007/s00344-019-09945-6

Catalá C, Rose JK, Bennett AB. 2000. Auxin-regulated genes encoding cell wall-modifying proteins are expressed during early tomato fruit growth. *Plant Physiol* **122:** 527–534. doi:10.1104/pp.122.2.527

Chae K, Isaacs CG, Reeves PH, Maloney GS, Muday GK, Nagpal P, Reed JW. 2012. *Arabidopsis* SMALL AUXIN UP RNA63 promotes hypocotyl and stamen filament elongation. *Plant J* **71:** 684–697. doi:10.1111/j.1365-313X.2012.05024.x

Chanclud E, Morel JB. 2016. Plant hormones: a fungal point of view. *Mol Plant Pathol* **17:** 1289–1297. doi:10.1111/mpp.12393

Cheng Y, Dai X, Zhao Y. 2007. Auxin synthesized by the YUCCA flavin monooxygenases is essential for embryogenesis and leaf formation in *Arabidopsis*. *Plant Cell* **19:** 2430–2439. doi:10.1105/tpc.107.053009

Cheng S, Xian W, Fu Y, Marin B, Keller J, Wu T, Sun W, Li X, Xu Y, Zhang Y, et al. 2019. Genomes of subaerial Zygnematophyceae provide insights into land plant evolution. *Cell* **179:** 1057–1067.e14. doi:10.1016/j.cell.2019.10.019

Chiu LW, Heckert MJ, You Y, Albanese N, Fenwick T, Siehl DL, Castle LA, Tao Y. 2018. Members of the GH3 family of proteins conjugate 2,4-D and dicamba with aspartate and glutamate. *Plant Cell Physiol* **59:** 2366–2380.

Choi HS, Seo M, Cho HT. 2018. Two TPL-binding motifs of ARF2 are involved in repression of auxin responses. *Front Plant Sci* **9:** 372. doi:10.3389/fpls.2018.00372

Cholodny N. 1924. Über die hormonale Wirkung der Organspitze bei der geotropischen Krümmung [About the hormonal effect of the organ tip in the geotropic curvature]. *Ber Dtsch Bot Ges* **42:** 356–362.

Cooke TJ, Poli D, Sztein AE, Cohen JD. 2002. Evolutionary patterns in auxin action. *Plant Mol Biol* **49:** 319–338. doi:10.1023/A:1015242627321

Cosgrove DJ. 1998. Molecular regulation of plant cell wall extensibility. *Gravit Space Biol Bull* **11:** 61–70.

Crawford BC, Sewell J, Golembeski G, Roshan C, Long JA, Yanofsky MF. 2015. Plant development. genetic control of distal stem cell fate within root and embryonic meristems. *Science* **347:** 655–659. doi:10.1126/science.aaa0196

Dai X, Mashiguchi K, Chen Q, Kasahara H, Kamiya Y, Ojha S, DuBois J, Ballou D, Zhao Y. 2013. The biochemical mechanism of auxin biosynthesis by an *Arabidopsis* YUCCA flavin-containing monooxygenase. *J Biol Chem* **288:** 1448–1457. doi:10.1074/jbc.M112.424077

Darwin CR. 1880. *The power of movement in plants*. Murray, London.

Davies RT, Goetz DH, Lasswell J, Anderson MN, Bartel B. 1999. *IAR3* encodes an auxin conjugate hydrolase from *Arabidopsis*. *Plant Cell* **11:** 365–376. doi:10.1105/tpc.11.3.365

De Smet I, Voß U, Lau S, Wilson M, Shao N, Timme RE, Swarup R, Kerr I, Hodgman C, Bock R, et al. 2011. Unraveling the evolution of auxin signaling. *Plant Physiol* **155:** 209–221. doi:10.1104/pp.110.168161

Dharmasiri N, Dharmasiri S, Estelle M. 2005. The F-box protein TIR1 is an auxin receptor. *Nature* **435:** 441–445. doi:10.1038/nature03543

Ding Z, Wang B, Moreno I, Dupláková N, Simon S, Carraro N, Reemmer J, Pěnčík A, Chen X, Tejos R, et al. 2012. ER-localized auxin transporter PIN8 regulates auxin homeostasis and male gametophyte development in *Arabidopsis*. *Nat Commun* **3:** 941. doi:10.1038/ncomms1941

Edgar RC. 2004. MUSCLE: multiple sequence alignment with high accuracy and high throughput. *Nucleic Acids Res* **32:** 1792–1797. doi:10.1093/nar/gkh340

Eklund DM, Ishizaki K, Flores-Sandoval E, Kikuchi S, Takebayashi Y, Tsukamoto S, Hirakawa Y, Nonomura M, Kato H, Kouno M, et al. 2015. Auxin produced by the indole-3-pyruvic acid pathway regulates development and gemmae dormancy in the liverwort *Marchantia polymorpha. Plant Cell* **27:** 1650–1669. doi:10.1105/tpc.15.00065

Esmon CA, Tinsley AG, Ljung K, Sandberg G, Hearne LB, Liscum E. 2006. A gradient of auxin and auxin-dependent transcription precedes tropic growth responses. *Proc Natl Acad Sci* **103:** 236–241. doi:10.1073/pnas.0507127103

Feraru E, Feraru MI, Barbez E, Waidmann S, Sun L, Gaidora A, Kleine-Vehn J. 2019. PILS6 is a temperature-sensitive regulator of nuclear auxin input and organ growth in *Arabidopsis thaliana. Proc Natl Acad Sci* **116:** 3893–3898. doi:10.1073/pnas.1814015116

Finet C, Berne-Dedieu A, Scutt CP, Marlétaz F. 2013. Evolution of the *ARF* gene family in land plants: old domains, new tricks. *Mol Biol Evol* **30:** 45–56. doi:10.1093/molbev/mss220

Fitting H. 1939. Untersuchungen über den einfluss von licht und dunkelheit auf die entwicklung von moosen. I. Die brutkörper der Marchantieen [Investigations into the influence of light and darkness on the development of mosses. I. The gemmae of the Marchantieen]. *Jahrb Wiss Bot* **88:** 633–722.

Flores-Sandoval E, Eklund DM, Bowman JL. 2015. A simple auxin transcriptional response system regulates multiple morphogenetic processes in the liverwort *Marchantia polymorpha. PLoS Genet* **11:** e1005207. doi:10.1371/journal.pgen.1005207

Flores-Sandoval E, Eklund DM, Hong SF, Alvarez JP, Fisher TJ, Lampugnani ER, Golz JF, Vázquez-Lobo A, Dierschke T, Lin SS, et al. 2018. Class C ARFs evolved before the origin of land plants and antagonize differentiation and developmental transitions in *Marchantia polymorpha. New Phytol* **218:** 1612–1630. doi:10.1111/nph.15090

Gray WM, Kepinski S, Rouse D, Leyser O, Estelle M. 2001. Auxin regulates SCF^{TIR1}-dependent degradation of AUX/IAA proteins. *Nature* **414:** 271–276. doi:10.1038/35104500

Hagen G, Martin G, Li Y, Guilfoyle TJ. 1991. Auxin-induced expression of the soybean GH3 promoter in transgenic tobacco plants. *Plant Mol Biol* **17:** 567–579. doi:10.1007/BF00040658

Halbsguth W, Kohlenbach HW. 1953. Einige versuche über die wirkung von heteroauxin auf die symmetrieentwicklung der brutkörperkeimlinge von *Marchanta polymorpha* L. [Some experiments on the effect of heteroauxin on polar development of the gemmae of *Marchanta polymorpha* L.]. *Planta* **42:** 349–366. doi:10.1007/BF01928191

Heitz E. 1940. Die Polaritat keimender Moossporen [The polarity of germinating moss spores]. *Verh Schweiz Naturforsch Ges* **120:** 168–170.

Hori K, Maruyama F, Fujisawa T, Togashi T, Yamamoto N, Seo M, Sato S, Yamada T, Mori H, Tajima N, et al. 2014. *Klebsormidium flaccidum* genome reveals primary factors for plant terrestrial adaptation. *Nat Commun* **5:** 3978. doi:10.1038/ncomms4978

Jackson RG, Lim EK, Li Y, Kowalczyk M, Sandberg G, Hoggett J, Ashford DA, Bowles DJ. 2001. Identification and biochemical characterization of an *Arabidopsis* indole-3-acetic acid glucosyltransferase. *J Biol Chem* **276:** 4350–4356. doi:10.1074/jbc.M006185200

Jiao C, Sorensen I, Sun X, Sun H, Behar H, Alseekh S, Philippe G, Palacio Lopez K, Sun L, Reed R, et al. 2020. The *Penium margaritaceum* genome: hallmarks of the origins of land plants. *Cell* **181:** 1097–1111.E12. doi:10.1016/j.cell.2020.04.019

Jin Q, Scherp P, Heimann K, Hasenstein KH. 2008. Auxin and cytoskeletal organization in algae. *Cell Biol Int* **32:** 542–545. doi:10.1016/j.cellbi.2007.11.005

Jin SH, Ma XM, Han P, Wang B, Sun YG, Zhang GZ, Li YJ, Hou BK. 2013. UGT74D1 is a novel auxin glycosyltransferase from *Arabidopsis thaliana. PLoS ONE* **8:** e61705. doi:10.1371/journal.pone.0061705

Jones VA, Dolan L. 2017. Mp*WIP* regulates air pore complex development in the liverwort *Marchantia polymorpha. Development* **144:** 1472–1476. doi:10.1242/dev.144287

Kamimoto Y, Terasaka K, Hamamoto M, Takanashi K, Fukuda S, Shitan N, Sugiyama A, Suzuki H, Shibata D, Wang B, et al. 2012. *Arabidopsis* ABCB21 is a facultative auxin importer/exporter regulated by cytoplasmic auxin concentration. *Plant Cell Physiol* **53:** 2090–2100. doi:10.1093/pcp/pcs149

Kaneko S, Cook SD, Aoi Y, Watanabe A, Hayashi KI, Kasahara H. 2020. An evolutionarily primitive and distinct auxin metabolism in the lycophyte *Selaginella moellendorffii. Plant Cell Physiol* **61:** 1724–1732. doi:10.1093/pcp/pcaa098

Kato H, Ishizaki K, Kouno M, Shirakawa M, Bowman JL, Nishihama R, Kohchi T. 2015. Auxin-mediated transcriptional system with a minimal set of components is critical for morphogenesis through the life cycle in *Marchantia polymorpha. PLoS Genet* **11:** e1005084. doi:10.1371/journal.pgen.1005084

Kato H, Kouno M, Takeda M, Suzuki H, Ishizaki K, Nishihama R, Kohchi T. 2017. The roles of the sole activator-type auxin response factor in pattern formation of *Marchantia polymorpha. Plant Cell Physiol* **58:** 1642–1651. doi:10.1093/pcp/pcx095

Kato H, Mutte SK, Suzuki H, Crespo I, Das S, Radoeva T, Fontana M, Yoshitake Y, Hainiwa E, van den Berg W, et al. 2020. Design principles of a minimal auxin response system. *Nat Plants* **6:** 473–482. doi:10.1038/s41477-020-0662-y

Kepinski S, Leyser O. 2005. The *Arabidopsis* F-box protein TIR1 is an auxin receptor. *Nature* **435:** 446–451. doi:10.1038/nature03542

Khasin M, Cahoon RR, Nickerson KW, Riekhof WR. 2018. Molecular machinery of auxin synthesis, secretion, and perception in the unicellular chlorophyte alga *Chlorella sorokiniana* UTEX 1230. *PLoS ONE* **13:** e0205227. doi:10.1371/journal.pone.0205227

Kim J, Harter K, Theologis A. 1997. Protein–protein interactions among the Aux/IAA proteins. *Proc Natl Acad Sci* **94:** 11786–11791. doi:10.1073/pnas.94.22.11786

Klämbt D, Knauth B, Dittmann I. 1992. Auxin dependent growth of rhizoids of *Chara globularis. Physiol Plant* **85:** 537–540. doi:10.1111/j.1399-3054.1992.tb05823.x

Kögl F, Kostermans GFR. 1934. Hetero-auxin als Stoffwechselprodukt niederer pflanzlicher Organismen. Isolierung aus Hefe. 13. Mitteilung über pflanzliche Wachstumsstoffe [Hetero-auxin as a metabolic product of lower plants. Isolation from yeast. 13. Communication on Vegetable Growth Substances]. *Hoppe-Seyler's Z Physiol Chem* **228:** 113–121. doi:10.1515/bchm2.1934.228.3-6.113

Korasick DA, Enders TA, Strader LC. 2013. Auxin biosynthesis and storage forms. *J Exp Bot* **64:** 2541–2555. doi:10.1093/jxb/ert080

Korasick DA, Westfall CS, Lee SG, Nanao MH, Dumas R, Hagen G, Guilfoyle TJ, Jez JM, Strader LC. 2014. Molecular basis for AUXIN RESPONSE FACTOR protein interaction and the control of auxin response repression. *Proc Natl Acad Sci* **111:** 5427–5432. doi:10.1073/pnas.1400074111

Kumar S, Stecher G, Li M, Knyaz C, Tamura K. 2018. MEGA X: molecular evolutionary genetics analysis across computing platforms. *Mol Biol Evol* **35:** 1547–1549. doi:10.1093/molbev/msy096

Kunkel BN, Harper CP. 2018. The roles of auxin during interactions between bacterial plant pathogens and their hosts. *J Exp Bot* **69:** 245–254. doi:10.1093/jxb/erx447

LaRue CD, Narayanaswami S. 1957. Auxin inhibition in the liverwort *Lunularia*. *New Phytol* **56:** 61–70. doi:10.1111/j.1469-8137.1957.tb07449.x

Lavy M, Prigge MJ, Tao S, Shain S, Kuo A, Kirchsteiger K, Estelle M. 2016. Constitutive auxin response in *Physcomitrella* reveals complex interactions between Aux/IAA and ARF proteins. *eLife* **5:** e13325. doi:10.7554/eLife.13325

Le Bail A, Billoud B, Kowalczyk N, Kowalczyk M, Gicquel M, Le Panse S, Stewart S, Scornet D, Cock JM, Ljung K, et al. 2010. Auxin metabolism and function in the multicellular brown alga *Ectocarpus siliculosus*. *Plant Physiol* **153:** 128–144. doi:10.1104/pp.109.149708

LeClere S, Tellez R, Rampey RA, Matsuda SP, Bartel B. 2002. Characterization of a family of IAA-amino acid conjugate hydrolases from *Arabidopsis*. *J Biol Chem* **277:** 20446–20452. doi:10.1074/jbc.M111955200

Li Y, Baldauf S, Lim EK, Bowles DJ. 2001. Phylogenetic analysis of the UDP-glycosyltransferase multigene family of *Arabidopsis thaliana*. *J Biol Chem* **276:** 4338–4343. doi:10.1074/jbc.M007447200

Li FW, Brouwer P, Carretero-Paulet L, Cheng S, de Vries J, Delaux PM, Eily A, Koppers N, Kuo LY, Li Z, et al. 2018. Fern genomes elucidate land plant evolution and cyanobacterial symbioses. *Nat Plants* **4:** 460–472. doi:10.1038/s41477-018-0188-8

Li FW, Nishiyama T, Waller M, Frangedakis E, Keller J, Li Z, Fernandez-Pozo N, Barker MS, Bennett T, Blázquez MA, et al. 2020. Anthoceros genomes illuminate the origin of land plants and the unique biology of hornworts. *Nat Plants* **6:** 259–272. doi:10.1038/s41477-020-0618-2

Ljung K. 2013. Auxin metabolism and homeostasis during plant development. *Development* **140:** 943–950. doi:10.1242/dev.086363

Ludwig-Müller J, Jülke S, Bierfreund NM, Decker EL, Reski R. 2009. Moss (*Physcomitrella patens*) GH3 proteins act in auxin homeostasis. *New Phytol* **181:** 323–338. doi:10.1111/j.1469-8137.2008.02677.x

Lv B, Yu Q, Liu J, Wen X, Yan Z, Hu K, Li H, Kong X, Li C, Tian H, et al. 2020. Non-canonical AUX/IAA protein IAA33 competes with canonical AUX/IAA repressor IAA5 to negatively regulate auxin signaling. *EMBO J* **39:** e101515.

Maher EP, Martindale SJ. 1980. Mutants of *Arabidopsis thaliana* with altered responses to auxins and gravity. *Biochem Genet* **18:** 1041–1053. doi:10.1007/BF00484337

Mallory AC, Bartel DP, Bartel B. 2005. MicroRNA-directed regulation of *Arabidopsis AUXIN RESPONSE FACTOR17* is essential for proper development and modulates expression of early auxin response genes. *Plant Cell* **17:** 1360–1375. doi:10.1105/tpc.105.031716

Maravolo NC. 1976. Polarity and localization of auxin movement in the hepatic, *Marchantia polymorpha*. *Am J Bot* **63:** 526–531. doi:10.1002/j.1537-2197.1976.tb11841.x

Martin-Arevalillo R, Nanao MH, Larrieu A, Vinos-Poyo T, Mast D, Galvan-Ampudia C, Brunoud G, Vernoux T, Dumas R, Parcy F. 2017. Structure of the *Arabidopsis* TOPLESS corepressor provides insight into the evolution of transcriptional repression. *Proc Natl Acad Sci* **114:** 8107–8112. doi:10.1073/pnas.1703054114

Martin-Arevalillo R, Thévenon E, Jégu F, Vinos-Poyo T, Vernoux T, Parcy F, Dumas R. 2019. Evolution of the auxin response factors from charophyte ancestors. *PLoS Genet* **15:** e1008400. doi:10.1371/journal.pgen.1008400

Mashiguchi K, Tanaka K, Sakai T, Sugawara S, Kawaide H, Natsume M, Hanada A, Yaeno T, Shirasu K, Yao H, et al. 2011. The main auxin biosynthesis pathway in *Arabidopsis*. *Proc Natl Acad Sci* **108:** 18512–18517. doi:10.1073/pnas.1108434108

Matsuo S, Miyatake K, Endo M, Urashimo S, Kawanishi T, Negoro S, Shimakoshi S, Fukuoka H. 2020. Loss of function of the *Pad-1* aminotransferase gene, which is involved in auxin homeostasis, induces parthenocarpy in Solanaceae plants. *Proc Natl Acad Sci* **117:** 12784–12790. doi:10.1073/pnas.2001211117

McClure BA, Hagen G, Brown CS, Gee MA, Guilfoyle TJ. 1989. Transcription, organization, and sequence of an auxin-regulated gene cluster in soybean. *Plant Cell* **1:** 229–239.

Métraux JP, Taiz L. 1977. Cell wall extension in *Nitella* as influenced by acids and ions. *Proc Natl Acad Sci* **74:** 1565–1569. doi:10.1073/pnas.74.4.1565

Métraux JP, Richmond PA, Taiz L. 1980. Control of cell elongation in *Nitella* by endogenous cell wall pH gradients: multiaxial extensibility and growth studies. *Plant Physiol* **65:** 204–210. doi:10.1104/pp.65.2.204

Monte I, Ishida S, Zamarreño AM, Hamberg M, Franco-Zorrilla JM, García-Casado G, Gouhier-Darimont C, Reymond P, Takahashi K, García-Mina JM, et al. 2018. Ligand-receptor co-evolution shaped the jasmonate pathway in land plants. *Nat Chem Biol* **14:** 480–488. doi:10.1038/s41589-018-0033-4

Morffy N, Strader LC. 2020. Old town roads: routes of auxin biosynthesis across kingdoms. *Curr Opin Plant Biol* **55:** 21–27. doi:10.1016/j.pbi.2020.02.002

Mutte SK, Kato H, Rothfels C, Melkonian M, Wong GK, Weijers D. 2018. Origin and evolution of the nuclear auxin response system. *eLife* **7:** e33399. doi:10.7554/eLife.33399

Nanao MH, Vinos-Poyo T, Brunoud G, Thévenon E, Mazzoleni M, Mast D, Lainé S, Wang S, Hagen G, Li H, et al. 2014. Structural basis for oligomerization of auxin transcriptional regulators. *Nat Commun* **5:** 3617. doi:10.1038/ncomms4617

Naramoto S. 2017. Polar transport in plants mediated by membrane transporters: focus on mechanisms of polar auxin transport. *Curr Opin Plant Biol* **40:** 8–14. doi:10.1016/j.pbi.2017.06.012

Nishiyama T, Sakayama H, de Vries J, Buschmann H, Saint-Marcoux D, Ullrich KK, Haas FB, Vanderstraeten L, Becker D, Lang D, et al. 2018. The *Chara* genome: secondary complexity and implications for plant terrestrialization. *Cell* **174:** 448–464.e24. doi:10.1016/j.cell.2018.06.033

Noh B, Murphy AS, Spalding EP. 2001. Multidrug resistance-like genes of *Arabidopsis* required for auxin transport and auxin-mediated development. *Plant Cell* **13:** 2441–2454.

Normanly J, Cohen JD, Fink GR. 1993. *Arabidopsis thaliana* auxotrophs reveal a tryptophan-independent biosynthetic pathway for indole-3-acetic acid. *Proc Natl Acad Sci* **90:** 10355–10359. doi:10.1073/pnas.90.21.10355

Ohtaka K, Hori K, Kanno Y, Seo M, Ohta H. 2017. Primitive auxin response without TIR1 and Aux/IAA in the charophyte alga *Klebsormidium nitens*. *Plant Physiol* **174:** 1621–1632. doi:10.1104/pp.17.00274

Okada K, Ueda J, Komaki MK, Bell CJ, Shimura Y. 1991. Requirement of the auxin polar transport system in early stages of *Arabidopsis* floral bud formation. *Plant Cell* **3:** 677–684. doi:10.2307/3869249

Okrent RA, Brooks MD, Wildermuth MC. 2009. *Arabidopsis* GH3.12 (PBS3) conjugates amino acids to 4-substituted benzoates and is inhibited by salicylate. *J Biol Chem* **284:** 9742–9754. doi:10.1074/jbc.M806662200

Östin A, Kowalyczk M, Bhalerao RP, Sandberg G. 1998. Metabolism of indole-3-acetic acid in *Arabidopsis*. *Plant Physiol* **118:** 285–296. doi:10.1104/pp.118.1.285

Ouyang J, Shao X, Li J. 2000. Indole-3-glycerol phosphate, a branchpoint of indole-3-acetic acid biosynthesis from the tryptophan biosynthetic pathway in *Arabidopsis thaliana*. *Plant J* **24:** 327–334. doi:10.1046/j.1365-313x.2000.00883.x

Paál A. 1919. Über phototropische reizleitung [About phototropic stimulus conduction]. *Jahrb Wiss Bot* **58:** 406–458.

Petrásek J, Mravec J, Bouchard R, Blakeslee JJ, Abas M, Seifertová D, Wisniewska J, Tadele Z, Kubes M, Covanová M, et al. 2006. PIN proteins perform a rate-limiting function in cellular auxin efflux. *Science* **312:** 914–918. doi:10.1126/science.1123542

Piya S, Shrestha SK, Binder B, Stewart CN Jr, Hewezi T. 2014. Protein–protein interaction and gene co-expression maps of ARFs and Aux/IAAs in *Arabidopsis*. *Front Plant Sci* **5:** 744. doi:10.3389/fpls.2014.00744

Poli D, Jacobs M, Cooke TJ. 2003. Auxin regulation of axial growth in bryophyte sporophytes: its potential significance for the evolution of early land plants. *Am J Bot* **90:** 1405–1415. doi:10.3732/ajb.90.10.1405

Porco S, Pěnčík A, Rashed A, Voß U, Casanova-Sáez R, Bishopp A, Golebiowska A, Bhosale R, Swarup R, Swarup K, et al. 2016. Dioxygenase-encoding *AtDAO1* gene controls IAA oxidation and homeostasis in *Arabidopsis*. *Proc Natl Acad Sci* **113:** 11016–11021. doi:10.1073/pnas.1604375113

Poulet A, Kriechbaumer V. 2017. Bioinformatics analysis of phylogeny and transcription of TAA/YUC auxin biosynthetic genes. *Int J Mol Sci* **18:** 1791. doi:10.3390/ijms18081791

Prigge MJ, Lavy M, Ashton NW, Estelle M. 2010. *Physcomitrella patens* auxin-resistant mutants affect conserved elements of an auxin-signaling pathway. *Curr Biol* **20:** 1907–1912. doi:10.1016/j.cub.2010.08.050

Qin G, Gu H, Zhao Y, Ma Z, Shi G, Yang Y, Pichersky E, Chen H, Liu M, Chen Z, et al. 2005. An indole-3-acetic acid carboxyl methyltransferase regulates *Arabidopsis* leaf development. *Plant Cell* **17:** 2693–2704. doi:10.1105/tpc.105.034959

Rampey RA, LeClere S, Kowalczyk M, Ljung K, Sandberg G, Bartel B. 2004. A family of auxin-conjugate hydrolases that contributes to free indole-3-acetic acid levels during *Arabidopsis* germination. *Plant Physiol* **135:** 978–988. doi:10.1104/pp.104.039677

Ren G, Li L, Huang Y, Wang Y, Zhang W, Zheng R, Zhong C, Wang X. 2018. GhWIP2, a WIP zinc finger protein, suppresses cell expansion in *Gerbera hybrida* by mediating crosstalk between gibberellin, abscisic acid, and auxin. *New Phytol* **219:** 728–742. doi:10.1111/nph.15175

Rensing SA, Lang D, Zimmer AD, Terry A, Salamov A, Shapiro H, Nishiyama T, Perroud PF, Lindquist EA, Kamisugi Y, et al. 2008. The Physcomitrella genome reveals evolutionary insights into the conquest of land by plants. *Science* **319:** 64–69. doi:10.1126/science.1150646

Romani F. 2017. Origin of *TAA* genes in charophytes: new insights into the controversy over the origin of auxin biosynthesis. *Front Plant Sci* **8:** 1616. doi:10.3389/fpls.2017.01616

Romani F, Reinheimer R, Florent SN, Bowman JL, Moreno JE. 2018. Evolutionary history of HOMEODOMAIN LEUCINE ZIPPER transcription factors during plant transition to land. *New Phytol* **219:** 408–421. doi:10.1111/nph.15133

Rousseau J. 1950. Action de l'acide indol beta-acétique sur les propagules de *Marchantia polymorpha* et *Lunularia cruciata* [Action of indol beta-acetic acid on the gemmae of *Marchantia polymorpha* and *Lunularia cruciata*]. *C R Hebd Séances Acad Sci* **230:** 675–676.

Sachs Jv. 1880. Stoff und Form der Pflanzenorgane. I [Substance and shape of plant organs. I]. *Arbeiten des Botanischen Instituts in Wurzburg* **1:** 452–488.

Sachs Jv. 1882. Stoff und Form der Pflanzenorgane. II [Substance and shape of plant organs. II]. *Arbeiten des Botanischen Instituts in Wurzburg* **2:** 689–718.

Sawa S, Ohgishi M, Goda H, Higuchi K, Shimada Y, Yoshida S, Koshiba T. 2002. The *HAT2* gene, a member of the *HD-Zip* gene family, isolated as an auxin inducible gene by DNA microarray screening, affects auxin response in *Arabidopsis*. *Plant J* **32:** 1011–1022. doi:10.1046/j.1365-313X.2002.01488.x

Skokan R, Medvecká E, Viaene T, Vosolsobě S, Zwiewka M, Müller K, Skůpa P, Karady M, Zhang Y, Janacek DP, et al. 2019. PIN-driven auxin transport emerged early in Streptophyte evolution. *Nat Plants* **5:** 1114–1119. doi:10.1038/s41477-019-0542-5

Spartz AK, Ren H, Park MY, Grandt KN, Lee SH, Murphy AS, Sussman MR, Overvoorde PJ, Gray WM. 2014. SAUR inhibition of PP2C-D phosphatases activates plasma membrane H^+-ATPases to promote cell expansion in *Arabidopsis*. *Plant Cell* **26:** 2129–2142. doi:10.1105/tpc.114.126037

Spartz AK, Lor VS, Ren H, Olszewski NE, Miller ND, Wu G, Spalding EP, Gray WM. 2017. Constitutive expression of *Arabidopsis SMALL AUXIN UP RNA19* (*SAUR19*) in tomato confers auxin-independent hypocotyl elongation. *Plant Physiol* **173:** 1453–1462. doi:10.1104/pp.16.01514

Staswick PE, Tiryaki I. 2004. The oxylipin signal jasmonic acid is activated by an enzyme that conjugates it to isoleucine in *Arabidopsis*. *Plant Cell* **16:** 2117–2127. doi:10.1105/tpc.104.023549

Staswick PE, Serban B, Rowe M, Tiryaki I, Maldonado MT, Maldonado MC, Suza W. 2005. Characterization of an *Arabidopsis* enzyme family that conjugates amino acids to indole-3-acetic acid. *Plant Cell* **17:** 616–627. doi:10.1105/tpc.104.026690

Steindler C, Matteucci A, Sessa G, Weimar T, Ohgishi M, Aoyama T, Morelli G, Ruberti I. 1999. Shade avoidance responses are mediated by the ATHB-2 HD-zip protein, a negative regulator of gene expression. *Development* **126:** 4235–4245.

Stepanova AN, Robertson-Hoyt J, Yun J, Benavente LM, Xie DY, Doležal K, Schlereth A, Jürgens G, Alonso JM. 2008. TAA1-mediated auxin biosynthesis is essential for hormone crosstalk and plant development. *Cell* **133:** 177–191. doi:10.1016/j.cell.2008.01.047

Stepanova AN, Yun J, Robles LM, Novak O, He W, Guo H, Ljung K, Alonso JM. 2011. The *Arabidopsis* YUCCA1 flavin monooxygenase functions in the indole-3-pyruvic acid branch of auxin biosynthesis. *Plant Cell* **23:** 3961–3973. doi:10.1105/tpc.111.088047

Stewart JL, Nemhauser JL. 2010. Do trees grow on money? Auxin as the currency of the cellular economy. *Cold Spring Harb Perspect Biol* **2:** a001420. doi:10.1101/cshperspect.a001420

Stirk WA, Ördög V, Novák O, Rolčík J, Strnad M, Bálint P, van Staden J. 2013. Auxin and cytokinin relationships in 24 microalgal strains. *J Phycol* **49:** 459–467. doi:10.1111/jpy.12061

Stortenbeker N, Bemer M. 2019. The *SAUR* gene family: the plant's toolbox for adaptation of growth and development. *J Exp Bot* **70:** 17–27. doi:10.1093/jxb/ery332

Strader LC, Wheeler DL, Christensen SE, Berens JC, Cohen JD, Rampey RA, Bartel B. 2011. Multiple facets of *Arabidopsis* seedling development require indole-3-butyric acid-derived auxin. *Plant Cell* **23:** 984–999. doi:10.1105/tpc.111.083071

Sun H, Basu S, Brady SR, Luciano RL, Muday GK. 2004. Interactions between auxin transport and the actin cytoskeleton in developmental polarity of *Fucus distichus* embryos in response to light and gravity. *Plant Physiol* **135:** 266–278. doi:10.1104/pp.103.034900

Swarup R, Kargul J, Marchant A, Zadik D, Rahman A, Mills R, Yemm A, May S, Williams L, Millner P, et al. 2004. Structure–function analysis of the presumptive *Arabidopsis* auxin permease AUX1. *Plant Cell* **16:** 3069–3083. doi:10.1105/tpc.104.024737

Szemenyei H, Hannon M, Long JA. 2008. TOPLESS mediates auxin-dependent transcriptional repression during *Arabidopsis* embryogenesis. *Science* **319:** 1384–1386. doi:10.1126/science.1151461

Sztein AE, Cohen JD, de la Fuente IG, Cooke TJ. 1999. Auxin metabolism in mosses and liverworts. *Am J Bot* **86:** 1544–1555. doi:10.2307/2656792

Sztein AE, Cohen JD, Cooke TJ. 2000. Evolutionary patterns in the auxin metabolism of green plants. *Int J Plant Sci* **161:** 849–859. doi:10.1086/317566

Takehara S, Sakuraba S, Mikami B, Yoshida H, Yoshimura H, Itoh A, Endo M, Watanabe N, Nagae T, Matsuoka M, et al. 2020. A common allosteric mechanism regulates homeostatic inactivation of auxin and gibberellin. *Nat Commun* **11:** 2143. doi:10.1038/s41467-020-16068-0

Tan X, Calderon-Villalobos LI, Sharon M, Zheng C, Robinson CV, Estelle M, Zheng N. 2007. Mechanism of auxin perception by the TIR1 ubiquitin ligase. *Nature* **446:** 640–645. doi:10.1038/nature05731

Tao Y, Ferrer JL, Ljung K, Pojer F, Hong F, Long JA, Li L, Moreno JE, Bowman ME, Ivans LJ, et al. 2008. Rapid synthesis of auxin via a new tryptophan-dependent pathway is required for shade avoidance in plants. *Cell* **133:** 164–176. doi:10.1016/j.cell.2008.01.049

Tivendale ND, Ross JJ, Cohen JD. 2014. The shifting paradigms of auxin biosynthesis. *Trends Plant Sci* **19:** 44–51. doi:10.1016/j.tplants.2013.09.012

Tiwari SB, Wang XJ, Hagen G, Guilfoyle TJ. 2001. AUX/IAA proteins are active repressors, and their stability and activity are modulated by auxin. *Plant Cell* **13:** 2809–2822. doi:10.1105/tpc.010289

Tiwari SB, Hagen G, Guilfoyle TJ. 2004. Aux/IAA proteins contain a potent transcriptional repression domain. *Plant Cell* **16:** 533–543. doi:10.1105/tpc.017384

Turnaev II, Gunbin KV, Afonnikov DA. 2015. Plant auxin biosynthesis did not originate in charophytes. *Trends Plant Sci* **20:** 463–465. doi:10.1016/j.tplants.2015.06.004

Turnaev II, Gunbin KV, Suslov VV, Akberdin IR, Kolchanov NA, Afonnikov DA. 2020. The phylogeny of class B flavoprotein monooxygenases and the origin of the YUCCA protein family. *Plants (Basel)* **9:** 1092. doi:10.3390/plants9091092

Ulmasov T, Hagen G, Guilfoyle TJ. 1997. ARF1, a transcription factor that binds to auxin response elements. *Science* **276:** 1865–1868. doi:10.1126/science.276.5320.1865

Ulmasov T, Hagen G, Guilfoyle TJ. 1999. Activation and repression of transcription by auxin-response factors. *Proc Natl Acad Sci* **96:** 5844–5849. doi:10.1073/pnas.96.10.5844

Vance BD. 1987. Phytohormone effects on cell division in *Chlorella pyrenoidosa* Chick (TX-7-11-05) (Chlorellaceae). *J Plant Growth Regulation* **5:** 169–173. doi:10.1007/BF02087185

von Witsch H. 1940. Versuch über den Einfluss von Wuchsstoff auf Wachstum und Entwicklung von *Calypogeia trichomanes* [Experiment on the influence of growth substances on the growth and development of *Calypogeia trichomanes*]. *Planta* **30:** 664–672. doi:10.1007/BF01917045

Vosolsobe S, Skokan R, Petrasek J. 2020. The evolutionary origins of auxin transport: what we know and what we need to know. *J Exp Bot* **71:** 3287–3295. doi:10.1093/jxb/eraa169

Wang C, Liu Y, Li SS, Han GZ. 2014. Origin of plant auxin biosynthesis in charophyte algae. *Trends Plant Sci* **19:** 741–743. doi:10.1016/j.tplants.2014.10.004

Wang C, Li SS, Han GZ. 2016. Commentary: plant auxin biosynthesis did not originate in charophytes. *Front Plant Sci* **7:** 158.

Wang S, Li L, Li H, Sahu SK, Wang H, Xu Y, Xian W, Song B, Liang H, Cheng S, et al. 2020. Genomes of early-diverging streptophyte algae shed light on plant terrestrialization. *Nat Plants* **6:** 95–106. doi:10.1038/s41477-019-0560-3

Went FW. 1926. On growth-accelerating substances in the coleoptile of *Avena sativa*. *Proc K Ned Akad Wet* **30:** 10–19.

Went FW. 1928. Wuchsstoff und Wachstum [Growth substances and development]. *Recueil Trav Bot Néerl* **24:** 1–116.

Went FW, Thimann KV. 1937. *Phytohormones*. Macmillan, New York.

Wilson AE, Tian L. 2019. Phylogenomic analysis of UDP-dependent glycosyltransferases provides insights into the evolutionary landscape of glycosylation in plant metabolism. *Plant J* **100:** 1273–1288. doi:10.1111/tpj.14514

Wong JH, Spartz AK, Park MY, Du M, Gray WM. 2019. Mutation of a conserved motif of PP2C.D phosphatases confers SAUR immunity and constitutive activity. *Plant Physiol* **181:** 353–366. doi:10.1104/pp.19.00496

Wood NL, Berliner MD. 1979. Effects of indoleacetic acid on the desmid *Micrasterias thomasiana*. *Plant Sci Lett* **16:** 285–289. doi:10.1016/0304-4211(79)90040-3

Wright AD, Sampson MB, Neuffer MG, Michalczuk L, Slovin JP, Cohen JD. 1991. Indole-3-acetic acid biosynthesis in the mutant maize orange pericarp, a tryptophan auxotroph. *Science* **254:** 998–1000. doi:10.1126/science.254.5034.998

Wu G, Cameron JN, Ljung K, Spalding EP. 2010. A role for ABCB19-mediated polar auxin transport in seedling photomorphogenesis mediated by cryptochrome 1 and phytochrome B. *Plant J* **62:** 179–191. doi:10.1111/j.1365-313X.2010.04137.x

Wu MF, Yamaguchi N, Xiao J, Bargmann B, Estelle M, Sang Y, Wagner D. 2015. Auxin-regulated chromatin switch directs acquisition of flower primordium founder fate. *eLife* **4:** e09269. doi:10.7554/eLife.09269

Yokoya NS, Handro W. 1996. Effects of auxins and cytokinins on tissue culture of *Grateloupia dichotoma* (Gigartinales, Rhodophyta). *Hydrobiologia* **326:** 393–400. doi:10.1007/BF00047837

Yue J, Hu X, Sun H, Yang Y, Huang J. 2012. Widespread impact of horizontal gene transfer on plant colonization of land. *Nat Commun* **3:** 1152. doi:10.1038/ncomms2148

Yue J, Hu X, Huang J. 2014. Origin of plant auxin biosynthesis. *Trends Plant Sci* **19:** 764–770. doi:10.1016/j.tplants.2014.07.004

Záveská Drábková L, Dobrev PI, Motyka V. 2015. Phytohormone profiling across the bryophytes. *PLoS ONE* **10:** e0125411. doi:10.1371/journal.pone.0125411

Zhang J, Lin JE, Harris C, Campos Mastrotti Pereira F, Wu F, Blakeslee JJ, Peer WA. 2016. DAO1 catalyzes temporal and tissue-specific oxidative inactivation of auxin in *Arabidopsis thaliana*. *Proc Natl Acad Sci* **113:** 11010–11015. doi:10.1073/pnas.1604769113

Zhang J, Fu XX, Li RQ, Zhao X, Liu Y, Li MH, Zwaenepoel A, Ma H, Goffinet B, Guan YL, et al. 2020. The hornwort genome and early land plant evolution. *Nat Plants* **6:** 107–118. doi:10.1038/s41477-019-0588-4

Zhao Y. 2018. Essential roles of local auxin biosynthesis in plant development and in adaptation to environmental changes. *Annu Rev Plant Biol* **69:** 417–435. doi:10.1146/annurev-arplant-042817-040226

Zhao Y, Christensen SK, Fankhauser C, Cashman JR, Cohen JD, Weigel D, Chory J. 2001. A role for flavin monooxygenase-like enzymes in auxin biosynthesis. *Science* **291:** 306–309. doi:10.1126/science.291.5502.306

Zhao Z, Zhang Y, Liu X, Zhang X, Liu S, Yu X, Ren Y, Zheng X, Zhou K, Jiang L, et al. 2013. A role for a dioxygenase in auxin metabolism and reproductive development in rice. *Dev Cell* **27:** 113–122. doi:10.1016/j.devcel.2013.09.005

Zheng Z, Guo Y, Novák O, Dai X, Zhao Y, Ljung K, Noel JP, Chory J. 2013. Coordination of auxin and ethylene biosynthesis by the aminotransferase VAS1. *Nat Chem Biol* **9:** 244–246. doi:10.1038/nchembio.1178

Žižková E, Kubeš M, Dobrev PI, Přibyl P, Šimura J, Zahajská L, Záveská Drábková L, Novák O, Motyka V. 2017. Control of cytokinin and auxin homeostasis in cyanobacteria and algae. *Ann Bot* **119:** 151–166. doi:10.1093/aob/mcw194

Zubieta C, Ross JR, Koscheski P, Yang Y, Pichersky E, Noel JP. 2003. Structural basis for substrate recognition in the salicylic acid carboxyl methyltransferase family. *Plant Cell* **15:** 1704–1716. doi:10.1105/tpc.014548

Auxin Biology in Bryophyta: A Simple Platform with Versatile Functions

Hidemasa Suzuki, Takayuki Kohchi, and Ryuichi Nishihama

Graduate School of Biostudies, Kyoto University, Kyoto 606-8502, Japan

Correspondence: nishihama@lif.kyoto-u.ac.jp

Bryophytes, including liverworts, mosses, and hornworts, are gametophyte-dominant land plants that are derived from a common ancestor and underwent independent evolution from the sporophyte-dominant vascular plants since their divergence. The plant hormone auxin has been shown to play pleiotropic roles in the haploid bodies of bryophytes. Pharmacological and chemical studies identified conserved auxin molecules, their inactivated forms, and auxin transport in bryophyte tissues. Recent genomic and molecular biological studies show deep conservation of components and their functions in auxin biosynthesis, inactivation, transport, and signaling in land plants. Low genetic redundancy in model bryophytes enable unique assays, which are elucidating the design principles of the auxin signaling pathway. In this article, the physiological roles of auxin and regulatory mechanisms of gene expression and development by auxin in Bryophyta are reviewed.

Land plants evolved from ancestral charophycean algae approximately 500 million years ago (Morris et al. 2018). Among the extant land plants, bryophytes, including liverworts, mosses, and hornworts, are the basalmost lineages that diverged from other land plants. Recent phylogenetic analyses support the monophyly of bryophytes (Nishiyama et al. 2004; Cox et al. 2014; Wickett et al. 2014; Puttick et al. 2018), and the synapomorphy of "Bryophyta" is the haploid gametophyte-dominant life cycle. Germinated spores develop filamentous protonemata or globular sporelings, which in turn give rise to three-dimensionally (3D) organized gametophores (in leafy liverworts and mosses) or thalli (in thalloid liverworts and hornworts), from which reproductive organs that contain germ cells develop. After fertilization, diploid sporophytes, consisting of a foot, seta or intercalary meristem, and sporogenous tissue, develop on the haploid bodies (Ligrone et al. 2012) and finally generate haploid spores via meiosis.

The plant hormone auxin is known to be involved in almost all aspects of development through regulation of cell division, expansion, and differentiation in sporophyte-dominant angiosperms (Demeulenaere and Beeckman 2014; Dresselhaus and Schneitz 2014; Landrein and Vernoux 2014; Rios et al. 2014). Classic auxin- and inhibitor-treatment experiments showed that in gametophyte-dominant bryophytes, auxin plays diverse roles in both gametophyte and sporophyte generations. Additionally, recent molecular approaches in model bryophytes and omics data revealed that the core compo-

Cite this article as *Cold Spring Harb Perspect Biol* doi: 10.1101/cshperspect.a040055

nents in auxin biology, including biosynthesis, metabolism, transport, and signaling, are highly conserved among land plants. At present, two model bryophytes, the moss *Physcomitrium* (*Physcomitrella*) *patens* and the liverwort *Marchantia polymorpha*, are widely used in part because of the ease of transformation and gene disruption, low genetic redundancy compared with angiosperms, and simple developmental traits (Rensing et al. 2008, 2020; Ishizaki et al. 2016; Bowman et al. 2017; Lang et al. 2018; Montgomery et al. 2020). These model bryophytes provide unique platforms for investigating auxin biology.

This article provides an overview of auxin biology in Bryophyta, including developmental roles revealed by pharmacological assays, homeostatic regulation revealed by chemical approaches, design principles of the nuclear auxin signaling (NAS) pathway to which genetic studies in model bryophytes contributed, and mechanisms underlying developmental regulation revealed by molecular biological studies. See also Bowman (2016), Kato et al. (2018), and Thelander et al. (2018) for reviews of classical auxin studies in gemma and gemmalings of *M. polymorpha*, recent auxin studies in *M. polymorpha*, and recent auxin studies in *P. patens*, respectively.

PHARMACOLOGICAL EFFECTS OF AUXINS AND THEIR INHIBITORS ON BRYOPHYTE DEVELOPMENT

Mosses spend a long time as protonemata, which consist of functionally distinct, interconvertible feeding (chloronema) and foraging (caulonema) filaments (Cove 2005). Some lateral branches on caulonema give rise to buds through several unique divisions, which in turn form 3D gametophores with stem-leaf structures (Harrison et al. 2009; Kofuji and Hasebe 2014). Exogenously supplied auxin was shown to inhibit protonema branching, induce caulonema differentiation, and promote bud initiation in caulonema together with cytokinin (Johri and Desai 1973; Ashton et al. 1979; Sood and Hackenberg 1979; Thelander et al. 2018). Because auxin transport inhibitors, *N*-1-naphthylphthalamic acid (NPA) and 3,4,5-triiodobenzoic acid, inhibit auxin-mediated caulonema differentiation, active transport seems necessary to accumulate sufficient auxin (Cove and Ashton 1984).

Moss gametophores, which harbor stem cell regions at their apices where leaves and reproductive organs arise, develop multicellular rhizoids from the basal epidermis (Sakakibara et al. 2003; Harrison et al. 2009; Kofuji and Hasebe 2014) and show various responses to exogenous auxin depending on the type and concentration of auxin used. Generally, low auxin enhances shoot development by elongating stems and increasing leaves, whereas high auxin arrests growth by stunting stalks and inhibiting leaf formation (Ashton et al. 1979; Bennett et al. 2014b; Lavy et al. 2016). Exogenous auxin generally elongates leaves because of longitudinal hyperelongation of cells (Barker 2011). Rhizoid development is also strikingly stimulated by auxin application (Ashton et al. 1979; Nyman and Cutter 1981; Chopra and Vashistha 1990; Sakakibara et al. 2003).

Various liverworts also show enhanced and/or ectopic rhizoid development in response to exogenous auxin, while excessive auxin sometimes suppresses rhizoid formation (LaRue and Narayanaswami 1957; Ilahi and Allsopp 1969; Allsopp and Ilahi 1970a,b). In the case of *M. polymorpha* (for its life cycle, see Fig. 1), auxin application enhances rhizoid development in sporelings and gemmalings (Rousseau 1950; Tarén 1958; Maravolo and Voth 1966; Allsopp et al. 1968; Ishizaki et al. 2012; Flores-Sandoval et al. 2015), whereas L-kynurenine (L-Kyn), an indole-3-acetic acid (IAA)-biosynthesis inhibitor, decreases rhizoids in growing thalli (Fig. 1; He et al. 2011; Flores-Sandoval et al. 2015). Under normal growth conditions, gemmae give rise to rhizoids from their ventral side after detaching from gemma cups as a germination process, which is regulated by light via phytochromes (Otto and Halbsguth 1976; Inoue et al. 2016). However, in the presence of moderate concentrations of auxin, rhizoids also emerge from the dorsal sides of the gemmae, even in the dark, suggesting that exogenous auxin compensates for light signals during gemma

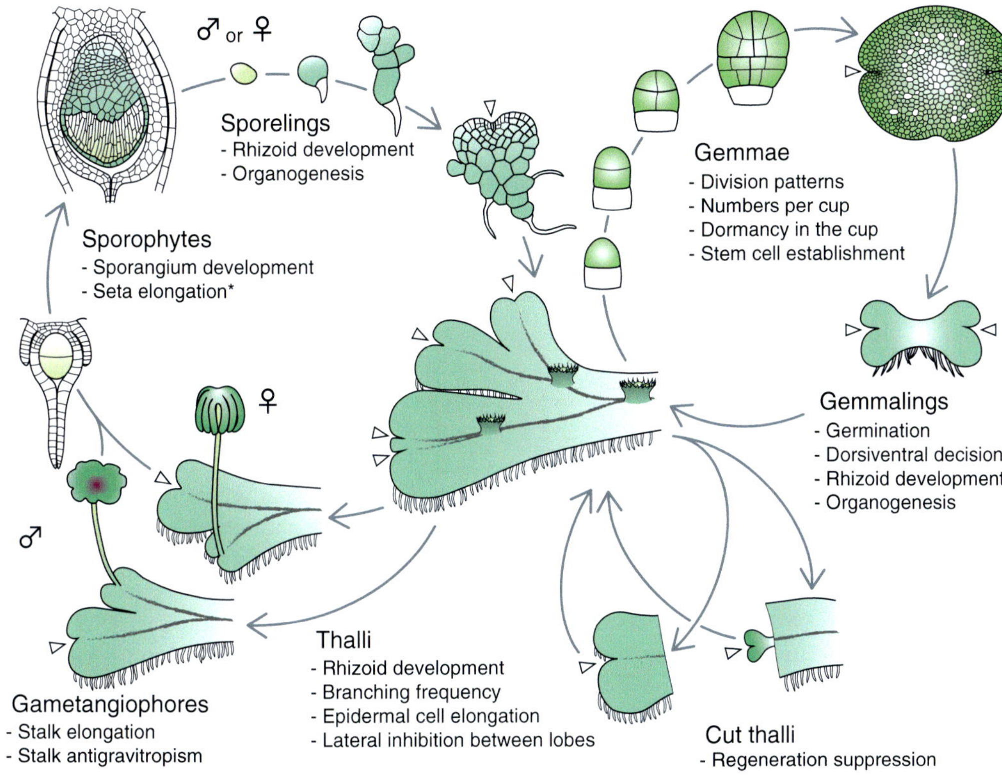

Figure 1. Auxin-regulated processes in thalloid liverwort throughout the life cycle. *Marchantia polymorpha* is depicted as a representative thalloid liverwort. Sporelings of *M. polymorpha* quickly establish 3D thalli, which have meristematic stem cell regions at the apical notches (arrowheads) and outgrow through repetitive bifurcations of the notches (Inoue 1960; Shimamura 2016; Solly et al. 2017). Thalli produce single-celled rhizoids on the ventral surface for substrate anchoring and develop the assimilatory organ air chamber with a single air pore and the asexual reproductive organ gemma cup on the dorsal surface, in which multicellular elliptic discoid propagules, gemmae, with notches at both longitudinal tips are produced and kept dormant (Barnes and Land 1908; Apostolakos et al. 1982; Shimamura 2016; Kato et al. 2020b; Suzuki et al. 2020). Outside the cup, gemmae germinate and develop into thalli via gemmalings. Upon thallus excision, new thalli regenerate only from the pieces without apical notches. Male and female gametangiophores, consisting of stalks and heads, emerge from notches of thalli, which in turn develop sporophytes through fertilization (Shimamura 2016). Auxin-regulated processes are listed below each developmental stage. Illustrations of the largest, the second-largest sporelings, and the mature gemma are modified from Kny (1890). See also Thelander et al. (2018) for auxin responses in the moss *Physcomitrella patens*. *Auxin responses not examined in *M. polymorpha*.

germination (Fig. 1; Rousseau 1954; Otto and Halbsguth 1976).

Auxin is also involved in gemma dormancy within gemma cups. In the thalloid liverwort *Lunularia cruciata*, surgical excision of apical notches caused gemma germination within gemma cups on the remaining thalli, which was suppressed by IAA applied to the cut faces (LaRue and Narayanaswami 1957), suggesting that mobile signals from apical notches keep gemmae dormant within gemma cups and that IAA is a potential candidate. *M. polymorpha* grown in the presence of L-Kyn showed an elevated ratio of gemma cups containing nondormant gemmae, supporting the concept that IAA facilitates gemma dormancy within gemma cups (Fig. 1; Eklund et al. 2015). Ectopic regeneration is also suppressed by apical notch–

derived mobile signals, probably IAA, because excision of apical notches caused regeneration from the cutting faces of the remaining thalli, which was suppressed by IAA application in *L. cruciata* and *M. polymorpha* (Fig. 1; Vöchting 1885; LaRue and Narayanaswami 1957; Rota and Maravolo 1975; Nishihama et al. 2015). Suppression of regeneration by auxin has also been reported in several phylogenetically distant liverworts (Allsopp and Ilahi 1970c; Ilahi and Allsopp 1970b). Thus, in general, asexual propagation is prevented by auxin.

Inhibitory effects of IAA on apical notches are also observed in intact thalli. Davidonis and Munroe (1972) showed that in *M. polymorpha*, notch bifurcations often resulted in two lobes with different growth rates and that surgical separation at an early stage after bifurcation released the smaller lobe from inhibition, which was suppressed by IAA application. Lateral inhibition seems to occur between the bifurcated lobes mediated by IAA (Fig. 1).

Moss gametophores also show inhibition by the apex: lateral branch primordia and basal stem regeneration were reactivated by removing apices, both of which were suppressed by IAA application to the decapitated stem tips (MacQuarrie and von Maltzahn 1959; Coudert et al. 2015). Auxin is also involved in regeneration in the moss *Splachnum ampullaceum* because regeneration of isolated leaves was promoted or inhibited by IAA application at low or high concentrations, respectively (MacQuarrie and von Maltzahn 1959).

Thallus shapes are also affected by auxin as L-Kyn caused a dose-dependent reduction of thallus growth in *M. polymorpha* (Solly et al. 2017). In general, exogenous auxin strongly inhibits growth at high concentrations, but slightly enhances growth at low concentrations in liverworts (Ilahi and Allsopp 1969, 1970a; Allsopp and Ilahi 1970b; Ishizaki et al. 2012), although the detailed effects on thallus growth vary according to plant species and auxin types. In *Riccardia multifida*, various auxins promoted and inhibited thallus elongation and branching at low and high concentrations, respectively. In *Moerckia flotowiana*, *Pellia epiphylla*, and *Makinoa crispata*, high auxin suppressed wing development and causes cylindrical thalli (Ilahi and Allsopp 1969). These results suggest that auxin greatly contributes to overall plant shapes.

In vascular plants, auxin is known to be a major player in tropic responses to suit environmental cues (Retzer et al. 2014). In *M. polymorpha*, the dorsiventrality of gemmalings is determined by sensing gravity soon after gemma germination, and exogenous auxin disturbs proper dorsiventral determination (Fig. 1; for review, see Bowman 2016). Auxin was also reported to elongate and bend stalks of gametangiophores, although the correlation with gravitropic or phototropic responses was not discussed (Fig. 1; Rousseau 1953). In *P. patens*, auxin-mediated negative gravitropism in the dark and positive phototropism were observed in gametophores (Cove et al. 1978; Bennett et al. 2014a; Bao et al. 2015). In the moss *Ceratodon purpureus*, the auxin efflux inhibitors NPA and pyrenoylbenzoic acid (PBA) inhibited the gravitropic curvature of protonemal cells (Schwuchow et al. 2001).

Auxin functions in sporophyte generation. Auxin promoted seta elongation in the liverwort *P. epiphylla* and the moss *Polytrichum ohioense* (Fig. 1; Schnepf et al. 1979; Thomas 1980; Poli et al. 2003). In the hornworts *Anthoceros laevis* and *Anaxyrus punctatus*, 2,4-dichlorophenoxyacetic acid (2,4-D) inhibited the elongation of sporophytes but increased their diameter (Rousseau 1958). In the hornwort *Phaeoceros personii*, although the application of IAA failed to promote or impede sporophyte elongation, an auxin antagonist, *p*-chlorophenoxyisobutyric acid (PCIB), reduced sporophyte elongation (Poli et al. 2003). In *P. patens*, NPA has been reported to cause branching of sporangia (Fujita et al. 2008).

AUXIN HOMEOSTASIS: BIOSYNTHESIS, METABOLISM, AND TRANSPORT

Auxin levels are controlled by biosynthesis, temporary inactivation, permanent degradation, intercellular transport, and isolation into several intracellular compartments (not mentioned here). In vascular plants, IAA is a major intrinsic auxin, which is also detected in bryophyte species and more basal algal sisters (Löbenberg 1959; Fries 1964; Schneider et al. 1967; Ashton et al. 1985; Atzorn et al. 1990; Cooke et al. 2002). In *Arabidopsis thaliana*, the indole-3-pyruvic acid (IPyA) pathway is known to be a major IAA biosynthesis route, where tryptophan is converted to IPyA and then to IAA. Each reaction is catalyzed by TRYPTOPHAN AMINOTRANSFERASE OF ARABIDOPSIS 1/TRYPTOPHAN AMINOTRANSFERASE-RELATED (TAA1/TAR) and YUCCA (YUC), respectively (for review, see Kasahara 2016), both of which are conserved in all three taxa of Bryophyta (Rensing et al. 2008; Eklund et al. 2015; Bowman et al. 2017; Li et al. 2020; Zhang et al. 2020). TAA1/TAR is inactivated by phosphorylation at an evolutionarily conserved Thr residue in an auxin-dependent manner. MpTAA, the sole TAA1 homolog in *M. polymorpha*, loses its enzymatic activity by a phosphorylation-mimicking mutation at the corresponding residue (Wang et al. 2020). The knockout of Mp*TAA* at the sporeling stage results in a cell mass phenotype without any organ differentiation, which is partially rescued by exogenous IAA (Eklund et al. 2015). Likewise, the IPyA pathway mediated by PpTARs and PpYUCs is the main IAA biosynthesis route in *P. patens* (Landberg et al. 2020), which is consistent with competitive effects on caulonema differentiation between a YUC inhibitor and exogenous auxin (Tsugafune et al. 2017). These results suggest that the IPyA pathway and its regulatory systems exist in the common ancestor of land plants and that the IPyA pathway is a major IAA source and plays pivotal roles in the development and environmental responses of bryophytes. Indeed, Mp*TAA* was reported to be regulated under the control of the circadian clock to regulate the nyctinastic movement of *M. polymorpha* thalli (Lagercrantz et al. 2020). However, other IAA biosynthetic pathways have not yet been the focus in bryophytes, although their existence is implied by the finding that Mp*taa* plants still contain detectable levels of IAA (Eklund et al. 2015).

In addition to IAA, phenylacetic acid (PAA) is detected in a broad range of plants, including angiosperms, *P. patens*, and *M. polymorpha* (Sugawara et al. 2015). In *Arabidopsis*, IAA and PAA show different biochemical features, such as affinity to coreceptors and transport profiles. Although PAA shows auxin activity in *P. patens*, physiological differences between IAA and PAA remain unclear in bryophytes (Sugawara et al. 2015).

In *Arabidopsis*, IAA is temporarily inactivated by GRETCHEN HAGEN 3 (GH3)-mediated conjugation with amino acids (Staswick et al. 2005) and by UDP-glycosyltransferase-mediated ester linking with glucose (Jackson et al. 2001), or permanently inactivated by DIOXYGENASE FOR AUXIN OXIDATION 1 (DAO1)-mediated conversion into 2-oxindole-3-acetic acid (ox[IAA]; Porco et al. 2016; Zhang et al. 2016). Both amide- and ester-linked IAA conjugates were detected in the three Bryophyta taxa. oxIAA has been detected in some liverwort and moss species (Sztein et al. 1995, 1999, 2000; Ludwig-Müller et al. 2009b; Záveská Drábková et al. 2015; Landberg et al. 2020). Sztein et al. proposed that free-IAA levels are primarily regulated via the balance between biosynthesis and degradation in liverworts, whereas they are regulated by competitive conjugation versus hydrolysis of conjugates in hornworts and mosses, based on their observation that hornwort and moss species exhibited higher conjugation rates than liverwort species (Sztein et al. 1995, 1999, 2000; for review, see Cooke et al. 2002). However, Záveská Drábková et al. (2015) reached the opposite conclusion that liverworts preferred conjugation, whereas mosses favored degradation to maintain IAA homeostasis because liverwort species exhibited higher amino acid conjugation rates than moss species in their analysis. Although these studies could not be simply compared because of the differences in plant species and growth conditions, the conservation of both conjugation and degradation strategies to sustain IAA homeostasis in Bryophyta could be determined. Indeed, two GH3 homologs in *P. patens* have been shown to mediate IAA conjugation to amino acids (Ludwig-Müller et al. 2009a,b) and are thought to regulate free IAA levels under unfavorable environmental conditions (Mittag et al. 2015). Further molecular studies of GH3 and other IAA metabolic enzymes, which are conserved in the genomes of bryophytes (Rensing et al. 2008; Bowman et al. 2017; Li et al. 2020; Zhang et al. 2020), are needed to clarify the strategies for IAA homeostasis.

In *Arabidopsis,* in addition to several active auxin transporters, including PIN-FORMED (PIN) (for review, see Grones and Friml 2015), a plasmodesmatal auxin transport controlled by callose deposition has been suggested, which occurs in *P. patens* (Han et al. 2014; Coudert et al. 2015). *P. patens* encodes polarly membrane-localized long-PIN homologs (PpPINA/B/C) and an endoplasmic reticulum-localized short-PIN homolog ([PpPIND]; Viaene et al. 2013, 2014; Bennett et al. 2014a,b). Generally, mutations of the former result in an auxin-excessive phenotype, whereas their overexpression results in auxin-repressive phenotypes, which are probably caused by trapping and overdrainage of IAA, respectively (Bennett et al. 2014b; Viaene et al. 2014). These facts suggest that the long PpPINs act as efflux carriers. Indeed, basipetal polar transport could be detected by tracer assays of radiolabeled IAA within multicellular rhizoids of the moss *Funaria hygrometrica* (Fig. 2A; Rose and Bopp 1983). Tracer assays did not demonstrate polar transport in stalks of gametophores along the longitudinal axis but did show basipetal polar transport in setae of several mosses (Fig. 2B; Poli et al. 2003; Fujita et al. 2008). Meanwhile, based on the velocity and direction of transport, IAA is thought to be carried by nonpolar active transport or simple diffusion in setae of liverworts or young sporophytes of hornworts, respectively (Fig. 2C), suggesting that auxin transport strategies in sporophytes were independently evolved in bryophytes. Radiolabeled IAA was transported basipetally in thalli of *M. polymorpha* (Binns and Maravolo 1972; Maravolo 1976; Gaal et al. 1982), which was inhibited by 2,3,5-triiodobenzoic acid, a transport inhibitor, suggesting that an active transporter(s) basipetally carries IAA (Fig. 2D; Maravolo 1976).

PRINCIPLE OF AUXIN SIGNALING

Auxin regulates gene expression through the NAS pathway, comprising TRANSPORT INHIBITOR RESPONSE 1/AUXIN SIGNALING

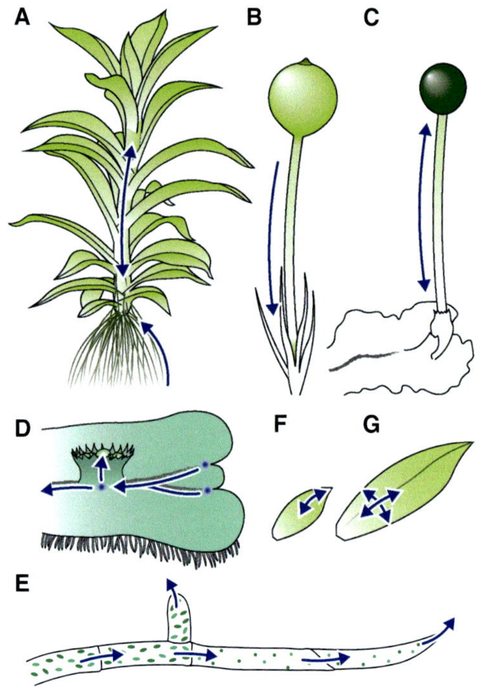

Figure 2. Auxin transport in bryophytes. Estimated auxin transport (blue arrows) in stalks (*A*), rhizoids (*A*), setae (*B*), protonemata (*E*), young leaves (*F*), and larger leaves (*G*) in mosses, and setae (*C*) and thalli (*D*) of liverworts.

F-BOX (TIR1/AFB) receptors, AUXIN/INDOLE-3-ACETIC ACID (AUX/IAA) transcriptional repressors, and AUXIN RESPONSE FACTOR (ARF) transcription factors (Fig. 3A; Table 1). TIR1/AFBs, an F-box-containing subunit of an SCF-type E3 ubiquitin ligase complex, form coreceptor complexes with AUX/IAAs via auxin, which in turn facilitate ubiquitination and subsequent degradation of AUX/IAAs (Gray et al. 1999, 2001; Dharmasiri et al. 2005; Kepinski and Leyser 2005; Tan et al. 2007). Canonical AUX/IAAs have three conserved domains: domain I (DI), domain II (DII), and the Phox and Bem1 domain ([PB1], also called domain III/IV). Each of these interacts with TOPLESS (TPL)-type corepressors, TIR1/AFBs, and PB1 domains of ARFs and AUX/IAAs, respectively (Kim et al. 1997; Dharmasiri et al. 2005; Kepinski and Leyser 2005; Szemenyei et al. 2008). DII is known as the degron region. ARFs consist of a DNA-binding domain (DBD), a middle region (MR), and a PB1 domain. Each of these binds to *cis*-elements, regulates transcriptional activity, and homo- and heterotypically interacts with ARFs and AUX/IAAs, respectively (Kim et al. 1997; Ulmasov et al. 1997, 1999a,b; Tiwari et al. 2003). Auxin releases ARFs from AUX/IAAs via TIR1/AFB-mediated degradation and hence facilitates ARF-mediated transcriptional regulation. Phylogenetic analyses demonstrated NAS establishment in the common ancestor of land plants (Bowman et al. 2017; Flores-Sandoval et al. 2018a; Mutte et al. 2018).

In *P. patens*, four TIR1/AFB and three AUX/IAA homologs interact with each other in an auxin-dependent manner (Prigge et al. 2010). *P. patens* expressing stabilized mutants of PpIAA, which had been originally forward-genetically isolated as auxin-insensitive mutants, and *PpAFB* knockdown plants showed auxin depletion phenotypes, confirming that PpAFB–PpIAA pairs act as auxin coreceptors (Ashton et al. 1979; Prigge et al. 2010). *M. polymorpha* encodes a single canonical AUX/IAA homolog, MpIAA. Overexpression of a DII-mutated MpIAA protein conferred

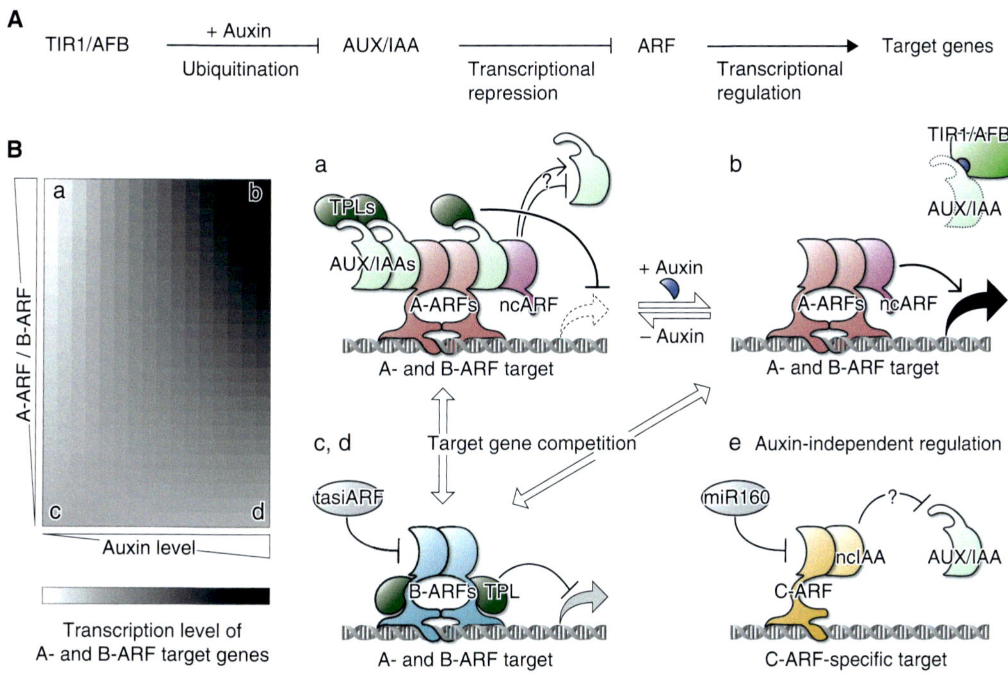

Figure 3. Deduced design principles of nuclear auxin signaling (NAS) from studies with bryophytes. (*A*) Components and basic scheme of the NAS pathway. (*B*) Heat map showing putative expression levels of A- and B-ARFs' common target genes (*left*) and functional models of the NAS components (*right*) at different A-ARF/B-ARF ratios and auxin levels (*a–d*). A-ARFs (and noncanonical ARFs [ncARFs]) recruit AUX/IAAs and thus repress target gene transcription at low auxin levels (*a*). A-ARFs (and ncARFs) activate target gene transcription at high auxin levels as TIR1/AFBs interact with AUX/IAAs via auxin and promote AUX/IAA degradation (*b*). B-ARFs buffer AUX/IAA-mediated intense repression at low auxin levels (*c*) and A-ARF-mediated activation at high auxin levels (*d*) by competing with A-ARFs for binding to their target DNA sites, which fine-tunes auxin responses. C-ARFs neither share target DNA sites with A- or B-ARFs, nor interact with AUX/IAAs, and thus regulate distinct target genes in an auxin-independent manner (*e*). ncIAA may prevent canonical AUX/IAAs from binding to C-ARFs. Note that the functions of ncARFs and ncIAAs are hypothetical. These models are based on the following papers and our deduction, in addition to the well-known NAS model established in angiosperms (*A*): Flores-Sandoval et al. (2018a,b), Kato et al. (2017, 2020a), Lavy et al. (2016), Lv et al. (2020), Mutte et al. (2018), and Tao and Estelle (2018).

Table 1. Copy numbers and IDs of the nuclear auxin signaling (NAS) component homologs in bryophytes

	Marchantia polymorpha	*Physcomitrium patens*	*Anthoceros angustus*	*Anthoceros agrestis* (Bonn)	*Anthoceros agrestis* (Oxford)	*Anthoceros punctatus*	*Arabidopsis thaliana*
TIR1/AFB	1 (MpTIR1; Mp6g02750)	4 (PpAFB1; Pp3c23_5870, PpAFB2; Pp3c20_17150, PpAFB3; Pp3c23_11300, PpAFB4; Pp3c24_9800/ Pp3c24_9814)	1 (AANG003364)	1 (AagrBONN344.691)	1 (AagrOXF000012l.465)	1 (Apun000117l.230)	6
AUX/IAA	1 (MpIAA; Mp6g05000)	3 (PpIAA1A; BAB71765.1, PpIAA1B; BAB71766.1/ Pp3c8_14720, PpIAA2; EDQ70046/ Pp3c24_6610)	1 (AANG009773)	1 (AagrBONN117.2250)	1 (AagrOXF000103l.142)	1 (Apun000023l.749)	28
ncIAA	1 (MpNCIAA/ MpAXI2; Mp8g07850)	0	1 (AANG012370)	1 (AagrBONN117.1187)	1 (AagrOXF000083l.132)	1 (Apun000058l.129)	1
TPL/TPL-RELATED	1 (MpTPL; Mp7g17410)	2 (PpTPL1; Pp3c15_9880, PpTPL2; Pp3c9_21250)	1 (AANG014214)	1 (AagrBONN340.1726)	1 (AagrOXF000065l.67)	1 (Apun000149l.87)	4

Cite this article as *Cold Spring Harb Perspect Biol* doi: 10.1101/cshperspect.a040055

A-ARF	1 (MpARF1; Mp1g12750)	8 (PpARFa1; Pp3c1_14480, PpARFa2; Pp3c1_14440, PpARFa3; Pp3c2_25890, PpARFa4; Pp3c13_4720, PpARFa5; Pp3c26_11550, PpARFa6; Pp3c17_19900, PpARFa7; Pp3c14_16990, PpARFa8; Pp3c1_40270)	1 (AANG012813)	1 (AagrBONN340.233)	1 (AagrOXF000126l.206)	1 (Apun000045l.202)	5
B-ARF	1 (MpARF2; Mp4g11820)	4 (PpARFb1; Pp3c27_60, PpARFb2; Pp3c16_6100, PpARFb3; Pp3c5_9420, PpARFb4; Pp3c6_21370)	0	0	0	0	15
C-ARF	1 (MpARF3; Mp1g07070)	3 (PpARFc1A; Pp3c4_12970, PpARFc1B; Pp3c4_13010, PpARFc2; Pp3c6_26890)	3 (AANG000742, AANG000743, AANG012642)	4 (AagrBONN340.199, AagrBONN344.1784, AagrBONN369.337, AagrBONN369.338)	4 (AagrOXF000006l.365, AagrOXF000069l.94, AagrOXF000069l.95, AagrOXF000126l.173)	4 (Apun000045l.242, Apun000089l.87, Apun000089l.88, Apun000144l.74)	3

Continued

Table 1. *Continued*

	Marchantia polymorpha	*Physcomitrium patens*	*Anthoceros angustus*	*Anthoceros agrestis* (Bonn)	*Anthoceros agrestis* (Oxford)	*Anthoceros punctatus*	*Arabidopsis thaliana*
Noncanonical (nc)ARF	1 (MpNCARF/ MpAXI1; Mp2g02890)	2 (PpARFd1; Pp3c6_26890, PpARFd2; Pp3c15_9710)	0	0	0	0	0

Homologs were obtained from MarpolBase (www.marchantia.info) for the liverwort *M. polymorpha*, and from Phytozome v13 (www. phytozome-next.jgi.doe.gov) and NCBI (www.ncbi.nlm.nih.gov) for the moss *P. patens* from DRYAD (www.datadryad.org/stash/dataset/doi:10.5061/dryad.msbcc2ftv) for hornwort *A. angustus*, from the University of Zurich (www.hornworts.uzh.ch/en.html) for the hornwort *A. agrestis* (Bonn), *A. agrestis* (Oxford), and *A. punctatus*. AagrOXF_evm.model.utg000008l.194.1 and Apun_evm.model.utg000039l.220.1 are omitted from TIR1/AFB candidates here, since they are closer to PpXFBs (Pp3c22_6020 and Pp3c18_16630), which are F-box/Leucine Rich Repeat proteins absent in angiosperms (Prigge et al. 2010) and are more homologous to jasmonate/oxo-phytodienoic acid receptors than TIR1/AFBs (see a phylogenetic tree in Mutte et al. 2018). Names: "AagrBONN_evm.model.Sc2ySwM_XXX.XXXX," "AagrOXF_evm.model.utgXXXXl.XX," and "Apun_evm.model.utgXXXXXXl.XXX" are modified as "AagrBONNXXX.XXXX," "AagrOXFXXXXl.XX," and "ApunXXXXXXl.XXX," respectively, to save space.

auxin tolerance, suggesting a conserved role of MpIAA in auxin perception (Kato et al. 2015).

Although several studies have used stabilized, dominant mutants of AUX/IAAs for functional analyses, reports of their loss-of-function phenotypes were limited. Lavy et al. (2016) performed a prominent mutational study with AUX/IAA-null mutants in *P. patens* in which all three *PpIAA* homologs were knocked out. The *Ppiaa* triple mutants showed no responses to exogenous auxin or L-Kyn, leading to an excessive auxin-response phenotype. Expression of a PpIAA1a mutant incapable of binding to TPL, which lacked an ethylene-responsive element binding factor-associated amphiphilic repression (EAR) motif (Causier et al. 2012) within DI and an EAR-like motif between DI and DII, still caused an excessive auxin phenotype, but slightly restored growth and auxin responsiveness, suggesting EAR-independent repressive functions of AUX/IAAs (Tao and Estelle 2018). Considering that *Arabidopsis* ARF5/MONOPTEROS interacts with chromatin remodeling ATPases to exert transcriptional regulation (Wu et al. 2015), simple binding of AUX/IAAs to ARFs could inhibit recruitment of other required factors, thus repressing ARF functions (Tao and Estelle 2018). A PB1 mutant of PpIAA1a, incapable of binding to itself nor PpARFs, failed to complement the null *Ppiaa* mutant. However, another PB1 mutant of PpIAA1a, incapable of self-binding but capable of binding to PpARFs, slightly rescued the mutant. Taken together, the PB1-mediated heterotypic interaction with ARFs is critical for the function of AUX/IAA, and their homotypic interactions and subsequent oligomerization contribute to, but are not essential for, their repressive functions (Tao and Estelle 2018).

ARF transcription factors are phylogenetically classified as A, B, and C (Finet et al. 2013). Comprehensive phylogenetic analysis in green plants showed that ancestral C-ARFs diverged from ancestral A/B-ARFs before the TIR1/AFB–AUX/IAA auxin coreceptor system (Flores-Sandoval et al. 2018a; Mutte et al. 2018). Based on sequence traits in the MR and results of transactivation assays, A- and B-ARFs are generally thought to be transcriptional activators and repressors, respectively (Ulmasov et al. 1999a; Tiwari et al. 2003; Kato et al. 2015). Overexpression of A- and/or B-ARF homologs in the *Ppiaa* triple mutant revealed that A- and B-ARFs share common target genes and that B-ARFs exert repressive functions in an AUX/IAA-independent manner (Lavy et al. 2016). Counterintuitively, knockout of B-ARFs reduced auxin sensitivity, and overexpression of A-ARFs down-regulated the transcription of some auxin-responsive genes. Based on these conflicting results, Lavy et al. (2016) proposed a model in which B-ARFs compete for target gene sites with A-ARFs, and affect the occupancy of A-ARFs and AUX/IAAs, providing a buffering capacity for fine-scale regulation of target genes.

M. polymorpha encodes three canonical ARFs, each belonging to class A, B, or C (Flores-Sandoval et al. 2015; Kato et al. 2015; Bowman et al. 2017). Knockout of Mp*ARF1*, the sole A-ARF, conferred auxin insensitivity and caused various developmental defects, including narrow twisted thalli, small tubular gemma cups, decreased gemmae per cups, nondormant gemmae, and disrupted patterning in gemma development (Kato et al. 2017). Interestingly, inhibition of NAS via DII-mutated nondegradable MpIAA resulted in very severe growth defects, such as formation of cell masses, ectopic air pores, ectopic serrated structures, and/or adventitious buds, and failure to develop thalli (Kato et al. 2015). Furthermore, the aforementioned Mp*taa* auxin biosynthesis mutant or overexpressors of a bacterial auxin-conjugating enzyme formed cell masses (Eklund et al. 2015). Mp*arf1* phenotypes are much milder than those in that they develop thalloid organs. One possible explanation is that AUX/IAA proteins are not recruited to target sites in the absence of the A-ARF, allowing its target genes to be transcribed at basal levels (Kato et al. 2017, 2018).

Kato et al. (2020a) investigated whether the conserved domains of MpARFs could be interchangeable by expressing various domain-swapped constructs in the Mp*arf1* mutant background. Chimeric MpARF1, which exchanged the DBD with MpARF2 but not MpARF3, restored development and auxin responsiveness, suggesting that MpARF1 and MpARF2 share target genes, while MpARF3 targets another group of genes. The MR of MpARF1 was interchangeable with that of neither MpARF2 nor MpARF3. Interestingly, MpARF2 was shown to interact directly with MpTPL through a conserved "LFG motif," and small fragments containing the LFG motif but not a mutated motif from all MpARFs exhibited repressor activities upon coexpression with MpTPL. Thus, the MR of MpARF2 (and probably that of MpARF3) exerts auxin-independent transcriptional repression via direct recruitment of MpTPL (Kato et al. 2020a).

All combinations of homo- and heterotypic interactions among the PB1 domains of MpARFs and MpIAA were demonstrated except for MpARF3–MpARF3 homotypic interactions (Kato et al. 2015). In the domain-swapping assay, the PB1 domain of MpARF1 was not interchangeable with that of MpARF3. Chimeric MpARF1 with its PB1 domain swapped with that of MpARF2 partially rescued developmental defects of Mp*arf1* but did not rescue responsiveness to exogenous auxin. In general, PB1 domains have positive and negative faces and oligomerize through consecutive interactions between the two faces (Korasick et al. 2014). Docking of structural models, however, predicted that both the positive and negative faces of MpARF2 and MpARF3 PB1 domains interact with one face of the MpIAA PB1 domain, while one face of the MpARF1 PB1 domain does so, allowing further oligomeric interactions, which seems critical for establishing sufficient auxin responsiveness. Chimeric MpARF1, which has a distinct oligomerization domain in place of the PB1 domain, rescued *Mparf1* growth defects but not auxin insensitivity, suggesting that MpARF1–MpARF1 homotypic interactions through the PB1 domain are critical for transcriptional activation by MpARF1 and that auxin response requires MpARF1–MpIAA oligomerization (Kato et al. 2020).

These results offer design principles of the auxin response system: in the absence of auxin, A-ARFs oligomerize with AUX/IAAs causing transcriptional repression, while in the presence of auxin, A-ARFs interact with each other and promote target transcription. B-ARFs compete for target sites with A-ARFs and A-ARF–AUX/IAA complexes to fine-tune auxin responsiveness (Kato et al. 2020a). Even though B-ARFs recruit TPLs, B-ARFs are speculated to exert weaker repression than A-ARF–AUX/IAA complexes, as the latter is capable of recruiting many more TPLs via oligomerized AUX/IAAs. Thus, B-ARFs could buffer both transcriptional activation and repression at high and low levels of auxin, respectively (Fig. 3B).

B-ARFs are known to be posttranscriptionally regulated by *trans*-acting small interfering RNA (tasiRNA) generated by miR390 from *TAS3* gene products in angiosperms (Allen et al. 2005; Williams et al. 2005). *P. patens* and *M. polymorpha* also have conserved miR390, suggesting the existence of a similar regulatory system (Axtell et al. 2007; Arif et al. 2012; Lin et al. 2016; Plavskin et al. 2016; Xia et al. 2017). In *P. patens*, it is demonstrated that *TAS3*-derived tasiRNAs target transcripts of B-ARF genes to restrict their expression at a random subset of protonemal tip cells, which increases sensitivity and robustness of auxin responses. The tasiARFs thus allow stochastic modulation of cell differentiation and developmental plasticity in response to environmental cues (Plavskin et al. 2016). In *M. polymorpha*, conditional knockout of Mp*ARF2* in gemmalings or thalli caused severe growth arrest, suggesting that B-ARF plays critical roles in development (Kato et al. 2020a). Interestingly, hornworts do not encode a B-ARF in their genome, which was probably lost after divergence from a liverwort/moss common ancestor (Flores-Sandoval et al. 2018a; Mutte et al. 2018; Li et al. 2020; Zhang et al. 2020), suggesting the acquisition of a distinct buffering mechanism for A-ARFs.

C-ARFs are posttranscriptionally regulated by miR160 (Mallory et al. 2005; Wang et al. 2005), whose target sites are conserved among land plants (Axtell et al. 2007; Lin et al. 2016; Tsuzuki et al. 2016; Flores-Sandoval et al. 2018a). In contrast to MpARF1 and MpARF2, neither knockout nor overexpression of MpARF3 affected auxin sensitivity, unless its miR160-resistant form is overexpressed, leading to auxin-insensitive cell masses (Flores-

Sandoval et al. 2018a). Additionally, Mp*ARF3* is not a part of the auxin-related coexpression module (Flores-Sandoval et al. 2018b). Together with the phylogenetic position, these studies suggest that C-ARFs are not under the control of auxin; rather, it is proposed that the TIR1/AFB–AUX/IAA- and miR160-mediated pathways evolved independently to regulate preexisting A/B- and C-ARF transcriptional networks, respectively (Flores-Sandoval et al. 2018a,b). Indeed, as the DBDs of MpARF1 and MpARF3 were unexchangeable, many, if not all, A- and B-ARFs seem to regulate gene sets distinct from target genes of C-ARFs (Kato et al. 2020a).

Phylogenetic analysis indicated the existence of conserved noncanonical AUX/IAAs (ncIAAs) lacking DI and DII (Flores-Sandoval et al. 2018a; Mutte et al. 2018), although ncIAA homologs are absent from *P. patens*. Liverworts, mosses, and lycophytes, but not hornworts and other vascular plants, have noncanonical ARFs (ncARFs) lacking DBD, which diverged from A-ARFs (Paponov et al. 2009; Mutte et al. 2018). *M. polymorpha* has one ncIAA, MpNCIAA/MpAXI2 and one ncARF, MpNCARF/MpAXI1 (Bowman et al. 2017). Although Mp*ncarf* mutants developed relatively normal thalli, they showed tolerance to exogenous auxin and reduced fold changes of some auxin-up-regulated genes (Mutte et al. 2018). Based on these results, Mutte et al. (2018) proposed two hypothetical roles of ncARFs: ncARFs protect canonical ARFs from AUX/IAAs and/or recruit cofactors to ARF-target loci via interaction with canonical ARFs. In contrast to the first model, given that oligomeric interactions of AUX/IAAs to A-ARFs are required for sufficient auxin responsiveness (Kato et al. 2020a; see above), ncARFs may assist the A-ARF–AUX/IAA interaction as spacers (Fig. 3B). Meanwhile, the roles of MpNCIAA in NAS are unclear, as Mp*nciaa* mutants showed comparable auxin responses to wild-type (Mutte et al. 2018). AtIAA33, the sole ncIAA ortholog in *Arabidopsis*, has recently been demonstrated to interact with C-ARFs, AtARF10/16, which protects AtARF10/16 from canonical AUX/IAA-mediated repression (Lv et al. 2020). Additionally, like Mp*ARF3*, Mp*NCIAA* is expressed independently of the auxin coexpression group (Flores-Sandoval et al. 2018b). Although further experiments are needed, ncIAAs may play a role in sequestering C-ARFs from NAS (Fig. 3B).

MOLECULAR MECHANISMS UNDERLYING DEVELOPMENTAL REGULATION

As seen in the previous sections, auxins and their inhibitors cause multiple responses in bryophytes. For these context-dependent, diverse auxin responses, its localization and responsiveness should be spatiotemporally controlled. Accordingly, reporters of the soybean *GmGH3* promoter (Hagen et al. 1991), which operate under the control of NAS also in a bryophyte (Kato et al. 2015, 2017), exhibit high activities in multiple tissues or organs throughout the life cycle in both *P. patens* (Bierfreund et al. 2003; Fujita et al. 2008; Landberg et al. 2013) and *M. polymorpha* (Ishizaki et al. 2012). Additionally, the synthetic auxin response reporter *DR5revV2* shows correlated expression patterns with those of *GmGH3* (Thelander et al. 2019). As an ARF-independent auxin-sensing reporter, a moss-optimized "ratiometric version of two D2s," PpR2D2, was recently developed. PpR2D2 semiquantitatively visualizes auxin sensitivities as AUX/IAA degradation rates and enables further fine monitoring in *P. patens* (Liao et al. 2015; Thelander et al. 2019).

In protonema colonies of *P. patens*, PpR2D2 shows a gradient along filaments where auxin-sensing levels are high in proximal cells and low in tip cells (Thelander et al. 2019). Both the *GmGH3* and *DR5revV2* reporters show stronger activity in proximal cells than in tip cells (Thelander et al. 2019). Thus, apical stem cells show relatively low auxin sensing and response levels. Polar localization of long PINs on the plasma membrane toward tips suggests acropetal auxin transport and its accumulation in and drainage from tip cells, which is needed for proper caulonema differentiation (Fig. 2E; Viaene et al. 2014). Although the auxin source is still unclear, PpSHI, a positive regulator of auxin biosynthesis homologous to *Arabidopsis* SHORT INTERNODES/STYLISH ([SHI/STY]; Sohlberg et al.

2006;Ståldal et al. 2008; Eklund et al. 2010a), is specifically expressed in caulonema (Eklund et al. 2010b). The above-mentioned B-ARF gradient generated by tasiARFs adds another tier of auxin response regulation for caulonema differentiation and bud formation (Plavskin et al. 2016). Under the control of auxin, homologs of ROOT HAIR DEFECTIVE 6-LIKEs and *Lotus japonicus* ROOTHAIRLESS LIKE (PpRSL1/2 and PpLRL1/2, respectively) function in caulonema differentiation (Jang and Dolan 2011; Tam et al. 2015). For gametophore development, auxin-responsive transcription factors, AINTEGUMENTA, PLETHORA, and BABY BOOM homologs (PpAPBs), act as master regulators of bud formation (Aoyama et al. 2012).

Consistent with the undetectable polar transport with radiolabeled IAA in gametophore stems (Fujita et al. 2008), *Pppina/b* double mutants, and NPA treatment have only a minor effect on branching patterns; rather, plasmodesmata are supposed to be a major transport route (Coudert et al. 2015). Nevertheless, since long *PpPIN* mutants exhibit a multifaceted phenotype in gametophores, such as stem elongation, leaf abnormalities, and disruption of tropic responses, PIN-mediated transport also plays pleiotropic roles in gametophores (Bennett et al. 2014b). During leaf development, PpPINA/B are expressed in apical regions and show bipolar localization on both apical and basal plasma membranes at the early stages. The expression zones move toward the basal region as leaves develop, where PpPINA/B localize on all sides. Such a shift in expression zones correlates with a similar shift in cell elongation (Viaene et al. 2014). The PpR2D2 reporter indicated that high auxin-sensing zones also expand from apex to base during leaf development (Fig. 2F,G; Thelander et al. 2019). Auxin application at a certain concentration transforms leaves into narrow, elongated shapes (Barker 2011; Bennett et al. 2014b). Knockout of long PINs also caused long narrow leaves because of enhanced longitudinal cell elongation and reduced mediolateral cell division, probably caused by auxin trapping (Bennett et al. 2014b). Taken together, long PINs affect auxin responsiveness via transport, which regulates the transition from cell division to expansion in leaf development (Bennett et al. 2014b; Viaene et al. 2014; Thelander et al. 2019).

Auxin promotes rhizoid development in both liverworts and mosses. In *P. patens*, the PpR2D2 reporter revealed that auxin-sensing levels are low in rhizoid precursors and show a gradient from proximal to tip after the three-cell stage (Thelander et al. 2019). Consistently, *GmGH3* and *DR5revV2* reporters show gradient patterns in multicelled stages, although they are also active in precursors (Thelander et al. 2019). On the other hand, *PpSHI* is highly expressed in tip cells (Eklund et al. 2010b). Considering the basipetal transport of radiolabeled IAA in *F. hygrometrica* rhizoids (Rose and Bopp 1983), it is suggested that IAA is biosynthesized in tip cells, is transported toward proximal regions, and then evokes signaling in rhizoids of mosses (Fig. 2A). In *P. patens*, mutations in the auxin-inducible *PpRSL1/2* or *PpLRL1/2* genes (see above) result in loss of rhizoid development regardless of auxin treatment, suggesting that these basic helix-loop-helix (bHLH) transcription factors regulate rhizoid development downstream of auxin (Menand et al. 2007; Jang et al. 2011; Tam et al. 2015). Their angiosperm homologs regulate root hair development (Masucci and Schiefelbein 1994; Menand et al. 2007; Ding et al. 2009; Karas et al. 2009). In particular, the class II RSL genes are regulated by auxin in *Arabidopsis* (Yi et al. 2010; Pires et al. 2013). It is known that their *M. polymorpha* homologs, Mp*RSL1* and Mp*LRL*, also regulate rhizoid development, but their association with auxin is unclear (Breuninger et al. 2016; Proust et al. 2016).

In the thalli of *M. polymorpha*, both Mp*TAA* and Mp*YUC2* promoters show strong activity at apical notches and the bottom of gemma cups (Eklund et al. 2015). In line with the observations that removal of apical notches broke gemma dormancy in the cup (LaRue and Narayanaswami 1957; Eklund et al. 2015), IAA biosynthesized at the bottom of gemma cups but also that supplied from apical notches (Maravolo 1976; Gaal et al. 1982) are suggested to play roles in maintaining gemmae dormancy. An increase in nondormant gemmae caused by Mp*YUC2* knockdown, overexpression of *iaaL*, and Mp*ARF1* knockout also supports the notion that auxin promotes gemma dormancy (Eklund et al. 2015; Kato et al. 2017). Regulation of dormancy, however, seems to be a complex process as direct IAA application broke gemma dormancy within gemma cups (Tarén 1958). Additionally, the question remains as to how auxin affects mature gemmae detached from mother plants but still in the cup. Recently, abscisic acid (ABA) was shown to function autonomously in gemmae to initiate and/or maintain dormancy within gemma cups (Eklund et al. 2018), but the underlying mechanisms bridging the two signaling pathways remain unclear.

Auxin also mediates cell elongation in *M. polymorpha*. Gametangiophore stem elongation was suppressed by high doses of nondegradable MpIAA and dominant-negative TPL mutants (Flores-Sandoval et al. 2015). The former also caused defects in the antigravitropic bending of stalks, supporting that the NAS pathway regulates tropic responses (Kato et al. 2015). Additionally, it suppressed auxin-induced protrusion of air pores and elongation of gemma cup cells, suggesting that the NAS pathway regulates cell elongation. Several EXPANSIN genes shown to be up-regulated in response to auxin treatment might be potential targets (Kato et al. 2017; Mutte et al. 2018).

Analysis of the NAS components revealed that auxin also plays a significant role in apical stem cell functions. Consistent with the fact that auxins promote branching in *R. multifida* (Ilahi and Allsopp 1969), overexpression of Mp*ARF1* and the dominant-negative Mp*TPL* mutant enhanced thallus branching (Flores-Sandoval et al. 2015). Conditional knockout of Mp*ARF2* terminated thallus growth, suggesting that coordinated regulation between A- and B-ARFs is essential for stem cell maintenance (Kato et al. 2020a). Mp*arf1* mutation causes abnormalities in division patterns and deviation of position and numbers of stem cell regions during gemma formation, suggesting that the establishment of developmental axes through formative divisions is a common trait of A-ARFs in land plants (Yoshida et al. 2014; Kato et al. 2017, 2018). A bHLH transcription factor, *M. polymorpha* TARGET OF MONOPTEROS 5 (MpTMO5), is also needed for proper patterning in gemma development (Lu et al. 2020). Given that *Arabidopsis TMO5* is a direct target of A-ARF (Schlereth et al. 2010), it would be worthwhile to test whether MpARF1 targets Mp*TMO5* in gemma development.

In sporophytes, strong *GmGH3* activities were observed in the whole or apical part of *M. polymorpha* and *P. patens*, respectively, at the immature stage, but limited to the basal side as they mature (Fujita et al. 2008; Ishizaki et al. 2012). In *M. polymorpha*, nondegradable AUX/IAA induction halts sporophyte development (Kato et al. 2015). In *P. patens*, transport disruption by NPA application or knockout of *PpPINB* occasionally caused the branching of sporangia, and knockout of both *PpPINA* and *PpPINB* caused severe or lethal growth arrest (Fujita et al. 2008; Bennett et al. 2014b). These studies highlight the importance of spatiotemporal auxin regulation in the sporophyte bodies of bryophytes. Although auxin response in sporophytic tissues plays an essential role in angiosperms, as shown by the embryonic lethal phenotype of At*TIR1/AFB* sextuple mutants (Prigge et al. 2020), little is known about the molecular mechanisms underlying sporophyte development of bryophytes. Future analyses are needed to determine whether conserved regulatory systems are used for sporophytes.

CONCLUDING REMARKS

Studies with bryophytes have contributed to our understanding of how the auxin system evolved in green plants. As described above, bryophytes share with vascular plants, not only the core mechanisms of auxin biosynthesis, polar transport, inactivation, and signaling, but also various auxin-regulated physiological processes, such as cell division, cell elongation, cell differentiation, tissue and organ development, and tropic responses. Thus, most of the auxin system was established early in land plant lineages, which reemphasizes the utility of bryophytes as models for auxin biology. In particular, the design principle of the functions of the NAS components has been determined effectively using simple platforms.

Auxin plays a role in reproductive organ differentiation in bryophytes. In *P. patens*, *PpTARs* regulate gamete precursor development by local auxin biosynthesis (Landberg et al. 2020); *PpSHI* regulates egg cell differentiation, sexual organ opening, and programmed canal cell death (Landberg et al. 2013). In *M. polymorpha*, Mp*IAA* is expressed in gametangia (Kato et al. 2015). In bryophytes, germline cells are formed directly on vegetative somatic tissues, distinct from seed plants where they are specified in meiosis-derived cells within several rounds of divisions (Hisanaga et al. 2019). Elucidation of the roles of auxin in reproductive development should provide insights into the reduction of gametophyte generation during land plant evolution. Further analyses are needed on how auxin biosynthesis, transport, signaling, and inactivation are used in various contexts in bryophytes.

Auxin is regarded as a signal that fuels programs predetermined by other mechanisms (Bennett and Leyser 2014) most likely via epigenetic regulation. However, its underlying biochemical basis has not been fully elucidated. For example, little is known regarding how A-ARFs reverse the deacetylated histone status. Additionally, there are still many unanswered questions on the functions of the NAS components, such as whether all or only a subset of A-ARF-target genes are coregulated by B-ARFs and whether and how the ratio of A- and B-ARFs varies for individual target genes. Simple auxin systems in bryophytes can make significant contributions to these central issues in auxin biology. So far, the majority of molecular studies on bryophytes have been reverse genetically performed. Given the low genetic redundancy, forward genetics would provide clues to explore novel molecular mechanisms for auxin biology.

ACKNOWLEDGMENTS

We thank Eduardo Flores-Sandoval for helpful information. Research in the authors' laboratory was/is funded by MEXT/JSPS KAKENHI, Grant Nos.: JP18J12698 to H.S., JP17H07424 to T.K., and JP20H04884 to R.N. R.N. was also supported by SPIRITS 2017 of Kyoto University.

REFERENCES

Allen E, Xie Z, Gustafson AM, Carrington JC. 2005. microRNA-directed phasing during *trans*-acting siRNA biogenesis in plants. *Cell* **121:** 207–221. doi:10.1016/j.cell.2005.04.004

Allsopp A, Ilahi I. 1970a. Studies in the Metzgeriales (Hepaticae). III: The effects of sugars and auxins on *Noteroclada confluens*. *Phytomorphology* **20:** 9–16.

Allsopp A, Ilahi I. 1970b. Studies in the Metzgeriales (Hepaticae). V: Investigation on *Blasia pusilla*. *Phytomorphology* **20:** 118–125.

Allsopp A, Ilahi I. 1970c. Studies in the Metzgeriales (Hepaticae). VII: Regeneration in *Noteroclada confluens* and *Blasia pusilla*. *Phytomorphology* **20:** 173–182.

Allsopp A, Pearman C, Rao AN. 1968. The effects of some growth substances and inhibitors on the development of *Marchantia* gemmae. *Phytomorphology* **18:** 84–94.

Aoyama T, Hiwatashi Y, Shigyo M, Kofuji R, Kubo M, Ito M, Hasebe M. 2012. AP2-type transcription factors determine stem cell identity in the moss *Physcomitrella patens*. *Development* **139:** 3120–3129. doi:10.1242/dev.076091

Apostolakos P, Galatis B, Mitrakos K. 1982. Studies on the development of the air pores and air chambers of *Marchantia paleacea* 1. Light microscopy. *Ann Bot* **49:** 377–396. doi:10.1093/oxfordjournals.aob.a086262

Arif MA, Fattash I, Ma Z, Cho SH, Beike AK, Reski R, Axtell MJ, Frank W. 2012. DICER-LIKE3 activity in *Physcomitrella patens DICER-LIKE4* mutants causes severe developmental dysfunction and sterility. *Mol Plant* **5:** 1281–1294. doi:10.1093/mp/sss036

Ashton NW, Grimsley NH, Cove DJ. 1979. Analysis of gametophytic development in the moss, *Physcomitrella patens*, using auxin and cytokinin resistant mutants. *Planta* **144:** 427–435. doi:10.1007/BF00380118

Ashton NW, Schulze A, Hall P, Bandurski RS. 1985. Estimation of indole-3-acetic acid in gametophytes of the moss, *Physcomitrella patens*. *Planta* **164:** 142–144. doi:10.1007/BF00391040

Atzorn R, Geier U, Sandberg G. 1990. The physiological role of indole acetic acid in the moss *Funaria hygrometrica* Hedw. I: Quantification of indole-3-acetic acid in tissue and protoplasts by enzyme immunoassay and gas chromatography-mass spectrometry. *J Plant Physiol* **135:** 522–525. doi:10.1016/S0176-1617(11)80628-0

Axtell MJ, Snyder JA, Bartel DP. 2007. Common functions for diverse small RNAs of land plants. *Plant Cell* **19:** 1750–1769. doi:10.1105/tpc.107.051706

Bao L, Yamamoto KT, Fujita T. 2015. Phototropism in gametophytic shoots of the moss *Physcomitrella patens*. *Plant Signal Behav* **10:** e1010900. doi:10.1080/15592324.2015.1010900

Barker EI. 2011. "An examination of leaf morphogenesis in the moss, *Physcomitrella patens*." PhD thesis, University of Regina, Regina, SK, Canada.

Barnes CR, Land WJG. 1908. Bryological papers. II: The origin of the cupule of *Marchantia*. *Bot Gaz* **46:** 401–409. doi:10.1086/329782

Bennett T, Leyser O. 2014. The auxin question: a philosophical overview. In *Auxin and its role in plant development* (ed. Zažímalová E, Petrasek J, Benková E), pp. 3–19. Springer, Vienna, Austria.

Bennett T, Brockington SF, Rothfels C, Graham SW, Stevenson D, Kutchan T, Rolf M, Thomas P, Wong GKS, Leyser O, et al. 2014a. Paralogous radiations of PIN proteins with multiple origins of noncanonical PIN structure. *Mol Biol Evol* **31:** 2042–2060. doi:10.1093/molbev/msu147

Bennett TA, Liu MM, Aoyama T, Bierfreund NM, Braun M, Coudert Y, Dennis RJ, O'Connor D, Wang XY, White CD, et al. 2014b. Plasma membrane-targeted PIN proteins drive shoot development in a moss. *Curr Biol* **24:** 2776–2785. doi:10.1016/j.cub.2014.09.054

Bierfreund NM, Reski R, Decker EL. 2003. Use of an inducible reporter gene system for the analysis of auxin distribution in the moss *Physcomitrella patens*. *Plant Cell Rep* **21:** 1143–1152. doi:10.1007/s00299-003-0646-1

Binns AN, Maravolo NC. 1972. Apical dominance, polarity, and adventitious growth in *Marchantia polymorpha*. *Am J Bot* **59:** 691–696. doi:10.1002/j.1537-2197.1972.tb10141.x

Bowman JL. 2016. A brief history of *Marchantia* from Greece to genomics. *Plant Cell Physiol* **57:** 210–229. doi:10.1093/pcp/pcv044

Bowman JL, Kohchi T, Yamato KT, Jenkins J, Shu S, Ishizaki K, Yamaoka S, Nishihama R, Nakamura Y, Berger F, et al. 2017. Insights into land plant evolution garnered from the *Marchantia polymorpha* genome. *Cell* **171:** 287–304.e15. doi:10.1016/j.cell.2017.09.030

Breuninger H, Thamm A, Streubel S, Sakayama H, Nishiyama T, Dolan L. 2016. Diversification of a transcription factor family led to the evolution of antagonistically acting genetic regulators of root hair growth. *Curr Biol* **26:** 1622–1628. doi:10.1016/j.cub.2016.04.060

Causier B, Ashworth M, Guo W, Davies B. 2012. The TOPLESS interactome: a framework for gene repression in *Arabidopsis*. *Plant Physiol* **158:** 423–438. doi:10.1104/pp.111.186999

Chopra RN, Vashistha BD. 1990. The effect of auxins and antiauxins on shoot-bud induction and morphology in the moss, *Bryum atrovirens* Will ex Brid. *Aust J Bot* **38:** 177–184. doi:10.1071/BT9900177

Cooke TJ, Poli D, Sztein AE, Cohen JD. 2002. Evolutionary patterns in auxin action. *Plant Mol Biol* **49:** 319–338. doi:10.1023/A:1015242627321

Coudert Y, Palubicki W, Ljung K, Novak O, Leyser O, Jill Harrison C. 2015. Three ancient hormonal cues co-ordinate shoot branching in a moss. *eLife* **4:** e06808. doi:10.7554/eLife.06808

Cove D. 2005. The moss *Physcomitrella patens*. *Annu Rev Genet* **39:** 339–358. doi:10.1146/annurev.genet.39.073003.110214

Cove DJ, Ashton WN. 1984. The hormonal regulation of gametophytic development in bryophytes. In *The experimental biology of bryophytes* (ed. Dyer AF, Duckett JG), pp. 177–201. Academic, London.

Cove DJ, Schild A, Ashton NW, Hartmann E. 1978. Genetic and physiological studies of the effect of light on the development of the moss, *Physcomitrella patens*. *Photochem Photobiol* **27:** 249–254. doi:10.1111/j.1751-1097.1978.tb07596.x

Cox CJ, Li B, Foster PG, Embley TM, Civáň P. 2014. Conflicting phylogenies for early land plants are caused by composition biases among synonymous substitutions. *Syst Biol* **63:** 272–279. doi:10.1093/sysbio/syt109

Davidonis GH, Munroe MH. 1972. Apical dominance in *Marchantia*: correlative inhibition of neighbor lobe growth. *Bot Gaz* **133:** 177–184. doi:10.1086/336631

Demeulenaere MJF, Beeckman T. 2014. The interplay between auxin and the cell cycle during plant development. In *Auxin and its role in plant development* (ed. Zažímalová E, Petrasek J, Benková E), pp. 119–141. Springer, Vienna, Austria.

Dharmasiri N, Dharmasiri S, Estelle M. 2005. The F-box protein TIR1 is an auxin receptor. *Nature* **435:** 441–445. doi:10.1038/nature03543

Ding W, Yu Z, Tong Y, Huang W, Chen H, Wu P. 2009. A transcription factor with a bHLH domain regulates root hair development in rice. *Cell Res* **19:** 1309–1311. doi:10.1038/cr.2009.109

Dresselhaus T, Schneitz K. 2014. The role of auxin for reproductive organ patterning and development. In *Auxin and its role in plant development* (ed. Zažímalová E, Petrasek J, Benková E), pp. 213–243. Springer, Vienna, Austria.

Eklund DM, Ståldal V, Valsecchi I, Cierlik I, Eriksson C, Hiratsu K, Ohme-Takagi M, Sundström JF, Thelander M, Ezcurra I, et al. 2010a. The *Arabidopsis thaliana* STYLISH1 protein acts as a transcriptional activator regulating auxin biosynthesis. *Plant Cell* **22:** 349–363. doi:10.1105/tpc.108.064816

Eklund DM, Thelander M, Landberg K, Ståldal V, Nilsson A, Johansson M, Valsecchi I, Pederson ERA, Kowalczyk M, Ljung K, et al. 2010b. Homologues of the *Arabidopsis thaliana SHI/STY/LRP1* genes control auxin biosynthesis and affect growth and development in the moss *Physcomitrella patens*. *Development* **137:** 1275–1284. doi:10.1242/dev.039594

Eklund DM, Ishizaki K, Flores-Sandoval E, Kikuchi S, Takebayashi Y, Tsukamoto S, Hirakawa Y, Nonomura M, Kato H, Kouno M, et al. 2015. Auxin produced by the indole-3-pyruvic acid pathway regulates development and gemmae dormancy in the liverwort *Marchantia polymorpha*. *Plant Cell* **27:** 1650–1669. doi:10.1105/tpc.15.00065

Eklund DM, Kanei M, Flores-Sandoval E, Ishizaki K, Nishihama R, Kohchi T, Lagercrantz U, Bhalerao RP, Sakata Y, Bowman JL. 2018. An evolutionarily conserved abscisic acid signaling pathway regulates dormancy in the liverwort *Marchantia polymorpha*. *Curr Biol* **28:** 3691–3699.e3. doi:10.1016/j.cub.2018.10.018

Finet C, Berne-Dedieu A, Scutt CP, Marlétaz F. 2013. Evolution of the ARF gene family in land plants: old domains, new tricks. *Mol Biol Evol* **30:** 45–56. doi:10.1093/molbev/mss220

Flores-Sandoval E, Eklund DM, Bowman JL. 2015. A simple auxin transcriptional response system regulates multiple morphogenetic processes in the liverwort *Marchantia polymorpha*. *PLoS Genet* **11:** e1005207. doi:10.1371/journal.pgen.1005207

Flores-Sandoval E, Eklund DM, Hong SF, Alvarez JP, Fisher TJ, Lampugnani ER, Golz JF, Vázquez-Lobo A, Dierschke T, Lin SS, et al. 2018a. Class C ARFs evolved before the origin of land plants and antagonize differentiation and developmental transitions in *Marchantia polymorpha*. *New Phytol* **218:** 1612–1630. doi:10.1111/nph.15090

Flores-Sandoval E, Romani F, Bowman JL. 2018b. Co-expression and transcriptome analysis of *Marchantia polymorpha* transcription factors supports class C ARFs as independent actors of an ancient auxin regulatory module. *Front Plant Sci* **9:** 1345. doi:10.3389/fpls.2018.01345

Fries K. 1964. Uber einen genuinen keimungs und streckungswachstumsaktiven hemmstoff bei *Marchantia polymorpha* L. [About genuine germination and extension growth inhibitor in *Marchantia polymorpha* L.]. *Beitr Biol Pflanz* **40:** 177–235.

Fujita T, Sakaguchi H, Hiwatashi Y, Wagstaff SJ, Ito M, Deguchi H, Sato T, Hasebe M. 2008. Convergent evolution of shoots in land plants: lack of auxin polar transport in moss shoots. *Evol Dev* **10:** 176–186. doi:10.1111/j.1525-142X.2008.00225.x

Gaal DJ, Dufresne SJ, Maravolo NC. 1982. Transport of ^{14}C-indoleacetic acid in the hepatic *Marchantia polymorpha*. *Bryologist* **85:** 410–418. doi:10.2307/3242908

Gray WM, del Pozo JC, Walker L, Hobbie L, Risseeuw E, Banks T, Crosby WL, Yang M, Ma H, Estelle M. 1999. Identification of an SCF ubiquitin-ligase complex required for auxin response in *Arabidopsis thaliana*. *Genes Dev* **13:** 1678–1691. doi:10.1101/gad.13.13.1678

Gray WM, Kepinski S, Rouse D, Leyser O, Estelle M. 2001. Auxin regulates SCF^{TIR1}-dependent degradation of AUX/IAA proteins. *Nature* **414:** 271–276. doi:10.1038/35104500

Grones P, Friml J. 2015. Auxin transporters and binding proteins at a glance. *J Cell Sci* **128:** 1–7. doi:10.1242/jcs.159418

Hagen G, Martin G, Li Y, Guilfoyle TJ. 1991. Auxin-induced expression of the soybean GH3 promoter in transgenic tobacco plants. *Plant Mol Biol* **17:** 567–579. doi:10.1007/BF00040658

Han X, Hyun TK, Zhang M, Kumar R, Koh E, Kang BH, Lucas WJ, Kim JY. 2014. Auxin-callose-mediated plasmodesmal gating is essential for tropic auxin gradient formation and signaling. *Dev Cell* **28:** 132–146. doi:10.1016/j.devcel.2013.12.008

Harrison CJ, Roeder AHK, Meyerowitz EM, Langdale JA. 2009. Local cues and asymmetric cell divisions underpin body plan transitions in the moss *Physcomitrella patens*. *Curr Biol* **19:** 461–471. doi:10.1016/j.cub.2009.02.050

He W, Brumos J, Li H, Ji Y, Ke M, Gong X, Zeng Q, Li W, Zhang X, An F, et al. 2011. A small-molecule screen identifies L-kynurenine as a competitive inhibitor of TAA1/TAR activity in ethylene-directed auxin biosynthesis and root growth in *Arabidopsis*. *Plant Cell* **23:** 3944–3960. doi:10.1105/tpc.111.089029

Hisanaga T, Yamaoka S, Kawashima T, Higo A, Nakajima K, Araki T, Kohchi T, Berger F. 2019. Building new insights in plant gametogenesis from an evolutionary perspective. *Nat Plants* **5:** 663–669. doi:10.1038/s41477-019-0466-0

Ilahi I, Allsopp A. 1969. Studies in the Metzgeriales (Hepaticae). II: The effects of certain auxins on the growth and morphology of some thalloid species. *Phytomorphology* **19:** 381–389.

Ilahi I, Allsopp A. 1970a. Studies in the Metzgeriales (Hepaticae). IV: Further investigations on the effects of various physiologically active substances on *Noteroclada confluens* and on some thalloid species. *Phytomorphology* **20:** 68–77.

Ilahi I, Allsopp A. 1970b. Studies in the Metzgeriales (Hepaticae). VI: Regeneration and callus formation in some thalloid species. *Phytomorphology* **20:** 126–136.

Inoue H. 1960. Studies in spore germination and the earlier stages of gametophyte development in the Marchantiales. *J Hattori Bot Lab* **23:** 148–191.

Inoue K, Nishihama R, Kataoka H, Hosaka M, Manabe R, Nomoto M, Tada Y, Ishizaki K, Kohchi T. 2016. Phytochrome signaling is mediated by PHYTOCHROME INTERACTING FACTOR in the liverwort *Marchantia polymorpha*. *Plant Cell* **28:** 1406–1421. doi:10.1105/tpc.15.01063

Ishizaki K, Nonomura M, Kato H, Yamato KT, Kohchi T. 2012. Visualization of auxin-mediated transcriptional activation using a common auxin-responsive reporter system in the liverwort *Marchantia polymorpha*. *J Plant Res* **125:** 643–651. doi:10.1007/s10265-012-0477-7

Ishizaki K, Nishihama R, Yamato KT, Kohchi T. 2016. Molecular genetic tools and techniques for *Marchantia polymorpha* research. *Plant Cell Physiol* **57:** 262–270. doi:10.1093/pcp/pcv097

Jackson RG, Lim EK, Li Y, Kowalczyk M, Sandberg G, Hogget J, Ashford DA, Bowles DJ. 2001. Identification and biochemical characterization of an *Arabidopsis* indole-3-acetic acid glucosyltransferase. *J Biol Chem* **276:** 4350–4356. doi:10.1074/jbc.M006185200

Jang G, Dolan L. 2011. Auxin promotes the transition from chloronema to caulonema in moss protonema by positively regulating *PpRSL1* and *PpRSL2* in *Physcomitrella patens*. *New Phytol* **192:** 319–327. doi:10.1111/j.1469-8137.2011.03805.x

Jang G, Yi K, Pires ND, Menand B, Dolan L. 2011. RSL genes are sufficient for rhizoid system development in early diverging land plants. *Development* **138:** 2273–2281. doi:10.1242/dev.060582

Johri MM, Desai S. 1973. Auxin regulation of caulonema formation in moss protonema. *Nat New Biol* **245:** 223–224. doi:10.1038/newbio245223a0

Karas B, Amyot L, Johansen C, Sato S, Tabata S, Kawaguchi M, Szczyglowski K. 2009. Conservation of *Lotus* and *Arabidopsis* basic helix-loop-helix proteins reveals new players in root hair development. *Plant Physiol* **151:** 1175–1185. doi:10.1104/pp.109.143867

Kasahara H. 2016. Current aspects of auxin biosynthesis in plants. *Biosci Biotechnol Biochem* **80:** 34–42. doi:10.1080/09168451.2015.1086259

Kato H, Ishizaki K, Kouno M, Shirakawa M, Bowman JL, Nishihama R, Kohchi T. 2015. Auxin-mediated transcriptional system with a minimal set of components is critical for morphogenesis through the life cycle in *Marchantia polymorpha*. *PLoS Genet* **11:** e1005084. doi:10.1371/journal.pgen.1005084

Kato H, Kouno M, Takeda M, Suzuki H, Ishizaki K, Nishihama R, Kohchi T. 2017. The roles of the sole activator-type auxin response factor in pattern formation of *Marchantia polymorpha*. *Plant Cell Physiol* **58:** 1642–1651. doi:10.1093/pcp/pcx095

Kato H, Nishihama R, Weijers D, Kohchi T. 2018. Evolution of nuclear auxin signaling: lessons from genetic studies with basal land plants. *J Exp Bot* **69:** 291–301. doi:10.1093/jxb/erx267

Kato H, Mutte SK, Suzuki H, Crespo I, Das S, Radoeva T, Fontana M, Yoshitake Y, Hainiwa E, van den Berg W, et al. 2020a. Design principles of a minimal auxin response system. *Nat Plants* **6:** 473–482. doi:10.1038/s41477-020-0662-y

Kato H, Yasui Y, Ishizaki K. 2020b. Gemma cup and gemma development in *Marchantia polymorpha. New Phytol* **228:** 459–465. doi:10.1111/nph.16655

Kepinski S, Leyser O. 2005. The *Arabidopsis* F-box protein TIR1 is an auxin receptor. *Nature* **435:** 446–451. doi:10.1038/nature03542

Kim J, Harter K, Theologis A. 1997. Protein–protein interactions among the Aux/IAA proteins. *Proc Natl Acad Sci* **94:** 11786–11791. doi:10.1073/pnas.94.22.11786

Kny L. 1890. Bau und Entwickelung von *Marchantia polymorpha* L. In *Botanische Wandtafeln*, pp. 364-401. VIII Abtheilung, Paul Parey, Berlin.

Kofuji R, Hasebe M. 2014. Eight types of stem cells in the life cycle of the moss *Physcomitrella patens. Curr Opin Plant Biol* **17:** 13–21. doi:10.1016/j.pbi.2013.10.007

Korasick DA, Westfall CS, Lee SG, Nanao MH, Dumas R, Hagen G, Guilfoyle TJ, Jez JM, Strader LC. 2014. Molecular basis for AUXIN RESPONSE FACTOR protein interaction and the control of auxin response repression. *Proc Natl Acad Sci* **111:** 5427–5432. doi:10.1073/pnas.1400074111

Lagercrantz U, Billhardt A, Rousku SN, Ljung K, Eklund DM. 2020. Nyctinastic thallus movement in the liverwort *Marchantia polymorpha* is regulated by a circadian clock. *Sci Rep* **10:** 8658. doi:10.1038/s41598-020-65372-8

Landberg K, Pederson ERA, Viaene T, Bozorg B, Friml J, Jönsson H, Thelander M, Sundberg E. 2013. The moss *Physcomitrella patens* reproductive organ development is highly organized, affected by the two *SHI/STY* genes and by the level of active auxin in the *SHI/STY* expression domain. *Plant Physiol* **162:** 1406–1419. doi:10.1104/pp.113.214023

Landberg K, Šimura J, Ljung K, Sundberg E, Thelander M. 2020. Studies of moss reproductive development indicate that auxin biosynthesis in apical stem cells may constitute an ancestral function for focal growth control. *New Phytol* doi: 10.1111/nph.16914

Landrein B, Vernoux T. 2014. Auxin, chief architect of the shoot apex. In *Auxin and its role in plant development* (ed. Zažímalová E, Petrasek J, Benková E), pp. 191–212. Springer, Vienna, Austria.

Lang D, Ullrich KK, Murat F, Fuchs J, Jenkins J, Haas FB, Piednoel M, Gundlach H, Van Bel M, Meyberg R, et al. 2018. The *Physcomitrella patens* chromosome-scale assembly reveals moss genome structure and evolution. *Plant J* **93:** 515–533. doi:10.1111/tpj.13801

LaRue CD, Narayanaswami S. 1957. Auxin inhibition in the liverwort *Lunularia. New Phytologist* **56:** 61–70. doi:10.1111/j.1469-8137.1957.tb07449.x

Lavy M, Prigge MJ, Tao S, Shain S, Kuo A, Kirchsteiger K, Estelle M. 2016. Constitutive auxin response in *Physcomitrella* reveals complex interactions between Aux/IAA and ARF proteins. *eLife* **5:** e13325. doi:10.7554/eLife.13325

Li FW, Nishiyama T, Waller M, Frangedakis E, Keller J, Li Z, Fernandez-Pozo N, Barker MS, Bennett T, Blázquez MA, et al. 2020. Anthoceros genomes illuminate the origin of land plants and the unique biology of hornworts. *Nat Plants* **6:** 259–272. doi:10.1038/s41477-020-0618-2

Liao C-Y, Smet W, Brunoud G, Yoshida S, Vernoux T, Weijers D. 2015. Reporters for sensitive and quantitative measurement of auxin response. *Nat Methods* **12:** 207–210. doi:10.1038/nmeth.3279

Ligrone R, Duckett JG, Renzaglia KS. 2012. The origin of the sporophyte shoot in land plants: a bryological perspective. *Ann Bot* **110:** 935–941. doi:10.1093/aob/mcs176

Lin PC, Lu CW, Shen BN, Lee GZ, Bowman JL, Arteaga-Vazquez MA, Liu LYD, Hong SF, Lo CF, Su GM, et al. 2016. Identification of miRNAs and their targets in the liverwort *Marchantia polymorpha* by integrating RNA-Seq and degradome analyses. *Plant Cell Physiol* **57:** 339–358. doi:10.1093/pcp/pcw020

Löbenberg E. 1959. "Über genuine wuchsstoffe bei *Marchantia polymorpha* L." [About genuine growth substances in *Marchantia polymorpha* L.]. PhD thesis, University of Frankfurt, Frankfurt, Germany.

Lu KJ, van't Wout Hofland N, Mor E, Mutte S, Abrahams P, Kato H, Vandepoele K, Weijers D, de Rybel B. 2020. Evolution of vascular plants through redeployment of ancient developmental regulators. *Proc Natl Acad Sci* **117:** 733–740. doi:10.1073/pnas.1912470117

Ludwig-Müller J, Decker EL, Reski R. 2009a. Dead end for auxin conjugates in *Physcomitrella*? *Plant Signal Behav* **4:** 116–118. doi:10.4161/psb.4.2.7536

Ludwig-Müller J, Jülke S, Bierfreund NM, Decker EL, Reski R. 2009b. Moss (*Physcomitrella patens*) GH3 proteins act in auxin homeostasis. *New Phytol* **181:** 323–338. doi:10.1111/j.1469-8137.2008.02677.x

Lv B, Yu Q, Liu J, Wen X, Yan Z, Hu K, Li H, Kong X, Li C, Tian H, et al. 2020. Non-canonical AUX/IAA protein IAA33 competes with canonical AUX/IAA repressor IAA5 to negatively regulate auxin signaling. *EMBO J* **39:** e101515. doi:10.15252/embj.2019101515

MacQuarrie IG, von Maltzahn KE. 1959. Correlations affecting regeneration and reactivation in *Splachnum ampullaceum* (L.) Hedw. *Can J Bot* **37:** 121–134. doi:10.1139/b59-011

Mallory AC, Bartel DP, Bartel B. 2005. MicroRNA-directed regulation of *Arabidopsis AUXIN RESPONSE FACTOR17* is essential for proper development and modulates expression of early auxin response genes. *Plant Cell* **17:** 1360–1375. doi:10.1105/tpc.105.031716

Maravolo NC. 1976. Polarity and localization of auxin movement in the hepatic, *Marchantia polymorpha. Am J Bot* **63:** 526–531. doi:10.1002/j.1537-2197.1976.tb11841.x

Maravolo NC, Voth PD. 1966. Morphogenic effects of three growth substances on *Marchantia gemmalings. Bot Gaz* **127:** 79–86. doi:10.1086/336346

Masucci JD, Schiefelbein JW. 1994. The *rhd6* mutation of *Arabidopsis thaliana* alters root-hair initiation through an auxin- and ethylene-associated process. *Plant Physiol* **106:** 1335–1346. doi:10.1104/pp.106.4.1335

Menand B, Keke Yi SJ, Hoffmann L, Ryan E, Linstead P, Schaefer DG, Dolan L. 2007. An ancient mechanism controls the development of cells with a rooting function in land plants. *Science* **316:** 1477–1480. doi:10.1126/science.1142618

Mittag J, Gabrielyan A, Ludwig-Müller J. 2015. Knockout of GH3 genes in the moss *Physcomitrella patens* leads to increased IAA levels at elevated temperature and in darkness. *Plant Physiol Biochem* **97:** 339–349. doi:10.1016/j.plaphy.2015.10.013

Montgomery SA, Tanizawa Y, Galik B, Wang N, Ito T, Mochizuki T, Akimcheva S, Bowman JL, Cognat V, Maréchal-Drouard L, et al. 2020. Chromatin organization in early land plants reveals an ancestral association between H3K27me3, transposons, and constitutive heterochromatin. *Curr Biol* **30:** 573–588.e7. doi:10.1016/j.cub.2019.12.015

Morris JL, Puttick MN, Clark JW, Edwards D, Kenrick P, Pressel S, Wellman CH, Yang Z, Schneider H, Donoghue PCJ. 2018. The timescale of early land plant evolution. *Proc Natl Acad Sci* **115:** E2274–E2283. doi:10.1073/pnas.1719588115

Mutte SK, Kato H, Rothfels C, Melkonian M, Wong GKS, Weijers D. 2018. Origin and evolution of the nuclear auxin response system. *eLife* **7:** e33399. doi:10.7554/eLife.33399

Nishihama R, Ishizaki K, Hosaka M, Matsuda Y, Kubota A, Kohchi T. 2015. Phytochrome-mediated regulation of cell division and growth during regeneration and sporeling development in the liverwort *Marchantia polymorpha*. *J Plant Res* **128:** 407–421. doi:10.1007/s10265-015-0724-9

Nishiyama T, Wolf PG, Kugita M, Sinclair RB, Sugita M, Sugiura C, Wakasugi T, Yamada K, Yoshinaga K, Yamaguchi K, et al. 2004. Chloroplast phylogeny indicates that bryophytes are monophyletic. *Mol Biol Evol* **21:** 1813–1819. doi:10.1093/molbev/msh203

Nyman LP, Cutter EG. 1981. Auxin–cytokinin interaction in the inhibition, release, and morphology of gametophore buds of *Plagiomnium cuspidatum* from apical dominance. *Can J Bot* **59:** 750–762. doi:10.1139/b81-106

Otto KR, Halbsguth W. 1976. Die förderung der bildung von primärrhizoiden an brutkörpern von *Marchantia polymorpha* L. durch licht und IES [The promotion of the formation of primary rhizoids on brood bodies of *Marchantia polymorpha* L.]. *Z Pflanzenphysiol* **80:** 197–205. doi:10.1016/S0044-328X(76)80021-9

Paponov IA, Teale W, Lang D, Paponov M, Reski R, Rensing SA, Palme K. 2009. The evolution of nuclear auxin signalling. *BMC Evol Biol* **9:** 126. doi:10.1186/1471-2148-9-126

Pires ND, Yi K, Breuninger H, Catarino B, Menand B, Dolan L. 2013. Recruitment and remodeling of an ancient gene regulatory network during land plant evolution. *Proc Natl Acad Sci* **110:** 9571–9576. doi:10.1073/pnas.1305457110

Plavskin Y, Nagashima A, Perroud PF, Hasebe M, Quatrano RS, Atwal GS, Timmermans MCP. 2016. Ancient *trans*-acting siRNAs confer robustness and sensitivity onto the auxin response. *Dev Cell* **36:** 276–289. doi:10.1016/j.devcel.2016.01.010

Poli D, Jacobs M, Cooke TJ. 2003. Auxin regulation of axial growth in bryophyte sporophytes: its potential significance for the evolution of early land plants. *Am J Bot* **90:** 1405–1415. doi:10.3732/ajb.90.10.1405

Porco S, Pěnčík A, Rashed A, Voß U, Casanova-Sáez R, Bishopp A, Golebiowska A, Bhosale R, Swarup R, Swarup K, et al. 2016. Dioxygenase-encoding At*DAO1* gene controls IAA oxidation and homeostasis in *Arabidopsis*. *Proc Natl Acad Sci* **113:** 11016–11021. doi:10.1073/pnas.1604375113

Prigge MJ, Lavy M, Ashton NW, Estelle M. 2010. *Physcomitrella patens* auxin-resistant mutants affect conserved elements of an auxin-signaling pathway. *Curr Biol* **20:** 1907–1912. doi:10.1016/j.cub.2010.08.050

Prigge MJ, Platre M, Kadakia N, Zhang Y, Greenham K, Szutu W, Pandey BK, Bhosale RA, Bennett MJ, Busch W, et al. 2020. Genetic analysis of the *Arabidopsis* TIR1/AFB auxin receptors reveals both overlapping and specialized functions. *eLife* **9:** e54740. doi:10.7554/eLife.54740

Proust H, Honkanen S, Jones VAS, Morieri G, Prescott H, Kelly S, Ishizaki K, Kohchi T, Dolan L. 2016. *RSL* class I genes controlled the development of epidermal structures in the common ancestor of land plants. *Curr Biol* **26:** 93–99. doi:10.1016/j.cub.2015.11.042

Puttick MN, Morris JL, Williams TA, Cox CJ, Edwards D, Kenrick P, Pressel S, Wellman CH, Schneider H, Pisani D, et al. 2018. The interrelationships of land plants and the nature of the ancestral embryophyte. *Curr Biol* **28:** 733–745.e2. doi:10.1016/j.cub.2018.01.063

Rensing SA, Lang D, Zimmer AD, Terry A, Salamov A, Shapiro H, Nishiyama T, Perroud PF, Lindquist EA, Kamisugi Y, et al. 2008. The *Physcomitrella* genome reveals evolutionary insights into the conquest of land by plants. *Science* **319:** 64–69. doi:10.1126/science.1150646

Rensing SA, Goffinet B, Meyberg R, Wu SZ, Bezanilla M. 2020. The moss *Physcomitrium (Physcomitrella) patens*: a model organism for non-seed plants. *Plant Cell* **32:** 1361–1376. doi:10.1105/tpc.19.00828

Retzer K, Korbei B, Luschnig C. 2014. Auxin and tropisms. In *Auxin and its role in plant development* (ed. Zažímalová E, et al.), pp. 361–387. Springer, Vienna, Austria.

Rios AF, Yoshida S, Weijers D. 2014. Auxin regulation of embryo development. In *Auxin and its role in plant development* (ed. Zažímalová E, Petrasek J, Benková E), pp. 171–189. Springer, Vienna, Austria.

Rose S, Bopp M. 1983. Uptake and polar transport of indoleacetic acid in moss rhizoids. *Physiol Plant* **58:** 57–61. doi:10.1111/j.1399-3054.1983.tb04143.x

Rota JA, Maravolo NC. 1975. Transport and mobilization of ^{14}C-sucrose during regeneration in the hepatic, *Marchantia polymorpha*. *Bot Gaz* **136:** 184–188. doi:10.1086/336800

Rousseau J. 1950. Action de l'acide indol beta-acétique sur les propagules de *Marchantia polymorpha* et *Lunularia cruciata* [Action of indole beta-acetic acid on the propagules of *Marchantia polymorpha* and *Lunularia cruciata*]. *C R Hebd Seances Acad Sci* **230:** 675–676.

Rousseau J. 1953. Action des hétéro-auxines sur les chapeaux du *Marchantia polymorpha* L. [Action of heteroauxins on the caps of *Marchantia polymorpha* L.]. *Bull Soc Bot Fr* **100:** 179–180. doi:10.1080/00378941.1953.10833187

Rousseau J. 1954. Action des hétéroauxines á l'obscurité sur les propagules de *Marchantia polymorpha* L. [Action of heteroauxins in the dark on the propagules of *Marchantia polymorpha* L.]. *C R Hebd Seances Acad Sci* **238:** 2111–2112.

Rousseau J. 1958. Morphoses du sporogone de deux *Anthoceros* par action de l'acide dichlorophénoxyacétique [Morphoses of the sporogonium of two *Anthoceros* by the action of dichlorophenoxyacetic acid]. *Bull Soc Bot Fr* **105:** 234–236. doi:10.1080/00378941.1958.10837870

Sakakibara K, Nishiyama T, Sumikawa N, Kofuji R, Murata T, Hasebe M. 2003. Involvement of auxin and a homeodomain-leucine zipper I gene in rhizoid development of the moss *Physcomitrella patens*. *Development* **130:** 4835–4846. doi:10.1242/dev.00644

Schlereth A, Möller B, Liu W, Kientz M, Flipse J, Rademacher EH, Schmid M, Jürgens G, Weijers D. 2010. MONOPTEROS controls embryonic root initiation by regulating a mobile transcription factor. *Nature* **464:** 913–916. doi:10.1038/nature08836

Schneider MJ, Voth PD, Troxler RF. 1967. Methods of propagating bryophyte plants, tissues, and propagules. *Bot Gaz* **128:** 169–174. doi:10.1086/336394

Schnepf E, Herth W, Morré DJ. 1979. Elongation growth of setae of *Pellia* (Bryophyta): effects of auxin and inhibitors. *Z Pflanzenphysiol* **94:** 211–217. doi:10.1016/S0044-328X(79)80160-9

Schwuchow J, Michalke W, Hertel R. 2001. An auxin transport inhibitor interferes with unicellular gravitropism in protonemata of the moss *Ceratodon purpureus*. *Plant Biol* **3:** 357–363. doi:10.1055/s-2001-16459

Shimamura M. 2016. *Marchantia polymorpha*: taxonomy, phylogeny and morphology of a model system. *Plant Cell Physiol* **57:** 230–256. doi:10.1093/pcp/pcv192

Sohlberg JJ, Myrenås M, Kuusk S, Lagercrantz U, Kowalczyk M, Sandberg G, Sundberg E. 2006. *STY1* regulates auxin homeostasis and affects apical–basal patterning of the *Arabidopsis gynoecium*. *Plant J* **47:** 112–123. doi:10.1111/j.1365-313X.2006.02775.x

Solly JE, Cunniffe NJ, Harrison CJ. 2017. Regional growth rate differences specified by apical notch activities regulate liverwort thallus shape. *Curr Biol* **27:** 16–26. doi:10.1016/j.cub.2016.10.056

Sood S, Hackenberg D. 1979. Interaction of auxin, antiauxin and cytokinin in relation to the formation of buds in moss protonema. *Z Pflanzenphysiol* **91:** 385–397. doi:10.1016/S0044-328X(79)80253-6

Ståldal V, Sohlberg JJ, Eklund DM, Ljung K, Sundberg E. 2008. Auxin can act independently of *CRC, LUG, SEU, SPT* and *STY1* in style development but not apical-basal patterning of the *Arabidopsis gynoecium*. *New Phytol* **180:** 798–808. doi:10.1111/j.1469-8137.2008.02625.x

Staswick PE, Serban B, Rowe M, Tiryaki I, Maldonado MT, Maldonado MC, Suza W. 2005. Characterization of an *Arabidopsis* enzyme family that conjugates amino acids to indole-3-acetic acid. *Plant Cell* **17:** 616–627. doi:10.1105/tpc.104.026690

Sugawara S, Mashiguchi K, Tanaka K, Hishiyama S, Sakai T, Hanada K, Kinoshita-Tsujimura K, Yu H, Dai X, Takebayashi Y, et al. 2015. Distinct characteristics of indole-3-acetic acid and phenylacetic acid, two common auxins in plants. *Plant Cell Physiol* **56:** 1641–1654. doi:10.1093/pcp/pcv088

Suzuki H, Harrison CJ, Shimamura M, Kohchi T, Nishihama R. 2020. Positional cues regulate dorsal organ formation in the liverwort *Marchantia polymorpha*. *J Plant Res* **133:** 311–321. doi:10.1007/s10265-020-01180-5

Szemenyei H, Hannon M, Long JA. 2008. TOPLESS mediates auxin-dependent transcriptional repression during *Arabidopsis* embryogenesis. *Science* **319:** 1384–1386. doi:10.1126/science.1151461

Sztein AE, Cohen JD, Slovin JP, Cooke TJ. 1995. Auxin metabolism in representative land plants. *Am J Bot* **82:** 1514–1521. doi:10.1002/j.1537-2197.1995.tb13853.x

Sztein AE, Cohen JD, García De La Fuente I, Cooke TJ. 1999. Auxin metabolism in mosses and liverworts. *Am J Bot* **86:** 1544–1555. doi:10.2307/2656792

Sztein AE, Cohen JD, Cooke TJ. 2000. Evolutionary patterns in the auxin metabolism of green plants. *Int J Plant Sci* **161:** 849–859. doi:10.1086/317566

Tam THY, Catarino B, Dolan L. 2015. Conserved regulatory mechanism controls the development of cells with rooting functions in land plants. *Proc Natl Acad Sci* **112:** E3959–E3968. doi:10.1073/pnas.1416324112

Tan X, Calderon-Villalobos LIA, Sharon M, Zheng C, Robinson CV, Estelle M, Zheng N. 2007. Mechanism of auxin perception by the TIR1 ubiquitin ligase. *Nature* **446:** 640–645. doi:10.1038/nature05731

Tao S, Estelle M. 2018. Mutational studies of the Aux/IAA proteins in *Physcomitrella* reveal novel insights into their function. *New Phytol* **218:** 1534–1542. doi:10.1111/nph.15039

Tarén N. 1958. Factors regulating the initial development of gemmae in *Marchantia polymorpha*. *Bryologist* **61:** 191–204. doi:10.1639/0007-2745(1958)61[191:FRTIDO]2.0.CO;2

Thelander M, Landberg K, Sundberg E. 2018. Auxin-mediated developmental control in the moss *Physcomitrella patens*. *J Exp Bot* **69:** 277–290. doi:10.1093/jxb/erx255

Thelander M, Landberg K, Sundberg E. 2019. Minimal auxin sensing levels in vegetative moss stem cells revealed by a ratiometric reporter. *New Phytol* **224:** 775–788. doi:10.1111/nph.16068

Thomas RJ. 1980. Cell elongation in hepatics: the seta system. *Bot Gaz* **107:** 229–245.

Tiwari SB, Hagen G, Guilfoyle T. 2003. The roles of auxin response factor domains in auxin-responsive transcription. *Plant Cell* **15:** 533–543. doi:10.1105/tpc.008417

Tsugafune S, Mashiguchi K, Fukui K, Takebayashi Y, Nishimura T, Sakai T, Shimada Y, Kasahara H, Koshiba T, Hayashi K. 2017. Yucasin DF, a potent and persistent inhibitor of auxin biosynthesis in plants. *Sci Rep* **7:** 13992. doi:10.1038/s41598-017-14332-w

Tsuzuki M, Nishihama R, Ishizaki K, Kurihara Y, Matsui M, Bowman JL, Kohchi T, Hamada T, Watanabe Y. 2016. Profiling and characterization of small RNAs in the liverwort, *Marchantia polymorpha*, belonging to the first diverged land plants. *Plant Cell Physiol* **57:** 359–372. doi:10.1093/pcp/pcv182

Ulmasov T, Hagen G, Guilfoyle TJ. 1997. ARF1, a transcription factor that binds to auxin response elements. *Science* **276:** 1865–1868. doi:10.1126/science.276.5320.1865

Ulmasov T, Hagen G, Guilfoyle TJ. 1999a. Activation and repression of transcription by auxin-response factors. *Proc Natl Acad Sci* **96:** 5844–5849. doi:10.1073/pnas.96.10.5844

Ulmasov T, Hagen G, Guilfoyle TJ. 1999b. Dimerization and DNA binding of auxin response factors. *Plant J* **19:** 309–319. doi:10.1046/j.1365-313X.1999.00538.x

Viaene T, Delwiche CF, Rensing SA, Friml J. 2013. Origin and evolution of PIN auxin transporters in the green lineage. *Trends Plant Sci* **18:** 5–10. doi:10.1016/j.tplants.2012.08.009

Viaene T, Landberg K, Thelander M, Medvecka E, Pederson E, Feraru E, Cooper ED, Karimi M, Delwiche CF, Ljung K, et al. 2014. Directional auxin transport mechanisms in early diverging land plants. *Curr Biol* **24:** 2786–2791. doi:10.1016/j.cub.2014.09.056

Vöchting H. 1885. Ueber die regeneration der Marchantieen [About the regeneration of Marchantieen]. *Jahrb Wiss Bot* **16:** 367–414.

Wang JW, Wang LJ, Mao YB, Cai WJ, Xue HW, Chen XY. 2005. Control of root cap formation by microRNA-targeted auxin response factors in *Arabidopsis*. *Plant Cell* **17:** 2204–2216. doi:10.1105/tpc.105.033076

Wang Q, Qin G, Cao M, Chen R, He Y, Yang L, Zeng Z, Yu Y, Gu Y, Xing W, et al. 2020. A phosphorylation-based switch controls TAA1-mediated auxin biosynthesis in plants. *Nat Commun* **11:** 679. doi:10.1038/s41467-020-14395-w

Wickett NJ, Mirarab S, Nguyen N, Warnow T, Carpenter E, Matasci N, Ayyampalayam S, Barker MS, Burleigh JG, Gitzendanner MA, et al. 2014. Phylotranscriptomic analysis of the origin and early diversification of land plants. *Proc Natl Acad Sci* **111:** E4859–E4868. doi:10.1073/pnas.1323926111

Williams L, Carles CC, Osmont KS, Fletcher JC. 2005. A database analysis method identifies an endogenous *trans*-acting short-interfering RNA that targets the *Arabidopsis ARF2, ARF3*, and *ARF4* genes. *Proc Natl Acad Sci* **102:** 9703–9708. doi:10.1073/pnas.0504029102

Wu MF, Yamaguchi N, Xiao J, Bargmann B, Estelle M, Sang Y, Wagner D. 2015. Auxin-regulated chromatin switch directs acquisition of flower primordium founder fate. *eLife* **4:** e09269. doi:10.7554/eLife.09269

Xia R, Xu J, Meyers BC. 2017. The emergence, evolution, and diversification of the miR390-*TAS3-ARF* pathway in land plants. *Plant Cell* **29:** 1232–1247. doi:10.1105/tpc.17.00185

Yi K, Menand B, Bell E, Dolan L. 2010. A basic helix-loop-helix transcription factor controls cell growth and size in root hairs. *Nat Genet* **42:** 264–267. doi:10.1038/ng.529

Yoshida S, BarbierdeReuille P, Lane B, Bassel GW, Prusinkiewicz P, Smith RS, Weijers D. 2014. Genetic control of plant development by overriding a geometric division rule. *Dev Cell* **29:** 75–87. doi:10.1016/j.devcel.2014.02.002

Záveská Drábková L, Dobrev PI, Motyka V. 2015. Phytohormone profiling across the bryophytes. *PLoS ONE* **10:** e0125411. doi:10.1371/journal.pone.0125411

Zhang J, Lin JE, Harris C, Campos Mastrotti Pereira F, Wu F, Blakeslee JJ, Peer WA. 2016. DAO1 catalyzes temporal and tissue-specific oxidative inactivation of auxin in *Arabidopsis thaliana*. *Proc Natl Acad Sci* **113:** 11010–11015. doi:10.1073/pnas.1604769113

Zhang J, Fu XX, Li RQ, Zhao X, Liu Y, Li MH, Zwaenepoel A, Ma H, Goffinet B, Guan YL, et al. 2020. The hornwort genome and early land plant evolution. *Nat Plants* **6:** 107–118. doi:10.1038/s41477-019-0588-4

The Systems and Synthetic Biology of Auxin

R. Clay Wright,[1] Britney L. Moss,[2] and Jennifer L. Nemhauser[3]

[1]Department of Biological Systems Engineering, Virginia Polytechnic Institute and State University (Virginia Tech), Blacksburg, Virginia 24061, USA

[2]Department of Biology, Whitman College, Walla Walla, Washington 99362, USA

[3]Department of Biology, University of Washington, Seattle, Washington 98195, USA

Correspondence: jn7@uw.edu

Auxin biology as a field has been at the forefront of advances in delineating the structures, dynamics, and control of plant growth networks. Advances have been enabled by combining the complementary fields of top-down, holistic systems biology and bottom-up, build-to-understand synthetic biology. Continued collaboration between these approaches will facilitate our understanding of and ability to engineer auxin's control of plant growth, development, and physiology. There is a need for the application of similar complementary approaches to improving equity and justice through analysis and redesign of the human systems in which this research is undertaken.

Nearly 20 years ago, Hiroaki Kitano published an article entitled "Systems Biology: A Brief Overview" (Kitano 2002). His conceptualization and articulation of the urgent need to "examine the structure and dynamics of cellular and organismal function" remains a major goal of molecular biology, biochemistry, development, and physiology. What would it mean to have "a continuous spectrum of knowledge" of plants from molecular/biochemical/biophysical details to signaling networks to intracellular communication to multicell phenomena to interorgan coordination to ecology? Kitano described four major domains of inquiry in pursuit of this awesome vision: (1) system structures, integrated maps that connect experimentally validated information like gene regulatory networks and biochemical pathways to the physical properties of individual cells and higher order assemblages; (2) system dynamics, comprehensive quantitative models of how system structures change over time in response to a variety of stimuli; (3) system control, a fully annotated dynamical model that is able to predict which interventions will be able to shift behavior of the system to meet distinct sets of specifications; and (4) system design, an automated process for constructing biological systems via well-articulated principles and simulation (Fig. 1). The intervening decades have seen remarkable achievements in the first two of these aims, largely under the umbrella of systems biology, but the second two have been more of a challenge. Synthetic biology, or engineering biology (Engineering Biology Research Consortium 2019), has in many ways emerged as a field to fill this gap. The decade ahead will require a framework that enables bridging between these two fields, a task that

Cite this article as *Cold Spring Harb Perspect Biol* doi: 10.1101/cshperspect.a040071

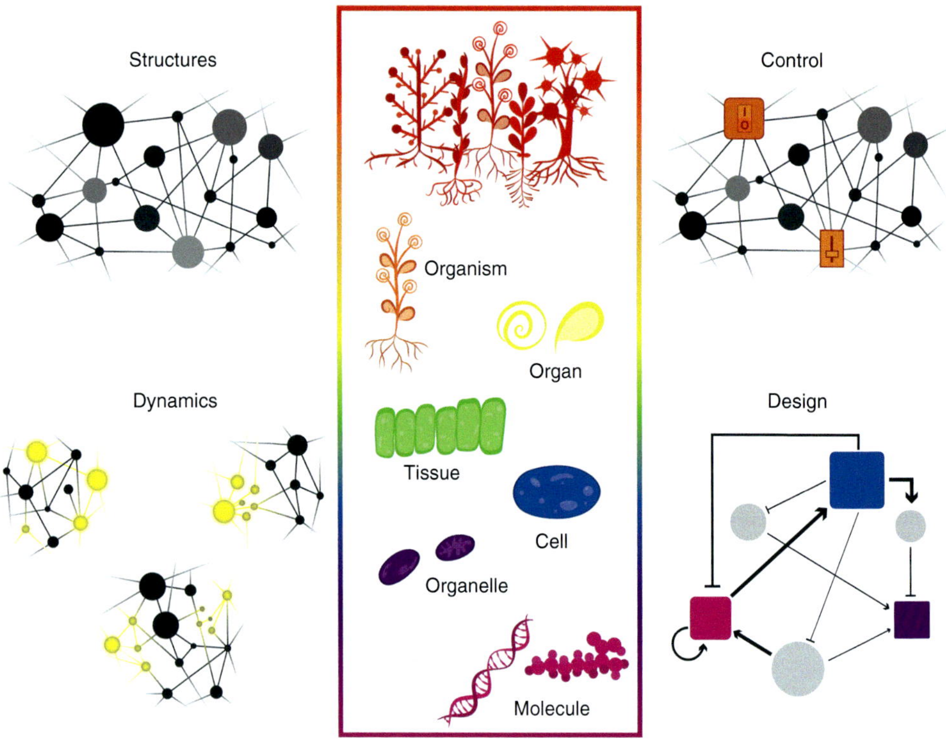

Figure 1. Kitano's four proposed domains of inquiry. The network maps shown represent the four paths of inquiry needed to build a continuous spectrum of knowledge from molecules to ecosystems. Structures represent experimentally validated connectivity maps of biological processes and pathways. Dynamics capture quantitatively how network behaviors change over time, often in response to stimuli. Control uses techniques like sensitivity analysis to identify which nodes in the network are best suited for engineering desired changes in network performance (phenotype). Design synthesizes knowledge arising from the other three domains to modify and construct novel biological systems with user-defined behaviors.

plant biologists are well positioned to pursue (Rhee et al. 2019; Henkhaus et al. 2020; Leydon et al. 2020).

In this review, we will lay out our operating definitions of systems and synthetic biology. In doing so, we will also provide examples of how auxin biology has been at the forefront in advancing both fields. Finally, we will position researchers themselves as a part of a system that needs attention and study.

DEFINITIONS

One challenge to building bridges between the disciplines of systems and synthetic biology is philosophical. Systems biology can be rooted in any one or all of three distinct lenses: methodological, ontological, and epistemological (Mazzocchi 2012). The first, methodological, is really about how data are collected and modeled. The ontological and epistemological lenses refer respectively to which level(s) the researcher believes encode system function, and whether the direction of causality for functions is unidirectional across levels of organization. Fulvio Mazzocchi coined the term "pragmatic systems-biologist" as someone who can build and use tools that prioritize capturing the function of the intact system, while retaining membership in the ontological and epistemological reductionist camps that hold that the principles of chemistry and physics can explain the phenomena observed at higher orders (Mazzocchi 2012). In contrast, a systems-theoretic biologist believes that the arrow of causality points in both directions. In other words, systems properties like emergence and self-construction make biological systems essentially non-Turing computable (Mazzocchi 2012), which is a potentially serious obstacle for a synthetic biologist to surmount. Fortunately, physicists and computer scientists have a head start on building reasonable conceptual and pragmatic work-arounds for exactly this problem (Cooper 2012; Grozinger et al. 2019). Perhaps the most important takeaway from full-on systems theory is the clarion call to explicitly state the assumptions underlying data acquisition and their interpretation and acknowledge what is missing or unmeasurable. Although Mazzocchi might not consider it a desirable state, we would argue that the "pragmatic systems-biologist" is an ideal vantage point from which to blend systems and synthetic biology approaches (Table 1).

Synthetic biology can be defined as the application of engineering design principles to biological systems, particularly the model-based design-build-test-learn cycle (Pouvreau et al. 2018). Synthetic biology is often associated with the translation of biological knowledge for applications of human value. However, the bottom-up, build-to-understand approach has perhaps made its greatest impact in improving our understanding of biology (Elowitz and Lim 2010; Church et al. 2014). The undeniably, unabashedly artificial methodology of synthetic biology thus plays multiple useful roles as a companion with systems approaches. Echoing the strengths of systems theory, synthetic biology approaches also force the researcher to articulate assumptions in the design of the experiment, and frequently reveal gaps in knowledge. Confronted with these gaps, the synthetic biologist can continue to mine the system for additional parts and interactions, thereby testing both methodological and epistemological reductionism. Attempts to engineer the intensity, placement, and mode of feedback, for example, in synthetic and natural systems has produced critical insights into highly desirable qualities of engineered systems like stability and multicell coordination (Chen 2013).

Table 1. A useful but oversimplified guide to differences between systems and synthetic biology

Biology type	Scope	Objective
Systems	Organismal, integration of multilevel and diverse types of networks, holistic	Behavior characterization, causal mechanisms, predictive modeling
Synthetic	Gene circuits, subnetworks, modules, reductionist	Parts characterization, engineered behavior, predictive modeling, model-guided design

USING SYSTEMS AND SYNTHETIC BIOLOGY TO IDENTIFY CONTROL AND LEARN TO DESIGN

Control is critical to biological systems. Plants must remain at least moderately in control of the chaos of thermodynamic equilibrium to extract work, organize and build their fractal structures, and operate molecular machinery to ensure their survival. A control system, or controller, can be defined as a mathematical function that calculates an output that drives the system closer to the set point or goal value of the system, based on an input value that is typically the difference between the current state of the system and the desired set point (Baetica et al. 2019). For example, the simple temperature controller in an oven turns on a heating element whenever the actual temperature falls below the set point temperature and turns the element off when the temperature rises above the set point. In a biological system example, we can think about the expression level of a gene critical for maintenance of cellular identity as being the system state (Fig. 2). In many cases, a biological controller composed of intertwined metabolic and gene regulatory networks will detect whether this gene deviates from the average expression level required for maintenance of cellular identity and correct its expression. To be a good controller that maintains the set point without significant oscilla-

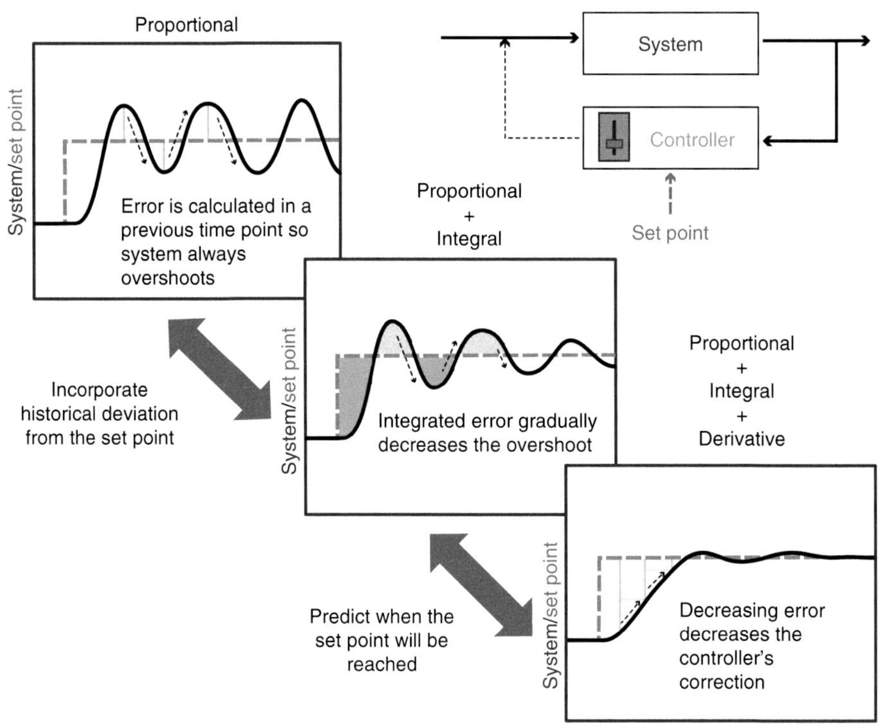

Figure 2. A useful but oversimplified introduction to control in biological systems. The dynamics of a system (dark, solid lines) are regulated by a controller with the aim of maintaining a set point (lighter, dashed lines). The controller modulates the system by applying corrections to the system input (thin, dark, dashed lines), which are calculated based on various functions of the difference between the system and set point values (lightest lines and fill).

tions, rapidly yet smoothly returning to the set point after perturbation, the driving function must integrate the previous behavior of the system and be an accurate predictor of the future behavior or dynamics of a system (Chevalier et al. 2019; Whitby et al. 2021).

So how do we go about understanding control in biological systems? We do it largely in the same way we tune engineered control systems, by tinkering. In tuning a slightly more complex controller than in a typical oven—a controller that regulates the intensity of the heating element instead of simply turning it on or off, or the controller for the cruise control in a car—there are generally three "tuning knobs" or parameters engineers tinker with to optimize a controller's ability to swiftly and accurately return to the set point temperature or speed in our examples after perturbation (Fig. 2). The first of these tuning knobs drives the heating element or accelerator proportionally to the difference between the current and the set point values. However, proportional control alone, similar to on–off logic, only enacts control when there is a difference between the set point and current values. Proportional control alone therefore leads to some consistent error around the set point. The second tuning knob, integral control, solves this problem of consistent error by summing the error between the actual and set point values over a period of time and adjusting accordingly. So small deviations from the set point over time gradually affect stronger responses in integral control. In this way, integral control takes into account the history of the system. The third mode of control in built systems, derivative control, takes into account the future of the system by including the rate of change of the system's deviation from the set point. Derivative control therefore provides information about how fast a system is returning to its set point and prevents the system from overshooting this set point by adjusting the control proportionally to the rate of approach.

Unlike the temperature controller in an oven, which has a clear, human-engineered goal of maintaining the set point temperature by implementing a simple on–off logic, controllers in biological systems often operate at vastly different temporal and spatial scales from the organismal "goal" of fitness, or in the case of domesticated organisms, their value to humans. At many scales across biology, it is unclear what the goal is, or more appropriately, the set point of these biological controllers. Even the most complex engineered systems are built upon logic that is orders of magnitude less complex than biological systems. After all, "evolution is a tinkerer, not an engineer" (Jacob 1977; Oster and Rocklin 1979). However, this comparison is still useful in introducing the concepts of control in biological systems (more mathematically detailed comparison can be found in Del Vecchio et al. 2016, Chevalier et al. 2019, and Whitby et al. 2021).

For auxin, "the specificity in the system is not in the signal but in the cells that perceive it" (Leyser 2018). It may be tempting to consider auxin itself as the system being controlled, but auxin is most often the signal communicating the difference between the current state and set point of gene expression (Roosjen et al. 2018). In the context of feedback, the lines between the controller and controlled are blurred. Auxin's ability to regulate its own transport and metabolism is an integral part of the controller's function by generating negative feedback to reject noise and maintain signaling and cellular homeostasis as well as positive feedback to drive changes in the set point, signaling behavior, and cellular identity (Casanova-Sáez and Voß 2019; Semeradova et al. 2020). So, to understand auxin signaling as a control system we need to answer the question, what is the set point of auxin signaling? What does auxin signaling control? The answer to this question depends upon the scope of the system we wish to understand, that is, the specific cells, tissues, or organs of interest at particular times in development or during their response to stimuli. At the cellular level, auxin controls cell size, shape, growth axis, and identity in coordination with other signals and control systems. At the tissue and organ levels, these cellular behaviors are integrated into morphology and physiology. The nuclear auxin controller calculates transcriptional activity of a significant fraction of the genomes of all plants (Mutte et al. 2018). The set point for auxin signaling may then be whatever rate of transcription is needed to coordinate cell growth and division or higher-order outputs.

CONTROL AND DESIGN OF AUXIN SIGNALING

Beyond its centrality to plant biology, auxin signaling has the attractive feature of at once embodying elegant simplicity (few parts comprise a minimal pathway) and expansive complexity (each component belongs to large gene families and has multiple tuning knobs) (Leyser 2018). We know more about how cells recognize and respond to auxin than almost any other signaling process. The nuclear auxin signaling pathway is as stripped down as nuclear receptors are in animals; yet, like nuclear receptors, it incorporates several highly conserved eukaryotic functions such as ubiquitin-mediated degradation and transcriptional de-repression/activation. As described in great detail in other articles in this collection, auxin-induced transcription is mediated by dimerization of transcription factors called auxin response factors (ARFs). In low auxin conditions, ARFs are repressed by interaction with proteins called Aux/IAAs that form a bridge between the ARFs and TOPLESS (TPL), a member of the Groucho/Tup1 family of corepressors. When auxin accumulates, it facilitates a high-affinity interaction between the Aux/IAA substrates and an auxin receptor F-box protein. Interaction with the F-box brings the substrate to an SCF-type E3 ubiquitin ligase, leading to polyubiquitination and subsequent degradation by the proteasome.

This simple architecture produces complex behaviors through expression of different complements of signaling components. In *Arabidopsis*, for example, there are six auxin F-box receptors, 29 Aux/IAA substrates, and 23 ARF transcription factors, each with its own production/decay rate and relative affinity for other components. Whereas it had been long hypothesized that specificity in auxin responses were conferred by different component combinations, it has been impossible to rigorously test this model in plants because of coexpression of many family members in the same cell, and multilevel feedback on pathway components and auxin itself. The number of accessory components that are required make in vitro assays impractical. An early plant synthetic biology application was porting the few plant-specific nuclear auxin signaling components into yeast (Havens et al. 2012; Pierre-Jerome et al. 2014). By tagging different components with fluorescent reporters, time-lapse flow cytometry can be used to quantify the dynamics of each part in user-defined circuits. In addition, the inducibility of the system with a small molecule that is readily taken up by diverse eukaryotic cells with low toxicity makes auxin-induced degradation (AID) and gene activation a desirable technology for applications in nonplant cells. A similar "plantified" yeast approach has now also been applied to abscisic acid signaling (Ruschhaupt et al. 2019).

The synthetic dissection of the auxin pathway in yeast is in many ways paralleled by the use of "humanized" yeast to dissect the design principles of the animal cell nuclear factor κB (NF-κB) signaling pathway (Zhang et al. 2017). Similar to auxin, the NF-κB pathway is a rapidly activated signaling system that functions during growth and development, and in animals has a particularly key role in immune response. Extensive live-cell imaging, biochemical studies, and computational modeling have revealed NF-κB signaling to be an oscillator circuit with time-delayed negative feedback. The negative feedback loop consists of a transcription factor (NF-κB family) and an inhibitor (IκB family). In response to a signal, IκB is ubiquitinated and degraded, releasing inhibition on NF-κB, which can then up-regulate new IκBs to inhibit NF-κBs, and the cycle goes on. Recently, Zhang et al. (2017) reconstituted a minimal IκB/NF-κB feedback loop in yeast cells and rewired it into the yeast-mating pathway with the α factor pheromone as an input. They successfully recapitulated oscillatory NF-κB dynamics in single cells, with a periodicity close to that observed in mammalian cells but with sustained rather than damped oscillations. Further, the simplified nature of the system enabled quantitative exploration of the relationship between oscillation waveforms and parameters such as IκB degradation rate, which was found to tune the frequency of oscillatory dynamics. NF-κB researchers are also making headway in understanding how signaling dynamics are "read-out" to generate specific transcriptional responses by using an integrated platform that enables live-cell imaging of NF-κB dynamics coupled with single-cell RNA-seq to measure downstream transcriptional profiles associated with specific outputs (Lane et al. 2017). These approaches forge a path for future auxin synthetic biology research seeking to connect circuit function with context dependency and evolutionary innovation.

Recently, the power of a systems biology approach to identifying control parameters in auxin signaling was highlighted in a study of how cell division and growth are organized in the shoot apical meristem of *Arabidopsis* (Galvan-Ampudia et al. 2020). Synthetic reporters of auxin flux and signaling output were deployed to generate a cellular resolution time course of auxin signaling during floral organ initiation. This work revealed a mismatch between the measurements of input (auxin flux) and output (auxin signaling), demonstrating that how cells perceive auxin is changing during the transition from stem cell to floral organ progenitor. This behavior is well known and makes sense in light of the highly reticulated auxin signaling network. An auxin stimulus may lead to degradation of extant Aux/IAAs and expression of other Aux/IAAs and ARFs resulting in a different response upon a second exposure to auxin. Thus, auxin signaling and specifically the milieu of auxin signaling proteins—TIR1/AFB receptors, Aux/IAA substrate/repressors, and ARF transcription factors—act as a cellular memory, recording the previous states of auxin within the cell. This approach also helped define the responsibility of WUSCHEL in modulating auxin responses in the SAM stem cell niche (Ma et al.

2019). WUS restrains the start of an auxin signaling cascade that integrates geometry via spatiotemporal differences in auxin flux and perception contingent upon cellular histories to organize floral organ initiation. In light of our control example, WUS changes the set point of the auxin controller. These works clearly demonstrate a tenant of developmental signaling: current and future signaling dynamics are predicted by cell location and history. Perhaps not surprisingly, these results also usher in nearly the identical question as the synthetic biology studies described above: what duration of, or changes in, auxin flux patterning are perceived by the signaling pathway, and how is this signal converted into transcriptional activation of specific genes? Are auxin signals encoded in the frequency domain, the amplitude domain, or both? How many channels are there on auxin radio?

CASE STUDY: SYSTEMS AND SYNTHETIC BIOLOGY APPROACHES APPLIED TO THE Aux/IAAs

The story of understanding and now engineering the auxin response is a fair encapsulation of the history of technical innovations in modern plant biology (Abel and Theologis 2010). Starting with the first wave of molecular biology studies in the mid-1980s, much of the focus has been on the earliest auxin-induced transcripts (Walker and Key 1982; Hagen and Guilfoyle 1985; Theologis et al. 1985). The Aux/IAAs were among these genes, and subsequent work led to the identification of *cis* elements in their promoters that conferred inducibility in response to auxin or to treatment with inhibitors of translation (Ballas et al. 1993). This second quality led to a model in which auxin regulated a short-lived repressor protein (Koshiba et al. 1995), a premise that was eventually supported by cloning the genes affected in some of the first mutants isolated in the then new model plant *Arabidopsis* (Rouse et al. 1998), and eventually in elegant in vitro biochemical assays and crystallography that proved that the Aux/IAAs themselves were coreceptors for auxin (Dharmasiri et al. 2005; Kepinski and Leyser 2005; Tan et al. 2007). Luciferase fusions to Aux/IAAs were used to delimit the auxin degron, and subsequently revealed a range of AID rates among family members (Zenser et al. 2001; Dreher et al. 2006). Aux/IAAs were also used in an early yeast two-hybrid study that identified sequences in the carboxy terminus (now known as a PB1 domain; Korasick et al. 2014) as interaction surfaces allowing for Aux/IAA and ARF homo- and heterodimerization (Kim et al. 1997), an experiment that has recently been repeated on a larger scale (Vernoux et al. 2011). Synthetic assays of Aux/IAA functions in yeast combined with dynamic modeling demonstrated that degradation rate is among the most effective tuning knobs for altering downstream transcriptional dynamics (Pierre-Jerome et al. 2014). This Aux/IAA degradation tuning knob can be adjusted by varying sequences within the Aux/IAAs (Dreher et al. 2006; Guseman et al. 2015; Moss et al. 2015). In addition, natural or induced variation in the TIR1/AFB receptors can also impact degradation rate, and thus tune auxin sensitivity (Havens et al. 2012; Yu et al. 2013; Dezfulian et al. 2016; Wright et al. 2017; Ramos Báez et al. 2020).

These observations collectively suggest a model where the rate of auxin-induced Aux/IAA turnover acts as a checkpoint or timer for auxin-regulated developmental events. Evidence in support of this model came from an experiment analyzing transgenic plants expressing degradation rate variants of *IAA14*, a crucial determinant of lateral root initiation (Guseman et al. 2015). Progression through the well-established stages of lateral root development was strongly correlated with the engineered rates of IAA14 turnover. This straightforward result is unexpected given the multiple levels of feedback within the auxin pathway, which might be predicted to provide a strong buffering effect. This work led to a hypothesis that Aux/IAAs are auxin-initiated timers that synchronize developmental transitions. Additional support for this idea came from work connecting light perception to lateral root initiation, where the authors present a quantitative model whereby AID of Aux/IAAs allows for on-the-fly environmental resetting of developmental timing (Kircher and Schopfer 2018).

AID of Aux/IAAs has also fueled the development of a number of new technologies. In recent years, there has been widespread adoption of AID using different portions of Aux/IAA degrons as fusion tags to allow inducible degradation of proteins of interest in a wide array of nonplant organisms (Nishimura et al. 2009, 2020; Zhang et al. 2015; Bence et al. 2017). Similarly, Aux/IAA degrons have also been incorporated into cell–cell communication systems. In yeast, auxin-responsive synthetic transcription factors were expressed in one strain and mixed with a second strain that was engineered to synthesize auxin (Khakhar et al. 2016). Recently, this idea was expanded to act as a way to penalize cheaters in a quorum-sensing circuit engineered into mammalian cells (Ma et al. 2020). Synthetic transcription factors that incorporate the Aux/IAA degron have also been used to semirationally design plant development (Khakhar et al. 2018). Finally, the engineering of an orthogonal auxin-TIR1 pair that interfaces with the endogenous auxin response has been a major breakthrough in resolving controversial areas of auxin biology, notably very rapid auxin-induced growth responses (Fendrych et al. 2018; Uchida et al. 2018). These types of approaches alter the control logic of endogenous nuclear auxin control systems and add exogenous control points.

THE NEXT SET OF QUESTIONS WELL SUITED FOR SYNTHETIC AND/OR SYSTEMS APPROACHES

Remarkably given all of the progress made to date, there remain a host of unanswered questions in auxin biology, many of which will likely benefit from incorporating synthetic and systems approaches. From the perspective of the molecular signaling system itself, there remain key gaps in knowledge in regard to fundamental questions like what components are working in the nucleus, in the cytoplasm, or associated with various membranes. This question has been thrust to the fore with new findings about the activity of TIR1/AFB1 in triggering rapid growth responses (Prigge et al. 2020), as well as with revelations about auxin sequestration in the endoplasmic reticulum (Barbez and Kleine-Vehn 2013) and regulated nucleocytoplasmic partitioning of at least some ARFs (Powers et al. 2019). These same findings also bring out questions about coordination or competition between different signaling events, as well as presenting new potential mechanisms for partitioning auxin action within the cell body during anisotropic growth responses.

There are also areas where we have seen great progress, but still lack complete understanding. For example, work on mosses and liverworts, in addition to an ever-growing number of angiosperms, has been critical for connecting the diversification of auxin signaling components to evolutionary innovations (Matthes et al. 2019; Blázquez et al. 2020; Israeli et al. 2020), and this rich source of information has yet to be fully incorporated into control models or leveraged in engineering efforts. There is also much to learn about how promoter architecture and different classes of ARFs feed into these dynamics (Lavy et al. 2016; Kato et al. 2017; Flores-Sandoval et al. 2018; Galli et al. 2018; Freire-Rios et al. 2020; Lanctot et al. 2020). The challenges in this regard call to mind the work on one of the most extensively studied enhancers regulating *eve2* in *Drosophila* embryogenesis. In their review article "The Appeasement of Doug: A Synthetic Approach to Enhancer Biology," Vincent et al. (2016) document several failures at reconstituting the function of *eve2* by minimizing, multimerizing, and even recoding spacer sequences between known transcription factor-binding sites. These authors argue, and we concur, that both successes and failures in enhancer reconstitution are informative and worthy of public report and advocate the use of iterative design-build-test-learn cycles to continue deciphering the design principles of regulatory logic in (auxin) biology.

From an engineering standpoint, we have yet to fully explore the natural features of transcription factors, such as modularity and cooperative binding. Bashor et al. (2019) recently demonstrated the ability to engineer cooperativity into synthetic transcription factors to generate nonlinear gene circuit behavior in yeast cells. They used a computational model-driven approach that enabled them to explore circuit configuration and behavior space for a large set of highly modular synthetic transcription factors. These highly tunable synthetic transcription factors enabled construction of circuits capable of such complex behaviors such as persistence filtering (activation only in response to an input of sufficiently long duration) and temporal decoding (bandpass filters that respond only to inputs of specified frequencies). How do the interactions between ARFs, and between ARFs and other transcription factors, impact native signaling dynamics? How could they be manipulated to change critical control parameters? The next generation of mathematical models, particularly if built with an eye to design, could be of great use in generating and evaluating experiments in this arena.

So how is information about development encoded in auxin signals? How can we learn to reprogram this controller to be better adapted to our changing climate to improve the lives of farmers and humanity more broadly? To answer these questions requires examination of a variety of tissues and how different auxin controllers are implemented in these systems. A beautiful example of such work examined auxin regulation of tomato leaf shape (Israeli et al. 2019). The *Aux/IAA* gene *Entire* regulates leaflet formation in tomatoes and is one of the rare examples of a clear phenotype associated with a loss of function in an *Aux/IAA*. Of course, auxin controllers are not operating in isolation, and many questions still remain as to how auxin and other signaling pathways interact. Other pathways modulate auxin responses and their effects on plant growth, development, and behavior. In addition to WUSCHEL and the shoot apical meristem cell network discussed above, interactions with the circadian clock are well known and have recently been dissected in circadian-regulated floral opening using systems approaches (Atamian and Harmer 2016; Ke et al. 2018). Hormone interactions in development are well founded, and our understanding of how auxin and cytokinin interact to control development and floral organ size have recently been advanced through a combination of synthetic signaling reporters and quantitative time-lapse microscopy (Zhu et al. 2020). Just as we have perturbed auxin controllers to study their behavior and logic, so too have other organisms sabotaged auxin signaling for their own gain (Zhang et al. 2020; Su et al. 2021). Over the past year, several auxin biosynthetic genes and gene circuits have been implicated in regulating plant–microbe interactions, pushing once again our system boundaries of auxin's realm of controls from organism to ecosystem (Kunkel and Harper 2018).

SCIENTISTS ARE THEMSELVES EMBEDDED IN A SYSTEM THAT NEEDS TO BE UNDERSTOOD AND INTENTIONALLY (SYNTHETICALLY) ENGINEERED FOR EQUITY AND JUSTICE

In 2016, the World Economic Forum brought together a group of young scientists from all over the world to discuss ethical conduct of research. The results of this workshop were published in 2018 as a code of conduct resting on seven principles: (1) engage with the public, (2) pursue the truth, (3) minimize harm, (4) engage with decision makers, (5) support diversity, (6) be a mentor, and (7) be accountable (World Economic Forum Young Scientists Community 2020). This expanded definition of responsible conduct of research is provocative and highlights the ways in which our research programs are themselves systems, embedded in a series of higher-order systems.

We are on the precipice of the next green revolution. With full knowledge of our past successes and failures, we must face this next transition with full attention to innovation in both the approaches we take and in which people participate in the work. In Kitano's domains, we might first focus on the individual stakeholders—what they can contribute, what they need, what success would look like for them—and then how they interact with one another. We can then use this information to build predictable models and guide rational design of our research systems (e.g., Byars-Winston et al. 2018). When we think of our mentees, this would necessitate enthusiastic adoption of best practices, especially those that make our groups truly welcoming and inclusive environments for all (Committee on Effective Mentoring in STEMM et al. 2019; Henkhaus et al. 2020). Professor Beronda Montgomery has cogently laid out a path for how we can use what we know from studying plants and microbes to inform how we approach our mentoring, our collaborative relationships, and our institutional structures (Montgomery 2018, 2020a,b). When we think of our students, we can develop course-based undergraduate research experiences (CUREs) that leverage the excitement around systems and synthetic approaches (Campbell et al. 2014; Murren et al. 2019; Govindan et al. 2020). This type of experience offers more equitable access to research and could substantially broaden the pool of future plant biologists (Lopatto et al. 2008; Bangera and Brownell 2014; Brownell et al. 2015).

Beyond the people directly engaged in the research, there are other stakeholders that historically have been marginalized, excluded, or disrespected. These include the public on whose behalf we are working, farmers working in an enormous range of socioeconomic and environmental conditions, indigenous communities with deep knowledge of specific plants and practices, agronomists, and decision makers in governmental and nongovernmental organizations, just to name a few. How would our science be different if we were regularly engaging with these audiences, not to "educate" them but to listen, to learn, and to build long-term relationships based on trust. There is a need for funding and leadership to foster scientific exchanges and collaborations between scientists in different parts of the world. Perhaps one of the most effective ways to minimize harm with technology is to transfer expertise to scientists who live in the communities where new varieties are needed, allowing for local conversations about values and priorities to guide the decision of which products are developed. We must reckon with how our choices as plant scientists can minimize harm and maximize accountability.

ACKNOWLEDGMENTS

We would like to thank Dr. Daria Chrobok of DC Sci-Art for collaborating with us to produce the figures. Research in the Wright Plant Synthetic Biology Laboratory is supported by the USDA National Institute of Food and Agriculture (Hatch Project 1021738). Research in the AuxSynBio Laboratory at Whitman College is supported by a National Science Foundation (IOS-1546873) award to Brit Moss. Systems and synthetic biology work in the Nemhauser Laboratory is supported by the National Institutes of Health (R01-GM107084), National Science Foundation (IOS-1546873), and the Howard Hughes Medical Institute Faculty Scholar Program.

REFERENCES

Abel S, Theologis A. 2010. Odyssey of auxin. *Cold Spring Harb Perspect Biol* **2:** a004572. doi:10.1101/cshperspect.a004572

Atamian HS, Harmer SL. 2016. Circadian regulation of hormone signaling and plant physiology. *Plant Mol Biol* **91:** 691–702. doi:10.1007/s11103-016-0477-4

Baetica AA, Westbrook A, El-Samad H. 2019. Control theoretical concepts for synthetic and systems biology. *Curr Opin Syst Biol* **14:** 50–57. doi:10.1016/j.coisb.2019.02.010

Ballas N, Wong LM, Theologis A. 1993. Identification of the auxin-responsive element, AuxRE, in the primary indoleacetic acid-inducible gene, PS-IAA4/5, of Pea (*Pisum sativum*). *J Mol Biol* **233:** 580–596. doi:10.1006/jmbi.1993.1537

Bangera G, Brownell SE. 2014. Course-based undergraduate research experiences can make scientific research more inclusive. *CBE Life Sci Educ* **13:** 602–606. doi:10.1187/cbe.14-06-0099

Barbez E, Kleine-Vehn J. 2013. *Divide Et Impera*—cellular auxin compartmentalization. *Curr Opin Plant Biol* **16:** 78–84. doi:10.1016/j.pbi.2012.10.005

Bashor CJ, Patel N, Choubey S, Beyzavi A, Kondev J, Collins JJ, Khalil AS. 2019. Complex signal processing in synthetic gene circuits using cooperative regulatory assemblies. *Science* **364:** 593–597.

Bence M, Jankovics F, Lukácsovich T, Erdélyi M. 2017. Combining the auxin-inducible degradation system with CRISPR/Cas9-based genome editing for the conditional depletion of endogenous *Drosophila melanogaster* proteins. *FEBS J* **284:** 1056–1069. doi:10.1111/febs.14042

Blázquez MA, Nelson DC, Weijers D. 2020. Evolution of plant hormone response pathways. *Annu Rev Plant Biol* **71:** 327–353. doi:10.1146/annurev-arplant-050718-100309

Brownell SE, Hekmat-Scafe DS, Singla V, Chandler Seawell P, Conklin Imam JF, Eddy SL, Stearns T, Cyert MS. 2015. A high-enrollment course-based undergraduate research experience improves student conceptions of scientific thinking and ability to interpret data. *Life Sci Educ* **14:** ar21. doi:10.1187/cbe.14-05-0092

Byars-Winston A, Womack VY, Butz AR, McGee R, Quinn SC, Utzerath E, Saetermoe CL, Thomas SB. 2018. Pilot study of an intervention to increase cultural awareness in research mentoring: implications for diversifying the scientific workforce. *J Clin Transl Sci* **2:** 86–94. doi:10.1017/cts.2018.25

Campbell AM, Eckdahl T, Cronk B, Andresen C, Frederick P, Huckuntod S, Shinneman C, Wacker A, Yuan J. 2014. Pclone: synthetic biology tool makes promoter research accessible to beginning biology students. *Life Sci Educ* **13:** 285–296. doi:10.1187/cbe.13-09-0189

Casanova-Sáez R, Voß U. 2019. Auxin metabolism controls developmental decisions in land plants. *Trends Plant Sci* **24:** 741–754. doi:10.1016/j.tplants.2019.05.006

Chen S, Harrigan P, Heineike B, Stewart-Ornstein J, El-Samad H. 2013. Building robust functionality in synthetic circuits using engineered feedback regulation. *Curr Opin Biotechnol* **24:** 790–796. doi:10.1016/j.copbio.2013.02.025

Chevalier M, Gómez-Schiavon M, Ng AH, El-Samad H. 2019. Design and analysis of a proportional-integral-derivative controller with biological molecules. *Cell Syst* **9:** 338–353.e10. doi:10.1016/j.cels.2019.08.010

Church GM, Elowitz MB, Smolke CD, Voigt CA, Weiss R. 2014. Realizing the potential of synthetic biology. *Nat Rev Mol Cell Biol* **15:** 289–294. doi:10.1038/nrm3767

Committee on Effective Mentoring in STEMM, Board on Higher Education and Workforce, Policy and Global Affairs, National Academies of Sciences, Engineering, and Medicine. 2019. *The science of effective mentorship in STEMM* (ed. Byars-Winston A, Dahlberg ML). National Academies Press, Washington, DC. www.nap.edu/catalog/25568

Cooper B. 2012. The incomputable reality. *Nature* **482:** 465. doi:10.1038/482465a

Del Vecchio D, Dy AJ, Qian Y. 2016. Control theory meets synthetic biology. *J R Soc Interface* **13:** 20160380. doi:10.1098/rsif.2016.0380

Dezfulian MH, Jalili E, Roberto DKA, Moss BL, Khoo K, Nemhauser JL, Crosby WL. 2016. Oligomerization of SCF^{TIR1} is essential for Aux/IAA degradation and auxin signaling in *Arabidopsis*. *PLOS Genet* **12:** e1006301. doi:10.1371/journal.pgen.1006301

Dharmasiri N, Dharmasiri S, Estelle M. 2005. The F-box protein TIR1 is an auxin receptor. *Nature* **435:** 441–445. doi:10.1038/nature03543

Dreher KA, Brown J, Saw RE, Callis J. 2006. The *Arabidopsis* Aux/IAA protein family has diversified in degradation and auxin responsiveness. *Plant Cell* **18:** 699–714. doi:10.1105/tpc.105.039172

Elowitz M, Lim WA. 2010. Build life to understand it. *Nature* **468:** 889–890. doi:10.1038/468889a

Engineering Biology Research Consortium. 2019. Engineering biology: A research roadmap for the next-generation bioeconomy. roadmap.ebrc.org/resources/2019roadmap

Fendrych M, Akhmanova M, Merrin J, Glanc M, Hagihara S, Takahashi K, Uchida N, Torii KU, Friml J. 2018. Rapid and reversible root growth inhibition by TIR1 auxin signalling. *Nat Plants* **4:** 453–459. doi:10.1038/s41477-018-0190-1

Flores-Sandoval E, Eklund DM, Hong SF, Alvarez JP, Fisher TJ, Lampugnani ER, Golz JF, Vázquez-Lobo A, Dierschke T, Lin SS, et al. 2018. Class C ARFs evolved before the origin of land plants and antagonize differentiation and developmental transitions in *Marchantia polymorpha*. *New Phytol* **218:** 1612–1630. doi:10.1111/nph.15090

Freire-Rios A, Tanaka K, Crespo I, van der Wijk E, Sizentsova Y, Levitsky V, Lindhoud S, Fontana M, Hohlbein J, Boer DR, et al. 2020. Architecture of DNA elements mediating ARF transcription factor binding and auxin-responsive gene expression in *Arabidopsis*. *Proc Natl Acad Sci* **117:** 24557–24566. doi:10.1073/pnas.2009554117

Galli M, Khakhar A, Lu Z, Chen Z, Sen S, Joshi T, Nemhauser JL, Schmitz RJ, Gallavotti A. 2018. The DNA binding landscape of the maize AUXIN RESPONSE FACTOR family. *Nat Commun* **9:** 4526. doi:10.1038/s41467-018-06977-6

Galvan-Ampudia CS, Cerutti G, Legrand J, Brunoud G, Martin-Arevalillo R, Azais R, Bayle V, Moussu S, Wenzl C, Jaillais Y, et al. 2020. Temporal integration of auxin information for the regulation of patterning. *eLife* **9:** e55832. doi:10.7554/eLife.55832

Govindan B, Pickett S, Riggs B. 2020. Fear of the CURE: a beginner's guide to overcoming barriers in creating a course-based undergraduate research experience. *J Microbiol Biol Educ* **21:** 21.2.48. doi:10.1128/jmbe.v21i2.2109, www.asmscience.org/content/journal/jmbe/10.1128/jmbe.v21i2.2109

Grozinger L, Amos M, Gorochowski TE, Carbonell P, Oyarzún DA, Stoof R, Fellermann H, Zuliani P, Tas H, Goñi-Moreno A. 2019. Pathways to cellular supremacy in biocomputing. *Nat Commun* **10:** 5250. doi:10.1038/s41467-019-13232-z

Guseman JM, Hellmuth A, Lanctot A, Feldman TP, Moss BL, Klavins E, Villalobos LIAC, Nemhauser JL. 2015. Auxin-induced degradation dynamics set the pace for lateral root development. *Development* **142:** 905–909. doi:10.1242/dev.117234

Hagen G, Guilfoyle TJ. 1985. Rapid induction of selective transcription by auxins. *Mol Cell Biol* **5:** 1197–1203. doi:10.1128/MCB.5.6.1197

Havens KA, Guseman JM, Jang SS, Pierre-Jerome E, Bolten N, Klavins E, Nemhauser JL. 2012. A synthetic approach reveals extensive tunability of auxin signaling. *Plant Physiol* **160:** 135–142. doi:10.1104/pp.112.202184

Henkhaus N, Bartlett M, Gang D, Grumet R, Jordon-Thaden I, Lorence A, Lyons E, Miller S, Murray S, Nelson A, et al. 2020. Plant science decadal vision 2020–2030: reimagining the potential of plants for a healthy and sustainable future. *Plant Direct* **4:** e00252. doi:10.1002/pld3.252

Israeli A, Capua Y, Shwartz I, Tal L, Meir Z, Levy M, Bar M, Efroni I, Ori N. 2019. Multiple auxin-response regulators enable stability and variability in leaf development. *Curr Biol* **29:** 1746–1759.e5. doi:10.1016/j.cub.2019.04.047

Israeli A, Reed JW, Ori N. 2020. Genetic dissection of the auxin response network. *Nat Plants* **6:** 1082–1090. doi:10.1038/s41477-020-0739-7

Jacob F. 1977. Evolution and tinkering. *Science* **196:** 1161–1166. doi:10.1126/science.860134

Kato H, Kouno M, Takeda M, Suzuki H, Ishizaki K, Nishihama R, Kohchi T. 2017. The roles of the sole activator-type auxin response factor in pattern formation of *Mar-*

chantia polymorpha. Plant Cell Physiol **58:** 1642–1651. doi:10.1093/pcp/pcx095

Ke M, Gao Z, Chen J, Qiu Y, Zhang L, Chen X. 2018. Auxin controls circadian flower opening and closure in the waterlily. *BMC Plant Biol* **18:** 143. doi:10.1186/s12870-018-1357-7

Kepinski S, Leyser O. 2005. The *Arabidopsis* F-box protein TIR1 is an auxin receptor. *Nature* **435:** 446–451. doi:10.1038/nature03542

Khakhar A, Bolten NJ, Nemhauser J, Klavins E. 2016. Cell–cell communication in yeast using auxin biosynthesis and auxin responsive CRISPR transcription factors. *ACS Synth Biol* **5:** 279–286. doi:10.1021/acssynbio.5b00064

Khakhar A, Leydon AR, Lemmex AC, Klavins E, Nemhauser JL. 2018. Synthetic hormone-responsive transcription factors can monitor and re-program plant development. *eLife* **7:** e34702. doi:10.7554/eLife.34702

Kim J, Harter K, Theologis A. 1997. Protein–protein interactions among the Aux/IAA proteins. *Proc Natl Acad Sci* **94:** 11786–11791. doi:10.1073/pnas.94.22.11786

Kircher S, Schopfer P. 2018. The plant hormone auxin beats the time for oscillating light-regulated lateral root induction. *Development* **145:** dev169839. doi:10.1242/dev.169839

Kitano H. 2002. Systems biology: a brief overview. *Science* **295:** 1662–1664. doi:10.1126/science.1069492

Korasick DA, Westfall CS, Lee SG, Nanao MH, Dumas R, Hagen G, Guilfoyle TJ, Jez JM, Strader LC. 2014. Molecular basis for AUXIN RESPONSE FACTOR protein interaction and the control of auxin response repression. *Proc Natl Acad Sci* **111:** 5427–5432. doi:10.1073/pnas.1400074111

Koshiba T, Ballas N, Wong LM, Theologis A. 1995. Transcriptional regulation of PS-IAA4/5 and PS-IAA6 early gene expression by indoleacetic acid and protein synthesis inhibitors in pea (*Pisum sativum*). *J Mol Biol* **253:** 396–413. doi:10.1006/jmbi.1995.0562

Kunkel BN, Harper CP. 2018. The roles of auxin during interactions between bacterial plant pathogens and their hosts. *J Exp Bot* **69:** 245–254. doi:10.1093/jxb/erx447

Lanctot A, Taylor-Teeples M, Oki EA, Nemhauser JL. 2020. Specificity in auxin responses is not explained by the promoter preferences of activator ARFs. *Plant Physiol* **182:** 1533–1536. doi:10.1104/pp.19.01474

Lane K, Van Valen D, DeFelice MM, Macklin DN, Kudo T, Jaimovich A, Carr A, Meyer T, Pe'er D, Boutet SC, et al. 2017. Measuring signaling and RNA-Seq in the same cell links gene expression to dynamic patterns of NF-κB activation. *Cell Syst* **4:** 458–469.e5. doi:10.1016/j.cels.2017.03.010

Lavy M, Prigge MJ, Tao S, Shain S, Kuo A, Kirchsteiger K, Estelle M. 2016. Constitutive auxin response in *Physcomitrella* reveals complex interactions between Aux/IAA and ARF proteins. *eLife* **5:** e13325. doi:10.7554/eLife.13325

Leydon AR, Gala HP, Guiziou S, Nemhauser JL. 2020. Engineering synthetic signaling in plants. *Annu Rev Plant Biol* **71:** 767–788. doi:10.1146/annurev-arplant-081519-035852

Leyser O. 2018. Auxin signaling. *Plant Physiol* **176:** 465–479. doi:10.1104/pp.17.00765

Lopatto D, Alvarez C, Barnard D, Chandrasekaran C, Chung HM, Du C, Eckdahl T, Goodman AL, Hauser C, Jones CJ, et al. 2008. Undergraduate research: genomics education partnership. *Science* **322:** 684–685. doi:10.1126/science.1165351

Ma Y, Miotk A, Šutiković Z, Ermakova O, Wenzl C, Medzihradszky A, Gaillochet C, Forner J, Utan G, Brackmann K, et al. 2019. WUSCHEL acts as an auxin response rheostat to maintain apical stem cells in *Arabidopsis*. *Nat Commun* **10:** 1–11.

Ma Y, Budde MW, Mayalu MN, Zhu J, Murray RM, Elowitz MB. 2020. Synthetic mammalian signaling circuits for robust cell population control. bioRxiv doi:10.1101/2020.09.02.278564

Matthes MS, Best NB, Robil JM, Malcomber S, Gallavotti A, McSteen P. 2019. Auxin EvoDevo: conservation and diversification of genes regulating auxin biosynthesis, transport, and signaling. *Mol Plant* **12:** 298–320. doi:10.1016/j.molp.2018.12.012

Mazzocchi F. 2012. Complexity and the reductionism–holism debate in systems biology. *Wiley Interdiscip Rev Syst Biol Med* **4:** 413–427. doi:10.1002/wsbm.1181

Montgomery B. 2018. From deficits to possibilities: mentoring lessons from plants on cultivating individual growth through environmental assessment and optimization. *Public Philos J* **1:** 1–12. osf.io/g83s9 [Accessed October 29, 2020].

Montgomery BL. 2020a. Lessons from microbes: what can we learn about equity from unculturable bacteria? *mSphere* **5:** e01046. msphere.asm.org/content/5/5/e01046-20, doi:10.1128/mSphere.01046-20

Montgomery BL. 2020b. Planting equity: using what we know to cultivate growth as a plant biology community. *Plant Cell* **32:** 3372–3375. doi:10.1105/tpc.20.00589

Moss BL, Mao H, Guseman JM, Hinds TR, Hellmuth A, Kovenock M, Noorassa A, Lanctot A, Villalobos LIAC, Zheng N, et al. 2015. Rate motifs tune auxin/indole-3-acetic acid degradation dynamics. *Plant Physiol* **169:** 803–813. doi:10.1104/pp.15.00587

Murren CJ, Wolyniak MJ, Rutter MT, Bisner AM, Callahan HS, Strand AE, Corwin LA. 2019. Undergraduates phenotyping *Arabidopsis* knockouts in a course-based undergraduate research experience: exploring plant fitness and vigor using quantitative phenotyping methods. *J Microbiol Biol Educ* **20:** 1–9. doi:10.1128/jmbe.v20i2.1650

Mutte SK, Kato H, Rothfels C, Melkonian M, Wong GKS, Weijers D. 2018. Origin and evolution of the nuclear auxin response system. *eLife* **7:** e33399. doi:10.7554/eLife.33399

Nishimura K, Fukagawa T, Takisawa H, Kakimoto T, Kanemaki M. 2009. An auxin-based degron system for the rapid depletion of proteins in nonplant cells. *Nat Methods* **6:** 917–922. doi:10.1038/nmeth.1401

Nishimura K, Yamada R, Hagihara S, Iwasaki R, Uchida N, Kamura T, Takahashi K, Torii KU, Fukagawa T. 2020. A super-sensitive auxin-inducible degron system with an engineered auxin-TIR1 pair. *Nucleic Acids Res* **48:** e108–e108. doi:10.1093/nar/gkaa748

Oster G, Rocklin S. 1979. Optimization models in evolutionary biology. In *Some mathematical questions in biology* (ed. Levin S), Vol. X, pp. 21–88. American Mathematical Society, Providence, RI.

Pierre-Jerome E, Jang SS, Havens KA, Nemhauser JL, Klavins E. 2014. Recapitulation of the forward nuclear auxin response pathway in yeast. *Proc Natl Acd Sci* **111:** 9407–9412. doi:10.1073/pnas.1324147111

Pouvreau B, Vanhercke T, Singh S. 2018. From plant metabolic engineering to plant synthetic biology: the evolution of the design/build/test/learn cycle. *Plant Sci* **273:** 3–12. doi:10.1016/j.plantsci.2018.03.035

Powers SK, Holehouse AS, Korasick DA, Schreiber KH, Clark NM, Jing H, Emenecker R, Han S, Tycksen E, Hwang I, et al. 2019. Nucleo-cytoplasmic partitioning of ARF proteins controls auxin responses in *Arabidopsis thaliana*. *Mol Cell* **76:** 177–190.e5. doi:10.1016/j.molcel.2019.06.044

Prigge MJ, Platre M, Kadakia N, Zhang Y, Greenham K, Szutu W, Pandey BK, Bhosale RA, Bennett MJ, Busch W, et al. 2020. Genetic analysis of the *Arabidopsis* TIR1/AFB auxin receptors reveals both overlapping and specialized functions. *eLife* **9:** e54740. doi:10.7554/eLife.54740

Ramos Báez R, Buckley Y, Yu H, Chen Z, Gallavotti A, Nemhauser JL, Moss BL. 2020. A synthetic approach allows rapid characterization of the maize nuclear auxin response circuit. *Plant Physiol* **182:** 1713–1722. doi:10.1104/pp.19.01475

Rhee SY, Birnbaum KD, Ehrhardt DW. 2019. Towards building a plant cell atlas. *Trends Plant Sci* **24:** 303–310. doi:10.1016/j.tplants.2019.01.006

Roosjen M, Paque S, Weijers D. 2018. Auxin response factors: output control in auxin biology. *J Exp Bot* **69:** 179–188. doi:10.1093/jxb/erx237

Rouse D, Mackay P, Stirnberg P, Estelle M, Leyser O. 1998. Changes in auxin response from mutations in an *AUX/IAA* gene. *Science* **279:** 1371–1373. doi:10.1126/science.279.5355.1371

Ruschhaupt M, Mergner J, Mucha S, Papacek M, Doch I, Tischer SV, Hemmler D, Chiasson D, Edel KH, Kudla J, et al. 2019. Rebuilding core abscisic acid signaling pathways of *Arabidopsis* in yeast. *EMBO J* **38:** e101859. doi:10.15252/embj.2019101859

Semeradova H, Montesinos JC, Benkova E. 2020. All roads lead to auxin: post-translational regulation of auxin transport by multiple hormonal pathways. *Plant Commun* **1:** 100048. doi:10.1016/j.xplc.2020.100048

Su P, Zhao L, Li W, Zhao J, Yan J, Ma X, Li A, Wang H, Kong L. 2021. Integrated metabolo-transcriptomics and functional characterization reveals that the wheat auxin receptor TIR1 negatively regulates defense against *Fusarium graminearum*. *J Integr Plant Biol* **63:** 340–352. doi:10.1111/jipb.12992

Tan X, Calderon-Villalobos LIA, Sharon M, Zheng C, Robinson CV, Estelle M, Zheng N. 2007. Mechanism of auxin perception by the TIR1 ubiquitin ligase. *Nature* **446:** 640–645. doi:10.1038/nature05731

Theologis A, Huynh TV, Davis RW. 1985. Rapid induction of specific mRNAs by auxin in pea epicotyl tissue. *J Mol Biol* **183:** 53–68. doi:10.1016/0022-2836(85)90280-3

Uchida N, Takahashi K, Iwasaki R, Yamada R, Yoshimura M, Endo TA, Kimura S, Zhang H, Nomoto M, Tada Y, et al. 2018. Chemical hijacking of auxin signaling with an engineered auxin–TIR1 pair. *Nat Chem Biol* **14:** 299–305. doi:10.1038/nchembio.2555

Vernoux T, Brunoud G, Farcot E, Morin V, Van den Daele H, Legrand J, Oliva M, Das P, Larrieu A, Wells D, et al. 2011. The auxin signalling network translates dynamic input into robust patterning at the shoot apex. *Mol Syst Biol* **7:** 508. doi:10.1038/msb.2011.39

Vincent BJ, Estrada J, DePace AH. 2016. The appeasement of Doug: a synthetic approach to enhancer biology. *Integr Biol (Camb)* **8:** 475–484.

Walker JC, Key JL. 1982. Isolation of cloned cDNAs to auxin-responsive poly(A)+RNAs of elongating soybean hypocotyl. *Proc Natl Acad Sci* **79:** 7185–7189. doi:10.1073/pnas.79.23.7185

Whitby M, Cardelli L, Kwiatkowska M, Laurenti L, Tribastone M, Tschaikowski M. 2021. PID control of biochemical reaction networks. *IEEE Trans Automat Contr* doi:10.1109/TAC.2021.3062544

World Economic Forum Young Scientists Community. 2020. World economic forum code of ethics for researchers. widgets.weforum.org/coe/#code

Wright RC, Zahler ML, Gerben SR, Nemhauser JL. 2017. Insights into the evolution and function of auxin signaling F-Box proteins in *Arabidopsis thaliana* through synthetic analysis of natural variants. *Genetics* **207:** 583–591. doi:10.1534/genetics.117.300092

Yu H, Moss BL, Jang SS, Prigge M, Klavins E, Nemhauser JL, Estelle M. 2013. Mutations in the TIR1 auxin receptor that increase affinity for auxin/indole-3-acetic acid proteins result in auxin hypersensitivity. *Plant Physiol* **162:** 295–303. doi:10.1104/pp.113.215582

Zenser N, Ellsmore A, Leasure C, Callis J. 2001. Auxin modulates the degradation rate of Aux/IAA proteins. *Proc Natl Acad Sci* **98:** 11795–11800. doi:10.1073/pnas.211312798

Zhang L, Ward JD, Cheng Z, Dernburg AF. 2015. The auxin-inducible degradation (AID) system enables versatile conditional protein depletion in *C. elegans*. *Development* **142:** 4374–4384. doi:10.1242/dev.129635

Zhang ZB, Wang QY, Ke YX, Liu SY, Ju JQ, Lim WA, Tang C, Wei P. 2017. Design of tunable oscillatory dynamics in a synthetic NF-κB signaling circuit. *Cell Syst* **5:** 460–470.e5. doi:10.1016/j.cels.2017.09.016

Zhang H, Li L, He Y, Qin Q, Chen C, Wei Z, Tan X, Xie K, Zhang R, Hong G, et al. 2020. Distinct modes of manipulation of rice auxin response factor OsARF17 by different plant RNA viruses for infection. *Proc Natl Acad Sci* **117:** 9112–9121. doi:10.1073/pnas.1918254117

Zhu M, Chen W, Mirabet V, Hong L, Bovio S, Strauss S, Schwarz EM, Tsugawa S, Wang Z, Smith RS, et al. 2020. Robust organ size requires robust timing of initiation orchestrated by focused auxin and cytokinin signalling. *Nat Plants* **6:** 686–698. doi:10.1038/s41477-020-0666-7

Modeling Auxin Signaling in Roots: Auxin Computations

Jaap Rutten,[1] Thea van den Berg,[1] and Kirsten ten Tusscher

Computational Developmental Biology Group, Utrecht University, Utrecht 3584 CH, The Netherlands

Correspondence: k.h.w.j.tentusscher@uu.nl

Auxin signaling and patterning is an inherently complex process, involving polarized auxin transport, metabolism, and signaling, its effect on developmental zones, as well as growth rates, and the feedback between all these different aspects. This complexity has led to an important role for computational modeling in unraveling the multifactorial roles of auxin in plant developmental and adaptive processes. Here we discuss the basic ingredients of auxin signaling and patterning models for root development as well as a series of key modeling studies in this area. These modeling studies have helped elucidate how plants use auxin signaling to compute the size of their root meristem, the direction in which to grow, and when and where to form lateral roots. Importantly, these models highlight how auxin, through patterning of and collaborating with other factors, can fulfill all these roles simultaneously.

Auxin signaling plays a key role in diverse processes governing the growth, development, and environmental adaptation of plant roots. Auxin maxima signal the presence of stem cell niches fueling the growth of individual root organs, with the auxin gradients originating at these maxima impacting the size and activity of the corresponding meristems (Sabatini et al. 1999; Petersson et al. 2009). Environmental factors such as nitrate presence or soil salinity impact these auxin patterns, either directly or via other hormones, resulting in the adaptation of root growth rates to environmental conditions (Krouk et al. 2010; Fu et al. 2019). At the same time, transient asymmetries in auxin signaling underlie asymmetries in elongation growth that enable tropic responses of roots toward gravity (Swarup et al. 2005) and away from salt (Galvan-Ampudia et al. 2013; van den Berg et al. 2016), while no such asymmetries are detected during hydrotropism and phototropism (Shkolnik et al. 2016; Kimura et al. 2018) and a potential role in nutrient tropisms remains to be investigated. In lateral roots (LRs), additional auxin-based processes influence gravitropism, resulting in nonvertical gravitropic set-point angles for LRs (Roychoudhry et al. 2013) to ensure the widening of the root system necessary to explore a larger soil surface area (Lynch and Brown 2001). Again, environmental conditions impact these auxin patterns and hence LR growth angles, with, for example, phosphate starvation inducing the shallow angles characteristic of topsoil foraging (Lynch and Brown 2001) and nitrate deprivation instead inducing steep root growth angles (Roychoudhry et al.

[1]These authors contributed equally to this work.

Cite this article as *Cold Spring Harb Perspect Biol* doi: 10.1101/cshperspect.a040089

2017). Additionally, temporal changes in auxin signaling confer on subsets of protoxylem-neighboring pericycle cells the competence for future LR formation (De Smet et al. 2007; Moreno-Risueno et al. 2010; Xuan et al. 2016). Whether or not these competent cells subsequently develop into LRs, and at which rate, is again heavily dependent on auxin signaling (Lavenus et al. 2013; Du and Scheres 2018; Santos Teixeira and ten Tusscher 2019). Finally, auxin signaling specifies the timing and location of cell differentiation dictating, for example, protophloem differentiation rate (Marhava et al. 2018) and the patterning of xylem and phloem poles (Mähönen et al. 2006; Bishopp et al. 2011). Auxin signaling thus enables plant roots to compute where to grow and at what rate, when and where, and at what angle to form LRs, and when and where to form which tissue types. Additionally, it allows plant roots to finely tune these decisions to the environmental conditions of individual root tips, the root system, and overall plant status. Despite the tremendous progress that has been and is being made in identifying the molecular players and networks involved in these processes, a complete mechanistic understanding often does not come easily.

Two major factors are complicating our understanding. First, all of these processes have an inherently multiscale and feedback-regulated nature, involving processes ranging from the rapid subcellular relocalization of auxin-exporting PIN proteins (Geldner et al. 2001; Kleine-Vehn and Friml 2008; Kleine-Vehn et al. 2009), cellular gene-expression changes (Mähönen et al. 2014; Santuari et al. 2016), tissue level growth, and hormone patterning (Sabatini et al. 1999; Dello Ioio et al. 2007; Mähönen et al. 2014; Salvi et al. 2020) to slow overall root system growth and architecture development (Nacry et al. 2005). While at widely different spatial and temporal scales, these processes are all influenced and being influenced by one another, making their mechanistic underpinnings hard to disentangle. As a further complication, the major role of cell division, cell elongation, and tissue growth in these processes gives rise to additional feedback and complicates interpretation (Rutten and ten Tusscher 2019). As an example, rapid cell growth may dilute cellular transcription factor or hormone levels (Band et al. 2012), while the combination of cell displacement with stability of a transcription factor may at the same time cause the presence of that factor well outside of its domain of transcription (Mähönen et al. 2014). Also, a phenotype in which cell differentiation occurs at a longer distance from the root tip may be interpreted as a slowdown in differentiation rate (Pacifici et al. 2018) but could potentially also arise from higher division rates enhancing the cumulative displacement of cells out of the meristem.

A second major issue revolves around the specificity of auxin signaling. Surely, tissue-specific differences in the expression of the auxin-response machinery enable the differential interpretation of similar auxin signals in different tissue contexts (Weijers et al. 2005; Rademacher et al. 2011). Likewise, simultaneously expressed response machinery components with different auxin affinity parameters enable the differential interpretation of quantitatively different auxin signals (Villalobos et al. 2012; Shimizu-Mitao and Kakimoto 2014). Still, this does not fully explain how auxin at similar times and tissues and within similar concentration ranges can control stable root developmental zonation, transient directional growth responses, and temporally periodic priming of LRs (van den Berg and ten Tusscher 2018). Additionally, it does not easily explain how auxin may control cell division, cell expansion, and cell differentiation rates (Evans et al. 1994) as well as control the sizes and locations of zones at which these different cellular processes occur (Ishida et al. 2010; Perrot-Rechenmann 2010). A major question is how, under these conditions, additional contextual information is generated and used for a cell to compute and decide what to do in response to the auxin signals it receives.

In this review, we discuss how computational models, through their capacity to integrate processes at various spatiotemporal scales as well as the feedback between them, have enabled us to gain insight in how auxin signaling controls root development, growth, and adaptation. Additionally, we show how in doing so, models have helped shed light on the issue of auxin signaling specificity.

A BRIEF PRIMER ON MULTISCALE ROOT MODELS

In this section, we provide a brief introduction to the different aspects involved in multiscale models of root growth, development, and adaptation.

Root Anatomy

As a first step, one needs to specify the root tissue for which one aims to model the dynamical behavior. Because of both computational efficiency and lack of full three-dimensional data to base the models on, most root models focus either on a longitudinal or transverse cross section of the root. Existing longitudinal root models range from one-dimensional models of a single strand of cells representing a single tissue type (Band et al. 2012; Muraro et al. 2013; Moret et al. 2020) to stylized rectangular models of the entire root tip (Grieneisen et al. 2007; Mironova et al. 2010, 2012; Mähönen et al. 2014; Tian et al. 2014; Moore et al. 2015; Hong et al. 2017; Retzer et al. 2019), to realistically shaped root tip models either idealized from (Cruz-Ramírez et al. 2012; van den Berg et al. 2016; Di Mambro et al. 2017; van den Berg and ten Tusscher 2018; Salvi et al. 2020), or based on, anatomical data (Fig. 1A; Band et al. 2014; Mellor et al. 2016a,b, 2020; Muraro et al. 2016; Liu et al. 2017). One-dimensional models represent a suitable approach if the research question focuses on a single tissue type. In other situations, based on the finding that the absence of a root cap and wedge-shaped root tip has a significant impact on root tip auxin patterning (Cruz-Ramírez et al. 2012; van den Berg et al. 2016), most researchers have moved from using rectangular simplifications to applying realistic root tip anatomies. In addition to incorporating overall root tip shape, two-dimensional models typically incorporate cell-type-specific differences in cell width, as well as the increase in cell length as cells mature from their birth in the meristem, through elongation and subsequent differentiation, creating the typical zoned patterns at the root tip (Fig. 1A,B). While model layouts directly taken from anatomical data may seem the most realistic and therefore preferable, idealized anatomies have the advantage of not introducing any asymmetries, which is important when, for example, studying the origin of auxin asymmetries during tropisms. Most transverse root models are two-dimensional (Péret et al. 2013; De Rybel et al. 2014; Muraro et al. 2014, 2016; el-Showk et al. 2015; Mellor et al. 2017), again either idealized from or directly taken from anatomical data, yet occasionally also one-dimensional, ring-shaped representations (Fàbregas et al. 2015; Mellor et al. 2019) or one-dimensional line sections (Miyashima et al. 2019) have been used.

Auxin Dynamics

To simulate auxin patterning, models incorporate auxin production and degradation, active PIN and ABCB/PGP-mediated auxin export, and passive as well as active AUX/LAX-mediated auxin import (Fig. 1C). Additionally, models incorporate auxin inflow into the vasculature from the (not explicitly modeled) shoot. Typically, cell-type and root-zone-specific levels and polarity of PIN proteins are superimposed on the root anatomy (for an exception, see Fig. 1D; Mironova et al. 2012). In most early models, auxin import was modeled as a constant, cell-type-, and developmental zone-type-independent process (Grieneisen et al. 2007; Mironova et al. 2010, 2012; Mähönen et al. 2014). In Swarup et al. (2005) as well as later models, cell-type- and zone-specific AUX/LAX expression levels were incorporated (Fig. 1D; Band et al. 2014; Di Mambro et al. 2017; Moore et al. 2017; Mellor et al. 2020). Recently, a modeling study also incorporated plasmodesmata, enabling gradient-dependent passive diffusion between neighboring cells (Mellor et al. 2020). On a similar note, early models assumed similar auxin production and degradation across all cells. More recent models incorporate the elevated production of auxin or auxin precursors in the region around the quiescent center and stem cell niche, the columella, and the LR cap (Fig. 1E;

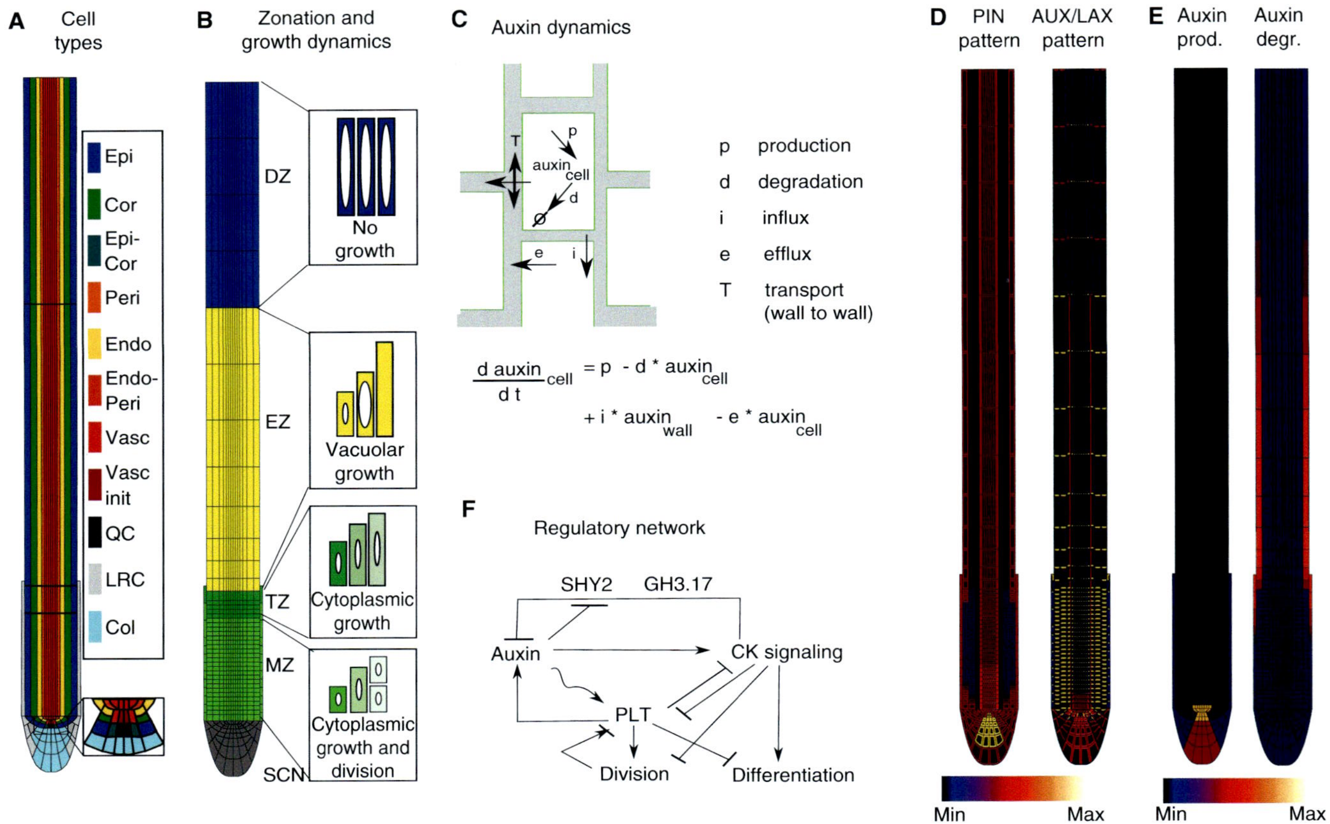

Figure 1. Overview of components of root tip models. (*A*) Root tip architecture indicating the different incorporated cell types. (Epi) Epidermis, (Cor) cortex, (Epi-Cor) epidermis-cortical initials, (Peri) pericycle, (Endo) endodermis, (Endo-Peri) endodermis-pericycle initials, (Vasc) vasculature, (QC) quiescent center, (Col) columella, (LRC) lateral root cap. (*B*) Layout of the root tip with the locations of the different developmental zones and the growth dynamics occurring inside these zones. (SCN) Stem cell niche, (COL) columella, (MZ) meristematic zone, (TZ) transition zone, (EZ) elongation zone, (DZ) differentiation zone. (*C*) Schematic depiction of auxin dynamics, with intracellular production and degradation, trans-membrane influx into and efflux out of cells, and wall-to-wall transport, and the (simplified) auxin dynamics model equation this results in. (*D*) Imposed pattern of auxin-exporting PIN and auxin-importing AUX/LAX proteins. (*E*) Imposed pattern of elevated auxin production and (CK-dependent) degradation. (*F*) Example network of regulatory interactions (based on data in Salvi et al. 2020).

Di Mambro et al. 2017; van den Berg and ten Tusscher 2018; Mellor et al. 2020; Salvi et al. 2020). Additionally, some recent models incorporate the GRETCHENHAGEN3.17 (GH3.17)-mediated elevated degradation of auxin in the LR cap and elongation zone epidermis (Fig. 1F; Di Mambro et al. 2017; Salvi et al. 2020).

Although similar in the types of processes considered, models can be divided into two groups based on whether individual cells and cell walls are assumed to have a single homogeneous auxin concentration (Band et al. 2014; Mellor et al. 2020) or whether subcellular and subwall compartment concentration differences are considered (Grieneisen et al. 2007; Moore et al. 2015; van den Berg et al. 2016; Di Mambro et al. 2017; van den Berg and ten Tusscher 2018; Salvi et al. 2020). In the latter type of models, the intracell and intrawall diffusion of auxin is also taken into account. It has thus far remained unclear to what extent such subcompartment auxin gradients exist and whether they are relevant for auxin patterning.

Regulatory Networks

To decipher how auxin signaling impacts growth, developmental, and adaptive decisions in plant roots, root models typically incorporate a regulatory network consisting of the genes and hormone signals considered relevant for the phenomenon under investigation (Fig. 1F). Models thus often incorporate regulation of the auxin-exporting PIN proteins, auxin-importing AUX/LAX proteins, auxin-biosynthesizing TAA and YUCCA enzymes, and auxin-degrading GH3 enzymes that together govern auxin patterns. Additionally, models may incorporate the regulation of the AUX/IAA and ARF factors involved in auxin signaling sensitivity; and downstream, cell-fate- and cell-behavior-controlling factors such as WOX5 and PLETHORAs are modeled (Mähönen et al. 2014; Tian et al. 2014). In addition to auxin dynamics, root models frequently incorporate the dynamics of other hormones such as gibberellic acid (GA) (Band et al. 2012; Muraro et al. 2016), ethylene (Moore et al. 2015), and cytokinin (CK) (signaling) (Muraro et al. 2013, 2016; Moore et al. 2015; Di Mambro et al. 2017; Salvi et al. 2020), which, by affecting auxin transporters, auxin signaling, and downstream factors modulate, antagonize, or synergize with auxin-dependent control of root growth, development, and adaptation.

Cell and Tissue Growth Dynamics

Only a subset of root models explicitly incorporate cell growth, division and expansion, and the resulting tissue growth dynamics (Grieneisen et al. 2007; Mironova et al. 2010; Band et al. 2012; Mähönen et al. 2014; Lebovka et al. 2020; Salvi et al. 2020). Importantly, auxin and gene-expression levels and patterns not only control these growth dynamics, but are subsequently also affected by them, giving rise to additional feedback that is ignored in static root models (Mähönen et al. 2014; Salvi et al. 2020).

So far, models combining hormone, gene expression, and growth dynamics have applied relatively simplistic descriptions of cell and tissue growth, and several aspects of growth dynamics can still be significantly improved. As an example, while mechanically more realistic, root growth models taking into account the symplastic and highly anisotropic growth of root tissue have been developed (Fozard et al. 2013, 2016; Vos et al. 2014; Weise and ten Tusscher 2019), and these have not yet been fully integrated with detailed models of auxin and gene-expression dynamics.

THE FRUITS OF MULTISCALE ROOT MODELING

Computing Zones and Rates of Growth

Mature Root Growth Control

It has long been known that the location of the quiescent center and stem cell niche coincide with an auxin maximum (Sabatini et al. 1999), with the auxin gradient emanating from it impacting root meristem size. Earlier modeling studies focused mostly on the mechanism underlying the formation of the auxin maximum and gradient in the root tip. These studies revealed the important role of the PIN exporter reflux loop (Grieneisen et al. 2007; Mironova et al. 2010) in generating an auxin maximum and gradient and the important modulatory functions of AUX/LAX importers (Band et al. 2014), ethylene, CK cross talk with auxin (Moore et al. 2015; Liu et al. 2017) and, more recently, plasmodesmatal connections (Mellor et al. 2020) for precise auxin patterning. They have been extensively reviewed in Rutten and ten Tusscher (2019).

More recent studies have focused on the interplay of auxin with other hormones, particularly its antagonist CK, and how this contributes to the control of root meristem size and root growth (Fig. 2). Because of the discrepancy in sites of CK production (root cap and meristem) (Antoniadi et al. 2015) and CK signaling (transition and elongation zone) (Mason et al. 2004) as well as the intricate feedback between auxin, CK, and other hormones (ethylene and gibberellin among others), reconstructing the auxin-CK patterning network has proven to be highly nontrivial (Moore et al. 2015). Using a model incorporating auxin, CK and GA signaling (Muraro et al. 2016) demonstrated that to correctly simulate CK signaling as occurring outside the meristem, an additional, not yet identified factor operating in the meristem and repressing CK signaling was necessary. Additionally, they demonstrated that CK signaling–dependent repression of PINs resulted in a secondary increase of auxin levels in the vasculature. Using a somewhat tangential approach in which CK signaling was superimposed to only occur in the correct location outside the meristem, Di Mambro and coworkers further investigated the impact of this domain on auxin patterning. They demonstrated that CK-mediated repression of PINs combined with the induction of the GH3.17 auxin-degrading enzyme generates an auxin minimum at the transition zone. In a recent study, Salvi et al. (2020) identify the PLETHORA transcription factors, together with auxin, major determinants of stem cell and meristem status (Galinha et al. 2007), as the factors responsible for repressing CK signaling inside the meristem. Importantly, auxin induces *PLT* expression (Mähönen et al. 2014), while PLTs through affecting auxin production, degradation, and transport positively affect auxin levels (Santuari et al. 2016), giving rise to a positive feedback loop. Salvi et al. (2020) showed that this positive feedback loop enables the activation and expansion of the root meristem after germination. Additionally, they demonstrated that meristem activation itself, through division-induced dilution of PLETHORA proteins in the distalmost meristem cells, induces the formation of the elongation zone and activation of CK signaling. This CK signaling eventually halts meristem size expansion. Finally, Salvi and coworkers demonstrate that CK not only antagonizes auxin, but also PLTs, either directly (ARR12) or through repressing divisions that contribute to the expansion of the PLT domain (ARR1). By performing hypothetical what-if experiments in which CK antagonizes only auxin or only PLETHORAs, the authors demonstrate that while PLETHORA antagonism enables CK to rapidly exert an effect on meristem size, auxin antagonism is responsible for stronger and longer-term effects on final meristem size.

Thus, the auxin-PLETHORA-CK regulatory network results in two distinct zones: the meristem in which auxin co-occurs with PLETHORAs, and the elongation zone in which auxin co-occurs with CK signaling. Thereby the necessary additional information is generated that enables auxin to control division rates in one domain and elongation, and differentiation rates in the other. Still, this leaves open the question of how auxin can control both the rates of cellular division, expansion, and differentiation processes, as well as—through its interaction with PLTs and CK—the location and size of the domains in which these processes occur. Rapid, transient changes in auxin patterns necessary to, for example, adjust elongation rates during gravitropism, may conflict with the need for long-term, robust control of developmental zonation processes that are also dependent on auxin. A combination of experiments and modeling (Mähönen et al. 2014) revealed critical PLT properties essential to resolve this conflict. The authors used their model to demonstrate that, because of the high auxin levels required for PLETHORA induction, combined with the long timescale for induction, PLETHORA transcription only occurs near the meristem. Additionally, they showed that the stable nature of the PLETHORA protein allows for its maintenance in cells that, as a result of division, are

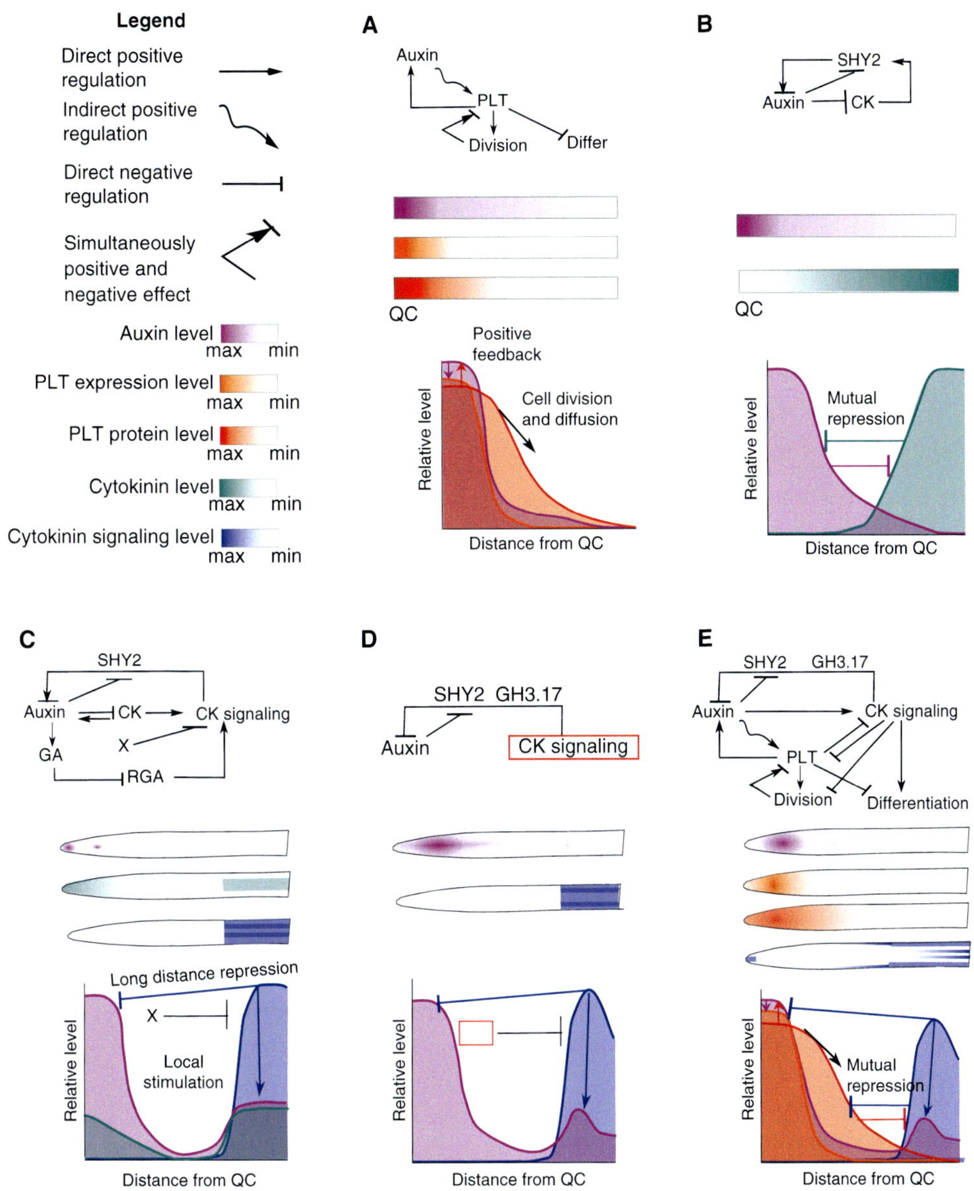

Figure 2. Models for auxin, PLT, and cytokinin (CK)-mediated root meristem patterning. (*Top* to *bottom*) Simplified cell-level regulatory network, resulting root level patterning of key factors, schematic depiction of tissue-level interactions underlying the tissue patterning. (*A*) Model for auxin-PLT-based meristem size patterning. (Based on data in Mähonen et al. 2014.) (*B*) Model for auxin-CK antagonism (additional ethylene interactions not shown). (Figure created from data in Moore et al. 2015.) (*C*) Model for auxin, gibberellic acid (GA), CK signaling-mediated meristem patterning. (Based on data in Muraro et al. 2016.) (*D*) Model for auxin and CK signaling–mediated meristem patterning. Red box indicates the superimposed patterning of CK signaling. Auxin-CK: auxin patterning in absence of the superimposed CK signaling. (Figure based on data in Di Mambro et al. 2017.) (*E*) Model for auxin-PLT-CK signaling–mediated meristem patterning. (Based on data in Salvi et al. 2020.) (QC) Quiescent center.

beyond this transcriptional domain. Combined with its cell-to-cell movement through plasmodesmata, this allows for the production of a PLETHORA protein gradient. The result is a slow and stable PLETHORA gradient immune to rapid, transient changes in auxin patterning and capable of maintaining stable root domains. Indeed, using a what-if simulation in which auxin directly controls both elongation rates and root zonation, the authors demonstrated that gravitropism-induced auxin asymmetries would result in a loss of radially coordinated development.

Computing Where to Grow

While it has long been known that many directional root growth responses arise from an asymmetry in auxin patterning, the complexities underlying the generation, maintenance, and modulation of these auxin asymmetries are still being unraveled (Fig. 3). A first study combining experiments with modeling to study root tropisms was conducted by Band et al. (2012). Band et al. used a modeling approach to derive auxin levels from DIIVenus measurements during gravitropism, enabling them to quantify gravitropic auxin asymmetry, demonstrating a twofold increase in auxin levels at the lower side.

In recent years, roots were also found to be capable of detecting small 5%–10% salt gradients and respond by growing away from the direction of highest salt levels (Galvan-Ampudia et al. 2013). A major question is how roots can translate such a weak and noisy directional signal into a robust directional response. Using a longitudinal root tip model (van den Berg et al. 2016) demonstrated that the originally reported asymmetry in PIN2 can initiate the process of auxin asymmetry formation yet is insufficient to generate a significant auxin asymmetry. Instead, the auxin-dependent expression of the auxin importer AUX1 was shown to be critical in spatially propagating and locally amplifying the auxin asymmetry, while a transient increase in PIN1 was shown to enhance the speed of asymmetry generation (van den Berg et al. 2016). In a more recent study, salt-induced polarization of AUX1 was shown to further contribute to the auxin asymmetry underlying halotropism (Korver et al. 2020).

Intriguingly, while during halotropism PIN2 asymmetry promotes auxin asymmetry, a recent study reported a dampening effect of PIN2 asymmetry on gravitropic auxin asymmetry (Retzer et al. 2019). Previous studies reported that during gravitropism a transient asymmetry in PIN2 levels occurs, with PIN2 levels increasing at the lower, high auxin side and decreasing at the upper, low auxin side (Baster et al. 2013). This PIN2 asymmetry was originally thought to contribute to gravitropic auxin asymmetry. Retzer et al. convincingly show that while a functional PIN2 is required to transport auxin to the elongation zone, the PIN2 asymmetry enhances auxin loss at the lower side while reducing it at the upper side, thereby damping the amplitude of the auxin asymmetry.

An at least partial explanation for the different roles of PIN2 asymmetry reported in the halotropism and gravitropism studies may lie in whether PIN2 changes occur on one or both sides of the root, and whether or not lateral PIN2 changes occur. In the gravitropism study, the combined increase of apical PIN2 on the lower side, reduces auxin accumulation, and decrease of apical PIN2 on the upper side, reduces auxin loss and dampens auxin asymmetry (Retzer et al. 2019). In contrast, in the halotropism study, PIN2 changes only occur on the salt-exposed side and involve the combined decrease of apical and increase of inward lateral PIN2 levels that result in the reshuffling of auxin to the nonsalt exposed side.

Computing Where to Branch

Lateral Root Priming

LR formation starts with the priming of subsets of pericycle cells that gain competence for the future formation of LRs. Priming involves periodic oscillations in auxin and/or auxin signaling at the end of the meristem and beginning of the elongation zone. Through growth, this temporal periodicity is transformed into a spatially repetitive pattern of prebranch sites. Several modeling studies have attempted to explain periodic auxin (signaling) oscillations (Lucas et al. 2008; Middleton et al. 2010; Mironova et al. 2010; Mellor et al. 2016b). Here we focus on the two models that take into account root tip morphol-

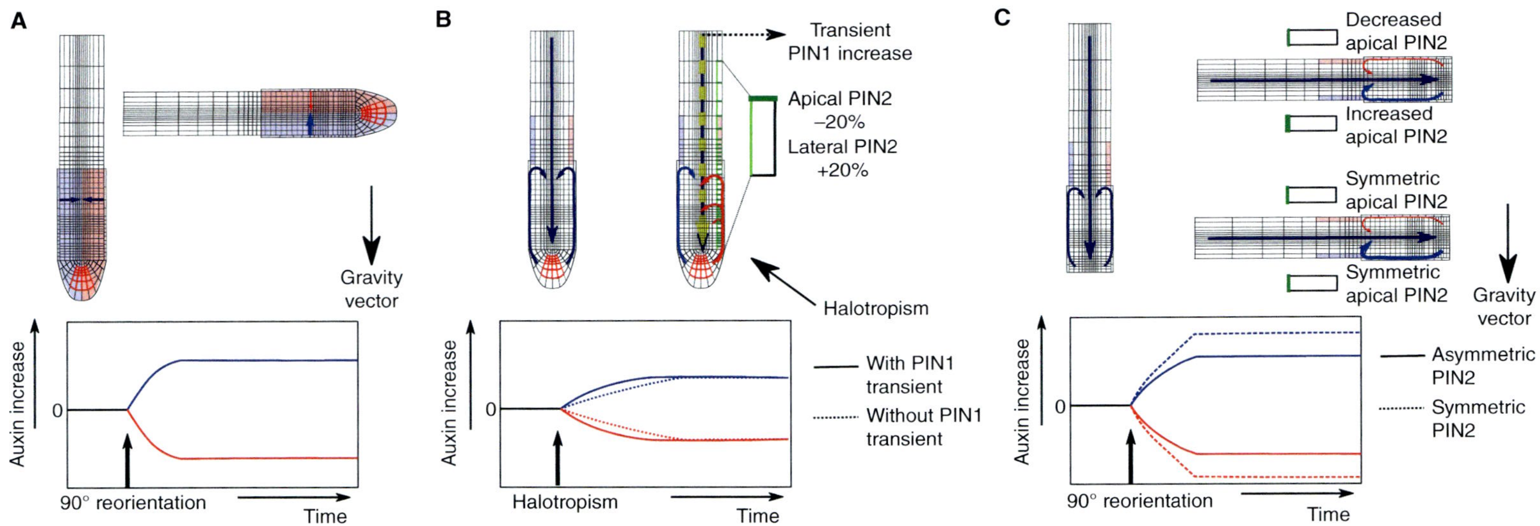

Figure 3. Models for root tropism-mediated auxin asymmetry dynamics. Blue = bottom (gravitropism) or nonsalt exposed (halotropism) root side. Red = top (gravitropism) or salt exposed (halotropism) root side. Yellow = change in PIN1 (halotropism only). Green = change in PIN2 membrane patterning. (*A*) Schematic depiction of the Band et al. (2012) model for root gravitropism. (*Top*) The two modeled root half compartments and the change in intercompartment auxin exchange upon gravitropic stimulation. (*Bottom*) Resulting buildup of auxin asymmetry. (*B*) Schematic depiction of the van den Berg et al. (2016) model for root halotropism. (*Top*) Change in PIN2 membrane patterning, PIN1-expression level, and the resulting changes in auxin reflux at the salt-exposed and non-exposed root sides. (*Bottom*) Resulting buildup of auxin asymmetry in presence and absence of transient increase in PIN1 levels. (*C*) Schematic depiction of the Retzer et al. (2019) model for root gravitropism. (*Top*) Changes in auxin reflux at the top and bottom root sides in the presence and absence of asymmetric changes in PIN2 membrane levels. (*Bottom*) Resulting buildup of auxin asymmetry in presence and absence of asymmetric changes in PIN2 membrane levels.

ogy and cellular resolution (Fig. 4; Xuan et al. 2016; van den Berg and ten Tusscher 2018). In the paper by Xuan et al. (2016), the authors investigated to what extent periodic apoptosis of the topmost LR cap cells, found to strongly correlate in space and time with pericycle priming, could drive auxin oscillations in the root vasculature. The authors modeled a nongrowing root tip with realistic anatomy and cellular layout, and assumed that, upon apoptosis, all auxin present in the root cap cells was passed on to the apoplast of directly neighboring epidermal cells. Additionally, the authors assumed apoptosis to occur synchronously across the root cap and performed three-dimensional axisymmetric simulations. The authors reported a 2- to 5-minute increase of 6%–12% in vascular auxin levels upon root cap apoptosis (Xuan et al. 2016). Although these auxin increases are lower and shorter than experimentally observed, the model highlights that significant amounts of auxin released from the root cap can be transported to the inner vascular tissues. Additionally, the model demonstrated that, as a result of PIN patterns, the highest auxin accumulation naturally occurs in the outermost vasculature cell file, the protoxylem, in which priming is known to start (De Smet et al. 2007).

A strong correlation of priming with root cap apoptosis has been observed, yet apoptosis is likely not driving priming given the persistence of priming in *smb* mutants. van den Berg and ten Tusscher (2018) investigated whether the combination of root growth with root tip auxin patterning could underlie auxin oscillations. They developed a two-dimensional root tip model in which cellular growth, division, expansion, and differentiation dynamics are incorporated. This dynamic model generated oscillations in auxin levels as an emergent outcome of root tip auxin reflux and root growth dynamics. van den Berg and ten Tusscher subsequently took apart their model, removing or modifying different aspects of auxin transport and growth dynamics in an effort to pin down the oscillatory mechanism. First, they demonstrated that the reflux of auxin in the root tip results in an auxin-loading zone at the transition and early elongation zone. Next, they showed that the potential for cells to load auxin increases with cell size, and largely arises from increased passive uptake of auxin through the membrane. Additionally, cellular auxin loading was shown to increase if cells are followed by a smaller cell below, resulting in reduced growth-dependent cumulative displacement and hence more time to grow and load auxin in the relevant domain. Furthermore, since the surface-to-volume ratio increase is largest for vascular cells, these cells load mostly auxin, explaining the preferential loading of auxin in the vasculature. Consistent with the results of Xuan et al. (2015), maximum auxin increase occurred in the vasculature cell file neighboring the pericycle. Finally, the authors demonstrate how the periodic formation of large- and small-cell pairs, generating peak auxin loading in the large cell of the pair, naturally arises from root growth dynamics. The model generated peak oscillation levels resulting in 50%–80% increases in vascular auxin levels that lasted 2–3 hours, which is more consistent with experimental data.

Computing When to Branch

Lateral Root Initiation

After priming, the initial steps in the formation of an actual LR involve the polarization and migration of nuclei in selected prebranch site cells, resulting in the asymmetric divisions with which LR primordium (LRP) development starts (De Smet 2012). Essential for these initial events to occur is the swelling of the pericycle prebranch site cells, for which the shrinkage of overlaying endodermal cells is required (Vermeer et al. 2014). Both this spatial accommodation process and the ensuing asymmetric divisions are auxin dependent (Vermeer et al. 2014; Moret et al. 2020), requiring a highly local elevation of auxin signaling in prebranch site cells (Dubrovsky et al. 2008; De Smet 2010). To investigate potential mechanisms for localized pericycle auxin accumulation, the Grieneisen group first developed a longitudinal root model (Laskowski et al. 2008) and, more recently, a transverse root model (el-Showk et al. 2015). The initial longitudinal model was used to investigate the mechanisms through which mechanical bending may induce LR initi-

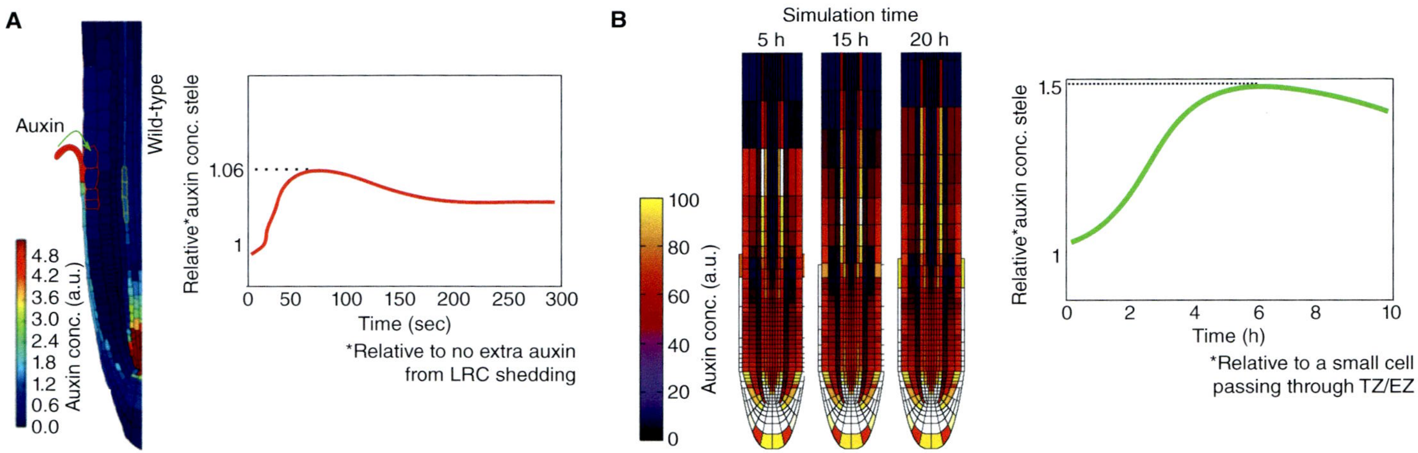

Figure 4. Models for the periodic auxin elevations underlying lateral root priming. (*A*) Panel created from data in the Xuan et al. (2016) model. (*Top*) Apoptosis of the uppermost lateral root cap (LRC) cell results in the auxin content of that cell being exported to the cell walls of the neighboring epidermal tissue. (*Bottom*) Resulting vascular auxin dynamics. (*B*) (*Top*) Synergy between root tip reflux loop and growth dynamics lead to emergent periodic auxin elevations. (Panel created from data in the Van den Berg et al. 2018 model.) (*Bottom*) Resulting vascular auxin dynamics. (TZ) Transition zone, (EZ) elongation zone.

ation. Laskowski et al. (2008) demonstrated that root bending, through local expansion of cell lengths in the outer curve of the bend, could initiate auxin accumulation. This potential for auxin accumulation upon bending was shown to be specific for the differentiation zone, which, due to its more apolar patterning of PIN exporters, enables a lateral reflux of auxin essential for auxin accumulation. Additionally, the authors demonstrated that auxin accumulation could be enhanced when incorporating the auxin-dependent induction of AUX1, giving rise to a local positive feedback loop. Finally, Laskowski et al. demonstrated that the decrease in auxin rootward of the bend, through decreased expression and polarity of auxin-dependent PIN3/PIN7, limits rootward while promoting lateral auxin transport, thereby further enhancing the auxin maximum. In their more recent study, Grieneisen and coworkers predominantly focused on vascular patterning. With regard to LR initiation, they demonstrated in this study that the polarity patterns of PIN1 and PIN7 that generate a transverse auxin flux circuit directed toward the xylem axis are critical for auxin-dependent AUX1 expression to result in auxin accumulation specifically in the pericycle (el-Showk et al. 2015).

Nodule development in legumes shares many parallels with LR development, both in terms of involved genes, as well as in the requirement for localized auxin accumulation to start the nodulation process (Franssen et al. 2015; Schiessl et al. 2019), although auxin accumulation occurs in the cortex rather than pericycle. Models of nodule initiation demonstrated that relative to an increase in AUX/LAX-mediated auxin uptake, reduction of PIN-mediated efflux may be even more effective in generating a robust auxin maximum (Deinum et al. 2012, 2016).

Lateral Root Emergence

A next critical step in the development of a new LR is its emergence from within the main root, requiring the buildup of turgor pressure of the LRP relative to its overlaying tissue as well as the separation of cells directly overlaying the LRP (Stoeckle et al. 2018). For this, a tightly localized elevation of auxin signaling resulting in the auxin-dependent modification of water transport and cell wall properties is essential. To investigate the mechanism underlying this localized auxin patterning (Péret et al. 2013), a model of a cross-section of the *Arabidopsis* root containing a stage 0/I primordium was used, and their modeling was combined with dedicated experiments. The authors first demonstrated that cortical, auxin-dependent expression of the auxin importer *LAX3*-enabled LRP auxin efflux results in cortical auxin accumulation. However, this auxin accumulation was not tightly constrained to cells overlaying the LRP. If also incorporating that auxin more rapidly induces the *PIN3* exporter as compared to *LAX3*, the more focused PIN3-mediated LRP auxin efflux produced a much more localized auxin and LAX3 activity pattern, consistent with experimental results. Thus, this study again underlines the importance of auxin dependence of auxin importers in generating and amplifying auxin patterns. Additionally, it shows the importance of timing. In an earlier study, using a highly simplified model of water fluxes between an LRP and its surroundings, the same authors had already demonstrated that through an auxin-dependent patterning of aquaporins, the necessary downstream changes in turgor pressure are achieved (Péret et al. 2012).

CONCLUDING REMARKS

In this article, we show how computational models are an invaluable tool in deciphering the complex, multiscale processes through which plant roots use auxin signaling to compute their developmental patterning, growth rate, and direction. Iterating between models and experiments has proven to be a powerful approach to delineate the necessary and sufficient ingredients underlying a particular developmental patterning or adaptation mechanism. This, for example, taught us the importance of PIN exporter polarity patterns, AUX/LAX importer tissue-specific patterns, plasmodesmatal connectivity, and the complex auxin-CK-PLT interaction network in determining root tip auxin pattern, root meristem size, and hence root growth rate (Grieneisen et al. 2007; Band et al. 2014; Mähönen et al. 2014; Di Mambro et al. 2017; Mellor et al. 2020; Salvi et al. 2020).

Through simulating what-if scenarios that are experimentally out of reach, these models additionally provide us with plausible explanations for the selective advantage of particular mechanisms and regulatory network architectures. As an example, what-if simulations assuming a direct control of both gravitropism and developmental zonation by auxin demonstrated that this would cause decorrelated differentiation (Mähönen et al. 2014). Similarly, what-if simulations for CK signaling only antagonizing auxin or PLETHORAs instead of both shows the importance of PLETHORA antagonism for early and auxin antagonism for later stages of meristem development (Salvi et al. 2020). These computational studies also underline that not all explanations should be sought for in genes or hormones, instead highlighting the importance of cell size differences and growth- or bending-induced changes therein for bootstrapping auxin elevations (Laskowski et al. 2008; van den Berg and ten Tusscher 2018). Additionally, by explicitly incorporating cellular and tissue growth processes, models have helped reveal the importance of growth and division-induced dilution for the resetting of cellular hormone and gene-expression patterns that is necessary for cells to switch to the next developmental stage (Salvi et al. 2020).

Taking a computational perspective, modeling studies have also enabled us to decipher exactly how plants use auxin signaling to compute their decision-making processes. In the case of halotropism, it was found that the auxin dependence of the auxin importer AUX1 plays an essential role in amplifying initially small auxin asymmetries, enabling plant roots to translate weak directional signals into a clear decision on the direction of growth (van den Berg et al. 2016). On a similar note, for root vascular patterning, a series of modeling studies have elucidated how the simultaneous auxin-dependent production of CK and AHP6-mediated antagonism of CK signaling modulate PIN-mediated auxin export such that initial small auxin differences are amplified into robustly patterned xylem and phloem poles (De Rybel et al. 2014; Muraro et al. 2014; el-Showk et al. 2015; Mellor et al. 2017). For gravitropism, modeling identified the auxin exporter PIN2 and, for LRs, CK signaling, both antagonizing auxin asymmetry as tuning devices for gravitropic response strength decisions (Retzer et al. 2019; Waidmann et al. 2019). In the case of LR emergence, the sequential auxin-dependent induction of auxin exporters and importers was shown to enable the computation of exactly which overlaying cells should elevate their auxin levels and give way to the growing LRP (Péret et al. 2013).

Finally, taking an information theoretic perspective, modeling studies have helped us understand auxin signaling specificity, by disentangling how auxin may simultaneously convey multiple messages. In root meristems, auxin (signaling) is known to control both meristem size, and hence where cells divide and where they stop dividing and start elongating and differentiating, as well as at which rates cells divide, elongate, and differentiate. Logically, a single auxin signal conveys insufficient information to decide both what processes to perform and at what rate to do this. First, modeling has helped to demonstrate that through directly controlling root elongation rate and, indirectly, via the more stable PLETHORA transcription factors, meristem patterning, auxin can control both rapid root tropisms and robust developmental zonation (Mähönen et al. 2014). Second, modeling highlights how auxin signaling, through interacting with both the PLETHORAs and CK signaling, self-organizes a meristematic PLETHORA and elongation/differentiation CK signaling domain (Salvi et al. 2020). Auxin signaling thus controls what happens where (division vs. expansion/differentiation). Additionally, by teaming up with either PLETHORAs or CK signaling, auxin signaling provides itself with the additional contextual information necessary to also control at which rates these processes take place.

A thus far mostly unexplored territory in the field of plant computational developmental biology is modeling beyond the scale of individual organs. Models simulating overall root system architecture (RSA) traditionally focus on the physiological processes of water and nutrient uptake (Ho et al. 2004). In these models, typically rates of LR formation, main and LR growth, and LR angles or simply final RSA are varied to investigate the consequences for foraging efficiency under different conditions (Rangarajan et al. 2018). Expanding these models to the realm of mechanistic, developmental modeling would enable us to decipher thus far undiscovered, potential feedback between processes such as LR angle and main root growth rate. However, such an expansion not only requires incorporating hormonal–genetic networks controlling different aspects of RSA that may not yet have been fully elucidated, but also plant level fluxes of auxin and other relevant systemic and local signals. A major factor limiting the development of these models is our current lack of detailed knowledge on auxin transport outside the well-studied root tips and forming LRP. In addition to active, PIN-mediated auxin transport (Blilou et al. 2005; Grieneisen et al. 2007), transport of auxin also occurs through the differentiated phloem and is unloaded in meristems (Goldsmith et al. 1974; Ross-Elliott et al. 2017). The relative importance of these two processes, and how this may differ in different regions of the root system and for different developmental stages is currently unclear. Another major limitation lies on the modeling side. While current RSA models are spatiotemporally too coarse grained to successfully incorporate developmental patterning processes, the fine-grained, cell-based models used to simulate individual root tips and LRP are computationally prohibitively complex. Development of root system models at an intermediate spatial resolution and their integration with experiments will likely be the next frontier in deciphering the role of auxin signaling in plant root development and adaptation.

ACKNOWLEDGMENTS

J.R., T.v.d.B., and K.t.T. were supported by Grant No. 864.14.003 of the Netherlands Scientific Organization (NWO).

REFERENCES

Antoniadi I, Plačková L, Simonovik B, Doležal K, Turnbull C, Ljung K, Novák O. 2015. Cell-type-specific cytokinin distribution within the *Arabidopsis* primary root apex. *Plant Cell* **27:** 1955–1967. doi:10.1105/tpc.15.00176

Band LR, Úbeda-Tomás S, Dyson RJ, Middleton AM, Hodgman TC, Owen MR, Jensen OE, Bennett MJ, King JR. 2012. Growth-induced hormone dilution can explain the dynamics of plant root cell elongation. *Proc Natl Acad Sci* **109:** 7577–7582. doi:10.1073/pnas.1113632109

Band LR, Wells DM, Fozard JA, Ghetiu T, French AP, Pound MP, Wilson MH, Yu L, Li W, Hijazi HI, et al. 2014. Systems analysis of auxin transport in the *Arabidopsis* root apex. *Plant Cell* **26:** 862–875. doi:10.1105/tpc.113.119495

Baster P, Robert S, Kleine-Vehn J, Vanneste S, Kania U, Grunewald W, De Rybel B, Beeckman T, Friml J. 2013. SCFTIR1/AFB-auxin signalling regulates PIN vacuolar trafficking and auxin fluxes during root gravitropism. *EMBO J* **32:** 260–274. doi:10.1038/emboj.2012.310

Bishopp A, Help H, El-Showk S, Weijers D, Scheres B, Friml J, Benková E, Mähönen AP, Helariutta Y. 2011. A mutually inhibitory interaction between auxin and cytokinin specifies vascular pattern in roots. *Curr Biol* **21:** 917–926. doi:10.1016/j.cub.2011.04.017

Blilou I, Xu J, Wildwater M, Willemsen V, Paponov I, Friml J, Heidstra R, Aida M, Palme K, Scheres B. 2005. The PIN auxin efflux facilitator network controls growth and patterning in *Arabidopsis* roots. *Nature* **433:** 39–44. doi:10.1038/nature03184

Cruz-Ramírez A, Díaz-Triviño S, Blilou I, Grieneisen VA, Sozzani R, Zamioudis C, Miskolczi P, Nieuwland J, Benjamins R, Dhonukshe P, et al. 2012. A bistable circuit involving SCARECROW-RETINOBLASTOMA integrates cues to inform asymmetric stem cell division. *Cell* **150:** 1002–1015. doi:10.1016/j.cell.2012.07.017

Deinum EE, Geurts R, Bisseling T, Mulder BM. 2012. Modeling a cortical auxin maximum for nodulation: different signatures of potential strategies. *Front Plant Sci* **3:** 96. doi:10.3389/fpls.2012.00096

Deinum EE, Kohlen W, Geurts R. 2016. Quantitative modelling of legume root nodule primordium induction by a diffusive signal of epidermal origin that inhibits auxin efflux. *BMC Plant Biol* **16:** 254. doi:10.1186/s12870-016-0935-9

Dello Ioio R, Linhares FS, Scacchi E, Casamitjana-Martinez E, Heidstra R, Costantino P, Sabatini S. 2007. Cytokinins determine *Arabidopsis* root-meristem size by controlling cell differentiation. *Curr Biol* **17:** 678–682. doi:10.1016/j.cub.2007.02.047

De Rybel B, Adibi M, Breda AS, Wendrich JR, Smit ME, Novák O, Yamaguchi N, Yoshida S, Van Isterdael G, Palovaara J, et al. 2014. Plant development. Integration of growth and patterning during vascular tissue formation in *Arabidopsis*. *Science* **345:** 1255215. doi:10.1126/science.1255215

De Smet I. 2010. Multimodular auxin response controls lateral root development in *Arabidopsis*. *Plant Signal Behav* **5:** 580–582. doi:10.4161/psb.11495

De Smet I. 2012. Lateral root initiation: one step at a time. *New Phytol* **193:** 867–873. doi:10.1111/j.1469-8137.2011.03996.x

De Smet I, Tetsumura T, Rybel BD, dit Frey NF, Laplaze L, Casimiro I, Swarup R, Naudts M, Vanneste S, Audenaert D, et al. 2007. Auxin-dependent regulation of lateral root positioning in the basal meristem of *Arabidopsis*. *Development* **134:** 681–690. doi:10.1242/dev.02753

Di Mambro R, De Ruvo M, Pacifici E, Salvi E, Sozzani R, Benfey PN, Busch W, Novak O, Ljung K, Di Paola L, et al. 2017. Auxin minimum triggers the developmental switch from cell division to cell differentiation in the *Arabidopsis* root. *Proc Natl Acad Sci* **114:** E7641–E7649. doi:10.1073/pnas.1705833114

Du Y, Scheres B. 2018. Lateral root formation and the multiple roles of auxin. *J Exp Bot* **69:** 155–167. doi:10.1093/jxb/erx223

Dubrovsky JG, Sauer M, Napsucialy-Mendivil S, Ivanchenko MG, Friml J, Shishkova S, Celenza J, Benková E. 2008. Auxin acts as a local morphogenetic trigger to specify lateral root founder cells. *Proc Natl Acad Sci* **105:** 8790–8794. doi:10.1073/pnas.0712307105

el-Showk S, Help-Rinta-Rahko H, Blomster T, Siligato R, Marée AFM, Mähönen AP, Grieneisen VA. 2015. Parsimonious model of vascular patterning links transverse hormone fluxes to lateral root initiation: auxin leads the way, while cytokinin levels out. *PLoS Comput Biol* **11:** e1004450. doi:10.1371/journal.pcbi.1004450

Evans ML, Ishikawa H, Estelle MA. 1994. Responses of *Arabidopsis* roots to auxin studied with high temporal resolution: comparison of wild type and auxin-response mutants. *Planta* **194:** 215–222. doi:10.1007/BF01101680

Fàbregas N, Formosa-Jordan P, Confraria A, Siligato R, Alonso JM, Swarup R, Bennett MJ, Mähönen AP, Caño-Delgado AI, Ibañes M. 2015. Auxin influx carriers control vascular patterning and xylem differentiation in *Arabidopsis thaliana*. *PLoS Genet* **11:** e1005183. doi:10.1371/journal.pgen.1005183

Fozard JA, Lucas M, King JR, Jensen OE. 2013. Vertex-element models for anisotropic growth of elongated plant organs. *Front Plant Sci* **4:** 233. doi:10.3389/fpls.2013.00233

Fozard JA, Bennett MJ, King JR, Jensen OE. 2016. Hybrid vertex-midline modelling of elongated plant organs. *Interface Focus* **6:** 20160043. doi:10.1098/rsfs.2016.0043

Franssen HJ, Xiao TT, Kulikova O, Wan X, Bisseling T, Scheres B, Heidstra R. 2015. Root developmental programs shape the *Medicago truncatula* nodule meristem. *Development* **142:** 2941–2950. doi:10.1242/dev.120774

Fu Y, Yang Y, Chen S, Ning N, Hu H. 2019. *Arabidopsis* IAR4 modulates primary root growth under salt stress through ROS-mediated modulation of auxin distribution. *Front Plant Sci* **10:** 522. doi:10.3389/fpls.2019.00522

Galinha C, Hofhuis H, Luijten M, Willemsen V, Blilou I, Heidstra R, Scheres B. 2007. PLETHORA proteins as dose-dependent master regulators of *Arabidopsis* root development. *Nature* **449:** 1053–1057. doi:10.1038/nature06206

Galvan-Ampudia CS, Julkowska MM, Darwish E, Gandullo J, Korver RA, Brunoud G, Haring MA, Munnik T, Vernoux T, Testerink C. 2013. Halotropism is a response of plant roots to avoid a saline environment. *Curr Biol* **23:** 2044–2050. doi:10.1016/j.cub.2013.08.042

Geldner N, Friml J, Stierhof YD, Jürgens G, Palme K. 2001. Auxin transport inhibitors block PIN1 cycling and vesicle trafficking. *Nature* **413:** 425–428. doi:10.1038/35096571

Goldsmith MH, Cataldo DA, Karn J, Brenneman T, Trip P. 1974. The rapid non-polar transport of auxin in the phloem of intact coleus plants. *Planta* **116:** 301–317. doi:10.1007/BF00390855

Grieneisen VA, Xu J, Marée AFM, Hogeweg P, Scheres B. 2007. Auxin transport is sufficient to generate a maximum and gradient guiding root growth. *Nature* **449:** 1008–1013. doi:10.1038/nature06215

Ho MD, McCannon BC, Lynch JP. 2004. Optimization modeling of plant root architecture for water and phosphorus acquisition. *J Theor Biol* **226:** 331–340. doi:10.1016/j.jtbi.2003.09.011

Hong JH, Savina M, Du J, Devendran A, Ramakanth KK, Tian X, Sim WS, Mironova VV, Xu J. 2017. A sacrifice-for-survival mechanism protects root stem cell niche from chilling stress. *Cell* **170:** 102–113.e14. doi:10.1016/j.cell.2017.06.002

Ishida T, Adachi S, Yoshimura M, Shimizu K, Umeda M, Sugimoto K. 2010. Auxin modulates the transition from the mitotic cycle to the endocycle in *Arabidopsis*. *Development* **137:** 63–71. doi:10.1242/dev.035840

Kimura T, Haga K, Shimizu-Mitao Y, Takebayashi Y, Kasahara H, Hayashi K, Kakimoto T, Sakai T. 2018. Asymmetric auxin distribution is not required to establish root phototropism in *Arabidopsis*. *Plant Cell Physiol* **59:** 828–840. doi:10.1093/pcp/pcy018

Kleine-Vehn J, Friml J. 2008. Polar targeting and endocytic recycling in auxin-dependent plant development. *Ann Rev Cell Dev Biol* **24:** 447–473. doi:10.1146/annurev.cellbio.24.110707.175254

Kleine-Vehn J, Huang F, Naramoto S, Zhang J, Michniewicz M, Offringa R, Friml J. 2009. PIN auxin efflux carrier polarity is regulated by PINOID kinase-mediated recruitment into GNOM-independent trafficking in *Arabidopsis*. *Plant Cell* **21:** 3839–3849. doi:10.1105/tpc.109.071639

Korver RA, van den Berg T, Meyer AJ, Galvan-Ampudia CS, ten Tusscher KHWJ, Testerink C. 2020. Halotropism requires phospholipase Dζ1-mediated modulation of cellular polarity of auxin transport carriers. *Plant Cell Environment* **43:** 143–158. doi:10.1111/pce.13646

Krouk G, Lacombe B, Bielach A, Perrine-Walker F, Malinska K, Mounier E, Hoyerova K, Tillard P, Leon S, Ljung K, et al. 2010. Nitrate-regulated auxin transport by NRT1.1 defines a mechanism for nutrient sensing in plants. *Dev Cell* **18:** 927–937. doi:10.1016/j.devcel.2010.05.008

Laskowski M, Grieneisen VA, Hofhuis H, Hove CAT, Hogeweg P, Marée AFM, Scheres B. 2008. Root system architecture from coupling cell shape to auxin transport. *PLoS Biol* **6:** e307. doi:10.1371/journal.pbio.0060307

Lavenus J, Goh T, Roberts I, Guyomarc'h S, Lucas M, De Smet I, Fukaki H, Beeckman T, Bennett M, Laplaze L. 2013. Lateral root development in *Arabidopsis*: fifty shades of auxin. *Trends Plant Sci* **18:** 450–458. doi:10.1016/j.tplants.2013.04.006

Lebovka I, Mele BH, Zakieva A, Gursanscky N, Merks R, Greb T. 2020. Computational modelling of cambium activity provides a regulatory framework for simulating radial plant growth. bioRxiv doi:10.1101/2020.01.16.908715

Liu J, Moore S, Chen C, Lindsey K. 2017. Crosstalk complexities between auxin, cytokinin, and ethylene in *Arabidopsis* root development: from experiments to systems modeling, and back again. *Mol Plant* **10:** 1480–1496. doi:10.1016/j.molp.2017.11.002

Lucas M, Guédon Y, Jay-Allemand C, Godin C, Laplaze L. 2008. An auxin transport-based model of root branching in *Arabidopsis thaliana*. *PLoS ONE* **3:** e3673. doi:10.1371/journal.pone.0003673

Lynch JP, Brown KM. 2001. Topsoil foraging—an architectural adaptation of plants to low phosphorus availability. *Plant Soil* **237:** 225–237. doi:10.1023/A:1013324727040

Mähönen AP, Bishopp A, Higuchi M, Nieminen KM, Kinoshita K, Törmäkangas K, Ikeda Y, Oka A, Kakimoto T, Helariutta Y. 2006. Cytokinin signaling and its inhibitor AHP6 regulate cell fate during vascular development. *Science* **311:** 94–98. doi:10.1126/science.1118875

Mähönen AP, ten Tusscher K, Siligato R, Smetana O, Díaz-Triviño S, Salojärvi J, Wachsman G, Prasad K, Heidstra R, Scheres B. 2014. PLETHORA gradient formation mechanism separates auxin responses. *Nature* **515:** 125–129. doi:10.1038/nature13663

Marhava P, Bassukas AEL, Zourelidou M, Kolb M, Moret B, Fastner A, Schulze WX, Cattaneo P, Hammes UZ, Schwechheimer C, et al. 2018. A molecular rheostat adjusts auxin flux to promote root protophloem differentiation. *Nature* **558:** 297–300. doi:10.1038/s41586-018-0186-z

Mason MG, Li J, Mathews DE, Kieber JJ, Schaller GE. 2004. Type-B response regulators display overlapping expression patterns in *Arabidopsis*. *Plant Physiol* **135:** 927–937. doi:10.1104/pp.103.038109

Mellor N, Band LR, Pěnčík A, Novák O, Rashed A, Holman T, Wilson MH, Voß U, Bishopp A, King JR, et al. 2016a. Dynamic regulation of auxin oxidase and conjugating enzymes AtDAO1 and GH3 modulates auxin homeostasis. *Proc Natl Acad Sci* **113:** 11022–11027. doi:10.1073/pnas.1604458113

Mellor N, Bennett MJ, King JR. 2016b. GH3-mediated auxin conjugation can result in either transient or oscillatory transcriptional auxin responses. *Bull Math Biol* **78:** 210–234. doi:10.1007/s11538-015-0137-x

Mellor N, Adibi M, El-Showk S, De Rybel B, King J, Mähönen AP, Weijers D, Bishopp A. 2017. Theoretical approaches to understanding root vascular patterning: a consensus between recent models. *J Exp Bot* **68:** 5–16. doi:10.1093/jxb/erw410

Mellor N, Vaughan-Hirsch J, Kümpers BMC, Help-Rinta-Rahko H, Miyashima S, Mähönen AP, Campilho A, King JR, Bishopp A. 2019. A core mechanism for specifying root vascular patterning can replicate the anatomical variation seen in diverse plant species. *Development* **146:** dev172411. doi:10.1242/dev.172411

Mellor NL, Voß U, Janes G, Bennett MJ, Wells DM, Band LR. 2020. Auxin fluxes through plasmodesmata modify root-tip auxin distribution. *Development* **147:** dev181669. doi:10.1242/dev.181669

Middleton AM, King JR, Bennett MJ, Owen MR. 2010. Mathematical modelling of the Aux/IAA negative feedback loop. *Bull Math Biol* **72:** 1383–1407. doi:10.1007/s11538-009-9497-4

Mironova VV, Omelyanchuk NA, Yosiphon G, Fadeev SI, Kolchanov NA, Mjolsness E, Likhoshvai VA. 2010. A plausible mechanism for auxin patterning along the developing root. *BMC Syst Biol* **4:** 98. doi:10.1186/1752-0509-4-98

Mironova VV, Omelyanchuk NA, Novoselova ES, Doroshkov AV, Kazantsev FV, Kochetov AV, Kolchanov NA, Mjolsness E, Likhoshvai VA. 2012. Combined in silico/in vivo analysis of mechanisms providing for root apical meristem self-organization and maintenance. *Ann Bot* **110:** 349–360. doi:10.1093/aob/mcs069

Miyashima S, Roszak P, Sevilem I, Toyokura K, Blob B, Heo J, Mellor N, Help-Rinta-Rahko H, Otero S, Smet W, et al. 2019. Mobile PEAR transcription factors integrate positional cues to prime cambial growth. *Nature* **565:** 490–494. doi:10.1038/s41586-018-0839-y

Moore S, Zhang X, Mudge A, Rowe JH, Topping JF, Liu J, Lindsey K. 2015. Spatiotemporal modelling of hormonal crosstalk explains the level and patterning of hormones and gene expression in *Arabidopsis thaliana* wild-type and mutant roots. *New Phytol* **207:** 1110–1122. doi:10.1111/nph.13421

Moore S, Liu J, Zhang X, Lindsey K. 2017. A recovery principle provides insight into auxin pattern control in the *Arabidopsis* root. *Sci Rep* **7:** 43004. doi:10.1038/srep43004

Moreno-Risueno MA, Norman JMV, Moreno A, Zhang J, Ahnert SE, Benfey PN. 2010. Oscillating gene expression determines competence for periodic *Arabidopsis* root branching. *Science* **329:** 1306–1311. doi:10.1126/science.1191937

Moret B, Marhava P, Aliaga Fandino AC, Hardtke CS, ten Tusscher KHW. 2020. Local auxin competition explains fragmented differentiation patterns. *Nat Commun* **11:** 2965. doi:10.1038/s41467-020-16803-7

Muraro D, Byrne H, King J, Bennett M. 2013. The role of auxin and cytokinin signalling in specifying the root architecture of *Arabidopsis thaliana*. *J Theor Biol* **317:** 71–86. doi:10.1016/j.jtbi.2012.08.032

Muraro D, Mellor N, Pound MP, Help H, Lucas M, Chopard J, Byrne HM, Godin C, Hodgman TC, King JR, et al. 2014. Integration of hormonal signaling networks and mobile microRNAs is required for vascular patterning in *Arabidopsis* roots. *Proc Natl Acad Sci* **111:** 857–862. doi:10.1073/pnas.1221766111

Muraro D, Larrieu A, Lucas M, Chopard J, Byrne H, Godin C, King J. 2016. A multi-scale model of the interplay between cell signalling and hormone transport in specifying the root meristem of *Arabidopsis thaliana*. *J Theor Biol* **404:** 182–205. doi:10.1016/j.jtbi.2016.04.036

Nacry P, Canivenc G, Muller B, Azmi A, Onckelen HV, Rossignol M, Doumas P. 2005. A role for auxin redistribution in the responses of the root system architecture to phosphate starvation in *Arabidopsis*. *Plant Physiol* **138:** 2061–2074. doi:10.1104/pp.105.060061

Pacifici E, Di Mambro R, Dello Ioio R, Costantino P, Sabatini S. 2018. Acidic cell elongation drives cell differentiation in the *Arabidopsis* root. *EMBO J* **37:** e99134. doi:10.15252/embj.201899134

Péret B, Li G, Zhao J, Band LR, Voß U, Postaire O, Luu DT, Da Ines O, Casimiro I, Lucas M, et al. 2012. Auxin regulates aquaporin function to facilitate lateral root emergence. *Nat Cell Biol* **14:** 991–998. doi:10.1038/ncb2573

Péret B, Middleton AM, French AP, Larrieu A, Bishopp A, Njo M, Wells DM, Porco S, Mellor N, Band LR, et al. 2013. Sequential induction of auxin efflux and influx carriers regulates lateral root emergence. *Mol Syst Biol* **9:** 699. doi:10.1038/msb.2013.43

Perrot-Rechenmann C. 2010. Cellular responses to auxin: division versus expansion. *Cold Spring Harb Perspect Biol* **2:** a001446. doi:10.1101/cshperspect.a001446

Petersson SV, Johansson AI, Kowalczyk M, Makoveychuk A, Wang JY, Moritz T, Grebe M, Benfey PN, Sandberg G, Ljung K. 2009. An auxin gradient and maximum in the

Arabidopsis root apex shown by high-resolution cell-specific analysis of IAA distribution and synthesis. *Plant Cell* **21:** 1659–1668. doi:10.1105/tpc.109.066480

Rademacher EH, Möller B, Lokerse AS, Llavata-Peris CI, van den Berg W, Weijers D. 2011. A cellular expression map of the *Arabidopsis* AUXIN RESPONSE FACTOR gene family. *Plant J* **68:** 597–606. doi:10.1111/j.1365-313X.2011.04710.x

Rangarajan H, Postma JA, Lynch JP. 2018. Co-optimization of axial root phenotypes for nitrogen and phosphorus acquisition in common bean. *Ann Bot* **122:** 485–499. doi:10.1093/aob/mcy092

Retzer K, Akhmanova M, Konstantinova N, Malínská K, Leitner J, Petrášek J, Luschnig C. 2019. Brassinosteroid signaling delimits root gravitropism via sorting of the *Arabidopsis* PIN2 auxin transporter. *Nat Commun* **10:** 5516. doi:10.1038/s41467-019-13543-1

Ross-Elliott TJ, Jensen KH, Haaning KS, Wager BM, Knoblauch J, Howell AH, Mullendore DL, Monteith AG, Paultre D, Yan D, et al. 2017. Phloem unloading in *Arabidopsis* roots is convective and regulated by the phloem-pole pericycle. *eLife* **6:** e24125. doi:10.7554/eLife.24125

Roychoudhry S, Del Bianco M, Kieffer M, Kepinski S. 2013. Auxin controls gravitropic setpoint angle in higher plant lateral branches. *Curr Biol* **23:** 1497–1504. doi:10.1016/j.cub.2013.06.034

Roychoudhry S, Kieffer M, Del Bianco M, Liao CY, Weijers D, Kepinski S. 2017. The developmental and environmental regulation of gravitropic setpoint angle in *Arabidopsis* and bean. *Sci Rep* **7:** 42664. doi:10.1038/srep42664

Rutten JP, ten Tusscher K. 2019. In silico roots: room for growth. *Trends Plant Sci* **24:** 250–262. doi:10.1016/j.tplants.2018.11.005

Sabatini S, Beis D, Wolkenfelt H, Murfett J, Guilfoyle T, Malamy J, Benfey P, Leyser O, Bechtold N, Weisbeek P, et al. 1999. An auxin-dependent distal organizer of pattern and polarity in the *Arabidopsis* root. *Cell* **99:** 463–472. doi:10.1016/S0092-8674(00)81535-4

Salvi E, Rutten JP, Di Mambro R, Polverari L, Licursi V, Negri R, Dello Ioio R, Sabatini S, ten Tusscher K. 2020. A self-organized PLT/auxin/ARR-B network controls the dynamics of root zonation development in *Arabidopsis thaliana*. *Dev Cell* **53:** 431–443.e23. doi:10.1016/j.devcel.2020.04.004

Santos Teixeira JA, ten Tusscher KH. 2019. The systems biology of lateral root formation: connecting the dots. *Mol Plant* **12:** 784–803. doi:10.1016/j.molp.2019.03.015

Santuari L, Sanchez-Perez GF, Luijten M, Rutjens B, Terpstra I, Berke L, Gorte M, Prasad K, Bao D, Timmermans-Hereijgers JLPM, et al. 2016. The PLETHORA gene regulatory network guides growth and cell differentiation in *Arabidopsis* roots. *Plant Cell Online* **28:** 2937–2951. doi:10.1105/tpc.16.00656

Schiessl K, Lilley JLS, Lee T, Tamvakis I, Kohlen W, Bailey PC, Thomas A, Luptak J, Ramakrishnan K, Carpenter MD, et al. 2019. Nodule inception recruits the lateral root developmental program for symbiotic nodule organogenesis in *Medicago truncatula*. *Curr Biol* **29:** 3657–3668.e5. doi:10.1016/j.cub.2019.09.005

Shimizu-Mitao Y, Kakimoto T. 2014. Auxin sensitivities of all *Arabidopsis* Aux/IAAs for degradation in the presence of every TIR1/AFB. *Plant Cell Physiol* **55:** 1450–1459. doi:10.1093/pcp/pcu077

Shkolnik D, Krieger G, Nuriel R, Fromm H. 2016. Hydrotropism: root bending does not require auxin redistribution. *Mol Plant* **9:** 757–759. doi:10.1016/j.molp.2016.02.001

Stoeckle D, Thellmann M, Vermeer JE. 2018. Breakout-lateral root emergence in *Arabidopsis thaliana*. *Curr Opin Plant Biol* **41:** 67–72. doi:10.1016/j.pbi.2017.09.005

Swarup R, Kramer EM, Perry P, Knox K, Leyser HMO, Haseloff J, Beemster GTS, Bhalerao R, Bennett MJ. 2005. Root gravitropism requires lateral root cap and epidermal cells for transport and response to a mobile auxin signal. *Nat Cell Biol* **7:** 1057–1065. doi:10.1038/ncb1316

Tian H, Wabnik K, Niu T, Li H, Yu Q, Pollmann S, Vanneste S, Govaerts W, Rolcík J, Geisler M, et al. 2014. WOX5-IAA17 feedback circuit-mediated cellular auxin response is crucial for the patterning of root stem cell niches in *Arabidopsis*. *Mol Plant* **7:** 277–289. doi:10.1093/mp/sst118

van den Berg T, Korver RA, Testerink C, ten Tusscher KHWJ. 2016. Modeling halotropism: a key role for root tip architecture and reflux loop remodeling in redistributing auxin. *Development* **143:** 3350–3362. doi:10.1242/dev.135111

van den Berg T, ten Tusscher KH. 2018. Lateral root priming synergystically arises from root growth and auxin transport dynamics. bioRxiv doi:10.1101/361709

Vermeer JEM, von Wangenheim D, Barberon M, Lee Y, Stelzer EHK, Maizel A, Geldner N. 2014. A spatial accommodation by neighboring cells is required for organ initiation in *Arabidopsis*. *Science* **343:** 178–183. doi:10.1126/science.1245871

Villalobos LIAC, Lee S, De Oliveira C, Ivetac A, Brandt W, Armitage L, Sheard LB, Tan X, Parry G, Mao H, et al. 2012. A combinatorial TIR1/AFB-Aux/IAA co-receptor system for differential sensing of auxin. *Nat Chem Biol* **8:** 477–485. doi:10.1038/nchembio.926

Vos DD, Vissenberg K, Broeckhove J, Beemster GTS. 2014. Putting theory to the test: which regulatory mechanisms can drive realistic growth of a root? *PLoS Comput Biol* **10:** e1003910. doi:10.1371/journal.pcbi.1003910

Waidmann S, Ruiz Rosquete M, Schöller M, Sarkel E, Lindner H, LaRue T, Petřík I, Dünser K, Martopawiro S, Sasidharan R, et al. 2019. Cytokinin functions as an asymmetric and anti-gravitropic signal in lateral roots. *Nat Commun* **10:** 3540. doi:10.1038/s41467-019-11483-4

Weijers D, Benkova E, Jäger KE, Schlereth A, Hamann T, Kientz M, Wilmoth JC, Reed JW, Jürgens G. 2005. Developmental specificity of auxin response by pairs of ARF and Aux/IAA transcriptional regulators. *EMBO J* **24:** 1874–1885. doi:10.1038/sj.emboj.7600659

Weise LD, ten Tusscher KHWJ. 2019. Discrete mechanical growth model for plant tissue. *PLoS ONE* **14:** e0221059. doi:10.1371/journal.pone.0221059

Xuan W, Audenaert D, Parizot B, Möller BK, Njo MF, De Rybel B, De Rop G, Van Isterdael G, Mähönen AP, Vanneste S, et al. 2015. Root cap-derived auxin pre-patterns the longitudinal axis of the *Arabidopsis* root. *Curr Biol* **25:** 1381–1388. doi:10.1016/j.cub.2015.03.046

Xuan W, Band LR, Kumpf RP, Damme DV, Parizot B, Rop GD, Opdenacker D, Möller BK, Skorzinski N, Njo MF, et al. 2016. Cyclic programmed cell death stimulates hormone signaling and root development in *Arabidopsis*. *Science* **351:** 384–387. doi:10.1126/science.aad2776

Computational Models of Auxin-Driven Patterning in Shoots

Mikolaj Cieslak, Andrew Owens, and Przemyslaw Prusinkiewicz

Department of Computer Science, University of Calgary, Calgary, Alberta T2N 1N4, Canada

Correspondence: pwp@ucalgary.ca

Auxin regulates many aspects of plant development and behavior, including the initiation of new outgrowth, patterning of vascular systems, control of branching, and responses to the environment. Computational models have complemented experimental studies of these processes. We review these models from two perspectives. First, we consider cellular and tissue-level models of interaction between auxin and its transporters in shoots. These models form a coherent body of results exploring different hypotheses pertinent to the patterning of new outgrowth and vascular strands. Second, we consider models operating at the level of plant organs and entire plants. We highlight techniques used to reduce the complexity of these models, which provide a path to capturing the essence of studied phenomena while running simulations efficiently.

Plant development is a multiscale self-organizing process. Self-organization—the emergence of higher-level processes and structures through interaction between lower-level components—occurs across all levels of plant organization, from molecular processes within cells to the patterning of tissues, plant organs, entire plants, and plant communities. An important common factor linking these phenomena is the plant hormone auxin (Sachs 2004; Vieten et al. 2007; Leyser 2011). An understanding of plant development, form, and function thus relies, in a fundamental way, on an understanding of the self-organizing processes involving auxin. Computational models and simulations provide useful insights because self-organizing processes are often difficult to grasp intuitively. Consequently, over the past 15 years they have become tightly integrated into experimental research on auxin-related patterning, as reflected in general reviews of the topic (e.g., Shi and Vernoux 2019; Ravichandran et al. 2020), and multiple reviews focused specifically on modeling (Rolland-Lagan and Prusinkiewicz 2005; Heisler and Jönsson 2006; Berleth et al. 2007; Kramer 2008; Smith and Bayer 2009; Krupinski and Jönsson 2010; Smith 2011; Prusinkiewicz and Runions 2012; van Berkel et al. 2013; Runions et al. 2014; Morales-Tapia and Cruz-Ramírez 2016).

Here we review models of auxin-driven self-organization in plant shoots from two perspectives. First, we consider models of mechanisms that control the distribution and flow of auxin between individual cells. These mechanisms form the foundation of auxin-driven morphogenesis in shoots and are directly involved in two key elements of shoot morphogenesis: the pat-

Cite this article as *Cold Spring Harb Perspect Biol* doi: 10.1101/cshperspect.a040097

terning of vascular strands and the initiation of new outgrowths. We then look at models formulated at the level of plant organs and entire plants. The size of these structures prevents effective modeling at the level of individual cells, which has led to the development of diverse techniques, abstractions, and approximations aimed at reducing the computational complexity of the models. We review and compare some of the proposed techniques using models of branching architecture and leaves as examples.

CELL AND TISSUE-LEVEL MODELS

Patterning Vascular Strands

The self-organization of auxin flow is the cornerstone of the canalization hypothesis put forward by Sachs (1969, 1981, 1989, 1991) to explain the patterning of vascular systems in plants. This hypothesis postulates a positive feedback loop between auxin transport and factors facilitating this transport, such that existing auxin flow increases the capacity for further flow in the same direction. The canalization hypothesis was the subject of the first computational models of auxin-driven patterning, devised by Mitchison (1980, 1981). Mitchison's models were ahead of their time (Jönsson 2019): their importance was only recognized two decades later (Rolland-Lagan and Prusinkiewicz 2005), when it became possible to compare the predictions of his models with experimental results enabled by new techniques of molecular genetics, including the molecular reporters of auxin (Ulmasov et al. 1997; Brunoud et al. 2012) and its efflux carrier, PIN-FORMED1 (PIN1) (Heisler et al. 2005). Mitchison considered two models of vascular patterning, differing in both the mode and regulation of auxin transport. According to the facilitated diffusion model, auxin transport is driven by the gradient of auxin concentration, with diffusivity depending on flux. In contrast, in the polar transport model there is a preferential direction to auxin transport, with transport coefficients depending on flux (Fig. 1A,B). Both models were shown to produce canals of high auxin flow connecting a source with high auxin concentration to a sink with low concentration, but closed paths with circular auxin flow, which also occur, could only be explained assuming polar transport (Mitchison 1980). The subsequent discovery of PIN1 proteins, facilitating the directional transport of auxin (Okada et al. 1991; Gälweiler et al. 1998), has lent further experimental support to the polar transport model. Nevertheless, several discrepancies between the models and biological reality emerged.

1. The original version of the model (Mitchison 1980, 1981) predicted a lower concentration of auxin in the high-flux streams compared to the surrounding tissues, yet experimental data indicated the opposite. For example, the expression of the *DR5::GUS* and *DR5::GFP* auxin response reporters is higher in provascular strands than in the surrounding tissue (Mattsson et al. 2003; Scarpella et al. 2003, 2006). This discrepancy was resolved by Feugier et al. (2005), who considered two different modes of allocating auxin transporters to the cell membrane. In the first mode, consistent with the polar transport model originally proposed by Mitchison, auxin transporters were generated locally in response to high auxin flux through different segments of the membrane. In the second mode, the transporters were allocated to cell membranes from a common pool within each cell, which introduced a competition for transporters between different membrane segments. Simulations showed that, in the second mode, auxin concentration in the stream was higher than in the surrounding tissue, because the competition for transporters restricted auxin export from the vein to these tissues. This result was consistent with studies showing that PIN1 proteins are indeed allocated to segments of the membrane from a common pool in the cell interior, through a dynamic equilibrium between competing processes of exo- and endocytosis (Steinmann et al. 1999; Geldner et al. 2001; Paciorek et al. 2005).

2. Biochemical analysis indicates that, because of a difference in pH inside and outside of the cells, the diffusion of auxin from the cells to the extracellular space is minimal (Rubery and Sheldrake 1974; Raven 1975; Goldsmith

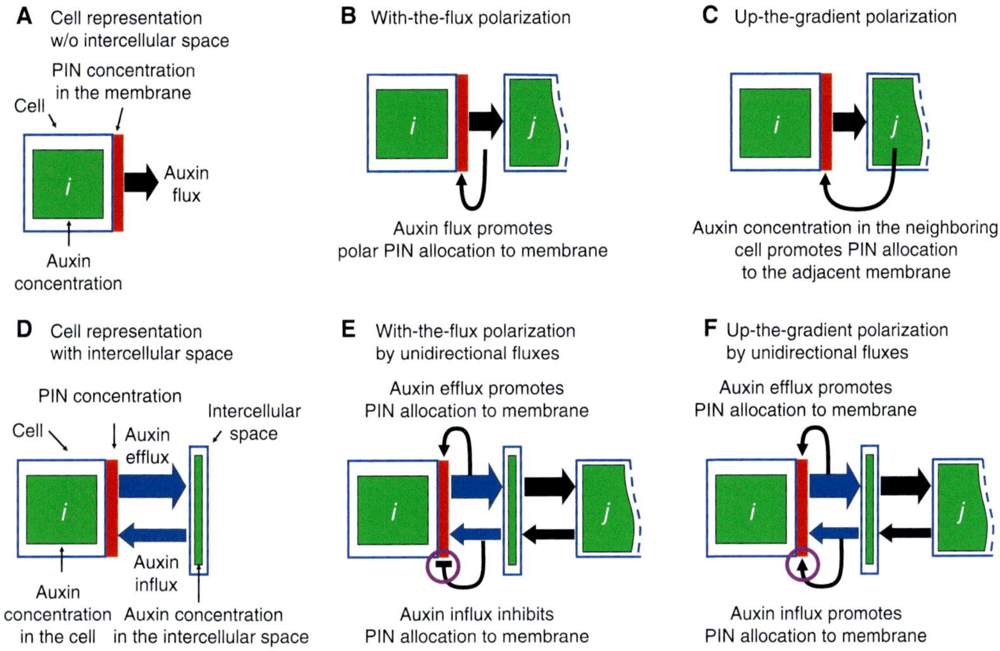

Figure 1. Schematic cell representations (*A*,*D*) and hypothetical types of feedback between auxin and its efflux carriers (PIN1 proteins) (*B*,*C*,*E*,*F*). (*B*) With-the-flux polarization underlying vascular strand formation. (*C*) Up-the-gradient polarization underlying the patterning of new outgrowth. (*E*,*F*) A model reconciling both types of feedback, based on separate measurements of auxin efflux and influx by the cell. Efflux always promotes allocation of PIN to the membrane, while influx either down-regulates (*E*) or up-regulates (*F*) PIN allocation, producing the effect of with-the-flux or up-the-gradient polarization, respectively.

1977; Mitchison 1981; see also Zažímalová et al. 2010 and Runions et al. 2014 as well as critical analysis by Mitchison 2015). This leads to the question of how the canalization process, driven by feedback between auxin transport and the location of auxin transporters, is initiated. One hypothetical possibility is a base-level presence of PIN1 or a different class of auxin exporters, such as ABCBs (Geisler et al. 2017) in all cell membranes, creating conditions mathematically equivalent to diffusion (Runions et al. 2014). Another possibility may be the transport of auxin through plasmodesmata, discussed below.

3. Simulations show that Mitchison's (1980, 1981) polar transport model, with or without the modification by Feugier et al. (2005), can stably maintain a high-flux stream connecting the source and the sink, but does not establish this connection robustly: a stream extending from the source may meander rather than form the shortest connection, or miss the sink altogether (Fig. 2A; Smith and Bayer 2009). O'Connor et al. (2014) found that in *Brachypodium*, the process of finding a sink is facilitated by the presence of two variants of the PIN1 protein, hypothesized to have different responses to auxin flux. Data and modeling support a hypothesis in which PIN1b, responding relatively weakly to flux (a linear dependence of PIN1b allocation on fluxes), is responsible for creating a broad field of auxin fluxes in the area surrounding the sink and the source. This field is then refined into a narrow stream by PIN1a, which strongly responds to auxin fluxes (a quadratic dependence). The generality of this model is unclear as many species do not have different PIN1s (although they do have multiple PIN family members). To simulate vascular patterning in the shoot api-

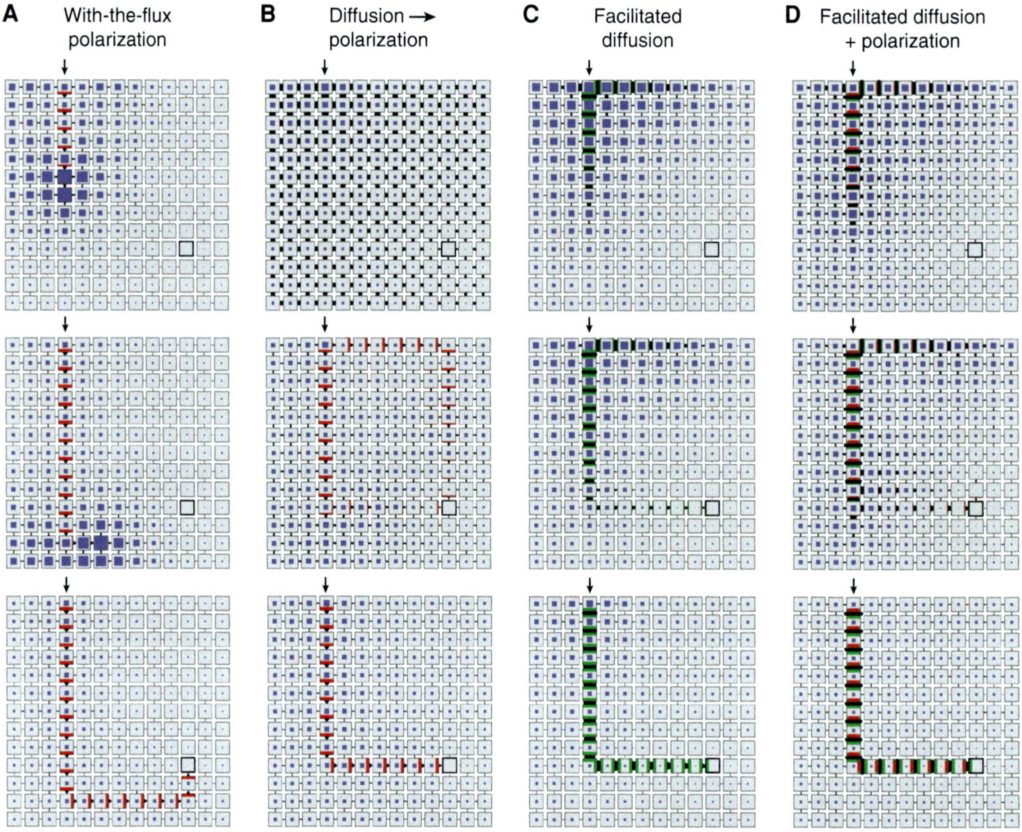

Figure 2. Development of a canal connecting auxin source (arrow) to sink (black square) according to variants of the canalization model. (*A*) With-the-flux polarization. (*B*) High nonfacilitated diffusion followed by with-the-flux polarization. (*C*) Facilitated diffusion. (*D*) Facilitated diffusion and with-the-flux polarization acting concurrently. In all simulations, each cell is initialized with a small amount of auxin. The auxin production and turnover rates are the same in all cells except for the source cell, which has a higher production rate, and the sink cell, which has a higher turnover rate. Model parameters common to two or more simulations have the same value, except for model *B*, which is initialized with a high diffusion coefficient. In the course of simulation, its value decreases while the polar transport gradually increases. The visualization of cells and fluxes is similar to Figure 1A, with auxin concentrations indicated by the size of blue squares and diffusivities in models (*C*,*D*) shown as green bars.

cal meristem (SAM) of plants with a single PIN1 variant, Bayer et al. (2009) and Hartmann et al. (2019) introduced a hypothetical "vein attraction factor": a separate substance that would create a gradient directing the emerging auxin stream toward the sink. At present, there is no experimental evidence of such a substance.

4. Mitchison and Feugier's models do not easily form loops, and thus do not readily explain reticulate venation patterns. Addressing this question, Feugier and Iwasa (2006) proposed a hypothetical "flux bifurcator" mechanism that modifies the allocation of auxin efflux carriers to the cell membrane such that a stable auxin efflux in more than one direction becomes possible. This model can create loops, but has not yet been shown to produce vein patterns with the geometry observed in nature. Moreover, the model predicts PIN1 polarity at the vein junctions opposite to the polarity observed experimentally (Scarpella et al. 2006).

5. Plant treatment (Mattsson et al. 1999) and mutant data (Verna et al. 2019; Krishna et al. 2020; Ravichandran et al. 2020) show that defects in PIN-mediated auxin transport distort vascular patterns but do not destroy them completely. A mechanism that patterns vascular strands in the absence of polar auxin transport thus appears to exist.

Mechanisms resolving some of these shortcomings may involve auxin transport through plasmodesmata (Rutschow et al. 2011; Han et al. 2014; Verna et al. 2019; Gao et al. 2020; Paterlini 2020; Ravichandran et al. 2020; Sager et al. 2020). Simulations show that efficient diffusion of auxin through (initially open) plasmodesmata could guide polar auxin transport and the emerging veins toward the auxin sinks without a separate attraction factor (Fig. 2B). Moreover, there is evidence that auxin signaling controls the size of the plasmodesmata aperture (Han et al. 2014; Sager et al. 2020), and thus auxin movement through plasmodesmata may create a feedback loop (Paterlini 2020; Ravichandran et al. 2020). The resulting facilitated diffusion through plasmodesmata—a possibility already considered by Mitchison (1980)—could explain vascular patterning in the absence of PIN proteins and polar transport (Fig. 2C). Connections between sinks and sources would also be reliably produced if facilitated diffusion and polar transport acted in concert (Fig. 2D). In addition, all of these models can produce simple patterns with loops, although they fall short of forming entire reticulate patterns. The emergence of reticulate patterns remains one of the most intriguing open questions of plant morphogenesis. Adding to the mystery is the ease with which plants transition between open and reticulate patterns as a result of mutations (Verna et al. 2019).

Patterning New Outgrowth

The patterning of vascular strands has played a prominent role in studies of auxin-driven shoot development in part for historical reasons. Sachs's canalization hypothesis was formulated over 50 years ago, and its model, devised by Mitchison, was the first computational model of auxin-driven morphogenesis in plants. Yet another self-organizing process was the subject of even earlier mathematical analysis and computational modeling: phyllotaxis, the patterning of new organs at the SAM. The cornerstone of the causal explanation of phyllotaxis is Hofmeister's (1868) hypothesis, augmented by Snow and Snow (1931, 1952) and termed by them as the "first available space" hypothesis. According to this hypothesis, new primordia are inserted where and when sufficient space becomes available within the organogenetic zone of a growing SAM. Schoute (1913) proposed that this process is controlled by a diffusing inhibitor produced by primordia, which prevents the initiation of new primordia in the proximity of existing ones. Computational models (e.g., Veen and Lindenmayer 1977) have demonstrated that simulations based on this assumption can indeed generate the observed patterns, but the existence of a diffusing inhibitor has not been confirmed (Sassi and Vernoux 2013). Following experiments showing a displacement (Snow and Snow 1937) and the initiation (Reinhardt et al. 2000) of primordia by locally applied auxin, Reinhardt et al. (2003) proposed a different conceptual model, in which low auxin concentration plays the inhibitory role. According to this model, new primordia arise at locations where the concentration of auxin is high, whereas polar auxin transport toward existing primordia depletes auxin in their proximity. Jönsson et al. (2006) and Smith et al. (2006a) provided support for this concept with computational models that assumed a preferential PIN polarization toward adjacent cells with the highest auxin concentration (Fig. 1C). The models generated a growing pattern of isolated primordia; moreover, with the additional assumptions of increased local production of auxin in the primordia and increased PIN polarization toward them, Smith et al. (2006a) simulated the spiral phyllotactic patterns observed in *Arabidopsis* (Fig. 3A). Subsequent experimental data (Cui et al. 2013; Pinon et al. 2013; reviewed by Shi and Vernoux 2019) showed that auxin biosynthesis in the SAM of *Arabidopsis* is indeed required for the correct formation of phyllotactic patterns; furthermore, auxin enhances PIN1

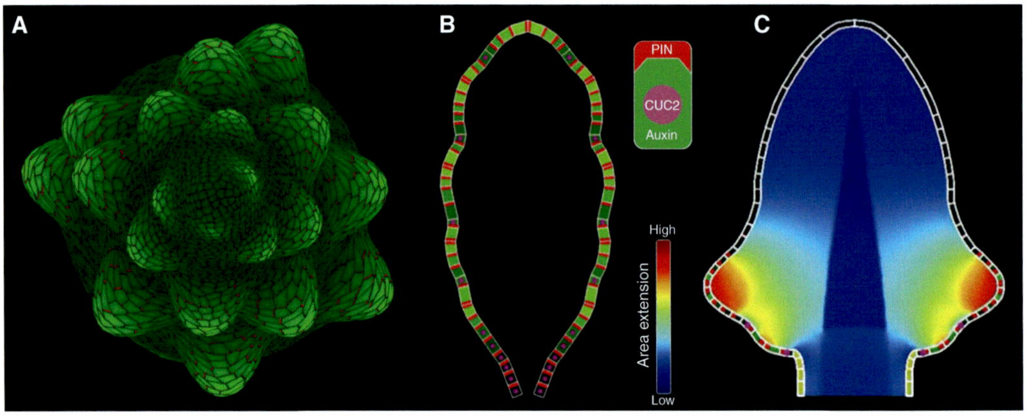

Figure 3. Sample models of outgrowth driven by auxin maxima. (*A*) A model of spiral phyllotaxis in the flowering shoot of *Arabidopsis thaliana*. Dark to bright green: low to high auxin concentration in the epidermis; red lines: location and concentration of PIN1 proteins on cell walls. The interaction between auxin and PIN in the growing shoot meristem produces new auxin maxima at locations consistent with the first available space paradigm. The maxima give rise to flower primordia. (Panel *A* is a variant of the model presented by Smith et al. 2006a.) (*B*) A model of *Arabidopsis* leaf serration. The interaction between auxin and PIN1 proteins has a similar character to that in the phyllotaxis model, but includes an additional factor, the CUC2 protein. Auxin maxima pattern new serrations, whereas CUC2 patterns the indentations. The suppression of CUC2 by auxin prevents auxin maxima from shifting. Symbolic representation of cells is shown in the *inset* (*top*). (Panel *B* is reprinted from Bilsborough et al. 2011 under the terms of the Creative Commons BY-NC license.) (*C*) Extension of the leaf model capturing the expansion rates of the whole leaf blade. Local growth rates are indicated by the heat map, as shown in the *inset* (*bottom*). (Panel *C* is reprinted from Kierzkowski et al. 2019 under the terms of the Creative Commons CC BY license.)

production (Vieten et al. 2005; Wenzel et al. 2007; Krogan et al. 2016), which may be related to the enhanced PIN1 polarization in the primordia postulated by Smith et al. (2006a). On the other hand, a modification of the auxin transport equations (Barbier de Reuille et al. 2006) makes it possible to generate spiral phyllotactic patterns without invoking additional assumptions (Smith 2011). This modification was employed in a model relating the chirality of phyllotaxis to the asymmetry of leaves (Chitwood et al. 2012).

In parallel with the discovery of the role of auxin in phyllotaxis by Reinhardt et al. (2000, 2003), Benková et al. (2003) observed that the primordia of lateral roots, cotyledons, and flower organs are patterned by auxin maxima as well, which suggested related mechanisms. Hay et al. (2006) extended this observation to leaf serrations. Adapting the phyllotaxis model of Smith et al. (2006a) to operate on the margin of a growing leaf, Bilsborough et al. (2011) proposed a plausible model of serration patterning in which new serrations are inserted "in the first available space," similarly to primordia in the phyllotaxis case (Fig. 3B). This model enhances the feedback between auxin and PIN1 with an additional factor, the CUC2 protein. CUC2 is required for the dynamic reallocation of PIN1 in the cells, leading to the formation of PIN1 convergence points and auxin maxima, but its expression is suppressed by auxin, which eliminates CUC2 from the convergence points. This elimination stabilizes the position of auxin maxima on the growing margin. In addition, CUC2 locally suppresses growth, thus contributing to the formation of the indentations between serrations. A subsequent extension (Kierzkowski et al. 2019) put this mechanism in the context of a biomechanical model capturing the expansion of the entire leaf blade, as opposed to only simulating processes taking place on the margin (Fig. 3C). Interestingly, Govindaraju et al. (2020) have constructed transgenic *Arabidopsis* plants in which PIN1 is eliminated from the leaf epidermis, but the shape of leaves, including serrations, remains preserved. It is thus possible that the analogy between the initiation of primordia in the SAM, where an instructive role for epidermal auxin is necessary for correct phyllotaxis (Kierzkowski et al. 2013), and the sculpting of leaf shape is more limited than previously thought.

Bilsborough et al. (2011) also noted that in a variant of the serration model without CUC2, the existing maxima shift their position as the simulated leaf grows, making room for the intercalary insertion of new maxima between those initiated previously. Such traveling maxima have not been observed in real leaves—which highlighted the role of CUC2 in the patterning of leaf serrations—but were subsequently found in the SAM of transformed *Arabidopsis* plants (Bhatia et al. 2016). Extending this observation, Galvan-Ampudia et al. (2020) observed that auxin maxima initiating new flower primordia in wild-type *Arabidopsis* travel radially by a distance of several cells, whereas Zhang et al. (2021) found that auxin maxima that pattern bracts in young gerbera flower heads are displaced in both the radial and lateral direction (along the head rim). Computational models reported by Zhang et al. (2021) suggest that the lateral component of this displacement plays a critical role in the formation of phyllotactic patterns with the high numbers of spirals typical of the Aster family. Simulated interactions between auxin and its transporters can also generate other spatiotemporal patterns, such as oscillations (Farcot and Yuan 2013) and "snaking": a pattern of auxin maxima progressing from the boundary of a tissue toward the center (Draelants et al. 2015). Whether such patterns also exist in nature remains an open question.

Reconciling Models of Vasculature Development and Outgrowth Positioning

The models outlined so far are capable of generating vascular patterns and new outgrowth positions, but, considered together, raise the question of the relationship between the "with-the-flux" polarization postulated by the canalization hypothesis and the "up-the-gradient" polarization model used to explain the emergence of maxima of auxin concentration. This link is particularly important because new outgrowth is intimately associated with the development of the vascular system (Reinhardt et al. 2003). Three scenarios have been proposed.

1. Stoma et al. (2008) attributed the initiation of both new primordia and vascular strands to with-the-flux polarization. The resulting model was capable of producing the auxin maxima and canals of auxin propagation, but predicted a transient drop in epidermal auxin concentration during the formation of PIN1 convergence sites, which was not confirmed experimentally (Heisler et al. 2005; Brunoud et al. 2012; see also Abley et al. 2016; Shi and Veroux 2019).
2. Merks et al. (2007) attributed the formation of both auxin maxima and veins to up-the-gradient polarization. This model was also capable of producing maxima of auxin concentration and vascular strands, but predicted transient PIN polarization toward the progressing tip of emerging veins, again contrary to experimental data.
3. Bayer et al. (2009) postulated a "dual-polarization" model, in which both polarization modes coexist, with the transition from up-the-gradient to with-the-flux mode controlled by increasing auxin concentration in individual cells.

The last model was consistent with experimental data, but highlighted further questions regarding the underlying molecular mechanisms: (1) what biochemical mechanism can respond to fluxes, as opposed to concentrations that obey the standard law of mass action; (2) how can a cell detect auxin concentration in the neighboring cell through local interactions only; and (3) what molecular implementation can explain the transition from up-the-gradient to with-the-flux polarization.

Addressing the first question, Coen et al. (2004) elaborated Mitchison's (1980) idea of a mechanism in which the passage of an auxin molecule through the cell membrane is associated with the production or turnover of a tally molecule. The concentration of tally molecules arising in this process would thus measure the flux. Extending this idea, Cieslak et al. (2015) postulated that the efflux and influx of auxin are measured separately (Fig. 1D–F), and proposed biochemically plausible reaction networks capable of generating both with-the-flux and up-the-gradient polarization on this basis. The switch between modes depended on reaction rates, which can potentially be controlled by auxin concentration within the cell. Hartmann et al. (2019) lent support to this model by employing it to show the emergence of organ primordia and vascular strands in the SAM.

Other models have also been proposed. Focusing on up-the-gradient polarization, Heisler et al. (2010) proposed that PIN1 localizes to membranes adjacent to cell walls that exhibit the highest stress. In the context of canalization, Kramer (2009) explored Mitchison's (1981) idea that auxin efflux carriers respond to the gradient of auxin concentration within cells, as opposed to directly measuring auxin fluxes. Wabnik et al. (2010) attributed a regulatory role to gradients of auxin concentration in the intercellular space. It is not clear, however, whether concentration gradients occurring at the relatively small distances in the cytoplasm or in the extracellular space suffice to play the proposed roles. Furthermore, the AUXIN-BINDING PROTEIN1 (ABP1), considered a potential molecular player in both models, was subsequently shown not to be critical, as mutant plants lacking it were not distinguishable from wild-type plants (Gao et al. 2015).

Abley et al. (2013) considered auxin-driven polarization in the broad context of polarization mechanisms in animals and plants. A distinctive feature of the proposed model was the assumption that cells are inherently polarized, in a manner similar to magnetic dipoles, and polarization at the tissue level results from the coordination of these individual polarities rather than their production de novo. The inherent polarity of PIN1 protein localization has not yet been demonstrated, although it has been shown for another polarity-inducing protein, BREAKING OF ASYMMETRY IN THE STOMATAL LINEAGE (BASL) (Chan et al. 2020).

ORGAN AND PLANT-LEVEL MODELS

Models accounting for individual cells make it possible to faithfully represent tissues consisting of tens or hundreds of cells, but become computationally impractical when modeling larger structures. Various methods have been used, separately or jointly, to mitigate computational complexity. The quest for these methods is at least as interesting as the biological insights obtained so far. Consequently, we review models of auxin-driven patterning at the organ and whole-plant levels from the methodological perspective, focusing on two aspects: the representation of space and the representation of auxin itself.

Representation of Space

1. *Reduction of dimensionality.* Model complexity is commonly reduced by considering spatial models in two or even only one dimension. Two-dimensional models often correspond to sections of a tissue or organ. For example, Bayer et al. (2009) and O'Connor et al. (2014) modeled the patterning of primordia and veins by considering longitudinal sections of the SAM only. Žádníková et al. (2016) modeled the formation of the apical hook (the bend in the hypocotyl that protects the SAM as it breaks through the soil during germination) by considering transverse and longitudinal sections of the hook separately. The transverse model captured the differential distribution of auxin in the upper and lower part of the stem. The longitudinal model employed this information to guide differential elongation and proliferation of cells on the upper and lower side of the hypocotyl, producing the curved hook. In a further reduction of dimensionality, Bilsborough et al. (2011) approximated the leaf margin as a sequence of cells: formally, a curve. Likewise, Ibañes et al. (2009) and Fàbregas et al. (2015) modeled the patterning of vascular bundles in *Arabidopsis* inflorescence stems by considering only transverse sections of the stem.

2. *Big cell models*. This class of models is based on the assumption—plausible, although formally not demonstrated—that groups of cells behave in a manner similar to individual cells. The resulting models are composed of a smaller number of larger cells, compared to those occurring in reality, which reduces their computational complexity. This simplification was employed, for instance, in the model of bud activation in *Arabidopsis* shown in Figure 4A (Prusinkiewicz et al. 2009). Another model in the same paper pushed the reduction even further: entire components of plant architecture—internodes and flowers—were viewed as large "cells," each characterized by the concentration of auxin in the interior and the allocation of PIN1 proteins to its boundary (Fig. 4B). Both models simulated apical dominance—the suppression of the outgrowth of lateral buds caused by auxin moving from the shoot apex down the stem—as the result of the interaction between auxin and its transporters inherent in the with-the-flux polarization

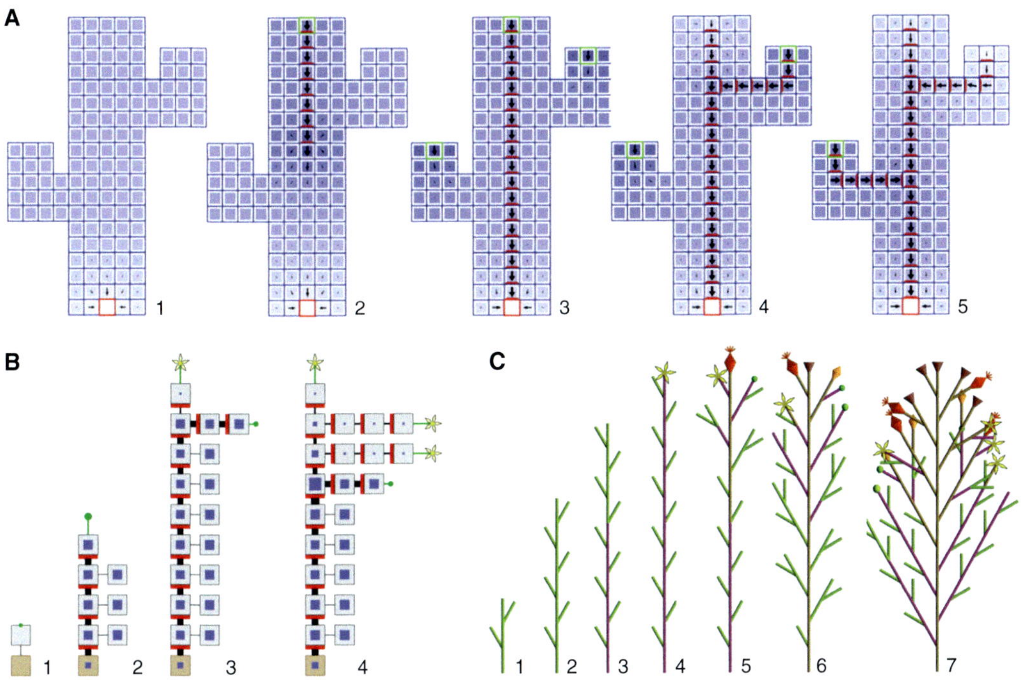

Figure 4. Sample models using approximate representations of space and auxin. (*A*) A model of bud activation and vascular strand emergence using big cell approximations of groups of cells in a longitudinal section of the stem. Auxin concentration is shown as the intensity of blue, PIN as the intensity of red, and the dominant direction and magnitude of auxin flux are indicated by black arrows. The auxin sink is the red square at the stem base. Panels 1–5 show selected stages of the simulation following the placement of auxin sources (green squares) first at the top of the stem, then in the lateral buds. Following the removal of the apical auxin source, vascular strands emerge in basipetal order due to interaction between auxin and PIN following Mitchison's polar transport canalization model. (*B*) A coarse version of the previous model, in which entire internodes are treated as big cells. The cells are visualized as in Figure 1A. (*C*) A further simplification of the shoot model, in which the flow of auxin (indicated by green internodes) and a flower-inducing signal (pink) is simulated kinematically, as signals moving through the stem and branches with given velocities. Simplification of the model components and signals facilitates the construction of increasingly complex, dynamic models of the whole plant. (Panels *A* and *B* are reprinted from Prusinkiewicz et al. 2009 under the terms of the Creative Commons CC BY license. Panel *C* was created using a variant of the model described by Janssen and Lindenmayer 1987 and Prusinkiewicz and Lindenmayer 1990. See the original publications for details of the operation of these models.)

model. They also provided a stepping stone for integrating apical dominance, carbon dynamics, and biomechanical factors within a single shoot model (Cieslak et al. 2011), explaining the interaction between auxin and another plant hormone affecting branching, strigolactone (Shinohara et al. 2013), and elucidating how organ numbers and auxin levels are maintained in the SAM (Shi et al. 2018). The interplay between hormones and sugars in the control of branching remains a topic of current interest (Barbier et al. 2019; Schneider et al. 2019; Bertheloot et al. 2020).

3. *Continuous-space models.* In this class of models, tissues and larger structures are considered as continua, thus bypassing the need to model individual cells (although the resulting partial differential equations may have to be discretized for the purpose of numerical simulations). This approach is widely used in models of phyllotaxis in which the interaction between primordia is mediated by abstract inhibitory fields (e.g., Douady and Couder 1996; Smith et al. 2006b). Newell et al. (2008) and Mirabet et al. (2012) showed that cell-level auxin-driven models of phyllotaxis could be distilled into continuous models as well. Likewise, Moulton et al. (2020) used a continuous model of polar auxin transport in a mathematical model of tropisms. These continuous models, however, have only been formulated in one dimension. An extension to two and three dimensions—a transport-driven counterpart of the reaction-diffusion equations of morphogenesis—is a topic of current studies (Haskovec et al. 2019).

Representation of Auxin

1. *Kinematic models.* At a more abstract level, auxin flow between components of a modeled structure can be simulated without accounting for the transport mechanism in detail. For example, Janssen and Lindenmayer (1987; see also Prusinkiewicz and Lindenmayer 1990) modeled the development and flowering sequence of the paniculate inflorescence of the Asteraceae plant *Mycelis muralis* in terms of a hypothetical interaction between two binary signals (Fig. 4C). The flow of both signals was defined kinematically by their propagation rates between adjacent internodes in a branching structure. A basipetal signal imparting apical dominance was interpreted as auxin. An acropetal flower-inducing signal may correspond to the subsequently characterized protein Flowering Locus T (FT) (for review, see Putterill and Varkonyi-Gasic 2016). The model provides a good example of the emergence of a complex, dynamic branching structure through the self-organizing interaction of simple components. To what extent it is consistent with the development of such structures in nature is an open question, as auxin and FT have not yet been considered jointly.

2. *Particle system models.* In the models considered so far, auxin concentration or presence has been described as a property of cells or regions of space. An alternative is to focus on the flow of auxin itself. The difference between these approaches corresponds to the distinction between Eulerian and Lagrangian viewpoints in fluid dynamics (Bridson 2015). According to the Eulerian viewpoint, one considers points in space and describes how the quantities associated with them change over time. In the context of auxin-driven patterning, these quantities typically represent auxin concentrations in cells or regions of intercellular space, and fluxes associated with cell walls. It is also possible to characterize auxin as particles representing individual molecules or molecule clusters. For example, Renton et al. (2012) considered a particle-based representation of auxin when modeling and analyzing long-distance auxin flow, and Cieslak and Prusinkiewicz (2019) considered a stochastic variation of particle numbers in a theoretical model of leaf shape diversity. This approach is consistent with the Lagrangian viewpoint, focusing on the particles themselves as they propagate through space and interact with each other. The idea of modeling auxin-driven patterning using particle systems and the Lagrangian

point of view has its origins in the work of Rodkaew et al. (2002, 2003), who represented leaf vasculature as trajectories of particles moving from the leaf margin toward the base. The connection to auxin was made explicit in a related method proposed by Runions et al. (2005). For instance, a cell-level model of *Brachypodium* shoots (O'Connor et al. 2014) is compared with its particle system counterpart in Figure 5A and B. In the particle system model, vascular bundles originate at auxin convergence points emerging periodically at the apical meristem and extend by the addition of new particles at the strand tip. The direction of these additions is a combination of three tendencies: an attraction toward previously formed veins, an attraction toward the base of the emerging stem, and a tendency to propagate the vein in the current direction. The three-dimensional extension of the same algorithm shown in Figure 5D captures the development of a vascular system in the inflorescence of *Arabidopsis*, consistent with observations by Kang et al. (2003) summarized in Figure 5C. Further extensions have led to a model of vasculature in fleshy fruit (tomato and peach) (Cieslak et al. 2016) and the compound inflorescence of yarrow (Fig. 5E–G; Owens et al. 2016). In all these examples, the particle system representation of auxin-driven patterning provides a means of controlling the complexity of the models.

3. *One-shot geometric models.* The particle system models described above simulate canalization using an iterative geometric construction that gradually extends a developing vein toward the sink. At an even higher level of abstraction and simplification, entire vein segments can be inserted into a structure in a single step, without simulating auxin propagation at all (Fig. 6). This approach has been employed in a model that aims at explaining the diversity of eudicot leaves in terms of variations in the underlying developmental process (Runions et al. 2017). According to this model, leaf development can be viewed as a feedback loop integrating three processes:

- the emergence of auxin concentration maxima that position new outgrowth (serrations, lobes, or entire leaflets) at the leaf margin;
- the patterning of veins connecting these maxima to the leaf base or to the veins formed earlier; and
- growth, locally aligned with these veins.

In an extension of the Hofmeister/Snow and Snow hypothesis to leaves, new outgrowth centers are inserted into morphogenetically competent zones of the leaf margin as soon as they are sufficiently distanced from the centers formed earlier. Each center is connected to the leaf base or a nearby vein by a new vein inserted instantaneously (the actual path is determined by minimizing an estimated resistance to the transport of water and sugars). In spite of representing the processes of convergence point formation and canalization at a high level of abstraction, this model captured the development of the shape and venation patterns of a wide range of eudicot leaves.

CONCLUSION

How organisms acquire their form is a fundamental question of developmental biology. Jacob (1998), Nobel Prize winner for the co-discovery of gene regulation, called it "the most intriguing question in the world" and qualified development as a miracle that calls for an explanation, but has been insufficiently studied because of the difficulty of the problem. Fox Keller (2002) wondered whether development could be comprehended in a manner comparable to physics at all. Refuting these concerns, dramatic progress in developmental plant biology has taken place over the last 20 years. Studies of auxin have played a prominent role in this progress, and models have contributed by showing the consequences of hypotheses based on experimental data and highlighting questions requiring further experiments. To date, they have been particularly helpful in the analysis of the different hypotheses related to the feedback between auxin and its transporters. In the years to come, we can expect methods for creating higher-level

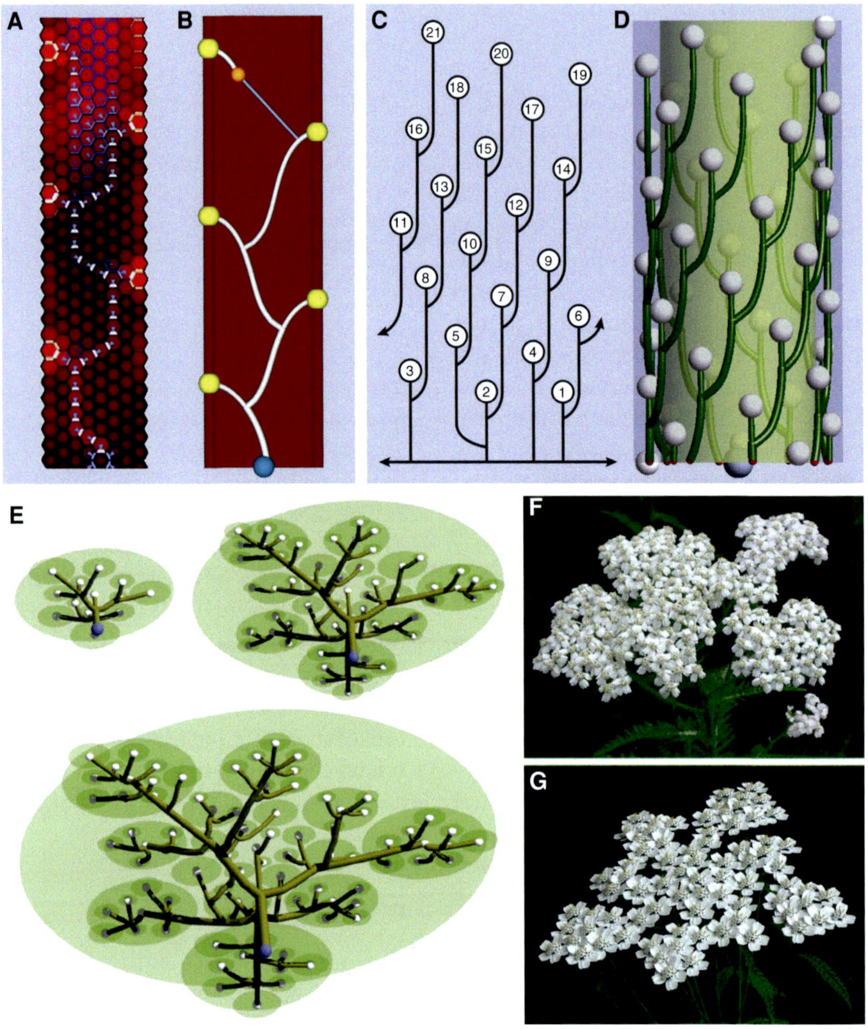

Figure 5. Particle system models of auxin-driven vascular patterning and growth. (*A*,*B*) A comparison of the cell-level and particle-system model of vascular patterning in a longitudinal section of a *Brachypodium* shoot. (*A*) In the cell-level model, SoPIN1 proteins (yellow), polarized up-the-gradient, create a pattern of auxin maxima (bright red) in the epidermis. New maxima emerge acropetally as the shoot grows. The flow of auxin is broadly oriented toward previously formed vascular strands by PIN1b proteins (blue) before becoming canalized by PIN1a (white). (Panel *A* is reprinted from O'Connor et al. 2014 under the Creative Commons CC0 public domain dedication.) (*B*) In the corresponding particle system model, a vascular strand emerges as a trace of an auxin front (orange particle), moving from an auxin source (yellow sphere) toward the sink (a previously formed vascular strand, white). The blue line indicates the direction of particle motion toward the nearest established vein. The shape of the emerging strand is also affected by the polarizing influence of the global sink (blue sphere), which biases the motion of the orange particle basipetally. (*C*,*D*) A comparison of the pattern of vascular strands connecting leaves and flowers along the $n + 5$ parastichies in a young reproductive shoot of *Arabidopsis thaliana* (based on Kang et al. 2003) with a model created as in panel *B*, operating on the cylindrical surface of the stem. (*E*) A developmental model of the branching structure of a yarrow inflorescence. Vascular strands (brown) connect auxin maxima (white spheres) emerging in a growing inflorescence meristem (green) to the structure base and previously formed strands as in *B* and *D*. The vascular strands define the branching structure of the inflorescence. The photograph (*F*) and model (*G*) of a fully developed inflorescence are similar, indicating the usefulness of the particle-system abstraction in simulating plant structures at the organ and whole plant levels. (Panels *E* and *G* are reprinted from Owens et al. 2016 with permission from the authors. Panel *F* was kindly provided by Frank L. Hoffman, www.all-creatures.org.)

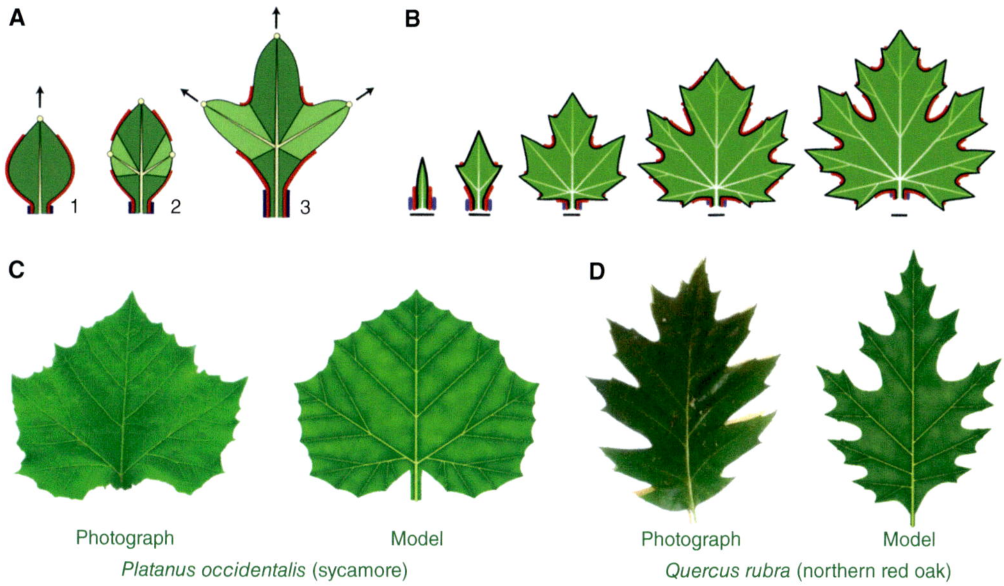

Figure 6. Leaf development and forms modeled with one-shot vein insertion. (*A*) Principle of model construction. (Stage 1) Auxin concentration maximum at the leaf margin (yellow circle) promotes leaf growth in the direction of the midvein (black arrow). (Stage 2) This growth creates space for new maxima, which are instantaneously connected to the main vein by lateral veins. (Stage 3) The new maxima promote growth in the direction of lateral veins. Additional factors include the red morphogen, which enables the formation of convergence points and defines sinuses (action similar to CUC2), and the blue morphogen defining the petiole. (*B*) Development of a maple leaf simulated by iterating the process shown in panel *A*. (*C*,*D*) A comparison of photographs and models of sample leaves. (All panels are reprinted from Runions et al. 2017 under the terms of the Creative Commons CC BY license.)

models to mature, with more clarification of their relation to models operating at the level of individual cells. The resulting methods will enhance the role of modeling in studies of the development of plant organs and entire plants.

ACKNOWLEDGMENTS

We thank Adam Runions and Richard Smith for sharing their models and insights pertinent to the sink-finding problems, Enrico Scarpella for sharing manuscripts of his papers prior to publication, and Ottoline Leyser, Adam Runions, and Enrico Scarpella for insightful comments on the draft manuscript. The support of our research by the Plant Phenotyping and Imaging Research Center/Canada First Research Excellence Fund (M.C. and P.P.) and the Natural Sciences and Engineering Research Council of Canada (Grant 2019-06279 to P.P.) is gratefully acknowledged. We apologize to authors whose work has not been included due to space limitations.

REFERENCES

Abley K, Barbier de Reuille P, Strutt D, Bangham A, Prusinkiewicz P, Marée A, Grieneisen V, Coen E. 2013. An intracellular partitioning-based framework for tissue cell polarity in plants and animals. *Development* **140:** 2061–2074. doi:10.1242/dev.062984

Abley K, Sauret-Güeto S, Marée AF, Coen E. 2016. Formation of polarity convergences underlying shoot outgrowths. *eLife* **5:** e18165. doi:10.7554/eLife.18165

Barbier FF, Dun EA, Kerr SC, Chabikwa TG, Beveridge CA. 2019. An update on the signals controlling shoot branching. *Trends Plant Sci* **24:** 220–236. doi:10.1016/j.tplants.2018.12.001

Bayer E, Smith R, Mandel T, Nakayama N, Sauer M, Prusinkiewicz P, Kuhlemeier C. 2009. Integration of transport-based models for phyllotaxis and midvein formation. *Genes Dev* **23:** 373–384. doi:10.1101/gad.497009

Benková E, Michniewicz M, Sauer M, Teichmann T, Seifertová D, Jürgens G, Friml J. 2003. Local, efflux-dependent auxin gradients as a common module for plant organ formation. *Cell* **115:** 591–602. doi:10.1016/S0092-8674(03)00924-3

Berleth T, Scarpella E, Prusinkiewicz P. 2007. Towards the systems biology of auxin-transport-mediated patterning. *Trends Plant Sci* **12:** 151–159. doi:10.1016/j.tplants.2007.03.005

Bertheloot J, Barbier F, Boudon F, Perez-Garcia MD, Péron T, Citerne S, Dun E, Beveridge C, Godin C, Sakr S. 2020. Sugar availability suppresses the auxin-induced strigolactone pathway to promote bud outgrowth. *New Phytol* **225:** 866–879. doi:10.1111/nph.16201

Bhatia N, Bozorg B, Larsson A, Ohno C, Jönsson H, Heisler MG. 2016. Auxin acts through MONOPTEROS to regulate plant cell polarity and pattern phyllotaxis. *Curr Biol* **26:** 3202–3208. doi:10.1016/j.cub.2016.09.044

Bilsborough GD, Runions A, Barkoulas M, Jenkins HW, Hasson A, Galinha C, Laufs P, Hay A, Prusinkiewicz P, Tsiantis M. 2011. Model for the regulation of *Arabidopsis thaliana* leaf margin development. *Proc Natl Acad Sci* **108:** 3424–3429. doi:10.1073/pnas.1015162108

Bridson R. 2015. *Fluid simulation for computer graphics.* CRC, Boca Raton, FL.

Brunoud G, Wells DM, Oliva M, Larrieu A, Mirabet V, Burrow AH, Beeckman T, Kepinski S, Traas J, Bennett MJ, et al. 2012. A novel sensor to map auxin response and distribution at high spatio-temporal resolution. *Nature* **482:** 103–106. doi:10.1038/nature10791

Chan J, Mansfield C, Clouet F, Dorussen D, Coen E. 2020. Intrinsic cell polarity coupled to growth axis formation in tobacco BY-2 cells. *Curr Biol* **30:** 4999–5006.e3. doi:10.1016/j.cub.2020.09.036

Chitwood DH, Headland LR, Ranjan A, Martinez CC, Braybrook SA, Koenig DP, Kuhlemeier C, Smith RS, Sinha NR. 2012. Leaf asymmetry as a developmental constraint imposed by auxin-dependent phyllotactic patterning. *Plant Cell* **24:** 2318–2327. doi:10.1105/tpc.112.098798

Cieslak M, Prusinkiewicz P. 2019. Gillespie–Lindenmayer systems for stochastic simulation of morphogenesis. *In Silico Plants* **1:** diz009. doi:10.1093/insilicoplants/diz009

Cieslak M, Seleznyova AN, Prusinkiewicz P, Hanan J. 2011. Towards aspect-oriented functional–structural plant modelling. *Ann Bot* **108:** 1025–1041. doi:10.1093/aob/mcr121

Cieslak M, Runions A, Prusinkiewicz P. 2015. Auxin-driven patterning with unidirectional fluxes. *J Exp Bot* **66:** 5083–5102. doi:10.1093/jxb/erv262

Cieslak M, Cheddadi I, Boudon F, Baldazzi V, Génard M, Godin C, Bertin N. 2016. Integrating physiology and architecture in models of fruit expansion. *Front Plant Sci* **7:** 1739. doi:10.3389/fpls.2016.01739

Coen E, Rolland-Lagan AG, Matthews M, Bangham JA, Prusinkiewicz P. 2004. The genetics of geometry. *Proc Natl Acad Sci* **101:** 4728–4735. doi:10.1073/pnas.0306308101

Cui D, Zhao J, Jing Y, Fan M, Liu J, Wang Z, Xin W, Hu Y. 2013. The *Arabidopsis* IDD14, IDD15, and IDD16 cooperatively regulate lateral organ morphogenesis and gravitropism by promoting auxin biosynthesis and transport. *PLoS Genet* **9:** e1003759. doi:10.1371/journal.pgen.1003759

de Reuille PB, Bohn-Courseau I, Ljung K, Morin H, Carraro N, Godin C, Traas J. 2006. Computer simulations reveal properties of the cell–cell signaling network at the shoot apex in *Arabidopsis. Proc Natl Acad Sci* **103:** 1627–1632. doi:10.1073/pnas.0510130103

Douady S, Couder Y. 1996. Phyllotaxis as a dynamical self-organizing process, parts I–III. *J Theor Biol* **178:** 255–312. doi:10.1006/jtbi.1996.0024

Draelants D, Avitabile D, Vanroose W. 2015. Localized auxin peaks in concentration-based transport models of the shoot apical meristem. *J R Soc Interface* **12:** 20141407. doi:10.1098/rsif.2014.1407

Fàbregas N, Formosa-Jordan P, Confraria A, Siligato R, Alonso JM, Swarup R, Bennett MJ, Mähönen AP, Caño-Delgado AI, Ibañes M. 2015. Auxin influx carriers control vascular patterning and xylem differentiation in *Arabidopsis thaliana. PLoS Genet* **11:** e1005183. doi:10.1371/journal.pgen.1005183

Farcot E, Yuan Y. 2013. Homogeneous auxin steady states and spontaneous oscillations in flux-based auxin transport models. *SIAM J Appl Dyn Syst* **12:** 1330–1353. doi:10.1137/120891538

Feugier FG, Iwasa Y. 2006. How canalization can make loops: a new model of reticulated leaf vascular pattern formation. *J Theor Biol* **243:** 235–244. doi:10.1016/j.jtbi.2006.05.022

Feugier FG, Mochizuki A, Iwasa Y. 2005. Self-organization of the vascular system in plant leaves: inter-dependent dynamics of auxin flux and carrier proteins. *J Theor Biol* **236:** 366–375. doi:10.1016/j.jtbi.2005.03.017

Fox Keller E. 2002. *Making sense of life: explaining biological development with models, metaphors, and machines.* Harvard University Press, Cambridge, MA.

Galvan-Ampudia CS, Cerutti G, Legrand J, Brunoud G, Martin-Arevalillo R, Azais R, Bayle V, Moussu S, Wenzl C, Jaillais Y, et al. 2020. Temporal integration of auxin information for the regulation of patterning. *eLife* **9:** e55832. doi:10.7554/eLife.55832

Gälweiler L, Guan C, Müller A, Wisman E, Mendgen K, Yephremov A, Palme K. 1998. Regulation of polar auxin transport by AtPIN1 in *Arabidopsis* vascular tissue. *Science* **282:** 2226–2230. doi:10.1126/science.282.5397.2226

Gao Y, Zhang Y, Zhang D, Dai X, Estelle M, Zhao Y. 2015. Auxin binding protein 1 (ABP1) is not required for either auxin signaling or *Arabidopsis* development. *Proc Natl Acad Sci* **112:** 2275–2280. doi:10.1073/pnas.1500365112

Gao C, Liu X, De Storme N, Jensen KH, Xu Q, Yang J, Liu X, Chen S, Martens HJ, Schulz A, et al. 2020. Directionality of plasmodesmata-mediated transport in *Arabidopsis* leaves supports auxin channeling. *Curr Biol* **30:** 1970–1977.e4. doi:10.1016/j.cub.2020.03.014

Geisler M, Aryal B, di Donato M, Hao P. 2017. A critical view on ABC transporters and their interacting partners in auxin transport. *Plant Cell Physiol* **58:** 1601–1614. doi:10.1093/pcp/pcx104

Geldner N, Friml J, Stierhof YD, Jürgens G, Palme K. 2001. Auxin transport inhibitors block PIN1 cycling and vesicle trafficking. *Nature* **413:** 425–428. doi:10.1038/35096571

Goldsmith MHM. 1977. The polar transport of auxin. *Annu Rev Plant Physiol* **28:** 439–478. doi:10.1146/annurev.pp.28.060177.002255

Govindaraju P, Verna C, Zhu T, Scarpella E. 2020. Vein patterning by tissue-specific auxin transport. *Development* **147:** dev187666. doi:10.1242/dev.187666

Han X, Hyun TK, Zhang M, Kumar R, Koh EJ, Kang BH, Lucas WJ, Kim JY. 2014. Auxin-callose-mediated plasmodesmal gating is essential for tropic auxin gradient formation and signaling. *Dev Cell* **28:** 132–146. doi:10.1016/j.devcel.2013.12.008

Hartmann FP, Barbier de Reuille P, Kuhlemeier C. 2019. Toward a 3D model of phyllotaxis based on a biochemically plausible auxin-transport mechanism. *PLoS Comput Biol* **15:** e1006896. doi:10.1371/journal.pcbi.1006896

Haskovec J, Jönsson H, Kreusser LM, Markowich P. 2019. Auxin transport model for leaf venation. *Proc R Soc Lond A Math Phys Sci* **475:** 20190015.

Hay A, Barkoulas M, Tsiantis M. 2006. ASYMMETRIC LEAVES1 and auxin activities converge to repress *BREVIPEDICELLUS* expression and promote leaf development in *Arabidopsis*. *Development* **133:** 3955–3961. doi:10.1242/dev.02545

Heisler MG, Jönsson H. 2006. Modeling auxin transport and plant development. *J Plant Growth Regul* **25:** 302–312. doi:10.1007/s00344-006-0066-x

Heisler MG, Ohno C, Das P, Sieber P, Reddy GV, Long JA, Meyerowitz EM. 2005. Patterns of auxin transport and gene expression during primordium development revealed by live imaging of the *Arabidopsis* inflorescence meristem. *Curr Biol* **15:** 1899–1911. doi:10.1016/j.cub.2005.09.052

Heisler MG, Hamant O, Krupinski P, Uyttewaal M, Ohno C, Jönsson H, Traas J, Meyerowitz EM. 2010. Alignment between PIN1 polarity and microtubule orientation in the shoot apical meristem reveals a tight coupling between morphogenesis and auxin transport. *PLoS Biol* **8:** e1000516. doi:10.1371/journal.pbio.1000516

Hofmeister W. 1868. Allgemeine Morphologie der Gewächse [General morphology of plants]. In *Handbuch der physiologischen Botanik* [*Manual of physiological botany*]. Engelmann, Leipzig, Germany.

Ibañes M, Fàbregas N, Chory J, Caño-Delgado AI. 2009. Brassinosteroid signaling and auxin transport are required to establish the periodic pattern of *Arabidopsis* shoot vascular bundles. *Proc Natl Acad Sci* **106:** 13630–13635. doi:10.1073/pnas.0906416106

Jacob F. 1998. *Of flies, mice, and men.* Harvard University Press, Cambridge, MA.

Janssen JM, Lindenmayer A. 1987. Models for the control of branch positions and flowering sequences of capitula in *Mycelis muralis* (L.) Dumont (Compositae). *New Phytol* **105:** 191–220. doi:10.1111/j.1469-8137.1987.tb00859.x

Jönsson H. 2019. The rise of a forgotten model. *Nat Rev Mol Cell Biol* **20:** 455–455. doi:10.1038/s41580-019-0144-0

Jönsson H, Heisler M, Shapiro B, Meyerowitz E, Mjolsness E. 2006. An auxin-driven polarized transport model for phyllotaxis. *Proc Natl Acad Sci* **103:** 1633–1638. doi:10.1073/pnas.0509839103

Kang J, Tang J, Donnelly P, Dengler N. 2003. Primary vascular pattern and expression of *ATHB-8* in shoots of *Arabidopsis*. *New Phytol* **158:** 443–454. doi:10.1046/j.1469-8137.2003.00769.x

Kierzkowski D, Lenhard M, Smith R, Kuhlemeier C. 2013. Interaction between meristem tissue layers controls phyllotaxis. *Dev Cell* **26:** 616–628. doi:10.1016/j.devcel.2013.08.017

Kierzkowski D, Runions A, Vuolo F, Strauss S, Lymbouridou R, Routier-Kierzkowska AL, Wilson-Sánchez D, Jenke H, Galinha C, Mosca G, et al. 2019. A growth-based framework for leaf shape development and diversity. *Cell* **177:** 1405–1418.e17. doi:10.1016/j.cell.2019.05.011

Kramer EM. 2008. Computer models of auxin transport: a review and commentary. *J Exp Bot* **59:** 45–53. doi:10.1093/jxb/erm060

Kramer EM. 2009. Auxin-regulated cell polarity: an inside job? *Trends Plant Sci* **14:** 242–247. doi:10.1016/j.tplants.2009.02.005

Krishna A, Gardiner J, Donner TJ, Scarpella E. 2020. Control of vein-forming, striped gene-expression by auxin signaling. bioRxiv doi:10.1101/2020.09.28.317644

Krogan NT, Marcos D, Weiner AI, Berleth T. 2016. The auxin response factor MONOPTEROS controls meristem function and organogenesis in both the shoot and root through the direct regulation of *PIN* genes. *New Phytol* **212:** 42–50. doi:10.1111/nph.14107

Krupinski P, Jönsson H. 2010. Modeling auxin-regulated development. *Cold Spring Harb Perspect Biol* **2:** a001560. doi:10.1101/cshperspect.a001560

Leyser O. 2011. Auxin, self-organization, and the colonial nature of plants. *Curr Biol* **21:** R331–R337. doi:10.1016/j.cub.2011.02.031

Mattsson J, Sung ZR, Berleth T. 1999. Responses of plant vascular systems to auxin transport inhibition. *Development* **126:** 2979–2991.

Mattsson J, Ckurshumova W, Berleth T. 2003. Auxin signaling in *Arabidopsis* leaf vascular development. *Plant Physiol* **131:** 1327–1339. doi:10.1104/pp.013623

Merks RM, Van de Peer Y, Inzé D, Beemster GT. 2007. Canalization without flux sensors: a traveling-wave hypothesis. *Trends Plant Sci* **12:** 384–390. doi:10.1016/j.tplants.2007.08.004

Mirabet V, Besnard F, Vernoux T, Boudaoud A. 2012. Noise and robustness in phyllotaxis. *PLoS Comput Biol* **8:** e1002389. doi:10.1371/journal.pcbi.1002389

Mitchison GJ. 1980. A model for vein formation in higher plants. *Proc R Soc Lond B Biol Sci* **207:** 79–109. doi:10.1098/rspb.1980.0015

Mitchison GJ. 1981. The polar transport of auxin and vein patterns in plants. *Philos Trans R Soc Lond B Biol Sci* **295:** 461–471. doi:10.1098/rstb.1981.0154

Mitchison G. 2015. The shape of an auxin pulse, and what it tells us about the transport mechanism. *PLoS Comput Biol* **11:** e1004487. doi:10.1371/journal.pcbi.1004487

Morales-Tapia A, Cruz-Ramírez A. 2016. Computational modeling of auxin: a foundation for plant engineering. *Front Plant Sci* **7:** 1881. doi:10.3389/fpls.2016.01881

Moulton DE, Oliveri H, Goriely A. 2020. Multiscale integration of environmental stimuli in plant tropism produces complex behaviors. *Proc Natl Acad Sci* **117:** 32226–32237. doi:10.1073/pnas.2016025117

Newell AC, Shipman PD, Sun Z. 2008. Phyllotaxis: cooperation and competition between mechanical and biochemical processes. *J Theor Biol* **251:** 421–439. doi:10.1016/j.jtbi.2007.11.036

O'Connor D, Runions A, Sluis A, Bragg J, Vogel J, Prusinkiewicz P, Hake S. 2014. A division in PIN-mediated auxin patterning during organ initiation in grasses. *PLoS Comput Biol* **10:** e1003447. doi:10.1371/journal.pcbi.1003447

Okada K, Ueda J, Komaki MK, Bell CJ, Shimura Y. 1991. Requirement of the auxin polar transport system in early stages of *Arabidopsis* floral bud formation. *Plant Cell* **3:** 677–684. doi:10.2307/3869249

Owens A, Cieslak M, Hart J, Classen-Bockhoff R, Prusinkiewicz P. 2016. Modeling dense inflorescences. *ACM Trans Graph* **35:** 1–14. doi:10.1145/2897824.2925982

Paciorek T, Zažímalová E, Ruthardt N, Petrášek J, Stierhof YD, Kleine-Vehn J, Morris DA, Emans N, Jürgens G, Geldner N, et al. 2005. Auxin inhibits endocytosis and promotes its own efflux from cells. *Nature* **435:** 1251–1256. doi:10.1038/nature03633

Paterlini A. 2020. Uncharted routes: exploring the relevance of auxin movement via plasmodesmata. *Biol Open* **9:** bio055541. doi:10.1242/bio.055541

Pinon V, Prasad K, Grigg SP, Sanchez-Perez GF, Scheres B. 2013. Local auxin biosynthesis regulation by PLETHORA transcription factors controls phyllotaxis in *Arabidopsis. Proc Natl Acad Sci* **110:** 1107–1112. doi:10.1073/pnas.1213497110

Prusinkiewicz P, Lindenmayer A. 1990. *The algorithmic beauty of plants* (with Hanan JS, Fracchia FD, Fowler D, de Boer MJM, Mercer L). Springer, New York.

Prusinkiewicz P, Runions A. 2012. Computational models of plant form and development. *New Phytol* **193:** 549–569. doi:10.1111/j.1469-8137.2011.04009.x

Prusinkiewicz P, Crawford S, Smith RS, Ljung K, Bennett T, Ongaro V, Leyser O. 2009. Control of bud activation by an auxin transport switch. *Proc Natl Acad Sci* **106:** 17431–17436. doi:10.1073/pnas.0906696106

Putterill J, Varkonyi-Gasic E. 2016. FT and florigen long-distance flowering control in plants. *Curr Opin Plant Biol* **33:** 77–82. doi:10.1016/j.pbi.2016.06.008

Raven JA. 1975. Transport of indoleacetic acid in plant cells in relation to pH and electrical potential gradients, and its significance for polar IAA transport. *New Phytol* **74:** 163–172. doi:10.1111/j.1469-8137.1975.tb02602.x

Ravichandran SJ, Linh NM, Scarpella E. 2020. The canalization hypothesis—challenges and alternatives. *New Phytol* **227:** 1051–1059. doi:10.1111/nph.16605

Reinhardt D, Mandel T, Kuhlemeier C. 2000. Auxin regulates the initiation and radial position of plant lateral organs. *Plant Cell* **12:** 507–518. doi:10.1105/tpc.12.4.507

Reinhardt D, Pesce E, Stieger P, Mandel T, Baltensperger K, Bennett M, Traas J, Friml J, Kuhlemeier C. 2003. Regulation of phyllotaxis by polar auxin transport. *Nature* **426:** 255–260. doi:10.1038/nature02081

Renton M, Hanan J, Ferguson BJ, Beveridge CA. 2012. Models of long-distance transport: how is carrier-dependent auxin transport regulated in the stem? *New Phytol* **194:** 704–715. doi:10.1111/j.1469-8137.2012.04093.x

Rodkaew Y, Siripant S, Lursinsap C, Chongstitvatana P. 2002. An algorithm for generating vein images for realistic modeling of a leaf. In *Proceedings of the International Conference on Computational Mathematics and Modeling*, Vol. 9, pp. 1–9.

Rodkaew Y, Chongstitvatana P, Siripant S, Lursinsap C. 2003. Particle systems for plant modeling. Plant growth modeling and applications. In *Proceedings of PMA03*, pp. 210–217.

Rolland-Lagan AG, Prusinkiewicz P. 2005. Reviewing models of auxin canalization in the context of leaf vein pattern formation in *Arabidopsis. Plant J* **44:** 854–865. doi:10.1111/j.1365-313X.2005.02581.x

Rubery PH, Sheldrake AR. 1974. Carrier-mediated auxin transport. *Planta* **118:** 101–121. doi:10.1007/BF00388387

Runions A, Fuhrer M, Lane B, Federl P, Rolland-Lagan AG, Prusinkiewicz P. 2005. Modeling and visualization of leaf venation patterns. In *ACM SIGGRAPH 2005 Papers*, pp. 702–711.

Runions A, Smith RS, Prusinkiewicz P. 2014. Computational models of auxin-driven development. In *Auxin and its role in plant development* (ed. Zažímalová E, Petrášek J, Benková E), pp. 315-357. Springer, Vienna.

Runions A, Tsiantis M, Prusinkiewicz P. 2017. A common developmental program can produce diverse leaf shapes. *New Phytol* **216:** 401–418. doi:10.1111/nph.14449

Rutschow HL, Baskin TI, Kramer EM. 2011. Regulation of solute flux through plasmodesmata in the root meristem. *Plant Physiol* **155:** 1817–1826. doi:10.1104/pp.110.168187

Sachs T. 1969. Polarity and the induction of organized vascular tissues. *Ann Bot* **33:** 263–275. doi:10.1093/oxfordjournals.aob.a084281

Sachs T. 1981. The control of the patterned differentiation of vascular tissues. *Adv Bot Res* **9:** 151–262. doi:10.1016/S0065-2296(08)60351-1

Sachs T. 1989. The development of vascular networks during leaf development. *Curr Top Plant Biochem Physiol* **8:** 168–183.

Sachs T. 1991. *Pattern formation in plant tissues.* Cambridge University Press, Cambridge.

Sachs T. 2004. Self-organization of tree form: a model for complex social systems. *J Theor Biol* **230:** 197–202. doi:10.1016/j.jtbi.2004.05.006

Sager R, Wang X, Hill K, Yoo BC, Caplan J, Nedo A, Tran T, Bennett MJ, Lee JY. 2020. Auxin-dependent control of a plasmodesmal regulator creates a negative feedback loop modulating lateral root emergence. *Nat Commun* **11:** 1–10. doi:10.1038/s41467-019-14226-7

Sassi M, Vernoux T. 2013. Auxin and self-organization at the shoot apical meristem. *J Exp Bot* **64:** 2579–2592. doi:10.1093/jxb/ert101

Scarpella E, Rueb S, Meijer AH. 2003. The *RADICLELESS1* gene is required for vascular pattern formation in rice. *Development* **130:** 645–658. doi:10.1242/dev.00243

Scarpella E, Marcos D, Friml J, Berleth T. 2006. Control of leaf vascular patterning by polar auxin transport. *Genes Dev* **20:** 1015–1027. doi:10.1101/gad.1402406

Schneider A, Godin C, Boudon F, Demotes-Mainard S, Sakr S, Bertheloot J. 2019. Light regulation of axillary bud outgrowth along plant axes: an overview of the roles of sugars and hormones. *Front Plant Sci* **10:** 1296. doi:10.3389/fpls.2019.01296

Schoute JC. 1913. Beiträge zur Blattstellungslehre. I: Die Theorie [Contributions to the theory of leaf position. I: The theory]. *Recl des Trav Bot Néerl* **10:** 153–325.

Shi B, Vernoux T. 2019. Patterning at the shoot apical meristem and phyllotaxis. *Curr Top Dev Biol* **131:** 81–107. doi:10.1016/bs.ctdb.2018.10.003

Shi B, Guo X, Wang Y, Xiong Y, Wang J, Hayashi KI, Lei J, Zhang L, Jiao Y. 2018. Feedback from lateral organs controls shoot apical meristem growth by modulating auxin transport. *Dev Cell* **44:** 204–216.e6. doi:10.1016/j.devcel.2017.12.021

Shinohara N, Taylor C, Leyser O. 2013. Strigolactone can promote or inhibit shoot branching by triggering rapid depletion of the auxin efflux protein PIN1 from the plasma membrane. *PLoS Biol* **11:** e1001474. doi:10.1371/journal.pbio.1001474

Smith RS. 2011. Modeling plant morphogenesis and growth. In *New trends in the physics and mechanics of biological systems: lecture notes of the Les Houches Summer School* (ed. Amar MB, Goriely A, Mueller MM, Cugliandolo LF), Vol. 92, pp. 301–336. Oxford University Press, Oxford.

Smith R, Bayer E. 2009. Auxin transport-feedback models of patterning in plants. *Plant Cell Environ* **32:** 1258–1271. doi:10.1111/j.1365-3040.2009.01997.x

Smith R, Guyomarc'h S, Mandel T, Reinhardt D, Kuhlemeier C, Prusinkiewicz P. 2006a. A plausible model of phyllotaxis. *Proc Natl Acad Sci* **103:** 1301–1306. doi:10.1073/pnas.0510457103

Smith RS, Kuhlemeier C, Prusinkiewicz P. 2006b. Inhibition fields for phyllotactic pattern formation: a simulation study. *Botany* **84:** 1635–1649.

Snow M, Snow R. 1931. Experiments in phyllotaxis. I: The effect of isolating a primordium. *Philos Trans R Soc Lond B Biol Sci* **221:** 1–43.

Snow M, Snow R. 1937. Auxin and leaf formation. *New Phytol* **36:** 1–18. doi:10.1111/j.1469-8137.1937.tb06899.x

Snow M, Snow R. 1952. Minimum areas and leaf determination. *Proc R Soc Lond B Biol Sci* **139:** 545–566. doi:10.1098/rspb.1952.0034

Steinmann T, Geldner N, Grebe M, Mangold S, Jackson CL, Paris S, Gälweiler L, Palme K, Jürgens G. 1999. Coordinated polar localization of auxin efflux carrier PIN1 by GNOM ARF GEF. *Science* **286:** 316–318. doi:10.1126/science.286.5438.316

Stoma S, Lucas M, Chopard J, Schaedel M, Traas J, Godin C. 2008. Flux-based transport enhancement as a plausible unifying mechanism for auxin transport in meristem development. *PLoS Comput Biol* **4:** e1000207. doi:10.1371/journal.pcbi.1000207

Ulmasov T, Murfett J, Hagen G, Guilfoyle TJ. 1997. Aux/IAA proteins repress expression of reporter genes containing natural and highly active synthetic auxin response elements. *Plant Cell* **9:** 1963–1971.

van Berkel K, de Boer R, Scheres B, ten Tusscher K. 2013. Polar auxin transport: models and mechanisms. *Development* **140:** 2253–2268. doi:10.1242/dev.079111

Veen AH, Lindenmayer A. 1977. Diffusion mechanism for phyllotaxis: theoretical physico-chemical and computer study. *Plant Physiol* **60:** 127–139. doi:10.1104/pp.60.1.127

Verna C, Ravichandran SJ, Sawchuk MG, Linh NM, Scarpella E. 2019. Coordination of tissue cell polarity by auxin transport and signaling. *eLife* **8:** e51061. doi:10.7554/eLife.51061

Vieten A, Vanneste S, Wiśniewska J, Benková E, Benjamins R, Beeckman T, Luschnig C, Friml J. 2005. Functional redundancy of PIN proteins is accompanied by auxin-dependent cross-regulation of PIN expression. *Development* **132:** 4521–4531. doi:10.1242/dev.02027

Vieten A, Sauer M, Brewer PB, Friml J. 2007. Molecular and cellular aspects of auxin-transport-mediated development. *Trends Plant Sci* **12:** 160–168. doi:10.1016/j.tplants.2007.03.006

Wabnik K, Kleine-Vehn J, Balla J, Sauer M, Naramoto S, Reinöhl V, Merks RM, Govaerts W, Friml J. 2010. Emergence of tissue polarization from synergy of intracellular and extracellular auxin signaling. *Mol Syst Biol* **6:** 447. doi:10.1038/msb.2010.103

Wenzel CL, Schuetz M, Yu Q, Mattsson J. 2007. Dynamics of *MONOPTEROS* and PIN-FORMED1 expression during leaf vein pattern formation in *Arabidopsis thaliana*. *Plant J* **49:** 387–398. doi:10.1111/j.1365-313X.2006.02977.x

Žádníková P, Wabnik K, Abuzeineh A, Gallemi M, Van Der Straeten D, Smith RS, Inzé D, Friml J, Prusinkiewicz P, Benková E. 2016. A model of differential growth-guided apical hook formation in plants. *Plant Cell* **28:** 2464–2477. doi:10.1105/tpc.15.00569

Zažímalová E, Murphy AS, Yang H, Hoyerová K, Hošek P. 2010. Auxin transporters—why so many? *Cold Spring Harb Perspect Biol* **2:** a001552.

Zhang T, Cieslak M, Owens A, Wang F, Broholm SH, Teeri TH, Elomaa P, Prusinkiewicz P. 2021. Phyllotactic patterning of gerbera flower heads. *Proc Natl Acad Sci* **118:** e2016304118. doi:10.1073/pnas.2016304118

Chemical Biology in Auxin Research

Ken-ichiro Hayashi

Department of Biochemistry, Okayama University of Science, Okayama City 700-0005, Japan

Correspondence: hayashi@dbc.ous.ac.jp

Molecular genetic and structural studies have revealed the mechanisms of fundamental components of key auxin regulatory pathways consisting of auxin biosynthesis, transport, and signaling. Chemical biology methods applied in auxin research have been greatly expanded through the understanding of auxin regulatory pathways. Many small-molecule modulators of auxin metabolism, transport, and signaling have been generated on the basis of the outcomes of genetic and structural studies on auxin regulatory pathways. These chemical modulators are now widely used as essential tools for dissecting auxin biology in diverse plants. This review covers the structures, primary targets, modes of action, and applications of chemical tools in auxin biosynthesis, transport, and signaling.

The plant hormone auxin plays a crucial role in almost every aspect of plant development and is implicated in cell division, elongation, and differentiation. At the whole-plant level, auxin regulates embryogenesis, apical dominance, lateral root formation, tropic responses to light and gravity, and vascular tissue differentiation (Woodward and Bartel 2005; Enders and Strader 2015). The term auxin was initially derived from the Greek word "auxein," meaning "to grow," and auxin was defined as a growth-promoting substance that promotes cell elongation in shoots before the identification of indole-3-acetic acid (IAA) as a genuine natural plant hormone (Abel and Theologis 2010). IAA is a simple small molecule that displays profound and pleiotropic effects on plant growth. During the 1940–1970s, many synthetic compounds were extensively screened for auxin-like activity using physiological bioassays, such as hypocotyl elongation and rooting promotion assays, leading to the identification of chemicals referred to as synthetic auxins (Abel and Theologis 2010; Quareshy et al. 2018b). Synthetic auxins, strictly defined as auxin receptor agonists, are now widely used as agrochemicals, especially as herbicides (Busi et al. 2018).

The diverse auxin responses can be modulated by three major pathways: auxin metabolism, polar auxin transport, and signal transduction. IAA is mainly biosynthesized from tryptophan via indole-3-pyruvic acid (IPA) through the IPA pathway (Zhao 2014; Kasahara 2016). IAA is then transported by the *AUXIN1/LIKE-AUX1* (*AUX1/LAX*) auxin influx transporters, *PIN-FORMED* (*PIN*) efflux facilitators, and *ATP-BINDING CASSETTE B* (*ABCB*) transporters to form auxin concentration gradients in plant tissues (Petrasek and Friml 2009; Peer et al. 2011). IAA is perceived by a nuclear-localized auxin coreceptor complex comprising TIR1/AFB F-box proteins and Aux/IAA tran-

Cite this article as *Cold Spring Harb Perspect Biol* doi: 10.1101/cshperspect.a040105

scriptional repressors, leading to the transactivation of auxin-responsive genes (Hayashi 2012; Salehin et al. 2015; Leyser 2018). IAA is metabolized by several inactivation pathways. GRETCHEN HAGEN3 (GH3) proteins synthesize IAA-amino acid conjugates, such as IAA-Asp and IAA-Glu, as inactive conjugates (Ludwig-Müller 2011). IAA is also oxidized to 2-oxo-indole-3-acetic acid (oxIAA) by DIOXYGENASE FOR AUXIN OXIDATION 1 (DAO1) (Porco et al. 2016; Zhang and Peer 2017). These pathways all contribute to the spatiotemporal dynamics of IAA distribution and allow the broad but subtle range of auxin responses to environmental and developmental cues.

In chemical biology, small-molecule modulators play a central role in dissecting biological functions in living systems (Dejonghe and Russinova 2017; Kinoshita et al. 2018). Small-molecule modulators used in auxin biology are frequently referred to as auxin chemical probes, and they can modulate all the redundant functions of cognate gene families in auxin biosynthesis, transport, and signaling, thus enabling us to dissect the physiological and developmental roles of redundant genes (De Rybel et al. 2009; Ma and Robert 2014). Additionally, fundamental components of auxin pathways have been conserved as pivotal genes over evolutionary history across diverse land plants such as liver-

Table 1. List of key modulators of auxin regulatory pathways

Compound name	Target protein(s)/ pathway	CAS registry number[a]	Molecular weight	Commercial source[b]	References
4-Cl-iMT	TIR1	2244499-31-6	233.6	N	Quareshy et al. 2018a
RN1	TIR1/AFBs	428446-82-6	370.3	B	Vain et al. 2019
RN2	TIR1/AFBs	2349373-16-4	426.7	B	Vain et al. 2019
RN3	TIR1/AFBs	425609-03-6	444.7	B	Vain et al. 2019
RN4	TIR1/AFBs	433950-87-9	364.2	B	Vain et al. 2019
Fipexide	Unknown	34161-24-5	388.8	B	Nakano et al. 2018
AAL1	Unknown	868215-48-9	320.4	B	Li et al. 2018
Auxinole	TIR1/AFBs	86445-22-9	321.4	A	Hayashi et al. 2008
PEO-IAA	TIR1/AFBs	6266-66-6	293.3	A	Hayashi et al. 2008
cvxIAA	ccvTIR1	2127037-35-6	281.3	A	Uchida et al. 2018
5-Adamantyl-IAA	ccvTIR1	2244426-40-0	309.4	A	Uchida et al. 2018
L-Kynurenine	TAA1/TARs	2922-83-0	208.2	A	He et al. 2011
KOK1169/AONP	TAA1/TARs	1394950-09-4	231.3	N	Narukawa-Nara et al. 2016
Pyruvamine 2031	TAA1/TARs	1394950-16-3	273.3	N	Kakei et al. 2017
BBo	YUCCA	5122-94-1	198.0	A	Kakei et al. 2015
PPBo	YUCCA	51067-38-0	214.0	A	Kakei et al. 2015
Yucasin	YUCCA	26028-65-9	211.7	A	Nishimura et al. 2014
Yucasin DF	YUCCA	1094690-87-5	213.2	N	Tsugafune et al. 2017
BUM	ABCB transporters	5809-23-4	313.4	A	Kim et al. 2010
Gravacin	ABCB19/PGP19	188438-05-3	283.1	A	Rojas-Pierce et al. 2007
Bz-IAA	AUX1, PIN, ABCB	4382-53-0	281.3	A	Tsuda et al. 2011
Bz-NAA	PIN, ABCB	1270002-60-2	292.3	N	Tsuda et al. 2011
CHPAA	AUX1/LAX	33697-81-3	186.6	A	Parry et al. 2001
1-NOA	AUX1/LAX	2976-75-2	202.2	A	Parry et al. 2001

[a]The CAS registry number is a unique numerical identifier assigned by the Chemical Abstracts Service to every chemical substance. The compounds in the online catalog of multiple chemical suppliers can be searched by the CAS registry number. The compound name and its abbreviation are not listed in some catalogs.

[b](A) commercially available from multiple suppliers, (B) commercially available as a rare chemical from a few suppliers, (N) commercially unavailable, contact author.

worts, mosses, ferns, and vascular plants (Eklund et al. 2015; Flores-Sandoval et al. 2015; Kato et al. 2018; Tao and Estelle 2018). Mutations in these pivotal genes cause severe pleiotropic defects in plant development. Auxin chemical probes can be applied as essential tools for the elucidation of auxin biology in diverse plants, as molecular genetic and genomic approaches are often not available for nonmodel plants.

This review introduces the recent list of representative synthetic auxins and chemical probes that target auxin signaling, biosynthesis, and transport. Additionally, I will summarize the mechanisms of action of some key chemical probes in auxin pathways (Table 1).

SYNTHETIC AUXINS AND AUXIN PRODRUGS

During the 1940–1970s, the potent pleiotropic hormonal activities of IAA led to a rush to explore synthetic chemicals exhibiting IAA-like activity. The synthetic chemical approach was initially applied to the chemical modification of the natural auxins phenylacetic acid (PAA) and IAA. Over time, many structurally divergent chemicals have been screened using phenotype-based bioassays based on auxin-like responses (Quareshy et al. 2018b). Synthetic chemicals with IAA-like activities are designated as synthetic auxins, many of which are widely used as agrochemicals, especially as herbicides to control weeds in arable crops. The synthetic auxins have been categorized into five groups (Ferro et al. 2010): (1) naphthalene-1-acetic acids (NAA); (2) phenoxyacetic acids (e.g., 2,4-dichlorophenoxyacetic acid) (2,4-D); (3) picolinate-type auxins (e.g., picloram); (4) benzoic acids (e.g., dicamba); and (5) indoles (e.g., halogenated IAA) (Fig. 1). The common structural features of synthetic auxins are an aromatic ring and the anion moiety attached to the aromatic ring, which are essential requirements for aux-

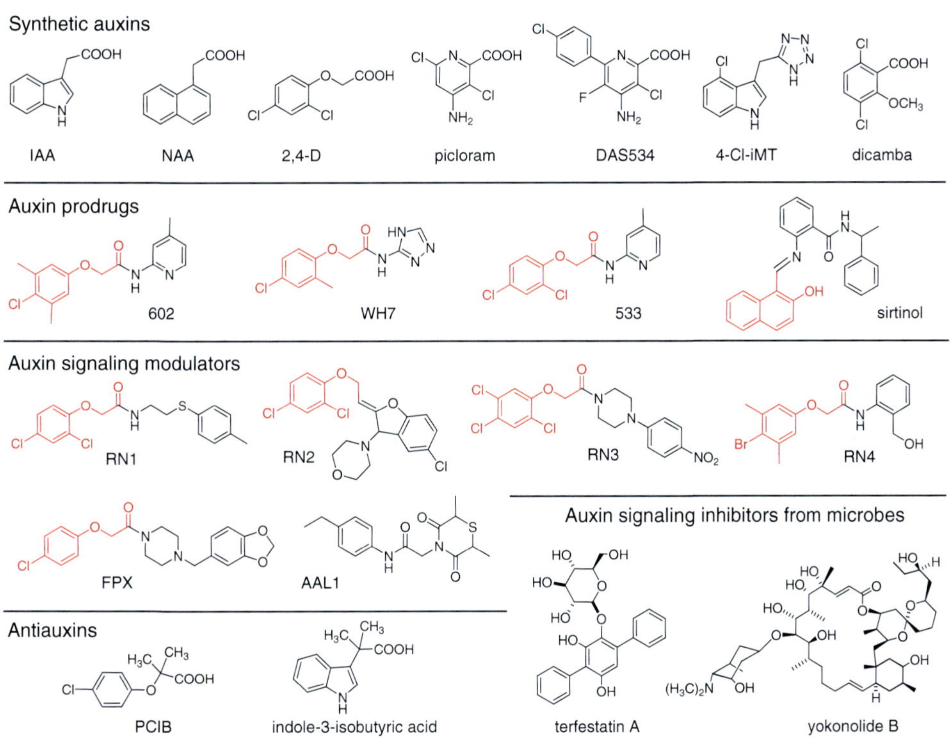

Figure 1. Structure of auxins and modulators of auxin signaling.

inic activity. These structurally divergent synthetic auxins are expected to bind to the TIR1/AFB-Aux/IAA coreceptor complex and consequently activate auxin signaling, but each synthetic auxin is likely to show different biological behaviors because they present distinct profiles of metabolic stability, polar transport in tissues, and affinity toward specific sets of auxin receptor complexes. Diverse synthetic auxins have been evaluated in the auxin responses of roots and hypocotyls, auxin-responsive gene expression, and auxin transport profiles (Simon et al. 2013), and this comprehensive analysis under identical assay conditions confirms that these synthetic auxins show distinct transport profiles and induce strikingly diverse phenotypes.

Two major synthetic auxins, NAA and 2,4-D, are widely used for plant research because both are chemically stable and show profound hormonal activities similar to IAA. Membrane-permeable NAA can passively move into cells and is actively exported by efflux transport proteins. NAA can rescue the impaired root phenotypes of *aux1* auxin influx mutants (Marchant et al. 1999). The AUX1 symporter takes up IAA and 2,4-D into the cell (Yang et al. 2006; Singh et al. 2018). PIN and ABCB efflux transporters export NAA as actively as IAA, but 2,4-D is a poor substrate for efflux transport proteins (Yang and Murphy 2009; Simon et al. 2013). Thus, NAA and 2,4-D can be useful diagnostic tools for dissecting the auxin polar transport machinery. For example, the *yuc* quintuple mutant (*yuc 3, 5, 7, 8, 9*) shows severe defects in root phenotypes (Chen et al. 2014). IAA and NAA can fully restore defects in both root growth and gravitropic responses. However, 2,4-D only restores the defects in root growth, not in root gravitropism, suggesting that root gravitropic responses require an asymmetric auxin concentration gradient formed by polar efflux transport (Sugawara et al. 2015). On the other hand, AUX1 does not mediate the uptake of other synthetic auxins, such as picloram and dicamba (Hoyerova et al. 2018).

IAA homeostasis is regulated by two major pathways involving GH3, IAA-amino acid conjugate synthases, and DAO, 2-oxoglutarate-dependent IAA oxidases. 2,4-D and dicamba have been shown to be very poor substrates of GH3 enzymes (Staswick et al. 2005; Chiu et al. 2018). Similarly, non-indole-type synthetic auxins are not oxidized by DAO1 oxidases. Thus, synthetic auxins induce a selective auxin response based on distinct transport and metabolic profiles (Busi et al. 2018; Hoyerova et al. 2018).

Precursors of a synthetic auxin (auxin prodrugs) release active auxin molecules via metabolic activation, for example, through hydrolysis by cellular enzymes. By employing auxin-related phenotype-based screening, auxinic compounds that selectively induce auxin responses were identified from a chemical library. The active compounds were found to be auxin prodrugs (WH-7, compounds 533 and 602 in Fig. 1) that release phenoxyacetic acid-type auxins after metabolic activation by plant amidohydrolases (Christian et al. 2008; Savaldi-Goldstein et al. 2008). These auxin prodrugs preferentially elicit auxin responses in the shoots, although 2,4-D shows auxin activity in both the roots and shoots. It is possible that these auxin prodrugs might be selectively activated in shoots where the hydrolases are expressed. Sirtinol was initially identified as a synthetic activator of auxin signaling because of its distinct structure from known synthetic auxins, but sirtinol was later found to be an auxin prodrug (Zhao et al. 2003). Sirtinol is metabolized to 2-hydroxy-1-naphthoic acid, which is an active auxin (Dai et al. 2005). Auxin prodrugs have important applications in agriculture. For example, auxin prodrugs IBA, 2,4-dichlorophenoxybutyric acid (2,4-BD), and naphthalene-1-acetamide (1-NAA amide) are widely used as agrochemicals.

SELECTIVE MODULATORS OF SCF$^{TIR1/AFB}$ SIGNALING

In canonical SCF^{TIR1} auxin signaling, IAA is perceived by TIR1 auxin receptors. The SCF^{TIR1} E3 ubiquitin ligase complex consisting of TIR1, Skp1 (ASK1), and Cullin (CUL1) catalyzes the ubiquitination of Aux/IAA repressor proteins. Auxin promotes the ubiquitination of Aux/IAAs by enhancing the interaction between Aux/IAA and TIR1 receptors (Leyser 2018). Consequently, auxin induces the degradation of Aux/IAA repressors via the 26S proteasome

 Cite this article as *Cold Spring Harb Perspect Biol* doi: 10.1101/cshperspect.a040105

pathway to activate ARF transcriptional activity regulating auxin-responsive gene expression. The crystal structure of the TIR1-auxin-Aux/IAA ternary complex shows that auxin acts as molecular glue through which the two proteins are tightly bound together. In the crystal structure, the domain II motif (WPPV) from the degron peptide of Aux/IAAs covers the auxin-bound cavity in TIR1. The tryptophan and second proline residues of the degron are involved in the hydrophobic interaction with IAA. In this scenario, the auxin-agonist activities of a small molecule are determined by the binding affinity to a hydrophobic cavity formed within the TIR1-Aux/IAA coreceptor.

This perception model suggests that auxinic compounds can be relatively simple small molecules such as IAA and PAA that fit within a small binding cavity. In *Arabidopsis*, TIR1 and five homologous proteins (AFB1–5) redundantly function as nuclear auxin receptors (Prigge et al. 2020). Interestingly, the *afb4* and *afb5* loss-of-function mutants are resistant to picolinate-type auxins, picloram and DAS534, but not to 2,4-D or IAA, suggesting that AFB4 and AFB5 are the major targets of picolinic acid-based herbicides (Walsh et al. 2006; Prigge et al. 2016). The binding affinity of synthetic auxins to TIR1/AFB-Aux/IAA coreceptors has been comprehensively investigated through surface-plasmon resonance analysis (Calderón Villalobos et al. 2012; Lee et al. 2014). NAA and 2,4-D show a lower affinity for both TIR1 and AFB5 receptors than IAA, and 2,4-D obviously prefers TIR1 to AFB5. Consistent with genetic evidence, picolinates display a higher affinity for AFB5 than the TIR1 receptor. This indicates that the receptor selectivity of synthetic auxins could be important for the design of auxinic herbicides. The tetrazole group functions as a bioisostere of carboxylic acid. The bioisostere analog of IAA, indole-3-methylene tetrazole was found to be a TIR1-selective agonist that is not recognized by AFB5 (Quareshy et al. 2018a). Through the optimization of indole-3-methylene tetrazole, 4-chloroindole-3-methylene tetrazole (4-Cl-iMT) was identified as a potent TIR1-selective agonist. Interestingly, the tetrazole group of 4-Cl-iMT is not conjugated with amino acids and glucose by the GH3 enzyme and UGT84B1 (IAA-glycoside synthase), suggesting that 4-Cl-iMT is more metabolically stable than IAA.

Depending on the auxin perception module, 6 TIR/AFB receptors and 24 Aux/IAAs harboring a domain II motif (WPPV) could establish specific subsets of TIR/AFB–Aux/IAA coreceptor complexes that could regulate specific ARFs (Parry et al. 2009). However, the physiological roles of each pair of TIR1/AFB-Aux/IAA coreceptor complexes cannot be easily dissected due to the redundant function of each Aux/IAA and TIR1/AFB. Vain et al. performed a chemical biology screen to identify auxin-like small molecules (Vain et al. 2019). A synthetic chemical library was screened based on the root and shoot phenotypes of wild-type (WT) and *auxin-resistant 1* (*axr1*) mutants. Four chemicals, RN1–RN4, were identified as active compounds that showed auxin-like activity specific to roots or shoots. RN compounds are derivatives of phenoxyacetic acids (Fig. 1), and physiological and biochemical analyses showed that RN1, RN3, and RN4 partially function as prodrugs of 2,4-D or 2,4,5-T in planta. Pull-down experiments with TIR1 and Aux/IAA demonstrated that RN3 and RN4 selectively induce the interaction of specific TIR1–Aux/IAA pairs. Therefore, RN3 and RN4 appear to act as selective molecular glues that can assemble specific subsets of TIR1 and Aux/IAA. These RN probes could be useful to dissect the physiological function of specific combinations of TIR1/AFB and Aux/IAAs.

Another phenotype-based systematic screen from a large chemical library identified two synthetic compounds exhibiting auxin-like activity. From screening 24,275 compounds, fipexide (FPX) was shown to inhibit hypocotyl elongation of etiolated *Arabidopsis* seedlings (Nakano et al. 2018). The inhibition of etiolated hypocotyl elongation is a typical auxin response, suggesting that FPX exhibits auxin-like activity. In general, efficient callus formation requires both auxin and cytokinin. Interestingly, FPX alone promotes potent callus formation and shoot regeneration. FPX contains a 4-chlorophenoxyacetyl amide moiety that can be converted by certain amidases to 4-chlorophenoxyacetic acid, a synthetic auxin. The complete molecular mechanism of FPX action has remained elusive. However, FPX shows structural similarity to RN3, implying that FPX might also be a selective agonist of particular sets of TIR1-Aux/IAA coreceptors. AAL1 was also identified as an inhibitor of etiolated hypocotyl extension from a set of 12,000 structurally diverse compounds (Li et al. 2018). A chemical genetic screen for an AAL1-resistant mutant identified a gain-of-function *iaa3/shy2 Aux/IAA* mutant. AAL1 induces *DR5::GUS* expression and promotes the degradation of DII-VENUS auxin-reporter proteins. Interestingly, *aux1* mutation and NPA treatment reduce hypocotyl inhibition by AAL1, suggesting that the auxin-like activity of AAL1 depends on polar auxin transport.

RATIONALLY DESIGNED AUXIN AGONISTS AND ANTAGONISTS OF THE TIR1/AFB RECEPTOR

Synthetic chemistry approaches have improved the selectivity, affinity, and stability of synthetic auxins and greatly contributed to auxin biology. Uchida et al. (2018) created a new engineered TIR1-auxin module referred to as "concave-TIR1-convexIAA." The modified IAA molecule and engineered TIR1 mutant receptor were designed according to a bump-and-hole strategy. The crystal structure of the TIR1-IAA complex suggests that phenylalanine 79 (F79) in the auxin-binding site is essential for the interaction with the aromatic ring of IAA. Indeed, the point mutation (F79G) resulted in a small hole in the TIR1 pocket, thereby disrupting the binding affinity for IAA. This TIR1 (F79G) mutant was designated concaveTIR1 (ccvTIR1), and it was also found to enable the binding of 5-aryl-substituted IAAs (Fig. 2A), including 5-methoxyphenyl IAA (convexIAA, cvxIAA). The large phenyl group (bump) of cvxIAA orients toward glycine 79 (G79) and fills the hole caused by G79 in ccvTIR1, thereby forming the ternary complex of ccvTIR1-cvxIAA-Aux/IAA. Thus, cvxIAA can selectively elicit auxin responses in transgenic plants expressing ccvTIR1 but can be completely inert in WT plants. ccvTIR1-cvxIAA can recover the impaired auxin responses of *tir1 afb2* mutants, suggesting that the ccvTIR1 and cvxIAA pair can function as a surrogate auxin signaling system. The highlights of the use of this system include its application for auxin-induced rapid hypocotyl elongation, referred to as acid growth. The ccvTIR1-cvxIAA system elegantly demonstrates that cvxIAA-induced acid growth is fully dependent on ccvTIR1, implying that TIR1 might mediate the rapid elongation response through a noncanonical pathway. Thus, the ccvTIR1-cvxIAA system can be an excellent tool for studying transcription-independent auxin signaling via TIR1 (Fendrych et al. 2016, 2018). The structural optimization of 5-substituted cvxIAA led to the generation of 5-adamantyl-IAA as a super-strong cvxIAA in a yeast two-hybrid system expressing another cvxTIR1 (F79A) with activity in the picomolar concentration range (Yamada et al. 2018). 5-Adamantyl-IAA and cvxTIR1 (F79A) are additional promising candidates for bump-and-hole synthetic auxin switches.

Detailed knowledge of the molecular mechanism of auxin perception has also allowed the structure-based design of TIR1/AFB auxin antagonists. A crystal structure showed that the formation of the ternary complex TIR1-IAA-Aux/IAA did not affect the conformation of TIR1, indicating that the IAA-binding site was a rigid and inflexible structure (Tan et al. 2007). Therefore, ligands specific to TIR1/AFB could be rationally designed according to the rigid structural model of the ternary complex. BH-IAA, *tert*-butoxycarbonylaminohexyl-IAA, was demonstrated as the first rationally designed TIR1 antagonist (Hayashi et al. 2008). BH-IAA inhibited almost all the typical auxin responses of *Arabidopsis* plants in a competitive manner with auxin. The crystal structure of the complex of BH-IAA and TIR1 demonstrated that the IAA moiety of BH-IAA sits in the auxin-binding site of TIR1 in the same binding position as IAA, and the long alkyl chain is oriented toward the Aux/IAA-binding cavity. This flexible long alkyl chain occupies the Aux/IAA-binding site to efficiently prevent access of the Aux/IAA WPPV degron motif (Fig. 2B). TIR1-Aux/IAA coreceptors tightly capture auxin within a small cavity in the coreceptor. On the

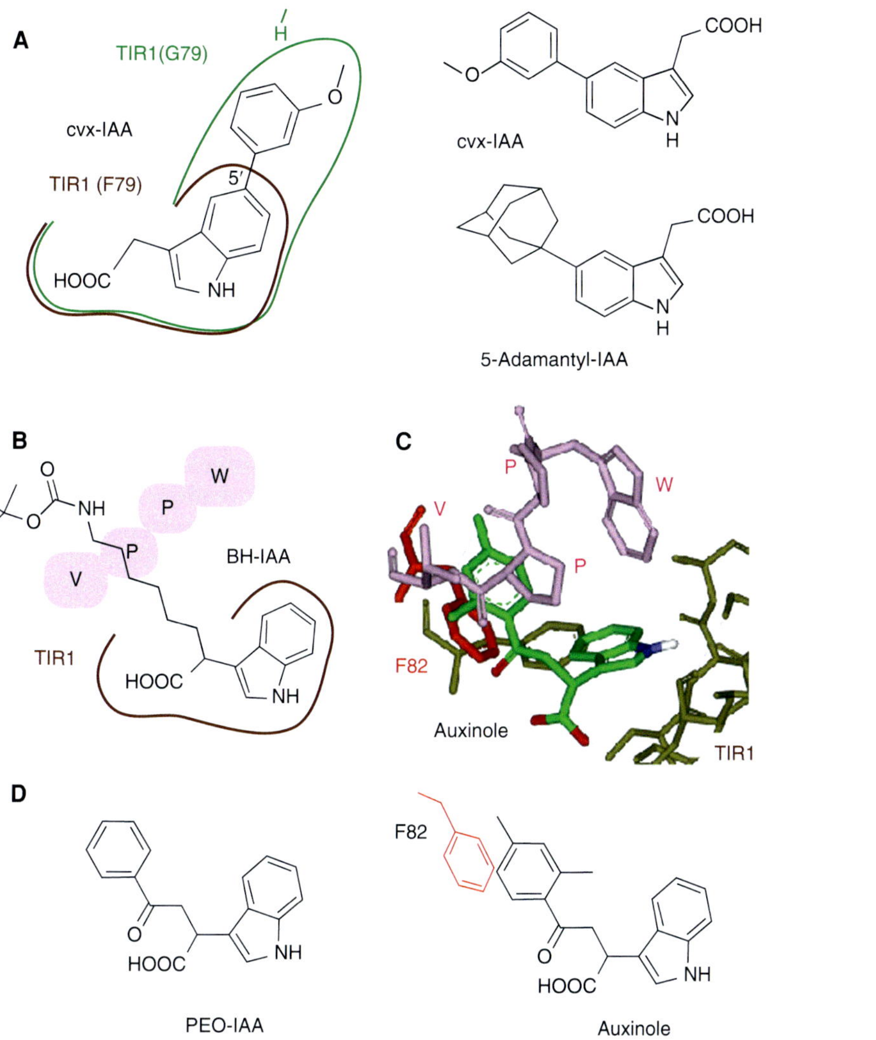

Figure 2. Examples of the structure-based design of auxin analogs. (*A*) cvxIAA was designed as a specific agonist of the engineered ccvTIR1 (F79G) receptor. The methoxyphenyl group can fit within a vacant pocket in the TIR1 mutant (F79G). 5-Adamantyl-IAA is a more potent analog of cvxIAA. (*B*) The IAA moiety in BH-IAA binds to the auxin-binding site of TIR1 in the same manner as IAA. The flexible long alkyl chain is oriented toward the Aux/IAA-binding site and blocks the access of the WPPV motif to the binding pocket. (*C,D*) Auxinole shows high affinity for the TIR1 auxin-binding site via a strong hydrophobic interaction between the dimethylphenyl ring and F82, which is essential for the interaction with the WPPV degron motif of Aux/IAAs. PEO-IAA is a lead compound of auxinole.

other hand, TIR1 receptors interact solely with the IAA moiety of BH-IAA and not with the flexible alkyl chain. Thus, BH-IAA is not a potent antagonist (Hayashi et al. 2008). To explore potent TIR1 antagonists, an in silico molecular docking screen was performed with an indole-focused chemical library, and α-(phenylethyl-2-oxo)-IAA (PEO-IAA) was identified as a lead compound (Fig. 2; Hayashi et al. 2012). Auxinole, the most potent derivative of PEO-IAA, was generated by the modification of the phenyl group of PEO-IAA. A phenyl ring at the α-position of auxinole interacts with the phenylalanine residue (F82) of the TIR1 auxin-binding site via a strong π-π stacking interaction. This F82 residue is conserved in TIR1/AFB members in diverse land plants and plays a crucial role in the interaction with the WPPV motif in domain II of Aux/IAAs. Thus, auxinole shows a high affinity to TIR1 and efficiently blocks Aux/IAA binding, explaining its strong antagonistic activity in auxin-responsive gene expression. Auxinole spatiotemporally blocks all the auxin responses mediated by redundant TIR1/AFB members in diverse land plants, including monocots, rice, and the moss *Physcomitrella patens,* and has been widely used as a TR1/AFB blocker in plant biology.

INHIBITORS OF AUXIN SIGNALING PATHWAYS

Several auxin derivatives, such as *p*-chlorophenoxyisobutylic acid (PCIB) (Oono et al. 2003), indole-3-isobutyric acid (ÅBerg 1951), and α-(5,7-dichloroindole-3-) isobutyric acid (Hatano et al. 1989), have been reported to be antiauxins that are expected to bind to a receptor in competition with auxin and thereby block the action of auxin (Fig. 1). The molecular mechanism of the antiauxin activity of PCIB has been extensively investigated in *Arabidopsis* plants. PCIB represses the IAA-induced degradation of Aux/IAAs, auxin-responsive gene expression, and typical physiological auxin responses, such as lateral root formation and root gravitropism. However, IAA-induced root phenotypes are not restored by PICB, suggesting that PCIB is not only an antagonist in auxin signaling but might also affect auxin transport and metabolism. Before the structural determination of the TIR1-IAA-Aux/IAA complex, auxin signaling inhibitors were screened from natural products using transgenic *Arabidopsis* auxin-responsive reporter systems (*BA3::GUS* and *DR5::GUS*) (Oono et al. 1998). These reporter assays are highly specific, rapid, and sensitive to auxin, making them very powerful for large library screens. Yokonolide A and B are macrolide-type compounds and were the first auxin signaling inhibitors isolated from the soil bacterium *Streptomyces diastatochromogenes* (Fig. 2; Hayashi et al. 2001, 2003). Similarly, microbial-origin terfestatin A and toyocamycin were identified by an auxin-responsive reporter screen (Fig. 1; Yamazoe et al. 2005). All these signaling inhibitors block Aux/IAA degradation and auxin-induced gene expression and repress various auxin responses in planta. However, the target proteins of these signaling inhibitors have not yet been determined.

TIR1/AFB are also thought to mediate nontranscriptional auxin signaling. This type of noncanonical auxin signaling is related to the very rapid inhibition of root elongation (Fendrych et al. 2018), vacuolar remodeling (Löfke et al. 2015), and auxin-induced Ca^{2+} signaling (Dindas et al. 2018; Li et al. 2019). However, little is known about the downstream pathway(s) of noncanonical TIR1/AFB signaling. A chemical screen for inhibitors of auxin-induced cytosolic Ca^{2+} dynamics was carried out using transgenic tobacco BY-2 cell lines expressing yellow fluorescent protein-apoaequorin as a reporter of Ca^{2+} signaling (De Vriese et al. 2019). Bepridil, a calcium channel blocker, was found to inhibit auxin-induced cytosolic Ca^{2+} accumulation in root cells and to affect auxin-induced vacuolar remodeling. Bepridil thus represents a valuable tool for exploring auxin-regulated Ca^{2+} signaling in plants.

AUXIN BIOSYNTHESIS INHIBITORS

The IPA pathway is the predominant IAA biosynthetic pathway and is composed of two sequential enzymatic steps (Mashiguchi et al. 2011; Won et al. 2011). In the initial step, TRYPTOPHAN AMINOTRANSFERASE OF ARABIDOPSIS1/TRYPTOPHAN AMINOTRANSFERASE RELATED (TAA1/TAR) converts Trp to IPA. The YUCCA (YUC) flavin monooxygenases then catalyzes the oxidative decarboxylation of IPA to IAA. YUC genes belong to a large redundant gene family including 11 genes in *Arabidopsis* and 14 genes in *Zea mays* and *Oryza sativa*. Therefore, auxin biosynthesis inhibitors would be powerful tools for studying auxin biology. Both TAA1 and YUC in the IPA pathway can be considered as targets for auxin biosynthesis inhibitors. L-Kynurenine (L-Kyn), which is the major metabolite of Trp in animals, was identified as the first competitive TAA1 inhibitor (He et al. 2011). L-Kyn was initially identified from a screen for an inhibitor of ethylene responses in *Arabidopsis* roots. Consistent with the structural similarity of L-Kyn and Trp, L-Kyn is regarded as an analog of Trp and shows competitive inhibition (Ki = 11 µM) of TAA1 activity. Intriguingly, other Trp derivatives, such as α-methyltryptophan and 5-methyltryptophan, have never been reported as IAA biosynthesis inhibitors, although TAA1 shows broad substrate specificity in vitro and will accept alternative amino acids such as Phe and Tyr. L-Kyn is commercially available and has been widely used to block IAA synthesis in various land plants, including the liverwort *Marchantia polymorpha* and the moss *Physcomitrella patens* (Eklund et al. 2015; Tsugafune et al. 2017). *L*-2-aminooxy-3-phenylpropionic acid (AOPP) was initially identified as an inhibitor of pyridoxyl-5-phosphate (PLP)-dependent enzymes, such as phenylalanine ammonia-lyase. Subsequently, AOPP was reported to inhibit the PLP-dependent enzymes TAA1/TAR (Soeno et al. 2010) and to reduce both endogenous IAA levels and root gravitropism in *Arabidopsis*. However, AOPP is not specific to TAA1/TAR enzymes and also represses other PLP-dependent enzymes, such as phenylalanine ammonia-lyase. Aminooxy-naphthylpropionic acid (AONP/KOK1169) was designed as a specific inhibitor of TAA1 (Narukawa-Nara et al. 2016). AONP confers typical auxin-deficient phenotypes and inhibits auxin-responsive gene expression. AONPs competitively inhibit TAA1 (Ki = 76.8 nM). Another aminooxy compound, pyruvamine 2031 (KOK2031), has also been identified as a potent inhibitor of rice OsTAR1 and represses endogenous IAA synthesis, resulting in apparent auxin-deficient phenotypes in rice (Kakei et al. 2017).

YUCs are flavin-containing monooxygenases that function at final step in the IPA pathway, making YUCs important targets of IAA biosynthesis inhibitors. Yucasin was discovered as a YUC inhibitor through a chemical library screen in maize coleoptiles (Nishimura et al. 2014). Yucasin has a similar structure to methimazole, another inhibitor of flavin-containing monooxygenases (Eswaramoorthy et al. 2006). Although yucasin counters the high-auxin phenotype of YUC overexpression lines, severe auxin-deficient phenotypes are not induced in WT plants by yucasin treatment. To improve the inhibitory activity of yucasin, the structure of yucasin was optimized by chemical modification (Tsugafune et al. 2017). Difluorinated yucasin, yucasin DF (YDF), causes severe auxin-deficient phenotypes in WT plants, but YDF shows slightly less inhibitory activity than yucasin against in vitro YUC1 enzyme. YDF was shown to be a chemically and metabolically stable compound, probably because of the difluorinated phenyl ring, suggesting that cellular YDF levels are sustained long enough to fully inhibit IAA biosynthesis. YDF was found to be effective in the moss *Physcomitrella patens* and the liverwort *Marchantia polymorpha*, demonstrating that YDF is a promising tool for modulating auxin biosynthesis in various land plants.

The phenylboronic acids 4-biphenylboronic acid (BBo) and 4-phenoxyphenylboronic acid (PPBo) have been reported to be potent IAA biosynthesis inhibitors targeting YUCs (Kakei et al. 2015). BBo and PPBo cause auxin-deficient phenotypes in *Arabidopsis* seedlings. These boronic acids competitively inhibit YUC enzyme activity against the substrate IPyA. Additionally, these boronic acids recover the extreme high-auxin phenotype of *35S::YUC1* seedlings and repress IAA biosynthesis and growth in the model monocot plant *Brachypodium distachyon*.

SYNTHETIC MODULATORS OF AUXIN TRANSPORT

N-1-naphthylphthalamic acid (NPA) and 2,3,5-triiodobenzoic acid (TIBA) are classical phytotropins and auxin polar transport inhibitors that have been widely used to study plant biology (Fig. 3). NPA blocks the auxin export activity of ABCB transporters expressed in both yeast and mammalian cells (Geisler et al. 2005; Blakeslee et al. 2007; Yang and Murphy 2009). Additionally, NPA represses PIN2-mediated IAA efflux in cultured tobacco cells (Barbez et al. 2013). However, the direct binding of

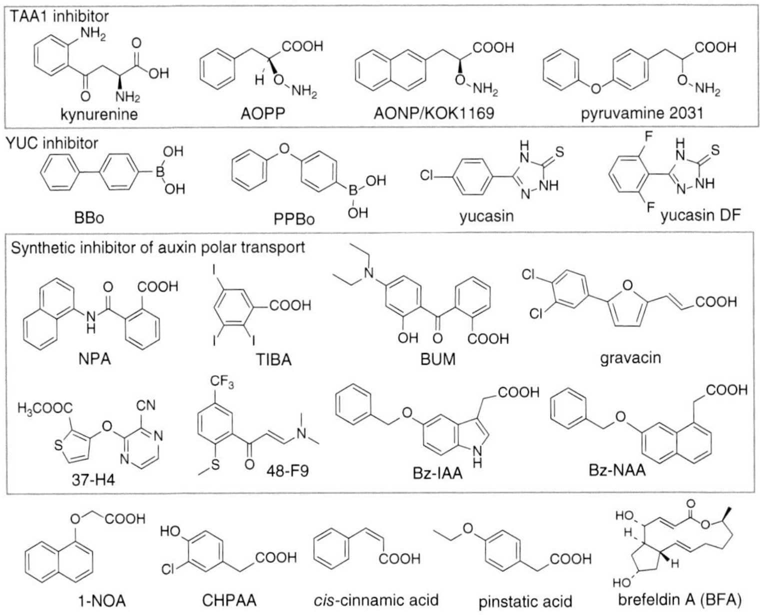

Figure 3. Structures of auxin biosynthesis inhibitors and auxin polar transport modulators.

NPA to PIN proteins has been unclear because of inconsistent results for NPA in relation to the activity of PINs in heterologous systems. The immunophilin-like protein TWD1 interacts with ABCBs to modulate auxin efflux. NPA binds to TWD1 to perturb the interaction of TWD1 with ABCB1 (Bailly et al. 2008; Wang et al. 2013). Additionally, NPA modulates the membrane localization of PINs and ABCBs by affecting the TWD1-ACTIN7 axis (Zhu et al. 2016). Thus, NPA efficiently inhibits polar auxin transport by modulating multiple targets of the polar auxin transport machinery. Interestingly, TIBA was also shown to disrupt PIN vesicle trafficking by perturbing the stability of actin (Dhonukshe et al. 2008). However, the molecular target of TIBA has remained elusive.

Gravacin (3-(5-[3,4-dichlorophenyl]-2-furyl) acrylic acid) and BUM (2-[4-(diethylamino)-2-hydroxybenzoyl] benzoic acid) (Fig. 3) were identified from a phenotype-based screen of synthetic chemicals. Gravacin perturbs root and shoot gravitropism in *Arabidopsis*. The primary target of gravacin was identified as ABCB19 by a chemical genetic approach (Rojas-Pierce et al. 2007), with gravacin requiring ABCB19 for its activity. BUM selectively targets ABCB transporters and causes pleiotropic defects in auxin-related phenotypes similar to what is observed in NPA-treated plants. BUM was demonstrated to compete at the NPA-binding sites of ABCB1 and ABCB19 (Kim et al. 2010). Two other auxin transport inhibitors, 37-H4 and 48-F9, were discovered by screening a chemical library using an IAA transport assay in maize coleoptiles (Nishimura et al. 2012). These two inhibitors exert NPA-like effects on *Arabidopsis* seedlings, but 37-H4 and 48-F9 are structurally distinct from the inhibitors NPA and BUM. All of these synthetic inhibitors inhibit polar auxin transport and affect the auxin-regulated response of plants. However, the molecular mechanisms of these inhibitors are still poorly understood due to the complicated regulation of IAA efflux and the interplay of PINs, TWD1, and ABCBs.

The auxin analogs 5-benzyloxy-IAA (Bz-IAA) and 7-benzyloxy-NAA (Bz-NAA) are inactive auxin analogs that cannot bind to the auxin-binding pocket of TIR1/AFB auxin receptors (Tsuda et al. 2011). The benzyloxy group of Bz-IAA and Bz-NAA (Bz-auxins) would likely compete with Phe79 in the binding pocket of TIR1/AFB in a similar way to cvxIAA (Tsuda et al. 2011). Bz-auxins do, however, show similar effects to NPA regarding auxin-regulated phenotypes in *Arabidopsis*, such as the inhibition of the gravitropic response and primary root growth. Additionally, Bz-auxins inhibited IAA movement in *Arabidopsis* and maize, but Bz-auxins do not affect the subcellular localization of PIN proteins in *Arabidopsis*. Bz-auxins inhibit the transport activity of PINs, ABCBs, and AUX1 expressed in yeast cells. Consistent with the transport profiles of the synthetic auxin NAA, Bz-NAA selectively acts on PINs and ABCBs, but not on AUX/LAX proteins. Therefore, Bz-auxins could be reorganized as active auxin analogs by auxin transport proteins and reduce endogenous IAA transport by competing with endogenous IAA movement, indicating that Bz-auxins are a new class of transport inhibitors.

1-Naphthoxyacetic acid (1-NOA) and 3-chloro-4-hydroxyphenylacetic acid (CHPAA) have been found to be selective inhibitors of auxin influx transport by the AUX1 importer (Hoyerova et al. 2018). 1-NOA and CHPAA repress the root gravitropic response without affecting root growth; phenocopying the *aux1* mutant (Parry et al. 2001) and 1-NOA can inhibit auxin uptake by AUX1 expressed in *Xenopus* oocytes (Yang et al. 2006).

Pinstatic acid, 4-ethoxyphenylacetic acid (PISA), was identified as another disruptor of auxin transport from a screen of modulators of PIN localization (Oochi et al. 2019). PISA shows auxin-like activity, such as the promotion of hypocotyl elongation and adventitious root formation. Interestingly, PISA does not directly activate the SCF^{TIR1} pathway as an active auxin. PISA inhibited gravitropic response and root hair formation in a manner similar to that observed when PIN1 is overexpressed in roots. It appears that PISA enhances the auxin flow by blocking PIN endocytosis and reorientation. PISA dramatically increases lateral root development under cotreatment with IAA, suggesting that PISA enhances the rate of shootward auxin transport, making PISA notably different from other known auxin chemical tools. Therefore, PISA represents a promising tool for studying the complicated regulation of PIN trafficking by auxin.

MODULATORS OF AUXIN POLAR TRANSPORT FROM NATURAL PRODUCTS

cis-Cinnamic acid has been reported as a natural auxin efflux inhibitor. *cis*-Cinnamic acid represses the efflux transport of NAA in tobacco BY-2 cells and inhibits root gravitropism in *Arabidopsis* (Steenackers et al. 2017). Intriguingly, *cis*-cinnamic acid promotes lateral root formation, in contrast to the effects of NPA and TIBA, probably because *cis*-cinnamic acid disrupts shootward auxin transport, leading to the accumulation of endogenous IAA in the root meristem. Flavonoid-deficient transparent *testa4* mutant shows enhanced basipetal auxin transport, demonstrating that flavonoids such as quercetin and naringenin are endogenous negative regulators of auxin transport (Peer and Murphy 2007). However, the mode of action and the primary targets of these natural inhibitors have remained obscure.

The polar localization of PIN proteins at the plasma membrane is crucial for the regulation of the rate and direction of cellular auxin export, which ultimately determines auxin gradients in the tissue. The polar distribution of PIN is maintained by the recycling of PIN between endosomal compartments and the plasma membrane. Clathrin-mediated endocytosis is involved in the internalization of PIN proteins from the plasma membrane (Adamowski and Friml 2015). Brefeldin A (BFA) is a fungal toxin that has been widely used to dissect auxin carrier trafficking pathways. BFA inhibits ADP ribosylation factor guanine nucleotide exchange factors (ARF-GEFs), which are modulators of vesicle formation. Thus, in tissues with BFA-sensitive ARF-GEFs, BFA represses the exocytosis of PIN by inhibiting ARF-GEFs, leading to the accumulation of PIN1 in artificial intracellular aggregates referred to as "BFA bodies" (Doyle et al. 2015). TIBA inhibits the accumulation of PINs in BFA bodies by disrupting the endocytosis of PINs (Geldner et al. 2001).

CONCLUDING REMARKS

Fundamental molecular mechanisms of auxin regulatory pathways involving metabolism, signaling, and polar transport have been unveiled by extensive genetic and structural biology studies in *Arabidopsis*. The core modules of all auxin regulatory pathways are well conserved in land plants. Small-molecule modulators of these auxin regulatory pathways have been generated as auxin chemical probes and are widely used in *Arabidopsis* biology. Given pathway conservation, auxin chemical probes can specifically modulate the auxin-related pathways of diverse land plants from the liverworts to vascular plants, making these probes valuable tools for research over a broad range of species, including in agriculture and horticulture. Furthermore, auxin chemical probes can be efficiently applied for the dissection of cross talk mechanisms between auxin and other hormones and environmental cues. The structural study of TIR1-Aux/IAA has led to the rational design of auxin agonists and antagonists, and structural information related to auxin biosynthesis (TAA1) (Tao et al. 2008) and metabolism (DAO, GH3, and ILL2) (Bitto et al. 2009; Peat et al. 2012; Takehara et al. 2020) could be applied to design more potent inhibitors of these pathways.

How rapid auxin responses are mediated by nontranscriptional auxin signaling remains one of the major questions in auxin biology (Kubeš and Napier 2019). By a chemical biology approach using the ccvTIR1-cvxIAA system, TIR1 was demonstrated to be involved in rapid auxin responses such as hypocotyl elongation and primary root growth inhibition, some of which are likely nontranscriptional (Fendrych et al. 2016, 2018; Uchida et al. 2018). In combination with new tools from synthetic biology, chemical biology continues to offer promising approaches for the elucidation of such new signaling pathways. Novel auxin chemical tools will continue to provide us with novel insights into auxin-regulated developmental processes.

ACKNOWLEDGMENTS

We thank Prof. Yunde Zhao (University of California San Diego), Prof. Richard Napier (University of Warwick), Dr. Kosuke Fukui (Okayama University of Science), and Prof. Hiroyuki Kasahara (Tokyo University of Agriculture and Technology) for helpful comments on the manuscript. This work was supported by a grant from JSPS KAKENHI (Grant No. JP19H03253).

REFERENCES

Abel S, Theologis A. 2010. Odyssey of auxin. *Cold Spring Harb Perspect Biol* **2:** a004572. doi:10.1101/cshperspect.a004572

ÅBerg B. 1951. The interaction of some auxin antagonists and 2,4-D in root growth. *Physiol Plant* **4:** 627–640. doi:10.1111/j.1399-3054.1951.tb07698.x

Adamowski M, Friml J. 2015. PIN-dependent auxin transport: action, regulation, and evolution. *Plant Cell* **27:** 20–32. doi:10.1105/tpc.114.134874

Bailly A, Sovero V, Vincenzetti V, Santelia D, Bartnik D, Koenig BW, Mancuso S, Martinoia E, Geisler M. 2008. Modulation of P-glycoproteins by auxin transport inhibitors is mediated by interaction with immunophilins. *J Biol Chem* **283:** 21817–21826. doi:10.1074/jbc.M709655200

Barbez E, Laňková M, Pařezová M, Maizel A, Zažímalová E, Petrášek J, Friml J, Kleine-Vehn J. 2013. Single-cell-based system to monitor carrier driven cellular auxin homeostasis. *BMC Plant Biol* **13:** 20. doi:10.1186/1471-2229-13-20

Bitto E, Bingman CA, Bittova L, Houston NL, Boston RS, Fox BG, Phillips GN Jr. 2009. X-ray structure of ILL2, an auxin-conjugate amidohydrolase from *Arabidopsis thaliana*. *Proteins* **74:** 61–71. doi:10.1002/prot.22124

Blakeslee JJ, Bandyopadhyay A, Lee OR, Mravec J, Titapiwatanakun B, Sauer M, Makam SN, Cheng Y, Bouchard R, Adamec J, et al. 2007. Interactions among PIN-FORMED and P-glycoprotein auxin transporters in *Arabidopsis*. *Plant Cell* **19:** 131–147. doi:10.1105/tpc.106.040782

Busi R, Goggin DE, Heap IM, Horak MJ, Jugulam M, Masters RA, Napier RM, Riar DS, Satchivi NM, Torra J, et al. 2018. Weed resistance to synthetic auxin herbicides. *Pest Manag Sci* **74:** 2265–2276. doi:10.1002/ps.4823

Calderón Villalobos LI, Lee S, De Oliveira C, Ivetac A, Brandt W, Armitage L, Sheard LB, Tan X, Parry G, Mao H, et al. 2012. A combinatorial TIR1/AFB-Aux/IAA co-receptor system for differential sensing of auxin. *Nat Chem Biol* **8:** 477–485. doi:10.1038/nchembio.926

Chen Q, Dai X, De-Paoli H, Cheng Y, Takebayashi Y, Kasahara H, Kamiya Y, Zhao Y. 2014. Auxin overproduction in shoots cannot rescue auxin deficiencies in *Arabidopsis* roots. *Plant Cell Physiol* **55:** 1072–1079. doi:10.1093/pcp/pcu039

Chiu LW, Heckert MJ, You Y, Albanese N, Fenwick T, Siehl DL, Castle LA, Tao Y. 2018. Members of the GH3 family of proteins conjugate 2,4-D and dicamba with aspartate and glutamate. *Plant Cell Physiol* **59:** 2366–2380.

Christian M, Hannah WB, Lüthen H, Jones AM. 2008. Identification of auxins by a chemical genomics approach. *J Exp Bot* **59:** 2757–2767. doi:10.1093/jxb/ern133

Dai X, Hayashi K, Nozaki H, Cheng Y, Zhao Y. 2005. Genetic and chemical analyses of the action mechanisms of sirtinol in *Arabidopsis*. *Proc Natl Acad Sci* **102:** 3129–3134. doi:10.1073/pnas.0500185102

Dejonghe W, Russinova E. 2017. Plant chemical genetics: from phenotype-based screens to synthetic biology. *Plant Physiol* **174:** 5–20. doi:10.1104/pp.16.01805

De Rybel B, Audenaert D, Beeckman T, Kepinski S. 2009. The past, present, and future of chemical biology in auxin research. *ACS Chem Biol* **4:** 987–998. doi:10.1021/cb9001624

De Vriese K, Himschoot E, Dünser K, Nguyen L, Drozdzecki A, Costa A, Nowack MK, Kleine-Vehn J, Audenaert D, Beeckman T, et al. 2019. Identification of novel inhibitors of auxin-induced Ca^{2+} signaling via a plant-based chemical screen. *Plant Physiol* **180:** 480–496. doi:10.1104/pp.18.01393

Dhonukshe P, Grigoriev I, Fischer R, Tominaga M, Robinson DG, Hasek J, Paciorek T, Petrasek J, Seifertova D, Tejos R, et al. 2008. Auxin transport inhibitors impair vesicle motility and actin cytoskeleton dynamics in diverse eukaryotes. *Proc Natl Acad Sci* **105:** 4489–4494. doi:10.1073/pnas.0711414105

Dindas J, Scherzer S, Roelfsema MRG, von Meyer K, Müller HM, Al-Rasheid KAS, Palme K, Dietrich P, Becker D, Bennett MJ, et al. 2018. AUX1-mediated root hair auxin influx governs $SCF^{TIR1/AFB}$-type Ca^{2+} signaling. *Nat Commun* **9:** 1174. doi:10.1038/s41467-018-03582-5

Doyle SM, Vain T, Robert S. 2015. Small molecules unravel complex interplay between auxin biology and endomembrane trafficking. *J Exp Bot* **66:** 4971–4982. doi:10.1093/jxb/erv179

Eklund DM, Ishizaki K, Flores-Sandoval E, Kikuchi S, Takebayashi Y, Tsukamoto S, Hirakawa Y, Nonomura M, Kato H, Kouno M, et al. 2015. Auxin produced by the indole-3-pyruvic acid pathway regulates development and gemmae dormancy in the liverwort *Marchantia polymorpha*. *Plant Cell* **27:** 1650–1669. doi:10.1105/tpc.15.00065

Enders TA, Strader LC. 2015. Auxin activity: past, present, and future. *Am J Bot* **102:** 180–196. doi:10.3732/ajb.1400285

Eswaramoorthy S, Bonanno JB, Burley SK, Swaminathan S. 2006. Mechanism of action of a flavin-containing monooxygenase. *Proc Natl Acad Sci* **103:** 9832–9837. doi:10.1073/pnas.0602398103

Fendrych M, Leung J, Friml J. 2016. TIR1/AFB-Aux/IAA auxin perception mediates rapid cell wall acidification and growth of *Arabidopsis* hypocotyls. *eLife* **5:** e19048. doi:10.7554/eLife.19048

Fendrych M, Akhmanova M, Merrin J, Glanc M, Hagihara S, Takahashi K, Uchida N, Torii KU, Friml J. 2018. Rapid and reversible root growth inhibition by TIR1 auxin signalling. *Nat Plants* **4:** 453–459. doi:10.1038/s41477-018-0190-1

Ferro N, Bredow T, Jacobsen HJ, Reinard T. 2010. Route to novel auxin: auxin chemical space toward biological correlation carriers. *Chem Rev* **110:** 4690–4708. doi:10.1021/cr800229s

Flores-Sandoval E, Eklund DM, Bowman JL. 2015. A simple auxin transcriptional response system regulates multiple morphogenetic processes in the liverwort *Marchantia polymorpha*. *PLoS Genet* **11:** e1005207. doi:10.1371/journal.pgen.1005207

Geisler M, Blakeslee JJ, Bouchard R, Lee OR, Vincenzetti V, Bandyopadhyay A, Titapiwatanakun B, Peer WA, Bailly A, Richards EL, et al. 2005. Cellular efflux of auxin catalyzed by the *Arabidopsis* MDR/PGP transporter AtPGP1. *Plant J* **44:** 179–194. doi:10.1111/j.1365-313X.2005.02519.x

Geldner N, Friml J, Stierhof YD, Jürgens G, Palme K. 2001. Auxin transport inhibitors block PIN1 cycling and vesicle trafficking. *Nature* **413:** 425–428. doi:10.1038/35096571

Hatano T, Kato Y, Katayama M, Marumo S. 1989. A new indolic antiauxin, α-(5,7-dichloroindole-3-)isobutyric acid: its chemical synthesis and biological activity. *Experientia* **45:** 400–402. doi:10.1007/BF01957494

Hayashi K. 2012. The interaction and integration of auxin signaling components. *Plant Cell Physiol* **53:** 965–975. doi:10.1093/pcp/pcs035

Hayashi K, Ogino K, Oono Y, Uchmiya H, Nozaki H. 2001. Yokonolide A, a new inhibitor of auxin signal transduction, from *Streptomyces diastatochromogenes* B59. *J Antibiot (Tokyo)* **54:** 573–581. doi:10.7164/antibiotics.54.573

Hayashi K, Jones AM, Ogino K, Yamazoe A, Oono Y, Inoguchi M, Kondo H, Nozaki H. 2003. Yokonolide B, a novel inhibitor of auxin action, blocks degradation of AUX/IAA factors. *J Biol Chem* **278:** 23797–23806. doi:10.1074/jbc.M300299200

Hayashi K, Tan X, Zheng N, Hatate T, Kimura Y, Kepinski S, Nozaki H. 2008. Small-molecule agonists and antagonists of F-box protein-substrate interactions in auxin perception and signaling. *Proc Natl Acad Sci* **105:** 5632–5637. doi:10.1073/pnas.0711146105

Hayashi K, Neve J, Hirose M, Kuboki A, Shimada Y, Kepinski S, Nozaki H. 2012. Rational design of an auxin antagonist of the SCF^{TIR1} auxin receptor complex. *ACS Chem Biol* **7:** 590–598. doi:10.1021/cb200404c

He W, Brumos J, Li H, Ji Y, Ke M, Gong X, Zeng Q, Li W, Zhang X, An F, et al. 2011. A small-molecule screen identifies L-kynurenine as a competitive inhibitor of TAA1/TAR activity in ethylene-directed auxin biosynthesis and root growth in *Arabidopsis*. *Plant Cell* **23:** 3944–3960. doi:10.1105/tpc.111.089029

Hoyerova K, Hosek P, Quareshy M, Li J, Klima P, Kubes M, Yemm AA, Neve P, Tripathi A, Bennett MJ, et al. 2018. Auxin molecular field maps define AUX1 selectivity: many auxin herbicides are not substrates. *New Phytol* **217:** 1625–1639. doi:10.1111/nph.14950

Kakei Y, Yamazaki C, Suzuki M, Nakamura A, Sato A, Ishida Y, Kikuchi R, Higashi S, Kokudo Y, Ishii T, et al. 2015. Small-molecule auxin inhibitors that target YUCCA are powerful tools for studying auxin function. *Plant J* **84:** 827–837. doi:10.1111/tpj.13032

Kakei Y, Nakamura A, Yamamoto M, Ishida Y, Yamazaki C, Sato A, Narukawa-Nara M, Soeno K, Shimada Y. 2017. Biochemical and chemical biology study of rice OsTAR1 revealed that tryptophan aminotransferase is involved in auxin biosynthesis: identification of a potent OsTAR1 inhibitor, Pyruvamine2031. *Plant Cell Physiol* **58:** 598–606.

Kasahara H. 2016. Current aspects of auxin biosynthesis in plants. *Biosci Biotechnol Biochem* **80:** 34–42. doi:10.1080/09168451.2015.1086259

Kato H, Nishihama R, Weijers D, Kohchi T. 2018. Evolution of nuclear auxin signaling: lessons from genetic studies with basal land plants. *J Exp Bot* **69:** 291–301. doi:10.1093/jxb/erx267

Kim JY, Henrichs S, Bailly A, Vincenzetti V, Sovero V, Mancuso S, Pollmann S, Kim D, Geisler M, Nam HG. 2010. Identification of an ABCB/P-glycoprotein-specific inhibitor of auxin transport by chemical genomics. *J Biol Chem* **285:** 23309–23317. doi:10.1074/jbc.M110.105981

Kinoshita T, McCourt P, Asami T, Torii KU. 2018. Plant chemical biology. *Plant Cell Physiol* **59:** 1483–1486. doi:10.1093/pcp/pcy142

Kubeš M, Napier R. 2019. Non-canonical auxin signalling: fast and curious. *J Exp Bot* **70:** 2609–2614. doi:10.1093/jxb/erz111

Lee S, Sundaram S, Armitage L, Evans JP, Hawkes T, Kepinski S, Ferro N, Napier RM. 2014. Defining binding efficiency and specificity of auxins for $SCF^{TIR1/AFB}$-Aux/IAA co-receptor complex formation. *ACS Chem Biol* **9:** 673–682. doi:10.1021/cb400618m

Leyser O. 2018. Auxin signaling. *Plant Physiol* **176:** 465–479. doi:10.1104/pp.17.00765

Li W, Li H, Xu P, Xie Z, Ye Y, Li L, Li D, Zhang Y, Li L, Zhao Y. 2018. Identification of Auxin Activity Like 1, a chemical with weak functions in auxin signaling pathway. *Plant Mol Biol* **98:** 275–287. doi:10.1007/s11103-018-0779-9

Li T, Yan A, Bhatia N, Altinok A, Afik E, Durand-Smet P, Tarr PT, Schroeder JI, Heisler MG, Meyerowitz EM. 2019. Calcium signals are necessary to establish auxin transporter polarity in a plant stem cell niche. *Nat Commun* **10:** 726. doi:10.1038/s41467-019-08575-6

Löfke C, Dünser K, Scheuring D, Kleine-Vehn J. 2015. Auxin regulates SNARE-dependent vacuolar morphology restricting cell size. *eLife* **4:** e05868. doi:10.7554/eLife.05868

Ludwig-Müller J. 2011. Auxin conjugates: their role for plant development and in the evolution of land plants. *J Exp Bot* **62:** 1757–1773. doi:10.1093/jxb/erq412

Ma Q, Robert S. 2014. Auxin biology revealed by small molecules. *Physiol Plant* **151:** 25–42. doi:10.1111/ppl.12128

Marchant A, Kargul J, May ST, Muller P, Delbarre A, Perrot-Rechenmann C, Bennett MJ. 1999. AUX1 regulates root gravitropism in *Arabidopsis* by facilitating auxin uptake within root apical tissues. *EMBO J* **18:** 2066–2073. doi:10.1093/emboj/18.8.2066

Mashiguchi K, Tanaka K, Sakai T, Sugawara S, Kawaide H, Natsume M, Hanada A, Yaeno T, Shirasu K, Yao H, et al. 2011. The main auxin biosynthesis pathway in *Arabidopsis*. *Proc Natl Acad Sci* **108:** 18512–18517. doi:10.1073/pnas.1108434108

Nakano T, Tanaka S, Ohtani M, Yamagami A, Takeno S, Hara N, Mori A, Nakano A, Hirose S, Himuro Y, et al. 2018. FPX is a novel chemical inducer that promotes callus formation and shoot regeneration in plants. *Plant Cell Physiol* **59:** 1555–1567. doi:10.1093/pcp/pcy139

Narukawa-Nara M, Nakamura A, Kikuzato K, Kakei Y, Sato A, Mitani Y, Yamasaki-Kokudo Y, Ishii T, Hayashi K, Asami T, et al. 2016. Aminooxy-naphthylpropionic acid and its derivatives are inhibitors of auxin biosynthesis targeting L-tryptophan aminotransferase: structure-activity relationships. *Plant J* **87:** 245–257. doi:10.1111/tpj.13197

Nishimura T, Matano N, Morishima T, Kakinuma C, Hayashi K, Komano T, Kubo M, Hasebe M, Kasahara H, Kamiya Y, et al. 2012. Identification of IAA transport inhibitors including compounds affecting cellular PIN trafficking by two chemical screening approaches using maize coleoptile systems. *Plant Cell Physiol* **53:** 1671–1682. doi:10.1093/pcp/pcs112

Nishimura T, Hayashi K, Suzuki H, Gyohda A, Takaoka C, Sakaguchi Y, Matsumoto S, Kasahara H, Sakai T, Kato J, et al. 2014. Yucasin is a potent inhibitor of YUCCA, a key enzyme in auxin biosynthesis. *Plant J* **77:** 352–366. doi:10.1111/tpj.12399

Oochi A, Hajny J, Fukui K, Nakao Y, Gallei M, Quareshy M, Takahashi K, Kinoshita T, Harborough SR, Kepinski, et al. 2019. Pinstatic acid promotes auxin transport by inhibiting PIN internalization. *Plant Physiol* **180:** 1152–1165. doi:10.1104/pp.19.00201

Oono Y, Chen QG, Overvoorde PJ, Köhler C, Theologis A. 1998. *age* Mutants of *Arabidopsis* exhibit altered auxin-regulated gene expression. *Plant Cell* **10:** 1649–1662. doi:10.1105/tpc.10.10.1649

Oono Y, Ooura C, Rahman A, Aspuria ET, Hayashi K, Tanaka A, Uchimiya H. 2003. *p*-Chlorophenoxyisobutyric acid impairs auxin response in *Arabidopsis* root. *Plant Physiol* **133:** 1135–1147. doi:10.1104/pp.103.027847

Parry G, Delbarre A, Marchant A, Swarup R, Napier R, Perrot-Rechenmann C, Bennett MJ. 2001. Novel auxin transport inhibitors phenocopy the auxin influx carrier mutation *aux1*. *Plant J* **25:** 399–406. doi:10.1046/j.1365-313x.2001.00970.x

Parry G, Calderon-Villalobos LI, Prigge M, Peret B, Dharmasiri S, Itoh H, Lechner E, Gray WM, Bennett M, Estelle M. 2009. Complex regulation of the TIR1/AFB family of auxin receptors. *Proc Natl Acad Sci* **106:** 22540–22545. doi:10.1073/pnas.0911967106

Peat TS, Böttcher C, Newman J, Lucent D, Cowieson N, Davies C. 2012. Crystal structure of an indole-3-acetic acid amido synthetase from grapevine involved in auxin homeostasis. *Plant Cell* **24:** 4525–4538. doi:10.1105/tpc.112.102921

Peer WA, Murphy AS. 2007. Flavonoids and auxin transport: modulators or regulators? *Trends Plant Sci* **12:** 556–563. doi:10.1016/j.tplants.2007.10.003

Peer WA, Blakeslee JJ, Yang H, Murphy AS. 2011. Seven things we think we know about auxin transport. *Mol Plant* **4:** 487–504. doi:10.1093/mp/ssr034

Petrasek J, Friml J. 2009. Auxin transport routes in plant development. *Development* **136:** 2675–2688. doi:10.1242/dev.030353

Porco S, Pěnčík A, Rashed A, Voß U, Casanova-Sáez R, Bishopp A, Golebiowska A, Bhosale R, Swarup R, Swarup K, et al. 2016. Dioxygenase-encoding *AtDAO1* gene controls IAA oxidation and homeostasis in *Arabidopsis*. *Proc Natl Acad Sci* **113:** 11016–11021. doi:10.1073/pnas.1604375113

Prigge MJ, Greenham K, Zhang Y, Santner A, Castillejo C, Mutka AM, O'Malley RC, Ecker JR, Kunkel BN, Estelle M. 2016. The *Arabidopsis* auxin receptor F-Box proteins AFB4 and AFB5 are required for response to the synthetic auxin picloram. *G3 (Bethesda)* **6:** 1383–1390. doi:10.1534/g3.115.025585

Prigge MJ, Platre M, Kadakia N, Zhang Y, Greenham K, Szutu W, Pandey BK, Bhosale RA, Bennett MJ, Busch W, et al. 2020. Genetic analysis of the *Arabidopsis* TIR1/AFB auxin receptors reveals both overlapping and specialized functions. *eLife* **9:** e54740. doi:10.7554/eLife.54740

Quareshy M, Prusinska J, Kieffer M, Fukui K, Pardal AJ, Lehmann S, Schafer P, Del Genio CI, Kepinski S, Hayashi K, et al. 2018a. The tetrazole analogue of the auxin indole-3-acetic acid binds preferentially to TIR1 and not AFB5. *ACS Chem Biol* **13:** 2585–2594. doi:10.1021/acschembio.8b00527

Quareshy M, Prusinska J, Li J, Napier R. 2018b. A cheminformatics review of auxins as herbicides. *J Exp Bot* **69:** 265–275. doi:10.1093/jxb/erx258

Rojas-Pierce M, Titapiwatanakun B, Sohn EJ, Fang F, Larive CK, Blakeslee J, Cheng Y, Cutler SR, Peer WA, Murphy AS, et al. 2007. *Arabidopsis* P-glycoprotein19 participates in the inhibition of gravitropism by gravacin. *Chem Biol* **14:** 1366–1376. doi:10.1016/j.chembiol.2007.10.014

Salehin M, Bagchi R, Estelle M. 2015. $SCF^{TIR1/AFB}$-based auxin perception: mechanism and role in plant growth and development. *Plant Cell* **27:** 9–19. doi:10.1105/tpc.114.133744

Savaldi-Goldstein S, Baiga TJ, Pojer F, Dabi T, Butterfield C, Parry G, Santner A, Dharmasiri N, Tao Y, Estelle M, et al. 2008. New auxin analogs with growth-promoting effects in intact plants reveal a chemical strategy to improve hormone delivery. *Proc Natl Acad Sci* **105:** 15190–15195. doi:10.1073/pnas.0806324105

Simon S, Kubeš M, Baster P, Robert S, Dobrev PI, Friml J, Petrášek J, Zažímalová E. 2013. Defining the selectivity of processes along the auxin response chain: a study using auxin analogues. *New Phytol* **200:** 1034–1048. doi:10.1111/nph.12437

Singh G, Retzer K, Vosolsobe S, Napier R. 2018. Advances in understanding the mechanism of action of the auxin permease AUX1. *Int J Mol Sci* **19:** 3391.

Soeno K, Goda H, Ishii T, Ogura T, Tachikawa T, Sasaki E, Yoshida S, Fujioka S, Asami T, Shimada Y. 2010. Auxin biosynthesis inhibitors, identified by a genomics-based approach, provide insights into auxin biosynthesis. *Plant Cell Physiol* **51:** 524–536. doi:10.1093/pcp/pcq032

Staswick PE, Serban B, Rowe M, Tiryaki I, Maldonado MT, Maldonado MC, Suza W. 2005. Characterization of an *Arabidopsis* enzyme family that conjugates amino acids to indole-3-acetic acid. *Plant Cell* **17:** 616–627. doi:10.1105/tpc.104.026690

Steenackers W, Klíma P, Quareshy M, Cesarino I, Kumpf RP, Corneillie S, Araújo P, Viaene T, Goeminne G, Nowack MK, et al. 2017. *cis*-Cinnamic acid is a novel, natural auxin efflux inhibitor that promotes lateral root formation. *Plant Physiol* **173:** 552–565. doi:10.1104/pp.16.00943

Sugawara S, Mashiguchi K, Tanaka K, Hishiyama S, Sakai T, Hanada K, Kinoshita-Tsujimura K, Yu H, Dai X, Takebayashi Y, et al. 2015. Distinct characteristics of indole-3-acetic acid and phenylacetic acid, two common auxins in plants. *Plant Cell Physiol* **56:** 1641–1654. doi:10.1093/pcp/pcv088

Takehara S, Sakuraba S, Mikami B, Yoshida H, Yoshimura H, Itoh A, Endo M, Watanabe N, Nagae T, Matsuoka M, et al. 2020. A common allosteric mechanism regulates homeostatic inactivation of auxin and gibberellin. *Nat Commun* **11:** 2143. doi:10.1038/s41467-020-16068-0

Tan X, Calderon-Villalobos LI, Sharon M, Zheng C, Robinson CV, Estelle M, Zheng N. 2007. Mechanism of auxin perception by the TIR1 ubiquitin ligase. *Nature* **446:** 640–645. doi:10.1038/nature05731

Tao S, Estelle M. 2018. Mutational studies of the Aux/IAA proteins in *Physcomitrella* reveal novel insights into their function. *New Phytol* **218:** 1534–1542. doi:10.1111/nph.15039

Tao Y, Ferrer JL, Ljung K, Pojer F, Hong F, Long JA, Li L, Moreno JE, Bowman ME, Ivans LJ, et al. 2008. Rapid synthesis of auxin via a new tryptophan-dependent pathway is required for shade avoidance in plants. *Cell* **133:** 164–176. doi:10.1016/j.cell.2008.01.049

Tsuda E, Yang H, Nishimura T, Uehara Y, Sakai T, Furutani M, Koshiba T, Hirose M, Nozaki H, Murphy AS, et al. 2011. Alkoxy-auxins are selective inhibitors of auxin transport mediated by PIN, ABCB, and AUX1 transporters. *J Biol Chem* **286:** 2354–2364. doi:10.1074/jbc.M110.171165

Tsugafune S, Mashiguchi K, Fukui K, Takebayashi Y, Nishimura T, Sakai T, Shimada Y, Kasahara H, Koshiba T, Hayashi K-i. 2017. Yucasin DF, a potent and persistent inhibitor of auxin biosynthesis in plants. *Sci Rep* **7:** 13992. doi:10.1038/s41598-017-14332-w

Uchida N, Takahashi K, Iwasaki R, Yamada R, Yoshimura M, Endo TA, Kimura S, Zhang H, Nomoto M, Tada Y, et al. 2018. Chemical hijacking of auxin signaling with an engineered auxin-TIR1 pair. *Nat Chem Biol* **14:** 299–305. doi:10.1038/nchembio.2555

Vain T, Raggi S, Ferro N, Barange DK, Kieffer M, Ma Q, Doyle SM, Thelander M, Pařízková B, Novák O, et al. 2019. Selective auxin agonists induce specific AUX/IAA protein degradation to modulate plant development. *Proc Natl Acad Sci* **116:** 6463–6472. doi:10.1073/pnas.1809037116

Walsh TA, Neal R, Merlo AO, Honma M, Hicks GR, Wolff K, Matsumura W, Davies JP. 2006. Mutations in an auxin receptor homolog AFB5 and in SGT1b confer resistance to synthetic picolinate auxins and not to 2,4-dichlorophenoxyacetic acid or indole-3-acetic acid in *Arabidopsis*. *Plant Physiol* **142:** 542–552. doi:10.1104/pp.106.085969

Wang B, Bailly A, Zwiewka M, Henrichs S, Azzarello E, Mancuso S, Maeshima M, Friml J, Schulz A, Geisler M. 2013. *Arabidopsis* TWISTED DWARF1 functionally interacts with auxin exporter ABCB1 on the root plasma membrane. *Plant Cell* **25:** 202–214. doi:10.1105/tpc.112.105999

Won C, Shen X, Mashiguchi K, Zheng Z, Dai X, Cheng Y, Kasahara H, Kamiya Y, Chory J, Zhao Y. 2011. Conversion of tryptophan to indole-3-acetic acid by TRYPTOPHAN AMINOTRANSFERASES OF ARABIDOPSIS and YUCCAs in *Arabidopsis*. *Proc Natl Acad Sci* **108:** 18518–18523. doi:10.1073/pnas.1108436108

Woodward AW, Bartel B. 2005. Auxin: regulation, action, and interaction. *Ann Bot* **95:** 707–735. doi:10.1093/aob/mci083

Yamada R, Murai K, Uchida N, Takahashi K, Iwasaki R, Tada Y, Kinoshita T, Itami K, Torii KU, Hagihara S.

2018. A super strong engineered auxin-TIR1 pair. *Plant Cell Physiol* **59:** 1538–1544. doi:10.1093/pcp/pcy127

Yamazoe A, Hayashi K, Kepinski S, Leyser O, Nozaki H. 2005. Characterization of terfestatin A, a new specific inhibitor for auxin signaling. *Plant Physiol* **139:** 779–789. doi:10.1104/pp.105.068924

Yang H, Murphy AS. 2009. Functional expression and characterization of *Arabidopsis* ABCB, AUX 1 and PIN auxin transporters in *Schizosaccharomyces pombe*. *Plant J* **59:** 179–191. doi:10.1111/j.1365-313X.2009.03856.x

Yang Y, Hammes UZ, Taylor CG, Schachtman DP, Nielsen E. 2006. High-affinity auxin transport by the AUX1 influx carrier protein. *Curr Biol* **16:** 1123–1127. doi:10.1016/j.cub.2006.04.029

Zhang J, Peer WA. 2017. Auxin homeostasis: the DAO of catabolism. *J Exp Bot* **68:** 3145–3154. doi:10.1093/jxb/erx221

Zhao Y. 2014. Auxin biosynthesis. *Arabidopsis Book* **12:** e0173. doi:10.1199/tab.0173

Zhao Y, Dai X, Blackwell HE, Schreiber SL, Chory J. 2003. SIR1, an upstream component in auxin signaling identified by chemical genetics. *Science* **301:** 1107–1110. doi:10.1126/science.1084161

Zhu J, Bailly A, Zwiewka M, Sovero V, Di Donato M, Ge P, Oehri J, Aryal B, Hao P, Linnert M, et al. 2016. TWISTED DWARF1 mediates the action of auxin transport inhibitors on actin cytoskeleton dynamics. *Plant Cell* **28:** 930–948. doi:10.1105/tpc.15.00726

Index

Index